HUMAN PHYSIOLOGY

HUMAN PHYSIOLOGY
THE MECHANISMS OF BODY FUNCTION

ARTHUR J. VANDER, M.D.
PROFESSOR OF PHYSIOLOGY
UNIVERSITY OF MICHIGAN

JAMES H. SHERMAN, Ph.D.
ASSOCIATE PROFESSOR OF PHYSIOLOGY
UNIVERSITY OF MICHIGAN

DOROTHY S. LUCIANO, Ph.D.
FORMERLY OF THE DEPARTMENT OF PHYSIOLOGY
UNIVERSITY OF MICHIGAN

THIRD EDITION

McGRAW-HILL BOOK COMPANY

NEW YORK   ST. LOUIS   SAN FRANCISCO   AUCKLAND   BOGOTÁ   HAMBURG
JOHANNESBURG   LONDON   MADRID   MEXICO   MONTREAL   NEW DELHI   PANAMA
PARIS   SÃO PAULO   SINGAPORE   SYDNEY   TOKYO   TORONTO

8 9 10 11 12 K P K P 8 9 8 7 6 5 4 3

This book was set in Primer by Ruttle, Shaw & Wetherill, Inc. The editors were
James E. Vastyan and Stephen Wagley; the designer was Hermann Strohbach; the pro-
duction supervisor was Charles Hess. New drawings were done by J & R Services, Inc.
Kingsport Press, Inc., was printer and binder.

The cover illustration is from the *Anatomical Drawings* by Leonardo da Vinci.

Library of Congress Cataloging in Publication Data

Vander, Arthur J date
Human physiology.

Includes index.
1. Human physiology. I. Sherman, James H.,
date   joint author. II. Luciano, Dorothy S.,
joint author. III. Title. [DNLM: 1. Physiology.
QT104.3 V228h]
QP34.5.V36 1980   612   79-17781
ISBN 0-07-066961-9

TO OUR PARENTS
AND TO JUDY, PEGGY, AND JOE
WITHOUT WHOSE UNDERSTANDING
IT WOULD HAVE BEEN IMPOSSIBLE.

CONTENTS

Preface   ix

1 A Framework for Human Physiology   1

PART ONE BASIC CELL FUNCTIONS

2 Cell Structure   13
3 Chemical Composition of the Body   21
4 Molecular Control Mechanisms – DNA and Protein   42
5 Energy and Cellular Metabolism   71
6 Movement of Molecules across Cell Membranes   99

PART TWO BIOLOGICAL CONTROL SYSTEMS

7 Homeostatic Mechanisms   127
8 Neural Control Mechanisms   144
9 Hormonal Control Mechanisms   191
10 Muscle   211

PART THREE COORDINATED BODY FUNCTIONS

11 Circulation   253
12 Respiration   327
13 Regulation of Water and Electrolyte Balance   366
14 The Digestion and Absorption of Food   402
15 Regulation of Organic Metabolism and Energy Balance   440
16 Reproduction   479
17 Defense Mechanisms of the Body: Immunology, Foreign Chemicals, Stress, and Aging   521
18 Processing Sensory Information   557
19 Control of Body Movement   589
20 Consciousness and Behavior   607

Appendix: English and Metric Units   631

Glossary   633

References   684

Index   697

PREFACE

The primary purpose of this book remains what it was in the first two editions: to present the fundamental mechanisms of human physiology. Our aim has been to tell a story, not to write an encyclopedia. The book is intended for undergraduate students, regardless of their scientific background. The physics and chemistry requisite for an understanding of the physiology are presented where relevant in the text. Students with little or no scientific training will find this material essential; while others, more sophisticated in the physical sciences, should profit from the review of basic science oriented toward specifically biological applications. Thus this book is suitable for most introductory courses in human physiology, including those taken by students in the health professions.

The overall organization and approach of the book is based upon a group of themes which are developed in Chapter 1 and which form the framework for our descriptions: (1) all phenomena of life, no matter how complex, are ultimately describable in terms of physical and chemical laws; (2) certain fundamental features of cell function are shared by virtually all cells and, in addition, constitute the foundation upon which specialization is built; (3) the body's various co-ordinated functions—circulation, respiration, etc. —result from the precise control and integration of specialized cellular activities, serve to maintain relatively constant the internal composition of the body, and can be described in terms of control systems similar to those designed by engineers.

In keeping with these themes, the book progresses from the cell to the total body, utilizing at each level of increasing complexity the information and principles developed previously. Part 1 is devoted to an analysis of basic cellular physiology and the essential physics and chemistry required for its understanding. Part 2 analyzes the concept of the body's internal environment, the nature of

biological control systems, and the properties of the major specialized cell types—nerve, muscle, and gland—which constitute these systems. Part 3 then analyzes the coordinated body functions in terms of the basic concepts and information developed in Parts 1 and 2. In this way we have tried to emphasize the underlying unity of biological processes.

This approach has resulted in several characteristics of the book: (1) Cell physiology has received extensive coverage. (2) We have been willing to spend a considerable number of pages logically developing a single cellular concept (such as the origin of membrane potentials) required for the understanding of total-body physiological processes, and which we have found to offer considerable difficulties for the student. (3) We have made every effort not to mention facts simply because they happen to be known, but rather to use facts as building blocks for general principles and concepts. In this last regard, we confess that even an introductory course in physiology must teach a rather frightening number of facts, and our book is no exception. We have tried to keep in mind, however, the words of John Hunter, the eighteenth-century British anatomist and surgeon: "Too much attention can not be paid to facts; yet too many facts crowd the memory without advantage, any further than they lead us to establish principles." (4) We have employed a very large number of figures as an aid to developing concepts and explanations. These figures also provide an excellent summary of the most important material in each chapter. (5) We have not shied away from pointing out the considerable gaps in current understanding. In summary, our book may be long in areas where other texts are brief, and vice versa; moreover, our approach requires the student to think rather than simply to memorize.

In this third edition, the level, scope, and emphasis remain unchanged. The text has been completely updated, a process involving changes too numerous to list here. We have also done considerable rewriting in order to improve clarity of organization and presentation. For example, the basic material on cell structure is now gathered together in a single chapter. This is also true for the descriptions of blood. We have also added a glossary to this edition.

We would like to thank the many physiologists who reviewed all or parts of this edition. We are grateful to Peggy Rogers for her superb typing of the manuscript.

Arthur J. Vander
James H. Sherman
Dorothy S. Luciano

1
A FRAMEWORK FOR HUMAN PHYSIOLOGY

A society of cells
    Cells: The basic units
    Tissues
    Organs and organ systems
The internal environment
    Body-fluid compartments
The importance of control
A rationale
Mechanism and causality

One cannot meaningfully analyze the complex activities of the human body without a framework upon which to build, a set of viewpoints to guide one's thinking. It is the purpose of this chapter to provide such an orientation to the subject of human physiology—the mechanisms by which the body functions.

A society of cells

Cells: The basic units

Individual cells are the basic units of both the structure and the function of living things. One of the crucial unifying generalizations of biology is that certain fundamental activities are common to almost all cells and represent the minimal requirements for maintaining the integrity and life of the cell. Thus, a human liver cell and an ameba are remarkably similar in their means of exchanging materials with their immediate environments, of obtaining energy from organic nutrients, of synthesizing complex proteins, and of duplicating themselves.

This is not to say that there are no significant differences between an ameba and a liver cell or between the liver cell and a nerve cell. However, a second crucial generalization is that these differences in cell function generally represent specializations of one or more of the fundamental common properties. For example, the excitability of nerve cells represents a specialization of electrical phenomena common to the membranes of virtually all cells; the secretion of protein hormones by certain gland cells of the body is a specialized form of the genetically controlled protein synthesis found in all cells; the transport of food molecules across the cells forming the intestinal wall results from a specialized orientation of transport mechanisms remarkably similar in most cells. These specializations, which arise during cellular differentiation, have all occurred as a part of evolution and have resulted in the adaptation of certain cells for specific roles.

1

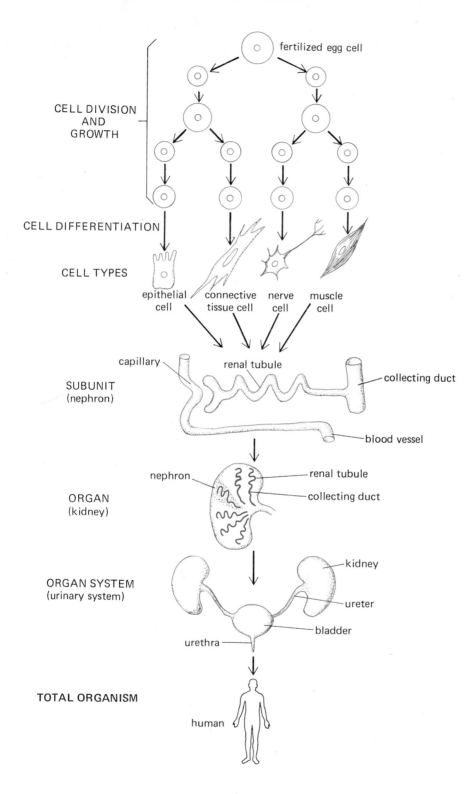

Figure 1-1. Levels of cellular organization.

The human organism begins as a single cell, the fertilized ovum, which divides to form two cells, each of which divides in turn, resulting in four cells, and so on as repeated cell divisions increase the number of cells in the developing embryo. If cell multiplication were the only event occurring, the end result would be a spherical mass of identical cells. However, other simultaneously occurring processes cause the embryo's rapidly dividing cells to become altered and arranged into a variety of configurations. *Cellular differentiation* is the process by which the initially identical cells present during the earliest stages of development not only undergo anatomical alteration but also acquire specialized functional properties, e.g., nerve cells' ability to conduct electric signals. Other processes in embryonic development are the migration of cells to new locations within the embryo and the selective adhesion of various types of cells to each other. In this manner the cells of the body are arranged in various combinations to form a hierarchy of organized structures. Differentiated cells of a similar type are organized into *tissues*, while different types of tissues are combined to form *organs* which are linked together to form *organ systems* (Fig. 1-1).

About 200 distinct kinds of cells can be identified in the human body. However, many of them perform similar general types of functions; for example, skeletal muscle cells, cardiac muscle cells, and smooth muscle cells, while distinguishable from each other, all have the special ability to generate forces. Accordingly, these three types of cells together comprise a functional category of specialized cells—muscle. Based upon the general type of function they perform, or in some cases upon their embryonic origins, four such broad categories of differentiated cell types are recognized, each category having a number of variations on a common theme: (1) muscle cells, (2) nerve cells, (3) epithelial cells, and (4) connective tissue cells.

Muscle cells generate forces and produce motion. They may be attached to bones and produce movements of the limbs or trunk, or they may enclose hollow cavities so that their contraction expels the contents of the cavity, as in the case of the pumping of the heart or the expulsion of urine by the bladder. Muscle cells are also found surrounding many of the hollow tubes in the body and their contraction changes the diameter of these tubes, as in the case of the smooth muscle cells surrounding blood vessels.

Nerve cells have the ability to initiate electric signals and to propagate these signals along their processes which extend from one area of the body to another. The electric signal may initiate a new electric signal in another nerve cell and thus pass information on from nerve cell to nerve cell, or it may initiate secretion from a gland cell or contraction of a muscle cell. Thus, nerve cells provide one of the major means of controlling the activities of other cells.

Epithelial cells are located on the surfaces that cover the body or individual organs, or line the walls of various tubular and hollow structures within the body. These cells form the boundaries between compartments and function as selective barriers regulating the exchange of molecules across these surfaces. For example, the epithelial cells at the surface of the skin form a barrier which prevents most substances in the external environment from entering the body through the skin; the epithelial linings of the lungs, gastrointestinal tract, and kidneys regulate the exchange of molecules between the blood and the external environment.

Connective tissue cells, as their name implies, have as their major function connecting, anchoring, and supporting the structures of the body. Some types of connective tissue cells form the loose meshwork of cells and fibers underlying most epithelial layers, but other cell types as diverse as fat-storing (adipose) cells, bone cells, and red and white blood cells also belong to this category. Many connective tissue cells secrete into the fluid surrounding them molecules which form a matrix consisting of various types of fibers embedded in a ground substance. This matrix may vary in consistency from a semifluid gel to the solid crystalline structure of bone. The extracellular fibers formed by these cells include the inelastic tendons and ligaments which link muscles to bone and bones to other bones, as well as rubber-band-like elastin fibers which provide the elastic qualities found in various organs. Connective tissue elements are found throughout all the organs of the body, often in the form of a fine meshwork of fibers which provides a supporting matrix to which other types of cells are anchored.

Tissues

Most of the differentiated cells in the body are organized into groups of cells of the same type. Such aggregates of similar cells form multicellular units which are known as *tissues*. Corresponding to the four general categories of differentiated cell types, there are four general classes of tissues: (1) *muscle tissue*, (2) *nerve tissue*, (3) *epithelial tissue*, and (4) *connective tissue*. It should be noted that the term "tissue" is frequently used in several different ways. It is formally defined as just described, i.e., as an aggregate of a single type of specialized cell; however, it is also commonly used to denote the general cellular fabric of any given organ or structure, e.g., kidney tissue, lung tissue, etc., each of which is in fact usually composed of all four specialized cell types.

Organs and organ systems

The organs of the body are composed of the four kinds of tissues arranged in sheets, tubes, layers, bundles, strips, etc. For example, the intestine is a long, hollow tubular organ whose inner surface is lined with a layer of epithelial cells. Underlying and attached to these cells is a thin layer of proteinaceous extracellular material known as basement membrane; under this is a layer of connective tissue through which run blood vessels and bundles of nerve fibers. Glands formed by the invagination of the epithelial surface layer are also located in this connective tissue layer and are connected by ducts to the luminal surface of the intestinal tract. Surrounding the connective tissue layer is a layer of smooth muscle cells whose contractile activity provides the forces for moving the luminal contents along the intestinal tract. Finally, surrounding the entire organ is connective tissue which anchors it to other structures.

Many organs are composed of a large number of similar subunits, the total function of the organ being the sum of the contributions made by each of the subunits. For example, the kidneys consist of 2 million similar functional units known as nephrons. Each nephron is a tuft of capillaries and a coiled tube of epithelial cells surrounded by a connective tissue layer which contains nerves and blood vessels. Urine is formed by the movement of molecules across the capillaries and epithelial cells. The total volume of urine excreted by the kidneys is the sum of the amounts formed by each of the individual nephrons.

Finally, the last order in classification is that of the *organ system*, a collection of organs which together subserve an overall function. For example, the kidneys, the urinary bladder, and the tubes leading from kidneys to bladder and from the bladder to the exterior constitute the urinary system. There are 10 organ systems in the body (Table 1-1).

The internal environment

In essence then, the human body can be viewed as a complex society of cells of many different types which are structurally and functionally combined and interrelated in a variety of ways to carry on the functions essential to the survival of the organism as a whole. Yet the fact remains that individual cells still constitute the basic units of this society and that almost all these cells individually exhibit the fundamental activities common to all forms of life. Indeed, many of the body's different cell types can be removed from the body and maintained in test tubes as free-living cells.

There is a definite paradox in this analysis. If each individual cell performs the fundamental activities required for its own survival, what contributions do the different organ systems make? How can we refer to a system's functions as being "essential to the survival of the organism as a whole" when each individual cell of the organism seems to be capable of performing its own fundamental activities? The resolution of this paradox is found in the isolation of most of the cells of a multicellular organism from the environment surrounding the body (*external environment*).

An ameba and a human liver cell both obtain their energy by the breakdown of certain organic nutrients; the chemical reactions involved in this intracellular process are remarkably similar in the two types of cells and involve the utilization of oxygen and the production of carbon dioxide. The ameba picks up oxygen directly from its environment and eliminates carbon dioxide into it. But how can the liver cell obtain its oxygen and elimi-

Table 1-1. Organ systems of the body

System	Major organs or tissues	Primary function(s)
Circulatory	Heart, blood vessels, blood	Rapid bulk flow of blood throughout the body's tissues.
Respiratory	Nose, throat, larynx, trachea, bronchi, bronchioles, lungs	Exchange of carbon dioxide and oxygen and regulation of hydrogen-ion concentration.
Digestive	Mouth, pharynx, esophagus, stomach, intestines, salivary glands, pancreas, liver, gallbladder	Digestion, absorption, and processing of nutrients.
Urinary	Kidneys, ureters, bladder, urethra	Regulation of plasma composition through excretion of organic wastes, salts, and water.
Musculo-skeletal	Cartilage, bone, ligaments, tendons, joints, skeletal muscle	Support, protect, and move the body.
Immune	White blood cells, lymph vessels and nodes, spleen, thymus, bone marrow, reticuloendothelial cells	Defense against foreign invaders. Also functions in the return of extracellular fluid to blood and the formation of red and white blood cells.
Nervous	Brain, spinal cord, peripheral nerves and ganglia, special sense organs	Regulates and coordinates many activities in the body. Detects changes in the internal and external environments. Is responsible for states of consciousness.
Endocrine	All glands secreting hormones into the blood; Pituitary, thyroid, parathyroid, adrenal, pancreas, testes, ovaries, intestinal glands, kidneys	Regulates and coordinates many activities in the body.
Reproductive	Male: Testes, penis, and associated ducts and glands Female: Ovaries, oviducts, uterus, vagina, mammary glands	Production of egg and sperm cells, transfer of male sperm to female, and provision of a nutritive environment for the developing embryo.
Integumentary	Skin	Protects against injury and dehydration, defends against foreign invaders, and regulates temperature.

nate the carbon dioxide when, unlike the ameba, it is not in direct contact with the external environment? Supplying oxygen to the liver is the function both of the respiratory system (comprising the lungs and the airways leading to them), which takes up oxygen from the external environment, and of the circulatory system, which distributes oxygen to all parts of the body. Conversely, the circulatory system carries the carbon dioxide generated by the liver cells and all the other cells of the body to the lungs, which eliminate it to the exterior. Similarly, the digestive and circulatory systems, working together, make nutrients from the external environment available to all the body's cells. Wastes other than carbon dioxide are carried by the circulatory system from the cells which produced them to the kidneys (and liver), which excrete them from the body (Fig. 1-2); the kidneys also regulate the concentrations of water and many essential minerals in the plasma.

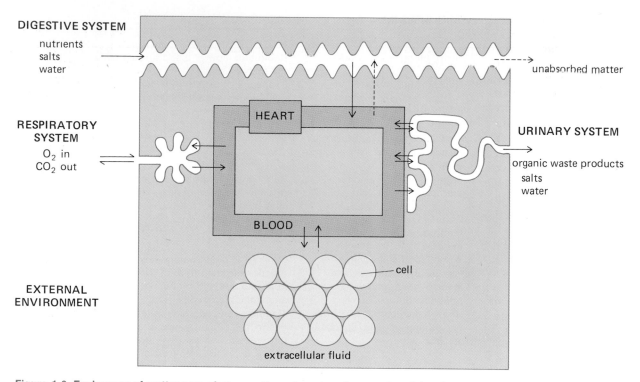

DIGESTIVE SYSTEM

nutrients
salts
water

unabsorbed matter

RESPIRATORY
SYSTEM

O_2 in
CO_2 out

HEART

URINARY SYSTEM

organic waste products
salts
water

BLOOD

cell

EXTERNAL
ENVIRONMENT

extracellular fluid

Figure 1-2. Exchanges of matter occur between the external environment and the circulatory system via the digestive system, respiratory system, and urinary system.

Figure 1-3. Extracellular fluid is the internal environment of the body.

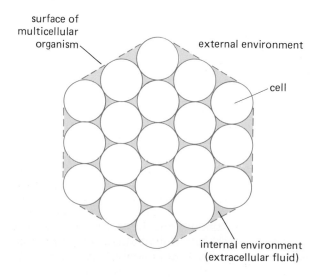

surface of
multicellular
organism

external environment

cell

internal environment
(extracellular fluid)

Thus the overall effect of the activities of organ systems is to create *within* the body the environment required for all cells to function; the last column of Table 1-1 summarizes the contribution each organ system makes to the regulation of the internal environment. The "internal environment" is not merely a theoretical physiological concept but can be identified quite specifically in anatomical terms: The body's *internal environment* is the *extracellular fluid* (literally, fluid "outside the cells"), which bathes each of the body's cells (Figs. 1-2 and 1-3). In other words, the environment in which each cell lives is not the "external environment" surrounding the entire body but the local extracellular fluid surrounding that cell. It is from this fluid that the cells receive nutrients and into which they excrete wastes. A multicellular organism can survive only as long as it is able to maintain the composition of its internal environment in a state compatible with the survival of its individual cells. The French physiologist Claude Bernard first clearly described in 1857 the central importance of the extracellular fluid of the body: "It is the fixity of the internal environment which is the condition of free and independent life. . . . All the vital mechanisms, however varied they may be, have only one object, that of preserving constant the conditions of life in the internal environment." This concept of an internal environment and the necessity for maintaining its composition relatively constant is the single most important unifying idea to be kept in mind while attempting to unravel and understand the functions of the body's organ systems and their interrelationships.

To summarize, the total activities of every individual cell in the body fall into two categories: (1) Each cell performs for itself all those fundamental basic cellular processes (movement of materials across its membrane, extraction of energy, protein synthesis, etc.) which represent the minimal requirements for maintaining its individual integrity and life; (2) each cell simultaneously performs one or more specialized activities which in concert with the other cells of its tissue or organ system contribute to the survival of the total organism by helping maintain the stable internal environment (extracellular fluid) required by all cells. These latter specialized activities together constitute the coordinated body processes (circulation, respiration, digestion, etc.) typical of multicellular organisms like human beings.

Clearly, the society of cells which constitutes the human body bears many striking similarities to a society of persons (although the analogy must not be pushed too far). Each person in a complex society must perform for himself a set of fundamental activities (eating, excreting, sleeping, etc.), which is virtually the same for all persons. In addition, because the complex organization of a society makes it virtually impossible for any individual within the society to raise his or her own food, arrange for the disposal of wastes, and so on, each individual participates in the performance of one of these supply-and-disposal operations required for the survival of all. A specialized activity, therefore, becomes an *additional* part of one's daily routine, but it never allows the individual to cease or to reduce the fundamental activities required for survival.

Body-fluid compartments

The extracellular fluid is divided between two so-called *compartments*, or general locations. Approximately 80 percent of the extracellular fluid surrounds all the body's cells, and because it lies between cells and tissues, it is known as intercellular or, more often, *interstitial fluid.* The remaining 20 percent of the extracellular fluid constitutes the fluid portion of the blood, the *plasma.* The blood plasma is continuously circulated by the action of the heart to all parts of the body and is the dynamic component of the extracellular fluid. As will be seen in Chap. 11, the plasma exchanges oxygen, nutrients, wastes, and other metabolic products with the interstitial fluid as the blood passes through the capillaries of the body. Because of the capillary exchanges between plasma and interstitial fluid, solute concentrations are virtually identical in the two fluids, except for protein. With this major exception, the entire extracellular fluid may be considered to have a homogeneous composition.

To generalize one step further, the human body can be viewed as containing three fluid compartments: (1) blood plasma, (2) interstitial fluid, and (3) intracellular fluid, the fluid in the cells of the body (Fig. 1-4). The major molecular component of all three compartments is water, which accounts for about 60 percent of the normal body weight, or 42 L, in an average man. Two-thirds of the total body water (28 L) is located inside the cells: the intracellular fluid. The remaining one-third of the total body water (14 L) comprises the two extracellular fluids: 80 percent as interstitial fluid (11 L) and 20 percent as plasma (3 L).

The above figures for the volumes of fluids in the various compartments of the body are those of a 70-kg man. Obviously, body measurements vary between different individuals. Some of these values may vary directly with body weight and can be expressed as a percentage of the total body weight, whereas others depend on age, sex, and state of health as well as body weight. The typical values for a healthy, 70-kg (154-lb), 21-year-old male have been used for years as representative of an average individual. This standard was chosen because a majority of the data collected over the years on normal healthy individuals has come from measurements made on students attending medi-

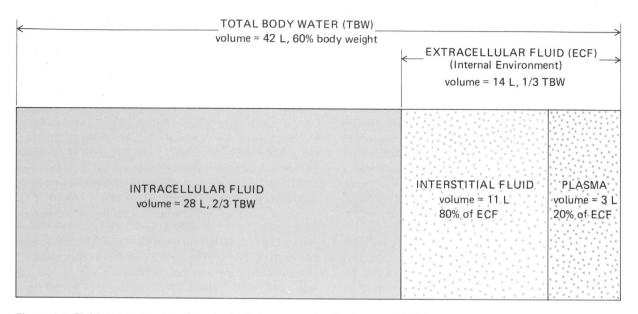

Figure 1-4. Fluid compartments of the body. Volumes are for the "average" 70-kg man.

cal school. The 21-year-old, 70-kg male represents the average body size, sex, and age of the medical students from which the data were obtained. Measurements on females of the same age and corrected for differences in total body weight give values of a similar magnitude, although slight sex-specific differences do exist. Thus, a female has a slightly lower total-body-water content than a male of the same weight because her body is composed of a higher percentage of fatty tissue containing less water than other tissues. Unless otherwise stated, quantitative values given in this text refer to a healthy, 70-kg, 21-year-old male.

The importance of control

Implicit in life is *control*. Regardless of its level of organizational complexity, no living system can exist without precise mechanisms for controlling its various activities.

Every one of the fundamental processes performed by any single cell (ameba or liver cell) must be carefully regulated. What determines how much sugar is to be transported across the cell's membrane? What proportion of this sugar, once inside the cell, is to be utilized for energy or transformed into fat or protein? How much protein of each type is to be synthesized and when? How large is the cell to grow, and when is it to divide? The list is almost endless. Thus, an understanding of cellular physiology requires not only a knowledge of the basic processes but also of the mechanisms which control them. Indeed, the two are inseparable.

In any multicellular organism like a human being, these basic intracellular regulators remain, but the existence of a multitude of different cells organized into specialized tissues and organs obviously imposes the need for overall regulatory mechanisms. Information about all important aspects of the external and internal environments must be monitored continuously; this information must be integrated, and on the basis of its content "instructions" must be sent to the various tissue and organ cells (particularly muscle and gland cells) directing them to increase or decrease their activities.

This transmission and integration of information is performed primarily by the nervous and hormonal systems. Thus these two systems contribute to the survival of the organism by controlling the activities of the various bodily components so that any change (or impending change) in the body's internal environment automatically initiates a chain of events wiping out the change or leading to a state of readiness should the impending change actually occur. For example, when (for any reason) the concentration of oxygen in the body significantly decreases below normal, the nervous system detects the change and increases its output to the skeletal muscles responsible for breathing movements; the result is a compensatory increase in oxygen uptake by the body and a restoration of normal internal oxygen concentration.

The above example is analogous to certain systems in engineering, say for maintaining constant oxygen concentration in a submarine. Similarly, the body temperature is regulated by a system whose underlying principles are nearly identical to the thermostatically controlled system which keeps a house at some specified temperature. These are known as *control systems.*

A rationale

With this framework in mind, the overall organization and approach of this book should easily be understood. Because the fundamental features of cell function are shared by virtually all cells and, in addition, constitute the foundation upon which specialization develops, we devote the first section of this book on the human animal to an analysis of basic cellular physiology. We also emphasize cell physiology at the start because the analysis of cellular phenomena in terms of physicochemical principles has proved very successful of late. Indeed, much of cell biology is now referred to as *molecular biology* in recognition of the ultimate goal of explaining all cellular processes in terms of interactions between molecules of known structure.

At the other end of the organizational spectrum, the third part of the book describes how the body's various coordinated functions (circulation, respiration, etc.) result from precisely controlled and integrated activities of specialized cells grouped together in tissues and organs. The theme of these descriptions is that each of these coordinated functions (with the obvious exception of reproduction) serves to keep some important aspect of the body's internal environment relatively constant and can be therefore described in terms of a control system similar to those familiar in engineering.

The second section of the book provides the principles and information required to bridge the gap between these two organizational levels, the cell and the body. First, the evolutionary origin and physicochemical composition of the internal environment are presented. Second, control systems are analyzed in general terms to emphasize that the basic principles governing virtually all control systems are the same. Finally, the bulk of this section is concerned with the major components of the body's control systems (nerve cells, muscle cells, and gland cells); the physicochemical basis for their specialized cellular activities and the interactions between them are emphasized. Once acquainted with the cast of major characters— nerve, muscle, and gland cells—and the theme— the maintenance of a stable internal environment through the interactions of these characters—the reader will be free to follow the specific plot lines— circulation, respiration, etc.—of Part 3.

Mechanism and causality

The *mechanist* view of life holds that all phenomena, no matter how complex, are ultimately describable in terms of physical and chemical laws and that no "vital force" distinct from matter and energy is required to explain life. Man is a machine—an enormously complex machine, but a machine nevertheless. This view has predominated in the twentieth century because virtually all information gathered from observation and experiment has agreed with it. But *vitalism,* the view that some force beyond physics and chemistry is required by living organisms, is not completely dead, nor is it surprising that it lingers specifically in brain physiology, where we are almost entirely lacking in hypotheses to explain such phenomena as thought and consciousness in physicochemical terms. We believe that even this area will ultimately yield to physicochemical analysis, but we also feel that it would be unscientific, on the basis of present knowledge, to dismiss the problem out of hand.

We have emphasized that a common denominator of physiological processes is their contribution to survival. Unfortunately, it is easy to misunderstand the nature of this relationship. Consider, for example, the statement that "during exercise a person sweats *because* the body *needs* to get rid of the excess heat generated." This type of statement is an example of *teleology*, the explanation of events in terms of purpose. But it is not an *explanation* at all in the scientific sense of the word. It is somewhat like saying, "The furnace is on because the house needs to be heated." Clearly, the furnace is on, not because it senses in some mystical manner the house's *needs*, but because the temperature has fallen below the thermostat's set point and the electric current in the connecting wires has turned on the heater and blower.

Is it not true to say that sweating actually serves a useful purpose because the excess heat, if not eliminated, might have caused sickness or even death? Correct, but this is totally different from stating that a "need" to avoid injury *caused* the sweating to occur. The *cause* of the sweating, in reality, was an automatically occurring sequence of events initiated by the increased heat generation: increased heat generation → increased blood temperature → increased activity of specific nerve cells in the brain → increased activity, in turn, of a series of nerve cells, the last of which stimulate the sweat glands → increased production of sweat by the sweat-gland cells. Each of these steps occurs by means of physicochemical changes in the cells involved. In science, to *explain* a phenomenon is to reduce it to a sequence of physicochemical events. This is the scientific meaning of causality, of the word "because."

On the other hand, that a phenomenon is beneficial to the person is of considerable interest and importance. It is attributable to evolutionary processes which result in the selecting out of those responses having survival value. Evolution is the key to understanding why most bodily activities do indeed appear to be purposeful. Throughout the book we emphasize how a particular process contributes to survival, but the reader must never confuse this survival value of a process with the explanation of the mechanisms by which the process occurs.

PART ONE

BASIC CELL
FUNCTIONS

2
CELL STRUCTURE

Microscopic observations of cells
Cell organelles

The cell is the simplest unit of biological structure into which an organism can be divided and still retain the characteristics we associate with life, yet each cell is so small that a microscope is required to observe its structure. Robert Hooke introduced the microscope into England and in 1663 demonstrated its remarkable powers before the Royal Society in London. As one of his demonstrations he placed a slice of cork (plant tissue) under a microscope, describing his observations as follows: "Our microscope informs us that the substance of cork is altogether filled with air and that the air is perfectly enclosed in little boxes or *cells* distinct from one another." It is from this description that the term "cell" was introduced into the biological literature. Hooke, however, did not realize that each of these air-filled boxes had once been occupied by a living cell. The concept that all living organisms are composed of microscopic compartments, cells, was not fully established until the early part of the last century.

Microscopic observations of cells

When a living cell is examined under a light microscope, numerous granules and particles, some of which appear to be long filamentous structures, can be seen. Many of these oscillate back and forth, as if floating in a fluid medium; this medium is water, which accounts for about 80 percent of the cell's weight. Such an abundance of water, which is highly transparent to light, makes a cell almost invisible under an ordinary light microscope, and many devices are used to make the cellular structure more visible. One is the application to the cell of dyes which combine with various cell structures, staining them so that they become visible.

Even with staining techniques, only a portion of the cell's structure can be seen through a light microscope. The smallest object the naked eye can resolve is about 0.1 mm in diameter. A light microscope can magnify an object about 2000 times, producing an image of a 10-μm[1] cell that is about 20 mm in diameter. But even at this magnification many cell structures are too small to be seen. For example, the outer membrane surrounding the cell is about 7.5 nm thick; after a magnification of 2000 times it would only be 0.015 mm thick and still invisible under the light microscope. The great diversity of structures within the cell began to be discovered only after the development of the *electron microscope* in the late 1940s.

A beam of electrons can be focused by an electron microscope to form an image of a specimen just as a beam of light can be focused by a light microscope. The magnifying power of the light microscope is limited by the wavelength of light.

[1] See Appendix I for a listing of metric units of measurement and their English equivalents.

13

The wavelength of a high-velocity electron may be of the order of 0.005 nm, which is about 100,000 times smaller than the wavelength of green light. The smaller wavelength of electrons increases the resolving power of the electron microscope some 200-fold over that of the light microscope, giving a total magnification of 400,000, which produces an image of the cell membrane that is 3 mm in diameter. This high magnification is sufficient to resolve very large individual molecules.

An electron microscope forms an image of a cell by accelerating electrons to very high velocities in an electric field, which focuses the electrons on the cell specimen. If an electron does not strike any of the atoms in the specimen, it passes through the cell and reacts with a photographic film placed underneath the specimen. If, on the other hand, the electron collides with a structure in the cell and is either captured or deflected to one side, it will not reach the photographic plate and a black spot or shadow is produced in that area. Since electrons are very easily scattered by atoms, most of the air inside the microscope must be pumped out, leaving a near vacuum surrounding the specimen.

Cells are so large, compared with electrons, that few electrons would manage to pass all the way through an entire cell. Therefore, the cell must be cut into many thin sections, some as thin as 20 nm, and these sections, rather than the whole living cell, are examined in the microscope. Because very thin sections of cells are used in electron microscopy, the final image represents only a very small portion of the whole cell, and the total three-dimensional image of a cell and its subcellular structures must be reconstructed from a number of sections taken through different portions of the cell and at different angles. When examining an electron micrograph, one must always remember that many structures which appear independent may actually be connected together through portions of the cell not included in the thin section. Thus, a section through a ball of string would appear as a series of lines and dots even though it was originally one continuous piece of string.

Cell organelles

Figure 2-1 is a diagrammatic view of a cell, its organelles and its association with adjacent cells, and Fig. 2-2 is an actual electron micrograph of a portion of a cell. Biologists once referred to the material found within a cell as *protoplasm* and studied it as if it were a substance having specific properties of its own and within which resided that most illusive quality—life. As techniques were developed for examining smaller and smaller portions of the cell, the concept of a cell as a semi-homogeneous protoplasmic matrix, much like a bag of thick soup, slowly disappeared, and the diversity of structure and function within a single cell became apparent. The interior of the cell, far from being a semiuniform medium, is structurally divided into a number of compartments, known as *cell organelles* (little organs), which have different chemical compositions and carry out specific functions. The combined interactions of these organelles contribute to the total quality we call life.

The outer surface of a cell is covered by a very thin structure known as the *plasma membrane.* It is this membrane that separates the intracellular fluid within a cell from the extracellular fluid (the internal environment of the body). The plasma membrane, acting as a selective barrier, regulates the flow of molecules into and out of a cell. This membrane, as well as the membranes which surround the internal cell organelles, is very flexible, much like a piece of cloth; it can be bent and folded, but it cannot be stretched without being torn. The plasma membrane is often invaginated into the cell, forming clefts, or is extended from the cell surface in fingerlike projections (Fig. 2-1).

The interior of a cell is composed of two large regions: (1) the *nucleus,* a spherical or oval body generally located near the center of the cell; (2) the *cytoplasm,* the region located outside the nucleus. Most of the cell organelles are in the cytoplasm, suspended in the intracellular fluid known as the *cytosol.*

The nucleus (Fig. 2-1) is the largest structure in the cell. It is surrounded by a barrier, the *nuclear envelope,* which consists of two membranes separated by a small space. At regular intervals along the surface of the nuclear envelope, the two nuclear membranes become joined, forming the rims of circular openings known as *nuclear pores.* These pores provide access to and from the nucleus for large molecules which cannot cross the other portions of the nuclear envelope. The most prominent structure within the nucleus is the densely staining *nucleolus,* a highly coiled filamentous structure associated with numerous granules, but

Figure 2-1. Diagram of the structures and organelles found in most cells of the body.

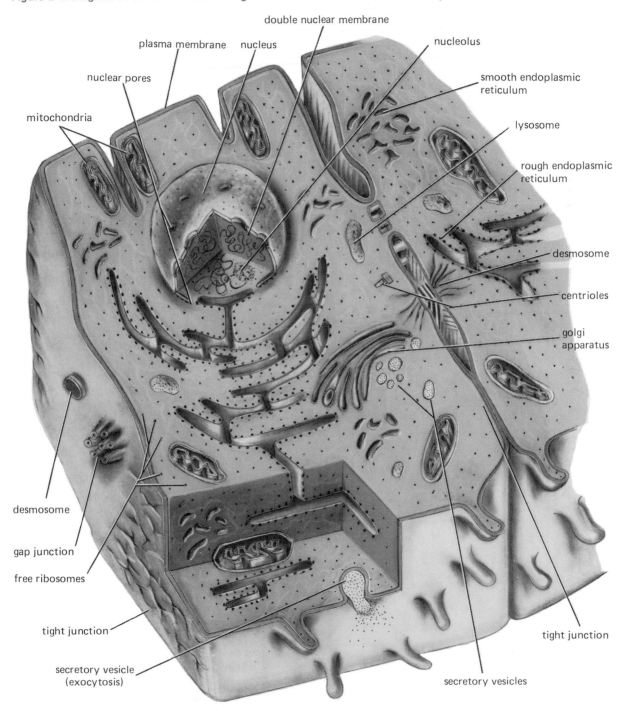

double nuclear membrane

plasma membrane nucleus nucleolus

nuclear pores smooth endoplasmic reticulum

mitochondria lysosome

rough endoplasmic reticulum

desmosome

centrioles

golgi apparatus

desmosome

gap junction

free ribosomes

tight junction

tight junction

secretory vesicle (exocytosis)

secretory vesicles

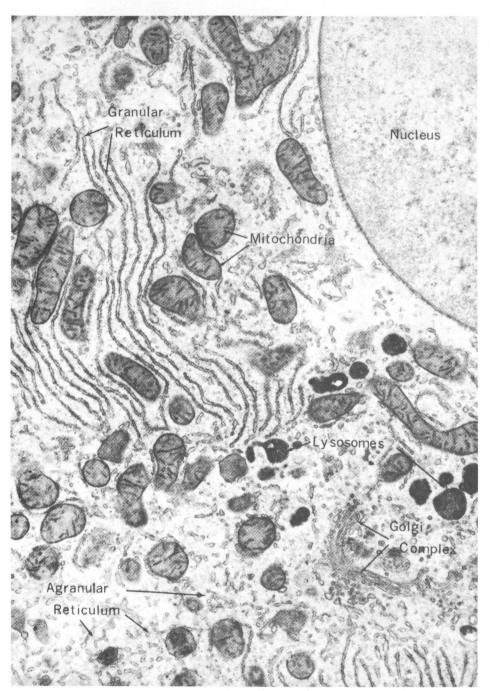

Figure 2-2. Electron micro-graph of a thin section through a portion of a rat liver cell. [*From K. R. Porter, in T. W. Goodwin and O. Lindberg* (eds.), *"Biological Structure and Function," vol. 1, Academic Press, Inc., New York, 1961.*]

not surrounded by a membrane. Fibrous threads, known as *chromatin*, are distributed throughout the remainder of the nucleus. These chromatin threads, which carry genetic information from parent cell to daughter cell and from parent to offspring, are coiled to a greater or lesser degree, producing a variation in the granular density of the nucleus.

All human cells contain a single nucleus during some stage of their life cycle. However, a few specialized types of cells, such as the red blood cell, lose their nucleus during the process of cell differentiation, while other cells formed by the fusion of many single-nucleated cells come to have more than one nucleus, as is the case with skeletal muscle cells.

The most extensive cytoplasmic cell organelle is the system of membranes which forms the *endoplasmic reticulum*. It consists of two opposing membranes separated by a small space. These membranes form a series of relatively flat sheets that are distributed throughout the cytoplasm and interconnect with each other. The small space between the membranes of the reticulum is continuous throughout this membranous network and with the space located between the two nuclear membranes (Fig. 2-1). Two types of endoplasmic reticulum can be distinguished: *granular* (rough-surfaced) and *agranular* (smooth-surfaced) *endoplasmic reticulum*. The granular endoplasmic reticulum has small particles, *ribosomes*, embedded in its membrane surface. Other ribosomal particles, *free ribosomes*, occur in the cytoplasm as free particles not attached to membranes. The agranular endoplasmic reticulum has no ribosomal particles on its surface and is more fragmented in appearance, being less likely to exist as extended sheets of membranes. Both granular and agranular endoplasmic reticulum can exist in the same cell and appear to be continuous with each other, but the relative amounts of the two types vary in different cells and even within the same cell with changes in cell activity.

The ribosomes are the sites at which proteins are synthesized. When the ribosomes are attached to the endoplasmic reticulum, the synthesized protein is released into the lumen (space) between the reticulum membranes, from which point it eventually is released to the exterior of the cell in the process of protein secretion. Proteins synthesized on free ribosomes are released into the cytosol for internal use by the cell. Agranular endoplasmic reticulum is involved in the synthesis of cell membranes and has other functions to be described in subsequent chapters.

The *Golgi apparatus*, named in honor of Camillio Golgi, who first described this cell organelle, consists of a series of closely opposed, flattened membranous sacs which are slightly curved, forming a cup-shaped structure. Associated with this organelle, particularly near its concave surface, are a number of membrane-enclosed vesicles. These vesicles, which are produced by the Golgi apparatus, contain proteins synthesized by the ribosomes attached to the endoplasmic reticulum. The Golgi apparatus concentrates these proteins in the vesicles, which then move to the periphery of the cell where they fuse with the plasma membrane and empty their contents to the outside of the cell during secretion of protein (Fig. 2-1). The vesicles are referred to either as *secretory granules* (when they contain densely staining material) or as *secretory vesicles* (when enclosing less dense material). Most cells have a single Golgi apparatus located near the nucleus, although some cells may have several.

The *mitochondria* (Greek *mitos*, thread; *chondros*, granule) are the cell organelles which produce most of the form of chemical energy used by cells. In doing so, they consume molecular oxygen and produce carbon dioxide. These organelles, which are usually rod- or oval-shaped (Fig. 2-3), are surrounded by two membranes, an inner and an outer. The outer membrane is smooth, whereas the inner membrane is folded into sheets or tubules, known as *cristae*, which extend into the internal space (*matrix*) of the mitochondrion. The mitochondria are found scattered throughout the cytoplasm. Large numbers of them are present in cells that use large amounts of energy; for example, a single liver cell may contain 1000 mitochondria. Less active cells contain fewer mitochondria.

Lysosomes are small, spherical or oval bodies, surrounded by a single membrane which encloses a densely staining, granular matrix. They often cannot be distinguished visually from the secretory vesicles derived from the Golgi apparatus and are, in fact, probably formed in the Golgi region of the cell. In contrast to the secretory vesicles which export material from the cell, the lysosomes function intracellularly to break down various complex structures, such as bacteria and cellular debris, that have been engulfed by the cell. They may also break down other intracellular organelles which have been damaged and are no longer functioning normally. The lysosomes are thus a highly specialized intracellular digestive system.

The cytoplasmic organelles we have described thus far—the endoplasmic reticulum, Golgi apparatus, secretory vesicles, mitochondria, and lysosomes—all have one structural element in common: they are surrounded by membranes. Membranes are a major structural element in cells, and we will discuss their structure and function in a later chapter.

Table 2-1. Cell structures*

Structure	Number per cell	Structural organization	Function
PLASMA MEMBRANE	1 cell surface	7.5 nm thick. Composed of a phospholipid bilayer and protein.	Selective barrier to movement of ions and molecules into and out of cell.
NUCLEUS			
Nuclear envelope	1 surrounds nucleus	Two opposed membranes separating small space. Nuclear pores, 50–70 nm diam.	Barrier to movement of most molecules. Messenger RNA passes to cytoplasm through pores.
Chromatin	46 strands per human cell nucleus	Coiled threads composed of DNA and protein.	DNA stores genetic information determining amino acid sequence of cell proteins. Chromatin condenses into chromosomes at time of cell division.
Nucleolus	Usually 1	Coiled filamentous structure associated with granules. Not surrounded by membrane.	Site of ribosomal RNA synthesis.
CYTOPLASM			
A. Membrane-bound cell organelles			
Endoplasmic reticulum (ER)	1 interconnected cell organelle	Two opposed membranes separating a space continuous throughout organelle and interconnecting with space of nuclear envelope.	
Granular ER		Ribosomal particles bound to ER membrane.	Synthesis of proteins to be secreted from cell.
Agranular ER		Smooth membrane; no ribosomes.	Fatty acid and steroid synthesis. Calcium storage and release in muscle cells.
Golgi apparatus	Usually 1 located near nucleus	Cup-shaped series of closely opposed membranous sacs and vesicles.	Concentration and modification of protein prior to secretion.
Secretory vesicles (granules)	Many	Membrane-bound sacs containing concentrated solution of proteins.	Protein secretion.
Mitochondria	Many	Rod- or oval-shaped bodies surrounded by two membranes. Inner membrane folds into inner matrix forming cristae.	Major site of ATP production, oxygen utilization, and CO_2 formation. Contains enzymes of Krebs cycle and oxidative phosphorylation.
Lysosomes	Several	Densely staining oval body surrounded by membrane, containing hydrolytic enzymes.	Digestive organelle, specialized for breakdown of engulfed bacteria and damaged cell organelles.

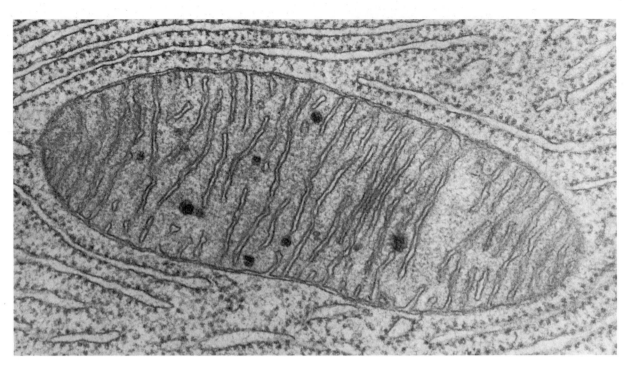

Figure 2-3. Electron micrograph of a mitochondrion surrounded by an extensive array of granular endoplasmic reticulum. *(Micrograph courtesy of Dr. K. R. Porter.)*

Table 2-1. *(Continued)*

Structure	Number per cell	Structural organization	Function
B. Nonmembranous structures			
Ribosomes	Many	20-nm-diameter particles composed of RNA and protein.	Site at which amino acids are assembled into proteins.
Free ribosomes		Not bound to membranes.	Site at which proteins to be used intracellularly are assembled.
Bound ribosomes		Bound to membranes of granular endoplasmic reticulum.	Site at which proteins to be secreted from cell are assembled.
Filaments	Many	5–15-nm-diameter protein threads of variable length.	Cell movements, especially in muscle cells. Also used to provide structural support at cell junctions.
Microtubules	Many	25-nm-diameter protein tubules with 15-nm-diameter hollow core.	Maintenance of cell shape (cytoskeleton). Associated with movements of cilia, flagella, and mitotic spindle.
Centrioles	2 located near nucleus	Two small cylindrical bodies composed of nine sets of three fused microtubules.	Formation of spindle apparatus at poles of cell during cell division. Also associated with formation and movement of cilia.
Granules	Few to many	Aggregates or crystals of chemical substances.	Storage of specialized end products of metabolism. Glycogen granules most common.
Fat droplets	Few	Spherical globule of triacylglycerol.	Storage of fat.

* Much of the information in this table will be described in later chapters.

In addition to these membranous organelles, several other structures are found in the cytoplasm. Most cells contain rodlike *filaments* and hollow *microtubules.* The microtubules appear to be more rigid than the filaments, and they provide structural support, a cytoskeleton, to which other organelles may be attached. These microtubules are often found along the inner surface of the plasma membrane where they appear to be involved in maintaining and changing the shape of the cell surface. Filaments and microtubules have both been implicated in cell processes which produce movement, whether it be movement of whole cells, the movement of organelles within cells, or the specialized movements of cell division. The *centrioles* are two small cylindrical bodies composed of nine sets of fused microtubules. They are generally located near the nucleus and participate in nuclear and cell division. *Granules,* which are aggregates or crystals of various chemical substances, and *fat droplets* make up the remainder of the cytoplasmic cell structures.

Table 2-1 provides a reference summary of the major cell structures and their functions. Later chapters will describe the mechanisms by which they perform these functions.

3
CHEMICAL COMPOSITION OF THE BODY

Atoms and molecules
  Atoms
    Chemical elements • *Atomic number* • *Atomic weight* • *Atomic composition of the body*
  Molecules
    Chemical bonds • *Molecular shape*
  Ions
  Polar molecules
    Water

Solutions
    Molecular solubility • *Concentration*
Classes of organic molecules
  Carbohydrates
  Lipids
    Triacylglycerols • *Phospholipids* • *Steroids*
  Proteins
  Nucleic acids
    DNA • *RNA*

Atoms and molecules are the chemical units of cell structure and function, just as cells are the units of multicellular organisms. In this chapter we describe the structures of the major classes of chemicals in the body and certain characteristics of atoms which underlie molecular structure in general. The specific functions and reactions of the various classes of molecules will be discussed in subsequent chapters. Thus, this chapter is, in essence, an expanded glossary of chemical terms and structures, and like a glossary, it should be consulted according to need.

Atoms and molecules

Atoms

Chemical elements Atoms are the smallest units of matter which have chemical characteristics. Over 100 different types of atoms have been identified, and all atoms of the same type are known as a specific *chemical element* (e.g., carbon or oxygen). Each atom is approximately spherical,

the smallest (hydrogen) being about 0.1 nm in diameter. Atoms are composed of even smaller units of matter, the *subatomic particles—electrons, protons,* and *neutrons.*

The subatomic particles differ in their mass, electric charge, and location in the atom. Neutrons and protons have approximately equal masses and are located in the *atomic nucleus* (Fig. 3-1), a very small volume at the atom's center. Electrons are much lighter particles, and revolve in orbits around the nucleus, much as the planets revolve around the sun. This miniature solar system model of an atom is an oversimplification, but it is sufficient to provide a conceptual framework for understanding the chemical properties of atoms.

Some of the most important properties of atoms depend upon the electric charge on the subatomic particles. Protons possess a positive electric charge, while electrons carry a negative electric charge, and neutrons have no electric charge. The magnitude of the electric charge on a proton and electron is the same, only the sign (positive or negative) is different. Since the total number of electrons in an atom is equal to the number of protons, an atom has no net electric charge.

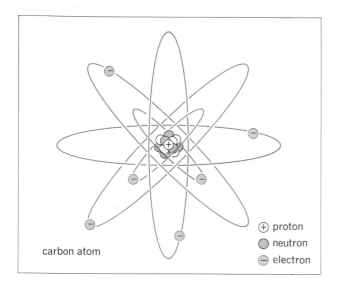

Figure 3-1. Atomic structure. Six electrons (negative) revolve around the nucleus of a carbon atom containing six protons (positive) and six neutrons (neutral).

Atomic number Each type of atom, i.e., chemical element, contains a characteristic number of protons which distinguishes it from other elements. This number is known as its *atomic number*. For example, hydrogen, the simplest atom, has an atomic number of 1, corresponding to its single proton (Table 3-1), whereas calcium has an atomic number of 20, corresponding to its 20 protons (since an atom is electrically neutral, the atomic number is also equal to the number of electrons surrounding the nucleus).

Atomic weight Since each of an atom's protons and neutrons has a mass that is about 2000 times the mass of an electron, the mass of an atom is approximately proportional to the number of protons and neutrons it contains. This sum is known as its *atomic weight.* Hydrogen, with an atomic weight of 1, is the lightest of all atoms, containing only a single proton in its nucleus. A carbon atom, atomic weight 12, is 12 times heavier than a hydrogen atom since its nucleus contains 12 heavy subatomic particles, 6 protons and 6 neutrons. Note that the atomic weight is actually a relative weight which indicates how much heavier a particular type of atom is than a hydrogen atom.

Atomic composition of the body Of the 100 or so known chemical elements, only 24 appear to be essential for maintaining the structure and function of the body (Table 3-2). In fact, just four of these elements—hydrogen, oxygen, carbon, and nitrogen—account for about 99 percent of the atoms in the body. Approximately 63 percent of the atoms are hydrogen, 26 percent oxygen, 9 percent carbon, and 1 percent nitrogen.

Most of the seven elements listed in Table 3-2 as *major minerals* are present as electrically charged atoms, known as *ions* (see below). These elements play important roles in the electrical properties of cells and in the transfer and utilization of chemical energy. Some of them also form a major part of the extracellular matrix formed by connective tissues, including the solid matrix of calcium and phosphorus atoms in bone tissue.

The 13 *trace elements* are present in extremely small quantities, but they are nonetheless essen-

Table 3-1. Atomic composition of the most abundant elements in the body

Element	Atomic number	Number of electrons	Number of protons	Number of neutrons	Atomic weight
Hydrogen	1	1	1	0	1
Carbon	6	6	6	6	12
Nitrogen	7	7	7	7	14
Oxygen	8	8	8	8	16
Sodium	11	11	11	12	23
Magnesium	12	12	12	12	24
Phosphorus	15	15	15	16	31
Sulfur	16	16	16	16	32
Chlorine	17	17	17	18	35
Potassium	19	19	19	20	39
Calcium	20	20	20	20	40

Table 3-2. Essential elements in the body

Symbol	Element
1 Major elements: 99.3% total atoms	
H	Hydrogen
O	Oxygen
C	Carbon
N	Nitrogen
2 Major minerals: 0.7% total atoms	
Ca	Calcium
P	Phosphorus
K (Latin, *kalium*)	Potassium
S	Sulfur
Na (Latin, *natrium*)	Sodium
Cl	Chlorine
Mg	Magnesium
3 Trace elements: less than 0.01% total atoms	
Fe (Latin, *ferrum*)	Iron
I	Iodine
Cu (Latin, *cuprum*)	Copper
Zn	Zinc
Mn	Manganese
Co	Cobalt
Cr	Chromium
Se	Selenium
Mo	Molybdenum
F	Fluorine
Sn (Latin, *stannum*)	Tin
Si	Silicon
V	Vanadium

tial for normal growth and function, e.g., the role of iodine in the functioning of the thyroid hormone and of iron in the transport of oxygen in the blood. It is likely that additional trace elements will be added to this list as the chemistry of the body becomes better understood.

Many other elements can be detected in the body in addition to the 24 elements listed in Table 3-2. These elements enter through the foods we eat and the air we breathe but do not have any known essential chemical function. Some elements, e.g., lead and mercury, not only are not required for normal function but can be extremely toxic.

Molecules

Molecules are formed by linking atoms together. The formula of a chemical compound (molecule) indicates the number and types of atoms it contains. Thus, the formula for a molecule of water, which contains two hydrogen atoms and one

oxygen, is H_2O, that for carbon dioxide is CO_2, and that for glucose, one ° the sugars found in the body, is $C_6H_{12}O_6$, indicating that it contains six carbon atoms, twelve hydrogen atoms, and six oxygen atoms. Such a formula, however, does not indicate which atoms are linked to which in the molecule.

Chemical bonds The atoms in molecules are linked together by chemical bonds, formed as a result of interactions between their electrons. One type of bond, a *covalent bond,* is formed between two atoms when each atom shares one of its electrons with the other atom. (An explanation of the physical forces involved in this interaction is beyond the scope of this book.) Since different elements have different numbers of electrons, they show differences in their ability to form chemical bonds. Some elements can form more than one chemical bond and thus become linked simultaneously to two or more atoms. Each element has a characteristic number of chemical bonds it can form (chemists refer to this number as the *valence* of an atom); the number of chemical bonds formed by the four major elements are: hydrogen, 1; oxygen, 2; nitrogen, 3; and carbon, 4.

When diagraming the structure of a molecule, each chemical bond is represented by a line. The chemical bonds of the four major elements can thus be represented as

$$\text{H—} \quad \text{—O—} \quad \text{—N—} \quad \text{—C—}$$

A molecule of water containing two hydrogen atoms linked to a single oxygen atom can be illustrated as

$$\text{H—O—H}$$

In some cases, two chemical bonds—a *double bond*—are formed between the same atoms; an example of this is carbon dioxide (CO_2):

$$\text{O=C=O}$$

Note that in this molecule, the carbon atom still forms four chemical bonds and each oxygen atom only two.

Molecular shape Although individual atoms are spherical, molecules have a variety of shapes (Fig. 3-2). When more than one chemical bond is formed

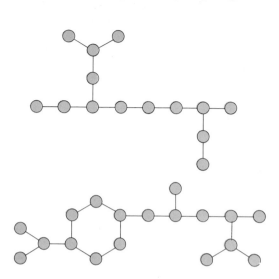

Figure 3-2. Combination of spherical atoms to form molecules of various shapes.

with a given atom, the bonds are distributed about the atom in a pattern which may or may not be symmetrical (Fig. 3-3). For example, the four bonds of a carbon atom occur at the four corners of a tetrahedron with the carbon atom at the center. The three bonds of a nitrogen atom form the three legs of a tripod and the two bonds of an oxygen atom form an angle of 105°.

Molecules are not necessarily rigid, inflexible structures. Within certain limits, the shape of a molecule can be changed without breaking the chemical bonds linking its atoms together. A chemical bond is like an axle, around which the atoms so joined can rotate. As illustrated in Fig. 3-4, a sequence of six carbon atoms can assume a number of different shapes as a result of rotations around various chemical bonds in the sequence. Such a sequence of atoms resembles a flexible piece of string more than a rigid rod. In contrast, cross-linking of various atoms in a molecule may cause its shape to become fixed in one conformation. Thus, a relatively inflexible ring of six carbon atoms is formed when the two atoms at the ends of the chain in Fig. 3-4 are linked by a chem-

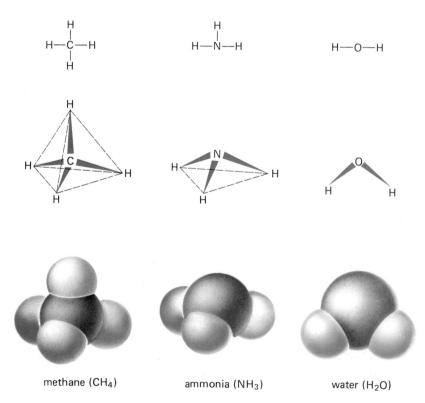

methane (CH₄)   ammonia (NH₃)   water (H₂O)

Figure 3-3. Geometrical configuration of chemical bonds around the carbon, nitrogen, and oxygen atoms.

Ions

Atoms or molecules which contain an unequal number of electrons and protons are electrically charged and are called *ions*. An ion is positively charged if it contains fewer electrons than protons; conversely, it is negatively charged if it contains an excess of electrons. The separation of a substance in solution into positive and negative ions is called *ionization*. For example, when NaCl (common table salt) dissolves in water, it separates into sodium ions (Na^+) and chloride ions (Cl^-). The ions formed by other minerals are listed in Table 3-3.

Minerals are not the only substances that can form ions. Many types of molecules can form ions by losing or gaining a proton, i.e., a hydrogen ion (H^+). For example, a common grouping of atoms in molecules is the *carboxyl group* (R–COOH, where R signifies the remaining portion of the molecule); ionization of this group occurs when the oxygen atom linked to the hydrogen captures the latter's electron, leading to the formation of H^+ and R–COO$^-$ (hydrogen ions and carboxyl ions) (Fig. 3-5). Another grouping, the *amino group* (R–NH$_2$) can combine with a hydrogen ion to become an amino ion (R–NH$_3^+$), Thus, portions of molecules which contain carboxyl or amino groups can acquire a net electric charge. Note that, in both cases, hydrogen ions were involved. The hydrogen ion plays a central role in the chemistry of the body as a result of its participation in ion formation.

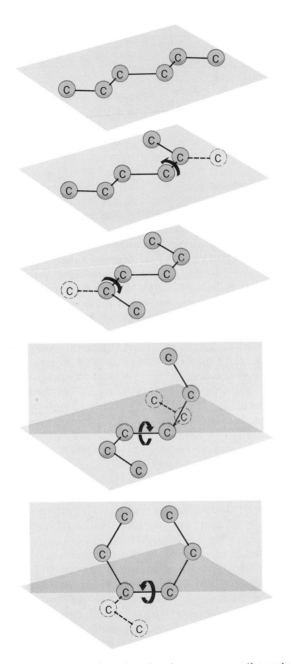

Figure 3-4. Changes in molecular shape occur as the molecule rotates around different carbon-to-carbon bonds, transforming it from a relatively straight chain into a ring.

ical bond. As we shall see, the three-dimensional shape of various molecules in the body is critical; cells can selectively react with molecules which have one specific shape but not with others.

Figure 3-5. Formation of ions from carboxyl and amino groups. (R represents the remainder of the molecule.)

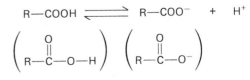

Table 3-3. Ions formed from mineral elements

Atom		Ion		Electrons gained or lost
Sodium	Na	Sodium ion	Na^+	1 lost
Potassium	K	Potassium ion	K^+	1 lost
Chlorine	Cl	Chloride ion	Cl^-	1 gained
Magnesium	Mg	Magnesium ion	Mg^{2+}	2 lost
Calcium	Ca	Calcium ion	Ca^{2+}	2 lost

The *acidity* of a solution is a measure of the concentration of hydrogen ions present; the higher the hydrogen ion concentration, the greater is the acidity. Molecules which release hydrogen ions are known as *acids*. For example, acetic acid (CH_3COOH) is an acid since ionization of its carboxyl group releases a hydrogen ion (and an acetate ion, CH_3COO^-). The ionized amino group, $R–NH_3^+$, is also an acid, since it can release a hydrogen ion (to form the neutral $R–NH_2$).

Figure 3-6. Distribution of electrons in nonpolar and polarized chemical bonds linking two atoms.

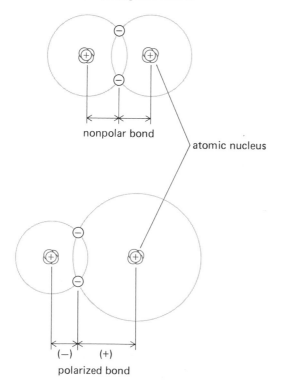

nonpolar bond

atomic nucleus

(–)   (+)

polarized bond

Polar molecules

When the electrons of two atoms interact, the result may vary from one extreme in which the two atoms equally share electrons (an electrically neutral covalent bond) to the opposite extreme, in which one of the atoms completely captures an electron from the other atom (ion formation). Between these two extremes there exist chemical bonds which are still covalent, but in which the shared electrons are located closer to one of the two atoms than the other; these are known as *polarized covalent bonds* (Fig. 3-6). For example, the bond between hydrogen and oxygen in a *hydroxyl group* (R–OH) is a polarized covalent bond in which the shared electrons reside closer to the oxygen atom than to the hydrogen atom. This causes the oxygen to become slightly negative and the hydrogen to be slightly positive. The electric charge associated with a polarized bond is considerably less than the charge on a fully ionized atom (for example, the oxygen in the polarized hydroxyl group has only about 13 percent of the negative charge associated with the oxygen in an ionized carboxyl group, $R–COO^-$).

Atoms of nitrogen and oxygen tend to form polar bonds with hydrogen atoms, whereas the bonds between carbon and hydrogen, or between two carbon atoms, are electrically neutral. Thus, molecules may have regions which are electrically neutral, polarized, or ionized. Molecules containing significant numbers of polar bonds are known as *polar molecules*, whereas molecules composed predominantly of electrically neutral chemical bonds are known as *nonpolar molecules*.

Water Whereas hydrogen is the most numerous atom in the body, water is the most numerous molecule; 99 out of every 100 molecules are water molecules. The bonds linking the two hydrogen atoms to oxygen are polarized bonds; therefore, the oxygen in water is slightly negative and each of the two hydrogen atoms has a slightly positive charge (Fig. 3-7). The positively polarized regions near the hydrogen atoms of one water molecule are electrically attracted to the negatively polarized regions of the oxygen atoms in adjacent water molecules (Fig. 3-7); thus, water molecules are held together by electrostatic bonds known as *hydrogen bonds.* A hydrogen bond is a very weak link, having only about 4 percent of the strength of the covalent bonds linking hydrogen and oxygen within a single water molecule. Therefore, clusters of hydrogen-bonded water molecules are continuously formed and broken, accounting for the fact that water is in a liquid state over a wide range of temperatures. As the temperature is lowered, larger and larger clusters of water molecules are formed until the water ultimately freezes into a continuous crystalline matrix—ice.

Solutions

Substances dissolved in a liquid are known as *solutes,* and the liquid in which they are dissolved is the *solvent.* Water is the solvent for the mineral ions and most of the organic molecules found in the body.

Molecular solubility In order for a substance to dissolve in water it must interact electrically with water molecules. For example, table salt (NaCl) is a solid crystalline substance because of the strong electric attraction between the positive sodium ions and the negative chloride ions, but when a crystal of sodium chloride is placed in water, the polarized water molecules are attracted to the charged sodium and chloride ions (Fig. 3-8); the ions become surrounded by clusters of water molecules which decrease the electric force of attraction between them, allowing them to separate into free sodium and chloride ions and go into solution, i.e., dissolve.

Molecules dissolve in water provided they contain a sufficient number of polarized covalent bonds or ionized groups electrically attracted to the polar water molecules. Thus, the presence of carboxyl, amino, or hydroxyl groups in a molecule promotes solubility in water. In contrast, molecules composed predominantly of carbon and

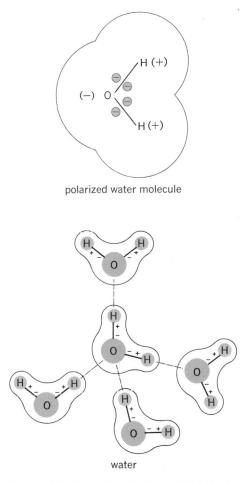

polarized water molecule

water

Figure 3-7. (A) Polarized chemical bonds link the hydrogen and oxygen atoms in a water molecule. (B) Hydrogen bonds between adjacent polarized water molecules forming the liquid state of water.

hydrogen are insoluble in water since their electrically neutral covalent bonds are not attracted to the polar water molecules.

Water is not the only possible solvent; e.g., oil and gasoline are solvents composed of molecules containing predominantly hydrogen and carbon atoms. Such nonpolar solvents will dissolve nonpolar solutes, whereas polar molecules and ions are insoluble in them. These examples of solute solubility in various types of liquids illustrate the general rule that "like dissolves like"; polar molecules dissolve in polar solvents and nonpolar molecules dissolve in nonpolar solvents.

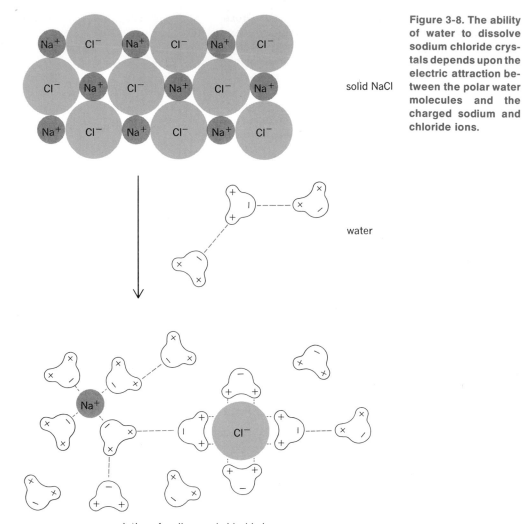

solid NaCl

water

solution of sodium and chloride ions

Some molecules contain polar or ionized groups at one end of the molecule and a nonpolar region at the opposite end; such molecules are called *amphipathic*. When mixed with water, amphipathic molecules form spherical clusters known as *micelles* in which the polar regions of the molecules are located at the surface of the micelle where they are attracted to the surrounding water molecules, and the nonpolar ends are oriented toward the center of the micelle (Fig. 3-9). As we shall see, this orientation of amphipathic molecules in aqueous environments plays an important role in the structure of cell membranes and the transport of nonpolar molecules in the blood.

Concentration *Concentration* is defined as the amount of material present per unit of volume. The standard unit of volume in the metric system is a *liter* (L). Smaller units are the milliliter (0.001 liter), or the microliter (0.001 milliliter) (see Appendix I). One way of expressing the amount of a chemical compound is by its mass in grams; its concentration then is expressed as the number of grams of compound present in a liter of solution.

However, a comparison of the concentrations of two different types of compounds on the basis of number of grams per liter of solution does not directly indicate how many molecules of each compound are present. For example, 10 g of compound X whose molecules are heavier than those of compound Y, will contain fewer molecules than 10 g of compound Y.

In contrast, a *mole* is the unit specifying the amount of a substance based on both the weight of the molecule and the number of molecules present. A mole is defined as the weight of a compound in grams equal to its molecular weight. Therefore:

$$\text{Number of moles} = \frac{\text{weight in grams}}{\text{molecular weight}}$$

The molecular weight is equal to the sum of the atomic weights of all the atoms in the molecule. For example, glucose ($C_6H_{12}O_6$) has a molecular weight of 180 (6-C $\times$ 12 + 12-H $\times$ 1 + 6-O $\times$ 16 = 180); thus 180 grams of glucose equals one mole of glucose. Water (H_2O) has a molecular weight of 18 (2-H $\times$ 1 + 1-O $\times$ 16 = 18); therefore, 18 grams of water is equal to one mole of water. Note that since one molecule of glucose is 10 times heavier than one molecule of water, 180 grams of glucose contain the same number of molecules as 18 grams of water. To generalize, one mole of any substance contains the same number of molecules; the actual number of molecules in one mole is 6×10^{23}. Thus, 1 liter of solution containing 180 grams of glucose will have a glucose concentration of one mole per liter and is called a 1 *molar* glucose solution. Such a solution will contain the same number of solute molecules as a 1 molar solution of any other type of solute.

Classes of organic molecules

Organic molecules denote molecules which contain carbon. Until the nineteenth century chemists had made little progress in understanding the chemical properties of carbon. It was known that molecules containing carbon were found primarily in living organisms, whereas nonliving matter was composed mostly of mineral elements. Mineral elements readily undergo chemical reactions with each other to form simple molecular structures consisting of relatively few atoms, but the carbon compounds present in living organisms were found to have very complex structures, containing many atoms (often thousands). These carbon compounds did not appear to undergo chemical reac-

tions as readily as the mineral elements, and it was thought that they were somehow basically different from the compounds found in the nonliving world and that they could be synthesized within living cells only under the influence of some mystical "vital force." In the early nineteenth century chemists finally succeeded in synthesizing carbon-containing molecules in a test tube in the absence of any living matter, and were able to show that these molecules obey the same basic chemical laws as do the mineral elements. No vital forces unique to living organisms were necessary to explain their chemical properties.

Because most of the naturally occurring carbon molecules are found in living organisms, the study of the chemistry of carbon compounds became known as *organic chemistry*, the chemistry of noncarbon molecules being known as *inorganic chemistry*. Once chemists were able to manipulate

Figure 3-9. Micellar organization of amphipathic molecules dissolved in water.

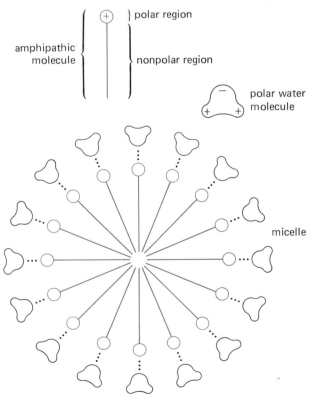

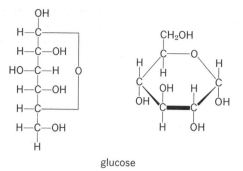

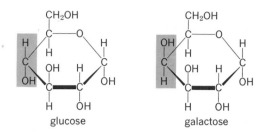

Figure 3-10. Two ways of diagraming the structure of the carbohydrate glucose.

Figure 3-11. The difference between the monosaccharides glucose and galactose depends upon the position of the hydroxyl group on the fourth carbon atom.

the reactions of carbon molecules in a test tube, the science of organic chemistry expanded rapidly. Although organic chemistry grew out of the study of molecules found in living organisms, the chemistry of living organisms, *biochemistry*, now forms only a portion of the broad field of organic chemistry.

One property of the carbon atom which makes life possible, with its variety of structures and functions, is the ability of carbon atoms to form four separate chemical bonds with other atoms, in particular with other carbon atoms. There is almost no limit to the number of carbon atoms that can be combined in this way. Since carbon atoms also combine with other atoms such as hydrogen, oxygen, nitrogen, and sulfur, the variety of molecular structures that can be formed with relatively few chemical elements is considerable. Moreover, very large organic molecules can be formed by linking together smaller molecular subunits to form long chains known as *polymers* (many

small parts). Most of the thousands of organic molecules in the body can be classified into four groups: *carbohydrates, lipids, proteins,* and *nucleic acids* (Table 3-4).

Carbohydrates

Although carbohydrates account for only 3 percent of the organic matter in the body, they play a central role in the chemical reactions which provide cells with energy. Carbohydrates are composed of atoms of carbon, hydrogen, and oxygen in the proportions represented by the general formula, $C_n(H_2O)_n$, where n is any whole number. The hydrogen and oxygen atoms equivalent to one molecule of water are linked to most of the carbon atoms according to the structural formula

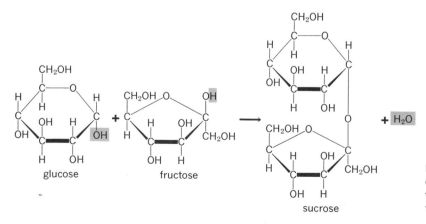

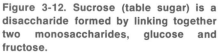

Figure 3-12. Sucrose (table sugar) is a disaccharide formed by linking together two monosaccharides, glucose and fructose.

The presence of the polar hydroxyl groups makes carbohydrates polar molecules; thus, they are readily soluble in water.

The simplest carbohydrates are the *monosaccharides* (Greek *sakcharon*, sugar), the most important of which is *glucose*, a 6-carbon molecule ($C_6H_{12}O_6$), often called "blood sugar." Two ways of representing the structure of glucose are shown in Fig. 3-10; the first is the conventional representation for organic molecules, but the second gives a better idea of the three-dimensional structure of glucose. Five carbon atoms are linked with an oxygen atom in a ring which lies in an essentially flat plane with the hydrogen and the hydroxyl groups lying above and below it. If one of the hydroxyl groups below the ring is shifted to a position above the ring, as shown in Fig. 3-11, a different mono-

saccharide results. Most monosaccharides in the body contain five or six carbon atoms.

Larger carbohydrate molecules can be formed by linking a number of monosaccharides together. Table sugar, *sucrose* (Fig. 3-12), is composed of two monosaccharides, glucose and fructose. Carbohydrate molecules composed of two monosaccharides are known as *disaccharides*. When many monosaccharides are linked together to form large polymers, the molecules are known as *polysaccharides*. Starch and cellulose, found in plant cells, and *glycogen*, present in animal cells (often called animal starch), are examples of polysaccharides. All three of these polysaccharides are composed of thousands of glucose molecules linked together in long chains (Fig. 3-13). The difference between them depends on how the glucose mole-

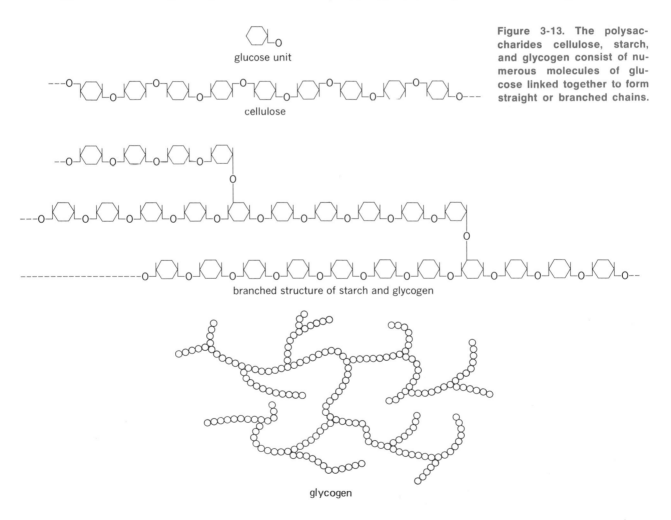

glucose unit

cellulose

branched structure of starch and glycogen

glycogen

Figure 3-13. The polysaccharides cellulose, starch, and glycogen consist of numerous molecules of glucose linked together to form straight or branched chains.

Table 3-4. Major categories of organic molecules in the body

Category	Percent body weight	Majority of atoms	Subclass	Subunits	Characteristics
Proteins	17	C, H, O, N		Amino acids	Large polymers of amino acids. Major source of nitrogen. Many functions.
Lipids	15	C, H ($-CH_2^-$)$_n$	Triacylglycerols	3 fatty acids + glycerol	Insoluble in water. Major store of reserve fuel which can be used by most cells to provide energy.
			Phospholipids	2 fatty acids + glycerol + phosphate + small charged nitrogen molecule	Molecule has a polar and nonpolar end. A major component of membrane structure.
			Steroids		Cholesterol and steroid hormones.
Carbohydrates	1	C, H, O $\left(-\overset{\displaystyle H}{\underset{\displaystyle OH}{C}}-\right)_n$	Monosaccharides (sugars)		Glucose is major fuel used by cells to provide energy.
			Polysaccharides	Sugars	Small amounts of glucose stored as glycogen.
Nucleic acids	2	C, H, O, N	DNA	Nucleotides containing the bases adenine, cytosine, guanine, *thymine*	Store genetic information.
			RNA	Nucleotides containing the bases adenine, cytosine, guanine, *uracil*	Translation of genetic information into protein synthesis.

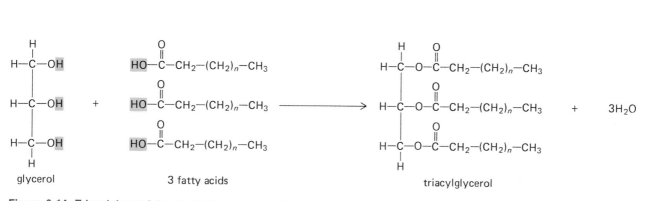

Figure 3-14. Triacylglycerol (neutral fat) consists of three fatty acids attached to the three hydroxyl groups of glycerol.

$$CH_3-CH_2-CH_2-(CH_2)_{12}-CH_2-CH_2-COOH$$
saturated fatty acid

$$CH_3-(CH_2)_3-CH_2-CH=CH-CH_2-CH=CH-CH_2-(CH_2)_5-CH_2-COOH$$
polyunsaturated fatty acid

Figure 3-15. Hydrocarbon structure of saturated and polyunsaturated fatty acids.

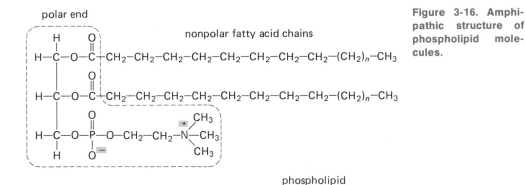

Figure 3-16. Amphi-pathic structure of phospholipid mole-cules.

phospholipid

cules are linked together and how frequently branches appear along the molecule. Cellulose is a straight-chain polysaccharide with no branches, whereas glycogen is highly branched and starch less so.

Lipids

Lipids (Greek *lipos*, fat) are molecules composed predominantly of hydrogen and carbon atoms. Since these atoms are linked by nonpolar covalent bonds, lipid molecules are insoluble in water. It is this physical property of insolubility in polar solvents (and solubility in nonpolar solvents) which characterizes this class of organic molecules. The lipids, which account for about 40 percent of the organic matter in the body, can be divided into three subclasses: *triacylglycerols, phospholipids,* and *steroids.*

Triacylglycerol The triacylglycerols[1] or neutral fats constitute the majority of the lipids in the body, and it is these molecules which are generally referred to simply as fat. Neutral fats are composed of two types of molecules linked together—*glycerol* and *fatty acids* (Fig. 3-14). Glycerol is a 3-carbon molecule which is actually a carbohydrate. A fatty acid consists of a chain of carbon atoms, with a carboxyl group at one end (Fig. 3-15). A fatty acid is attached to each of the three hydroxyl groups of glycerol, the linkage occurring at the carboxyl end of the fatty acid. Because fatty acids are synthesized in the body by linking together 2-carbon fragments, most fatty acids have an even

Figure 3-17. Ring structure of various steroids, including the female and male sex hormones estrogen and testo-sterone.

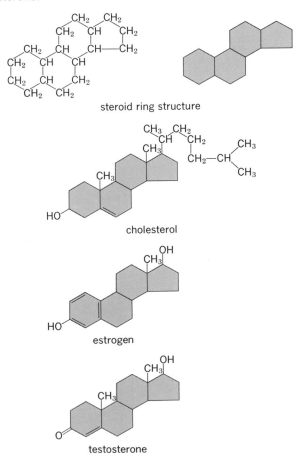

[1] These substances were called triglycerides until recently.

charge on side chain   side chain   amino acid

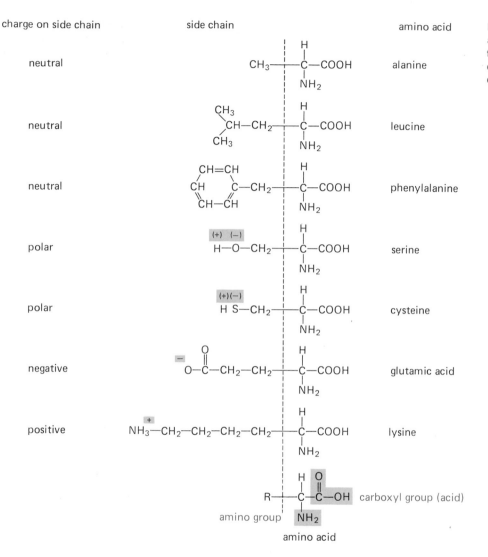

neutral CH_3——C—COOH   alanine

neutral CH_3 \\ CH—CH_2——C—COOH   leucine
                           CH_3 /

neutral   CH=CH \\ CH   C—CH_2——C—COOH   phenylalanine
                           CH—CH /

polar   (+) (−) \\ H—O—CH_2——C—COOH   serine

polar   (+)(−) \\ H S—CH_2——C—COOH   cysteine

negative   O \\ O—C—CH_2—CH_2——C—COOH   glutamic acid

positive   + \\ NH_3—CH_2—CH_2—CH_2—CH_2——C—COOH   lysine

                           R——C—C—OH   carboxyl group (acid)
                           amino group NH_2
                           amino acid

Figure 3-18. Structure of amino acids. R stands for the remainder of the molecule (the amino acid side chain).

number of carbon atoms. Sixteen- and eighteen-carbon fatty acids are the most common, although both shorter and longer fatty acids are also present. When each of the interior carbons of the fatty acid chain is linked by a single bond to the carbon atoms adjacent to it, the remaining bonds being with hydrogen atoms, the fatty acid is said to be *saturated*. Some fatty acids contain double bonds linking certain of the carbons in the chain, and these are known as *unsaturated* fatty acids. If more than one double bond is present the fatty acid is said to be *polyunsaturated* (Fig. 3-15). Animal fats generally contain a high proportion of saturated fatty acids, whereas vegetable fats contain more polyunsaturated fatty acids. Since the three fatty acids in a molecule of triacylglycerol need not be identical, a variety of neutral fats can be formed with fatty acids of different chain lengths and degrees of saturation.

Phospholipids Phospholipids are similar in overall structure to neutral fats with one important difference: the third hydroxyl group of glycerol, rather than being attached to a molecule of fatty acid, is linked to a phosphate group ($-PO_4^-$), to which in turn is usually attached a small polar or ionized nitrogen-containing molecule (Fig. 3-16). Both the phosphate and nitrogen groups are usually electrically charged, and these groups constitute a charged, polar region at one end of the phospholipid molecule, the fatty acid chains providing a nonpolar region at the opposite end. Therefore, these molecules are amphipathic and, when mixed in water, they become organized into micelles, the polar ends of the molecules associating with the polar water molecules.

Steroids Steroids have a distinctly different structure from the other two subclasses of lipid molecules. Four interconnected rings of carbon atoms form the basic skeleton of all steroids (Fig. 3-17). A few polar groups may be attached to this ring

structure, but they are not numerous enough to make the steroids water-soluble. The steroid family includes, among others, cholesterol and the male and female sex hormones, testosterone and estrogen, respectively.

Proteins

The term "protein" comes from the Greek *proteios* (of the first rank) which aptly describes their importance. These molecules, which account for about 50 percent of the organic material in the body, are components of most of the body's structures and also play critical roles in almost all

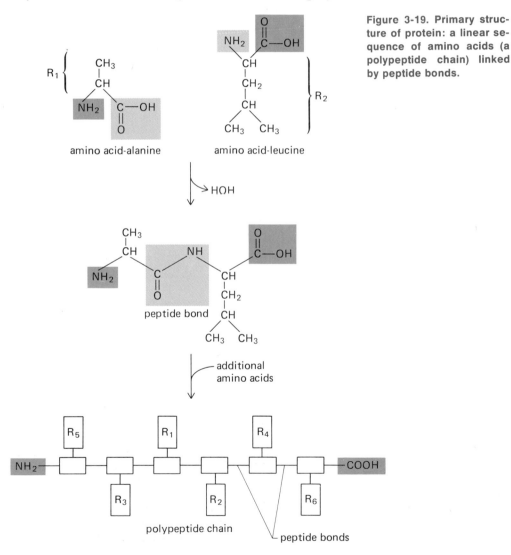

Figure 3-19. Primary structure of protein: a linear sequence of amino acids (a polypeptide chain) linked by peptide bonds.

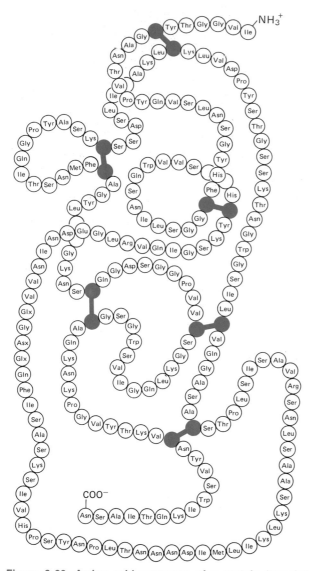

Figure 3-20. Amino acid sequence of a protein (trypsin) illustrating its three-dimensional shape that is formed by the folding of the linear sequence of amino acids (circles). The bars cross-linking various segments of the protein chain represent chemical bonds between the side chains of particular amino acids. *(Redrawn from Sidney Bernhard, "The Structure and Function of Enzymes," W. A. Benjamin, Inc., New York, 1968.)*

chemical interactions. Proteins are composed of carbon, hydrogen, oxygen, nitrogen, and small amounts of sulfur. Most proteins are very large molecules, containing thousands of atoms, and like most very large molecules, they are formed by link-

ing together a large number of small subunits. The subunit of protein structure is an *amino acid;* thus, proteins are polymers of amino acids.

The structural property all amino acids share is the presence of an amino (−NH₂) and a carboxyl (−COOH) group, both attached to the terminal carbon of the molecule (Fig. 3-18). The third bond of this terminal carbon is usually to hydrogen and the fourth to the remainder of the molecule, indicated by R. In the human body there are 20 different amino acids and 20 different chemical structures corresponding to R. Figure 3-18 illustrates only seven of them. Some of these side chain structures are nonpolar, others are polar but not ionized, and a few contain ionized groups.

Protein molecules are linear sequences of amino acids in which the carboxyl group of one amino acid is linked by a *peptide bond* to the amino group of the next amino acid in the sequence (Fig. 3-19). During the formation of this bond, a molecule of water is generated by removal of one of the hydrogen atoms on the amino group of one amino acid and the OH⁻ portion of the carboxyl group of the second amino acid. Note that when two amino acids are linked together, one end of the resulting molecule has a free amino group and the other has a free carboxyl group which can become linked to another amino acid, and so on, thereby extending the length of the molecule. The peptide bonds form the backbone of the polypeptide chain while the different R groups of the amino acids stick out to the sides of the chain. For this reason the R groups are called the amino acid side chains.

The sequence of amino acids along a polypeptide chain constitutes the *primary structure* of the protein molecule. There are two variables in this primary structure: (1) the total number of amino acids in the sequence (polypeptide chains can vary greatly in length, the very short chains being known as *peptides*); (2) the specific types of amino acids at each position along the sequence. Each position along a polypeptide chain may be occupied by any one of the 20 different amino acids. Let us consider the number of different peptides that can be formed that are only three amino acids long: Any one of 20 different amino acids may occupy the first position in the sequence, 20 the second position, and 20 the third position, for a total of $20 \times 20 \times 20 = 8000$ possible sequences! If the peptide is six amino acids in length, $20^6 = 64,000,000$ possible combinations can be formed. Thus, starting with 20 different amino acids an almost unlimited variety of polypeptide chains can be formed by altering both the amino acid sequence and the total number of amino acids in the sequence.

The primary structure of a polypeptide chain is analogous to a string of beads, each bead representing a single amino acid (Fig. 3-20). Since atoms can rotate around their chemical bonds, a polypeptide chain is a flexible structure that can be bent into a number of three-dimensional conformations, a phenomenon which provides the key to understanding the functions of proteins in the body, as we shall see in the next chapter. Three factors determine the total three-dimensional shape (so-called *tertiary structure*) of the polypeptide chain once the amino acid sequence has been formed: (1) electric attractions between various polar and ionized regions along the polypeptide chain; (2) electric attractions between these regions and surrounding water molecules; (3) covalent bonds linking the side chains of two amino acids. Let us look at each of these in turn.

An important example of the first factor is the interaction between peptide bonds. Because individual peptide bonds are polar, they interact with each other through hydrogen bonds (Fig. 3-21), and because peptide bonds are located at regular intervals along the polypeptide chain, these hydrogen bonds produce a coiling of the chain into the helical conformation known as an *alpha helix*. The peptide bonds are, of course, not the only polar or ionized sites in a protein; some of the amino acid side chains also contain polar bonds or ionized groups which are attracted to oppositely charged regions of the polypeptide chain. These electric interactions tend to disrupt portions of the regular alpha-helix coil, producing bends and irregular, random conformations.

In addition to forming bonds with each other, the various polarized and ionized regions of the molecule interact with the surrounding polar water molecules in which the protein is dissolved. Because of these forces of attraction, the polar side chains come to be located along the surfaces of the molecule (since these come into contact with water). In contrast, the nonpolar side chains cannot interact with water and become located in the

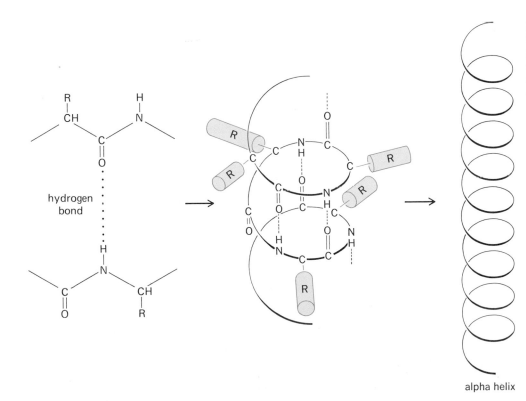

Figure 3-21. Hydrogen bonds between polarized peptide bonds produce the alpha-helical structure of polypeptide chains.

hydrogen bond

alpha helix

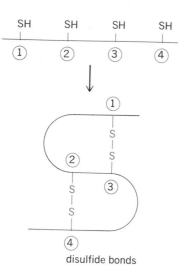

single polypeptide chain

Figure 3-22. The sulfhydryl groups of cysteine side chains form disulfide bonds cross-linking portions of a polypeptide chain.

oligomeric protein

Figure 3-24. Oligomeric protein composed of four polypeptide chains (protomers). *(Redrawn from Albert L. Lehninger, "Biochemistry," Worth Publishers, Inc., 1970.)*

Figure 3-23. Tertiary structure of a protein molecule (myoglobin) containing regions of alpha-helical and random-coil conformations. *(Redrawn from Albert L. Lehninger," Biochemistry," Worth Publishers, Inc., 1970.)*

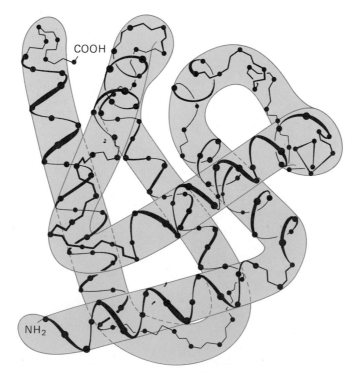

interior of the molecule, removed from contact with water.

Finally, the third factor—covalent binding between side chains—is provided by the amino acid cysteine. Because it contains a sulfhydryl group (R-SH), it can undergo a chemical reaction with another cysteine side chain to produce a disulfide bond (R-S-S-R), linking the two amino acids together (Fig. 3-22).

To recapitulate, a protein consists of a hierarchy of structure (Fig. 3-23) beginning with its linear sequence (primary structure), which is forced into regions of α-helical and random-coil configurations (secondary structure), and additional bends and folds which complete its total three-dimensional shape (tertiary structure). Moreover, yet another level of structural organization (quatenary structure) exists in oligomeric proteins (Fig. 3-24), i.e., proteins which contain more than one polypeptide chain. Each chain is known as a protomer. Protomers are sometimes held together by electrostatic interactions (an example is the hemoglobin molecule which consists of four polypeptide chains). In other cases the chains may be linked together by disulfide bonds as in the hormone insulin, which consists of two polypeptide chains.

In the next chapter we shall consider how the three-dimensional shape of proteins is related to their functions and how a cell goes about assembling these large molecules from amino acid subunits.

Nucleic acids

The nucleic acids contribute very little to the body's weight, but these molecules are the largest and most specialized of all. They are responsible for the storage of genetic information and its passage from parent to offspring and from cell to cell. It is the nucleic acids which determine whether one is a man or a mouse, or whether a cell is a muscle cell or a nerve cell. There are two types of nucleic acids: deoxyribonucleic acid (DNA) and ribonucleic acid (RNA). The DNA molecules store genetic information coded within their structures, whereas RNA molecules are involved in the decoding of this information into a form that can be utilized by a cell, specifically, into instructions for protein synthesis.

Both types of nucleic acids are polymers composed of linear sequences of repeating subunits.

Each subunit, known as a *nucleotide*, has three components: a phosphate group, a sugar, and a ring of carbon and nitrogen atoms known as a *base* (Fig. 3-25).

Figure 3-25. Structures of the four gases found in DNA nucleotides.

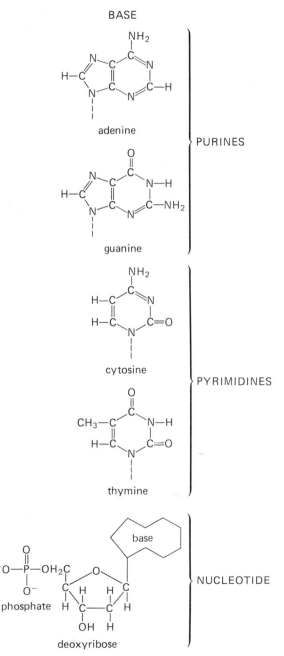

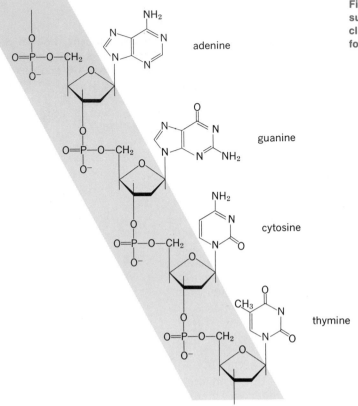

adenine

guanine

cytosine

thymine

Figure 3-26. Phosphate sugar bonds link nucleotides in sequence to form nucleic acids.

Figure 3-27. Base pairings between the two nucleotide chains form the double-helical structure of DNA.

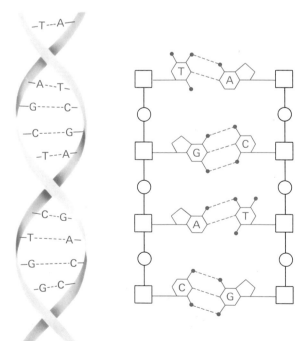

DNA The nucleotides in DNA contain the sugar *deoxyribose*, thus the name deoxyribonucleic acid. Four different nucleotides are present in DNA, corresponding to the four different bases that may be attached to deoxyribose. The four bases are divided into two classes: (1) the *purine* bases *adenine* and *guanine*, which have two rings of nitrogen and carbon atoms; (2) the *pyrimidine* bases *cytosine* and *thymine*, which have only a single ring. The phosphate group of one nucleotide is linked to the sugar of the adjacent nucleotide to form a chain with the bases sticking out to the side of the phosphate-sugar backbone (Fig. 3-26).

A DNA molecule consists of not one, but two chains of nucleotides coiled around each other in the form of a double helix (Fig. 3-27). Polar groups along the ring of a base on one chain form hydrogen bonds with those of a base on the second chain

to link the two chains of DNA together. The structure of the four bases is such that adenine (A) always pairs with thymine (T) and guanine (G) with cytosine (C). As we shall see in the next chapter, this specificity in base pairings provides the mechanism for duplicating and transfering genetic information.

RNA The general structure of RNA molecules differs in only a few respects from that of DNA: (1) RNA consists of a single, rather than a double chain of nucleotides; (2) the sugar in each nucleotide is *ribose* rather than deoxyribose; (3) the base thymine in DNA is replaced by the base *uracil* (U) (Fig. 3-28). Although RNA contains only a single chain of nucleotides, portions of this chain may bend back upon itself to undergo base pairings with other nucleotides in the same chain.

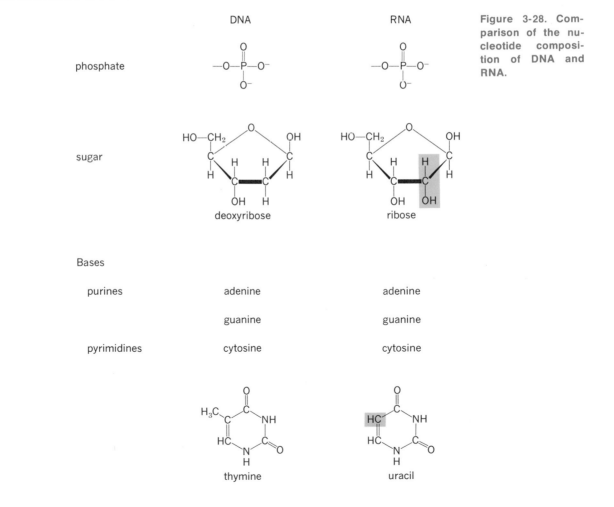

Figure 3-28. Comparison of the nucleotide composition of DNA and RNA.

4

MOLECULAR CONTROL MECHANISMS—DNA AND PROTEIN

Section A: Protein binding sites

Characteristics of protein binding sites
  Chemical specificity
  Affinity
  Saturation
  Competition
Functions of protein binding sites
  Aggregation of proteins into organized structures
  Enzymes
  Detection of chemical messengers
Regulation of binding-site characteristics
  Allosteric regulation
  Chemical alteration of
    protein structure

Section B: Genetic information and protein synthesis

Genetic information
  DNA and the genetic code
Protein synthesis
  Transcription: Formation of messenger RNA
  Translation
    Ribosomes • *Transfer RNA* • *Protein assembly*
  Regulation of protein synthesis
    Possible sites of regulation • *Induction and repression* • *Role of DNA-associated nuclear proteins*
Replication and expression of genetic information
  Replication of DNA
  Cell division
  Cell differentiation
  Mutation
  Recombinant DNA

The outstanding accomplishment of twentieth-century biology has been the elucidation of the chemical basis of heredity and its relationship to protein synthesis. Whether an organism is a man or a mouse, has blue eyes or black, has light skin or dark, is directly determined by the proteins it possesses. Moreover, within an individual organism, muscle cells differ from nerve cells or epithelial cells or any other cell types solely because they have different proteins. This is true because proteins not only function as structural elements in cells but regulate the rates at which almost all cellular chemical reactions occur.

Crucial for an understanding of protein functions is the fact that each type of protein consists of a unique sequence of amino acids, which confer upon it a unique three-dimensional shape. In the first section of this chapter we describe how these shapes underlie protein function.

So preeminent is the role of proteins in cellular function that the primary hereditary information passed from cell to cell is a set of specifications for the amino acid sequences of all proteins theoretically producible by the cell. This information is coded into the molecular structure of DNA molecules, and the middle section of this chapter describes the nature of this code and the manner in which it is passed from cell to cell.

Given that different cell types have different proteins and that the specifications for these proteins are coded into DNA molecules, one might be led to conclude that different cell types have received different sets of DNA molecules. However, such is not the case; all cells in the body (with the exception of sperm or ova) possess precisely the same sets of DNA molecules. The solution to this seeming paradox is that the DNA molecules do not transcribe all their coded instructions in all cells, nor even in a single cell at all times. The regulation of DNA function is described in the last section of this chapter.

Section A
Protein binding sites

Characteristics of protein binding sites

Proteins differ from other classes of molecules in their extraordinary ability to bind organic molecules and ions selectively. This binding may be so specific that one type of protein may bind only one type of organic molecule and no other. Such selectivity allows a protein to "identify" (by binding) the presence of one particular type of molecule in a solution containing hundreds of different kinds of molecules.

A *ligand* is any molecule or ion which is bound to the surface of another molecule, such as a protein, by forces other than covalent chemical bonds. These forces are either electrical attractions between oppositely charged, ionized (or polarized) groups on the two molecules, or very weak attractions between the electric fields surrounding electrically neutral atoms in them. On the protein surface, the chemical groups of only a small region, known as a *binding site*, interact with any given ligand, and the protein may contain different binding sites, each specific for different ligand types. The reversible binding of a ligand (L) to a protein binding site (P_b) to form a bound complex ($P_b \cdot L$), the dot indicating the absence of a covalent bond between P_b and L, is written:

$$L + P_b \rightleftharpoons P_b \cdot L$$

Chemical specificity

The force of electrical attraction between opposite charges or polarized regions of molecules decreases markedly as the distance between the charges increases. Therefore, in order for a ligand to bind to the surface of a protein, its atoms must

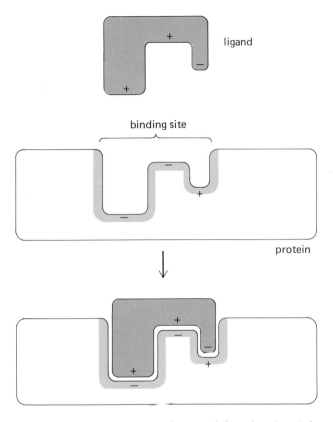

Figure 4-1. Complementary shapes of ligand and protein binding site determine the chemical specificity of binding.

be close enough to those of the protein for adequate electrical attraction to occur. This happens only when the shape of the surface of the ligand molecule is complementary to the shape of the surface of the protein binding site, such that the two molecules fit together like pieces of a jigsaw puzzle (Fig. 4-1). Therefore, a protein binding site with a specific shape can bind only those ligands having a complementary shape; thus, the *chemical specificity* of a protein binding site is determined by the shape of the protein molecule. In the last chapter we described how the three-dimensional shape of a protein molecule results from interactions ultimately ascribable to the location of the different amino acids along its polypeptide chain. Accordingly, proteins with different amino acid sequences have different shapes and, therefore, different binding sites.

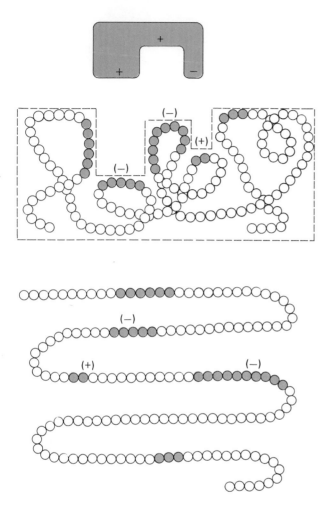

Figure 4-2. Amino acids that interact with the ligand at a binding site may be at far-removed sites along the polypeptide chain, as indicated in this model of the three-dimensional folding of a protein and the unfolded polypeptide chain shown below it.

Polar and ionized amino-acid side chains in the region of the binding site provide the major electrical links to the complementary polar and ionized regions of the ligand. As illustrated in Fig. 4-2, the amino acids in the region of the binding site need not be adjacent to each other in the primary structure, since folding of the polypeptide chain may bring various segments of the molecule into juxtaposition. Thus, amino acids from widely separated regions of the molecule may contribute to the specificity of the binding site.

Some binding sites have such a high degree of chemical specificity that they can bind only one type of ligand, whereas other binding sites may have a broader range of specificity, allowing them to bind a number of slightly differing ligands, all of which contain chemical regions similar enough to be complementary to the binding site. For example, protein X in Fig. 4-3 has a binding site that can combine with three different ligands, whereas protein Y has a greater (i.e., more limited) specificity and can combine only with a single ligand. Clearly, the degree of chemical specificity depends on the shape of the binding site; the more exactly its shape conforms to the shape of the ligand, the greater is its chemical specificity.

Figure 4-3. Protein X is able to bind all three ligands, a, b, and c, which have similar chemical structures, whereas protein Y can bind only ligand c.

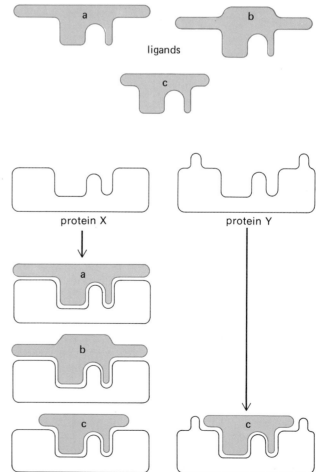

Affinity

Binding sites which bind a ligand tightly are said to be *high-affinity* sites, while those that bind weakly are *low-affinity* sites. Affinity and chemical specificity are two distinct properties of binding sites; therefore, two different proteins (Fig. 4-4) may be able to bind the same type of ligand, i.e., have the same chemical specificity, but may have different affinities for it. Specificity depends upon shape, whereas the strength of binding (affinity) depends upon both the number of interacting loci on the ligand and binding site and the strength of the electrical attraction at each of the loci; for example, the low-affinity binding site in Fig. 4-4 has few charged groups which correspond to the oppositely charged groups on the ligand so that the total electrical attraction between the two molecules is weaker than at the high-affinity site.

Saturation

At any one time, a single binding site on a protein is either occupied by a ligand or it is not. In a solution containing a number of identical protein molecules and complementary ligands, some of the sites may be occupied by ligands and others not. When all are occupied, then the binding sites are said to be fully *saturated*—there are no unoccupied sites available to bind additional ligands. When half of the available sites are occupied, the system is 50 percent saturated, and so on.

The probability that a given binding site will be occupied by a ligand depends upon two factors: (1) the concentration of free ligand in the solution; (2) the affinity of the binding site for the ligand. The importance of the first factor—concentration—is straightforward: In order for a ligand to be bound, it must approach near enough to the binding site so that the electrical forces holding it to the site become greater than the forces producing random molecular motion of the ligand in solution; the greater the ligand concentration, the greater the probability of a ligand encountering an unoccupied binding site and becoming bound (as illustrated by the so-called saturation curve of Fig. 4-5). The physiological implications of this dependence on concentration can be shown by a simple example. If the ligand were a drug which exerts its biological effect by binding to a specific protein, the effect of the drug would increase with

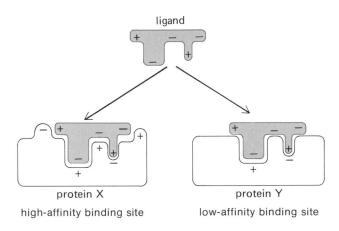

Figure 4-4. Protein X has five sites at which there are electrical attractions between the ligand and the binding site and, therefore, binds the ligand more tightly, i.e., has a higher affinity for the ligand, than has protein Y, which only has two sites of electrical attraction.

increasing drug concentration until all the binding sites become saturated. Further increases in the drug concentration would produce no further increase in the magnitude of the effect since there would be no additional sites to be occupied. To generalize, a continuous increase in the magnitude of a chemical stimulus (ligand concentration) which exerts its effects by binding to specific sites on proteins does not produce a continuous increase in the biological response to the stimulus.

The second factor which determines saturation is the affinity of the binding site for the ligand. This is because collisions between molecules in a solution and those on the binding site could dislodge a loosely bound ligand, much as a football player may fumble when tackled if he does not have a tight grip on the ball. If a binding site has a high affinity for a ligand, even a low concentration of ligand will result in a large proportion of the binding sites being occupied, since once bound to the site, the ligand will remain bound for a long time. A low-affinity site, on the other hand, requires a much higher concentration of ligand to achieve the same degree of saturation, since with the shorter duration of occupancy of low-affinity sites, more frequent encounters are required to maintain a given degree of saturation. The affinity can be expressed in terms of the concentration of ligand required to produce 50 percent saturation of the available binding sites.

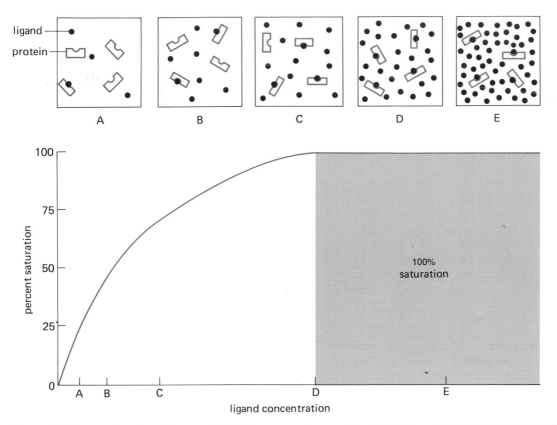

Figure 4-5. Increasing ligand concentration increases the number of binding sites that are occupied (percent saturation). At 100 percent saturation, all the binding sites have been occupied and further increases in concentration of ligand will not increase the amount bound.

Figure 4-6 reemphasizes that the degree of saturation of a population of binding sites depends upon both the concentration of the ligand and the affinity of the binding sites for the ligand.

Competition

As we have seen, more than one ligand may be able to bind to some binding sites. Consider the situation in which two ligands, A and B, are able to bind to the same site. Obviously only one ligand, either A or B, can occupy a given site at one time, and therefore, they compete with each other for the available sites. The number of sites occupied by each type of ligand will depend on the relative affinities of the binding site for the different ligands and on the concentrations of the ligands.

As a result of competition, the biological effects of one ligand may be markedly diminished by the presence of a competing ligand. For example, many drugs produce their effects by competing with the body's natural ligands for binding sites; by occupying the binding site, the drug prevents the natural ligand from binding and producing its response.

Functions of protein binding sites

Having defined the physical characteristics of ligand binding to protein binding sites—chemical specificity, affinity, saturation, and competition— we now ask what roles such binding reactions play in cell function. A detailed answer to this question is given in the remaining chapters of this book, since practically every cell function and its regula-

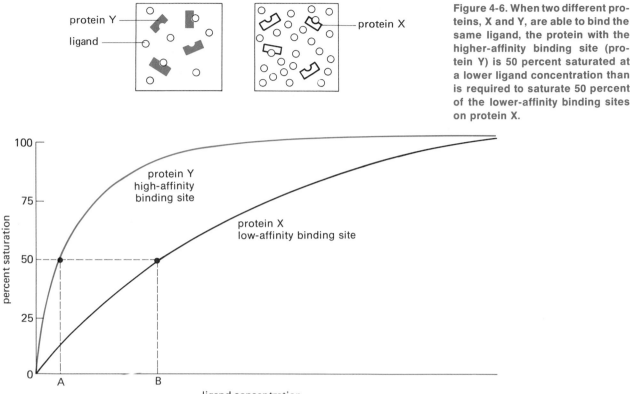

Figure 4-6. When two different proteins, X and Y, are able to bind the same ligand, the protein with the higher-affinity binding site (protein Y) is 50 percent saturated at a lower ligand concentration than is required to saturate 50 percent of the lower-affinity binding sites on protein X.

tion is mediated by specific interactions between ligands and their corresponding binding sites on proteins. In this section, a few examples will suffice to illustrate general principles.

Aggregation of proteins into organized structures

An obvious result of binding is the holding together of two molecules—the ligand and the protein molecule containing the binding site. Multimolecular aggregates arranged in specific three-dimensional patterns (for example, ribosomes and protein filaments) can be constructed by this type of binding of the individual component molecules. Each protein molecule in these structures may contain one or more binding sites for nonprotein ligands, but, in addition, it has binding sites specific for binding sites on another protein molecule. In other words, such proteins act both as binding sites and ligands. When the various individual protein constituents of a multimolecular structure are placed into a solution, the structure will actually assemble itself, since the location of each molecule in the structure is determined by the various binding sites and corresponding ligands. If one were to take the component pieces of a clock and shake them up in a basket, you would not expect them to fall into the positions that they occupy in the functioning clock, but this is, in large part, the way a cell goes about forming its structures. It synthesizes a number of different proteins and mixes them together in solution: Because of the specificity of their binding-site–ligand interactions, they aggregate into specific structures. Thus, the only information required by the cell is the specification for the specific sequences of amino acids to go into each protein; it does not require instructions on how to combine the proteins into more complex structures.

The assembly during embryonic development of multicellular tissues and organs from individual cells provides another example of selective binding. Different types of cells have in their plasma membranes proteins which contain specific binding sites for a class of proteins known as *aggrega-*

tion factors, each of which itself contains multiple sites. Therefore, one site on the aggregation factor binds to the surface of one cell while another binds to the surface of an adjacent cell, thereby linking the two cells together. Given a variety of surface binding sites on different cells, and corresponding aggregation factors, cells are arranged during embryonic growth into the various multicellular tissues and organs.

Enzymes

Substances which accelerate chemical reactions but do not themselves undergo any net chemical change during the reaction are known as *catalysts.* Certain proteins are the catalysts for the chemical reactions in the body and are called *enzymes.* The property of enzymes which enables them to be specific in the chemical reactions they accelerate is the specificity of their binding sites. The ligands which bind to the enzymes' binding sites are known as *substrates* and the molecules that are ultimately formed from these ligands in the chemical reaction catalyzed by the enzyme are known as *products.* The mechanism by which the binding of a substrate to the enzyme leads to an accelerated chemical reaction will be discussed in the next chapter.

Practically every chemical reaction in the body is catalyzed by a specific enzyme, and without such catalysis, the reaction usually proceeds at a negligible rate. Therefore, the types of chemical reactions that occur within a particular cell depend upon which enzymes are present in that cell. Furthermore, particular types of reactions are localized to particular regions of the cell because the specific sets of enzymes required for their catalysis are localized to those regions; for example, the chemical reactions that occur in the mitochondria are different from those that occur in the nucleus or in the endoplasmic reticulum because each of these cell organelles contains a different set of enzymes.

Certain enzymes are present in almost all cells, whether they be muscle, nerve, etc., and to that extent these cells are all similar (although not identical since the concentrations of these common enzymes may differ from cell to cell). In contrast, other enzymes may be totally absent from certain cells so that the reactions they catalyze occur so slowly in those cells as to be negligible.

The crucial point is that cells differ from each other almost entirely because they contain differing amounts of the various enzymes. Moreover, a given cell may undergo dramatic changes in its activity solely as a result of a change in its rate of enzyme production. Again we see that a cell is what it is because of the instructions it receives concerning the types of proteins (enzymes, in this case) it should synthesize and the rates at which it should synthesize them.

Detection of chemical messengers

In addition to holding molecules and cells together and catalyzing the chemical reactions of metabolism, protein binding sites play a key role in the communications systems of the body. Chemical signals that are transmitted between cells or between one area of a cell and another are detected by specific binding sites for the chemical messengers. For example, all cells of the body are exposed to the same hormones (chemical messengers secreted by endocrine glands into the blood) reaching them by way of the circulatory system, but only those cells that contain binding sites for a particular hormone will respond to it. Thus, hormone binding sites provide the basis for the selective action of hormones.

There are many other types of chemical signals that depend on specific binding sites. For example, the excitation of any cell by a nerve cell depends on a chemical agent which is released from the nerve cell and binds to specific binding sites for it on the responding cell. To take another example, chemical signals within a cell influence the activity of enzyme molecules by binding to them at specific protein binding sites.

Regulation of binding-site characteristics

Because a cell is what its protein binding sites make it, molecular control mechanisms center upon alteration of these sites. One mechanism for achieving either more or fewer sites is to control the synthesis of the protein molecules which contain them, and this process is discussed later in this chapter. The other mechanisms, the subject of this section, operate not by altering the number of protein molecules but rather by changing the characteristics of their binding sites.

Allosteric regulation

When a ligand binds to the surface of a protein, the electrical forces between the ligand and the protein may produce a change in the shape of the protein in the region of the binding site; this may in turn alter the electrical interactions between other regions of the protein thereby producing a change in the total conformation of the folded polypeptide chain, as illustrated in Fig. 4-7. Such changes in protein shape resulting from the ligand–binding-site interaction may cause new binding sites to appear at regions of the molecule where none existed previously, or they may modify the shapes of already existing binding sites. In the latter case, the change in shape may alter the chemical specificity of the site or its affinity for ligand. If, for example, this other site whose shape was changed were an enzymatic binding site, its enzymatic activity could be increased or decreased by the change in shape. Thus, these conformational changes provide a mechanism for regulating enzyme activity by way of chemical messengers; such regulation is known as *allosteric* (other site) regulation, and the chemical messenger whose binding at one site (the regulator site) on the protein regulates the activity of another binding site on the same protein is known as a *modulator*. It should be emphasized that allosteric modulation is not limited to enzymatic binding sites, for any binding site can be so influenced.

To reiterate, allosteric modulation of a binding site may alter the chemical specificity of the site or its affinity for ligand. Alteration of chemical specificity usually leads to an *on-off* type of control; i.e., it permits the ligand specific for that site to bind to the site or prevents any binding at all. In contrast, when the affinity of the binding site is changed, the result is not an *on-off* phenomenon but a change in the percent saturation of the binding sites. If we assume that the magnitude of the biological response produced by binding is proportional to the number of binding sites occupied, then increasing the affinity of the sites will increase the response at any given ligand concentration. This type of control mechanism therefore allows for a much finer regulation than does an *on-off* response. For example, a number of different types of cells may have binding sites for a particular hormone molecule; all of these cells are exposed to the same concentration of the hormone,

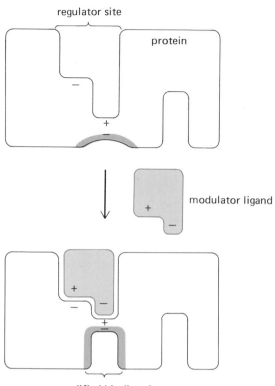

Figure 4-7. Allosteric modulation. Binding of a modulator ligand at the regulatory site alters the conformation of a second binding site.

yet the response of one type of cell may be increased by allosterically increasing the affinity of its binding sites for the hormone, thereby increasing the number of sites on the cell occupied by hormone molecules.

The binding of a chemical messenger may set in motion a chain reaction of allosteric modulations, as illustrated in Fig. 4-8. The initial binding of the chemical messenger to the regulator site on protein A produces a conformational change at a second site on protein A which allows protein B to bind to this site. The binding of protein B in turn produces a conformational change in protein B which allows it to bind protein C, and so on. As we shall see, as in the case of blood coagulation, for example, such cascades of modulation are frequently encountered in control systems.

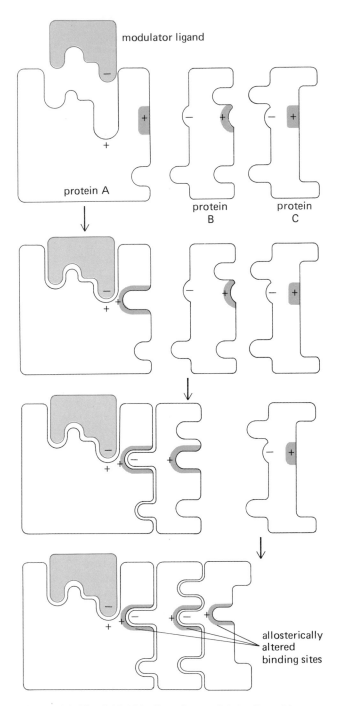

Figure 4-8. The initial binding of a modulator ligand to protein A triggers the successive bindings of proteins B and C. The binding of each ligand alters a second site on each protein, allowing it to bind the next protein.

Chemical alteration of protein structure

Allosteric modulation involves changing the shape of protein molecules but not their chemical composition. In contrast, another means of altering binding-site configuration is to covalently link various chemical groups to the amino acid side chains of the protein. For example, the addition of a phosphate group to the side chain of the amino acid serine converts a slightly polar side chain into a highly ionized side chain carrying a net negative charge. The negative charge on the phosphate group interacts with other charged groups in the protein, altering the shape of the polypeptide chain and thereby, its binding sites. The addition or removal of such chemical groups is normally mediated by specific enzymes, and the activity of these enzymes may be controlled through allosteric regulation by a variety of chemical messengers. Thus, nonallosteric modification of specific binding sites may be the end result of a sequence triggered by allosteric modulation of another protein.

In addition to the specific mechanisms mentioned above, there are a number of nonspecific factors in the environment of a protein which have the potential for altering its binding sites. Two of the most important are temperature and acidity, but normally these factors are not critical because they are maintained relatively constant in the body's fluids. Should large changes occur, as in disease, marked alterations of binding sites can occur. The body has a number of mechanisms for maintaining the acidity of the internal environment within a very narrow range. Diseases which interfere with the ability of the body to regulate its hydrogen-ion concentration produce life-threatening consequences because of the resulting alterations in the conformation of protein binding sites and the biological activities that are controlled by these sites.

Section B
Genetic information and protein synthesis

Genetic information

The transmission of hereditary information from cell to cell is the function of deoxyribonucleic acid (DNA), which contains instructions for the synthesis of proteins coded into its molecular structure. As described in the previous sections, the protein composition of a cell then determines its structure and functional activity. The portion of a DNA molecule which contains the information required to determine the amino acid sequence of a single polypeptide chain is known as a *gene*. A single DNA molecule contains many genes—the units of hereditary information.

As an example of the expression of genetic information, consider eye color in man. The color is due to the presence of particular molecules (pigments) in cells of the eye. A sequence of enzyme-mediated reactions is required to synthesize these pigments, and the genes determining eye color control the synthesis of specific enzymes in this pathway. Note that the genes for eye color do not contain information about the chemical structure of the eye pigments themselves; rather they contain the information required to synthesize the enzymes which mediate the formation of the pigments.

Although DNA molecules contain the information necessary for the synthesis of specific proteins, the DNA molecule does not itself participate *directly* in the assembly of a protein molecule. The DNA of a cell is located in the nucleus, whereas most protein synthesis occurs in the cytoplasm. The transfer of information from the DNA genes to the site of protein synthesis is the function of molecules of RNA (ribonucleic acid) synthesized on the surface of the DNA. This mechanism of expressing genetic information occurs in all living organisms and has led to what is now called the "central dogma" of molecular biology: In all organisms, genetic information flows from DNA to RNA and then to protein:

$$\text{DNA} \longrightarrow \text{RNA} \longrightarrow \text{PROTEIN}$$

Figure 4-9 summarizes the general pathway by which the information stored in DNA influences cell activity via the process of protein synthesis.

DNA and the genetic code

As described in the previous chapter, DNA consists of two polynucleotide chains coiled around each other to form a double helix. Each chain is a sequence of nucleotides (Fig. 4-10) joined together by phosphate-sugar linkages. Each nucleotide

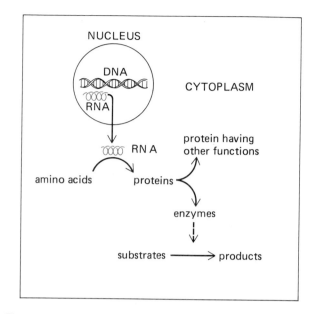

Figure 4-9. The expression of genetic information in a cell occurs through its transcription from DNA to RNA in the nucleus followed by the translation of the RNA information into protein synthesis in the cytoplasm.

contains one of four different bases (adenine [A], guanine [G], cytosine [C], or thymine [T]), and each of these bases is specifically paired, A to T and G to C, with a base on the corresponding polynucleotide chain of the double helix (Fig. 3-27). Thus, both chains contain a precisely ordered sequence of bases, one chain being complementary to the other. It is the sequence of these bases along the polynucleotide chains of the helix which provides the code ultimately specifying the sequence of amino acids in proteins.

The genetic language is very similar in principle to a written language, like English, which consists of a set of symbols forming the letters of an alphabet. The letters are arranged in specific sequences to form words, and the words are often arranged in linear sequences to form sentences. The genetic language contains only four letters, corresponding to the four bases A, G, C, and T. The words are three-base sequences which specify particular amino acids (thus, each word in the genetic language is only three letters long). These words are arranged in a linear sequence along the chains of DNA, and the sequence of words specifying the structure in a single protein comprises a gene (Fig. 4-11); thus a gene is equivalent to a sentence, and the entire collection of genes in a cell is equivalent to a book.

How are the four "letters" of the DNA alphabet, A, G, C, and T, arranged to form at least 20 different three-letter code words? Since the four bases can be arranged in 64 different combinations ($4 \times 4 \times 4 = 64$), a triplet code actually provides more than enough code words. It turns out that not just 20, but 61 out of the 64 possible triplet combinations are actually used to specify amino acids. This means that a given amino acid is usually specified by more than one code word (of the 20 different amino acids, 18 are represented by more than one code word).

The three code words which do not specify amino acids are known as *termination* code words; they perform the same function as does a period at the end of a sentence—they specify that the end of a gene has been reached. This is crucial since a single DNA molecule contains many genes in linear sequence and the translation process must "recognize" where a particular gene ends and another begins.

The genetic code is a universal language used by all living cells; for example, the code word for the amino acid tryptophan is the same in the DNA molecules of a bacterium, an ameba, a plant, and a human being. Although the same code words are used by all living cells, the messages they spell out—the sequence of code words that determines the amino acid sequence in proteins—are of course, different in each organism. The universal nature of the genetic code is one piece of evidence which strongly supports the concept that all forms of life on earth evolved from a common ancestor.

The giant molecules of DNA are located in the nucleus of a cell and are unable to pass through the nuclear membrane into the cytoplasm; yet it is in the cytoplasm on the surface of ribosomes that proteins are synthesized. Since DNA cannot leave

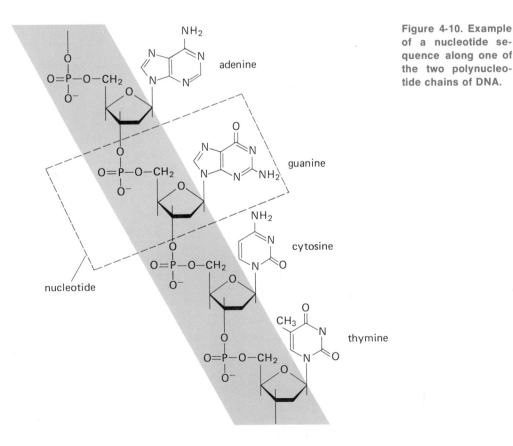

Figure 4-10. Example of a nucleotide sequence along one of the two polynucleotide chains of DNA.

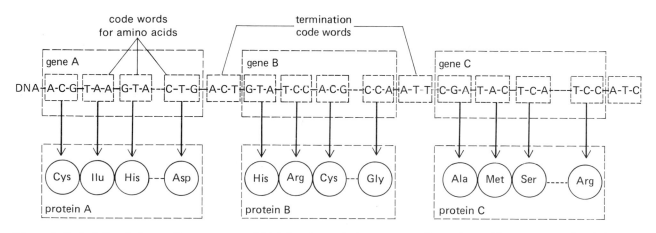

Figure 4-11. Relation between the sequence of triplet code words in genes and the amino acid sequences of proteins corresponding to them. The names of the individual amino acids are abbreviated. Note that all three genes are part of the same DNA molecule, which may contain hundreds of genes in linear sequence.

the nucleus, a message carrying its genetic information must pass from the nucleus to the cytoplasm. This message is in the form of RNA known as *messenger RNA* (mRNA). The transfer of genetic information from DNA into protein thus occurs in two stages: first, the genetic message must be passed from DNA to messenger RNA *(transcription)*; then the message in mRNA is used to direct the assembly of the proper sequence of amino acids to form a protein *(translation)*.

Protein synthesis

Transcription: Formation of messenger RNA

The essence of transcription is that the information in DNA is passed on to the linear sequence of nucleotides in messenger RNA. Recall from Chap. 3 that ribonucleic acids are single-chain polynucleotides, whose nucleotides differ from DNA in that they contain the sugar ribose (rather than deoxyribose) and the base uracil (rather than thymine). The other three bases, adenine, guanine, and cytosine, occur in both DNA and RNA.

The sugars and bases of the individual nucleotides are synthesized by enzyme-mediated reactions in cells and these components are then combined with phosphate. Each nucleotide consists of one of the four bases—A, U, C, or G—the sugar ribose, and a sequence of three phosphate groups

attached to ribose. It is from this pool of free (uncombined) nucleotides that molecules of RNA are formed. The four types of free nucleotides (corresponding to the four different bases) must be positioned in linear sequence so that the coded information of the genes is retained and chemical bonds must be formed between them. This is what occurs during transcription.

Recall that, in DNA, the two polynucleotide chains are linked together by hydrogen bonds between specific pairs of bases—a pairing of A with T, and G with C. Transcription begins with breakage of these weak hydrogen bonds so that a portion of the DNA double helix uncoils. Free RNA nucleotides are then able to pair with their exposed DNA counterparts in the two DNA chains. Free RNA nucleotides containing adenine will pair with any exposed thymine base in DNA; likewise, RNA nucleotides containing G, C, and U will pair with exposed C, G, and A bases in DNA, respectively (uracil, which replaces thymine in RNA, pairs with adenine). In this way, the nucleotide sequence in DNA (the genetic code) determines the sequence of nucleotides along a molecule of messenger RNA.

Once the appropriate nucleotides have been lined up properly by pairing with bases in DNA, they are joined together by an enzyme known as *DNA-dependent–RNA-polymerase* (also known as transcriptase); this enzyme catalyzes the splitting off of two of the three phosphate groups from each nucleotide and the linking of each nucleotide to the next one in sequence (Fig. 4-12). As its name im-

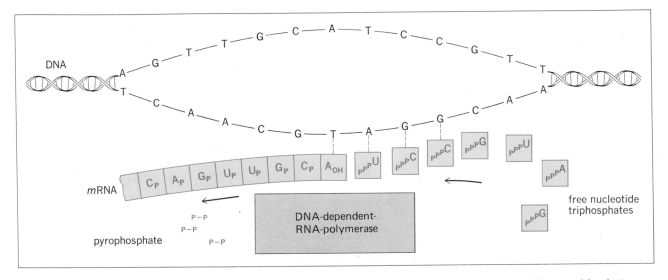

Figure 4-12. Synthesis of mRNA on the surface of one of the two strands of a DNA molecule through base pairing between free ribonucleotides and the bases in DNA.

plies, it is active only in the presence of DNA and will, therefore, not link free nucleotides together (in what would be a random sequence) when they are not base-paired with DNA.

Since DNA consists of two strands of polynucleotide, both of which are exposed during transcription, it theoretically is possible to form two different messenger RNA molecules, one from each strand (Fig. 4-13). These potential messenger RNA molecules would have different nucleotide sequences because the DNA chains are not identical, and thus, would code for two entirely different proteins. However, in reality only one of the two potential molecules of mRNA is ever formed. Which mRNA is formed depends on a specific sequence of nucleotides located at the beginning of the gene; this sequence is present in only one of the two DNA strands and can combine with a binding site on DNA-dependent–RNA-polymerase. Since this binding is required for activation of the enzyme, the result is that only the strand of DNA containing the initial sequence will be transcribed; the nucleotides paired with the other strand are not linked and are simply released as free nucleotides.

Figure 4-13. Transcription of a gene from a single strand of DNA to messenger RNA.

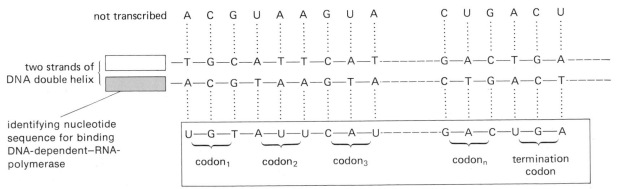

transcribed messenger RNA

Because the nucleotide sequence to which DNA-dependent–RNA-polymerase must bind may be located in either strand of the DNA double helix, some genes are transcribed from one strand while other genes in the same DNA molecule are transcribed from the opposite strand (note that a single gene includes both strands of the helix even though only one strand is transcribed).

It must be emphasized that the nucleotide sequence in a molecule of mRNA is not identical to that in the corresponding strand of DNA, since its formation depended on the pairings between complementary (not identical) bases (Fig. 4-13). Thus, during transcription, the RNA code words which are formed are complementary to the code words in DNA, and these triplet base sequences in mRNA corresponding to code words in DNA are called *codons*. It is the sequence of these codons in mRNA that constitutes the message carried to the cytoplasm and translated there into a specific sequence of amino acids. From the termination code word at the end of the gene in DNA is formed a complementary *termination codon* in mRNA.

One molecule of mRNA contains the genetic information corresponding to a single gene (or, at most, a few genes). In a given cell, only a few of the thousands of genes present in DNA are transcribed into mRNA at any given time. Which genes are transcribed depends on the presence or absence of diverse chemical signals. Such regulation of gene transcription provides a major control over the synthesis of specific proteins and, therefore, over the activities of cells and will be considered later in this chapter.

Once transcribed, a molecule of mRNA must move into the cytoplasm where its message can be translated into the synthesis of a protein. Although molecules of mRNA are considerably smaller than DNA molecules, they are still very large. Since a typical protein may consist of a sequence of 100 amino acids, the mRNA molecule coding for it must contain 100 codons, i.e., 300 nucleotides. Such large molecules do not readily pass through the nuclear membranes which separate the nucleus from the cytoplasm. However, the nucleus is actually enclosed by two membranes (Fig. 4-14) traversed at intervals by pore-like structures large enough to allow passage of mRNA molecules. These apparent openings are not simply holes through which any molecules can move, but

are filled with proteins which selectively mediate the movements of RNA molecules. The mechanisms responsible for these movements are still unknown.

Translation

Having reached the cytoplasm, the messenger RNA binds to a ribosome, the cell organelle which mediates the actual protein assembly. Before describing this final assembly, we must describe these ribosomes and two other forms of RNA involved.

Ribosomes Ribosomes are small particles (about 23 nm in diameter) found throughout the cytoplasm, either suspended in the cytosol or attached to the surface of the endoplasmic reticulum. The ribosomes attached to the endoplasmic reticulum are associated with the synthesis of proteins eventually to be secreted from the cell (Chap. 6), whereas the ribosomes suspended in the cytosol synthesize the proteins that remain within the cell (for example, enzymes and structural proteins). The number of ribosomes in a cell and their distribution between free and bound forms varies during different periods of cell activity, being very much increased during cell growth or when large amounts of protein are being produced.

Each ribosome consists of over 70 protein molecules associated with large quantities of the type of RNA known as *ribosomal RNA* (rRNA). Ribosomal RNA is synthesized in the nucleus, DNA serving as a template for the positioning of its nucleotides in the same manner described for synthesis of messenger RNA (thus, transcription of certain segments of DNA yields messenger RNA, whereas transcription of other segments yields other forms of RNA which assist in protein synthesis but do not themselves carry coded information about amino acid sequences in proteins). The portions of DNA which code for rRNA are associated with the *nucleolus*, a densely staining nuclear organelle (Fig. 4-14) consisting of filamentous and granular elements. The granular elements appear to be formed by a combination of rRNA and proteins, the latter having been synthesized in the cytoplasm and then transferred to the nucleus. These nucleoprotein granules move into the cytoplasm (probably through the nuclear

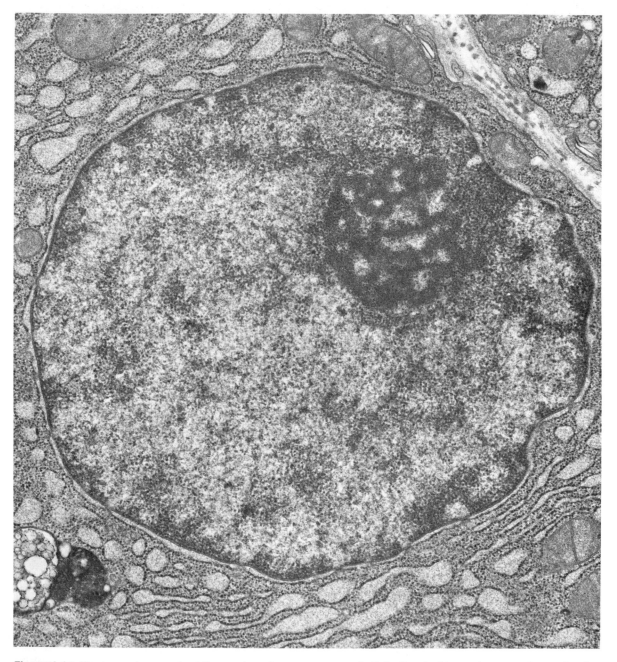

Figure 4-14. Electron micrograph of the nucleus in a pancreas cell. Note the double nuclear membrane that is interrupted at intervals by nuclear pores. The densely stained nucleolus is located in the upper right portion of the nucleus. *(Courtesy of Keith R. Porter.)*

pores) where they combine with other proteins to form the two subunits of a ribosome—a large so-called 60S subunit and a smaller 40S subunit.

When a messenger RNA molecule arrives in the cytoplasm, it becomes bound to the 40S subunit, which then binds to the 60S subunit to form a fully functional ribosome (Fig. 4-15). The messenger RNA molecule appears to lie in a groove between the two subunits of the ribosome. The numerous proteins of the ribosomal particle provide the enzymes and other proteins required for the translation of this mRNA molecule into protein. The exact function of the rRNA is still unclear.

Transfer RNA How do individual amino acids become arranged in the proper sequence to form a protein molecule? By themselves free amino acids do not have the ability to bind to the bases in RNA. Orientation of the amino acids on the mRNA molecule involves yet a third class of RNA known as *transfer RNA* (tRNA). Transfer RNA molecules are the smallest (about 80 nucleotides long) of the three classes of RNA. Like mRNA and rRNA, they are synthesized by base pairing with nucleotides in DNA and then enter the cytoplasm. The key to the functioning of tRNA is that each of these molecules can combine both with a specific amino acid and with a codon in mRNA specific for that amino acid. This permits tRNA to act as a linkage between the amino acid and the mRNA codon (Fig. 4-16).

The tRNA combines covalently with its specific amino acid under the influence of an enzyme known as amino acyl synthetase. There are 20 different amino acyl synthetase enzymes, each of which catalyzes the linkage of only one type of amino acid to its particular type of tRNA. The next

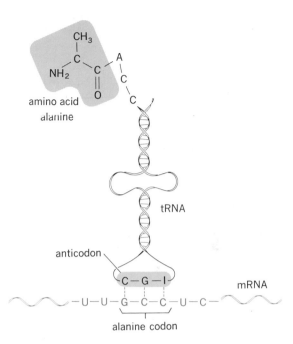

Figure 4-16. Base pairing between the anticodon region of a tRNA molecule with the corresponding codon region of an mRNA molecule. The exposed loop in the central region of tRNA is the site of interaction with amino acyl synthetase.

step is to link the tRNA molecule (bearing its attached amino acid) to the codon for that amino acid in mRNA. As one might have predicted, this is achieved by base-pairing between tRNA and mRNA. Of the 80-odd nucleotides in a given tRNA molecule, one three-nucleotide sequence is exposed and can base-pair with particular codons in mRNA; it is appropriately termed an *anticodon.* Since there are 61 different codons for the 20 different amino acids, one might expect to find 61 different molecules of tRNA, each with a different anticodon. There are, however, fewer than 61 types of tRNA molecules because some of the anticodons are able to pair with several of the different codons specifying the same type of amino acid.

Figure 4-17 illustrates the functioning of a tRNA molecule, in this case a tRNA for alanine. Note that it is covalently linked to alanine at one end and base-paired with G-C-C in mRNA at the other end. Its anticodon for this base-pair is C-G-I (tRNA has several unusual bases, the I representing *inosine,* which base-pairs with cytosine).

Figure 4-15. mRNA attaches to the 40S portion of the ribosomal particle.

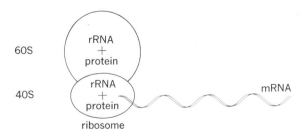

Figure 4-17. Sequence of events during the movement of a strand of mRNA
along a ribosome.

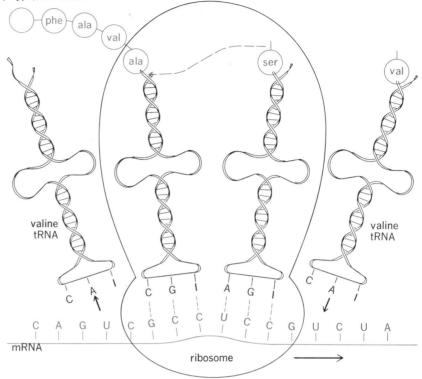

Protein assembly All the different parts involved in the assembly of protein have now been described. The final interactions between mRNA, ribosomes, and tRNA take place on the surface of the ribosomal particles. Synthesis is initiated by the attachment of one end of the mRNA to a ribosomal particle. This initial reaction appears to be quite specific since the coded message in mRNA is read in one direction, always beginning at the same end of the mRNA molecule, just as in English one reads a sentence from left to right, never from right to left. Protein synthesis begins at the free-amino end of the protein chain and proceeds toward the free-carboxyl end.

The various tRNAs, with their specific amino acids, base-pair through their anticodons with the corresponding codons in mRNA. Enzymes within the ribosomal particle then catalyze the formation of a peptide bond between two adjacent amino acids while they are still attached to their tRNA molecules. Thus, one end of the growing protein chain is always attached to a molecule of tRNA. The large 60S subunit of the ribosome has two binding sites for tRNA; one holds the tRNA that is also attached to the growing peptide chain, and the second holds the tRNA containing the next amino acid to be added to the peptide chain. With the formation of a peptide bond between the protein chain and the next amino acid, the initial tRNA is released from the mRNA and the peptide chain is transferred to the next tRNA. The mRNA now moves one codon space down the ribosome, making room for the binding of the next amino acid–tRNA molecule (Fig. 4-17). This process is repeated over and over as each amino acid is added in succession to the growing peptide chain (at a rate of 2 to 3 amino acids per second). When the ribosome reaches a termination codon specifying the end of the polypeptide chain, the completed protein is released from the ribosome.

The same strand of mRNA can be used more than once to synthesize several molecules of the protein because the message in mRNA is not destroyed during protein assembly. In fact, the same strand of mRNA can simultaneously synthesize several molecules of the protein. As mRNA is moved across the ribosome during the process of assembling a protein, it may become attached to a second ribosome and begin the synthesis of a second protein. Thus, a number of ribosomes may be attached to the same strand of mRNA forming what is known as a *polyribosome*. Polyribosomes containing as many as 70 ribosomes attached to a single strand of mRNA have been observed. Each ribosome has a growing peptide chain attached to it. Ribosomes near the beginning of the chain are associated with very short peptide chains representing the first few amino acids in the protein; ribosomes near the end are associated with a protein chain which is almost completed (Fig. 4-18). Once formed, messenger RNA does not remain in the cytoplasm indefinitely. Eventually it is broken down by enzymes which break the links between its nucleotides. Therefore, if a gene corresponding to a particular protein ceases to be transcribed into messenger RNA, the synthesis of that protein will eventually slow down and cease as the existing molecules of messenger RNA are broken down.

The steps leading from DNA to a completed protein are summarized in Table 4-1.

Regulation of protein synthesis

Possible sites of regulation The different types of cells in the body synthesize different proteins, and no one type of cell synthesizes all its proteins at the same rate. Cells thus have mechanisms which regulate both the types and rates of protein synthesis.

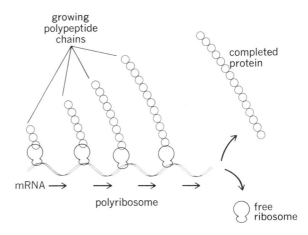

Figure 4-18. A polyribosome, showing the different stages of protein assembly as each ribosome moves along the strand of messenger RNA.

From our previous discussion we can identify a number of different stages at which protein synthesis might be regulated (Fig. 4-19):

1 If the ability to synthesize a specific mRNA on the surface of DNA were blocked, the protein corresponding to that particular gene would not be synthesized.
2 Regulation can modify the ability of mRNA to combine with the ribosome, the movement of the ribosome along mRNA, or the activity of the various ribosomal enzymes. Drugs are known which block protein synthesis at the ribosomal

Figure 4-19. Possible sites at which the rate of protein synthesis may be regulated.

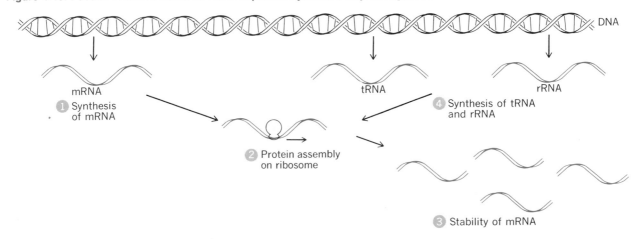

Table 4-1. Sequence of events leading from DNA to protein synthesis

Transcription

1 The two strands of the DNA double helix separate in the region of the gene to be transcribed.
2 Free nucleotides base-pair with the nucleotide bases in DNA.
3 The nucleotide triphosphates paired with one strand of DNA are linked together by DNA-dependent–RNA-polymerase to form mRNA containing a sequence of bases complementary to the DNA base sequence.

Translation

4 mRNA passes from the nucleus to the cytoplasm where one end of the mRNA binds to a ribosome.
5 Free amino acids combine with their corresponding tRNAs in the presence of specific amino acyl synthetase enzymes in the cytoplasm.
6 Amino acid–tRNA complexes bind to sites on the ribosome and the three base anticodons in tRNA pair with the corresponding codons in mRNA.
7 Each amino acid is then transferred from its tRNA to the growing peptide chain, which is attached to the adjacent tRNA.
8 The tRNA freed of its amino acid is released from the ribosome.
9 A new amino acid–tRNA complex is attached to the vacated site on the ribosome.
10 mRNA moves one codon step along the ribosome.
11 Steps 6 to 10 are repeated over and over.
12 The completed protein chain is released from the ribosome when the termination codon in mRNA is reached.

level or which disrupt the structure of the ribosome so that the message in mRNA is read incorrectly, giving rise to abnormal inactive proteins.

3 Although the mRNA molecule can be used over and over again to synthesize many molecules of the same protein, it is eventually destroyed and must be replaced with newly synthesized mRNA. Thus, by controlling the rate at which mRNA is destroyed relative to the rate at which it is synthesized, it is possible to regulate the amount of a particular protein being synthesized.

4 In addition to coding the sequence of amino acids in proteins, the DNA molecule also codes the base sequences in tRNA and rRNA. Thus, a mechanism which regulates the rates at which tRNA and rRNA are synthesized might also affect the process of protein synthesis.

We have listed the sites at which the regulation of protein synthesis *may* occur. A brief discussion of enzyme induction and repression will serve to illustrate one class of these regulatory mechanisms.

Induction and repression The functions of DNA and its relation to protein synthesis are better understood in bacteria than in any other type of cell because they are a relatively simple type of cell that can be subjected to a wide variety of experimental techniques more easily than cells from higher organisms.

The aspect of protein synthesis that has been most extensively studied is the mechanism by which the synthesis of particular proteins can be turned on and off. If the type of bacterium known as *Escherichia coli* is grown in a nutrient medium containing the disaccharide lactose, an enzyme, galactosidase, can be isolated which catalyzes the splitting of lactose into a molecule of glucose and a molecule of galactose:

$$\text{Lactose} \xrightarrow{\text{galactosidase}} \text{glucose} + \text{galactose}$$

Glucose and galactose are metabolized by the cell to provide energy and a source of carbon for the synthesis of other types of molecules, but in the absence of galactosidase the molecules of lactose are not broken down and cannot be used. If *E. coli* are grown in a nutrient medium that does not contain lactose, the cells are found to have very few molecules of galactosidase, about three molecules of enzyme per cell. If lactose is then added to the medium, a rapid synthesis of galactosidase takes place. In the presence of lactose, each cell has about 3000 molecules of the enzyme. The presence of lactose has *induced* the synthesis of the enzyme required for its metabolism (Fig. 4-20). Such a system has many advantages since the cell need not expend energy synthesizing an enzyme unless there is a substrate present for the enzyme to act upon.

Inhibition of enzyme synthesis in the presence of substrate is also observed in bacteria. The synthesis of the amino acid histidine, which is required for protein synthesis, involves a series of reactions catalyzed by 10 different enzymes. If *E. coli* are grown in a medium to which histidine has been added, the cells do not synthesize the 10 enzymes needed for histidine synthesis but rather utilize the histidine in the medium. When histidine is removed from the medium, the 10 enzymes are rapidly synthesized to provide the histidine needed for growth. In this example, it appears that the presence of histidine has *repressed* the synthesis of the enzymes utilized in its synthesis (Fig. 4-20). This type of system is useful to a cell because the end product of a biochemical pathway may be able to regulate the synthesis of the enzymes in that pathway. As the concentration of the end product is increased, the synthesis of the enzymes in the pathway is reduced and the rate of product formation is decreased.

In general, both induction and repression involve a number of enzymes simultaneously. Thus, in histidine repression, 10 different enzymes are simultaneously repressed by histidine. Genetic mapping of the genes which code for each of these different enzymes has shown that most of the genes are located next to each other in the DNA molecule and that one strand of mRNA is formed at this site which contains the information for the synthesis of all 10 enzymes. Repression of enzyme synthesis involves blocking the synthesis of this mRNA molecule. A collection of genes which can be induced or repressed as a unit is known as an *operon*.

Figure 4-21 diagrams the mechanism of enzyme induction found in bacterial cells. A repressor gene contains the coded information for the synthesis of a repressor protein molecule. The repressor protein is synthesized in the standard manner on the surface of a ribosome from mRNA synthesized on DNA at the site of the repressor gene. Once formed, the repressor protein can bind to the initial gene of an operon, known as the *operator gene.* The combination of the repressor protein with the operator gene prevents synthesis of the operon's mRNA. As long as the repressor is bound to the operator gene, enzyme synthesis is prevented since no mRNA can be formed.

The induction of enzyme synthesis is believed to involve the combination of the inducer molecule, such as lactose in the example above, with the repressor protein, thereby leading to inactivation of the repressor so that it is unable to bind to the operator gene. Since the operator gene is no longer inhibited, it can initiate the synthesis of the mRNA. This mRNA then can initiate enzyme synthesis on the ribosomes. Thus, induction of

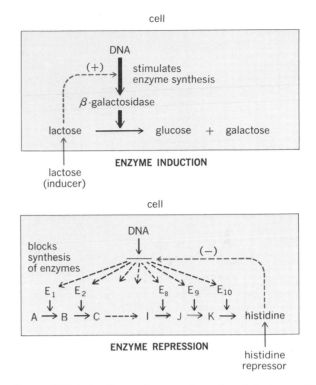

Figure 4-20. Induction and repression of protein (enzyme) synthesis. A plus sign indicates stimulation, and a minus sign indicates inhibition.

enzyme synthesis really involves the inhibition of an inhibitor (the repressor molecule) usually present in the cell.

Enzyme repression by the end product of a biochemical pathway involves a set of interactions similar to enzyme induction except that, in this case, the repressor molecule synthesized by the repressor gene is normally inactive. The end-product molecule, histidine in the example above, combines with the inactive repressor molecule, forming an active repressor molecule which is able to block the synthesis of mRNA at the site of the operator gene and thus block the synthesis of the corresponding enzymes.

Not all genes in DNA are subject to induction and repression; many appear to lack operator genes and are thus continuously active, independent of the presence or absence of inducers or repressors. However, the proteins synthesized by these genes may be subject to other types of control.

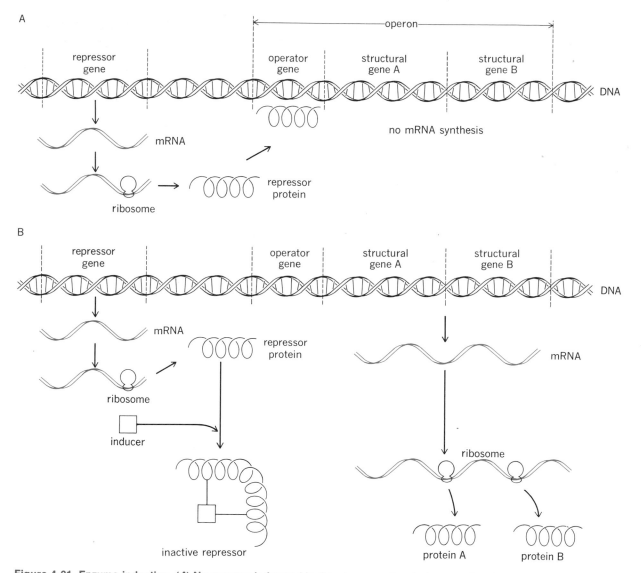

Figure 4-21. Enzyme induction. (*A*) No enzyme is formed in the absence of an inducer molecule. (*B*) Enzyme is formed in the presence of an inducer molecule due to the inactivation of the repressor protein.

The basic patterns of induction and repression described above were worked out mainly in studies on bacteria. However, it is clear that many of the genes in human cells also undergo induction and repression in the presence or absence of certain substrate molecules. For example, one of the mechanisms by which hormones influence the metabolism of cells is by interactions with DNA that alter protein synthesis by the cell.

Role of DNA-associated nuclear proteins. DNA molecules do not exist as isolated molecules in the nucleus but are combined with several types of protein and small amounts of RNA, the entire complex being known as *chromatin*. The chromatin proteins are intimately involved in the control of gene activity described in the previous section. For example, certain of these proteins provide specific binding sites for repressor molecules and hormones which regulate genetic transcription; i.e., they position the repressor or hormone at the right place along the DNA chain for affecting a particular gene. Other proteins may, themselves, block genetic transcription by preventing access of nucleotides to the DNA surface.

A group of positively charged chromatin proteins, known as *histones*, appear to be responsible for the various conformational states of the chromatin threads. The positive charges on the histone proteins interact with the negative charges of the phosphate groups in the nucleotides of DNA to produce a supercoiling of the chromatin threads. The result is that large regions of chromatin remain tightly coiled, forming regions of *heterochromatin* which stain more intensely than the uncoiled regions of *euchromatin*. Most gene transcription into mRNA appears to occur in the regions of euchromatin, whereas the heterochromatin regions are inactive.

Replication and expression of genetic information

The development of the human body from a single fertilized egg cell involves cell growth, cell replication, and cell differentiation. Each of these processes depends on the expression of genetic information into protein synthesis.

Replication of DNA

When a cell divides, forming two new independent cells, the genetic information stored in its DNA must be duplicated and a copy passed on to each of the daughter cells. DNA is the only molecule in a cell which is able to form a duplicate copy of itself without requiring a set of instructions from some other component in the cell. Thus, mRNA molecules can be formed only in the presence of DNA, which provides the information for the ordering of its base sequence. Likewise, protein can only be formed if mRNA is present to provide the information necessary for ordering the sequence of amino acids in the protein, and the other molecules of a cell are formed by the action of enzymes resulting from protein synthesis.

The replication of DNA is, in principle, very similar to the process whereby mRNA is synthe-

sized on the surface of DNA through base pairings between the bases in DNA and a collection of free nucleotides. During DNA replication (Fig. 4-22), the two strands of DNA separate and the exposed bases in each strand base pair with free nucleotides, forming a complementary strand. Once base-pairing with triphosphates is complete, the enzyme DNA polymerase joins the nucleotides together in a reaction which is similar to the reaction which linked the nucleotides in RNA together. In contrast to the synthesis of mRNA, however, both strands of DNA act as templates for the synthesis of new DNA molecules. The end result is two identical molecules of DNA, each containing one strand of nucleotides present in the DNA molecule before duplication and one strand newly synthesized using the old strand as a template. When two identical molecules of DNA have been formed, one copy is passed on to each of the two daughter cells during cell division. Thus, each daughter cell receives the same set of genetic instructions as was originally present in the parent cell.

Cell division

Starting with a single fertilized egg cell, the first division produces two cells. When these daughter cells divide, they produce two cells, giving a total of four cells. When these four cells divide, they produce a total of eight cells, and these eight produce sixteen cells, ... Thus, starting from a single cell, four division cycles will produce 16 cells (2^4), 10 division cycles will produce $2^{10} = 1024$ cells, and 20 division cycles will produce $2^{20} = 1,048,576$ cells. If the development of the human body proceeded at a constant rate by the repeated cycle of cell division and growth, it would require only about 46 division cycles to produce all of the cells in the adult body, starting from a single cell.

The ability of a cell to divide and thereby reproduce itself is a major characteristic of most (but not all) living cells. We have described above how DNA is able to reproduce itself. A cell, however, is much more than a single molecule; it is a complex organized collection of membrane structures and molecules. The division of such a complex structure into two roughly equivalent parts involves marked alterations in the structure and metabolism of the cell. Aside from the morphological description of the various stages of the division

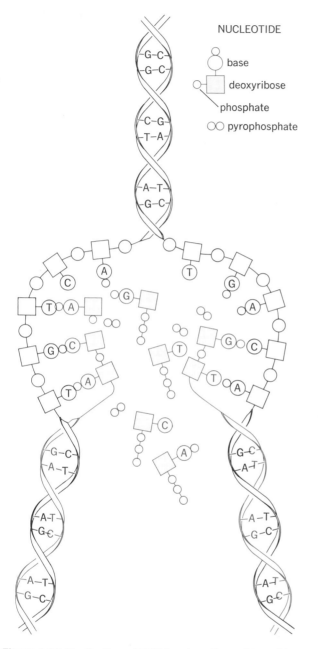

NUCLEOTIDE

base

deoxyribose

phosphate

pyrophosphate

Figure 4-22. Replication of DNA involves the pairing of free nucleotides with the bases of each DNA strand, giving rise to two new DNA molecules, each containing one old and one new nucleotide strand.

process as visualized with a light or electron microscope, we know surprisingly little about the underlying molecular events which occur during division or how the process is initiated and regulated. We therefore limit our discussion of cell division to a brief description of the various stages of the process and focus our attention on the net result, which is the duplication, packaging, and distribution of DNA molecules to the daughter cells.

Although the time between cell divisions varies considerably in different types of cells and some differentiated cells do not divide at all, the more rapidly growing cells divide about once every 24 h. During most of this period there is no visible evidence that the cell will undergo a division at some later time. For example, in a "24-h" cell, visible changes in cell structure begin to appear 23 h after the last division. The period between the end of one division and the beginning of the next is known as *interphase.* Since the actual process of cell division lasts only about 1 h, the cell spends most of its time in the interphase, and most of the cell properties described in this book are properties of interphase cells. However, one very important event related to the subsequent cell division does occur during the interphase, namely, the replication of DNA. A few hours after cell division, the process of DNA replication begins and proceeds over a period which may last 10 to 12 h, at which time each DNA molecule has been duplicated.

Just prior to cell division the duplicated DNA-protein threads (chromatin) become highly coiled and condense to form rod-shaped bodies known as *chromosomes.* The total amount of DNA found in a single human cell during interphase (when the chromatin threads are in the uncoiled state) could form a thread about 180 cm long, a distance which is about 100,000 times the diameter of a typical cell; thus, the coiling and condensation of chromatin to form chromosomes prior to cell division provide a way of transferring these long threads to daughter cells.

The first sign that a cell is going to divide is the appearance of the chromosomes in the nucleus. Each chromosome consists of two identical chromatin threads (recall that each DNA molecule had already replicated early in interphase) attached at a single point known as the *centromere.* Human cells, with the exception of the male and female reproductive cells, contain 46 chromosomes.

As the duplicated chromatin threads begin to coil, forming the chromosomes, the nuclear membrane breaks down and a new structure appears in the cell, the *spindle apparatus,* which consists of a number of microtubules. Some of these microtubules extend from one side of the cell to the other between two small cylindrical bodies known as *centrioles* (Fig. 4-23C). Other fibers pass from the centrioles and are attached to the centromere region of a chromosome.

As division proceeds, each chromosome separates at the centromere, and the two identical segments move toward the opposed centrioles (Fig. 4-23D). The spindle fibers act as if they were pulling the chromosome segments toward the poles although the actual mechanism of chromosome movement is still unknown. This process of nuclear division, which leads to the separation of identical sets of chromosomes, is known as *mitosis* (Greek *mitos,* thread). As the chromosome segments move toward opposite poles of the cell, the cell begins to constrict along a plane perpendicular to the spindle apparatus and constriction continues until the cell has been completely separated into two halves. It should be noted that mitosis is usually, but not always, followed by the division of the cell into daughter cells, each daughter cell receiving an identical set of chromosomes. Following division the spindle elements dissolve, the chromosomes become uncoiled, and a new nuclear membrane is formed in each daughter cell (Fig. 4-23E).

Cell differentiation

Since identical sets of chromosomes pass to each of the daughter cells during cell division, each cell in the body (with the exception of the reproductive cells) contains an identical set of DNA molecules. How, then, is it possible for one cell to become a muscle cell (and synthesize muscle proteins) while another cell containing the same molecules of DNA differentiates into a nerve cell (and synthesizes a different set of proteins)? We have already given the answer in earlier sections—different combinations of genes are active in the different cells. Thus, the genes which contain the information responsible for the synthesis of muscle proteins are able to synthesize mRNA in muscle cells. These same genes are also present in nerve cells

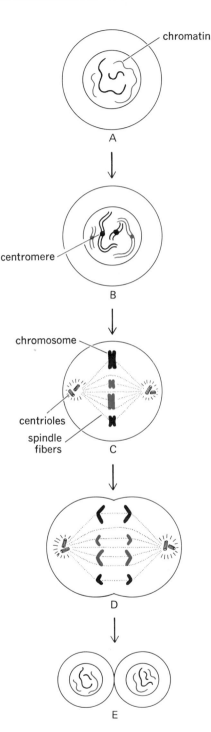

Figure 4-23. Chromosome condensation and separation during mitosis and cell division.

Figure 4-24. Cell differentiation depends upon the selective transcription of those genes associated with the specialized functions of the differentiated cell.

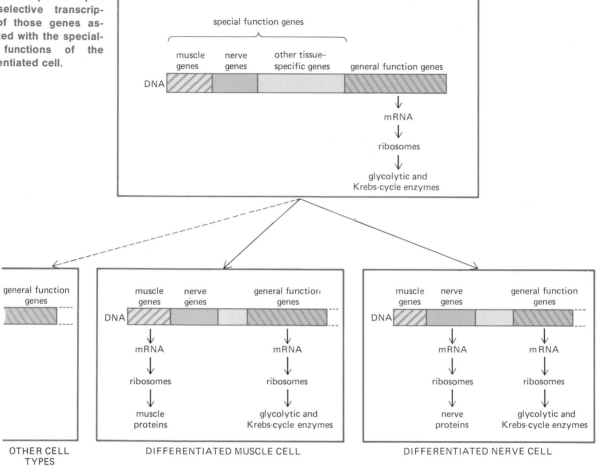

but are unable to form the corresponding mRNA and thus do not synthesize muscle proteins. Other genes are active in the nerve cell which are not active in muscle cells (Fig. 4-24). The problem of cell differentiation is thus related to the general problem of regulating protein synthesis. Certain genes appear to be "turned on" or "turned off" during cell differentiation. Mechanisms similar to those employed in enzyme induction and repression may be involved. To understand the complex process of cell differentiation a great deal more must be learned about the processes which regulate protein synthesis in the cells of higher organisms.

Mutation

In order to form the approximately 40 trillion cells of the adult human body a minimum of 40 trillion *individual* cell divisions must occur, and the DNA molecules in the original fertilized egg cell must be replicated at least 40 trillion times. If a secretary typed the same letter 40 trillion times, one would expect to find some typing errors. It is also not surprising to find that errors occur during the duplication of DNA, which result in an altered sequence of bases and a change in the genetic message. What is surprising is that DNA can be duplicated so many times with relatively few errors.

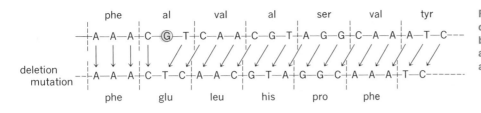

Figure 4-25. A deletion mutation caused by the loss of a single base, G, in the DNA sequence and the resulting misreading of a large number of code words.

Any alteration in the genetic message carried by DNA is known as a *mutation.*

Mutations may result from a multitude of factors, all of which ultimately lead to an alteration in the base sequence in DNA and thus a change in the genetic information. The simplest type of mutation occurs when a single base is incorrectly inserted at the wrong position in DNA. For example, the base sequence C–G–T forms the DNA code word for the amino acid alanine; if the guanine (G) is replaced by adenine (A) in this sequence, it becomes C–A–T, which is the code word for valine.

It is during the replication of DNA, when free nucleotides are being incorporated into new strands of DNA, that incorrect base substitution is most likely to occur. Even in the absence of specific agents which increase the likelihood of mutations, mistakes in copying DNA occur, so that the mutation rate is never zero. The major factors in the environment which increase the mutation rate are certain chemicals and various forms of ionizing radiation, such as x-rays, cosmic rays, and atomic radiation. Most of these factors cause the breakage of chemical bonds, so that incorrect pairings between bases occur or the wrong base is incorporated when the broken bonds are reformed.

Cells possess mechanisms for protecting themselves against certain types of mutation. For example, if an abnormal base pairing occurs in DNA, such as C with T (C normally pairs with G), a special set of enzymes will cut out the segment of one strand of DNA containing the abnormal base T, allowing the normal strand to resynthesize the deleted segment by normal base pairing. Certain diseases in which the cells of the skin tend to become cancerous when exposed to the ultraviolet radiation in sunlight appear to result from a lack of the DNA repair enzymes in these cells.

Mutations can also occur in which large sections of DNA are deleted from the molecule or in which single bases are added or deleted. Such mutations may cause the loss of an entire gene or group of genes or may cause the misreading of a large sequence of bases. Figure 4-25 shows the effect of removing a single base on the reading of the genetic code. Since the code is read in sequences of three bases, the removal of one base alters not only the code word containing that base but causes a misreading of all subsequent bases by shifting the reading sequence. Addition of an extra base would cause a similar phase-shift misreading.

When a single base is incorrectly substituted in a code word, the single amino acid in the protein coded by the gene containing the mutation is affected. However, because the genetic code has several different code words representing the same amino acid, a mutation in a single code word does not always alter the type of amino acid coded. We saw that substituting A for G in the alanine code word C–G–T produced the valine code word C–A–T. If, however, the mutation caused cytosine (C) to be substituted for thymine (T), the new codon C–G–C is one of several that correspond to alanine, and the protein formed by the mutant gene will not be altered in amino acid sequence in spite of the change in base sequence.

Assume that a mutation has altered the code word so that it now codes for a different amino acid, say the alanine C–G–T to valine C–A–T mutation. What effect does such a mutation have upon the cell? The effect depends upon both the type of gene and where in the gene the mutation has occurred. Although proteins are composed of many amino acids, the properties of a protein often depend upon a very small region of the total molecule, such as the binding site of an enzyme. If the mutation alters an amino acid not in the region of a binding site, there may be little or no change in the properties of the protein. On the other hand, if the mutation alters the binding site, a marked change in the properties of the protein may occur. If the protein is an enzyme, a mutation may render it totally inactive or change its specificity for substrate. The mutated enzyme may even catalyze an entirely different type of reaction. Thus, the alteration of a single amino acid may lead to a protein whose activity is unchanged, increased, or decreased or to a protein with entirely new properties.

Let us assume that the mutation leads to an enzyme that is totally inactive. If the enzyme is in a pathway supplying most of the cell's chemical energy, the loss of the enzyme may lead to the death of the cell. On the other hand, the enzyme may be involved in the synthesis of a particular amino acid, and if the cell exists in a medium containing that amino acid, the cell's functioning will not be impaired by the loss of the ability to synthesize its own amino acid from raw materials. In this case the mutation has led to a totally inactive enzyme, but since even the normal enzyme carries out no essential function in the cell (in that environment), its loss does not affect the life of the cell. The effect of such a mutation, however, becomes apparent if the cell is deprived of preformed amino acids; in such an environment, the enzyme performs an essential function, and the loss of its activity by mutation may lead to the death of the cell.

To generalize, mutation may have any one of three effects upon a cell: (1) It may cause no noticeable change in the cell's functioning; (2) it may modify cell function but still be compatible with cell growth and reproduction; (3) it may lead to the death of the cell. Any one of these three types of mutation can occur in any of the cells in the human body, but the overall consequences of mutation depend upon the type of cell in which it occurs. If a liver cell in an adult undergoes a mutation that causes it to function abnormally or even die, the effect upon the total organism is usually negligible since there are thousands of similar liver cells performing similar functions and the loss of one cell does not affect the overall functioning of the liver or the total organism. In contrast, since all the cells in the body are descended from a fertilized egg cell through repeated cell division, any mutation occurring during the early stages of development which does not lead to the immediate death of the cell is passed on to most of the cells in the developing organism. If the mutation is in an egg or sperm cell, all the cells descended from them inherit the mutation. Thus, mutations in the reproductive cells of the body do not affect the individual in which they occur but do affect the children produced by that individual.

Many types of abnormal functions in the body are the result of such genetic mutations which are passed on from generation to generation. Inherited diseases of this type are often due to a single gene, which either fails to produce an active protein or produces an abnormally active protein, and have been termed *inborn errors in metabolism.* An example is phenylketonuria, a disorder that can lead to one form of mental retardation in children. Because of a single abnormal enzyme, the victim is unable to convert the amino acid phenylalanine into the amino acid tyrosine at a normal rate. Phenylalanine is therefore diverted into other biochemical pathways in large amounts, giving rise to products which interfere with the normal activity of the nervous system. These products, excreted in large amounts in the urine, account for the name of the disease. The symptoms of the disease are prevented if the content of phenylalanine in the diet is restricted during childhood, thus preventing the accumulation of the products formed from phenylalanine. As our knowledge of biochemistry, genetics, and disease increases, more and more diseases are being found that are passed on from generation to generation and are the result of abnormally functioning enzymes.

Another category of diseases that is now thought to result, in many cases, from mutation is *cancer.* In some manner, presently not understood, the resulting abnormality in protein synthesis leads to the uncontrolled growth and cell division typical of cancer cells. As cancer cells continue to grow and divide, they form a mass known as a *malignant tumor,* which invades the surrounding tissues, disrupts the structure and function of organs, and eventually leads to the death of the organism. If cancer is detected in the early stages of its growth, the cancer cells can often be removed by surgery. Another property of cancer cells is their lack of adhesiveness to other cells, which allows them to break away from the parent tumor and spread by way of the circulatory system to other parts of the body (metastasis), where they continue to grow, forming multiple tumor sites. After the cancer cells have spread to different parts of the body, surgery is impossible as a means of removing all the cancer cells; thus, it is important to detect cancer in its early stages, when it can be effectively treated.

Although mutations may alter the properties of the protein coded by the mutant gene, all such mutations are not necessarily harmful to the cell in which they occur. In fact, mutation is the mechanism by which evolution occurs. Mutations may alter the activity of an enzyme in such a way that it is more rather than less active, or they may introduce an entirely new type of enzyme activity into a cell. If an organism carrying such a mutant gene is able to perform some function more effectively than an organism lacking the mutant gene, it has a better chance of surviving and passing the mutant gene on to its descendants. If the mutation produces an organism which functions less effectively than organisms lacking the mutation, the organism is less likely to survive and pass on the mutant gene. This is the principle underlying *natural selection.* Although any one mutation, if it is able to survive in the population, may cause only a very slight alteration in the properties of a cell, given enough time, a large number of small changes can accumulate to produce very large changes in the structure and function of an organism. The evolution of life on earth from the first cell to human beings has proceeded over a period of some 3 billion years. During this long period of time, the environment has selected those mutations which best enable the cells carrying them to survive and propagate.

Recombinant DNA

In 1972 a bacterial enzyme known as a *restriction enzyme* was discovered that splits molecules of DNA into a number of smaller fragments in a unique manner. Rather than splitting both strands of the DNA double helix at the same site, this enzyme cuts the two strands at two different locations, resulting in fragments having exposed segments of single-stranded DNA at each end (Fig. 4-26). The enzyme only acts at locations which have a particular sequence of nucleotides, and the resulting single-stranded ends therefore have exposed sets of complementary bases. These ends can undergo complementary base pairings with the ends of other DNA fragments produced in the

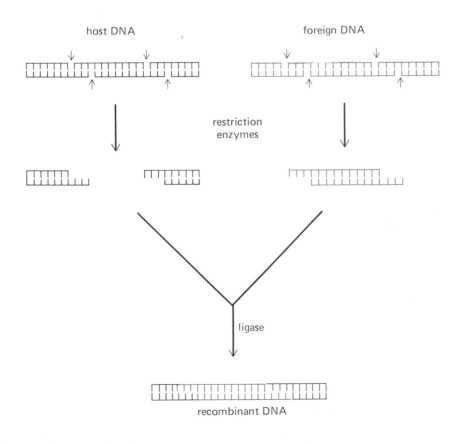

Figure 4-26. Formation of recombinant DNA by the use of bacterial enzymes.

host DNA   foreign DNA

restriction enzymes

ligase

recombinant DNA

same way. Therefore, when DNA molecules from two different organisms are treated with restriction enzyme, the resulting fragments can be recombined through base pairings between their single stranded ends (the ends of the fragments are covalently linked by another enzyme known as a *ligase*) to form a new molecule of DNA containing a mixture of genes from the two organisms. When this *recombinant DNA* is then inserted into a living cell, both sets of genes may be transcribed into mRNA and translated into proteins.

The potential benefits and hazards of this technique are many. For example, splicing the gene that codes for human insulin onto a bacterial DNA and inserting it into a bacterium would allow the organism to synthesize insulin; this hormone could then be extracted from the bacterium and used to treat diabetic patients who are unable to synthesize this hormone. On the other hand, it would be equally possible theoretically to insert a gene that codes for a toxic protein into a harmless bacterium that commonly inhabits the human gastrointestinal tract and thereby produce a widespread epidemic. Shortly after this technique was discovered, biologists placed a voluntary moratorium on further research utilizing it until its potential long-range effects could be evaluated and guidelines set. Research is once again proceeding under these guidelines, but only the future will tell whether actual benefits will outweigh the hazards of being able to manipulate the genetic constitution of organisms.

5

ENERGY AND CELLULAR METABOLISM

Energy
Activation energy
Determinants of biochemical reaction rates
  Chemical concentration: Law of mass action
  Enzymes
    Mechanisms of enzyme action • *Cofactors* • *Vitamins*
  Regulation of enzyme-mediated reactions
  Regulation of multienzyme metabolic pathways
ATP and cellular energy transfer
  The role of ATP

Oxidative phosphorylation
The Krebs cycle
Oxygen-independent ATP synthesis: Glycolysis
Photosynthesis
Metabolic pathways of carbohydrate, lipid, and protein
  Carbohydrate metabolism
    Carbohydrate catabolism • *Carbohydrate storage* • *Glucose synthesis*
  Fat metabolism
  Protein metabolism
  Summary

Metabolism (Greek: change) refers to the total collection of chemical reactions that occur within a living organism. Reactions which result in the fragmentation of a molecule into smaller and smaller parts are known as *degradative* or *catabolic reactions,* and those which put small molecular fragments together to form larger molecules are known as *synthetic* or *anabolic reactions.*

The molecules of a cell undergo continuous transformation as some molecules are broken down while others of the same type are being synthesized. Some molecules are replaced every few minutes, whereas other replacements may take years. Chemically, no person is the same at noon as at 8 o'clock in the morning, since during even this short period much of the body's structure has been torn apart and replaced with newly synthesized molecules. In adults the composition of the body is in a *dynamic steady state*[1] in which

[1] This has in the past been termed "dynamic equilibrium"; although "steady state" and "equilibrium" denote situations in which no net change is occurring within a system, a steady state requires continual energy input to the system to prevent change while a system in equilibrium does not.

the rates of chemical synthesis and degradation are in balance. Any factor which disturbs the balance between catabolism and anabolism can lead on the one hand to the destruction of the cell or on the other, to abnormal growth.

In addition to the overall dynamic steady state between catabolism and anabolism within a cell, there is a continuous input to and loss from the cell of organic molecules. The new molecules provide the raw materials used to synthesize those molecules that have been broken down and lost from the cell, and most important, they are ultimately the source of chemical energy used to maintain the structure and functioning of the cell.

In this chapter we focus attention on three basic aspects of metabolism: (1) The mechanisms for achieving highly specific types of chemical transformation in a cell. Most organic molecules do not readily undergo chemical reactions in a test tube, yet in a cell hundreds of different chemical reactions proceed simultaneously in an orderly fashion. How are these transformations achieved? (2) The mechanisms for releasing energy from molecules and making it available for the performance of cellular functions. (3) The mechanisms for regulating the chemical reactions within a cell. How are the rates of anabolism and catabolism kept in balance? How are the rates of energy production increased or decreased to meet the varying energy requirements of the cell during different states of activity?

Energy

The molecules taking part in a chemical reaction are composed of many atoms joined by chemical bonds. Several ways of breaking these bonds and rearranging them to form new combinations are possible. Which bonds will be broken and what types of new linkages will be formed are determined by the geometry of the molecules and the distribution of energy in the molecules at the time of their reaction. The concept of energy is the key to understanding the properties of chemical reactions and ultimately the properties of living cells. Energy is defined in dynamic terms as the ability to produce change, or more precisely, as the ability to perform work.

All physical and chemical change involves a redistribution of energy. Indeed, the presence of energy is revealed only when change is occurring. Whenever a change occurs in a system, energy must be transferred from one part of the system to another, but the total energy content of the system remains constant. That energy is neither created nor destroyed during any physical or chemical process is one of the most important axioms of science, the *first law of thermodynamics* or, more generally, the *law of the conservation of energy.*

The total energy content of any physical object consists of two components: the energy associated with the object because of its motion, called *kinetic energy,* and the energy associated with the object because of its position or internal structure, called *potential energy.* Movement is, therefore, a form of energy, and energy must be transferred to an object in order to produce movement. The amount of kinetic energy a moving object has is determined by its mass and its velocity. The faster an object moves, the greater its kinetic energy, and the larger the mass of an object, the greater is its kinetic energy at any given velocity. Kinetic energy is also associated with the motion of individual molecules, not just large objects. This kinetic energy of molecular motion is manifested as heat;

the hotter an object is, the faster its molecules move and the greater their kinetic energy. Thus, heat too is a form of energy—the kinetic energy of molecular motion.

Potential energy has the potential of becoming kinetic energy when it is released. For example, the energy transferred to a book by lifting it is present in the book as potential energy. When the book is dropped, this potential energy is converted into kinetic energy as it falls. When the book strikes the floor, the kinetic energy of its motion is converted into sound and heat energy, both of which are phenomena associated with increased molecular motion and are thus, themselves, forms of kinetic energy.

Chemical energy is a form of potential energy that is locked within the structure of molecules. It can be released during a chemical reaction in which chemical bonds are broken or formed. Since energy can neither be created nor destroyed during a chemical reaction, the difference in potential energy content between the reactant molecules and the product molecules must equal the amount of energy added or released during the reaction. For example, the reaction between hydrogen and oxygen to form water proceeds with the release of a considerable amount of heat energy, measured in units of *calories* (a calorie [cal] is the amount of heat energy required to raise the temperature of one gram of water one degree centigrade). Energies associated with chemical reactions are generally of the order of several thousand calories and are reported as kilocalories (1 kcal = 1000 cal). The formation of 1 mol of water from hydrogen and oxygen releases 68 kcal of heat energy, the full reaction being

$$H_2 + O \longrightarrow H_2O + 68 \text{ kcal/mol}$$

In other words, the chemical potential energy stored in 1 mol of water molecules is less by 68 kcal than that originally present in the hydrogen and oxygen molecules.

In order for a chemical bond (such as that between hydrogen and oxygen in the above equation) to be formed between two molecules, they must come close enough to each other for their electrons to interact. This happens when the random motion of the molecules brings them together in a collision. However, although merely the occurrence of the collision is a necessary event for bond formation to occur, it is not a sufficient one; in addition, an adequate amount of energy must be transferred during the collision. For example, if hydrogen and oxygen are mixed together at room temperature, water is formed at an extremely slow rate, but if the mixture is heated, water is formed rapidly; the explanation for this phenomenon is that heating the mixture increases the rate of molecular motion (kinetic energy) so that not only is the frequency of collisions increased but so is the magnitude of the kinetic energy transferred during each collision. The importance of this latter effect leads us to the concept of activation energy.

Activation energy

In the formation of new chemical bonds, the electric forces holding the atoms of a molecule together in a stable configuration must be disrupted before the altered arrangement of electrons in the new chemical bond can occur. To upset the balance of forces in the molecule, a quantity of energy known as *activation energy* must be added. The activation energy can be acquired through collisions with other molecules, in which kinetic energy transferred to the molecule increases the potential energy stored in its structure. Heating the mixture of hydrogen and oxygen in our example above increased the magnitude of the kinetic energy transferred during each collision by an amount sufficient to achieve the activation energy necessary for the reaction to occur.

As a mechanical analogy to a chemical reaction, consider the potential energy of a ball resting in a depression on top of a hill as representing the chemical potential energy stored in a molecule (Fig. 5-1). If the ball rolls down the hill, potential energy is converted into kinetic energy, just as chemical potential energy can be released as kinetic energy (heat) during a reaction. However, the ball will not roll down the hill spontaneously; it must be given a slight push. This push represents the activation energy necessary to initiate a chemical reaction. The magnitude of the activation energy can be represented by the size of the hump which the ball must roll over before it can roll down the hill. It is an important determinant of the rate of the reaction since only molecules which have acquired this amount of energy are able to react. At any given temperature the larger the activation energy, the fewer the number of mole-

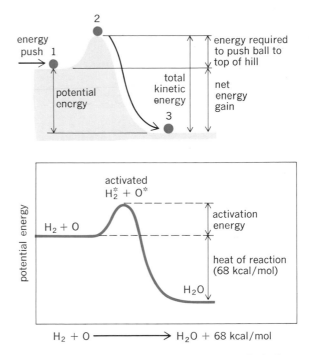

Figure 5-1. Changes in the potential energy of a ball moved from the top to the bottom of a hill compared with changes in chemical potential energy during a chemical reaction.

cules in the population that have this amount of energy and thus the slower the reaction.

Determinants of biochemical reaction rates

The factors which determine the rate of a chemical reaction are understandable in terms of the basic requirements for molecular interaction described above—the frequency of collision and the achievement of the required activation energy. In the previous sections, on several occasions, we made reference to heating a solution as the method for increasing collision frequency and supplying activation energy. The body temperature of 37°C (98.6°F) provides a warm environment in which molecules have a fairly high kinetic energy, but since it remains nearly constant, *changes* in chemical reaction rates in cells are not normally due to *changes* in temperature. Rather the major factors influencing reaction rates are molecular concentrations (which alter collision frequency) and the class of protein molecules known as enzymes, the function of which is to facilitate the acquisition of the required activation energy.

Chemical concentration: Law of mass action

In a reversible chemical reaction such as

$$A + B \underset{(2)}{\overset{(1)}{\rightleftharpoons}} C + D$$

Reactants   Products

no further change in the concentration of reactants or products will occur when the rates of the forward and reverse reactions become equal. Under these conditions of *chemical equilibrium*, molecules of A and B still react to form C and D, but since the latter molecules also react to form A and B at the same rate, no net change in the concentrations of any of the molecules occurs. If, after chemical equilibrium has been reached, the concentration of A is increased by adding more A to the solution, the rate of the forward reaction (1) increases since the larger number of A molecules increases the probability of a collision with a B molecule. The higher rate of the forward reaction leads to greater formation of C and D. As their concentrations rise, the rate of the reverse reaction (2) also rises, until a new equilibrium is achieved. At equilibrium the concentration of B will be decreased. It should be evident that conversion of A molecules into product molecules can also be achieved by increasing the concentration of B, rather than A, in the solution. However, note that in this case, at equilibrium, the concentration of A will be decreased. If the concentration of B molecules is made very large, almost all the A molecules present can be converted into product molecules.

The conversion of A molecules into product molecules can also be influenced by initially changing the concentration of product, rather than reactant, molecules. If the concentration of C is lowered by removing it from the reaction mixture, the rate of the reverse reaction (2) is lowered. Therefore, A and B molecules are not replaced as rapidly as they react to form C and D. The net effect is a decrease in the concentration of A and B. Thus, the conversion of A molecules into product molecules can be increased either by raising the concentration of A or B, or by lowering the concentration of one of the product molecules. This effect of molecular concentration on chemical reactions, known as the *law of mass action*, follows directly

from the effect of concentration on the frequency of molecular collisions.

The products formed in one reaction may become the reactant molecules necessary for another reaction, which in turn forms products taking part in further reactions, so that many reactions are interconnected by having certain molecules in common. In such a situation, altering the concentration of one of the molecules can affect a great many reactions, leading to increases or decreases in the concentrations of certain key molecules and shifting the balance between the rates of degradative and synthetic reactions.

Enzymes

During the nineteenth century chemists began to discover ways of accelerating the rates of chemical reactions which did not involve raising the temperature or changing initial reactant or product concentrations. Adding small amounts of certain substances, such as powdered platinum, was found to accelerate some reactions. Such substances are called *catalysts*. A catalyst does not undergo any net chemical alteration during the reaction and can be used over and over again; thus only small amounts of a catalyst are required to transform large amounts of reactants into products. A catalyst only accelerates the rates of reactions that would still occur spontaneously in its absence, although at a much slower rate.

The behavior of enzymes fits this description of a catalyst, and since all enzymes are proteins, an enzyme can be defined as a protein catalyst. (Although all enzymes are proteins, not all proteins are enzymes.) In order to accelerate a reaction an enzyme must come into contact with the reactant molecules, known as the *substrates* of the enzyme. The reaction between enzyme and substrate can be written

The enzyme combines with substrate to form an enzyme-substrate complex, which breaks down to release product molecules and enzyme. At the end of the reaction, the enzyme molecule is free to undergo the same reactions with additional substrate molecules. The overall effect is to accelerate the conversion of substrate molecules into product molecules with the enzyme acting as a catalyst:

$$\text{Substrates} \xrightarrow{\text{enzyme}} \text{products}$$

Mechanisms of enzyme action How does the formation of an enzyme-substrate complex result in the acceleration of the overall reaction? There are several proposed mechanisms at the molecular level, all of which have as a common denominator the transformation of a single-step reaction, with a high activation energy, into multiple steps, each having smaller activation energies. For example (Fig. 5-2), binding of the substrate to the enzyme surface (step 1) may distort the substrate structure, thereby producing a strain on some of its chemical bonds (like bending a stick to the point where it almost breaks), thus increasing the probability of its breakdown into products (step 2). When the bond is broken, the products may form transient linkages with the enzyme, but are soon released from the enzyme surface (step 3). Each of these steps requires a smaller activation energy than the direct conversion of substrate to products in the absence of enzyme.

The interaction between substrate and enzyme has all the characteristics described in Chap. 4 for the binding of a ligand to a protein binding site — specificity, affinity, and saturation. The region of the enzyme molecule to which the ligand (substrate) binds is known as the enzyme's *active site* (a term equivalent to "binding site"), and the shape of the enzyme molecule in the region of the active site provides the basis for its chemical specificity, since the shape of the active site must be complementary to the shape of the substrate (how *strongly* the substrate is bound to the active site determines not the specificity, but the *affinity* of the enzyme for its substrate). Enzymes are highly specific for the type of substrate molecule they act upon. Thousands of different reactions occur in a cell, and almost all of them are catalyzed by different enzymes. Some enzymes are so specific that they interact with only one particular type of substrate and no other. At the other extreme, some enzymes interact with a wide range of different substrates, all of which contain a particular type of chemical bond or grouping. Enzymes are generally named by adding the suffix *-ase* to the name of the substrate or to the type of reaction catalyzed by the enzyme. For example, the enzyme lactic dehydrogenase catalyzes the removal of hydrogen atoms from lactic acid.

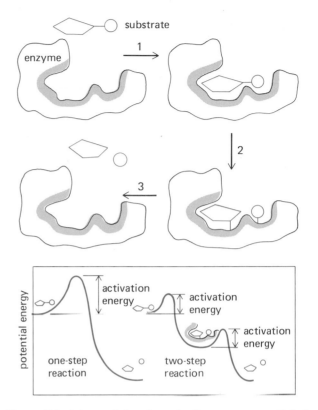

Figure 5-2. Intermediate steps in the enzyme-mediated conversion of substrate to product. Each intermediate step has a lower activation energy than the overall reaction. For simplicity, the activation energies of only two intermolecular steps are shown.

The catalytic activity of a single enzyme molecule is the rate at which it converts substrate to product, and each type of enzyme has a different catalytic activity. Because enzymes have a high catalytic activity at very low concentrations in cells, in some cases only a few molecules of enzyme per cell are sufficient.

Cofactors Many enzymes are inactive in the absence of certain nonprotein substances known as *cofactors*. Some cofactors are metal ions (such as magnesium, iron, zinc, or copper) which bind to specific regions on the enzyme and thereby either maintain the appropriate shape of the active site or participate directly in the binding of the substrate to the active site (by linking a charged region of the substrate to a similar region on the active site) (Fig. 5-3). In some cases, such as with iron,

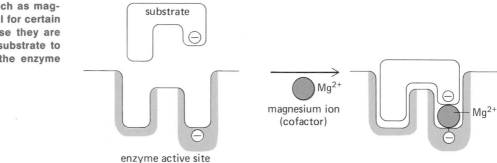

the metal directly participates in the electron exchanges occurring during the catalyzed reaction. Many of the essential trace elements (Chap. 3) appear to function as cofactors for specific enzymes.

In addition to metal ions, a number of nonprotein organic molecules, known as *coenzymes*, also function as cofactors. Coenzymes, most of which are derived from vitamins, may be covalently linked to the enzyme in the region of the active site or they may be loosely bound. They generally serve as carrier molecules which transfer atoms (such as hydrogen) or small molecular fragments containing one or two carbon atoms from one reaction to another. In the process of donating these atoms, the original coenzyme is reformed and can repeat the process. As an example, the enzyme lactic dehydrogenase requires the coenzyme nicotinamide adenine dinucleotide (NAD). In this reaction two atoms of hydrogen are removed from lactic acid.

$$CH_3\!-\!\overset{\displaystyle \overset{OH}{|}}{CH}\!-\!COOH + NAD \xrightarrow{\text{lactic}\atop \text{dehydrogenase}}$$

Lactic acid

$$CH_3\!-\!\overset{\displaystyle \overset{O}{\|}}{C}\!-\!COOH + NADH + H^+$$

Pyruvic acid

In one sense NAD acts as a substrate for the enzyme along with lactic acid, but the chemical fate of the NADH is quite different from that of pyruvic acid; pyruvic acid is metabolized through a number of reactions into various other products, whereas NADH is converted back into NAD through a second reaction, in which it donates hydrogen to another molecule, X:

$$X + NADH + H^+ \longrightarrow XH_2 + NAD$$

Thus, the NAD molecule has acted as a temporary carrier for hydrogen as it is transferred from one molecule to another. Only small amounts of the coenzyme NAD are necessary to maintain the enzyme's activity because of the continuous regeneration of NAD.

Vitamins Around the turn of the century, biologists studying the dietary requirements of animals found that trace amounts of certain organic molecules must be present in the diet to maintain normal health and growth. The Polish biochemist Funk, in 1912, believing these substances to belong to a group of organic molecules known as amines, gave them the name *vitamins* (Latin *vita*, life). A young animal placed on a diet lacking vitamins ceases to grow and eventually sickens and dies. Very small amounts of these vitamins will restore the animal to health. Over 20 different such substances have been shown to be essential in small amounts for the normal growth and health of various animal species.

The exact chemical structures of the first vitamins to be observed were unknown. Therefore, they were simply identified by letters of the alphabet (vitamin A, B, C, D, E, and K) and only later were their specific chemical structures identified (vitamin A = retinal, vitamin C = ascorbic acid, etc.). Vitamin B turned out to be composed of eight different substances now known as the vitamin B complex (B_1, thiamine; B_2, riboflavin; niacin; B_6, pyridoxine; pantothenic acid; folic acid; biotin; and B_{12}, cobalamine). The vitamins as a class have no particular chemical structure in common but they can be divided into the water-soluble vitamins—the B complex and vitamin C—and the fat-soluble (water-insoluble) vitamins—A, D, E, and K. The water-soluble vitamins function as coenzymes. For example, the essential ingredient in the coenzyme NAD is the B vitamin niacin. The fat-soluble vitamins in general do not function as coenzymes, and their specific roles in the body will be described in later chapters.

Vitamins are required in the diet because the body is unable to synthesize them, or is unable to synthesize adequate amounts of them. However, plants and bacteria have the enzyme machinery necessary for vitamin synthesis, and it is from these sources that we ultimately obtain vitamins. Since coenzymes can be used over and over again in a cycle of reactions, only small quantities of vitamins are required in the diet to replace those destroyed or excreted from the body. By themselves vitamins do not provide chemical energy, although they may participate in chemical reactions which release energy from other molecules. Increasing the amount of vitamins in the diet does not necessarily increase the activity of those enzymes for which the vitamin functions as a coenzyme. Only very small quantities of coenzymes are necessary to saturate their binding sites on the enzyme molecules, and increasing the concentration above this level does not increase the enzyme's activity. Obviously, a lack of vitamins in the diet causes ill effects since they are essential for the activity of many enzymes, especially those involved in providing chemical energy for cellular functions.

Quantities of vitamins in the diet that exceed the small quantities required for the functioning of enzymes have not yet been proven to have beneficial effects, but this possibility is being investigated. On the other hand, in the case of the fat-soluble vitamins, large quantities are known to have harmful effects.

Regulation of enzyme-mediated reactions

The rate of an enzyme-mediated reaction depends on three factors: substrate concentration, enzyme concentration, and enzyme activity (Fig. 5-4).

Increasing the concentration of substrate will, by mass action, increase the rate of the reaction. The concentration of a given substrate in a cell may vary for a variety of reasons. It may increase because the circulation delivers more substrate to the cell, because of changes in membrane transport which allow more substrate to enter the cell, or because other reactions in the cell are producing the substrate at a faster rate. However, there is a limit to the extent to which mass action will increase the rate of the reaction and this limit is reached when the enzyme becomes saturated with substrate. Enzymes with a high affinity for substrate become saturated at lower substrate concentrations.

The second way of altering the reaction rate is to change the concentration of enzyme; this has the effect of providing more or less sites with which substrate can react. Thus, for any given substrate concentration, the rate of a reaction will depend on the amount of enzyme present. Certain reactions proceed faster in some cells than in others because more enzyme is present. The amount of enzyme depends on the rate at which enzyme is synthesized and its rate of degradation; for most enzymes these rates are fairly constant and

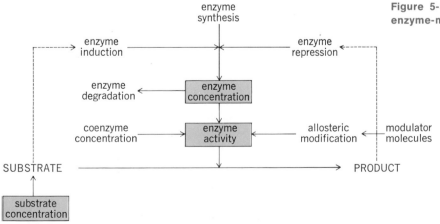

Figure 5-4. Factors which affect the rate of enzyme-mediated reactions.

thus their concentrations do not change appreciably during different states of cell activity. The concentrations of certain enzymes, however, are altered by controlling, through the mechanisms of enzyme induction and repression, transcription of their genes from DNA to messenger RNA (Chap. 4). Through enzyme induction, certain enzymes are synthesized only when substrate is available; through repression, the accumulation of product molecules turns off the synthesis of the enzyme responsible for forming the product. Both enzyme induction and repression are highly selective regulatory processes which act upon specific enzymes rather than affecting all of the enzymes in a cell.

The third factor determining the rate of an enzyme-mediated reaction is the catalytic activity of the individual enzyme molecules. For certain enzymes, the concentration of cofactors directly affects the activity of the enzyme, and as we shall see, the concentrations of coenzymes change during different states of metabolic activity. A completely different cause of changes in enzyme activity is allosteric regulation, the process described in Chap. 4 whereby the binding of a modulator molecule to one site on a protein alters the shape (and, thereby, the activity) of a second binding site on the same molecule. The products of certain reactions function as such modulator molecules, binding to allosteric enzymes and thereby producing either an increase or decrease in the activity of the enzyme. Note that such allosteric regulations produce changes in the activity of already existing enzyme molecules without altering the enzyme concentration, whereas enzyme induction and repression alter the enzyme's concentration without altering the activity of individual enzyme molecules.

Regulation of multienzyme metabolic pathways

The sequence of enzyme-mediated reactions leading to the formation of a particular product is known as a *metabolic pathway*. For example, the 19 reactions which convert glucose to carbon dioxide and water constitute the metabolic pathway for glucose degradation. Another set of enzymes make up the metabolic pathway for fatty acid synthesis. Moreover, since glucose can be converted into fat, there must be metabolic connections between these two pathways and some

mechanism for determining whether glucose will be degraded to carbon dioxide and water or converted into fat. Ultimately the regulation of these pathways depends upon the interaction of the factors discussed in the previous section which determine the rates of individual reactions. In this section we apply these general principles to the regulation of the flow of material through multienzyme metabolic pathways.

Consider a general metabolic pathway consisting of four enzymes (e), leading from an initial substrate A to the end product E, through a series of intermediates, B, C, and D

$$A \xrightarrow{e_1} B \xrightarrow{e_2} C \xrightarrow{e_3} D \xrightarrow{e_4} E$$

By mass action, increasing the concentration of A will lead to an increase in the concentration of B and so on until eventually there is an increase in the concentration of the end product E. If, as a result of the increasing concentration of A and the intermediates, one of the enzymes in the pathway becomes saturated with substrate, the rate of the overall pathway will no longer increase with increasing concentration of A. Such an enzyme becomes rate-limiting for the flow of material through the entire pathway. Therefore, by regulating the activity of this one rate-limiting enzyme, the rate of flow through the whole pathway can be increased or decreased. Thus it is not necessary to alter the activities of all the enzymes in a metabolic pathway to be able to control the overall rate at which the final end product is produced. These rate-limiting enzymes are usually the sites of allosteric regulation in metabolic pathways. Let us assume that enzyme e_2 is the rate-limiting enzyme in this pathway. The end product E may interact with enzyme e_2 to produce an allosteric inhibition of its activity (Fig. 5-5), and thus E becomes the modulator molecule for this enzyme. This form of *end-product inhibition* is quite common in many synthetic pathways. It effectively prevents an excessive accumulation of end product when the end product is not being utilized.

Let us now add another level of complexity. Suppose the reaction

$$A \xrightarrow{e_1} B$$

is reversible:

$$A \xrightleftharpoons{e_1} B$$

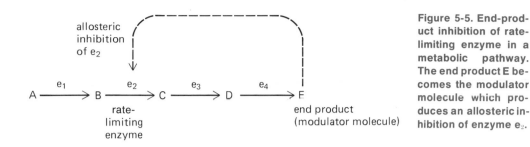

Figure 5-5. End-product inhibition of rate-limiting enzyme in a metabolic pathway. The end product E becomes the modulator molecule which produces an allosteric inhibition of enzyme e_2.

Note that the same enzyme (e_1) catalyzes the reaction in both directions. This is a very important point: An enzyme speeds up a reaction but does not determine the *direction* in which the reaction proceeds. Therefore, in a reversible reaction, increasing the concentration or activity of the mediating enzyme will speed the reaction in both directions.

In theory, all chemical reactions are reversible. However, whether any given reaction is spontaneously reversible to any extent, given a finite amount of time, depends upon the amount of chemical energy exchanged during the reaction. If there has been little change in the energy content of the substrate and product molecules, the reaction is readily reversible. However, if large exchanges of energy have occurred, this energy must be made available if the reaction is to be reversed. Therefore, in practice, reactions can be classified into two categories, those that are reversible because they involve only small exchanges of energy and those which are irreversible because of the release of large quantities of energy. (It should be emphasized that it is the quantity of energy released and not the property of the enzymes which determines the reversibility of a given reaction.)

Reversible reaction: $A \xrightleftharpoons{e_1} B$
   + small amount of energy

Irreversible reaction: $C \xrightarrow{e_3} D$
   + large amount of energy

The product D in the irreversible reaction above can, however, be converted into C, provided a source of sufficient energy is added to the system. Often this energy is provided by coupling the reaction to the simultaneous breakdown of another substrate which releases large quantities of

energy. Thus, an irreversible step can be "reversed" through an alternate route, using a second enzyme and an additional substrate to provide the required energy. Two such high-energy irreversible reactions catalyzed by separate enzymes are indicated by bowed arrows to emphasize that two distinct enzymes are involved in the two directions:

$$C \underset{e_5}{\overset{e_3}{\rightleftharpoons}} D$$

In summary, consider the following modification of our general metabolic pathway:

$$A \xrightleftharpoons{e_1} B \xrightleftharpoons{e_2} C \underset{e_5}{\overset{e_3}{\rightleftharpoons}} D \xrightarrow{e_4} E$$

The first two reactions are low-energy reversible reactions, each catalyzed by a single enzyme. The third reaction is a high-energy reaction having separate enzymes catalyzing the reactions in each direction. The fourth reaction is a high-energy irreversible reaction catalyzed by a single enzyme. In this metabolic pathway there is no route by which E can be converted back into A because of the irreversible last step. However, D and other intermediates can all lead to the formation of A by reversal of the pathway. The third step in the pathway is mediated by two separate enzymes and provides an example of how the direction of flow in a metabolic pathway can be regulated. If the two separate enzymes e_3 and e_5 are subject to separate allosteric control, it is possible to inhibit e_5, allowing the reactions to proceed in the direction of product E or alternatively to inhibit e_3 and allow D to be converted into A.

The combination of reversible and irreversible reactions, together with the variety of mechanisms for controlling enzyme activities, provides the cell with a wide range of controls for regulating metabolic activity. When one considers the thousands of reactions that occur in the body, the permutations and combinations of possible control points, the overall result is indeed staggering, and the details of metabolic regulation at the enzymatic level are beyond the scope of this book. In the remainder of the chapter we shall consider the general pathways by which cells obtain energy and metabolize carbohydrates, fats, and proteins, and indicate some of the key control points in these pathways.

ATP and cellular energy transfer

The role of ATP

The functioning of a cell depends upon its ability to extract and use the chemical potential energy locked within the structure of organic molecules. When a mole of glucose is broken down in the presence of oxygen into carbon dioxide and water, 686 kcal of energy is released. In a test tube all this energy is released as heat which can be used to perform work, as in a steam engine. However, a cell cannot transform heat energy into work. Rather, in a cell some of the chemical energy released is captured by transferring chemical potential energy from one molecule to another; i.e., energy lost by one molecule is transferred to the chemical structure of another and thus does not appear as heat.

In all cells, from those of bacteria to those of human beings, the major such energy-carrier molecule is *adenosine triphosphate* (ATP) (Fig. 5-6). (Note that the monophosphate derivative of ATP [AMP] is found in one of the four nucleotides found in DNA and RNA.) ATP is synthesized from adenosine diphosphate (ADP) and inorganic phosphate, and 7 kcal/mol must be added to achieve this reaction:

$$ADP + P_i + 7 \text{ kcal/mol} \rightarrow ATP + H_2O$$

The energy (7 kcal/mol) is transferred to ATP as a result of the catabolism of carbohydrates, fats, and protein. It is then released during subsequent breakdown of ATP:

$$ATP + H_2O \rightarrow ADP + P_i + 7 \text{ kcal/mol}$$

and is used to perform work by the cell (muscle contraction, the active transport of molecules across cell membranes, and the synthesis of organic molecules).

Energy is constantly cycled through ATP molecules in the cell. A typical ATP molecule may exist for only a few seconds before its energy is transferred to another molecule, and the ADP then formed is rapidly reconverted into ATP through coupling to energy-releasing reactions, i.e., the breakdown of carbohydrates, lipids, or proteins. Although energy is present in the structure of the ATP molecule, its function is not to *store* energy but to *transfer* it from one molecule to another. Thus the total energy stored in all the ATP molecules of the cell can supply the cell's energy requirements for only a fraction of a minute. The

Figure 5-6. Breaking the chemical bond that joins the terminal phosphorus atom to the ATP molecule releases 7 kcal/mol of energy and forms ADP and inorganic phosphate.

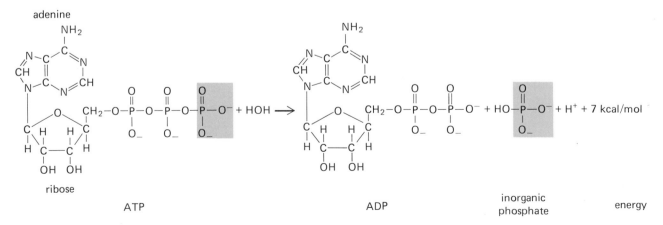

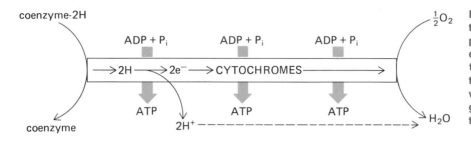

Figure 5-7. Energy is coupled to the formation of ATP at three points in the cytochrome chain during oxidative phosphorylation. Some coenzyme-2H enters this chain beyond the point at which the first ATP is formed, giving rise to two rather than three ATP.

actual storage of energy in the body is performed by the molecules of carbohydrate, lipid, and protein which transfer energy to ATP.

There are two general mechanisms by which this transfer occurs—*substrate phosphorylation* and *oxidative phosphorylation.* Substrate phosphorylation occurs in the cytoplasm when one of the phosphate groups on a catabolic intermediate is transferred directly to ADP to form ATP. In contrast to substrate phosphorylation, oxidative phosphorylation takes place in the mitochondria and occurs through the highly specialized coupling process to be described in the next secton. It is quantitatively far more important than substrate phosphorylation in that it accounts for 95 percent of the ATP molecules synthesized from glucose breakdown and 100 percent of those from fatty acid breakdown. For this reason, the mitochondrion has often been called the powerhouse of the cell.

Oxidative phosphorylation

In the breakdown of carbohydrates, fats, and proteins many of the chemical reactions involve the removal of hydrogen atoms from various intermediates formed during degradation. These reactions usually require a coenzyme, such as NAD, to which the hydrogen atoms are transferred during the reaction. Since coenzymes other than NAD also transfer hydrogen we shall use the general reaction

$$R\text{-}H_2 + coenzyme \rightleftharpoons R + coenzyme\text{--}2H$$

to represent the transfer of two hydrogens.[1] As a result of such hydrogen transfers, some of the chemical potential energy in molecules of carbo-

hydrate, fat, and protein is transferred to the co-enzyme molecules. The process of oxidative phosphorylation then uses this energy to synthesize ATP. Since the hydrogen atoms can be obtained from a wide variety of organic molecules, the process illustrates how energy obtained from many sources can be funneled into the common energy carrier, ATP.

How does the energy transferred to the hydrogen-coenzyme molecules actually get passed on to ATP? This involves a complex series of reactions which take place on the inner membranes of the mitochondria. These membranes contain a group of iron-containing proteins known as *cytochromes* (because they are brightly colored) which accept a pair of electrons from two hydrogens (Fig. 5-7) and, after passing these electrons through a sequence of different cytochromes, eventually donate them to molecular oxygen, which then reacts with the hydrogen ions formed at the beginning of the sequence of reactions to form water. The overall reaction can be written

$$coenzyme\text{--}2H + \tfrac{1}{2}O_2 \xrightarrow{cytochromes} coenzyme + H_2O + 52\ kcal/mol$$

The total result is energetically similar to the reaction between hydrogen and oxygen to form water with the release of heat energy.

Why must the cell proceed through a complex sequence of reactions to achieve what is essentially the formation of water from hydrogen and oxygen? If coenzyme–2H reacted directly with oxygen, all 52 kcal of energy would be released, but since only 7 kcal of energy is necessary to form ATP from ADP and inorganic phosphate, the remaining 45 kcal of energy would appear as wasted heat. In contrast, by passing through the cytochrome chain, the same 52 kcal of energy is released but in small steps along with each elec-

[1] For simplicity this convention ignores the fact that for some coenzymes the actual form of coenzyme-2H is coenzyme-H + H+.

tron transfer between successive cytochromes. The energy released at a number of these steps can be coupled to ATP synthesis, resulting in the formation of either two or three molecules of ATP from each reaction between coenzyme–2H and oxygen (Fig. 5-7) rather than only one. The overall efficiency of the process is 40 percent, the remaining 60 percent of the energy appearing as heat.

It should be emphasized that, aside from the supply of energy, an important result of oxidative phosphorylation is the removal of hydrogen from the coenzyme, which then becomes available again to pick up more hydrogen and transfer it to the cytochrome chain. Without the return of the coenzymes to their original state, the enzymes which require this form of the coenzymes would become inactive, and the metabolic pathways containing them would cease to function.

Note also that if oxygen is not available to the cytochrome system, ATP will not be formed by the mitochondria. As we shall see, an alternate oxygen-independent pathway does exist for non-mitochondrial synthesis of ATP, but it is not adequate to provide enough ATP to maintain structure and function of most cells for long. The cause of death from cyanide poisoning is the same as that from a lack of oxygen. Cyanide reacts with the final cytochrome in the chain, preventing electron transfer to oxygen and blocking the production of ATP by the mitochondria.

The Krebs cycle

As we have just seen, most ATP is formed by the process of oxidative phosphorylation in which hydrogen atoms derived from the catabolism of fuel molecules are transferred by way of coenzymes to the mitochondria and ultimately combined with molecular oxygen. The most important catabolic pathway giving rise to these hydrogen atoms is a pathway known as the *Krebs cycle*. Various intermediates derived from the breakdown of carbohydrates, fats, and proteins can enter the Krebs cycle, but the most common one is the 2-carbon *acetyl* group, linked to a molecule known as *coenzyme A*. Just as hydrogen is transferred from one reaction to another by way of a coenzyme carrier-molecule, 2-carbon acetyl fragments are transferred from one reaction to another by way of coenzyme A. Coenzyme A is a derivative of the B vitamin pantothenic acid and has a terminal

sulfhydryl group (R—SH) to which the acetyl fragment is covalently linked:

$$CH_3-\overset{\overset{\displaystyle O}{\|}}{C}-R + CoA-SH \rightleftharpoons CH_3-\overset{\overset{\displaystyle O}{\|}}{C}-S-CoA + R-H$$

coenzyme A   acetyl coenzyme A

We shall encounter a number of metabolic reactions in which coenzyme A functions in the transfer of acetyl groups.

The Krebs-cycle sequence of seven reactions is diagramed in Fig. 5-8. The enzymes for this pathway are in the mitochondria along with the enzymes for oxidative phosphorylation, the two pathways being linked by way of the coenzyme hydrogen-carriers. In the first reaction of the Krebs cycle, a 2-carbon acetyl fragment is transferred from acetyl coenzyme A to a 4-carbon molecule, forming the 6-carbon molecule citric acid. Following an internal rearrangement (reaction 2) one of the carboxyl groups (—COOH) is split off (reaction 3) giving rise to the first molecule of carbon dioxide formed in the Krebs cycle. A similar reaction occurs at step (4), releasing a second molecule of carbon dioxide. At this point, the 6-carbon citric acid has been degraded to a 4-carbon molecule through the loss of two molecules of carbon dioxide. This is equivalent to degrading each of the two carbon atoms in the acetyl fragment to carbon dioxide (although the carbon atoms which appear in carbon dioxide are not the actual carbon atoms that were donated to the pathway by acetyl coenzyme A). Note that the oxygen that appears in carbon dioxide is not derived from molecular oxygen but from the oxygen in the carboxyl groups of the Krebs-cycle intermediates.

Following the removal of the two carbon dioxide molecules, the remaining 4-carbon intermediate undergoes three additional reactions and ends up as the 4-carbon molecule to which the acetyl group was donated at the beginning of the cycle. Thus, the series of reactions operates in a cycle; a 2-carbon fragment derived from the degradation of fuel molecules is combined with a 4-carbon intermediate and two molecules of carbon dioxide are released during one turn of the cycle, thereby reforming the original 4-carbon intermediate which can then repeat the cycle. In 1952 Hans Krebs received the Nobel Prize for discovering this metabolic pathway named in his honor. It is also

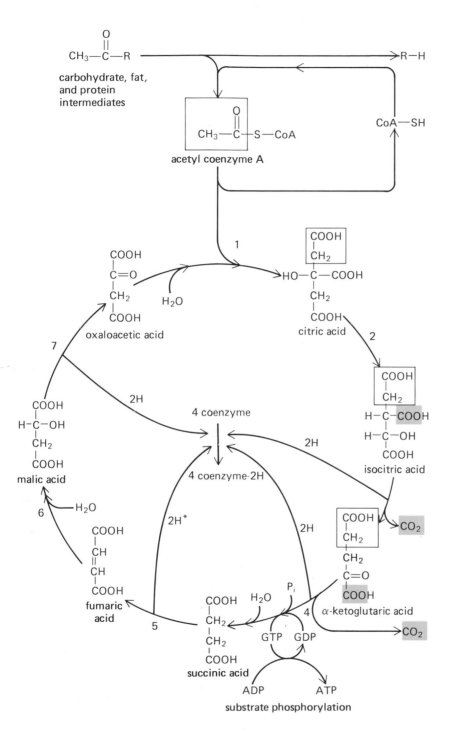

Figure 5-8. Krebs-cycle pathway by which fuel molecules are catabolized to CO_2, and hydrogen is transferred to coenzyme carrier molecules. The two hydrogens ($2H^*$) transferred to coenzyme in reaction 5 produce only two ATP during oxidative phosphorylation, the other three pairs of hydrogen each give rise to three ATP.

known as the citric acid or tricarboxylic acid cycle because of the presence of these intermediates.

Note that in the Krebs-cycle pathway, no molecular oxygen is consumed and only one ATP is formed; i.e., the Krebs cycle reactions, themselves, generate almost no ATP. Rather, it is in the process of oxidative phosphorylation acting upon the eight hydrogens released to the coenzymes in the process of the Krebs cycle reactions that oxygen is consumed and ATP formed. To reiterate, the Krebs cycle is *not* identical to oxidative phosphorylation; it is a totally distinct pathway which generates coenzyme–2H molecules which can then enter the oxidative phosphorylation pathway (by the same token, note that CO_2 is produced by the Krebs cycle reactions but not by oxidative phosphorylation).

We can now describe the origin and fate of the hydrogens passed to the coenzymes in the Krebs cycle. Pairs of hydrogen atoms are transferred from intermediates to coenzyme molecules at reactions (3), (4), (5), and (7), producing a total of four coenzyme–2H molecules for each 2-carbon fragment that passes through the cycle. These hydrogens are then passed on to oxygen by way of oxidative phosphorylation. As we have seen, there are three sites along the oxidative phosphorylation pathways at which ATP can be formed from ADP and inorganic phosphate. Thus, for each pair of hydrogen atoms delivered to the oxidative phosphorylation pathway it might be expected that three molecules of ATP would be formed (Fig. 5-7), and this is true for three of the coenzyme–2H molecules. However, the type of coenzyme hydrogen involved in reaction (5) enters the oxidative phosphorylation pathway beyond the site where the first ATP is formed, and therefore, only two molecules of ATP are produced from its hydrogens.

Reaction (4) is very complex, actually involving three enzymes and five coenzymes. During this reaction a molecule of inorganic phosphate is linked by substrate phosphorylation to guanosine diphosphate (GDP) to form the high-energy molecule guanosine triphosphate (GTP) (recall that substrate phosphorylation is the direct transfer of phosphate from a substrate). The energy stored in GTP can be transferred to ATP by the following enzymatic reaction:

$$GTP + ADP \rightarrow GDP + ATP$$

The overall result of reactions (4) and (5) is to form one molecule of ATP indirectly by way of substrate phosphorylation and two molecules of ATP by oxidative phosphorylation. Thus, the net formation of ATP linked to reactions in the Krebs cycle is a total of 12 ATP from each 2-carbon acetyl fragment that enters the cycle:

$$CH_3-\overset{\overset{\displaystyle O}{\|}}{C}-R + 2O_2 + 12ADP + 12P_i \rightarrow$$
$$2CO_2 + R-H + H_2O + 12ATP$$

This completes our description of the Krebs cycle is a total of 12 ATP from each 2-carbon carbohydrates, lipids, and proteins are catabolized to acetyl CoA (and other substances which can enter the cycle).

Oxygen-independent ATP synthesis: Glycolysis

As discussed above, the formation of ATP by oxidative phosphorylation is dependent on the transfer of hydrogen to oxygen and is thus dependent on the availability of oxygen. ATP can also be formed in the absence of oxygen by substrate phosphorylation in a series of reactions that is linked to the breakdown of carbohydrate.

When the first cells were formed some 3 to 4 billion years ago, there was no molecular oxygen in the earth's atmosphere (since atmospheric oxygen is a product of photosynthesis occurring in plant cells, which did not exist then). These first cells obtained their energy from the anaerobic ("without oxygen") breakdown of organic molecules which had accumulated in the oceans over millions of years. Anaerobic pathways for energy release have been retained throughout evolution, and some unicellular organisms today are still able to grow and multiply in the absence of oxygen, using only these pathways to provide the energy for ATP synthesis. As cells evolved, some modifications did occur in the pathway with regard to the final end products formed, but the first 10 enzymatic steps in the pathway (from glucose to pyruvic acid) remained the same. In *alcoholic fermentation*, which occurs in some species of bacteria and yeasts, the final end products formed from pyruvic acid are alcohol and carbon dioxide. In other microorganisms and in all higher organisms, including human beings, the final end product formed from pyruvic acid is lactic acid. This pathway from glucose to lactic acid is known as

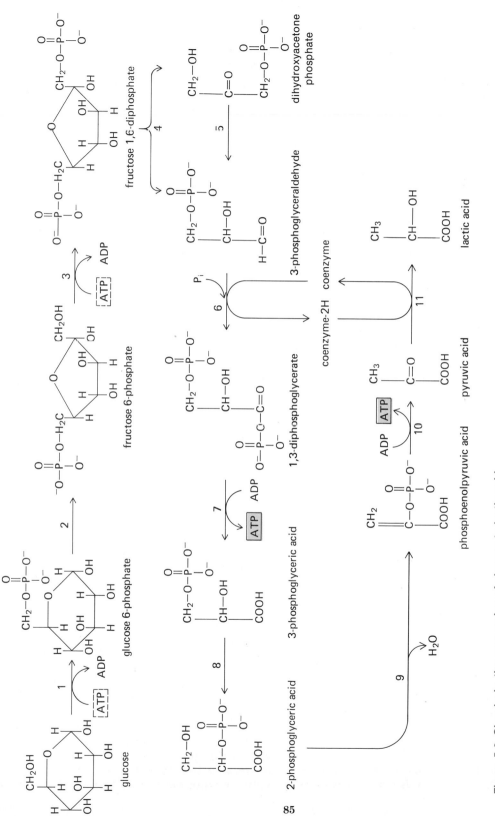

Figure 5-9. Glycolysis, the conversion of glucose to lactic acid.

85

glycolysis. Unlike the Krebs cycle, which can be entered by any fuel type—carbohydrate, lipid, or protein—capable of generating acetyl CoA (or another entry molecule) glycolysis is strictly a carbohydrate pathway, beginning with glucose or a glucose intermediate.

Each of the 11 reactions of glycolysis (Fig. 5-9) is catalyzed by a separate enzyme which has chemical specificity for its particular type of substrate (these 11 enzymes are dissolved in the fluid portion of the cytoplasm and are not associated with any particular cell organelle). Each reaction produces only a small change in the structure of the substrate, usually involving the breaking of a single chemical bond followed by the formation of a new bond with a different group. By such a sequence of small steps, one molecule of glucose is transformed into two molecules of lactic acid. The chemical transformation in glycolysis (Fig. 5-10), involving ADP, ATP, inorganic phosphate, and the coenzyme involved in hydrogen transfer, can be written:

Glucose + 2ATP + 4ADP + 2P$_i$ + 2 coenzyme → 2 lactic acid + 2ADP + 4ATP + 2 coenzyme + 2H$_2$O

In the early stage of glycolysis (reactions [1] and [3]), two molecules of ATP are used to add phosphate groups to glucose, but this energy debt will be repaid with interest when four molecules of ATP are formed later in the pathway. The addition of these ionized phosphate groups to glucose causes the glycolytic intermediates to become trapped in the cell since ionized molecules do not readily cross cell membranes (for reasons that will be discussed in the next chapter). The last two molecules in the pathway, pyruvic acid and lactic acid, lack ionized phosphate groups and are able to cross cell membranes. Therefore, whenever there is an increased production of lactic acid, as for example during intense exercise, the blood concentration of lactic acid increases due to its leakage across skeletal muscle membranes.

The first synthesis of ATP in glycolysis occurs by substrate phosphorylation during reaction (7) in which one of the phosphate groups on a glycolytic intermediate is transferred to ADP to form ATP. (Recall that substrate phosphorylation is the direct transfer of phosphate from a substrate to ADP, in contrast to oxidative phosphorylation which uses the energy released from the combination of hydrogen and oxygen to couple free inorganic

Figure 5-10. Glycolytic pathway leading to the net synthesis of two molecules of ATP.

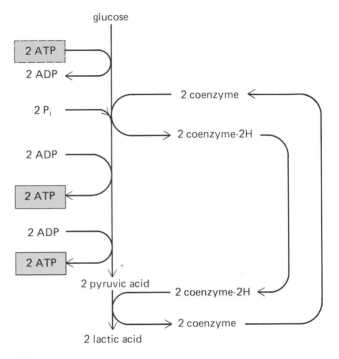

phosphate to ADP in the mitochondria.) A similar substrate phosphorylation occurs at reaction (10). Since reactions (4) and (5) convert one 6-carbon molecule into two 3-carbon molecules, and each 3-carbon intermediate forms a molecule of ATP at reaction (7) and another at reaction (10), each glucose yields a total of 4 new molecules of ATP during its breakdown to lactic acid. However, since two ATP were used in the early stages, there is an overall net gain of only two molecules of ATP (4ATP produced − 2ATP consumed).

In higher organisms, the amount of ATP produced by glycolysis is very small compared to that formed by oxidative phosphorylation. Despite this, there are special cases in which glycolysis plays the major role in ATP generation: (1) several cell types derive all of their limited energy from glycolysis either because they possess no mitochondria (the red blood cell, for example) or because they always receive a very low supply of oxygen; (2) skeletal muscle cells, as mentioned above, may produce large quantities of ATP by glycolysis during periods of intense activity when oxygen supply does not keep pace with energy utilization; (3) most, if not all, cells are capable of withstanding very short periods of inadequate oxygen supply by using glycolysis.

If glucose is to be broken down to lactic acid, not only must the glycolytic enzymes be present, but ATP, ADP, and inorganic phosphate must be available to participate as substrates in their appropriate reactions. In addition, a coenzyme to which hydrogen atoms are transferred in reaction (6) is required. Without this coenzyme, reaction (6) would not proceed and the subsequent formation at reactions (7) and (10) would not occur. Once the coenzyme has received hydrogen it must be able to transfer it to another reaction and thus regenerate its original form. Such a mechanism exists within the glycolytic pathway itself: Note that the last reaction, the conversion of pyruvic to lactic acid, requires hydrogen, and the source of this hydrogen is the same coenzyme that picked up hydrogen in reaction (6); thus, by donating the hydrogen to pyruvic acid, the coenzyme regenerates its original form which can accept once again another hydrogen in reaction (6); i.e., the coenzyme is not consumed during glycolysis, but merely shuttles hydrogen atoms between reactions (6) and (11).

We have emphasized that glycolysis is a biochemical pathway that can form ATP in the absence of oxygen, but in fact the first 10 reactions from glucose to pyruvic acid also occur when oxygen is present. Under these aerobic conditions the coenzyme–2H that is formed by reaction (6) is regenerated by oxidative phosphorylation in the mitochondria, rather than by transferring its hydrogen to pyruvic acid to form lactic acid. Instead of being converted to lactic acid, pyruvic acid is broken down to acetyl CoA, which then enters the Krebs cycle. Whether oxygen is present or not, a net synthesis of two molecules of ATP will still result from the breakdown of glucose to pyruvic acid. To reiterate, in the absence of oxygen the end product of glucose metabolism will be lactic acid, while in the presence of oxygen, glucose will be completely metabolized to carbon dioxide and water.

Photosynthesis

Since energy is released during the breakdown of carbohydrates, fats, and proteins, energy must have been supplied to these molecules during the course of their synthesis. A cell can synthesize a molecule of fatty acid, for example, which contains a high energy content, by using energy (ATP) derived from the breakdown of carbohydrate which is supplied to the cell in the form of food. But, ultimately there must be some source of energy that is used to synthesize the organic molecules which

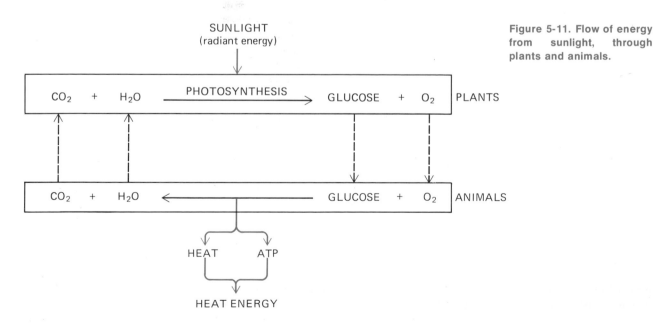

Figure 5-11. Flow of energy from sunlight, through plants and animals.

provide the food that is used by animal cells. The source of this energy is the sun.

Plants are able to capture the energy in sunlight and use it to synthesize glucose by a chemical process known as *photosynthesis* (Fig. 5-11). During photosynthesis, carbon dioxide, water, and the energy of sunlight are combined to form glucose and release molecular oxygen. The oxygen currently present in the earth's atmosphere is derived from this process of photosynthesis. Plants can also break down this newly synthesized glucose to release energy that can be used to synthesize lipids and proteins. Animals eat plants or other animals that have eaten plants and reverse the process, breaking down glucose (fats and proteins) in the presence of oxygen to release the stored energy, transferring some of it to ATP and forming carbon dioxide and water (which, of course, are the initial substrates for the process of photosynthesis). Thus, there is a continuous flow of energy from the sun through all living organisms that is coupled to the cyclic transformations between glucose and carbon dioxide. All life is ultimately dependent upon the energy of sunlight.

Metabolic pathways of carbohydrate, lipid, and protein

In this section, we shall focus upon the major pathways through which carbohydrates, fats, and proteins are catabolized or synthesized. Although these catabolic and anabolic pathways are present in almost all cells, a particular type of cell may contain higher concentrations of the enzymes in one pathway than do other cells or be subject to different intracellular and extracellular control mechanisms.

Carbohydrate metabolism

Carbohydrate catabolism Carbohydrates function primarily as fuel molecules which can be catabolized to release chemical energy coupled to the formation of ATP. We have actually already described, in the last sections, the major pathways of carbohydrate catabolism, namely the breakdown of glucose to lactic acid by way of the glycolytic pathway and the breakdown of carbohydrate intermediates to carbon dioxide and hydrogen by way of the Krebs cycle. We have only

to identify the link between these two pathways to provide the complete pathway for the catabolism of carbohydrate. Pyruvic acid, the carbohydrate intermediate preceding lactic acid in the glycolytic pathway, provides this link (Fig. 5-12).

We have emphasized that the full glycolytic pathway from glucose to lactic acid occurs under anaerobic conditions (since in the absence of oxygen, the conversion of pyruvic acid to lactic acid provides the only means of regenerating hydrogen-carrying coenzyme molecules). However, when the oxygen supply is adequate (the far more common condition), pyruvic acid does not go to lactic acid to any great extent; instead it enters the mitochondria where enzymes transfer the 2-carbon acetyl fragment of pyruvic acid to coenzyme A and release a molecule of carbon dioxide from the carboxyl group. (These reactions are also associated with the transfer of two hydrogens to a coenzyme molecule, which will pass them on to the oxidative phosphorylation pathway, thereby regenerating the original coenzyme.) The acetyl group can now enter the Krebs cycle in which its two carbon atoms are degraded to carbon dioxide, and its hydrogens passed on to the oxidative phosphorylation pathway.

We have illustrated the catabolism of carbohydrates in terms of the breakdown of glucose because glucose is the most important carbohydrate in the body. Other sugars, such as fructose (found in table sugar) and galactose (in milk), are catabolized by the same pathway as glucose after first being converted into glucose or to one of the early intermediates in the glycolytic pathway.

The total amount of chemical energy that is stored in the structure of glucose and released during its aerobic catabolism to carbon dioxide and water equals 686 kcal per mol of glucose:

$$C_6H_{12}O_6 + 6O_2 \rightarrow 6H_2O + 686 \text{ kcal/mol}$$

How much of this energy can be transferred to ATP? Figure 5-13 illustrates the overall pathway of glucose catabolism and the sites at which a total of 38 ATP are formed. As we have seen, a net formation of two molecules of ATP occurs by substrate phosphorylation during the breakdown of glucose to pyruvic acid. Two coenzyme–2H molecules are formed during the initial stages of glucose breakdown, two more during the coupling between pyruvic acid and the Krebs cycle (one for each of the two molecules of pyruvic acid formed

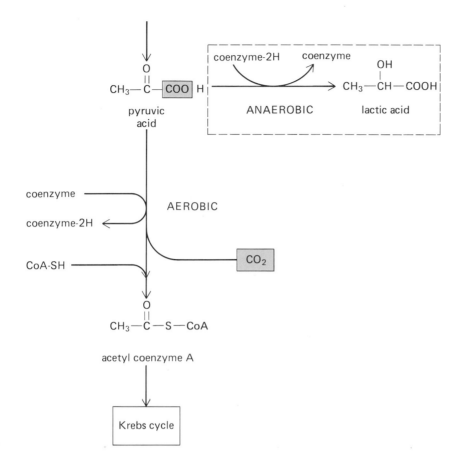

Figure 5-12. Link between glycolytic pathway and Krebs cycle under aerobic conditions.

from one molecule of glucose), and eight more during the Krebs cycle reactions. Thus, a total of 12 coenzyme–2H molecules are formed, which transfer their hydrogens to oxygen in the oxidative phosphorylation pathway, forming 34 molecules of ATP. Finally two additional ATP are formed by substrate phosphorylation in the Krebs cycle. Since 7 kcal are required to form one mole of ATP, and 38 moles of ATP are formed per mole of glucose catabolized, a total of 266 kcal (7×38) are transferred from glucose to ATP:

$$266 \text{ kcal} + 38\text{ADP} + 38\text{P}_i \rightarrow 38\text{ATP}$$

and the remaining 420 kcal ($686 - 266$) is released as heat. Thus, 39 percent ($266/686 \times 100$) of the energy in glucose is transferred to ATP under aerobic conditions.

In the absence of oxygen, only two molecules of ATP can be formed by glycolysis, which is only 2 percent of the energy that is stored in glucose. Thus, the evolution of aerobic metabolic pathways greatly increased the amount of energy that could be made available to a cell from glucose catabolism. For example, if a muscle consumed 38 molecules of ATP during the process of contraction, this amount of ATP could be supplied by the breakdown of one molecule of glucose in the presence of oxygen but would require the catabolism of 19 molecules of glucose to lactic acid under anaerobic conditions to supply the same 38 molecules of ATP. This example also illustrates that, although only two molecules of ATP are formed per molecule of glucose under anaerobic conditions, large amounts of ATP can still be supplied by this glycolytic pathway if large amounts of glucose are broken down to lactic acid.

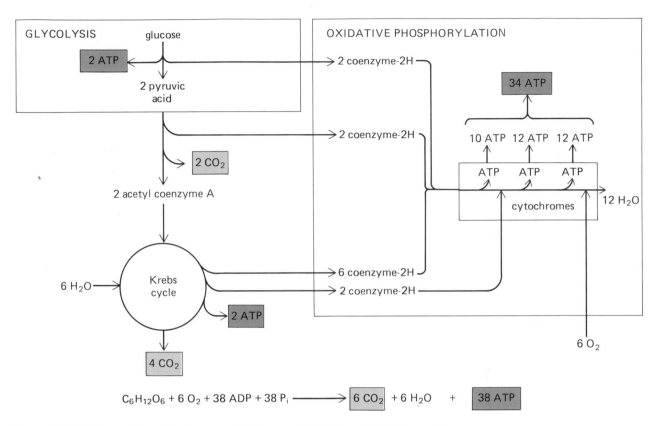

Figure 5-13. Pathway of aerobic glucose catabolism and its linkage to ATP formation.

Figure 5-14. Pathways for the formation and breakdown of glycogen.

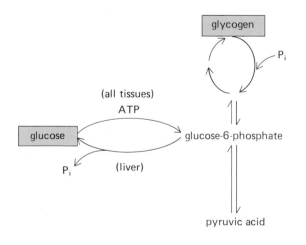

Carbohydrate storage Although the major role of carbohydrate is to provide energy for ATP formation, very little carbohydrate is stored as a reserve supply of fuel (most fuel is stored in the form of fat, as we shall see). The small amount of carbohydrate that is stored, about 430 g in a 70-kg man, is in the form of the branched-chain glucose-polysaccharide known as *glycogen*, which is similar in structure to *starch*, the primary storage form of carbohydrate in plants. Most of the glycogen is stored in skeletal muscles and in the liver, with smaller amounts found in most other tissues.

Glycogen is synthesized from glucose by the pathway illustrated in Fig. 5-14, the enzymes of which are in the cytosol. The first step, the transfer of phosphate to glucose from a molecule of ATP, forming glucose 6-phosphate, is the same as the first step in glycolysis (thus, glucose 6-phosphate can either be broken down to pyruvic acid or be incorporated into glycogen). Note that (as indicated by the bowed arrows) different enzymes are required to split a molecule of glucose 6-phosphate from glycogen than to incorporate it into glycogen. The existence of these different enzymes provides a mechanism for regulating the flow of glucose to and from glycogen. These enzymes are allosteric enzymes whose activity can be stimulated or inhibited by modulator molecules. When an excess of glucose is available to a liver or muscle cell, the enzymes in the glycogen-synthetic pathway are activated and the enzyme breaking down glycogen is simultaneously inhibited, leading to the net storage of glucose in the form of glycogen. When less glucose is available, the reverse combination of enzyme stimulation and inhibition occurs and there is a net breakdown of glycogen into glucose intermediates.

Most cells, including skeletal muscle cells, do not have the enzyme that catalyzes the removal of phosphate from glucose 6-phosphate to form free glucose, which can cross the plasma membrane. Thus, the glucose that is stored as glycogen in these cells cannot be released to provide glucose for other cells; the major function of the relatively large quantities of glycogen stored in muscle cells is to provide an immediately available source of energy (by way of glycogen → glucose 6-phosphate → glycolysis) for the cells themselves during periods of intense contractile activity when their consumption of glucose may greatly exceed the rate at which glucose is delivered to them by way of the blood.

Liver cells, on the other hand, do contain the enzyme which converts glucose 6-phosphate to glucose, thus allowing phosphorylated glucose intermediates formed from glycogen breakdown or other synthetic pathways to be converted into free glucose and be released from the cells into the blood. Because of this ability to release glucose into the blood, the liver plays a very important role in maintaining blood glucose concentration.

Glucose synthesis Liver cells can generate glucose not only from glycogen but can synthesize it from other sources. Figure 5-15 illustrates the pathways used in the synthesis of glucose. The first important point to note is that the pathway leading upward to glucose 6-phosphate uses most, but not all, of the same cytoplasmic enzymes used in the glycolytic breakdown of glucose to pyruvic acid, since these enzymes catalyze reversible reactions. However, two reactions (3 and 10) in the glycolytic pathway are irreversible. Reaction (3) is a major control point in the carbohydrate pathway because it involves two distinct allosterically regulated enzymes, one mediating the flow of metabolites from glucose to pyruvic acid and the other the flow in the opposite direction. By regulating the activities of these two enzymes, carbohydrate metabolism can be shifted between net glucose breakdown and net synthesis depending on the metabolic state of the cell and its energy requirements.

The second irreversible reaction in the glycolytic pathway, reaction (10), does not have a corresponding enzyme that will mediate the direct conversion of pyruvic acid back to phosphoenolpyruvic acid; however, two additional enzymes accomplish this conversion by an important indirect route: First, a molecule of carbon dioxide is combined with pyruvic acid to form a 4-carbon intermediate of the Krebs cycle, oxaloacetic acid; the second enzyme catalyzes the removal of carbon dioxide from oxaloacetic acid to form phosphoenolpyruvic acid. Note, that as a result of these two enzymes, any organic molecule whose carbon atoms can be converted to pyruvic or oxaloacetic acid can give rise to glucose 6-phosphate. Regulation of these enzymes, therefore, provides an important mechanism for altering glucose synthesis.

The conversion of pyruvic acid to oxaloacetic acid not only provides a pathway that can eventually lead to glucose synthesis but it also provides a means of forming this very important Krebs-cycle intermediate whenever it is being removed by transformation into some other product.

Having identified the enzymes and the pathways leading to glucose synthesis we can now ask what types of molecules can be fed into this synthetic pathway to form glucose. One of the most important sources of carbon atoms comes from amino acids which, as we shall see, can be converted into either pyruvic acid or one of the intermediates in the Krebs cycle which leads to oxaloacetic acid. Thus, proteins can be converted into carbohydrates. Secondly, since lactic acid can be converted into pyruvic acid, lactic acid released into the blood from muscle cells can be taken up by liver cells and converted into glucose which can then be released into the blood. There is one

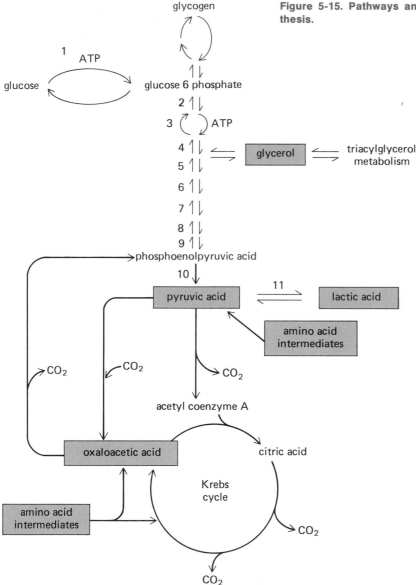

Figure 5-15. Pathways and intermediates leading to glucose synthesis.

other important source of material for glucose synthesis and that is glycerol, the 3-carbon backbone to which fatty acids are attached to form triacylglycerols. Glycerol can be converted into one of the 3-carbon intermediates in glycolysis formed by reaction (4), from which point it can be converted into glucose 6-phosphate.

Fat metabolism

Triacylglycerols (neutral fats) consist of three fatty acids linked to glycerol (Chap. 3) and constitute the majority of stored fuel molecules (Table 5-1). Under resting conditions, approximately half of the energy used by such tissues as muscle, liver,

Table 5-1. Fuel content of a 70-kg individual

	Total body content	Energy per gram	Total body energy content	
	(kg)	(kcal/g)	(kcal)	(%)
Triacylglycerols	15.6	9	140,000	78
Proteins	9.5	4	38,000	21
Carbohydrates	0.5	4	2,000	1

and kidney is derived from the catabolism of fatty acids. Although most cells store small amounts of fat in the form of fat droplets in their cytosol, most of the body's fat is stored in specialized cells known as *adipocytes*. Clusters of adipocytes form *adipose tissue*, most of which is located in deposits underlying the skin. The function of adipocytes is to synthesize and store triacylglycerols during periods of food uptake. Between meals, these fat stores are broken down into glycerol and fatty acids which are released into the blood and delivered to other cells which catabolize them to provide the energy for ATP formation.

The glycerol portion of triacylglycerol is a carbohydrate and, as we have seen, can be converted into one of the 3-carbon intermediates in the glycolytic pathway by which it can generate acetyl coenzyme A. The pathway for the catabolism of fatty acids to acetyl coenzyme A is quite different and is illustrated in Fig. 5-16. (Its enzymes are in the mitochondria so that the acetyl coenzyme A, once generated, can proceed directly through the Krebs cycle.) In the first step of fatty acid breakdown, a molecule of fatty acid is combined with coenzyme A, a reaction that consumes (rather than generates) one molecule of ATP. A series of reactions then follows in which an atom of oxygen derived from water is added to the third carbon atom of the fatty acid, two pairs of hydrogens are transferred to coenzyme molecules, and the bond between the second and third carbon atoms is broken, releasing a molecule of acetyl coenzyme A; simultaneously a second molecule of coenzyme A is added to the remaining fatty acid chain. This chain, now two carbon atoms shorter than the original fatty acid molecule, undergoes the same series of reactions, thereby forming another molecule of acetyl coenzyme A, and so on. Thus, each passage through this sequence shortens the fatty acid chain by two carbons until all of the carbons have been transferred to acetyl coenzyme A. These

can then enter the Krebs cycle, each 2-carbon fragment producing two molecules of CO_2 and, mainly by oxidative phosphorylation, 12 molecules of ATP. The two coenzyme–2H molecules produced directly during the formation of each acetyl coenzyme A also enter the oxidative phosphorylation pathway, one forming three molecules of ATP and the other two.

Most fatty acids in the body contain 14 to 22 carbons, 16 and 18 being most common. The catabolism of an 18-carbon saturated fatty acid (molecular weight = 284) yields 147 molecules of ATP, while the catabolism of one glucose molecule (molecular weight 180) yields 38 molecules of ATP. Thus, the catabolism of a gram of fatty acid yields about two and one-half times as much ATP as does a gram of carbohydrate. If an average man stored most of his fuel as carbohydrate rather than as fat, his body weight would be approximately 30 percent greater (89 versus 70 kg), and he would consume more energy moving this extra weight around in search of food. Thus, a major step in fuel economy occurred when animals evolved the ability to store fuel as fat; in contrast, plants store almost all of their fuel as carbohydrate (starch).

The synthesis of fatty acids occurs by reactions which are almost the reverse of those that degrade them. The enzymes in the synthetic pathway are located in the cytosol, whereas as we have just seen, the enzymes catalyzing the breakdown of fatty acids are located in the mitochondria. Fatty acid synthesis begins with acetyl coenzyme A, which through a series of reactions requiring ATP and coenzyme–2H, is combined with another molecule of acetyl coenzyme A to form a 4-carbon chain. By repetition of this process, long-chain fatty acids are built up, two carbons at a time, which accounts for the fact that almost all of the fatty acids in the body contain an even number of carbon atoms. The overall reaction for the synthesis of an 18-carbon fatty acid can be written:

$$9CH_3\overset{\overset{\text{O}}{\|}}{C}\text{—S—CoA} + 8ATP + 16 \text{ coenzyme–2H} \rightarrow$$
$$CH_3(CH_2)_{16}COOH + 8ADP + 8P_i + 16 \text{ coenzyme} +$$
$$7H_2O + 9CoA\text{—SH}$$

Finally, triacylglycerols are formed by combining three fatty acids with glycerol.

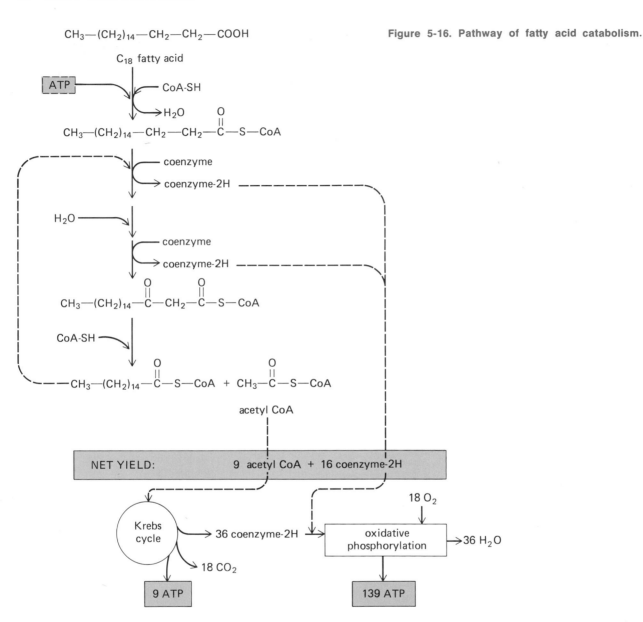

Figure 5-16. Pathway of fatty acid catabolism.

A comparison of the molecules required for fatty acid and glycerol synthesis and those produced by glucose catabolism should reveal why glucose is so easily converted to triacylglycerols. First, note that acetyl coenzyme A, the starting material for fatty acid synthesis, is formed from glucose. Second, the other ingredients required for fatty acid synthesis—coenzyme–2H and ATP—are formed during carbohydrate catabolism. Third,

glycerol can be formed from glucose. It should not be surprising, therefore, that much of the carbohydrate in food is converted into fat and stored, at least transiently, in adipose tissue shortly after its absorption from the gastrointestinal tract.

In contrast, fatty acids, or more specifically, the acetyl coenzyme A derived from fatty acid breakdown, cannot be used to synthesize new molecules of glucose. The reasons for this can be seen by

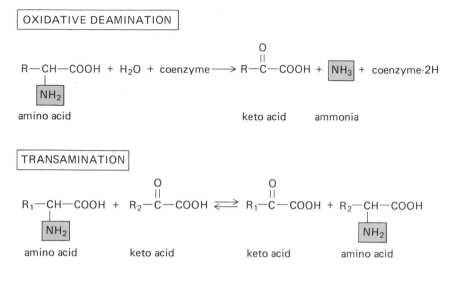

Figure 5-17. Oxidative deamination and transamination of amino acids.

examining the pathways for glucose synthesis (Fig. 5-15). First, because the reaction in which pyruvic acid is broken down to acetyl coenzyme A and carbon dioxide is irreversible, acetyl coenzyme A cannot be converted into pyruvic acid, via which "reverse glycolysis" could yield glucose. Second, the equivalent of the two carbon atoms in acetyl coenzyme A are converted into two molecules of carbon dioxide during their passage through the Krebs cycle before reaching oxaloacetic acid, another takeoff point for glucose synthesis.

Protein metabolism

In the last chapter we described how amino acids are linked together to form proteins, and how DNA and RNA are involved in directing the process. In contrast to the complexities of this synthetic pathway, protein catabolism requires only a few enzymes (*proteases*) to break the peptide bonds between amino acids. Some of these enzymes split off single amino acids, one at a time, from the ends of the protein chain, while others break peptide bonds between specific amino acids within the chain, thereby forming shorter peptides; these are then broken down to individual amino acids by the proteases. The overall pathways for protein synthesis and degradation can thus be summarized as

$$\text{amino acids} \underset{\text{proteases}}{\overset{\text{nucleic acids + ribosomes}}{\rightleftarrows}} \text{protein}$$

In addition to their incorporation into proteins, amino acids undergo many other metabolic reactions, but a few basic types of reactions common to most of these pathways should provide an overview of their general metabolism. Unlike carbohydrates and fats, amino acids consist of nitrogen-containing amino groups in addition to carbon, hydrogen, and oxygen. Once this amino group is removed, the remainder of the molecule (with a few exceptions) can be metabolized to intermediates capable of entering the carbohydrate pathway. The two types of reactions by which the amino group is removed are illustrated in Fig. 5-17. In the first type, known as *oxidative deamination,* the amino group is released as a molecule of ammonia (NH_3) and replaced by an oxygen atom (derived from water) to form a *keto acid* (during this reaction the two hydrogens in water are transferred to a coenzyme). The second type of reaction, known as *transamination,* involves the transfer of the amino group from an amino acid to a keto acid. Note that the keto acid to which the amino group is transferred becomes an amino acid. Thus, by means of transamination, new amino acids can be formed from carbohydrate intermediates (keto acids).

Figure 5-18 shows the oxidative deamination and transamination of two amino acids, glutamic acid and alanine. Note that the keto acids formed are intermediates either in the Krebs cycle (α-ketoglutaric acid) or glycolytic pathway (pyruvic acid). Once formed, these keto acids can therefore be metabolized to produce carbon dioxide and form ATP, or they can be used as intermediates in the synthetic pathway leading to the formation of glucose. As a third alternative, after their conversion to acetyl coenzyme A, they can be used to synthesize fatty acids. Thus, amino acids can be used as a source of energy, and they can be converted into both carbohydrate and fat. However, it must be recognized that, unlike fat (and to a lesser extent carbohydrate), excess protein is not really "stored" in cells; i.e., the proteins of the body directly participate in the ongoing functions of cells. Therefore, when endogenous protein is catabolized to provide energy or to serve as a precursor for carbohydrate and fat, this occurs at the expense of degrading some portion of a cell. During starvation, for example, muscle cells become progressively smaller as the contractile proteins are degraded to provide amino acids used by the liver to form glucose.

When carbohydrate or fat is catabolized to pro-vide energy for ATP formation, the carbon, hydrogen, and oxygen atoms they contain end up in carbon dioxide and water, which can, therefore, be considered the waste products of their metabolism. The carbon dioxide is eliminated from the body by the lungs and water is excreted primarily in the urine. As we have seen, the catabolism of amino acids leads, in addition, to the nitrogen-containing product ammonia. (The nitrogen derived from amino groups can also be used by cells to synthesize other important nitrogen-containing molecules, such as the purine and pyrimidine bases found in nucleic acids.) Ammonia, which can be highly toxic to cells if allowed to accumulate, readily leaks into the blood and is carried to the liver (Fig. 5-19), where in a series of reactions, two molecules of ammonia are linked with carbon dioxide to form *urea*. Thus, urea, which is not toxic to cells, is the major nitrogenous waste product of protein catabolism; it leaves the liver and is excreted by the kidneys into the urine. Two of the

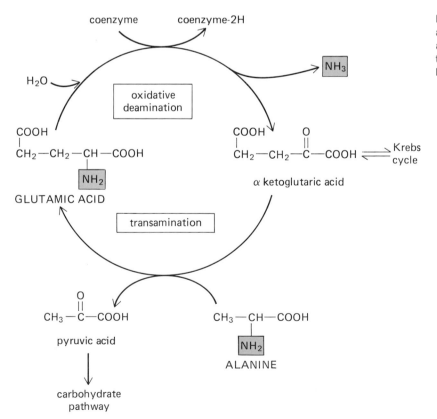

Figure 5-18. Oxidative deamination and transamination of the amino acids glutamic acid and alanine lead to keto acids that can enter the carbohydrate pathways.

20 amino acids also contain atoms of sulfur which can be converted to sulfate, SO_4^{-2}, and excreted in the urine.

Thus far, we have discussed mainly amino acid catabolism. Figure 5-18 also demonstrates how amino acids can be formed. The keto acids, such as α-ketoglutaric and pyruvic acid, can be formed from the breakdown of glucose and then be transaminated, as described above, to form glutamic acid and alanine; thus glucose can be used to form certain amino acids, provided other amino acids are available in the diet to supply amino groups for transamination. However, not all amino acids can be formed by the body, and the so-called *essential amino acids* must be obtained from the diet.

Summary

Having discussed the metabolism of the three major classes of organic molecules, we can now briefly review how each class is related to the others and to the process of synthesizing ATP. Figure 5-20 shows the major pathways we have discussed and the relations of the common intermediates. All three classes of molecules can enter the Krebs cycle through some intermediate, and thus all three can be used as a source of chemical potential energy for the synthesis of ATP by oxidative phosphorylation. Hydrogen and oxygen provide the substrates for this oxidative phosphorylation; some of these hydrogens are derived from the Krebs-cycle reaction directly; others may come from glycolysis or the breakdown of fatty acids. Glucose can be converted into fat or into amino acids by way of the common intermediates such as pyruvic acid, α-ketoglutaric acid, and acetyl CoA. Similarly, amino acids can be converted into glucose and fat. Fatty acids cannot be converted into glucose because of the irreversibility of the pyruvic acid-to-acetyl CoA reaction, but the glycerol portion of triacylglycerols can be converted into glucose. Metabolism is, thus, a highly integrated process in which all classes of molecules can be used, if necessary, to provide energy for the cell through ATP synthesis and in which each class of molecule can, to a large extent, provide the raw materials required to synthesize members of other classes.

We have specifically noted in the glycolytic path-

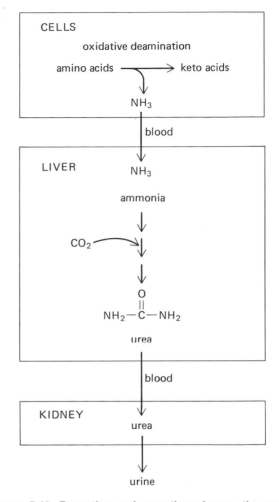

Figure 5-19. Formation and excretion of urea, the major nitrogenous waste product of protein catabolism.

way a number of allosteric enzymes whose activity can be altered by various modulator molecules, thereby regulating the net anabolism or catabolism of glucose. Other allosterically regulated enzymes occur in the Krebs cycle and in the pathways for fatty acid and amino acid metabolism. For example, ATP and its products, ADP and P_i, are modulator molecules able to activate or inhibit various allosteric enzymes. If a cell is utilizing large amounts of energy, the rate of ATP breakdown to ADP and P_i will be rapid, leading to a fall in ATP concentration and a rise in the concentrations of ADP and P_i. The changes in the concen-

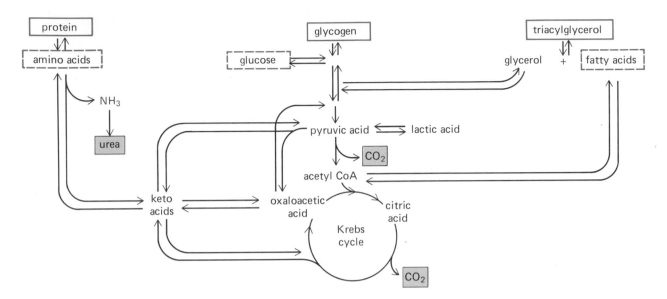

Figure 5-20. Pathways linking carbohydrate, fat, and protein metabolism. For the sake of clarity, the convention of bowed arrows for coupled irreversible reactions has not been used.

trations of these molecules allosterically activate or inhibit various enzymes so that the overall rates of glucose and fat catabolism are increased, thereby providing the energy necessary to form more ATP. Likewise, when the cell is using less energy, the rise in ATP and the fall in ADP and P_i inhibit these catabolic enzymes and activate the anabolic enzymes, leading to a net storage of fuel as glycogen and fat. Similar changes in the concentrations of the hydrogen-carrying coenzymes (high concentration of coenzyme–2H when energy consumption is low and low concentration of coenzyme–2H when energy consumption is high) can also influence various allosteric enzymes by acting as modulator molecules. Other intermediates, including glucose 6-phosphate, citric acid, and acetyl coenzyme A, also function as modulator molecules for allosteric enzymes, thereby providing the cell with a variety of interacting control systems.

6
MOVEMENT OF MOLECULES ACROSS CELL MEMBRANES

Membrane structure
Diffusion
   Factors determining the net rate of diffusion
   Diffusion through lipid portions of the membrane
   Diffusion through membrane pores
   Regulation of membrane permeability
Mediated-transport systems
   Properties characteristic of mediated transport
     Specificity • *Saturation* • *Competition*
   The carrier hypothesis

Facilitated diffusion
Active transport
Epithelial transport
Osmosis
Endocytosis and exocytosis
   Endocytosis
   Exocytosis
     Protein secretion
Membrane junctions

The contents of a cell are separated from the surrounding extracellular medium by a thin layer of lipids and protein—the *plasma membrane.* In most cells there are numerous other membranes which are part of the cell organelles—mitochondria, endoplasmic reticulum, lysosomes, the Golgi apparatus, and the nucleus—which divide the intracellular fluid into separate compartments. The term "membrane" refers nonspecifically to any of these cell membranes, whereas plasma membrane denotes specifically the surface membrane.

If all the membranes in 1 g of liver were unfolded and pieced together, they would cover an area of approximately 30 m² (250 ft²). The contributions of this array of membranes to cell function fall into two general categories: (1) they provide barriers to the movements of molecules and ions both between the various compartments within the cell and between the cell and the extracellular fluid; (2) they provide a scaffolding to which various cell components are anchored.

The chemical compositions of the intracellular and extracellular fluids are quite different. This fact alone demonstrates that the plasma membrane prevents free mixing of the molecules on the two sides of the membrane. However, cells must be able to take in nutrients from the extracellular environment and give off metabolic end products to it, which requires that certain molecules be able to cross the plasma membrane to some degree. Thus, the plasma membrane is selectively permeable. Similarly, selectivity of organelle membranes provides an important means for isolating various chemical reactions within different cell organelles. Moreover, the rates of movement of molecules through all the various cell membranes are not necessarily constant but can be altered, thereby providing one of the major mechanisms for regulating cell functions; for example, as will be seen in a subsequent chapter, the underlying basis of excitation in nerve and muscle cells is a selective change in their membrane permeability to ions.

Now we turn to examples of the second general category of membrane functions—its availability as a scaffolding. For one thing, a variety of specific binding sites are located on the outer surface of the plasma membrane and function as "recognition sites" (receptors) for hormones and other

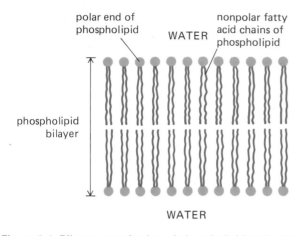

polar end of phospholipid   WATER   nonpolar fatty acid chains of phospholipid

phospholipid bilayer

WATER

Figure 6-1. Bilayer organization of phospholipid molecules in water.

chemical messengers; thus the plasma membrane serves as a signal-receiving device for the detection of chemical signals which regulate the cell's activity. Surface binding sites also participate in the arrangement of cells into tissues during growth and development by providing sites for specific cell-to-cell adhesions. Other specialized junctions which occur between certain cell types not only strengthen the adhesion between cells, but, in some cases, provide special channels for direct cell-to-cell communication. Fibrous elements are attached to the inner surface of the plasma membrane and function both in the maintenance of cell shape and in the changes in the surface configuration of cells during certain states of activity. In muscle cells, the contractile protein filaments are anchored to the plasma membrane, enabling them to transmit force to the outside of the cell. A variety of enzymes are anchored to cell membranes rather than being dissolved in the intracellular fluid; therefore, certain types of chemical reactions occur only in association with particular cell organelles because the enzymes mediating these reactions are either bound to the membranes of the organelles or are located within the organelle and are unable to penetrate the surrounding membrane (for example, the reactions of oxidative phosphorylation occur on the inner membrane of the mitochondria because the enzymes mediating them are arranged in a specific spatial configuration in that membrane).

Membrane structure

In spite of the large number of distinct functions associated with membranes in different types of cells or within the same cell, the general structures of all cell membranes appear to be similar. They are composed of a bimolecular lipid matrix in which proteins are embedded. Lipids and proteins are present in approximately equal proportions by weight. The lipids form the barrier to movements of molecules through the membrane; the proteins provide selective means for the transfer of certain molecules through the barrier, and they also constitute the binding sites and enzymes associated with the membrane.

The majority of the membrane lipids are phospholipids. As described in Chap. 3, these are amphipathic molecules, having a charged region at one end (due to the presence of a negatively charged phosphate group, which is usually attached to another small charged group), whereas the remainder of the molecule, consisting of two long-chain fatty acids, is electrically neutral. Whenever phospholipids are mixed with water they associate with each other to form biomolecular layers in which the polar ends of the phospholipids are positioned at the surfaces of the bilayer, due to their electrical attraction to polar water molecules; the nonpolar fatty acid chains are perpendicular to the surfaces and form a nonpolar region within the bilayer (Fig. 6-1). The phospholipids in cell membranes are also organized into a bimolecular layer similar to those which form spontaneously in phospholipid-water mixtures; their polar ends are located at the inner and outer surfaces of the membrane. The individual phospholipids in the membrane have considerable freedom of movement in that the fatty acid chains are flexible and can wiggle back and forth, and the entire molecule can move laterally in the membrane. Thus, the lipid phase of the membrane is more like a fluid than a rigid crystalline matrix. This fluidity makes the entire membrane quite flexible; it can easily be bent and folded, although horizontal stretching may rupture its loose phospholipid associations.

In addition to phospholipids, cell membranes contain another lipid—the steroid cholesterol. The plasma membrane contains about one molecule of cholesterol for each molecule of phospholipid, but intracellular membranes contain relatively much

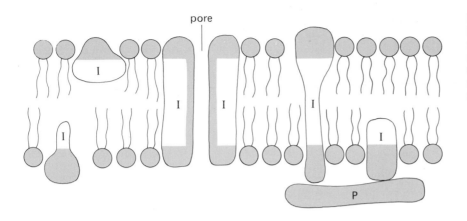

pore

Figure 6-2. Arrangement of integral (I) and peripheral (P) proteins in membrane in association with bimolecular layer of phospholipids. Shaded areas at the membrane surfaces indicate the polar regions of the proteins and phospholipids.

smaller quantities. The cholesterol molecules are located at the polar surfaces of the membrane, the nonpolar ring portion of the cholesterol inserted into the membrane. The presence of the smaller cholesterol molecule between the phospholipids increases the space available for movement of the fatty acid chains and, thus, increases the fluid state of the membrane.

The membrane proteins fall into two classes, *peripheral proteins* and *integral proteins*. Peripheral proteins are water soluble and can be removed from the membrane by solutions which alter the electric charge on proteins, suggesting that they are weakly bound to the surface of the membrane (Fig. 6-2). In contrast, the integral proteins are insoluble in water and cannot be removed without disrupting the lipid components of the membrane; this is because the integral proteins are closely associated with the nonpolar regions of the membrane lipids. The polypeptide chain of an integral protein is folded in such a way that certain regions contain mainly ionized and polar amino acid side chains, whereas the remaining portion of the protein has primarily nonpolar side chains. This creates an amphipathic molecule, which like the amphipathic phospholipids orients within the membrane so that its polar regions are at the surface of the membrane, whereas the nonpolar region is in the middle. In some cases, the integral proteins are located on only one side of the membrane; in others, the protein has a polar region at each of its ends, separated by a nonpolar region, this structure allowing it to span the entire membrane (Fig. 6-2). Thus, the distribution of proteins

in the membrane is highly asymmetrical in that different proteins are exposed on the inner and outer surfaces and different regions of the proteins that span the membrane are present on the two surfaces. Many of the membrane proteins are free to move laterally in the plane of the membrane, the lipid bilayer providing the fluid medium in which they move. Other membrane proteins are less mobile, possibly because they are bound to peripheral proteins.

Membrane proteins perform a variety of functions. For example, some of the peripheral proteins have contractile properties, and their activation has been implicated in the rearrangement of membrane proteins which occurs in response to various signals. To take another example, some of the integral proteins which span the membrane seem to be organized in clusters to form aqueous channels (pores) through which small water-soluble molecules and ions can pass, thereby bypassing the nonpolar lipid regions of the membrane; as we shall see, they also provide specialized mechanisms for the movement of certain substances across the membrane.

Integral proteins which span the membrane may also participate in transmitting signals from one side of the membrane to the other. An example of this is the following sequence: The surface portion of the protein provides a binding site for a chemical messenger; binding of the messenger produces a change in the conformation of the protein at the inner surface of the membrane; this latter region of the protein may be an enzyme that is thereby activated. In some cases, two integral proteins,

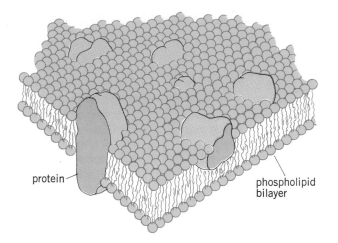

protein — — phospholipid bilayer

Figure 6-3. Fluid mosaic model of cell membrane structure. *(Redrawn from S. J. Singer and G. L. Nicholson,* Science, **175***:723. Copyright 1972 by the American Association for the Advancement of Science.)*

neither of which spans the membrane, may be involved in transmitting the message across the membrane, the protein on one surface having the binding site and the protein on the opposite surface being an inactive enzyme; binding of the messenger to the outer protein leads to a change in its conformation so that it can then bind to the inactive enzyme (the inner protein) and thereby activate it. The mobility of proteins in the membrane makes possible the bringing together of the two proteins so that they can bind.

The *fluid-mosaic* structure of membranes (Fig. 6-3) described above emphasizes the dynamic nature of membranes—a mosaic of proteins floating in a sea of lipid. Components of the membrane are asymmetrically arranged and their configuration within the membrane can be altered by

Figure 6-4. Electron micrograph of a human red cell plasma membrane. *(From J. D. Robertson in Michael Locke (ed.), "Cell Membranes in Development," Academic Press, Inc., New York, 1964.)*

various chemical and physical signals. Although all membranes appear to have this general structure, specific membranes contain different proteins and lipids and are subject to different modifying signals. In addition to lipids and proteins, the plasma membrane contains small amounts of carbohydrate covalently linked to lipids and proteins to form glycolipids and glycoproteins. The carbohydrate portions of these molecules are asymmetrically distributed across the membrane, all of the carbohydrate being located at the extracellular surface.

Cell membranes are 6 to 10 nm thick, intracellular membranes being thinner than plasma membranes. This variation reflects the types of proteins that are present. An electron micrograph of the plasma membrane of a red blood cell is shown in Fig. 6-4. The membrane appears as two dark lines separated by a light interspace. The dark lines correspond to the staining of the polar regions of the proteins and phospholipids, whereas the light interspace corresponds to the nonpolar regions in the membrane.

This completes our survey of the structure of cell membranes. The mechanisms which govern the movements of molecules and ions across them is the subject of the remainder of this chapter.

Diffusion

Diffusion is the movement of molecules from one location to another by random molecular motion. All molecules undergo continuous motion, and the warmer an object is, the faster its molecules move. This thermal motion enables molecules to move from one region to another, the velocity at which they travel depending upon the temperature and the mass of the molecule. At body temperature, an average molecule of water moves about 2500 km/h (1500 mi/h), whereas a molecule of glucose, which is 10 times heavier, moves about 850 km/h. In water, where individual molecules are separated from each other by about 0.3 nm, such rapidly moving molecules cannot travel very far before colliding with other molecules. They bounce off each other like rubber balls, undergoing millions of collisions every second. Each collision alters the direction of the molecule's movement, so that the path of any one molecule through a solution may look like that shown in Fig. 6-5. Such

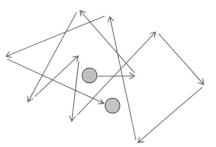

Figure 6-5. Path traveled by a single molecule in a solution. Changes in direction result from collisions with other molecules in the solution.

movement is said to be *random*, since a molecule moving in one direction may in the next instant be moving in the opposite direction, all directions being equally probable.

Many properties of living organisms are closely associated with the process of diffusion: oxygen and nutrients enter the blood by diffusion; molecules leave the blood and enter the extracellular compartments of the tissues by diffusion; the exchange of molecules between the cell and the extracellular environment and between compartments within cells occurs, at least in part, by diffusion. The rate of diffusion is thus an important factor in determining the rate at which nutrients can reach a cell (and the rate at which metabolic end products can leave it and gain entry to the blood).

Although individual molecules travel at high velocities, the number of collisions they undergo prevents them from traveling very far in a straight line. Figure 6-6 illustrates the diffusion of glucose from the blood, where it is maintained at a constant concentration, into the extracellular space surrounding the cells of a tissue. The question we wish to answer is: How long will it take the concentration of glucose to reach 90 percent of the glucose concentration in the blood at points 10 μm and 10 cm away from the blood vessel? The answer is 3.5 s and 11 years, respectively. Thus, although diffusion can distribute molecules rapidly over short distances (for example, within cells or the extracellular regions surrounding a few layers of cells), diffusion is an extremely slow process when distances of a few centimeters or more are involved. For an organism as large as a person, the diffusion of oxygen and nutrients from the body surface to the interior tissues of the body would be

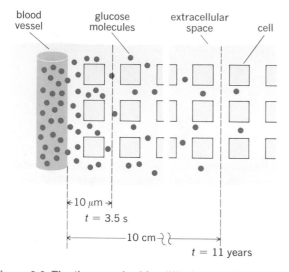

Figure 6-6. The time required for diffusion to raise the concentration of glucose at a point 10 μm (about one cell diameter) away from a blood vessel to 90 percent of the blood glucose concentration is about 3.5 s, whereas it would take over 11 years for glucose to reach the same concentration at a point 10 cm (3.9 in) away.

far too slow to provide adequate nourishment. Accordingly the circulatory system (utilizing a transport process known as bulk flow [Chap. 11]) provides the mechanism for rapidly transporting materials (blood) over large distances, with diffusion providing the means for entry and exit from the blood and movement through the extracellular fluid. Because cells depend upon the diffusion of materials from the blood, all the cells in the body must be close to a blood vessel. Moreover, the rate at which diffusion is able to move molecules within a cell is one of the major factors limiting the size to which a cell can grow and function. The larger the cell, the greater is the volume of cytoplasm that must be supplied by diffusion from a given unit of cell surface area. A cell does not have to be very large before diffusion becomes too slow to provide sufficient nutrients to the central portions. For example, the center of a 20-μm cell becomes equilibrated with oxygen in about 15 ms, whereas it would require 256 days for oxygen to reach the center of a cell the size of a basketball. Some cells which have a fairly large cell volume have solved this problem by assuming a shape that still provides a small volume/surface area ratio. For example, muscle cells are often

more than 10 cm long but only 10 to 100 μm thick. Thus, a molecule has to diffuse only a short distance from the surface to reach the center of the muscle cell.

Factors determining the net rate of diffusion

We begin our discussion of the factors which determine the net rate of diffusion with a consideration of diffusion within a solution (or gas) and will apply these basic principles to diffusion across cell membranes in a later section.

When billions of molecules are undergoing random motion (as is the case for the aqueous medium surrounding cells and inside them), at any instant there will be about as many molecules moving to the right in a given volume as there are molecules moving to the left, as many moving up as down, etc. Any volume of solution can be looked upon as being composed of a number of small volume elements joined together at their surfaces (Fig. 6-7). A *flux* is the amount of material (in this case the number of molecules) which crosses one of these imaginary surfaces moving from one side to the other in a unit of time. Thus, the fluxes of molecules across the imaginary surfaces of these volume elements represent the exchange of molecules between the different portions of the total solution. If the number of molecules in a unit of volume were doubled, the flux of molecules at each surface would be doubled since twice as many molecules would be moving in any one direction at a given time. To generalize, the concentra-

Figure 6-7. Collection of molecules undergoing random motion in a small volume element (A) results in approximately equal numbers of molecules moving up and down, right and left, etc. A large volume of solution is equivalent to a number of small volume elements (B) in which the fluxes of the molecules between the volume elements depend upon the concentration of molecules in each volume element.

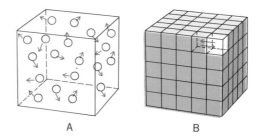

A   B

tion of molecules (number of molecules in a unit of volume) in any region of a solution determines the magnitude of the flux across the surfaces of this region.

We can formulate this relationship in an equation where f is the flux,

$$f = k_D C$$

i.e., number of molecules crossing a unit area in a unit time, C is the concentration, and k_D is a proportionality constant known as the *diffusion coefficient*. The magnitude of the diffusion coefficient depends upon the molecular weight and chemical structure of the molecule and upon the nature of the solution (water versus oil, for example) and its k_D increases as molecular size decreases because (at any given temperature) small molecules move more rapidly than large ones; for example, the magnitude of k_D for glucose is less than that for water. Increasing the temperature of a solution increases the velocity of molecular motion and thus increases the magnitude of k_D.

The diffusion of glucose between two compartments (not separated by a barrier) is illustrated in Fig. 6-8. Initially, glucose is present in compart-ment 1 at a concentration of 20 mmol/L, and there is no glucose in compartment 2. The random movements of the glucose molecules in compartment 1 carry some of them into compartment 2:

$$f_{1\text{-}2} = k_D C_1$$

where C_1 is the initial concentration of glucose in compartment 1. Some of the glucose molecules that have entered compartment 2 will randomly move back into compartment 1. The magnitude of the glucose flux from 2 to 1 depends upon the concentration of glucose in compartment 2, C_2, at any given time:

$$f_{2\text{-}1} = k_D C_2$$

Therefore, as the concentration of glucose in 2 increases, the flux of glucose from 2 to 1 also increases. The net increase at any instant in the number of glucose molecules in compartment 2 is given by the difference between $f_{1\text{-}2}$ and $f_{2\text{-}1}$. This difference between these two so-called one-way fluxes is known as the *net flux*, F. The net flux of glucose from compartment 1 to 2 causes the concentration in 1 to decrease and the concentration in 2 to increase. Eventually the concentrations in

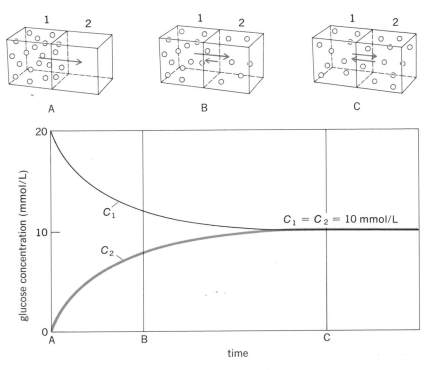

Figure 6-8. Diffusion of glucose between two compartments of equal volume. (A) Initial conditions: No glucose is present in compartment 2. (B) Some glucose molecules have moved into compartment 2, and some of them are moving at random back into compartment 1. (C) Diffusion equilibrium has been reached, the flux of glucose between the two compartments being equal in the two directions.

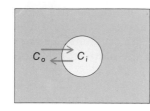

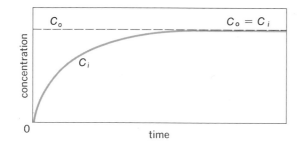

Figure 6-9. Change in intracellular concentration C_i as material diffuses from a constant extracellular concentration C_o across the plasma membrane into the cell.

the two compartments become equal, the glucose fluxes in the two directions become equal, and the net flux of glucose becomes zero. The system has now reached *diffusion equilibrium,* and no further change in the concentration of the two compartments occurs.

Several important properties of the diffusion process should be emphasized from this example. In the overall diffusion process three fluxes can be identified, the 2 one-way fluxes occurring in opposite directions from one compartment to the other and the net flux, which is the difference between them. It is the net flux, F, which is the most important component in the diffusion process since it is the net amount of material moving from one location to another. This net flux always proceeds from a high to a low concentration, i.e., down a concentration gradient. Since the magnitude of the net flux is determined by the difference in concentration (the concentration gradient between two compartments), the basic diffusion equation becomes

$$F = k_D(C_1 - C_2)$$

Let us now apply this analysis to diffusion across cell membranes. The rate at which a substance diffuses across the plasma membrane can be

measured as the rate at which the intracellular concentration, C_i, increases when the cell is placed in a solution containing the substance. We shall set the volume of solution outside the cell so large that the concentration of the substance in this solution, C_o, undergoes very little change as the material diffuses into the small intracellular volume (Fig. 6-9). As with all diffusion processes the net diffusional flux of material across the membrane is from the region of high concentration to the region of low concentration, the equation having the same form as the diffusion equation in free solution

$$F = k_p(C_o - C_i)$$

The magnitude of the net flux is proportional to the concentration gradient $(C_o - C_i)$ across the membrane and the *membrane permeability constant,* k_p. Like the diffusion coefficient, k_D, the magnitude of k_p depends on the type of molecule, its molecular weight, and the temperature, but in addition, it also depends on the characteristics of the membrane through which the molecule is diffusing.

Net diffusion of material into the cell continues until the intracellular and extracellular concentrations become equal, $C_i = C_o$. The rates at which a number of different molecules diffuse into a cell, as evidenced by the times for equilibrium to be

Figure 6-10. Comparison of the different rates at which various molecules A, B, . . . , G are able to diffuse across the cell membrane from an extracellular medium of constant concentration C_o. Molecule A enters the cell rapidly and has a high permeability constant k_p. Molecule F enters the cell very slowly, and molecule G is unable to cross the membrane at all.

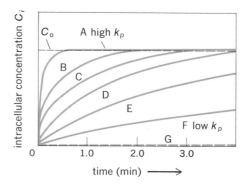

reached, are shown in Fig. 6-10. The more permeable the membrane (the larger the value of the membrane permeability constant, k_p), the faster a substance enters the cell. Such measurements reveal several properties of the membrane. First, the permeability constants of the membrane are found to be a thousand to a million times smaller than the diffusion coefficients of the same molecules in water; i.e., the molecules are entering the cell at a much slower rate than if there were no barrier at the surface. Second, not only is the membrane a barrier which slows down the entry of molecules, it is also a highly selective barrier which allows some molecules to enter rapidly while others, which may have a similar diffusion coefficient in free solution, enter more slowly.

Diffusion through lipid portions of the membrane

When the permeability constants of a number of different organic molecules are examined in relation to their molecular structure, a correlation emerges. Whereas most polar molecules enter very slowly or not at all, molecules with few polar groups enter cells rapidly (i.e., they have large permeability constants). This observation is explained by the fact that nonpolar molecules can dissolve in the nonpolar regions of the membrane occupied by the fatty acid chains of the membrane phospholipids. In other words, uncharged molecules rapidly cross the membrane by directly passing through its lipid regions. The more lipid-soluble the substance (the fewer its polar or ionized groups), the greater will be the number of molecules dissolved in the membrane lipids and thus the greater will be its flux across the membrane. Oxygen, carbon dioxide, and the steroid hormones are examples of nonpolar molecules which diffuse rapidly through the lipid portions of membranes.

That the phospholipid bilayer of the membrane actually provides this type of passive selective barrier, allowing nonpolar molecules to pass but excluding polar or ionized substances, is documented by experiments using artificial membranes consisting of a bimolecular layer of phospholipids; the permeability characteristics of these lipid bilayers to lipid-soluble molecules are very similar to those of real cell membranes.

Diffusion through membrane pores

Although ions and polar molecules diffuse across cell membranes more slowly than lipid-soluble substances, many diffuse more rapidly than expected from their lipid solubility. The permeability constants of these substances vary with the size of the ion or molecule, the constant being higher when the size is small. This relation between molecular size and rate of diffusion suggests that small charged ions and molecules may be diffusing through water-filled pores in the membrane. Polar molecules larger than the pore would be unable to diffuse across the membrane except by passing through the lipid portions of the membrane, where their rate of entry would be low because of their low lipid solubility. The estimated pore diameter of about 0.8 nm, based on the largest polar molecules that can diffuse through the membrane (Fig. 6-11), is just about at the limit of resolution of the electron microscope, and these postulated pores have, therefore, not been directly observed.

Artificial phospholipid bilayers have, compared to cell membranes, a very low permeability to small ions, suggesting that it is the protein components in the membrane that form the pores. As we have seen, some membrane proteins completely span the thickness of the membrane. Clusters of these proteins could be arranged to form the walls of water-filled pores through the membrane. It has been estimated that pores occupying less than 1

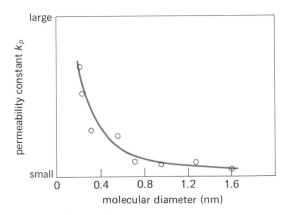

Figure 6-11. Relationship between the diameter of polar molecules and the rate at which the molecules diffuse across the plasma membrane.

percent of the total membrane surface area would account for the observed permeabilities to small polar molecules.

Regulation of membrane permeability

Different membranes have different permeabilities to the same molecule, and in some membranes the permeability may be altered during certain states of activity. Such changes in the permeability constant are achieved by an alteration in the structure or composition of the membrane.

Differences in the permeability of membranes must result from differences in the two pathways for diffusion through the membrane—the lipid pathway or the pores. There are a number of types of phospholipids in membranes, and different membranes contain varying proportions of these lipids, thereby producing slightly different permeability characteristics. Furthermore, the density of packing of the phospholipids may vary between membranes, and this can be influenced by the arrangement of the proteins. The permeability characteristics of the lipid portion of the membrane do not appear to be subject to rapid regulation, although slow changes in permeability may occur when particular membrane phospholipids are altered due to changes in the lipid composition of the diet. Certain drugs may affect the permeability of the lipid portion of the membrane by altering the packing of the lipids and their interaction with membrane proteins.

The number and size of pores can also vary in different membranes. Furthermore, the movements of ions, such as sodium and potassium, through pores, are highly specific. Sodium ions pass through a so-called sodium pore which does not allow the passage of potassium; similarly, potassium passes through its own pore which will not allow the passage of sodium. Thus, the permeability of a membrane to sodium and potassium depends on the number of separate sodium and potassium pores present in the membrane. The permeability of a given membrane may be regulated by controlling the conformation of the proteins that form the pores. Various signals may lead to the closing or opening of existing pores or to the formation or dissociation of the proteins that form them. In later chapters we shall encounter several factors which alter membrane permeability by affecting membrane pores.

Mediated-transport systems

Thus far we have considered the physical factors that influence the diffusion of molecules through the membrane. Although diffusion through pores may account for some of the transmembrane movement of small ions, it does not account for all; moreover, diffusion through pores cannot be responsible for the observed movements of large polar molecules such as the sugars and amino acids, yet the cell depends upon the entry of large quantities of these molecules to provide a source of chemical energy and the raw materials for protein synthesis. Moving such molecules (and many others) across the cell membrane requires special mechanisms, which are built into the membrane structure. These molecules bind to specific proteins on the membrane surface, which in some manner leads to their movement through the membrane. This process is called *mediated transport*. The properties of chemical specificity, competition, and saturation, which are characteristics of binding sites, distinguish molecules which cross by mediated transport from those which cross by diffusion.

Properties characteristic of mediated transport

Specificity. There are a number of different mediated-transport systems, and each is able to react with only a very select group of chemical substances. For example, although both amino acids and sugars undergo mediated transport, the system which transports amino acids does not transport sugars, and vice versa.

Saturation. In diffusion the one-way flux of molecules entering a cell at any instant is proportional to the concentration of the substance outside the cell, C_o, and is given by $f = k_p C_o$. If C_o is doubled, the flux into the cell is doubled (Fig. 6-12A). No matter how large C_o, the flux into the cell is always proportional to the concentration. In mediated-transport systems, however, the flux increases with extracellular concentration only up to a certain point (Fig. 6-12B). Thus, there is a maximum rate at which material can enter a cell through a mediated-transport system. When the maximum flux is reached, the system is said to be *saturated*. A transport system is saturated when all the specific sites on the membrane are occupied and operating at their maximum capacity. Some trans-

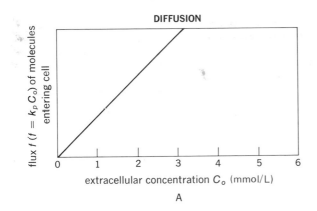

DIFFUSION

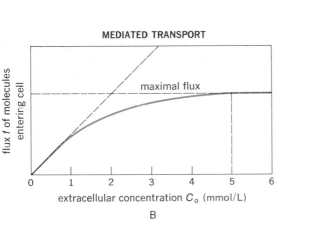

MEDIATED TRANSPORT

maximal flux

Figure 6-12. Relationship between the one-way flux of molecules into a cell and the extracellular concentration for: (A) a molecule entering by simple diffusion; (B) a molecule entering by a mediated-transport system, showing a saturation of the transport flux when the extracellular concentration C_o is 5 mmol/L or greater. The dotted line in B represents the flux that would be expected if the molecule entered by simple diffusion.

port systems become saturated at very low concentrations, whereas others become saturated only at concentrations far above those normally found in the body. The saturation of mediated-transport systems has important consequences in various organs in the body, particularly in the kidney where mediated-transport systems regulate the excretion of solutes in the urine.

Competition. A third characteristic of mediated-transport systems is the competition that occurs between similar molecules which enter the cell by interacting with the same binding site. If two

molecules A and B (for example, the amino acids glycine and alanine), which both enter the cell by the same mediated-transport system, are present simultaneously outside the cell, they must compete with each other for the available binding sites on the membrane (Fig. 6-13). The presence of B

Figure 6-13. Entry of molecule A into a cell either by diffusion or by a mediated-transport system. The presence of a similar molecule B does not alter the diffusion of A into the cell but, by competition with A for the available binding sites, B decreases the entry of A via the mediated-transport system.

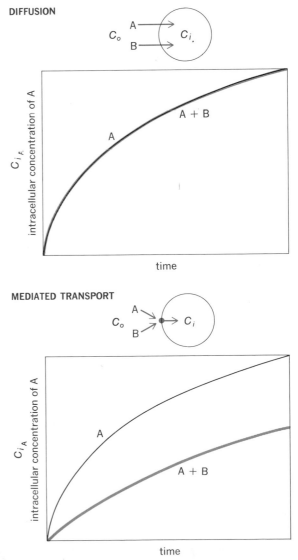

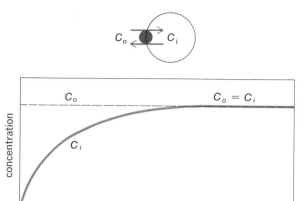

Figure 6-14. Changes in intracellular concentration C_i with time, resulting from the entry of molecules by facilitated diffusion.

decreases the rate at which A enters the cell since B occupies some of the binding sites that would otherwise be available to A. Molecules entering by simple diffusion do not compete with each other, since their entry is not restricted to a limited number of sites.

The carrier hypothesis

To reiterate, the properties of specificity, saturation, and competition are due to the fact that specific protein binding sites in the membrane combine with the transported molecule. These protein binding sites have been called *carriers* because it was once thought that the proteins actually "carried" the bound solute molecules across the membrane. However, we now know that, although protein molecules can move laterally in the plane of the membrane, they cannot readily move across it from one side to the other. Therefore, the mechanism by which a carrier protein enables a solute to pass through the membrane is still unknown but is presumed to involve a conformational change in the carrier molecule brought about by the solute's binding to it. The term "carrier" has been retained despite the fact that these proteins do not actually move across the membrane.

The behavior of carrier-mediated transport systems can be further divided into two classes: *facilitated-diffusion systems*, which lead to equal concentrations of solute across the membrane, and *active-transport systems*, which can move solutes "uphill" from a low-concentration region on one side of a membrane to a high-concentration region on the other side.

Facilitated diffusion

When the intracellular concentration C_i is measured as a function of time in a facilitated-diffusion system, the results appear very much like a diffusion process, since the final intracellular concentration equals the extracellular concentration, $C_i = C_o$ (Fig. 6-14). However, when the concentration of the medium is varied or competing molecules are added to the system, specificity, saturation, and competition appear.

In facilitated diffusion, binding sites on carrier proteins are accessible to solute molecules in both the extracellular and intracellular fluids. This permits the solute to move in either direction across the membrane. Let us apply this to the entry of solute from the extracellular fluid. If extracellular concentration initially exceeds intracellular concentrations, there is net movement into the cell. As entry proceeds, the concentration of solute increases within the cell; some of the intracellular solute molecules react with unoccupied carriers and move out of the cell. When the intracellular concentration becomes equal to the extracellular concentration, molecules move out of the cell as rapidly as they enter and no further change in the intracellular concentration occurs. Thus, in facilitated diffusion all net movement of molecules across the membrane is from a region of high concentration to a region of low concentration. It is unfortunate that the term "facilitated diffusion" is used to identify this process, since the process is not really one of diffusion but involves a chemically specific binding reaction. The only similarity to diffusion is the net movement from high to low concentration until the concentrations on the two sides of the membrane become equal.

The most important example of facilitated diffusion in the body is the movement of glucose across the membranes of most cells. This relatively large polar molecule cannot diffuse readily through the lipid or pore regions of the membrane, but facilitated diffusion through the membrane provides a mechanism for supplying cells with it.

Active transport

Not only do some molecules cross membranes more rapidly than expected, but some are even moved from a region of low concentration into a region of higher concentration. Such movement is quite contrary to that expected for either simple diffusion or facilitated diffusion since these alone can never lead to the net movement of molecules uphill against a concentration gradient. In order for molecules to move against a concentration gradient, energy must be supplied. The movement of molecules across a cell membrane from a region of low to high concentration at the expense of energy provided by metabolism is known as *active transport* (Fig. 6-15).

For a higher concentration of molecules to be achieved on one side of a membrane than on the other, molecules must be able to move more readily in that direction through the membrane than in the other. In a facilitated-diffusion system we saw that the reactions between solute molecule and carrier were the same on both sides of the membrane. In contrast, for active transport to occur, these reactions must be asymmetric, allowing the extracellular solute molecules to combine more readily with the carrier than do the intracellular (or vice versa, depending on the direction of the transport system); i.e., the affinities of the carrier for solute molecules must be different on the two sides. If the extracellular and intracellular concentrations of molecules on both sides of the membrane are initially equal, more molecules combine with the high-affinity carriers which are on one side (let us say, the extracellular side), than with low-affinity carriers, which are on the other side. More solute will therefore be moving into the cell than will be moving out; i.e., there will be a net flux of solute into the cell even though the solute concentrations on the two sides of the membrane are initially equal. As the intracellular concentration continues to increase, more of the low-affinity carrier sites will become occupied, thereby increasing the flux out of the cell until an intracellular concentration is reached at which the flux into the cell is equal to the flux out. At this point a steady state will have been reached at which the intracellular concentration exceeds the extracellular, and no further change in these concentrations occurs. Such active-transport systems can produce concentrations 30 to 50 times higher on one side of the membrane than on the other for many solutes; in a few cases, concentrations a millionfold higher are achieved.

Energy provided by cellular metabolism is required to maintain the differences in carrier affinity. Inhibiting the source of energy provided by the cell prevents an active-transport system from moving molecules against a concentration gradient.

One of the most important active-transport systems in cells moves sodium and potassium ions across plasma membranes. The extracellular fluids normally have a sodium concentration of 144 mmol/L and a potassium concentration of 4.4 mmol/L. Inside the cell the respective concentrations are 15 and 150 mmol/L. Thus, the concentration of sodium outside the cell is about 10 times greater than inside, and the potassium concentration inside the cell is about 35 times greater than outside. These concentration gradients are maintained by the active transport of potassium into and sodium out of the cell. Both sodium and potassium ions may interact with the same carrier molecule in crossing the membrane, both move up their concentration gradients, and the movement of one ion is linked through the common carrier to the movement of the other in the opposite direction. What is the nature of this particular carrier? An enzyme known as Na-K ATPase is located in the

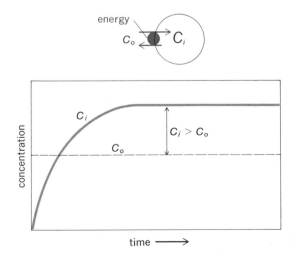

Figure 6-15. Active transport. Time course of changes in intracellular concentration C_i during active transport, resulting in an intracellular concentration of solute higher than the extracellular concentration C_o.

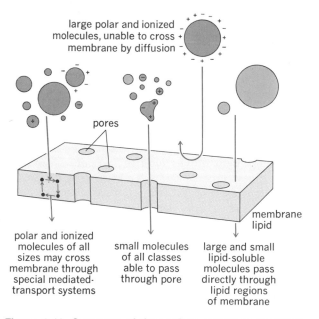

large polar and ionized molecules, unable to cross membrane by diffusion

pores

polar and ionized molecules of all sizes may cross membrane through special mediated-transport systems

small molecules of all classes able to pass through pore

large and small lipid-soluble molecules pass directly through lipid regions of membrane

membrane lipid

Figure 6-16. Summary of the various pathways by which molecules can cross cell membranes.

plasma membranes of most cells. It splits ATP in the presence of sodium and potassium ions and releases energy. Both types of ions are required for enzyme activity and, furthermore, the affinities of the enzyme for sodium and potassium are similar to the affinities of the membrane carrier which transports them. Finally, tissues which transport large quantities of sodium and potassium contain large quantities of this enzyme in their membranes. Thus, it is very likely that the Na-K ATPase not only supplies the chemical energy (in the form of ATP splitting) for ion transport but may well be the carrier molecule itself. It has been estimated that as much as 10 to 40 percent of the total energy produced by a cell may be utilized for the transport of sodium and potassium ions across the membrane. Since these ions can also cross the membrane by diffusion through pores, two routes are available to them, an active-transport pump and a diffusion leak through pores.

Amino acids are also actively transported into cells, attaining intracellular concentrations 2 to 20 times higher than the extracellular concentrations. Sugars enter most cells by facilitated diffusion but are actively transported by the epithelial cells of the intestine and in the kidneys. Unlike the case for active transport of ions, the energy used to move sugars and amino acids up concentration gradients is not *directly* linked to the splitting of ATP or any other source of chemical energy. Rather it is the sodium gradient across the membrane, itself established by active transport out of the cells, which provides the energy for organic solute transport. Let us take amino acid transport into cells as an example. The carrier molecule for the amino acid also has a binding site for sodium and the binding of sodium to the carrier increases the affinity of the amino acid binding site. Since the intracellular concentration of sodium is low, the probability of a sodium ion binding to the intracellular amino acid carrier is low; just the opposite is true on the outer surface of the membrane. Thus, the sodium gradient across the membrane leads to an asymmetry in the affinities of the amino acid carrier, resulting in the active transport of amino acid into the cell. Note that the energy for this process comes *indirectly* from ATP, which was required by the sodium pump to maintain a low intracellular sodium concentration, i.e., the sodium gradient which *directly* drives the uphill amino acid transport.

Mediated-transport systems are subject to regulation. Several hormones have as one of their effects upon cells an alteration in the mediated-transport activity of specific solutes. In some cases the effect of the hormone is to alter the affinity of the carrier for the transported solute. In some cases the hormone stimulates the synthesis of new carrier molecules which are inserted into the membrane, thereby increasing the maximal transport capacity of the membrane.

Figure 6-16 summarizes the various pathways by which molecules can cross cell membranes: (1) lipid-soluble molecules can diffuse through the lipid bilayer of the membrane; (2) small polar molecules, including ions, can diffuse through pores; (3) some ions and large polar molecules can enter through chemically specific carrier-mediated transport systems which include both facilitated-diffusion and energy-requiring active-transport systems. Very large polar molecules, including proteins, polysaccharides, and nucleic acids, are lipid insoluble, too large to diffuse through pores, and possess no specific carrier systems; they therefore do not move directly through membranes (it is for this reason that many enzymes remain localized within various membrane-bound organelles). However, as we shall see in subsequent sections, they are capable of entering or leaving cells by mechanisms which circumvent this inability to move directly through membranes.

Epithelial transport

The previous discussion centered on the transport of solutes across individual cell membranes. Another level of complexity emerges when one considers so-called transcellular transport, the movement of molecules from an extracellular compartment into an epithelial cell (across the plasma membrane at one pole of the cell) and out of the cell (across the plasma membrane at its opposite pole) into a second extracellular compartment. Such transport systems, for example, absorb nutrients from the intestinal tract into the blood and transport solutes and water between the blood and the tubules of the kidney during the formation of urine. For some substances, diffusion alone can account for transcellular movement, but in many cases an active-transport process is involved.

Active transport may be operating at one surface of the cell to move molecules up a concentration gradient, while at the other end of the cell the molecules leave by diffusion or facilitated diffusion down a concentration gradient (Fig. 6-17). Alternatively, the active transport may move the molecule out of the cell against a concentration gradient, thereby lowering its intracellular concentration. This creates a gradient for diffusion or facilitated diffusion into the cell across the opposite plasma membrane. In either case, the net result is the movement of molecules from one extracellular compartment having a low concentration across the cell into a second extracellular compartment containing the molecules at a higher concentration.

Osmosis

Water, which makes up the bulk of the fluid inside and outside of cells, is a small, polar molecule about 0.3 nm in diameter and crosses most cell membranes very rapidly. One might expect that, because of its polar structure and thus its low lipid solubility, water would not penetrate through the lipid regions of the membrane, but would be able to pass through the 0.8-nm pores because of its

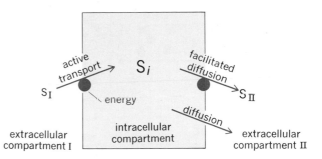

Figure 6-17. Transepithelial transport between two extracellular compartments I and II. Solute S enters the cell by active transport from compartment I and leaves the cell by diffusion or facilitated diffusion from the high intracellular concentration S_i into the lower concentration S_{II} in compartment II.

small size. However, artificial membranes containing only phospholipid bilayers also have a high permeability to water, which suggests that this small molecule may be able to pass between the fatty acid chains of the membrane. There is no evidence to suggest that the movement of water occurs by way of mediated-transport systems. Whatever the route, the fact remains that most membranes are highly permeable to water. Exceptions are certain epithelial membranes, such as the skin and portions of the kidney tubules, which have a relatively low permeability to water.

A net movement of water across a cell membrane occurs whenever there is a difference between the water concentrations on the two sides of the membrane; this is known as *osmosis*. If water happens to be the only molecule in the solutions on the two sides of the membrane able to move through the membrane, osmosis into or out of the cell will lead to a change in cell volume.

How can a difference in water concentration be established across a membrane? One usually thinks of the concentration of a substance in a solution as referring to the amount of solute dissolved in a given volume of solvent; e.g., a one molar (1 M) solution of glucose contains one mole of glucose dissolved in enough water to form one liter of solution. However, the water in this same solution also has a concentration. A liter of pure water weighs about 1000 g. The molecular weight of water is 18; thus, the concentration of pure water is $1000/18 = 55.5$ mol/L. However, if a solute molecule such as glucose is dissolved in water, the concentration of water in the resulting solution is

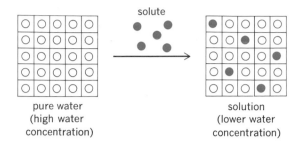

solute

pure water
(high water
concentration)

solution
(lower water
concentration)

Figure 6-18. Decrease in water concentration resulting from the addition of solute molecules.

less than that of pure water. Each molecule of solute added to a solution occupies an element of volume formerly occupied by water molecules. The more solute molecules added, the greater the number of water molecules displaced. Figure 6-18 illustrates this decrease in water concentration that results from the addition of solute.

It is essential to recognize that the degree to which the concentration of water is decreased by the addition of solute depends upon the number of particles of solute in solution and not upon the chemical nature of the solute because one molecule of solute will displace one molecule of water.[1] For example, one mole of glucose added to one liter of water decreases the water concentration to approximately the same extent as one mole of an amino acid, or one mole of a protein, or one mole of any molecule that exists as a single particle in solution. A molecule which ionizes in solution, such as sodium chloride, which forms the two ions Na^+ and Cl^-, decreases the water concentration in proportion to the number of ions formed. Hence, one mole of sodium chloride lowers the water concentration twice as much as one mole of glucose because sodium chloride forms two ions ("particles") in solution for every molecule added.

Since the concentration of water depends upon the number of solute particles in solution rather than their chemical properties, it is useful to have

[1] In reality all particles do not have an identical effect upon lowering the water concentration of a solution. This is because they differ in size and electric charge which produces slight differences in their interaction with water molecules. These effects are, however, so small, they do not significantly alter the generalization that it is the total number of solute particles in solution and not their chemical nature or size that is the major determinant of the concentration of water.

a concentration term which refers to the total concentration of all solute particles in a solution. The total solute concentration of a solution is known as the *osmolarity*. One osmole (osmol) is equal to one mole of an ideal nonionizing molecule. Thus, a 1 M solution of glucose has a concentration of 1 osmol/L, but a 1 M solution of sodium chloride contains 2 osmol of particles per liter of solution. A liter of solution containing 1 mol of glucose and 1 mol of sodium chloride has an osmolarity of 3 osmol/L since it would contain 3 mol of solute particles. The concentration of water in any two solutions having the same osmolarity is always the same, since the total number of solute particles per unit volume is the same. A 3-osmol solution may contain 1 mol of glucose and 1 mol of sodium chloride, or 3 mol of glucose, or $1\frac{1}{2}$ mol of sodium chloride, or any other combination of solutes so long as the total number of solute particles is equal to the number of solute particles in a 3 M solution of an ideal nonionizing molecule. A crucial point is that although osmolarity refers to the concentration of solute particles in solution, it is indirectly a measure of the *water concentration* in the solution; the *higher* the osmolarity of a solution, the *lower* is the water concentration.

Figure 6-19 shows two compartments separated by a membrane. The concentration of solute (4 osmol/L) in compartment 2 is higher than the concentration of solute in compartment 1 (2 osmol/L). Thus, there is a solute concentration gradient forcing net diffusion of solute from 2 to 1. This difference in solute concentration also means there is a difference in water concentration across the membrane, with the water in compartment 1 having a higher concentration than the water in compartment 2. Therefore, water undergoes a net diffusion from compartment 1 to 2, in the opposite direction from the net solute movement. If both water and solute can cross the membrane, their respective net movements proceed in opposite directions until the concentrations of both water and solute are equal in the two compartments. The diffusion of water from compartment 1 to compartment 2 not only lowers the concentration of water in 1 and raises it in 2, but also has the effect of concentrating the solute in compartment 1 and diluting it in compartment 2. While the water is moving between the two compartments, solute is diffusing in the opposite direction and also tends to bring the water and solute concentrations in the

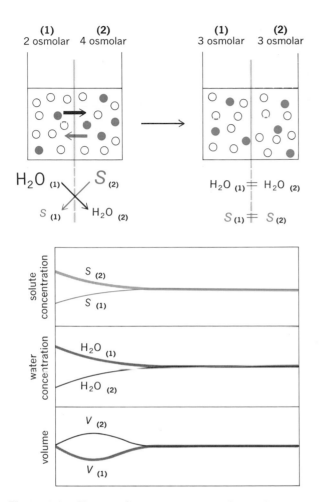

Figure 6-19. Changes in water concentration, solute concentration, and compartment volume when both water and solute are able to diffuse across a membrane separating two compartments that initially contained different concentrations of solute. (Osmolar = osmol/L)

two compartments into equilibrium. The net result is the redistribution of water and solute between the two compartments so that the final concentration of solute in both compartments is 3 osmol/L. However, the final volumes of the two compartments remain unchanged since some of the water and solute molecules have merely exchanged positions between the two compartments (transitory changes in volume may occur because of differences in the rates at which solute and water move between the two compartments).

If we make one alteration in this system, namely replacing the membrane with one permeable to water molecules but impermeable to the solute, the final result is different (Fig. 6-20). Initially, the same concentration gradients are found across the membrane as in the previous example. Water moves from its high concentration in compartment 1 into compartment 2, which has a lower water concentration, but since there can be no movement of solute in the opposite direction to compensate for the water movement, the volume of compartment 2 increases. Since solute molecules cannot leave compartment 1, the loss of water

Figure 6-20. Changes in water concentration, solute concentration, and compartment volume when only water is able to diffuse across a membrane separating two compartments that initially contained different concentrations of solute. In this example the two compartments are treated as if they were infinitely expandable and thus there is no pressure gradient created across the membrane.

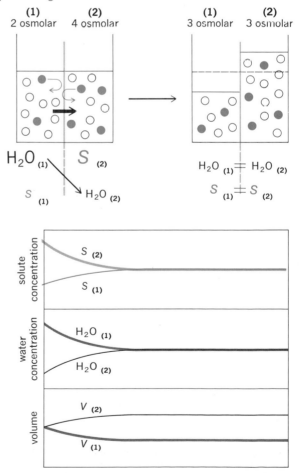

from this compartment has the effect of increasing the concentration of solute there, and the water entering compartment 2 dilutes the solute concentration there. Therefore, this system also comes to equilibrium when the concentrations of water and solute in the two compartments become equal, but, in this case, the equilibrium has resulted from the transfer of water alone and has led to a change in the final volume of the two compartments. (We have treated the two compartments as if they were infinitely expandable and thus the net transfer of water will not create a difference in pressure across the membrane; this is essentially the situation that occurs across cell membranes; in contrast, if the walls of compartment 2 could not expand, the movement of water into compartment 2 would raise the pressure in this compartment, which would oppose further water entry.) This example illustrates the following generalization: In order for a volume change to occur, the movement of solute molecules must be restricted by a semipermeable membrane. During osmosis the water always moves from a region of low solute concentration (high water concentration) to a region of high solute concentration (low water concentration).

The cell behaves like such an osmotic system; it is surrounded by a membrane that is permeable to water, and its cytoplasm contains many molecules which are unable to cross the cell membrane. The permeability constant of the cell membrane to water is larger than to almost any other substance so that any change in the water concentration surrounding a cell rapidly leads to a change in cell volume. Since changes in cell volume are completed very rapidly, for all practical purposes the concentration of water within a cell may be assumed to always be equal to the concentration of water in the extracellular fluids.

When a cell is in osmotic equilibrium, the water concentration is the same on the two sides of the membrane. Thus, the total solute (osmolar) concentrations on the two sides of the membrane must also be the same. If a solute cannot cross the cell membrane, any change in its concentration produces a water concentration gradient across the membrane, leading to a net flux of water into or out of the cell and a change in cell volume. Thus, the volume of a cell ultimately depends upon the concentration of nonpenetrating solutes on the two sides of the cell membrane.

About 85 percent of the solutes in the extracellular fluids of the body are sodium and chloride ions, which can diffuse into the cell through the pores in the plasma membrane. However, as we have seen, the membrane contains an active-transport system for pumping sodium ions out of the cell, the result being that there is no net movement of sodium into the cell. Therefore, it is as if sodium ions were unable to cross the membrane, since for every sodium ion entering the cell, another sodium ion is returned to the outside by the active-transport system. Since chloride ions follow the sodium because of the electric attraction between them (sodium is a positive ion and chloride a negative ion), both sodium and chloride ions act as if they were nonpenetrating solutes and any changes in their concentration outside the cell alter the concentration of extracellular water and leads to a change in cell volume.

Inside the cell the major solute particles are potassium ions and a number of organic solutes. Most of the latter are unable to cross the membrane, partly because they contain polar or ionized groups (which limit their diffusion through the lipid portions of the membrane) and partly because they are too large to pass through the pores. Although potassium ions can leak out of the cell, they are actively transported back into the cell, and the net effect is the same as if potassium could not cross the membrane. Thus, sodium chloride outside the cell and potassium and organic solutes inside the cell represent the major effective nonpenetrating solutes determining the water concentrations on the two sides of the membrane. In contrast, solutes which are not actively transported across the membrane but which can cross by diffusion have no net effect upon the cell volume since they eventually reach the same concentration on both sides of the membrane and thus have the same effect on water concentration of the two sides. It is only the nonpenetrating or effectively nonpenetrating solutes which can alter cell volume by producing a change in the water concentration across the membrane.

Let us take some examples. Red blood cells placed in solutions containing different concentrations of sodium chloride change their volume in each solution according to the water concentration inside relative to the concentration outside the cell (Fig. 6-21). If the outside water concentration is greater, water moves into the cell, which

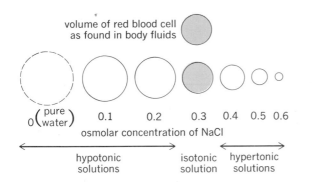

Figure 6-21. Comparison of the volume of a red blood cell in solutions of different sodium chloride concentrations with the volume of a red blood cell surrounded by body fluids.

swells. The cell shrinks if the concentration of water outside is less than inside. By comparing the volume of the cell as it exists in the blood with the volume in each of the sodium chloride solutions, one can determine that sodium chloride solution in which the cell volume remains the same as when surrounded by body fluids. This solution must have the same number of nonpenetrating solute particles as do the red cells. Such a solution is said to be *isotonic*. An isotonic solution is defined as any solution in which the cell volume remains the same as when surrounded by body fluids. Solutions in which the cell swells are called *hypotonic*, and solutions in which the cell shrinks are *hypertonic* solutions (Fig. 6-22). The terms isotonic, hypotonic, and hypertonic refer to the effects of solutions on the volume of any cell as compared with its volume in normal body fluids.

In the example above, a solution of 0.15 *M* sodium chloride was found to be isotonic for the human red blood cell; its osmolar concentration is 0.3 osmol/L. Any other solution containing 0.3 osmol/L of *nonpenetrating* solutes will be isotonic for the same cell. But if a cell is placed in a 0.3 osmolar solution of *penetrating* solute, the cell will swell and burst just as if it had been placed in pure water (see below), since the solute is able to enter the cell and will ultimately have the same effect on lowering the intracellular water concentration as it does on the extracellular water. The difference in water concentration that leads to the cell swelling is due to the presence of the usual nonpenetrating solutes within the cell and their absence in the bathing solution.

A cell placed in pure water, the most hypotonic of all solutions, swells until its plasma membrane ruptures. Such a cell can never reach a stable volume because there are no solute molecules in the extracellular medium, and therefore the water concentration outside the cell is always greater than that in the cell, leading to the continuous flow of water into the cell. Because the membranes are quite flexible, small osmotically induced changes in cell volume lead to a folding or unfolding of the surface membrane, changing the contours of the surface but not the surface area or the permeability properties of the membrane;

Figure 6-22. Changes in cell volume that result from the diffusion of water when a cell is placed in a hypotonic, isotonic, or hypertonic solution of sodium chloride.

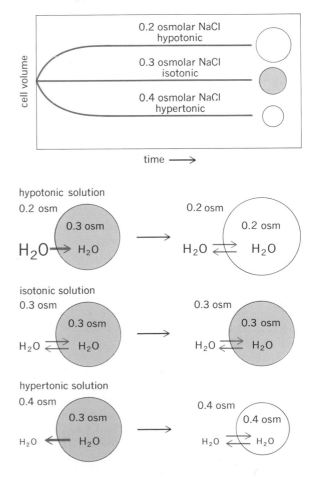

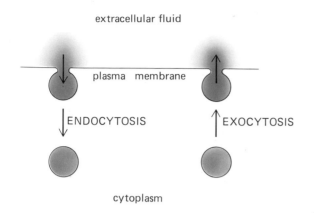

extracellular fluid

plasma   membrane

↓ENDOCYTOSIS   ↑EXOCYTOSIS

cytoplasm

Figure 6-23. Endocytosis and exocytosis.

however, when an increase in cell volume reaches the point at which the membrane begins to be stretched significantly, it ruptures producing large holes and destroying its selective permeability. When red blood cells are placed in pure water, the membrane ruptures (osmotic hemolysis), allowing the large protein molecules of hemoglobin to leak out. Since red cells would be hemolyzed if water were injected into a blood vessel, solutions to be injected must be nearly isotonic.

As will be described in Chap. 13, the kidneys, by regulating the excretion of salts and water in the urine, maintain the osmolarity of the extracellular fluids nearly constant, thereby preventing large changes in cell volume despite variable dietary intakes of salts and water. A hypotonic urine is excreted when there has been an excess water intake and a hypertonic urine is excreted when there has been an excess salt intake or when the body has lost water.

Endocytosis and exocytosis

Earlier in this chapter we described the various pathways by which molecules pass through cell membranes. There is an additional pathway which does not actually require the molecule to cross the structural matrix of the membrane. When living cells are observed under a light microscope, small regions of the plasma membrane can be seen to invaginate into the cell. These invaginations then become pinched off, forming small intracellular membrane-bound vesicles which enclose a small

volume of extracellular fluid; this process is known as *endocytosis* (Fig. 6-23). A similar process in the reverse direction, known as *exocytosis*, occurs when membrane-bound vesicles in the cytoplasm fuse with the plasma membrane and release their contents to the outside of the cell.

Endocytosis

Several varieties of endocytosis can be identified. When the vesicle encloses a small volume of extracellular fluid, the process is known as *fluid endocytosis*. The composition of the vesicle contents is the same as that of the extracellular fluid and there is no selective concentration of any particular component. In other cases, specific molecules bind to sites on the membrane and are carried into the cell when the membrane invaginates; this is known as *adsorptive endocytosis* and, in addition to taking in trapped extracellular fluid, leads to the selective concentration in the vesicle of the material bound to the membrane. Both fluid and adsorptive endocytosis are often referred to as *pinocytosis*, meaning cell drinking. A third type of endocytosis occurs when large multimolecular particles, such as bacteria, are engulfed by the plasma membrane and enter the cell. In this case, the membrane binds around the surface of the particle so that little extracellular fluid is enclosed within the vesicle; this form of endocytosis is known as *phagocytosis* (meaning cell eating).

The molecular mechanisms regulating endocytosis are still unclear. In some cells, fluid endocytosis occurs continuously, whereas adsorptive endocytosis and phagocytosis are stimulated by the binding of specific molecules or particles to sites on the cell surface; this binding, itself, appears to trigger the endocytosis locally, since adjacent regions of the membrane which lack bound molecules or particles do not increase their rate of endocytosis. The process of endocytosis requires metabolic energy, and contractile proteins associated with the plasma membrane have been implicated in this process.

What is the fate of the endocytotic vesicles once they have entered the cell? In some cases, such as the capillary endothelium, the vesicle passes through the cytoplasm and fuses with the opposite plasma membrane, releasing its contents to the extracellular space on the opposite side of the cell (exocytosis) (Fig. 6-24). This provides a pathway for transferring large molecules, which cannot cross cell membranes by other means, from one side of an epithelial layer of cells to the other. However, in most cells, endocytotic vesicles do not cross the cell but rather fuse with the membranes of lysosomes (Fig. 6-24), cell organelles which contain the digestive enzymes which break down large molecules such as protein, polysaccharides, and nucleic acids. The fusion of the vesicle with the lysosomal membrane exposes the contents of the vesicle to these digestive enzymes. The small molecular products of digestion, such as amino acids and sugars, then cross the membrane of the lysosome and enter the cytosol. There, they can be utilized by the cell or transported out of it by the usual transport pathways for such molecules. The endocytosis of bacteria and their digestion in the lysosomes is one of the body's major defense mechanisms against microorganisms and will be discussed in Chap. 17.

Obviously, the process of endocytosis removes a small portion of the membrane from the surface of the cell. If this membrane were not replaced, the surface area of the cell would decrease. In cells which have a large amount of endocytotic activity, more than 100 percent of the plasma membrane may be internalized in an hour, yet the cell volume and membrane surface area remain constant. This is because the membrane is replaced at about the same rate that it is removed. Much of this replacement seems to occur through the fusion of intracellular vesicles with the membrane via exocytosis.

Exocytosis

Exocytosis performs two functions for cells: (1) It provides a way to replace portions of the plasma membrane that have been removed by endocytosis or to add new membrane during cell growth; and (2) it provides a route by which certain types of molecules that are synthesized by cells can be released into the extracellular fluid. Many of the specialized end products of metabolism secreted by certain cells into the extracellular space are unable to cross plasma membranes because they are large polar molecules (proteins, for example). By being enclosed in secretory vesicles, they can be released from the cell by exocytosis.

The secretion of specialized end products in most

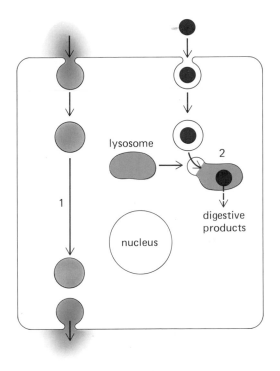

Figure 6-24. Fate of endocyotic vesicles. Pathway 1 transfers extracellular materials from one side of the cell to the other. Pathway 2 leads to fusion with lysosomes and the digestion of the vesicle contents.

cells is triggered by specific stimuli and does not occur continuously; for example, nerve cells release their chemical messengers by the process of exocytosis in response to a nerve impulse. Whenever the release of a specific end product involves exocytosis, calcium ions appear to be the essential coupler of the stimulus to the release mechanism. The concentration of free calcium ions in cells is very low, compared to its extracellular concentration, for two reasons: (1) Active transport systems in the plasma membrane and the membranes of cell organelles (particularly the mitochondria and the endoplasmic reticulum) "pump" calcium out of the cell or into these organelles, respectively; (2) several cell proteins have high-affinity binding sites for calcium, and therefore bind most of the available free calcium; for example, the plasma membrane itself binds a considerable amount of calcium. As we shall see in subsequent chapters, various stimuli cause entry of calcium across the plasma membrane or release of bound calcium from the membrane and cell organelles; it is the resulting increase in the concentration of free intracellular calcium which triggers the fusion of intracellular vesicles with the plasma membrane. Exactly how it does so is still unknown. One hypothesis is that calcium activates contractile proteins in the membrane and surrounding cytoplasm which act upon the vesicle.

How are molecules packaged into vesicles? In some cases, where relatively small organic end products are sequestered in vesicles, the enzymes for their synthetic pathways are located within the vesicles. In other cases, the vesicle membrane has special mediated-transport systems which concentrate the product within the vesicle, after it has been synthesized in the cytosol. Very high concentrations of product can be achieved within the vesicle by a combination of mediated transport followed by specific binding to protein sites within the vesicle. By being stored in large quantities in secretory vesicles, the product is immediately available for rapid secretion in response to a stimulus, and no delay is required for stimulus-induced synthesis.

Protein secretion. The packaging of protein end products into vesicles creates special problems for the cell because of the complex enzymatic machinery associated with protein synthesis and the inability of these large molecules to pass through membranes. As described in Chap. 4, ribosomes are the sites at which amino acids are assembled into proteins; proteins that are to be secreted from the cell are synthesized on ribosomes attached to the surface of the endoplasmic reticulum. As the protein is assembled from amino acids, it passes through the ribosome and the underlying membrane of the endoplasmic reticulum and enters the lumen of the reticulum (Fig. 6-25). This is the

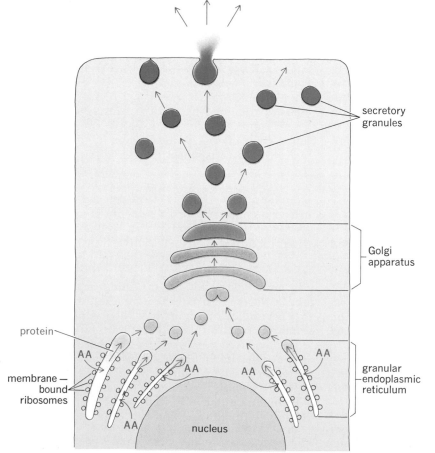

secretory granules

Golgi apparatus

granular endoplasmic reticulum

protein

AA

AA

AA

AA

AA

membrane bound ribosomes

nucleus

Figure 6-25. Pathway leading to the release of protein from a secretory cell. (AA signifies free amino acids which are assembled into proteins on ribosomes.)

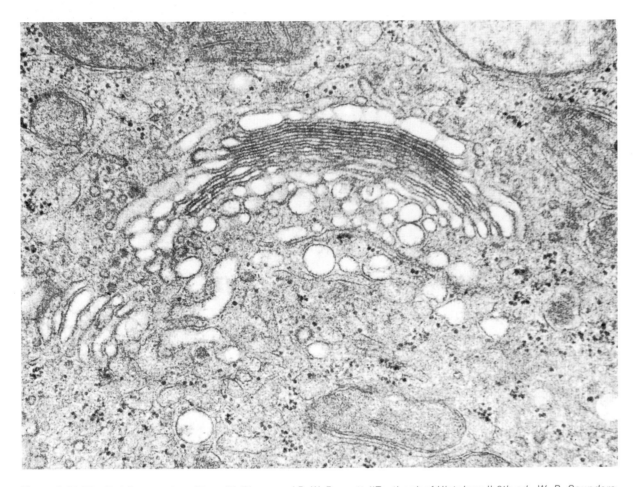

Figure 6-26. The Golgi apparatus. *(From W. Bloom and D. W. Fawcett, "Textbook of Histology," 9th ed., W. B. Saunders Company, Philadelphia, 1968.)*

only stage along its route to the outside of the cell at which the protein passes through a membrane. Portions of the ends of the reticulum then pinch off, forming small vesicles containing the newly synthesized protein. These vesicles migrate to the region of the Golgi apparatus (Fig. 6-26) where they fuse with the Golgi membranes. Within the Golgi apparatus, the protein solution becomes progressively more concentrated as a result of fluid removal from the vesicles. In addition, carbohydrate groups are linked covalently to many of the proteins by enzymes located in the Golgi apparatus. Following their concentration and enzymatic modification in the Golgi apparatus, the secretory proteins are present in small vesicles now known as *secretory (zymogen) granules* which have pinched off from the stack of Golgi membranes and migrated toward the plasma membrane where they can be released by exocytosis in response to the proper stimulus. The movement of secretory granules to only one end of the cell may be guided by the network of cellular microtubules and contractile filaments.

The endoplasmic reticulum, in addition to sequestering newly synthesized secretory proteins, also appears to be the site at which membranes are formed. The enzymes for the synthesis of membrane phospholipids are associated with the endoplasmic reticulum, and membrane proteins, following their synthesis on the ribosomes, can be added to the phospholipid bilayer. The newly synthesized membrane then passes to the Golgi apparatus and ultimately to the plasma membrane by a series of vesicle-membrane fusions.

Membrane junctions

In addition to providing a barrier to the movements of molecules between the intracellular and extracellular media, plasma membranes are involved in interactions between cells to form organized tissues. Some cells, particularly those of the blood, do not associate with other cells but remain as independent cells suspended in a fluid (the blood plasma). Most cells, however, are packaged into tissues and organs and are not free to move around the body. But even the cells in an organized tissue are not packaged so tightly that the adjacent cell surfaces are in direct contact with each other. There usually exists a space of at least 20 nm between the opposing membranes of adjacent cells; this space is filled with extracellular fluid and provides the pathway for the extracellular diffusion of substances within the tissue.

The forces that bring about the organization of cells into tissues and organs are poorly understood but appear to depend, at least in part, on the ability of binding sites on the cell surface to recognize and associate with specific protein aggregation factors which bind cells together (Chap. 4).

Some tissues can be separated into individual cells by fairly gentle procedures such as shaking the tissue in a medium from which calcium ions have been removed. The absence of calcium ions decreases the ability of these cells to stick together, suggesting that the doubly charged calcium ion (Ca^{2+}) may act as a link between cells by combining with negatively charged groups on adjacent cell surfaces. Many types of cells, however, require more drastic chemical procedures to separate them from their neighbors.

The electron microscope has revealed that a variety of cells are physically joined by specialized types of junctions. One type, known as a *desmosome* (Greek *desmos*, binding), is illustrated in Fig. 6-27A. The desmosome consists of two opposed membranes that remain separated by about 20 nm but show a dense accumulation of matter at each membrane surface and between the two membranes. In addition, fibers extend from the inner surface of the desmosome into the cytoplasm and appear to be linked to other desmosomes of the cell. The function of the desmosome is that of holding adjacent cells together in areas that are subject to considerable stretching, such as in the skin and heart muscle. Desmosomes are usually disk-shaped and thus could be likened to rivets or spot-welds as a means of linking cells together.

Another type of membrane junction, the *tight junction,* is found between epithelial cells. Such epithelial layers usually separate two compartments having differing chemical compositions; for example, the intestinal epithelium lies between the lumen of the intestinal tract containing the products of food digestion and the blood vessels that pass beneath the epithelial layer. These epithelial layers generally mediate the passage of molecules between the two compartments. It was noted earlier that most cells are separated from each other by a space of at least 20 nm which is filled with extracellular fluid. However, such spaces would make for a very leaky barrier that would allow even large protein molecules to diffuse between the two compartments by passing between adjacent cells. In fact, however, the intestinal epithelium in an adult is practically impermeable to protein, and the reason is the presence of tight junctions joining the epithelial cells near their luminal border. Tight junctions (Fig. 6-27B) are an actual fusing of the two adjacent plasma membranes so that there is no space between adjacent cells in the region of the tight junction. This type of fusion extends around the circumference of the cell and greatly reduces (but does not eliminate) the extracellular route for the passage of molecules between the epithelial cells. Therefore, for the most part, in order to cross the epithelium, a molecule must first cross the plasma membrane of an epithelial cell, pass through the cytoplasm, and exit through the membrane on the opposite side of the cell. Thus, the tight junction, in addition to helping to hold cells together, also seals the passageway between adjacent cells. Figure 6-27D shows both a tight junction and a desmosome located near the luminal border between two epithelial cells.

When a molecule to which the plasma membrane is impermeable is injected into most cell types through a micropipette inserted into the cell, it will usually remain within the cell and not pass into adjacent cells. When this experiment is performed on certain types of cells, however, the marker molecule is found to appear in the cell adjacent to the injected cell but not in the extracellular medium, suggesting that there is a direct channel linking the cytoplasms of the two cells. The electron microscope has revealed a structure

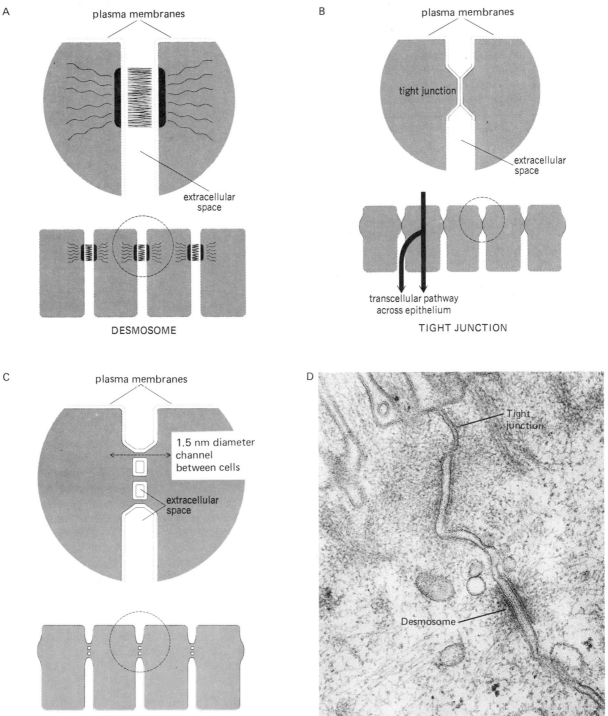

Figure 6-27. Schematic diagram of three types of specialized cell junctions: (A) desmosome, (B) tight junction, (C) gap junction, and (D) electron micrograph of two adjacent intestinal epithelial cells joined by a tight junction near the luminal surface and a desmosome a short distance below the tight junction. [*Electron micrograph from M. Farquhar and G. E. Palade,* J. Cell Biol., *17:375–412 (1963).*]

in these types of cells known as a *gap junction* (Fig. 6-27C). In the region of the gap junction the two opposing plasma membranes come within 2.0 to 4.0 nm of each other. Small channels about 1.5 nm in diameter extend across this gap and directly link the cytoplasm of the two cells. The small diameter of these channels limits the size of molecules that can pass between the connected cells to small molecules and ions, such as sodium and potassium, and excludes the exchange of large protein molecules. A variety of cell types possess gap junctions, including the muscle cells of the heart and smooth muscle cells. As we shall see, the gap junctions in these cells play a very important role in the transmission of electrical activity between adjacent muscle cells. In other cases, gap junctions are thought to coordinate the activities of adjacent cells by allowing chemical messengers to move from one cell to the next.

PART TWO

BIOLOGICAL
CONTROL
SYSTEMS

7
HOMEOSTATIC MECHANISMS

General characteristics of control systems
Components of living control systems
  Reflexes
    Some semantic problems
  Local homeostatic responses
  Chemicals as intercellular transmitters in homeostatic
    systems
    Prostaglandins

Receptors
Second messengers
  Calcium
  Cyclic AMP
  Relationship between calcium and cyclic nucleotides
The balance concept and chemical homeostasis
  Essential nutrients: An example
Biological rhythms

In Chap. 1, we described how each cell of a complex multicellular organism is bathed by extracellular fluid, which constitutes the body's internal environment. We also described how the activities of the various organ systems were directed toward the maintenance of relative constancy of this fluid's physical and chemical characteristics—its temperature and its concentrations of oxygen and carbon dioxide, organic nutrients, wastes, and inorganic ions. The liver acts as a metabolic factory, adding or removing organic molecules as needed; the lungs take in oxygen and eliminate carbon dioxide at rates required to keep the pressures of these gases constant; the gastrointestinal tract takes into the body ingested water, nutrients, and salts; the kidney eliminates just the right amount of wastes, water, and salts; and so on down the long list of bodily activities. Each cell of the body contributes, in its own way, to the survival of the total organism by helping to maintain the stable conditions required for life in the fluid bathing them all.

This concept of the maintenance of a stable internal environment was further elaborated and supported by the American physiologist W. B. Cannon, who emphasized that such stability could be achieved only through the operation of carefully coordinated physiological processes which he termed *homeostatic*. The activities of tissues and organs must be regulated and integrated with each other in such a way that any change in the internal environment automatically initiates a reaction to minimize the change. *Homeostasis* denotes the stable conditions which result from these compensating regulatory responses. Some changes in the composition of the internal environment do occur, of course, but the fluctuations are minimal and are kept within narrow limits through the multiple coordinated homeostatic processes, descriptions of which constitute the bulk of the remaining chapters.

Concepts of regulation and relative constancy have already been introduced in the context of a single cell. Thus, we described in Chap. 5 how metabolic pathways in a cell are regulated by the principle of mass action and by changes in enzyme activity so as to maintain the concentrations of the various metabolites in the cell, and in Chap. 4 how the control of protein synthesis is regulated by the genetic apparatus. In people, these basic intracellular regulators still remain, so that each individual cell exhibits some degree of self-regulation, but the existence of a multitude of different cells organized into specialized tissues, which are

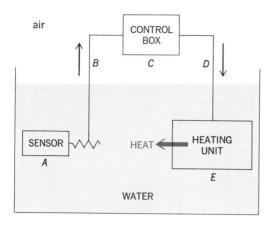

air

Figure 7-1. Components of a control system for regulating the temperature of a water bath.

further combined to form organs, obviously imposes the need for overall regulatory mechanisms to coordinate and integrate the activities of all cells. For this, intercellular communication, often over relatively long distances, is essential. Such communication is accomplished in large part by means of nerves and the blood-borne chemical messengers known as hormones.

The mechanisms by which these two communications systems operate are the subject of the next two chapters, but their overall role and the general characteristics of homeostatic processes can be appreciated only in terms of control systems, to which we now turn.

General characteristics of control systems

We shall define a homeostatic system as a control system consisting of a collection of interconnected components which functions to keep a physical or chemical parameter of the body relatively constant. Let us first analyze a nonbiological control system (Fig. 7-1), one designed to maintain the temperature of a water bath at approximately 30°C despite fluctuations in room temperature from 25 to 10°C. Since the water temperature is always to be higher than the room temperature, there is a continuous loss of heat to the room from the water. Moreover, the lower the room temperature, the greater this heat loss. Accordingly, the water must be continuously heated in order to offset the loss,

and the degree of heating must be altered whenever the room temperature changes. This adjustment of heat input to heat loss so that water temperature remains approximately 30°C is the job of the control system.

The first component of the system is known as a sensor. This particular sensor is sensitive to temperature, and it generates an electric current the magnitude of which increases as water temperature decreases. The current flows through wire B into the control box C constructed in such a way that the amount of current flowing out of the box in wire D is proportional to that entering along B. The current from the box flows to the heating unit E in the water bath. Its activity and therefore the amount of heat it produces per unit time is proportional to the magnitude of the current flow in D.

We fill the bath with water at room temperature, close the switches, and allow the system to operate (Fig. 7-2). Current generated in the sensor A controls the output of the heating unit by way of the control box. Because of the initially low water temperature, the magnitude of current flow from A is large and the heating unit is running full blast. This heats the water rapidly, but as the water temperature rises, two opposing events occur: (1) More heat is lost from the bath to the room; (2) the signal from A decreases and results in a decreased input to the heating unit, thereby decreasing the amount of heat it produces. The system ultimately stabilizes at a particular water

Figure 7-2. Effects of filling a water bath with water at room temperature and allowing the system to operate. Shown are the initial state, i.e., just after the water was added, and the ultimate steady state achieved. Note the differences in output of the sensor and the heating unit in the two states. The operating point is 30°C.

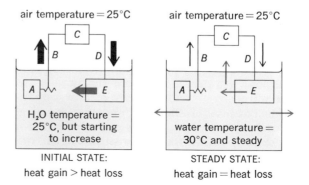

temperature when heat loss to the room exactly equals heat gain from the unit. At this point, the system is said to be in a *steady state;* input equals output, and the temperature remains steady. Actually, as shown in Fig. 7-3, there is always some oscillation around the steady-state temperature because of the time required for the heating unit to heat up or cool down.

The steady-state temperature is determined, in large part, by the characteristics of the sensor and the control box *C*, since these components are what determine the output of the heating unit. If we had chosen a heat-sensitive sensor which generates only half as much current at any given temperature as *A* does, the input to the control box would always be less, the output from the control box would be less, and the heating unit would always be generating less heat. Therefore, the steady-state temperature of the bath would be lower. Similarly, by altering the transforming function, i.e., the relationship between current in and current out, of the control box, we alter the steady-state temperature ultimately reached. In any control system, the actual steady state, or so-called *operating point* of the system, depends upon the characteristics of the individual components of the system. Our components were chosen to achieve an operating point of 30°C when the air temperature is 25°C.

Most important is that this type of system resists any changes from the operating point; thus, the control system in our example will automatically prevent any large deviation of the water temperature from its 30°C operating point. Suppose that after 30°C has been reached, the room temperature is suddenly lowered (and kept low) so that the loss of heat from water to room is increased. This loss unbalances heat loss and heat gain, and the water temperature falls. But the decrease in water temperature immediately increases the current generated in *A;* therefore, more current flows out of the control box *C* and the heating unit increases its activity, thereby raising the water temperature back toward its original value. A new steady state will be reached when heat loss once again equals heat gain, both having been increased by a proportionate amount. What is the new steady-state temperature at this point? If the system is extremely sensitive to change, the temperature will be only very slightly below what it was before the room temperature was lowered *but it cannot be precisely the same.* Compensation is incomplete be-

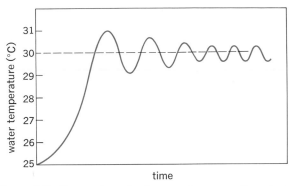

Figure 7-3. Change in water temperature over time during operation of the system shown in Fig. 7-2.

cause the new steady state depends upon the maintenance of an increased heat production to balance the increased heat loss, and this increased heat production is due to the increased signal coming from *A*. The reason *A* generates more current is the slightly lower water temperature. If the temperature actually returned completely to 30°C, the signal from *A* would return to its value for that temperature, heat production would return to the original value, and the water temperature would immediately decrease as heat loss again became greater than heat gain (recall that the room temperature is being maintained lower than the bath temperature). Thus, control systems of this type cannot absolutely prevent changes from occurring in the physical or chemical variable being regulated, but they do keep such changes within very narrow limits, dependent upon the sensitivity of the system. A crucial generalization emerges: The operating point for a particular control system is not a single number but rather a narrow range of values (say 29–31°C in our example); the precise value within that range existing at any time is dependent on conditions in the external environment (the room temperature in our example).

We can now summarize the basic characteristics of a control system in general terms. There must be a component which is sensitive to the variable being regulated and which changes its signal rate as the variable changes. There must be a continuous flow of this information from this sensor to an integrating control box, from which, in turn, the so-called command signal flows to an apparatus which responds to the signal by altering its rate of output (heat, in our example).

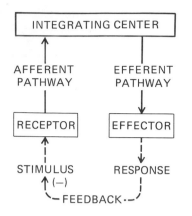

Figure 7-4. General components of a biological control system, the reflex arc. The response of the system has the effect of counteracting or eliminating the original stimulus. This phenomenon of negative feedback is emphasized by the minus sign in the feedback loop.

There remains one more concept which we have described but not named: *feedback*. The ultimate effects of the change in output by the system must somehow be made known (fed back) to the sensor which initiated the sequence of events. Our example illustrated perhaps the commonest form of feedback: When the water temperature is reduced, the sensor A detects this change and relays the information to the control box, which in turn signals the heating unit to increase its output; sensor A is "informed" of this change in output by the resulting rise in water temperature; accordingly, its current generation is again altered, which, in turn, results in an alteration of signal to the control box and thereby the heating unit. The water temperature acts as a continuous link between the sensor and the heating unit (without feedback, the sensor's signal would be unrelated to the heating-unit output and the system would be unable to maintain constancy). This type of feedback, in which an increase in the output of the system results in a decrease in the input, is known as *negative feedback*. It clearly leads to stability of a system and is crucial to the efficient operation of homeostatic mechanisms.

Note that a major characteristic of negative-feedback control systems is that they *restore* the regulated variable toward its operating point after its initial displacement, but they cannot *prevent* the initial displacement. Suppose, however, that we add another component to our temperature-control system of Fig. 7-1, namely, a thermosensor on the *outside* of the water bath which can detect changes in room temperature. Now, when the room temperature is lowered, as in our example, this information is immediately relayed to the control box which causes the heating unit to increase its output. In this manner, additional heat can be supplied to the water bath *before* the water temperature begins to fall. Thus, the use of the external sensor permits the system to *anticipate* a pending fall in water temperature and begin to take action to counteract the change before it occurs. This provides *feedforward* information which has the net effect of minimizing *fluctuations* in the level of the parameter being regulated. We shall see that the body frequently makes use of feedforward control in conjunction with negative-feedback systems.

There is, however, another type of feedback known as *positive feedback* in which an initial disturbance in a system sets off a train of events which increases the disturbance even further. Generally, such cycles occur explosively; they usually lead to instability and are quite uncommon. However, several important positive-feedback relationships occur in the body, blood clotting being an example.

Components of living control systems

Reflexes

Homeostatic control systems in living organisms manifest characteristics very similar to those just described (although some of the terminology is different). Many belong to the general category of stimulus-response sequences known as reflexes. A *reflex* is the sequence of events elicited by a stimulus. We may be aware of only the final event in the sequence, the *reflex response* (pulling one's hand away from a hot stove, for example). The entire reflex, including the response, often occurs without any awareness on the part of the person. The pathway mediating the reflex is known as the *reflex arc*, and its components are shown in Fig. 7-4; note that they are completely analogous to those of Fig. 7-1.

A *stimulus* is defined as a detectable change in the environment, such as a change in temperature, potassium concentration, pressure, etc. A *receptor* is the component which detects the environmental change (it is identical to the sensor of the previous section). The stimulus acts upon the receptor to alter the signal emitted by the receptor, and this signal is the information relayed to the control box, or *integrating center*. The pathway between the receptor and the integrating center is known as the *afferent pathway*.

The integrating center usually receives input from many receptors, some of which may be responding to quite different types of stimuli. Thus, the output of the integrating center reflects the net effect of the total afferent input; i.e., it represents an integration of numerous and frequently conflicting bits of information.

The output of the integrating center is then relayed to the last component of the system, the device whose change in activity constitutes the overall response of the system. This component, known as an *effector*, is analogous to the heating unit of our previous example. The information going from the integrating center to the effector is like a command directing the effector to alter its activity. The pathway along which this information travels is known as the *efferent pathway*.

As a result of the effector's response, the original stimulus (environmental change) which triggered this entire sequence of events may be counteracted (at least in part), just as in our example the increased heat production by the heating unit caused the lowered water temperature to return toward its operating point (30°C). As the stimulus is diminished by the effector's response, the activity of the receptor is diminished, so that the flow of information from receptor to integrating center returns toward the original level and, in turn, the effector's activity is turned toward its previous rate. Thus, the counteracting of the stimulus by the effector's response constitutes the *negative feedback*.

In summary, the components of a typical reflex arc are:

1 Receptor
2 Afferent pathway
3 Integrating center
4 Efferent pathway
5 Effector

To illustrate, let us again take a thermoregulatory example, this time one of the several reflexes which maintain body temperature relatively constant. The receptors are the endings of certain nerve cells in various parts of the body which generate electrical signals at a rate determined by their temperature. This information is relayed by the nerve fibers (the afferent pathway) to a specific part of the brain (which acts as integrating center) which in turn influences, via a chain of nerve cells, the rate of firing of those nerve cells which stimulate skeletal muscle to contract. These latter nerve fibers are the efferent pathway and the muscles they innervate are the effectors. The amount of muscle contraction is altered so as to raise or lower heat production (heat is produced during muscle contraction). Thus, when external temperature decreases, thereby enhancing heat loss from the body, this reflex automatically increases heat production by increasing muscle contraction—familiar to us as shivering—so that heat loss and gain can remain equal and body temperature remain relatively unchanged.

Traditionally, the term reflex is restricted to situations in which the first four of the components listed above are all parts of the nervous system, as in the thermoregulatory reflex just described. However, present usage is not so restrictive and recognizes that the principles are essentially the same when a hormone, rather than a nerve fiber, serves as the afferent or (more commonly) efferent pathway, or when a hormone-secreting gland serves as integrating center. Thus, in our example above, the integrating brain centers not only trigger off increased activity of nerve fibers to skeletal muscle but also cause release of several hormones which travel by the blood to cells and increase their heat production by altering the rates of various heat-producing chemical reactions; these hormones therefore also serve as an efferent pathway in thermoregulatory reflexes.

Accordingly, in our use of the term "reflex" we include hormones as reflex components, so that afferent or efferent information can be carried by either nerve fibers or hormones. In any case, two different components must serve as afferent and efferent pathways; thus the input to, or output from, the integrating center may both be neural, but they must be two different nerve fibers. Of course, they can also both be hormonal, or one neural and the other hormonal. Depending on the specific nature of the reflex, the integrating center may reside in either the nervous system or an endocrine gland.

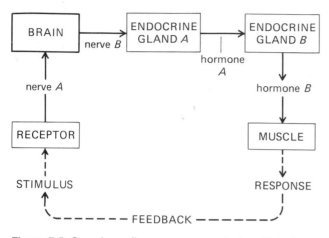

Figure 7-5. Complex reflex arc composed of multiple hormones and nerve fibers.

Finally, we must identify the effectors, the cells whose outputs constitute the ultimate responses of the reflexes. Actually, most cells of the body act as effectors in that their activity is subject to control by nerves or hormones. There are, however, two specialized tissues, muscle and gland, which comprise the major effectors of biological control systems. Muscle cells are specialized for the generation of force and movement. Glands consist of epithelial cells specialized for the function of secretion.

Glands are classified into two groups on the basis of the site into which their secretion is released. *Endocrine glands* have no ducts and secrete their products into the blood stream—more precisely, into the extracellular space around the gland cell, from which they then diffuse into the blood vessels. *Exocrine glands* secrete into ducts leading to a specific compartment or surface, such as the lumen of the gastrointestinal tract or the skin surface.

To summarize, most biological control systems function to keep a physical or chemical parameter of the body relatively constant. One may analyze any such system by answering a series of questions: (1) What is the parameter (blood glucose, body temperature, blood pressure, etc.) which is being maintained constant in the face of changing conditions? (2) Where are the receptors which detect changes in the state of this parameter? (3) Where is the integrating center to which these receptors send information and from which information is sent out to the effectors, and what is

the nature of these afferent and efferent pathways? (4) What are the effectors and how do they alter their activities so as to maintain the regulated variable at the operating point of the system?

Some semantic problems. There are problems which arise when one attempts to categorize the components of some complex reflex arcs according to the five terms listed above. For example, in the reflex arc shown in Fig. 7-5, the stimulus, receptor, and final effector (muscle) are quite clear, as is the designation of the first nerve (A) in the chain and the last hormone (B) as afferent and efferent pathways. But what about the brain, nerve B, hormone A, and the two endocrine glands? Actually, one is dealing here with a chain of small reflex arcs all subserving the overall reflex arc. Thus, endocrine gland A may be considered to be an effector whose response is the secretion of hormone A, or it may be viewed as an integrating center; in the latter case, nerve B becomes its afferent input and hormone A becomes its efferent output. It is fruitless to assign rigid terms to the interior components of such complex reflex chains. It is far more important for the reader to appreciate the sequence of events rather than worry about the labels.

Another problem is that some reflex arcs seem to lack a receptor and afferent pathway. This may seem puzzling, but the following example may help (Fig. 7-6). Parathormone is a hormone which is secreted by the parathyroid glands and which

Figure 7-6. Homeostatic reflex by which plasma calcium concentration is controlled. Note the absence of an afferent pathway.

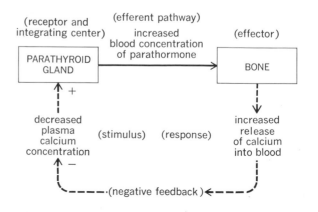

pon bone, causing it to increase its release
ium into the blood. When the blood calcium
tration is decreased for any reason, an in-
d amount of parathormone is secreted into
ood; the blood-borne parathormone reaches
hroughout the body and induces it to release
alcium into the blood. Clearly, in this reflex,
orne parathormone acts as the efferent
ay and bone is the effector. But we have
no mention yet of the receptors or the affer-
thway involved. Actually, the parathyroid
cells are themselves sensitive to the calcium
tration of the blood supplying them; thus,
me cells which produce the hormone act as
ceptors for this reflex. Clearly, there is no
t pathway in such a situation since the re-
s and integrating center (i.e., hormone-
ing cells) are one and the same cell; differ-
mponents in the cell may be involved in
ing the stimulus and integrating the output.
lly, we must point out a problem to which we
turn in several subsequent chapters. In the
narrow sense of the word, a reflex is an in-
ary, unpremeditated, unlearned response to
ulus; the pathway over which this chain of
occurs is "built in" to all members of a
s. Examples of such basic reflexes would be
g one's hand away from a hot object or
ng one's eyes as an object rapidly approaches
ce. However, there are also many responses
appear to be automatic and stereotyped but
actually are the result of learning and prac-
or example, an experienced driver performs
complicated acts in operating his car; to him
motions are, in large part, automatic, stereo-
and unpremeditated, but they occur only
se a great deal of conscious effort was spent
n them. We shall refer to such acts as *learned*
uired. In general, most reflexes, no matter
asic they may appear to be, are subject to
tion by learning; i.e., there is often no clear
ction between a basic reflex and one with a
d component.

homeostatic responses

es reflexes, another group of biological re-
es is of immense importance for homeostasis.
hall call them *local responses.* Local re-
es are initiated by a change in the external
ernal environment, i.e., a stimulus, which

acts upon cells in the immediate vicinity of the
stimulus inducing an alteration of cell activity
with the net effect of counteracting the stimulus.
Thus, a local response is, like a reflex, a sequence
of events proceeding from stimulus to response,
but, unlike a reflex, the entire sequence occurs
only in the area of the stimulus, no hormones or
nerves being involved.

Two examples should help clarify the nature and
significance of local responses: (1) Damage to an
area of skin causes the release of certain chemicals from cells in the damaged area which help
the local defense against further damage; (2) an
exercising muscle liberates chemicals into the
extracellular fluid which act locally to dilate the
blood vessels in the area, thereby permitting the
required inflow of additional blood to the muscle.
The great significance of such local responses is
that they provide individual areas of the body with
mechanisms for local self-regulation.

Chemicals as intercellular transmitters in homeostatic systems

It should be evident that the *sine qua non* of
reflexes and local responses is the ability of cells
to communicate with one another, i.e., the capacity
of one cell to alter the activity of another. When
a hormone is involved in a reflex, it is clear that the
communication between cells, i.e., between endocrine gland cell and effector, is accomplished by a
chemical agent, the hormone (the blood, of course,
acting as the delivery service). What has not been
said, however, is that virtually all nerve cells in
the body also communicate with each other or with
effectors by means of chemical agents known as
neurotransmitters. Thus, one neuron alters the
activity of the next neuron in a reflex chain by releasing from its ending a neurotransmitter which
diffuses across the very narrow space separating
the two neurons and acts upon the second, altering
its activity. Similarly, neurotransmitters released
from the ends of the neurons going to effectors
constitute the immediate signal, or input, to the
effector cells. The detailed physiology of these
neurotransmitters and of neuron-neuron or
neuron-effector communication will be described
in Chaps. 8 and 9. We mention them here to
emphasize that chemicals, whether they are secreted by endocrine gland cells or released from
neuronal endings, constitute the ultimate mes-

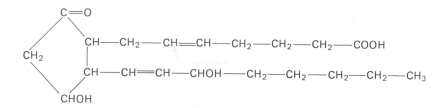

Figure 7-7. Structure of prosta
PGE$_2$.

sages by which one cell signals another to alter its activity.

This is true not only for reflexes but for local responses (as the examples above illustrate). The chemical messengers involved in local responses are referred to as *paracrine agents* to distinguish them from the hormones. Many paracrines are synthesized by local cells and released, given the appropriate stimulus, into the extracellular fluid surrounding the cell; there, they exert their particular effects. Perhaps the best known example of such an agent is histamine, which is produced and stored in several cell types and plays important roles in the body's responses to injury (Chap. 17). Other paracrines are not released by local cells, but are generated locally from inactive precursors circulating in the plasma; this generation is initiated by a change in the local environment.

As we shall see, a given chemical messenger may be synthesized by a number of different cell types so that a single molecular type may serve as a neurotransmitter (released from neuron terminals), as a hormone (released from endocrine gland cells), and as a paracrine agent. For example, epinephrine functions both as a hormone and as a neurotransmitter, and estrogen is both a hormone and a paracrine.

Prostaglandins. In recent years, an enormous amount of work has been devoted to the study of one particular group of chemical messengers, the family of unsaturated lipid acids containing a five-membered (cyclopentane) ring and known collectively as *prostaglandins*. Prostaglandins were originally discovered in semen but they are now known to be synthesized in most, possibly all, organs in the body. They are divided into main groups, each designated by a letter—PGA, PGE, PGF, etc.—on the basis of the configuration of their cyclopentane ring, and within each group, further subdivision occurs according to the num-

ber of double bonds in the side chains; thus, (Fig. 7-7) has two double bonds and this is de by the subscript in its name. At present, at le different prostaglandins as well as a num closely related substances called *thrombo* have been isolated from tissues; they are a thesized from arachidonic acid and other ess unsaturated fatty acids (it is because the precursors for the prostaglandins that these acids are essential nutrients).

The prostaglandins are not stored in tissu any great extent but are synthesized and rel immediately in response to a wide varie stimuli; they then act locally in the tissu which they are formed and are quickly metab to inactive forms. Accordingly, the prostagla fit best into the category of paracrine a (whether blood-borne prostaglandins have important effects on sites distant from their ti of origin remains unsettled).

The prostaglandins and thromboxanes e: bewildering array of effects and it is prov difficult task to elucidate their precise physiol roles. They almost certainly are important in clotting, regulation of smooth muscle tone, lation of neurotransmitter release and a multiple processes in the reproductive system trol of hormone secretion, and the body's def against injury and infection. Clearly, this v distributed and all-encompassing system of ch cal messengers is of enormous importance agents such as aspirin, which inhibit one or of the enzymes involved in their synthesis, cause many alterations of body function.

Receptors

The first step in the action of any chemical me ger on its target cell is its chemical combin with certain specific molecules of that cell; molecules are known as *receptors*. This tern be the source of confusion because the same is used to denote the "sensors" (stimulation of which initiate reflexes) described earlier in this chapter; the reader must keep in mind the fact that "receptor" has two totally distinct meanings, but the context in which it is used usually makes it quite clear as to which is meant.

What is the nature of these receptors with which chemical messengers combine? Present evidence suggests that most, if not all, are segments of proteins located either in the cell's plasma membrane or cytoplasm (in the general language of Chap. 4, the messenger is a "ligand" and the receptor a "binding site"). It is the combination of chemical messenger and receptor which initiates the cellular events leading to the cell's response. This response takes the ultimate form of a change in membrane structure, permeability, or transport, in the rate at which a particular substance is synthesized or secreted by the cell, or in the rate or strength of muscle contraction.

One extremely important characteristic of chemical mediation that is understandable in terms of membrane receptors is *specificity*. A chemical messenger—hormone, neurotransmitter, or locally released paracrine—influences only certain cells and not others. The explanation is that the membranes or cytoplasm of different cell types differ in the types of receptors they contain. Accordingly, only certain cell types, frequently just one, possess the precise protein receptor required for combination with a given chemical messenger. Here is another example of how protein structure confers specificity upon a biological process.

We have pointed out that receptors may be located in the cell's plasma membrane or cytoplasm. In fact, the former location is much more common, the major important exception being the cytoplasmic receptors for steroid hormones. Thus, the cell surface membrane not only maintains the physical and chemical integrity of the cell but serves as a signal-receiving device for the detection of chemical signals which regulate the cell's activity. A single cell usually contains many different receptor types, each capable of combining with only one chemical messenger.

It should be emphasized that receptors are, themselves, subject to physiological regulation. As we shall see in subsequent chapters, the number of receptors on a cell and the affinity of the receptors for their specific messenger can be increased or decreased, at least in certain systems.

Such alterations provide an additional mechanism for feedback control.

How does the messenger-receptor (i.e., ligand-binding site) combination elicit the cell's responses? In some cases, the alteration in protein structure produced by the combination of messenger with receptor may be the immediate cause of the altered function observed; for example, if the receptor were a membrane-bound protein involved in membrane permeability to sodium, the alteration of its structure by the chemical messenger might alter the passive flux of sodium into the cell. However, in most cases, the path from messenger-receptor combination to cell response is far more complicated and requires the participation of so-called second messengers, to be described in the next section.

Second messengers

"Second messengers" are substances which are either generated within the cell or which enter its cytosol (either across the cell membrane or by release from cell organelles) as a result of the original messenger-receptor combination, and which then act to alter the cell's enzymes, membranes, contractile proteins, etc. (i.e., to trigger the cell's overall response). In other words, the only function of the "first messenger"—hormone, neurotransmitter, or paracrine—is to combine with the receptors and bring a second messenger into play; the latter is then responsible for eliciting the cell's response. One of the most exciting developments in physiology has been the identification and elucidation of the second messengers and the recognition that the same ones may operate in many areas of the body and in different types of cells. The two best understood second messengers at present are calcium and cyclic AMP.

Calcium

In many cells, the combination of first chemical messenger with membrane receptor leads to an increase in cytosolic-calcium second messenger, which triggers, in turn, the cell's overall response. The cytosolic concentration of calcium is extremely low in virtually all cells, because of the active transport of calcium either out of the cell across the plasma membrane or into cellular or-

Table 7-1. Calcium in excitation-response coupling

1 Contraction in all forms of muscle
2 Secretion
  A. Exocrine
  B. Endocrine
  C. Neurotransmitter release
  D. Lysosomal enzyme release
  E. Histamine release from mast cells
3 Photoreceptor activation in rods
4 Cell division
5 Fertilization
6 Metabolic reactions
  A. Thermogenesis
  B. Gluconeogenesis
  C. Glycogenolysis
  D. Steroid synthesis
  E. Cyclic nucleotide metabolism
7 Membrane transport

Adopted from H. Rasmussen and D. B. P. Goodman, *Physiol. Rev.,* **57:**421–509 (1977).

ganelles, particularly endoplasmic reticulum and mitochondria. Because of the low cytosolic calcium and the fact that the inside of the cell is negatively charged relative to the outside, there always exists a large electrochemical gradient for the net diffusion of calcium into the cell, but, in the unstimulated state, this net inward movement is exactly counterbalanced by net active transport of calcium out of the cell (or into organelles). However, should the membrane permeability to calcium increase, the net inward diffusion would increase and a transient elevation of cytosolic calcium would occur. This is precisely what occurs in certain cells as a result of membrane changes induced by the combination of first chemical messenger and membrane receptor. An example of this was given in Chap. 6 when we described how release of secretory vesicles is triggered by an influx of calcium across the cell's plasma membrane.

In other cells, notably muscle, the source of at least a fraction of the increased cytosolic calcium triggered by messenger-receptor combination is not influx of extracellular calcium but rather release of calcium from cell organelles into the cytosol. The end result is, however, the same: The increased calcium acts as second messenger to elicit the response.

Calcium acts as a second messenger in many different cells, including muscle, nerve, and glands

(Table 7-1) (each of these is described in a subsequent chapter). Its role as a second messenger is frequently expressed by the term "coupler"; thus calcium is an "excitation-contraction coupler" in muscle (meaning that it carries the "message" from the "excited" membrane to the contractile apparatus inside the muscle. Similarly, it is termed an "excitation-secretion coupler" in those cells in which it is the second messenger between the membrane events and the secretory apparatus.

Cyclic AMP

The organic molecule *cyclic AMP* was first thought to be a second messenger only for hormones. However, it is now recognized to serve this function in the responses to certain neurotransmitters and paracrine agents, as well.

Most nonsteroid first messengers combine with receptors on the outer surface of the plasma membrane. The next step is identical for a large number of these messengers: The receptor molecule, when combined with the messenger, is able to bind to a second protein on the inner surface of the membrane (Fig. 7-8). This latter protein is an enzyme, *adenylate cyclase,* which becomes activated by the binding of receptor to it. The activated adenylate cyclase then catalyzes the transformation of cell ATP to another molecule known as cyclic 3',5'-adenosine monophosphate or simply *cyclic AMP* (cAMP), which then acts within the cell as a second messenger to trigger the intracellular sequence of events leading ultimately to the cell's overall response. The action of cAMP is terminated by its breakdown to noncyclic AMP, a reaction catalyzed by the enzyme, phosphodiesterase.

Let us take as an example of the cAMP system the action of the hormone epinephrine to stimulate glycogen breakdown to glucose in the liver (Fig. 7-9). Once activated by the epinephrine-receptor combination, adenylate cyclase catalyzes the synthesis of cAMP which, in turn, activates another enzyme known as protein kinase. The function of protein kinase is to catalyze the phosphorylation (addition of a phosphate group) to yet another enzyme, phosphorylase kinase. This results in activation of the latter enzyme which, in turn, phosphorylates (and, thereby, activates) yet another enzyme; this last enzyme then catalyzes the rate-limiting step in the catabolism of glycogen to glucose. Thus, the original activation of adenyl-

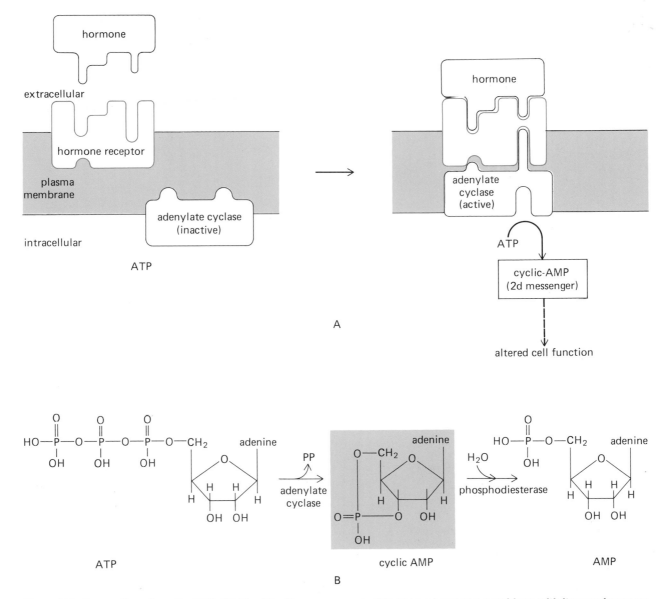

Figure 7-8. Generation of cyclic AMP. (A) The "first" messenger, in this case a hormone, combines with its membrane receptor, changing its configuration and permitting it to activate adenylate cyclase (the allosteric nature of this activation pictured in the figure is hypothetical). **(B)** Structures of ATP, cyclic AMP, and AMP, the latter resulting from enzymatic alteration of cyclic AMP.

ate cyclase initiates an elaborate "cascade" in which enzymes are converted in sequence from inactive to active forms. The following analysis describes the benefit of such a cascade. A single enzyme molecule is capable of transforming into product not just one but many substrate molecules, let us say 100; therefore, adenylate cyclase may catalyze the generation of 100 cAMP molecules. At each of three subsequent enzymatic steps (steps 4, 5, and 6 in Fig. 7-9), another 100-fold amplification occurs. Therefore, the end result is that a single molecule of epinephrine (the original

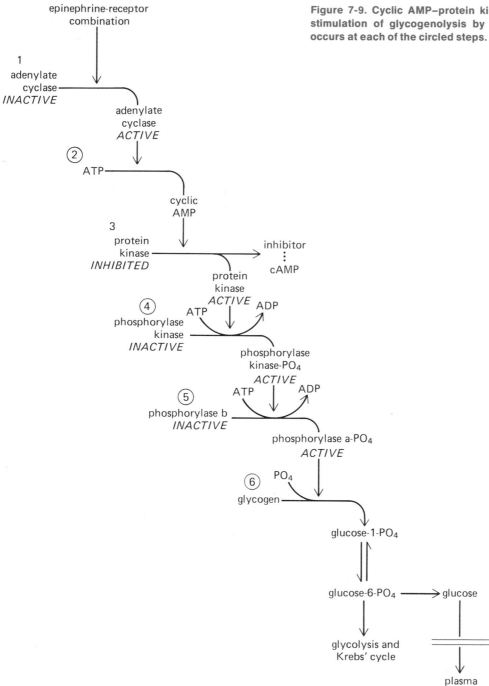

Figure 7-9. Cyclic AMP–protein kinase system mediates the stimulation of glycogenolysis by epinephrine. Amplification occurs at each of the circled steps.

stimulus) could theoretically generate 100 million glucose molecules (100^4). This fact helps to explain how hormones and other messengers can be effective at extremely low extracellular concentrations.

It is likely that activation of a protein kinase by cAMP is an event common to the biochemical sequences initiated by all cAMP-generating messengers, not just epinephrine. Protein kinase is not just a single enzyme but a group of enzymes having distinct protein substrates and existing in different target cells and within different organelles of the same target cell. These different protein kinases therefore activate (by phosphorylation) different proteins and so can trigger a variety of cascade activations simultaneously; for example, this could explain how a fat cell responds to epinephrine not only with glycogen breakdown (mediated by a particular enzyme) but also with triacylglycerol breakdown (mediated by a different enzyme). Moreover, whereas phosphorylation activates certain enzymes, it inhibits others; this is true, for example, of the enzyme catalyzing the rate-limiting step in glycogen synthesis and explains how epinephrine inhibits this process at the same time it stimulates glycogen breakdown.

In summary (Fig. 7-10), the biochemistry of protein kinase can explain how activation of a single molecule, cAMP, can produce so many different effects. It must be reemphasized that the ability of a cell to respond at all to a first messenger depends upon the presence of specific receptors for that messenger in the cell membrane. Given the presence of these receptors, cAMP will be generated and the actual response of that particular type of cell will depend upon its types of protein kinases and enzymes capable of being phosphorylated.

Of particular interest is the mechanism of action of certain (first) messengers, the cell responses to which are opposite to those produced by other (first) messengers known to generate cAMP. For example, the actions of the hormone insulin are opposite to those of epinephrine, the latter known to be mediated by cAMP, and this has led to the suggestion that insulin acts by *inhibiting* adenylate cyclase and thereby lowering intracellular cAMP. An alternative explanation invokes a role for a second cyclic nucleotide—cyclic 3′,5′-guanosine monophosphate (cGMP)—present in many cells. This theory proposes that, since cGMP exerts

actions opposed to those of cAMP, hormones like insulin may cause the generation of cGMP within their target cells. Thus, according to this theory, many cells may be subject to bidirectional control by opposing second messengers.

Relationship between calcium and cyclic nucleotides

The preceding sections may well have given the impression that, when a known second messenger is involved in a cell's response to a neurotransmitter, hormone, or paracrine, it is *either* calcium *or* a cyclic nucleotide. However, the fact is that *both* may be involved. For example, there are situations in which the receptor-messenger combination triggers, via adenylate cyclase, an increased cytoplasmic cAMP, which, in turn, leads to release of calcium from cell organelles; this increased calcium then elicits the cell's response. (The semantic complexity is as bad as the physiological—should we speak of calcium in this example as a "third messenger"?) These types of interactions between calcium and the cyclic nucleotides are the subject of much current investigation.

Finally, it would be incorrect to leave the impression that cAMP (along with cGMP) and calcium are the only possible second messengers. Others almost certainly remain to be discovered. It should also be noted that although any given cell may possess different types of receptors, each specific for a particular first messenger, there need not be a distinct second-messenger system for each receptor type; for example, a fat (adipose-tissue) cell has on its surface membrane distinct receptors for several different hormones, but activation of these receptors leads to changes in the same second messenger (cyclic AMP, in this case).

The balance concept and chemical homeostasis

One of the most important concepts in the physiology of control systems is that of balance. Our example of the water bath was really a study of heat balance within the water bath; i.e., the control system functioned to maintain a precise balance between the rates at which heat was added to and left the bath. Almost every homeostatic system

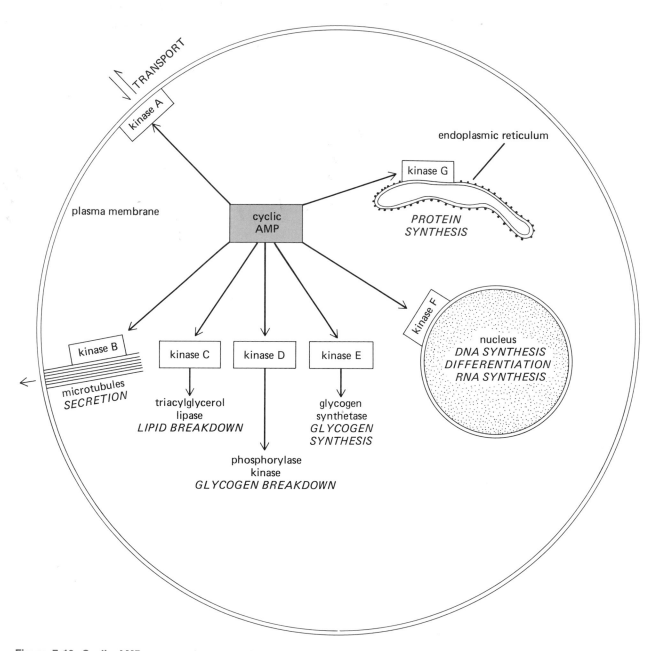

Figure 7-10. Cyclic AMP may exert many actions within a single cell by activating different protein kinases. The letters designating the different kinases are arbitrary. It is extremely unlikely that any one cell would exhibit all these responses to cyclic AMP. The sequence leading from first messenger to activation of adenylate cyclase and, thereby, to formation of the cyclic AMP, is not shown in the figure. *(Adapted from Goldberg.)*

NET GAIN TO BODY   DISTRIBUTION WITHIN   NET LOSS FROM BODY

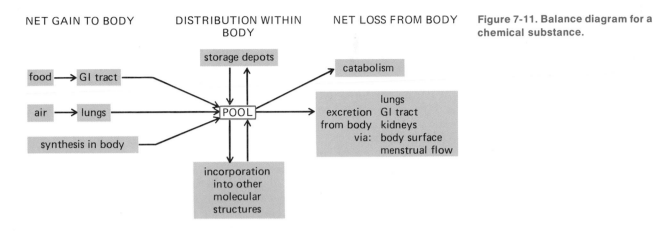

Figure 7-11. Balance diagram for a chemical substance.

in the body can be studied in terms of balance; some regulate the balance of a physical parameter (heat, pressure, flow, etc.), but most are concerned with the balance of a chemical component of the body. This section is intended to provide a foundation of general principles upon which can be built the study of any specific chemical substance.

Figure 7-11 is a generalized schema of the possible pathways involved in the balance of a chemical substance. The *pool* occupies a position of central importance in the balance sheet; it is the body's readily available quantity of the particular substance and is frequently identical to the amount present in the extracellular fluid. The pool functions as "middleman," receiving from and contributing to all the other pathways.

The pathways on the left of the figure are sources of *net gain* to the body. A substance may be ingested and then absorbed from the gastrointestinal (GI) tract. It is important to realize that all that is ingested may not be absorbed; some may either fail to be absorbed or be consumed by the bacteria residing in the gut. The lungs offer another site of entry to the body for gases (O_2) and airborne chemicals. Finally, the substance may be synthesized by cells within the body itself.

The pathways to the right of the figure are sources of *net loss* from the body. A substance may be excreted in the urine, feces, expired air, and menstrual flow, as well as from the surface of the body (skin, hair, nails, sweat, tears, etc.). The substance may be catabolized or transformed within the body to some other chemical; this fate—the opposite of synthesis—represents net loss of the substance.

The central portion of the figure illustrates the *distribution* of the substance within the body. From the readily available pool, it may be taken up by storage depots; conversely, material may leave the storage depots to reenter the pool. Finally, the chemical may be incorporated into some other molecular structure (fatty acids into membranes, iodine into thyroxine, etc.). This process is reversible in that the substance is liberated again whenever the more complex molecule is broken down. For this reason, this pathway differs from catabolism or transformation by which the substance is irretrievably lost. This pathway is also distinguished from storage in that the latter has no function other than the passive one of storage, whereas the incorporation of the substance into other molecules is done to fulfill an active function of the substance (e.g., the function of iodine in the body is to provide an essential component of the thyroxine molecule).

Of course, it should be recognized that every pathway of this generalized schema is not applicable to every substance; for example, the mineral electrolytes cannot be synthesized or catabolized by the body.

The orientation of the figure illustrates two important generalizations concerning the balance concept: (1) The total body balance depends upon the rates of total body net gain and net loss; (2) the pool concentration depends not only upon total body losses and gains but upon exchanges of the substance within the body.

It should be apparent that, with regard to total body balance, three states are possible: (1) Loss exceeds gain, the total amount of the substance in the body decreases, and the person is said to be in *negative balance*; (2) gain exceeds loss, the total amount of the substance in the body increases, and the person is said to be in *positive balance*; (3) gain equals loss, and balance is stable. Physiology is, in large part, concerned with the homeostatic mechanisms which match gain with loss to achieve a stable balance. Pool size, too, tends to be maintained relatively constant as a result of these homeostatic mechanisms as well as those operating on the pathways for internal exchange. Clearly a stable balance can be upset by alteration of the magnitude of any single pathway in the schema, e.g., severe negative water balance can occur in the presence of increased sweating. Conversely, stable balance can be restored by homeostatic control of the pathways. Therefore, much research has been concerned with determining which are the key homeostatically controlled pathways for each substance, what are the specific mechanisms involved, and what are the limits of intake and output beyond which balance cannot be achieved.

Essential nutrients: An example

There are many substances which are required for normal or optimal body function but which are synthesized by the body either not at all or in amounts inadequate to achieve balance. They are known as *essential nutrients.* Because they are all excreted or catabolized at some finite rate, a continuous new supply must be provided by the diet. Approximately 50 in number, they are water, 9 amino acids, several unsaturated fatty acids, approximately 20 vitamins, and a similar number of inorganic minerals.

It should be reemphasized that the term essential nutrient is reserved for substances that fulfill *two* criteria: They not only must be essential for good health but must not be synthesized by the body in adequate amounts. Thus, glucose, although "essential" for normal metabolism, is not classified as an essential nutrient because the body normally can synthesize all it needs.

The physiology of each essential nutrient can be studied in terms of the schema of Fig. 7-11, i.e., by analyzing each pathway relevant for that nutrient. There is considerable variation in the relative importances of the pathways for the homeostatic regulation of the different nutrients. Thus, in Chap.

13 we shall see that ingestion and urinary excretion are the main controlled variables for water. In contrast, control of iron balance, as described in Chap. 11, is dependent largely upon the control of iron absorption by the gastrointestinal tract. Balance of essential amino acids is achieved in still another way; the rate of catabolism of these amino acids (specifically the loss of the NH_2 group from the amino acid) is reduced in the presence of amino acid deficiency (and increased in the presence of excess).

The reason for placing so much emphasis on the pathways which are homeostatically controlled is that alteration of the nutrient flow via these pathways constitutes the mechanism by which balance is achieved whenever a primary change occurs in any of the other pathways. For example, iron balance can be upset either by a primary change in intake or by excretion; in either case, balance can be reestablished by a compensating change in the rate of absorption.

The key event in triggering off the homeostatic response is a change in the body content of the nutrient, manifested frequently as a change in pool content, i.e., a change in the internal environment. Therefore, it is important to realize that nothing in the body is maintained *absolutely* constant; relatively small deflections from normal are continuously occurring and constitute the signals for the control systems which then operate to limit the extent of deviation.

Let us look more closely at the spectrum of conditions possible as we alter primary intake of an essential nutrient, assuming all the other pathways are functioning normally. At zero intake, balance is impossible since excretion or catabolism, or both, cannot be reduced to zero; accordingly negative balance persists as the body stores of the nutrient are progressively depleted. Thus the minimal combined rate of excretion and catabolism sets the lower limit for achievement of balance. At the other end of the spectrum, the maximal combined rate of excretion and catabolism sets the upper limit for achievement of balance, i.e., the maximal intake of nutrient compatible with balance. If intake is greater than this, body stores will continuously increase, with the potential for toxic effects. This occurs for certain of the fat-soluble vitamins (A, D, and K) and is a problem for other nutrients as well.

Between the extremes there obviously exists a

range of intakes at which balance can be maintained without overt manifestations of either deficiency or toxicity. How wide this range is depends upon how much the rates of excretion or catabolism can be altered. (For example, sodium balance can be achieved readily at intakes between 0.3 and 25 g/day.) It must be reemphasized that, despite the fact that balance is stable at all intakes in this range, there are small differences in body content and pool size over the range, since these differences are required to drive the homeostatic control systems.

The ultimate aim of nutrition is to determine which intake in this range is the *optimal* intake for each nutrient. The critical question may be stated as follows: Does ingestion of more than the minimal amount of a nutrient produce greater health, growth, intelligence, etc.; i.e., is there an amount of nutrient which is *optimal* for health rather than merely adequate for avoiding disease? But this question raises the further question: Optimal for what? For example, most Americans ingest very large quantities of protein (which supplies essential amino acids), and this almost certainly has contributed to our increased body height. This would seem desirable, but in experimental animals it has been found that protein intakes which are optimal for maximal total body growth enhance the development of cancers and atherosclerosis as well. Obviously, "optimal for what?" cannot be answered without a clearly definable "what" and a careful consideration of many factors, including existing environmental influences and life style.

Biological rhythms

A striking characteristic of many bodily functions is the rhythmical changes they manifest. Body temperature, for example, fluctuates considerably during a normal 24-h period, as do the concentrations of many hormones. Such daily rhythms are called *circadian* (literally, around the day). Most likely, they occur as a result of changes in the operating points of the control systems regulating them, but we generally are uncertain as to their precise mechanisms or significance. We do know that modern life, with its stepped-up pace, rapid changes, and creation of artificial environments, may frequently disrupt them with, as yet, unknown results. Many of the body's rhythms are much longer than 24 h. The menstrual cycle is the best-known longer cycle, but there may well be others with even greater time spans.

Central to this field is the problem of "biological clocks" within the body which set the rhythms. This question has received particular attention by physiologists of reproduction, but it is now recognized as a critical area in the biology of growth, aging, and many other fields relevant to human physiology.

8

NEURAL CONTROL MECHANISMS

Section A: Membrane potentials

Basic principles of electricity
Electric and chemical properties of a resting cell
  Diffusion potentials
    Equilibrium potentials • *The resting cell membrane potential* • *The membrane pump* • *Summary*
Signals of the nervous system: Graded potentials and action potentials
  Graded potentials
  Action potentials
    Ionic basis of the action potential • *Mechanism of permeability changes* • *Threshold* • *Refractory periods* • *Action-potential propagation*

Section B: Physiological activities of neurons

Functional anatomy of neurons
Receptors
  Receptor activation: The receptor potential
  Amplitude of the receptor potential
    Intensity and velocity of stimulus application • *Summation of receptor potentials* • *Adaptation* • *Summary*
  Intensity coding
Synapses
  Functional anatomy
    Excitatory synapse • *Inhibitory synapse*

Activation of the postsynaptic cell: Neural integration
Neurotransmitters
  Dopamine, norepinephrine, and epinephrine • *Acetylcholine* • *Serotonin* • *Other possible neurotransmitters*
Modification of synaptic transmission by drugs and disease
Presynaptic inhibition
Neuromodulators
Neuroeffector communication
Patterns of neural activity
  The flexion reflex
  Swallowing
  Respiratory movements

Section C: Divisions of the nervous system

Central nervous system
  Spinal cord
  Brain
    Brainstem • *Cerebellum* • *Forebrain*
Peripheral nervous system
  Peripheral nervous system: Afferent division
  Peripheral nervous system: Efferent division
    Somatic nervous system • *Autonomic nervous system*
Blood supply, blood-brain barrier phenomena, and cerebrospinal fluid

One of the greatest achievements of biological evolution has been the development of the human nervous system. It is made up of the brain and spinal cord as well as the many nerve processes which pass between these two structures and the muscles, glands, or receptors which they innervate.

In Chaps. 18 to 20 we shall consider specific aspects of neural activity concerned with the special senses (sight, hearing, etc.), the coordination of muscle activity in posture and movement, and the relation between brain activity and consciousness, memory, emotions, etc. In this chapter we are concerned with the basic components common to all neural mechanisms. We shall review the principles of electricity and describe the electric signals (*graded potentials* and *action potentials*) generated in nerve cells, the functional anatomy of the individual nerve cell *(neuron)*, the processes by which these electrical signals are normally initiated in the body, the passage of information from one neuron to another, and some basic patterns of neural interaction. Finally, we shall introduce the basic organization and major divisions of the nervous system.

Section A
Membrane potentials

Basic principles of electricity

All chemical reactions are basically electric in nature since they involve exchanging or sharing negatively charged electrons between atoms to form ions or bonds. Most chemical reactions result in neutral molecules, containing equal numbers of electrons and protons, but in some cases ions are formed, which have a net electric charge, e.g., the ionized groups in organic molecules, such as the negative carboxyl group $RCOO^-$ and the positive amino group RNH_3^+, or the inorganic ions such as sodium, potassium, and chloride (Na^+, K^+, and Cl^-). With the exception of water, the major chemical components of the extracellular fluid are the sodium and chloride ions, whereas the intracellular fluid contains high concentrations of potassium ions and organic molecules containing ionized groups, particularly proteins and phosphate compounds. Since the environment of the cell contains many charged particles, it is not surprising to discover that electric phenomena resulting from the interaction of these charged particles play a significant role in cell function.

According to the laws of physics, all physical and chemical phenomena can be described using only five fundamental properties of matter: length, time, mass, temperature, and electric charge. Each of these properties is independent in the sense that none can be defined in terms of a combination of the others. For example, velocity is not a fundamental property since it can be defined in terms of a length moved in a given period of time. Force is not a fundamental property since it is defined by Newton's law as $F = ma$, force equals mass times acceleration, and thus is defined in terms of mass, length, and time. In contrast, electric charge is a fundamental property of matter and cannot be

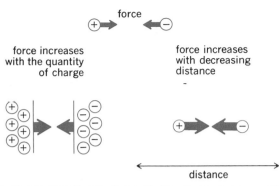

Figure 8-1. The electric force of attraction between positive and negative charges increases with the quantity of charge and with decreasing distance between charges.

defined in terms of length, time, mass, or temperature.

There are two types of charge, arbitrarily called *positive* and *negative*. This labeling occurred before it was known that atoms consist of protons and electrons. When electrons were discovered, they behaved like the charged particles which had been labeled negative. Protons, on the other hand, behaved like the electric charges labeled positive. If the original labeling had been the opposite, we would call electrons positive and protons negative.

When positive and negative charges are separated, an electric force draws the opposite charges together; yet positive charge repels positive charge and negative charge repels negative charge. Why there is such a force cannot be answered in terms of other physical properties of matter since electric charge is a fundamental property of matter. However, the force can be measured, and the relation between the amount of force, the quantity of charge, and the distance separating the charges can be studied. The amount of force acting between electric charges increases when the charged particles are moved closer together and with increasing quantity of charge (Fig. 8-1).

Energy E is measured by its ability to do work, and work W is defined as the product of force F and distance X: $W = FX$. When oppositely charged particles come together as a result of the attracting force between them, work is done by these moving particles. Conversely, to separate oppositely charged particles, work must be done, i.e., energy added (Fig. 8-2). The amount of work depends

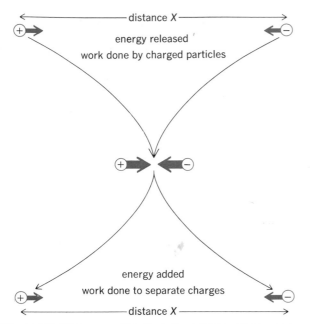

Figure 8-2. When opposite charges come together, energy is released and can be used to perform work. Work must be done (energy added) in order to separate opposite charges.

upon both the total number of charges involved and the distance between them. Thus, when electric charges are separated, they have the "potential" of doing work if they are allowed to come together again. *Voltage* is a measure of the potential of separated electric charge to do work and is defined as the amount of work done by an electric charge when moving from one point in a system to another. Voltage is always measured with respect to two points; thus, one refers to the potential difference, or simply the *potential*, between two points, and the units of measurement are known as *volts*. Since the total amount of charge that can be separated in most biological systems is very small, the potential differences are small. For example, the voltage measured across a nerve cell membrane is approximately 70 millivolts (mV). A millivolt is 0.001 V.

The movement of electric charge is known as *current*. If electric charge is separated between two points, there is a potential between these points, and the electric force tends to make charges flow, producing a current. The amount of charge that does move, the current, depends upon the voltage and the frictional interactions between the moving charges and the material through which

they are moving. The hindrance to the movement of electric charge through a particular material is known as *resistance*. The relationship between current (I), voltage (E), and resistance (R) is given by Ohm's law, $I = \dfrac{E}{R}$. The higher the resistance of the material, the lower the amount of current flow for any given voltage. Some materials, like glass and rubber, which have relatively few charge carriers, have such high electric resistance that the amount of current flow through them is very small, even when high voltages are applied. Such materials, known as *insulators*, are used to prevent the flow of current. Thus, the rubber insulation around electric wires prevents the flow of current from the wire to areas outside it. Materials having low resistance to current flow, due to the relatively large number of mobile charge carriers, are known as *conductors.*

Pure water is a relatively poor conductor because it contains very few charged particles, but when sodium chloride is added, the sodium and chloride ions provide charges that can move and carry the current, and the solution becomes a relatively good conductor with a low resistance. The water compartments inside and outside the cells in the body contain numerous charged particles (ions) which are able to move between areas of charge separation and thus carry current. Lipids contain very few charged groups and thus have a high electric resistance. The lipid components of the cell membrane provide a region of high electric resistance separating two water compartments of low resistance.

Electric and chemical properties of a resting cell

One can determine the presence of an electric potential (voltage) difference across cell membranes by inserting a very fine electrode into the cell and another into the extracellular fluid surrounding the cell and connecting the two to a voltmeter (Fig. 8-3). In this manner it has been found that all cells exhibit a membrane potential oriented so that the inside of the cell is negatively charged with respect to the outside. (As we shall see, this is true for nerve and muscle cells except when they are undergoing action potentials.) This potential difference is called the *resting membrane poten-*

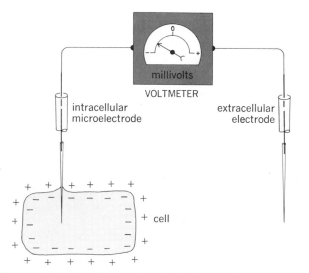

Figure 8-3. Intracellular microelectrode used to measure electric-potential difference across the cell membrane.

tial; its magnitude varies from 5 to 100 mV, depending upon the type of cell and its chemical environment.

The normal ionic composition of the fluid bathing the cells (the extracellular fluid) is approximately that listed in Table 8-1. Note that the chemical composition of the intracellular fluid is entirely different. There are many other substances in both fluid compartments, such as Mg^{2+}, Ca^{2+}, HCO_3^-, PO_4^{2-}, SO_4^{2-}, glucose, urea, amino acids, and hormones, but sodium and potassium play the most important roles in the generation of the resting membrane potential.

Table 8-1. Distribution of ions across the cell membrane of a cat nerve cell

Ion	Extracellular concentration (mM/L H_2O)	Intracellular concentration (mM/L H_2O)
Na	150	15
Cl	125	10
K	5	150

Diffusion potentials

Given that we are dealing with two solutions of different ionic composition and, for the moment, ignoring the nature of the membrane which sepa-

rates them, consider how the solutions interact. Figure 8-4 depicts two dilute solutions of sodium chloride, the solution on side 1 at a concentration of 0.1 M and that on side 2 at 0.01 M. The barrier separating the solutions is permeable to all ion species. Both sodium and chloride are more concentrated on side 1, and they will diffuse down their concentration gradients, moving from side 1 to side 2. However, the mobility of chloride, i.e., the ease with which it can move through the solution, is about $1\frac{1}{2}$ times greater than that of sodium; thus, chloride will move to side 2 more rapidly than sodium and side 2 will, at least transiently, become slightly negatively charged with respect to side 1. This electric gradient is due to the differential diffusion of charged particles in solution and is called a *diffusion potential.* The potential developed by a system such as that in Fig. 8-4 will disappear over time as the concentrations of sodium and chloride in sides 1 and 2 become equal.

What would happen at the junction between sides 1 and 2 if side 1 contained a solution of 0.15 M NaCl (similar to extracellular fluid) and side 2 contained 0.15 M KCl (similar to intracellular fluid) (Fig. 8-5)? Again, we ignore the membrane that separates the two compartments. Chloride concentrations on the two sides are equal, but those of sodium and potassium are not. The mobility of potassium, like that of chloride, is about $1\frac{1}{2}$ times greater than that of sodium; thus, potassium will diffuse down its concentration gradient faster than sodium; i.e., positive charge will initially leave side 2 faster than it will enter, and side 2

Figure 8-4. Generation of a diffusion potential by differential ion movement through a completely permeable membrane. Arrows represent ion movements.

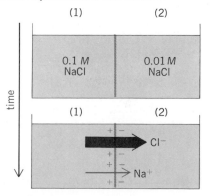

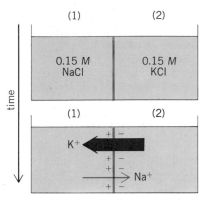

Figure 8-5. Generation of a diffusion potential by potassium movement through a completely permeable membrane. Arrows represent ion movements.

will become electronegative with respect to side 1. Again, with time, equilibrium will be reached and the potential difference will disappear.

Now consider the situation of Fig. 8-6 but assume that a selectively permeable membrane separates the two compartments, such that potassium can pass through but sodium and chloride cannot. In this case, all the sodium will remain on side 1, and some of the potassium, diffusing down its concentration gradient, will be added to it; side 1 will become relatively positive. In contrast to the previous examples, this diffusion potential will not disappear with time and will be of greater magnitude, the actual magnitude depending upon the concentration gradient for potassium. Thus, the diffusion potential across the membrane is determined by both the permeability properties of the membrane and the orientation (direction) of the concentration gradients.

Equilibrium potentials. The concentration gradient, which causes net diffusion of particles from a region of higher to a region of lower concentration, is called the *concentration force*. As this force moves potassium from side 2 to side 1 and side 1 becomes increasingly positive (Fig. 8-6), the electric potential difference itself begins to influence the movement of the positively charged potassium particles; they are attracted by the relatively negative charge of side 2 and repulsed by the positive charge of side 1. This attraction because of a difference in electric charge (or repulsion because of a similarity of charge) is the *electric force*. As long

as the concentration force driving potassium from side 2 to 1 is greater than the electric force driving it in the opposite direction, there will be net movement of potassium from side 2 to 1 and the potential difference will increase. Side 1 will become more and more positive until the electric force opposing the entry of potassium into that compartment equals the concentration force favoring entry. The membrane potential at which the electric force is equal in magnitude but opposite in direction to the concentration force is called the *equilibrium potential* for that ion. At the equilibrium potential there is no net movement of the ion because the opposing forces acting upon it are exactly balanced.

It can be seen that the value of the equilibrium

Figure 8-6. Generation of a diffusion potential across a membrane permeable only to potassium. Arrows represent ion movements.

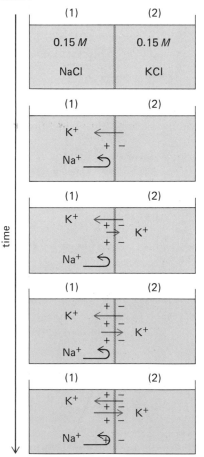

potential for any ion depends upon the concentration gradient for that ion across the membrane; if the concentrations on the two sides were equal, the concentration force would be zero and the electric potential required to oppose it would also be zero. The larger the concentration gradient, the larger is the equilibrium potential. Using potassium concentrations typical for neurons and extracellular fluid, the equilibrium potential for potassium is close to 90 mV, the inside of the cell being negative with respect to the outside.

If the membrane separating sides 1 and 2 is replaced with one permeable only to sodium, the initial net flow of the positively charged sodium will be from side 1 to 2 and side 2 will become positive (Fig. 8-7). This movement of sodium down its concentration gradient is opposed by the electric force generated by that movement. A sodium equilibrium potential will be established with side 2 positive with respect to side 1, at which point net movement will cease. For most neurons, the sodium equilibrium potential is about 60 mV, inside positive. Thus, the equilibrium potential for each ion species is different in magnitude and, possibly, even in direction, from those for other ion species since the concentration gradients and membrane permeabilities are different.

The resting cell membrane potential. It is not difficult to move from these hypothetical experiments to a nerve cell at rest where (1) the potassium concentration is much greater inside the cell than outside and the sodium concentration gradient is the opposite and (2) the cell membrane is some 50 to 75 times more permeable to potassium than to sodium. Given these characteristics, it should be evident that a diffusion potential will be generated across the membrane largely because of the movement of potassium down its concentration gradient so that the inside of the cell is negative with respect to the outside. The experimentally measured membrane potential is not, however, equal to the potassium equilibrium potential because the membrane is not perfectly impermeable to sodium and some sodium continually diffuses down its electric and concentration gradients, adding a small amount to the inside of the cell. Thus the measured membrane potential, of a nerve cell at least, is closer to −70 mV than to the potassium equilibrium potential of −90 mV. An important result of this fact is that, since

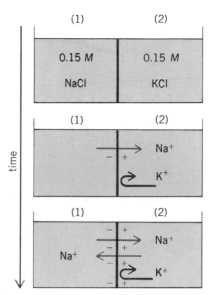

Figure 8-7. Generation of a diffusion potential across a membrane permeable only to sodium. Arrows represent ion movements.

the membrane is not at the potassium equilibrium potential, there is a continual net diffusion of potassium out of the cell. Realize that the values given for the various potentials are only average values since they vary from cell to cell with differences in ion concentration and membrane permeability.

The membrane pump. If there is net movement of sodium into and potassium out of the cell, why do the concentration gradients not run down? The reason is that active-transport mechanisms in the membrane utilize energy derived from cellular metabolism to pump the sodium back out of the cell and the potassium back in. Actually, the pumping of these ions is linked because they are both transported by the Na-K ATPase in the membrane (Chap. 6). If the linkage were to result in the exchange of sodium and potassium on a one-to-one basis, the pump would not directly separate charge. Whenever the exchange is not one-to-one, the pump does directly separate charge and is known as an *electrogenic pump*. In most cells (but by no means all), the electrogenic contribution of the pump to the membrane potential is quite small. However, the pump does make an essential *indirect* contribution, because it maintains the

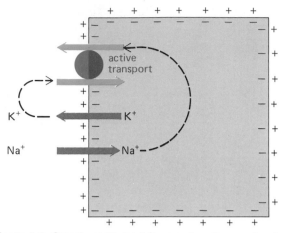

Figure 8-8. Steady-state total fluxes of sodium and potassium ions across the cell membrane. The net flux of sodium and potassium ions by diffusion is balanced by the active transport of these ions in the opposite direction across the membrane (i.e., potassium in and sodium out).

concentration gradients down which the ions diffuse to establish a membrane potential. The membrane potential is then due directly to the diffusion of these ions as described above.

Summary. Some potassium ions diffuse out of the cell down their concentration gradient and some move in down the electric gradient, but because the resting membrane potential is not as negative as the potassium equilibrium potential (-70 mV rather than -90 mV), less potassium enters passively than leaves. The difference is relatively small and is made up by active transport via the membrane pump; therefore the *total* potassium entering equals that leaving, and the resting cell neither gains nor loses potassium (Fig. 8-8). Sodium diffuses passively into the cell because of both electric and concentration forces but, because the membrane permeability of a resting cell to sodium is very low, the amount entering is small. There is no passive force to remove sodium from the cell, and that which enters must be actively transported out by the membrane pump. The amount of sodium pumped out equals, in most cases, the amount of potassium pumped in.

When more than one ion species can diffuse across the membrane, the membrane permeability properties as well as the concentration gradient of each species must be considered when accounting

for the membrane potential. If the membrane is impermeable to a given ion species, no ion of that species can cross the membrane and contribute to a diffusion potential, regardless of the electric and concentration gradients that may exist. And, for a given concentration gradient, the greater the membrane permeability to an ion species is, the greater the influence that ion species will have on the membrane potential. Since the resting membrane is much more permeable to potassium than to sodium, the resting membrane potential is much closer to the potassium equilibrium potential than to that of the sodium.

Signals of the nervous system: Graded potentials and action potentials

Changes in the membrane potential from its resting level can convey meaningful information to the cell. Nerve and muscle cells in particular use such voltage changes as signals in receiving, integrating, and transmitting information. These signals occur in two forms: *graded potentials* and *action potentials.* Graded potentials are extremely important in signaling over short distances; action potentials, on the other hand, are the long-distance signals of nerve and muscle membrane.

Figure 8-9. As potentials become more positive inside than the resting potential they are said to be *depolarizing;* those returning toward the resting membrane potential are said to be *repolarizing,* and those becoming more negative inside than the resting membrane potential are *hyperpolarizing.*

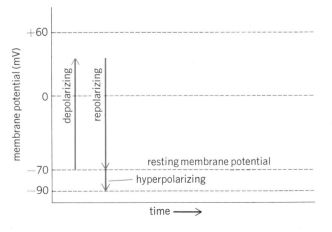

Recall that the word "potential" means a voltage difference between two points. Adjectives such as *membrane* potential, *resting* potential, *diffusion* potential, *graded* potential, etc. define the conditions under which the potential is measured or the way it developed (Table 8-2). The terms *depolarize,* *hyperpolarize,* and *repolarize* will be used frequently. The membrane is said to be *depolarized* when the membrane potential is less negative than the resting membrane potential, i.e., closer to zero, and *hyperpolarized* when it is more negative than the resting level. When the membrane potential is changing so that it moves toward or even above zero, it is *depolarizing,* and when it moves away from zero back toward its resting level, it is *repolarizing* (Fig. 8-9). When it becomes more negative than its resting level it is *hyperpolarizing.*

Graded potentials

Graded potentials are local changes in membrane potential in either a depolarizing or hyperpolarizing direction (Fig. 8-10 part 1). They are usually produced by some specific change in the cell's environment acting on a specialized region of the membrane. They are called graded potentials because the amplitude of the potential change is variable and related to the magnitude of the external event (Fig. 8-10 part 2). If the event is some change in ex-

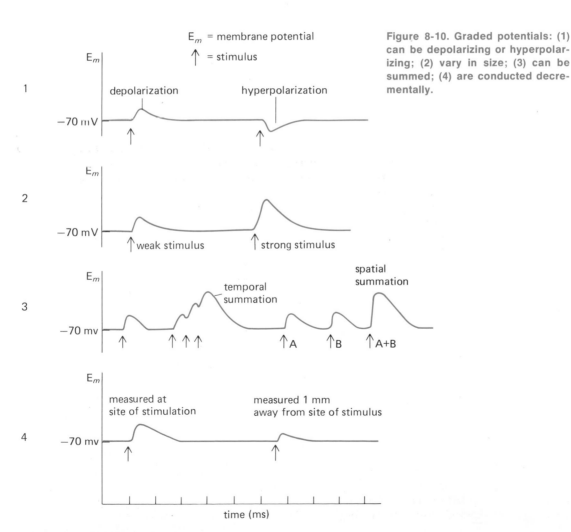

Figure 8-10. Graded potentials: (1) can be depolarizing or hyperpolarizing; (2) vary in size; (3) can be summed; (4) are conducted decrementally.

E_m = membrane potential

↑ = stimulus

1   E_m depolarization   hyperpolarization   −70 mV

2   E_m weak stimulus   strong stimulus   −70 mV

3   E_m temporal summation   spatial summation   −70 mv   A   B   A+B

4   E_m measured at site of stimulation   measured 1 mm away from site of stimulus   −70 mv

time (ms)

Table 8-2. A miniglossary

Potential = potential difference the voltage difference between two points
Membrane potential = transmembrane potential the voltage difference between the inside and outside of a cell
Diffusion potential a voltage difference created by the different rates of movement of ions
Equilibrium potential the voltage difference across a membrane which is equal in force but opposite in direction to the concentration force affecting a given ion species.
Resting membrane potential = resting potential the steady transmembrane potential of a cell that is not producing an electrical signal.
Graded potential a potential change of variable amplitude and duration that is conducted decrementally; has no threshold or refractory period.
Action potential a brief, all-or-none reversal of the membrane potential polarity; has a threshold, refractory period, and is conducted without decrement.
Postsynaptic potential a graded potential change produced in the postsynaptic neuron in response to release of transmitter substance by a presynaptic terminal; may be depolarizing (an *excitatory postsynaptic potential* or EPSP) or hyperpolarizing (an *inhibitory postsynaptic potential* or IPSP)
Receptor potential also called **generator potential** a graded potential change produced at the peripheral endings of afferent neurons (or in separate receptor cells) in response to an external stimulus.
Pacemaker potential spontaneously occurring graded potential change; occurs only in certain specialized cells.

ternal energy (a *stimulus*, such as heat, light, or mechanical deformation), the graded signal produced is called a *receptor* (or *generator*) *potential*. If the event is a signal from another neuron, the resulting graded potential is called a *synaptic potential* because it is produced at a *synapse*, the specialized junction between two nerve cells. In a few specialized types of cells graded potential changes may occur spontaneously (not as a result of events outside the cell); these are called *pacemaker potentials*. How these graded potentials are initiated will be discussed in detail later.

Graded potentials cause local current flows— the greater the potential change, the greater the current flow. The intracellular and extracellular fluids are fairly good conductors, and current flows through them whenever a voltage difference occurs. In Fig. 8-11, a stimulus has caused a graded potential that slightly depolarizes the membrane. Accordingly, this area of the membrane has an electric potential different from

that of adjacent areas. Because unlike charges attract and like charges repel, inside the cell current flows away from the activated membrane region through the cytoplasm, and outside the cell it flows toward the activated region through the extracellular fluid. Note that, as in all biological systems, the current is carried by ions such as K^+, Na^+, Cl^-, and HCO_3^-. By convention the direction of movement of the positive ions is designated the direction of current flow, but negatively charged particles simultaneously move in the opposite direction. This current flow changes the membrane potential of the adjacent regions.

The change in electric charge or potential, however, passes from one point to another much faster than the ions themselves move between these same two points. This can best be understood by considering water movement through a water-filled pipe. If the pressure is changed at one end of the pipe, e.g., by putting in more water, the water flow at the opposite end is changed long before

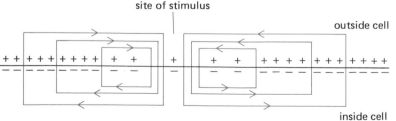

site of stimulus

outside cell

inside cell

Figure 8-11. Ion current flow in the cytoplasm of a nerve fiber and in the extracellular fluid surrounding it. Membrane is slightly depolarized (less negative inside) at the site of stimulus.

those molecules that were put in could have traveled to the outflow end. This occurs because the influence of the molecules (in this example the increase in water pressure) is transmitted much faster than the molecules themselves.

In much the same way, the influence of electric charge is transmitted from one area to another much faster than the ions could move between the same two points. But here the analogy between water flow in the fluid-filled pipe and current flow in the cytoplasm or extracellular fluid begins to break down. In the case of the pipe, the flow out one end equals the input. But current flow in the cytoplasm is more like water flow through a leaky hose; charge is lost across the membrane because the membrane is permeable to ions, with the result that the flow at the end is less than at the beginning. In fact, when conducting current, cells are so leaky that the current almost completely dies out within a few millimeters of its point of origin. For this reason, local current flow is *decremental;* i.e., its amplitude decreases with increasing distance (Fig. 8-10 part 4).

Because of decremental conduction, graded potentials (and the current flows they generate) can function as signals only over very short distances. Nevertheless, graded potentials play very important roles in information processing by nerve cells, as we shall see.

Action potentials

Action potentials are very different from graded potentials (Table 8-3). They are rapid alterations

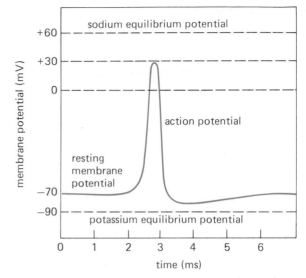

Figure 8-12. Changes in membrane potential during an action potential. (The sodium equilibrium potential is +60 mV; the potassium equilibrium potential is −90 mV.)

in membrane potential which may last only 1 ms; during this time the membrane changes from −70 to +30 mV and then returns to its original value (Fig. 8-12). Of all the types of cells in the body, only nerve and muscle cells are capable of producing action potentials; this property is an example of *excitability.*

How is an excitable membrane able to make rapid changes in its membrane potential? How does a change in the environment (a stimulus) interact with an excitable membrane to bring

Table 8-3. The differences between graded potentials and action potentials

Graded potentials	Action potentials
1 Graded response; amplitude reflects conditions of the initiating event.	All-or-none response; once membrane is depolarized to threshold, amplitude is independent of initiating event.
2 Graded response; can be summed.	All-or-none response; cannot be summed.
3 Has no threshold.	Has a threshold that is usually 10–15 mV depolarized relative to the resting potential.
4 Has no refractory period.	Has a refractory period.
5 Is conducted decrementally; i.e., amplitude decreases with distance.	Is conducted without decrement; the amplitude is constant.
6 Duration varies with initiating conditions.	Duration is constant.
7 Can be a depolarization or hyperpolarization.	Is a depolarization (with overshoot).
8 Initiated by stimulus, neurotransmitter, or spontaneously.	Initiated by membrane depolarization.

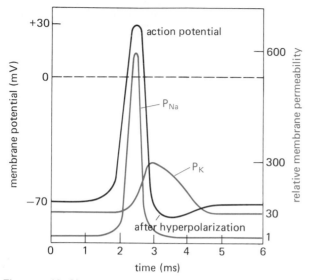

Figure 8-13. Changes in membrane permeability to sodium and potassium ions during an action potential.

about an action-potential response? How is an action potential propagated along the surface of an excitable membrane? These questions will be discussed in the following sections.

Ionic basis of the action potential. Action potentials can be explained by the concepts already developed for the origins of resting membrane potentials. This explanation, known as the *ionic hypothesis*, was developed mainly by the English scientists A. L. Hodgkin and A. F. Huxley, who received the Nobel Prize in 1963. We have seen that the magnitude of the resting membrane potential depends upon the concentration gradients of and membrane permeabilities to different ions, particularly sodium and potassium. This situation is true for the period of the action potential as well; the action potential results from a transient change in the membrane permeabilities. In the resting state the membrane is 50 to 75 times more permeable to potassium than to sodium ions, and the resting potential is much closer to the potassium equilibrium potential than to the sodium equilibrium potential. During an action potential, however, the permeability of the membrane to sodium and potassium ions is markedly altered. In the rising phase of the action potential the membrane permeability to sodium ions undergoes

a several hundredfold increase and sodium ions rush into the cell. During this period more positive charge is entering the cell in the form of sodium ions than is leaving in the form of potassium ions, and thus the membrane potential becomes much closer to the sodium equilibrium potential, becoming positive on the inside and negative on the outside of the membrane. In this phase the membrane potential approaches but does not quite reach the sodium equilibrium potential.

Action potentials in neurons last about 1 ms (0.001 s). What causes the membrane to return so rapidly to its resting level? The answer to this question is twofold: (1) The increased sodium permeability is rapidly turned off (*sodium inactivation*); (2) the membrane permeability to potassium increases relative to its resting level. The timing of these two events can be seen in Fig. 8-13. As the membrane becomes more positive inside, sodium inactivation causes the sodium permeability to decrease toward its resting value, and sodium entry rapidly decreases. This alone would restore the potential to its resting level. However, the repolarization process is speeded up by a simultaneous increase in potassium permeability which causes more potassium ions to move out of the cell down their concentration gradient. These two events, sodium inactivation and increased potassium permeability, allow potassium diffusion to regain predominance over sodium diffusion, and the membrane potential rapidly returns to its resting level. In fact, after sodium permeability has returned to its low resting level and the potassium permeability is still greater than normal, there is generally a small hyperpolarizing overshoot of the membrane potential (*after hyperpolarization,* Fig. 8-13).

In our description it may have seemed as though the sodium and potassium fluxes across the membrane involved large numbers of ions. Actually, only one of every 100,000 potassium ions within the cell need diffuse out to charge the membrane potential to its resting value, and very few sodium ions need enter the cell to cause the depolarization during an action potential. Thus, there is virtually no change in the concentration gradients during an action potential. Yet if the tiny number of additional ions crossing the membrane with each action potential were not eventually compensated for, the concentration gradients for sodium and potassium across the cell membrane would gradually disappear. As might be expected, an accumulation of sodium and loss of potassium are prevented by the continuous action of the membrane active-transport system for sodium and potassium (this restoration occurs mainly after the action potential is over). The number of ions that cross the membrane during an action potential is so small, however, that the pump need not keep up with the action potential fluxes, and hundreds of action potentials can occur even if the pump is stopped experimentally. Action potentials would eventually disappear in such circumstances as the sodium and potassium gradients ran down. Note that the pump plays no direct role in the generation of the action potential itself.

Mechanism of permeability changes. The cause of the permeability changes which underlie action potentials has been elucidated by experiments in which membrane permeability and ion fluxes are measured as the membrane potentials are changed. The changes are accomplished by electrodes which can be made to add positive charge to the outside of the membrane while simultaneously removing positive charge from the inside of the cell, thereby causing the membrane potential to become hyperpolarized. If the settings are reversed, the electrodes remove positive charge from the outside while adding it to the inside, thereby depolarizing the cell membrane.

Measurement of membrane permeability changes during these experiments revealed that the permeability to sodium was altered whenever the membrane potential was changed; specifically, hyperpolarization of the membrane caused a decrease in sodium permeability whereas depolarization caused an increase in sodium permeability. In light of our previous discussion of the ionic basis of membrane potentials, it is very easy to confuse the cause-and-effect relationships of the statement just made. Earlier we pointed out that an increase in sodium permeability *causes* membrane depolarization; now we are saying that depolarization *causes* an increase in sodium permeability. Combining these two distinct causal relationships yields the positive-feedback cycle (Fig. 8-14) responsible for the rising phase of the action potential: Depolarization alters the cell membrane structure so that its

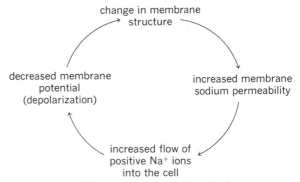

Figure 8-14. **Positive-feedback relationship between membrane depolarization and increased sodium permeability which leads to the rapid rising phase of the action potential.**

permeability to sodium increases; because of increased sodium permeability, sodium diffuses into the cell; this addition of net positive charge to the cell further depolarizes the membrane, which, in turn, produces a still greater increase in sodium permeability, which, in turn, causes. . . .

The sodium ions thus move through membrane channels which respond to changes in the membrane potential as though governed by "gates," so that the channels are open under some conditions but closed at other times, e.g., the membrane at rest. The channels are thought to consist of assemblies of proteins which stretch through the membrane, and their gating functions are probably achieved by changes in the conformation of certain of these proteins. For example, the "gates" to the channels could be closed when the proteins are arranged so that parts of the molecules effectively block the channel; under the influence of a reduced voltage across the membrane (depolarization) a rearrangement of the proteins could occur and open the channel. Sodium inactivation may be another change in protein structure, which blocks the channels again, thus keeping them closed even if the membrane remains depolarized.

The generation of action potentials is prevented by local anesthetics such as novocaine and xylocaine because they prevent the increases in membrane permeability to sodium and potassium caused by depolarization. Without action potentials, there is no way to transmit information over distances greater than a few millimeters. Different nerve fibers are differentially sensitive to local

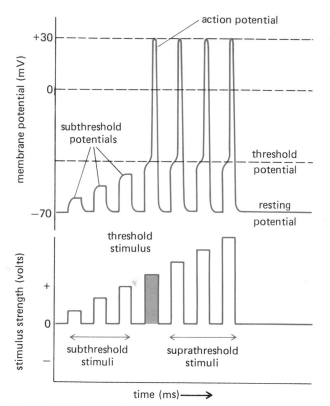

Figure 8-15. Decreasing membrane potential with increasing strength of depolarizing stimulus. When the membrane potential reaches the threshold potential, action potentials are generated. Increasing the stimulus strength above threshold levels does not alter the action-potential response. (The after-hyperpolarization has been omitted from this figure and from Figs. 8-16 and 8-17.)

anesthetics; for example, small fibers are generally more sensitive to them than are large fibers.

Threshold. Not all depolarizations trigger the positive-feedback relationship that leads to an action potential. Action potentials occur only when the membrane is depolarized enough (and, therefore, sodium permeability is increased enough) so that sodium entry exceeds potassium efflux. In other words, action potentials occur only when the *net* movement of positive charge is inward. The membrane potential at which this happens is called the *threshold potential*, and stimuli strong enough to depolarize the membrane to this level are called *threshold stimuli* (Fig. 8-15). The threshold potential of most excitable membranes

is 5 to 15 mV more depolarized than the resting membrane potential. Thus, if the resting potential of a neuron is -70 mV, the threshold potential may be -60 mV; in order to initiate an action potential in such a membrane, the potential must be depolarized by at least 10 mV. Thus, at depolarizations less than threshold, potassium movement still dominates and the positive-feedback cycle cannot get started despite the increase in sodium entry. In such cases, the membrane returns to its resting level as soon as the stimulus is removed, and no action potentials are generated. These weak depolarizations are *subthreshold potentials*, and the stimuli that cause them are *subthreshold stimuli*.

Stimuli of more than threshold magnitude *(suprathreshold stimuli)* elicit action potentials, but as can be seen in Fig. 8-15, the action potentials resulting from such stimuli are exactly the same as those caused by threshold stimuli. This is explained by the fact that, once threshold is reached, membrane events are no longer dependent upon stimulus strength. The depolarization continues on to become an action potential because the positive-feedback cycle is operating. Action potentials either occur maximally as determined by the electrochemical conditions across the membrane or they do not occur at all. Another way of saying this is that action potentials are *all or none.*

Because of this all-or-none nature, a single action potential cannot convey information about the magnitude of the stimulus which initiated it, since a threshold-strength stimulus and one of twice threshold strength give the same response. Since one function of nerve cells is to transmit information, one may ask how a system utilizing all-or-none signals can convey information about the strength of a stimulus. How can one distinguish between a loud noise and a whisper, a light touch and a pinch? This information, as we shall see later, depends upon the number of action potentials transmitted per unit time, i.e., the frequency of action potentials, and not upon their size. It also depends upon the number of neurons activated.

Refractory periods. How soon after firing an action potential can an excitable membrane fire a second one? If we apply a threshold-strength stimulus to a membrane and then stimulate the

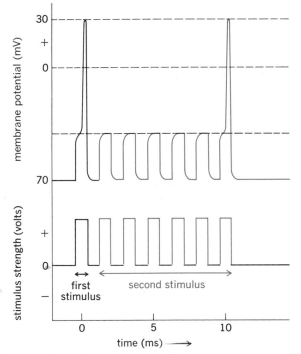

Figure 8-16. Following generation of an action potential, the membrane remains refractory to a second threshold-level stimulus for several milliseconds.

The mechanisms responsible for the refractory periods are related to the membrane mechanisms that alter the sodium and potassium permeability. The absolute refractory period corresponds roughly with the period of sodium permeability changes and the relative refractory period with the period of increased potassium permeability. Following an action potential, time is required to return the membrane structure to its original resting state, i.e., the sodium inactivation must be removed and the potassium permeability returned to normal.

The absolute refractory period limits the number of action potentials that can be produced by an excitable membrane in a given period of time. Recordings made from nerve cells in the intact organism that are responding to physiological stimuli indicate that most nerve cells respond at fre-

Figure 8-17. The magnitude of a single second stimulus necessary to generate a second action potential during the refractory period is greater than the initial stimulus and decreases in magnitude as the time between the first and second stimulus increases. Immediately following an action potential, the membrane is absolutely refractory to all stimulus strengths.

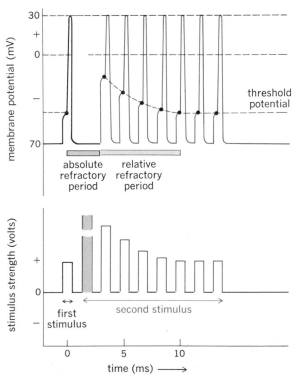

membrane with a second threshold-strength stimulus at various time intervals following the first, the membrane does not always respond to the second stimulus (Fig. 8-16). Even though identical stimuli are applied, the membrane appears unresponsive for a certain time. The membrane during this period is said to be *refractory* to a second stimulus.

Instead of applying the second stimulus at threshold strength, if we increase it to suprathreshold levels, we can distinguish two separate refractory periods associated with an action potential (Fig. 8-17). During the action-potential depolarization a second stimulus will not produce a second action-potential response no matter how strong it is. The membrane is said to be in its *absolute refractory period.* Following the absolute refractory period there is an interval during which a second action potential can be produced but only if the stimulus strength is considerably greater than the usual threshold level. This is known as the *relative refractory period* and can last some 10 to 15 ms or longer.

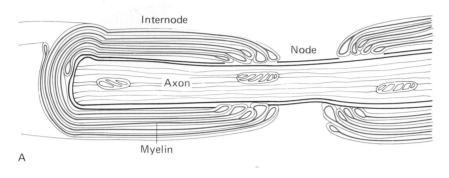

Internode

Node

Axon

Myelin

A

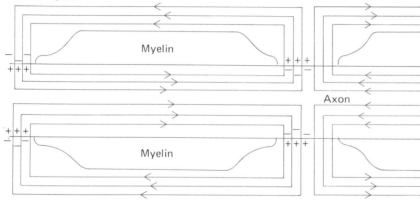

Active region

Myelin

Myelin

Axon

B

Figure 8-18. (A) Structure of a myelinated axon. (B) Ion current flows during an action potential in a myelinated axon.

quencies up to 100 action potentials per second although some may produce much higher frequencies for brief periods of time.

Action-potential propagation. In describing the generation of action potentials, we used stimulating electrodes to bring the membrane to threshold. In the body the threshold stimulus is provided by the graded potentials described earlier. How this occurs will be presented later. Once generated, how does the action potential travel from one point to another?

One particular action potential does not itself travel along the membrane; rather, each action potential triggers, by local current flow, a new one at an adjacent area of membrane. The local current flow is great enough to depolarize the new membrane site to just past its threshold potential, the sodium positive feedback cycle at the new membrane site takes over, and an action potential occurs there. Once the new site is depolarized to threshold, the action potential generated there is solely dependent upon the electrochemical gradi-

ents and membrane permeability properties at the new site. Since these factors are usually identical to those involved in the generation of the old one, the new action potential is virtually identical to the old. This is an important point because it means that no distortion occurs as the signal passes along the membrane; the action potential arriving at the end of the membrane is precisely identical to the initial one. Meanwhile, at the original membrane site, sodium inactivation is occurring and potassium permeability is increasing so that the membrane is repolarizing. These processes repeat themselves until the end of the membrane is reached, a wave of depolarization and repolarization traveling smoothly along the membrane as each active area in turn stimulates by local current flow the area immediately ahead of it.

Local current flow occurs wherever there is an electric gradient (meaning, of course, that it also flows toward the original site of stimulation); however, the membrane areas which have just undergone an action potential are refractory and cannot undergo another; thus, the only direction of action-

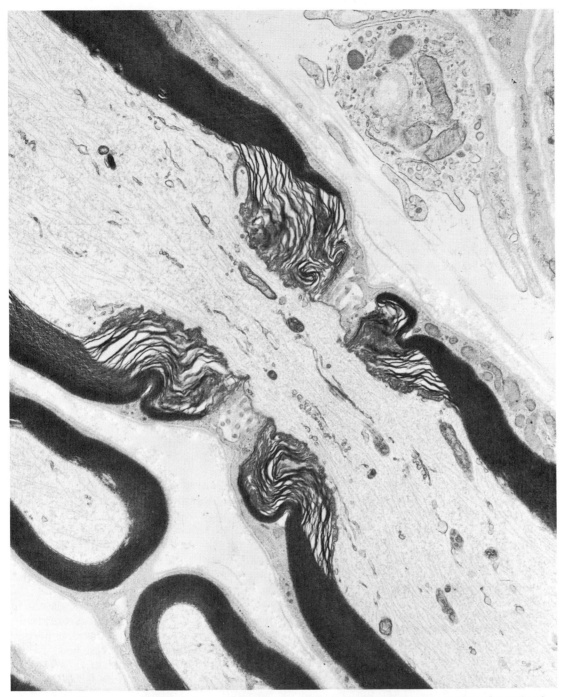

Figure 8-19. Electron micrograph of a myelinated axon (mouse sciatic nerve) in the region of a node of Ranvier. The axon, bordered by dark areas of myelin, is oriented diagonally in this micrograph (top left and bottom right). The myelin appears to fray on both sides of the node. *(From Keith R. Porter and Mary A. Bonneville, "An Introduction to the Fine Structure of Cells and Tissues," 2d ed., Lea & Febiger, Philadelphia, 1964.)*

potential propagation is away from the stimulation site.

The action potentials in skeletal muscle membrane are initiated near the middle of the cell and propagate from this region toward the two ends, but in most nerve-cell membranes action potentials are normally initiated at one end of the cell and propagate in only one direction toward the other end of the cell. Realize, though, that this unidirectional propagation of action potentials is determined by the stimulus location rather than an intrinsic inability to conduct in the opposite direction, and if action potentials are artificially triggered at any location on an excitable membrane, they travel away from that site in both directions.

The velocity with which an action potential is transmitted down the membrane depends upon fiber diameter and whether or not the fiber is myelinated. The larger the fiber diameter, the faster is the action-potential propagation, because a large fiber offers less resistance to local current flow, and adjacent regions of the membrane are brought to threshold faster.

Myelinization is the second factor influencing propagation velocity. *Myelin* is a fatty material covering the axons of many neurons. Myelin electrically insulates the membrane, making it more difficult for current to flow between intra- and extracellular fluid compartments. In effect, it "reduces the leak in the hose"; less current passes out through the myelin-covered section of the membrane during local current flow so that there is a lesser change in the voltage gradient along the fiber (Fig. 8-18). Action potentials do not occur along the sections of membrane protected by myelin; they occur only where the myelin coating is interrupted and the membrane is exposed to the extracellular fluid (Figs. 8-18 and 8-19). These interruptions, called the nodes of Ranvier, occur at regular intervals along the axon. Thus the action potential appears to jump from one node to the next as it propagates along a myelinated fiber, and for this reason this method of propagation is called *saltatory conduction,* from the Latin *saltare,* to leap. The membrane of nodes adjacent to the active node is brought to threshold faster and undergoes an action potential sooner than if myelin were not present. The velocity of action-potential propagation in large myelinated fibers can approach 120 m/s (250 mi/h).

Section B
Physiological activities of neurons

Functional anatomy of neurons

The basic unit of the nervous system is the individual nerve cell, or *neuron.* Only about 10 percent of the cells in the nervous system are neurons; the remainder are *glial cells*, which sustain the neurons metabolically, support them physically, and help regulate the ionic concentrations in the extracellular space. (The glia do not, however, branch as extensively as the neurons do, so that neurons occupy 50 percent of the volume of the central nervous system.) The human nervous system is thus composed of the neurons and glial cells which make up the brain and spinal cord as well as the many nerve processes which pass between these two structures and the receptors, muscles, or glands which they innervate.

Nerve cells occur in a wide variety of sizes and shapes (Fig. 8-20); nevertheless, they can be considered as consisting of three basic parts (Fig. 8-21): (1) the dendrites and cell body; (2) the axon; (3) the axon terminals. The *dendrites* form a series of highly branched cell outgrowths connected to the cell body and may be looked upon as an extension of its membrane. The dendrites and cell body are the site of most of the specialized junctions where signals are received from other neurons. As in other types of cells, the cell body contains the nucleus and many of the organelles involved in metabolic processes and is responsible for maintaining the metabolism of the neuron and for its growth and repair.

The *axon,* or *nerve fiber,* is a single process extending from the cell body. It is long (sometimes more than a meter) in neurons which connect with distant parts of the nervous system or with muscles and glands; such cells function as *projection neurons.* Axons are short or even nonexistent in *local circuit neurons,* which connect only with cells in their immediate vicinity. The first portion of the axon plus the part of the cell body where the axon is joined is known as the *initial segment.* The axon may give off branches called *collaterals* along its course, and near the end it undergoes considerable branching into numerous *axon terminals,* which are responsible for transmitting signals from the neuron to the cells contacted by the axon terminals.

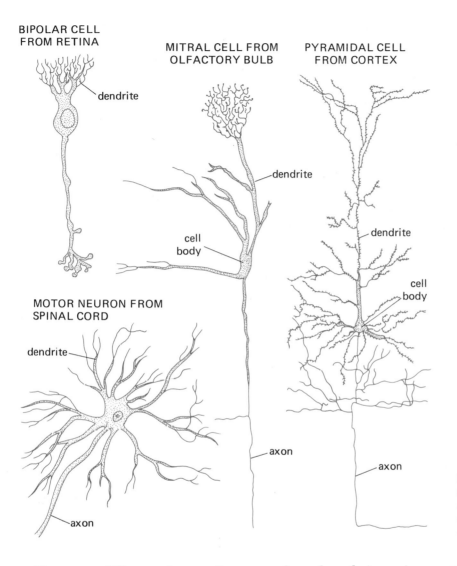

BIPOLAR CELL
FROM RETINA

dendrite

MITRAL CELL FROM
OLFACTORY BULB

dendrite

cell
body

PYRAMIDAL CELL
FROM CORTEX

dendrite

dendrite

cell
body

MOTOR NEURON FROM
SPINAL CORD

dendrite

axon

axon

axon

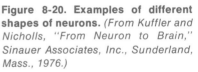

Figure 8-20. Examples of different shapes of neurons. *(From Kuffler and Nicholls, "From Neuron to Brain," Sinauer Associates, Inc., Sunderland, Mass., 1976.)*

The many different shapes of neurons depend on their location and function (Fig. 8-20). Yet, regardless of their shape, neurons can be divided into three classes: afferent neurons, efferent neurons, and interneurons (Fig. 8-22). Afferent and efferent neurons lie largely outside the skull or vertebral column, and interneurons lie within the central nervous system.[1] At their peripheral endings afferent neurons have *receptors*, which, in response to various physical or chemical changes in their environment, cause action potentials to be generated in the afferent neuron. The afferent neurons carry information from the receptors *into* the brain or spinal cord. After transmission to the central nervous system, some of this afferent information may be perceived as a conscious sensation. Efferent neurons transmit the final integrated information *from* the central nervous system out to the effector organs (muscles or glands). Efferent neurons which innervate skeletal muscle are also called *motor neurons*. The third group of nerve cells, the *interneurons*, both originate and terminate within the central nervous system, and 99 percent of all nerve cells belong to this group.

[1] Some equally acceptable systems of classification use the term "interneuron" for only the local circuit neurons in the central nervous system.

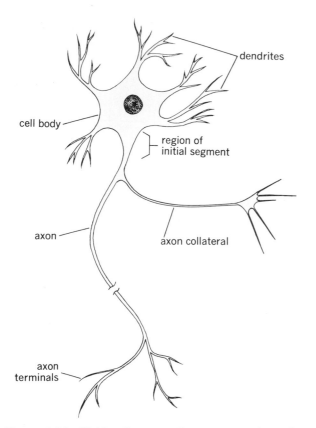

Figure 8-21. Highly diagrammatic representation of a neuron.

and thus appears to be an inherent property of these neurons; such neurons are thought to play important roles in rhythmic events such as breathing.

Receptors

Information about the external world and internal environment exists in different energy forms— pressure, temperature, light, sound waves, etc.— but only receptors can deal with these energy forms. The rest of the nervous system can extract meaning only from action potentials or, over very short distances, from graded potentials. Regardless of its original energy form, information from peripheral receptors must be translated into the language of action potentials.

The devices which do this are receptors. These are either specialized peripheral endings of afferent neurons or separate cells synaptically connected to afferent neurons. In the former case, the afferent neuron is activated directly when the stimulus impinges on the specialized membrane of the neuron's receptor ending (Fig. 8-23A); in the latter case, the separate receptor cell contains the specialized membrane which is activated by the stimulus (Fig. 8-23B), and upon stimulation, the receptor cell releases a chemical transmitter. This

The number of interneurons in the pathway between afferent and efferent neurons varies according to the complexity of the action. One type of basic reflex, the stretch reflex, has no interneurons, the afferent neurons synapsing directly upon the efferent neurons, whereas stimuli invoking memory or language may involve millions of interneurons.

Information is relayed along neurons in the form of action potentials, which are initiated physiologically by the three types of graded potentials: receptor, synaptic, and pacemaker potentials. The frequency of action potentials can be altered to signal different types of information. The mechanisms of action-potential initiation at receptors and synapses and the determinants of action-potential frequency are the subject of this section. The spontaneous generation of action potentials by pacemaker potentials in certain neurons occurs in the absence of any identifiable external stimulus

Figure 8-22. Three classes of neurons. Note that interneurons are entirely within the central nervous system.

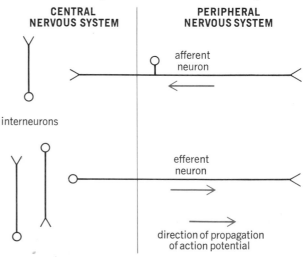

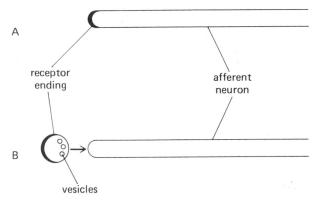

Figure 8-23. The sensitive membrane that responds to the stimulus is either an ending of the afferent neuron itself (A), or on a separate cell adjacent to the afferent neuron (B).

transmitter diffuses across the extracellular cleft separating the receptor cell from the afferent neuron and activates specific sites on the afferent neuron. In both cases, the result is an electrical change (a graded potential) in the afferent neuron which leads to the generation of action potentials in it.

There are several types of receptors, each of which is specific; i.e., each responds much more readily to one form of energy than to others, although virtually all receptors can be activated by several different forms of energy if the intensity is sufficient. (For example, the receptors of the eye normally respond to light, but they can be activated by intense mechanical stimuli like a poke in the eye.) Usually much more energy is required to excite a receptor by energy forms to which it is not specific. On the other hand, most receptors are exquisitely sensitive to their specific energy forms. For example, olfactory receptors can respond to as few as three or four odorous molecules in the inspired air, and visual receptors can respond to a single photon, the smallest known quantity of light.

Receptor activation: The receptor potential

The basic mechanism of receptor activation, believed to be the same for all types of receptors, is that the stimulus acts on a specialized receptor membrane to increase its permeability and thereby produce a graded potential, called a *receptor potential* (sometimes called a *generator potential*). The membrane mechanisms which generate the

potential seem to be the same regardless of whether the receptor membrane is on the afferent neuron itself or on a separate receptor cell and the afferent neuron is activated by transmitters diffusing to it from the receptor cell.

In describing here the general mechanisms for receptor activation we will use the simple example in which the receptor is the peripheral ending of the afferent neuron itself and responds to pressure or mechanical deformation (Fig. 8-24). In subsequent chapters, we shall describe receptors which respond to stimuli such as light, sound, and heat.

In the mechanoreceptor, stimuli such as pressure may bend, stretch, or press upon the receptor membrane, and somehow, perhaps by opening up pores in the membrane, increase its permeability to certain ions. Because of the increased permeability, ions move across the membrane down their electric and concentration gradients. Although these ion movements have not been worked out as clearly as those associated with action potentials, the permeability increases in mechanoreceptors appear to be nonselective and to apply to all small ions.

Remember that the intracellular fluid of all nerve cells has a higher concentration of potassium and a lower concentration of sodium than the extracellular fluid and that the inside of a resting neuron is about 70 mV negative with respect to the outside. The effect of a nonselective increase in membrane permeability at a stimulated

Figure 8-24. *(Top)* **A mechanoreceptor (pacinian corpuscle) in which the nerve ending is modified by cellular structures.** *(Bottom)* **The naked nerve ending of the same mechanoreceptor. The receptor potential arises at the nerve ending** *A,* **and the action potential arises at the first node in the myelin sheath** *B. (Adapted from Loewenstein.)*

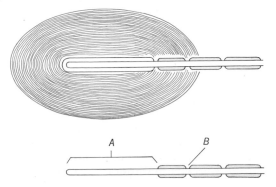

receptor is a net outward diffusion of a small number of potassium ions and the simultaneous movement in of a larger number of sodium ions. The result is net movement of positive charge into the cell, leading to a decrease in membrane potential (depolarization). The movement of sodium into the nerve fiber thus plays the major role in producing the receptor potential. In the receptor region the cell membrane has a very high threshold, so that the receptor potential cannot depolarize this region enough to cause an action potential. Instead, this depolarization (i.e., the receptor potential) is conducted by local current flow a short distance from the nerve ending to a region where the membrane's threshold is lower. In myelinated afferent neurons, this is usually at the first node of the myelin sheath. As is true of all graded potentials, the magnitude of the receptor potential decreases with distance from its site of origin but, if the amount of depolarization which reaches the first node is large enough to bring the membrane there to threshold, action potentials are initiated. The action potentials then propagate along the nerve fiber; the only function of the receptor potential is to trigger action potentials.

Receptor potentials, being graded potentials, are not all-or-none; their amplitude and duration vary with stimulus strength and other variables to be discussed shortly. A change in the amplitude of the receptor potential causes, via local current flow, a similar but smaller change in the degree of depolarization at the first node. Because of this relationship, the amplitude and duration of the receptor potential determine action-potential frequency in the afferent neuron. As long as the first node is depolarized to threshold, action potentials continue to fire and propagate along the afferent neuron. In fact, it is much more common for stimuli to cause trains, or bursts, of action potentials than single ones.

Amplitude of the receptor potential

Since the amplitude and duration of the receptor potential determine the number of action potentials initiated in the afferent neuron, the factors controlling these parameters are important. They include the stimulus intensity, rate of change of stimulus intensity, summation of successive receptor potentials, and adaptation. Note that receptor-potential amplitude determines action-

potential *frequency*, i.e., number of action potentials fired per unit time; it does *not* determine action-potential magnitude. Since the action potential is all-or-none, its amplitude is always the same regardless of the magnitude of the stimulus.

Intensity and velocity of stimulus application. Receptor potentials become larger with greater intensity of the stimulus (Fig. 8-10 part 2) because the permeability changes increase and the transmembrane ion movements are greater. The explanation for this phenomenon may be an increase in the number of channels which permit transmembrane ion flow. The amplitude of the receptor potential also rises with a greater rate of change of stimulus application, e.g., the faster the stimulus is applied, the larger the receptor potential.

Summation of receptor potentials. Another way of varying the amplitude of the receptor potential is by adding two or more together. This is possible because they are graded and because they may last as long as the stimulus is applied. If the receptor membrane is stimulated again before the receptor potential from a preceding stimulus has died away, the two potentials sum and make a larger depolarization (Fig. 8-10 part 3).

Adaptation. *Adaptation* is a decrease in frequency of action potentials in the afferent neuron despite a constant stimulus magnitude. It occurs for any or all of the following reasons: (1) The stimulus energy can be dissipated in the tissues as it passes through them to reach the receptor. As the energy loss gradually increases, the amount of energy reaching the receptor decreases. Thus, the membrane permeability change and receptor potential amplitude decrease with time. (2) The responsiveness of the receptor membrane can decrease with time so that the receptor-potential amplitude drops even though the energy reaching the receptor remains unchanged. These two reasons for adaptation are probably the most frequent. In both the lower action-potential frequency is due to lower amplitude of the receptor potential. (3) Even if the receptor-potential amplitude remains unchanged, there can be a decreased frequency of action potentials in response to a constant stimulus because of changes at the first node. The actual reasons for adaptation vary in different receptor types. Adaptation of an afferent

Figure 8-25. Action potentials in a single afferent nerve fiber showing adaptation.

neuron in response to the constant stimulation of its mechanoreceptor ending can be seen in Fig. 8-25. Some receptors adapt completely so that in spite of a constantly maintained stimulus, the generation of action potentials stops. In some extreme cases, the receptors fire only once at the onset of the stimulus. In contrast to these rapidly adapting receptors, slowly adapting types merely drop from an initial high action-potential frequency to a lower level, which is then maintained for the duration of the stimulus.

Summary. Because the generator potential is a graded potential, its magnitude can vary with stimulus intensity, rate of change of stimulus intensity, summation, and adaptation. The amplitude of the receptor potential in turn determines the frequency of the action potentials in the afferent neuron.

Intensity coding

We are certainly aware of different stimulus intensities. How is information about stimulus strength relayed by action potentials of constant amplitude? One way is related to the frequency of action potentials; increased stimulus strength means a larger receptor potential and higher frequency of firing of action potentials. A record of an experiment in which increased stimulus intensity is reflected in increased action-potential frequency in a single afferent nerve fiber is shown in Fig. 8-26.

There is an upper limit to this positive correlation between stimulus intensity and action-potential frequency. When stimulus strength becomes very great, the receptor potential reaches a maximum and a further increase in rate of firing action potentials by that receptor cannot occur. However, even though that particular receptor cannot generate a higher frequency of action potentials, receptors at other branches of the same neuron can be stimulated. Most afferent neurons have many branches, each with a receptor at its ending, but

the receptors at different branches do not respond with equal ease to a given stimulus; some are less easily excited and respond only to stronger stimuli. Thus, as stimulus strength increases, more and more receptors begin to respond. Action potentials generated by these receptors propagate along the branch to the main afferent nerve fiber and increase the frequency of action potentials there. The maximal firing frequency of the afferent fiber is limited only by the membrane refractory period.

In addition to the increased frequency of firing in a single neuron, similar receptors on the endings of other afferent neurons are also activated as stimulus strength increases, because stronger stimuli usually affect a larger area. For example, when one touches a surface lightly with a finger, the area of skin in contact with the surface is small

Figure 8-26. Action potentials recorded from a blood-pressure receptor afferent nerve fiber as the receptor was subjected to pressures of different magnitudes.

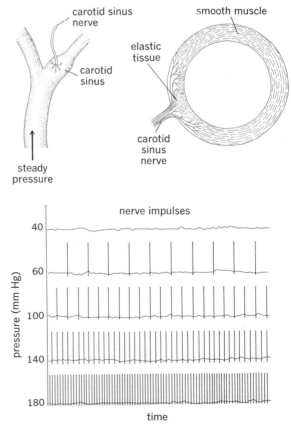

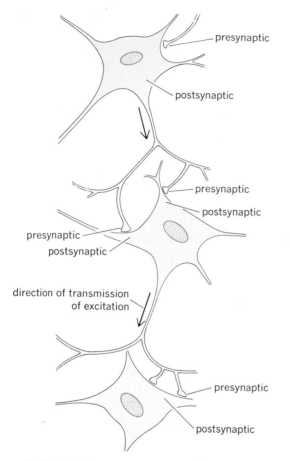

presynaptic

postsynaptic

direction of transmission
of excitation

presynaptic

postsynaptic

presynaptic

postsynaptic

presynaptic

postsynaptic

Figure 8-27. A single neuron postsynaptic to one group of cells and presynaptic to another.

and only receptors in that area of skin are stimulated; pressing down firmly increases the area of skin stimulated. This "calling in" of receptors on additional nerve cells is known as *recruitment.* These generalizations are true of virtually all afferent systems: Increased stimulus intensity is signaled both by an increased firing rate of action potentials in a single nerve fiber and by recruitment of additional receptors on other afferent neurons.

By the mechanisms discussed in this section different energy forms activate specific receptors to supply information about the kind of stimulus and its duration and intensity. This information is then transmitted in the form of action potentials along the afferent neuron. Next to be considered is the mechanism by which the activity is transferred from one neuron to another.

Synapses

A synapse is an anatomically specialized junction between two neurons where the electrical activity in one neuron influences the activity of the second. Most synapses occur between the axon terminals of one neuron and the cell body or dendrites of a second; however, in certain areas, synapses also occur between dendrite and dendrite, dendrite and cell body, and axon and axon. The neurons conducting information toward synapses are called *presynaptic neurons,* and those conducting information away are *postsynaptic neurons.* Figure 8-27 shows how, in a multineuronal pathway, a single neuron can be postsynaptic to one group of cells and, at the same time, presynaptic to another.

A postsynaptic neuron may have thousands of synaptic junctions on the surface of its dendrites or cell body so that information from hundreds or even thousands of presynaptic nerve cells converges upon it (Fig. 8-28). A single motor neuron in the spinal cord probably receives some 15,000 synaptic endings, and it has been calculated that certain neurons in the brain receive more than 100,000. Each activated synapse produces a brief, small electric signal, either excitatory or inhibitory, in the postsynaptic cell. The picture we are left with is one of thousands of synapses from many different presynaptic cells *converging* upon a single postsynaptic cell. The level of excitability of this cell at any moment, i.e., how close the membrane potential is to threshold, depends upon the number of synapses active at any one time and how many are excitatory or inhibitory. If the postsynaptic neuron reaches threshold it will generate action potentials. These are transmitted out along its axon to the terminal branches, which *diverge* to influence the excitability of many other cells. Figure 8-29 demonstrates the neuronal relationships of convergence and divergence.

In this manner, postsynaptic neurons function as neural *integrators;* i.e., their output reflects the sum of all the incoming bits of information arriving in the form of excitatory and inhibitory synaptic inputs.

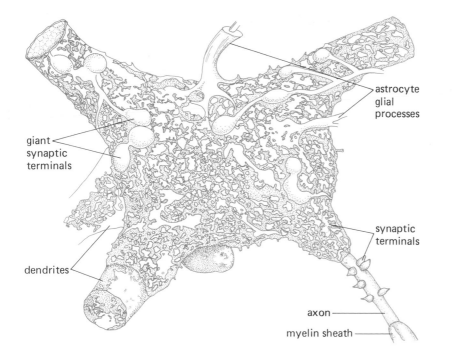

Figure 8-28. Synaptic endings on a nerve cell body. (There are also some glial processes.) *(Adapted from Poritsky.)*

astrocyte glial processes

giant synaptic terminals

synaptic terminals

dendrites

axon

myelin sheath

Functional anatomy

Synapses are of two types: *electrical* and *chemical.* At electrical synapses, the local currents resulting from action potentials in the presynaptic neuron flow through gap junctions (see Chap. 6) directly into the postsynaptic cell. Chemical synapses are much more common in the mammalian nervous system, and it is this type of synapse that is discussed in the following sections.

Figure 8-30 shows the anatomy of a single chemical synaptic junction. The axon terminal of the presynaptic neuron ends in a slight swelling, the *synaptic terminal.* A narrow extracellular space, the *synaptic cleft,* separating the pre- and postsynaptic neurons, prevents direct propagation of the action potential from the presynaptic neuron to the postsynaptic cell. Information is transmitted across the synaptic cleft by means of a chemical *neurotransmitter* substance stored in small, membrane-enclosed vesicles in the synaptic terminal. When an action potential in the presynaptic neuron reaches the end of the axon and depolarizes the terminal, small quantities of the neurotransmitter are released from the synaptic

terminal into the synaptic cleft. The link between membrane depolarization and neurotransmitter release is calcium. Depolarization of the synaptic terminal causes an increase in the permeability of the terminal membrane to calcium ions. Calcium enters the presynaptic terminal during the action potentials and causes some of the vesicles to fuse with the cell membrane and liberate their contents into the synaptic cleft. Once released from the vesicles, the transmitter molecules diffuse across the cleft and bind to receptor sites[1] on the membrane of the postsynaptic cell lying right under the synaptic terminal *(subsynaptic membrane).* The combination of the transmitter with the receptor site causes changes in the permeability of the subsynaptic membrane and thereby in the membrane potential of the postsynaptic cell.

Because all the transmitter is stored on one side

---

[1] Again, the reader is cautioned not to confuse this use of "receptors" (signifying a membrane molecular configuration with which chemical transmitters combine) with the afferent receptors described in the previous section.

A

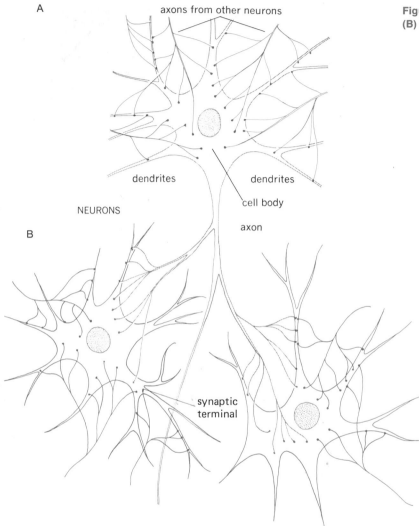

axons from other neurons

dendrites   dendrites

NEURONS

B

cell body

axon

synaptic
terminal

Figure 8-29. (A) Convergence of neural input.
(B) Divergence of neural output.

of the synaptic cleft and all the receptor sites are on the other side, chemical synapses operate in only one direction. Because of this one-way conduction across synapses, action potentials are transmitted along a given multineuronal pathway in only one direction.

There is a delay between excitation of the synaptic terminal and membrane-potential changes in the postsynaptic cell; it lasts less than a thousandth of a second and is called the *synaptic delay.* The delay is due to the mechanism which releases transmitter substance from the synaptic terminal,

since the time required for the transmitter to diffuse across the synaptic cleft is negligible. The postsynaptic permeability change is then terminated when the transmitter is removed from the subsynaptic membrane. At different synapses this may be caused by (1) its chemical transformation into an ineffective substance, (2) diffusion away from the receptor site, or (3) reuptake by the synaptic terminal.

Excitatory synapse. The two kinds of chemical synapses, excitatory and inhibitory, are differenti-

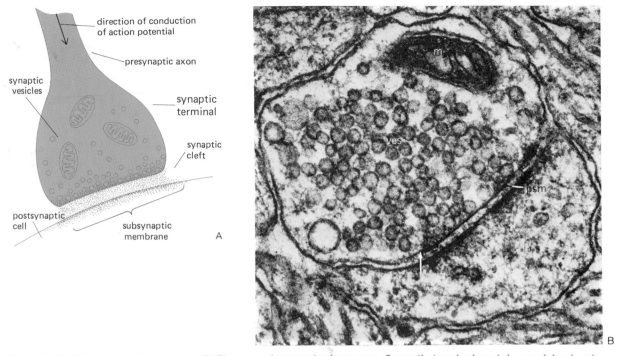

Figure 8-30. (A) Diagram of synapse. (B) Electron micrograph of synapse. Synaptic terminal contains vesicles (ves) and a mitochondrion (m), and dendrite (den) shows thickening of postsynaptic membrane (psm). Arrow marks beginning of synaptic cleft. *(From W. F. Windle, "Textbook of Histology," 5th ed., McGraw-Hill, New York, 1976.)*

ated by the effects which the neurotransmitter has on the postsynaptic cell. Whether the effect of the transmitter is inhibitory or excitatory depends on the structure of the subsynaptic membrane. An excitatory synapse, when activated, increases the likelihood that the membrane potential of the postsynaptic cell will reach threshold and generate action potentials. Here the effect of the neurotransmitter-receptor combination is to increase the permeability of the subsynaptic membrane to sodium and potassium ions so that they are free to move according to the electric and chemical forces acting upon them. Thus, at the subsynaptic membrane of an excitatory synapse there occurs the simultaneous movement of a relatively small number of potassium ions out of the cell and a larger number of sodium ions into the cell. The *net* movement of positive ions is into the neuron, which slightly depolarizes the postsynaptic cell. This potential change, called the *excitatory postsynaptic potential* (EPSP), brings the membrane closer to threshold (Fig. 8-31). The EPSP, like the receptor potential, is a graded potential which

spreads decrementally, by local current flow; its only function is to help trigger action potentials.

Inhibitory synapse. Activation of an inhibitory synapse produces changes in the postsynaptic cell which lessen the likelihood that the cell will generate an action potential. At inhibitory synapses the combination of the neurotransmitter molecules with the receptor sites on the subsynaptic membrane also changes the permeability of the membrane, but only the permeabilities to potassium or chloride ions are increased; sodium permeability is not affected. At some inhibitory synapses the transmitter acts to increase potassium permeability; at others (e.g., motor neurons) it increases chloride permeability, and at some it may increase both. Earlier it was noted that if a cell membrane were permeable only to potassium ions the resting membrane potential would equal the potassium equilibrium potential; i.e., the resting membrane potential would be −90 mV instead of −70 mV. The increased potassium permeability at an activated inhibitory synapse

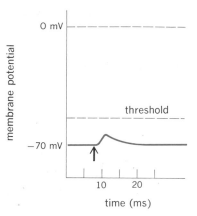

Figure 8-31. Excitatory postsynaptic potential (EPSP). Stimulation of the presynaptic neuron is marked by the arrow. Note the short synaptic delay before the postsynaptic cell responds.

makes the postsynaptic cell more like the hypothetical cell that is permeable only to potassium ions. Consequently, the membrane potential becomes closer to the potassium equilibrium potential. This increased negativity (hyperpolarization) is an *inhibitory postsynaptic potential* (IPSP) (Fig. 8-32). Thus, when an inhibitory synapse is activated, the postsynaptic neuron's membrane potential is moved farther away from the threshold level. In many cells the chloride equilibrium potential is also more negative (e.g., -80 mV) than the resting potential so that an increase in chloride permeability would have a similar hyperpolarizing effect. Even in cells in which the equilibrium potential for chloride is very close to the resting membrane potential, a rise in chloride-ion permeability will still lessen the likelihood that the cell will reach threshold, the reason being that it increases the tendency of the membrane to stay at the resting potential.

Activation of the postsynaptic cell: Neural integration

A feature that makes postsynaptic integration possible is that, in most neurons, one excitatory synaptic event is not enough by itself to change the membrane potential of the postsynaptic neuron from its resting level to threshold; e.g., a single EPSP in a motor neuron is estimated to be only 0.5 mV whereas changes of up to 25 mV are necessary to depolarize the membrane from its resting level

to threshold. Since a single synaptic event does not bring the postsynaptic membrane to its threshold level, an action potential can be initiated only by the combined effects of many synapses. Of the thousands of synapses on any one neuron, probably hundreds are active simultaneously (or at least close enough in time that the effects can summate), and the membrane potential of the neuron at any one moment is the resultant of all the synaptic activity affecting it at that time. There is a general depolarization of the membrane toward threshold when excitatory synaptic activity predominates (this is known as *facilitation*) and a hyperpolarization when inhibition predominates (Fig. 8-33).

Let us perform a simplified experiment to see how two EPSPs, two IPSPs, or an EPSP plus an IPSP interact (Fig. 8-34). Let us assume that there are three synaptic inputs to the postsynaptic cell; A and B are excitatory and C is an inhibitory synapse. There are stimulators on the axons to A, B, and C so that each of the three inputs can be activated individually. A very fine electrode is placed in the cell body of the postsynaptic neuron and connected to record the membrane potential. In part I of the experiment we shall test the interaction of two EPSPs by stimulating A and then, a short while later, stimulating A again. Part I of Fig. 8-34 shows that no interaction occurs between the two EPSPs. The reason is that the change in membrane potential associated with an EPSP is fairly short-lived. Within a few ms (by the time axon A is restimulated) the postsynaptic cell

Figure 8-32. Inhibitory postsynaptic potential (IPSP). Stimulation of the presynaptic neuron is marked by the arrow. Note the short synaptic delay.

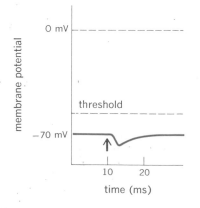

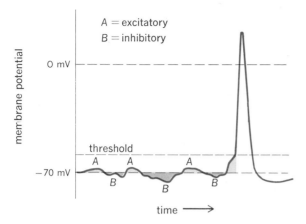

Figure 8-33. Intracellular recording from a postsynaptic cell during episodes when (A) excitatory synaptic activity predominates and the cell is facilitated and (B) inhibitory synaptic activity dominates.

has returned to its resting condition. In part II of the experiment, axon *A* is restimulated before the first EPSP has died away, and so-called *temporal summation* occurs. In part III, axons *A* and *B* are stimulated simultaneously and the two EPSPs that result also summate in the postsynaptic neuron; this phenomenon is called *spatial summation.* The summation of EPSPs can bring the membrane to its threshold so that an action potential is initiated. So far we have tested only the patterns of interaction of excitatory synapses. What happens if an excitatory and inhibitory synapse are activated so that their effects occur at the postsynaptic cell simultaneously? Since EPSPs and IPSPs are due

to local currents flowing in opposite directions, they tend to cancel each other, and the resulting potential change is smaller (Fig. 8-34, part IV). Inhibitory potentials can also show temporal and spatial summation.

As described above, the subsynaptic membrane is depolarized at an activated excitatory synapse and hyperpolarized at an activated inhibitory synapse. By the mechanisms of local current flow described earlier, current flows through the cytoplasm away from an excitatory synapse and toward an inhibitory synapse (Fig. 8-35). Thus, the entire cell body, including the initial segment, becomes slightly depolarized during activation of excitatory synapses and slightly hyperpolarized during activation of inhibitory synapses.

In the above examples we referred to the threshold of the postsynaptic neuron. However, the fact is that different parts of a neuron have different thresholds. The cell body and larger dendritic branches reach threshold when their membrane is depolarized about 25 mV from the resting level, but in many cells the initial segment has a threshold which is much closer to the resting potential. In cells whose initial-segment threshold is lower than that of their dendrites and cell body, the initial segment is activated first whenever enough EPSPs summate, and the action potential originating there is propagated down the axon.

Individual postsynaptic potentials last much longer than action potentials do. In the event that the initial segment is still depolarized above threshold after an action potential has been fired and the refractory period is over, a second action

Figure 8-34. Interaction of EPSPs and IPSPs at the postsynaptic neuron.

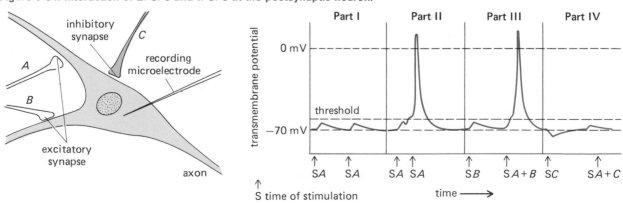

A

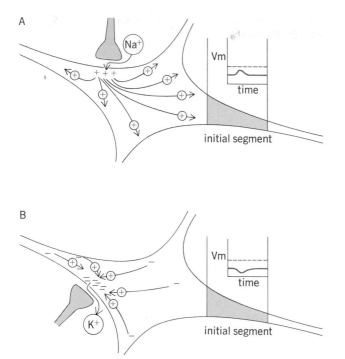

B

Figure 8-35. Comparison of excitatory (A) and inhibitory (B) synapses showing direction of current flow through the postsynaptic cell following synaptic activation. (A) Current flows through the cytoplasm of the postsynaptic cell away from the excitatory synapse, depolarizing the cell. (B) Current flows through the cytoplasm of the postsynaptic cell toward an inhibitory synapse, hyperpolarizing the cell. Arrows show direction of positive ion flow.

potential will occur. In fact, the greater the summed postsynaptic depolarization, the greater is the number of action potentials fired (up to the limit imposed by the duration of the absolute refractory period). Neuronal responses are almost always in the form of so-called bursts or trains of action potentials.

The significance of the lower threshold of the initial segment can best be demonstrated if we suppose for a moment that the threshold were the same over the entire neuron so that an action potential could be initiated with equal ease at any point of the cell body or dendrites. In Fig. 8-36 the greatest input to the cell is clearly inhibitory, but in the upper left corner three active excitatory synapses are clustered together. At this one point there is sodium-ion movement into a relatively small portion of the cell. If this depolari-

zation is great enough, an action potential could be initiated at this site and conducted over the entire cell membrane despite the fact that most of the input is inhibitory. This possible bias of the cell's activity by synapse grouping is greatly lessened by the fact that the initial segment acts to average all the synaptic input.

On the other side of the coin, those synapses right next to the initial segment have a greater influence upon cell activity than those at the ends of the dendrites, and thus synaptic placement provides a mechanism for giving different inputs a greater or lesser influence on the postsynaptic cell's output. Finally, it should be noted that, in some cells, action potentials can be initiated in regions other than the initial segment.

This discussion has no doubt left the impression that one neuron can influence another only at a synapse. However, this view requires qualification. We have emphasized how local currents influence the membrane of the cell in which they are generated (as, for example, in action potential propagation and synaptic excitation). Under certain circumstances, local currents in one neuron may directly affect the membrane potential of other

Figure 8-36. Synaptic input to a neuron.

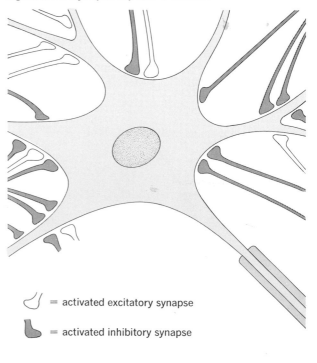

⌐ = activated excitatory synapse

◣ = activated inhibitory synapse

nearby neurons. This phenomenon is particularly important in areas of the central nervous system which contain a high density of unmyelinated neuronal processes.

Neurotransmitters

As described above, presynaptic neurons influence postsynaptic neurons by means of chemical transmitters. This section lists some of the most important generalizations relating to these neurotransmitters as well as specific facts about those chemicals most generally accepted as transmitters.

There are many different suspected synaptic transmitters used by different types of cells in the nervous system (Table 8-4); some of these are shown in Fig. 8-37. All synaptic terminals from a given cell probably liberate the same transmitter. Several criteria must be met before the status of a chemical is changed from suspected to positively identified neurotransmitter and only four of the

Table 8-4. Some chemicals known or presumed to be neurotransmitters

Dopamine ⎫
Norepinephrine ⎬ ← Catecholamines
Epinephrine ⎭
Acetylcholine
Serotonin
Gamma-aminobutyric acid
Glycine
Aspartic acid
Glutamic acid
Enkephalins
Endorphins
Substance P
Somatostatin
Histamine

compounds shown in Fig. 8-37 (acetylcholine, norepinephrine, epinephrine, and GABA) are known for certain to be transmitters, although many of the others are also widely accepted. It

Figure 8-37. Chemical structures of some neurotransmitter substances. Other suspected transmitters not shown are listed in Table 8-4.

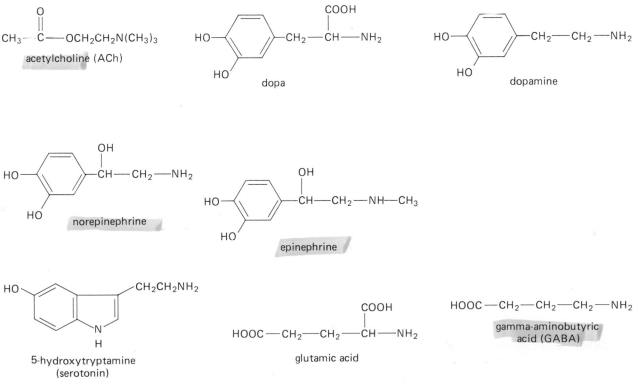

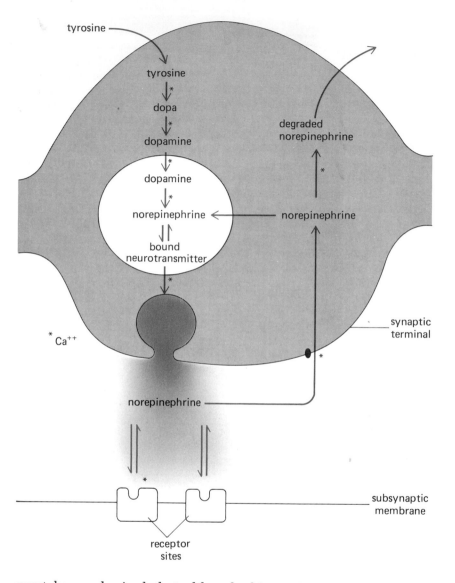

Figure 8-38. Site of catecholamine synthesis, storage, release, action, reuptake, and degradation. The asterisks indicate steps that can be modified by currently available drugs.

must be emphasized that although this section deals with these substances as neurotransmitters, some of them are found in nonneural tissue and function as hormones and paracrine agents. In other words, evolution has selected the same chemical messengers for use in widely differing circumstances.

Dopamine, norepinephrine, and epinephrine. Dopamine, norepinephrine (NE), and epinephrine (E) all contain a catechol ring (a six-sided carbon ring with two adjacent hydroxyl groups) and an NH₂ (amine) group; thus, they are *catecholamines.* They share the synthetic pathway shown in Fig. 8-38, which begins with the uptake of the amino acid tyrosine by synaptic terminals, and, depending on the enzymes present in the terminals, the neurotransmitter may be any one of the catecholamines. Transmitter activity ceases when the catecholamine concentration in the synaptic cleft declines because of catecholamine transport back into the cytoplasm of the synaptic terminal or when it is broken down by an enzyme in the synaptic cleft. (There is also an enzyme in the synaptic terminal that breaks down catecholamines; its action helps to maintain constant levels of neurotransmitter there.)

Acetylcholine. Acetylcholine (ACh) is synthesized from choline in the cytoplasm of synaptic terminals. It is transported into synaptic vesicles for storage, and in response to depolarization of the synaptic terminal and the influx of calcium, the vesicles empty into the synaptic cleft. The transmitter diffuses across the cleft and, if it is not first enzymatically destroyed, activates receptors on the postsynaptic membrane. Thus far the story is much like that for the catecholamines, but there is an important difference in the way the concentration of acetylcholine is reduced so that receptor activation is stopped. Instead of being taken back into the synaptic terminal, free acetylcholine is rapidly destroyed by the enzyme *acetylcholinesterase* located near the receptors on the postsynaptic membrane. The choline released during this reaction is taken up by the synaptic terminals to be reused in the synthesis of new ACh.

ACh is the transmitter for only a few pathways in the brain and spinal cord, and relatively little is known about these ACh-secreting neurons in the central nervous system. In contrast, ACh is an important synaptic transmitter in the autonomic nervous system (see below), and it is the transmitter at the junction between motor nerve terminals and skeletal muscle cells.

Serotonin. Serotonin (5-hydroxytryptamine, 5-HT, Fig. 8-37) is found in greatest concentration in blood platelets and specialized cells lining the digestive tract, but it also most probably functions as a neurotransmitter. It is produced in axon terminals from tryptophan, an essential amino acid, and changes in the blood levels of tryptophan and in the rate of tryptophan transport from the blood into the extracellular space of the nervous system and into the synaptic terminals can affect the rate at which serotonin is produced.

Serotonin is important in the neural pathways controlling mood, and several substances chemically related to it, e.g., psilocybin, a hallucinogenic agent found in some mushrooms, have potent psychic effects. Moreover, the drug LSD, a hallucinogen which induces a state resembling schizophrenia, is a serotonin antagonist.

Other possible neurotransmitters. There is good evidence that some amino acids act as synaptic neurotransmitter gamma-aminobutyric acids; (GABA, Fig. 8-37) and glycine function at inhibitory synapses whereas aspartic and glutamic acid (Fig. 8-37) are among the most potent excitatory chemicals in the brain.

A group of peptides, the *endorphins* ("endogenous morphines") and the *enkephalins*, was recently discovered by following an interesting process of reasoning. Since the opiate drugs (e.g., morphine and codeine) act specifically in the brain, there must be receptors for them and such receptors have indeed been found in discrete locations in the central nervous system (for example, in areas containing pathways which convey pain information). Then the next question was asked: Why are there receptors for opiates in the brains of people? They certainly wouldn't have evolved just so they would be available when the benefits of the opium poppy were discovered; there must be normally occurring opiate-like substances in the brain, and so these "endogenous morphines" were sought and found. Most of these substances, when injected into the brain, have a pain-killing potential about equal to that of morphine. This fact, along with many others, suggests that the enkephalins and endorphins play a role in sensory systems associated with pain. However, opiate receptors, endorphins, and enkephalins have also been found in parts of the brain that have nothing to do with pain but are involved in mood and emotions.

Modification of synaptic transmission by drugs and disease

The great majority of drugs which act upon the nervous system do so by altering synaptic mechanisms, and all the synaptic mechanisms labeled in Fig. 8-39 are vulnerable.

Disease processes, too, often affect synaptic mechanisms. For example, the toxin produced by the tetanus bacillus acts at inhibitory synapses on neurons supplying skeletal muscles, presumably by blocking the receptors. This eliminates inhibitory input to the neurons and permits unchecked influence of the excitatory inputs, leading to muscle spasticity and seizures; the spasms of the jaw

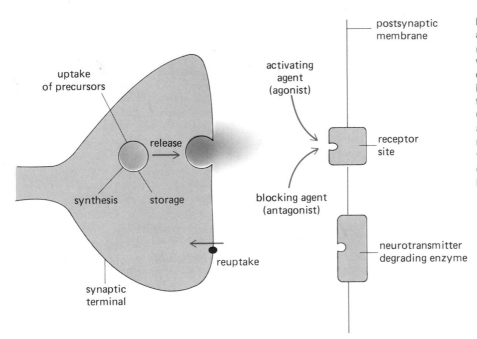

Figure 8-39. Sites of drug action at synapses. Activating agents *(agonists)* include those drugs which, upon binding to the receptor, produce a response similar to that caused by the neurotransmitter released at synapse. On the other hand, blocking agents *(antagonists)* bind to the receptors but are unable to activate them. By occupying the receptors, however, they prevent binding by the neurotransmitter.

muscles appearing early in the disease are responsible for the common name, lockjaw.

Presynaptic inhibition

As discussed above, the activation of an inhibitory synapse hyperpolarizes the postsynaptic cell, thereby decreasing the effect of all excitatory inputs to that cell. A second type of inhibitory action, called *presynaptic inhibition*, provides a means by which only certain excitatory inputs are depressed. Presynaptic inhibition works by affecting the transmission at a single excitatory synapse. The structures underlying presynaptic inhibition consist of a synaptic junction between the terminal of the inhibitory fiber (*A* in Fig. 8-40) and the synaptic terminal of the excitatory neuron (neuron *B*). When activated, this synapse reduces the amount of the transmitter released by neuron *B*. Therefore, the size of the EPSP in the postsynaptic cell (neuron *C*) will be less influenced by input from neuron *B*. The mechanisms of presynaptic inhibition are unknown.

Figure 8-40. Presynaptic inhibition.

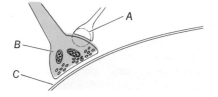

Neuromodulators

The concept of neuromodulation is new and still not clearly delineated. Neuromodulators do not, themselves, function as specific neurotransmitters but rather amplify or dampen neuronal activity by altering neurons directly or by influencing the effectiveness of a neurotransmitter. They might do the latter by affecting the neurotransmitter's synthesis, release, interactions with receptor sites, reuptake, or metabolism. For example, the hormone cortisol, by influencing the steady-state levels of the rate-limiting enzyme in the synthesis of norepinephrine, modulates the activity of norepinephrine -mediated synapses.

From this example, it is clear that neuromodulators may be hormones, i.e., substances which are secreted by endocrine glands into the blood via which they reach their distant target cells (Chap. 9). However, not all hormones act as neuromodulators, nor are all neuromodulators hormones. The others are paracrine agents which are synthesized and released locally. In this last regard we do not know whether it is possible for a neuron to secrete both neurotransmitters and neuromodulators. It is also unclear whether a given substance may act both as a neurotransmitter in one area of the nervous system and as a neuromodulator in the same or different area.

Table 8-5 lists some of those substances presently thought to function as neuromodulators. Trying to memorize such a table is of no value. We have presented it only to emphasize a particular point: Comparison of this table with Table 8-4 (neurotransmitters) and Table 9-1 (hormones) emphasizes the overlapping functions of neuromodulators. In this regard, each of the neuromodulators shown in the table will be encountered in a nonneuromodulator context elsewhere in the book; i.e., all presently known neuromodulators also exert nonneural effects. All this emphasizes one of the most exciting developments in physiology: A given chemical messenger may be synthesized in differing sites and may serve as a neurotransmitter, paracrine agent, or hormone. The brain and gastrointestinal tract, in particular, seem to produce many of the same peptide messengers.

Neuroeffector communication

To complete our analysis of the steps involved in neural control, we must at least mention neuroeffector communication. Efferent neurons innervate muscle or gland cells. Information is transmitted from axons to these effector cells by means of chemical transmitters. When an action potential reaches the terminal of an axon, it causes the release of transmitter which diffuses to the effector cell and alters its activity. The basic mechanisms are essentially the same as at synapses. The structures of these axon terminals and their anatomical relationships to the effector cells vary, depending upon the effector cell type; they will be described in detail in subsequent chapters. The neuroeffector transmitters are well

Table 8-5. Some possible neuromodulators

Histamine
Prostaglandins
Enkephalins
Endorphins
Substance P
Somatostatin
Aldosterone
Cortisol
Estrogens
Testosterone
Thyroid hormone
Gastrin
Vasoactive intestinal peptide (VIP)
Adrenocorticotropic hormone (ACTH)
Thyrotropin-releasing hormone (TRH)
Gonadotropin-releasing hormone (Gn-RH)
Vasopressin (antidiuretic hormone, ADH)
Angiotensin
Oxytocin

characterized; they are either acetylcholine or norepinephrine.

Patterns of neural activity

The purpose of this section is to illustrate through a few specific examples some of the varying degrees of complexity exhibited by neural control systems and to review the neural mechanisms presented in the previous sections of this chapter. We start with one of the simpler reflexes—the flexion reflex.

The flexion reflex

The sequence of events elicited by a painful stimulus to the toe leads to withdrawal of the foot from the source of injury: this is an example of the flexion, or withdrawal, reflex (Fig. 8-41). Receptors in the toe are stimulated and, via receptor potentials, transform the energy of the stimulus into the electrochemical energy of action potentials in afferent neurons. Information about the intensity of the stimulus is coded both by the frequency of action potentials in single afferent fibers and by the number of afferent fibers activated. The afferent fibers branch after entering the spinal cord, each branch terminating at a synaptic junction with another neuron. In the flexion reflex, the

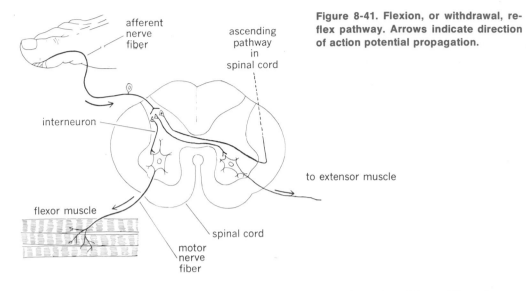

Figure 8-41. Flexion, or withdrawal, reflex pathway. Arrows indicate direction of action potential propagation.

second neurons in the pathway are interneurons. Because interneurons are interspersed between the afferent and efferent limbs of the reflex arc, the flexion reflex is one of the very large class of *polysynaptic* reflexes.

The afferent-neuron branches serve different functions. Some branches synapse with interneurons whose processes carry the information to the brain; it is only after the information transmitted in these pathways reaches the brain that the conscious correlate of the stimulus, i.e., the sensation of pain, is experienced. Other branches synapse with interneurons which, in turn, synapse upon the efferent neurons innervating flexor muscles. These muscles, when activated, cause flexion (bending) of the ankle and withdrawal of the foot from the stimulus. If the stimulus is very intense and the afferent discharge of very high frequency, the number of motor neurons which fires action potentials is large, and muscles of other joints are activated so that the knee and thigh also flex.

Still other branches of the afferent neurons activate interneurons which inhibit the motor neurons of antagonistic muscles whose activity would oppose the reflex flexion. This inhibition of antagonistic motor neurons is called *reciprocal innervation*. In this case, the motor neurons innervating the foot and leg extensor muscles (which straighten the ankle and leg) are inhibited. Other interneurons excite extensor and inhibit flexor motor neurons controlling movement of the op-

posite leg so that as the injured leg is flexed away from the stimulus, the opposite leg is extended more strongly to support the added share of the body's weight. This involvement of the opposite limb is called the *crossed-extensor reflex*.

The neural networks activated in the flexion reflex are among the simplest in humans, yet the reflex response is purposeful and coordinated. The foot flexion away from the stimulus removes the receptors from the stimulus source, and further damage is prevented (an example of negative feedback). The purposefulness of the reflex response is independent of the sensation of pain, for the withdrawal occurs before the sensation of pain is experienced. In fact, the reflex response can still be elicited after the spinal cord pathways which would normally transmit the afferent information up to the brain have been severed. An important characteristic of this type of control system is that the observed response varies with stimulus strength (between threshold and maximal stimulus strength). In general, low levels of stimulation lead to responses localized to the area of stimulus application whereas higher levels of stimulation cause more widespread responses. Thus, to take another example, the hotter an object, the more intense is the response to touching it; first one simply withdraws the finger; if it is hotter, the wrist and arm are pulled away; if hotter still, one jumps away with appropriate exclamations.

Swallowing

A swallow is brought about by the coordinated interaction of some 20 muscles whose efferent neurons' cell bodies lie in the lower part of the brain. If food is to be moved from the mouth to the stomach, the muscles of the throat and esophagus must contract in a precise sequence over time, the upper muscles contracting first to push the food downward while the lower regions of the esophagus are relaxed to allow the food to enter. After the food enters these lower regions, the muscles there must contract to push the food still farther down. The contractions of the upper regions must be maintained during this time to prevent the food from moving back up. And, during a swallow, the respiratory passages must be protected so that food does not enter them. In order for each of these muscles to contract, they must be activated by the motor neurons specific to them. Accordingly, swallowing requires an orderly sequence of excitation of many neurons.

Thus, a single brief stimulus triggers off a chain of responses which may last more than 10 s. Moreover, once initiated, the course of the swallow is largely unaffected by further sensory input. Contrast this to the flexion reflex in which the stimulus induces a very brief and nearly simultaneous response. Clearly, there must be a far more complex network of interacting neurons built into the swallowing reflex to ensure the timing of the response.

Respiratory movements

The control system which mediates respiratory movements has one component which is qualitatively different from any present in the flexion reflex and swallowing, namely, a group of neurons whose membranes manifest rhythmic spontaneous depolarization. In the absence of any external input, the membranes of these neurons undergo rhythmic depolarizations (pacemaker potentials) which cause periodic bursts of action potentials. This probably occurs because of a gradual change in membrane permeabilities, either an increasing sodium or decreasing potassium permeability. These neurons, which are the key neurons in controlling respiration, are strongly influenced by afferent input, but this input only plays upon a basic rhythm which is generated spontaneously.

Spontaneously depolarizing neurons such as these provide, as it were, "biological clocks" for the nervous system and may be of considerable importance in other rhythmic activities, including the reproductive cycle, daily variations in hormone secretion, etc. Above all, the question of spontaneous activity of neurons may be basic to one's view of voluntary behavior and will be discussed again in Chap. 20.

This completes our analysis of the functional characteristics of the nervous system, and for purposes of general orientation we now turn to a brief description of its anatomy. The functions of each component of the nervous system, merely stated in the coming section, will all be discussed more fully in future chapters.

Section C
Divisions of the nervous system

The various parts of the nervous system are interconnected, but for convenience they can be divided into the *central nervous system*, composed of the brain and spinal cord, and the *peripheral nervous system*, consisting of the nerves extending from the brain and spinal cord (Fig. 8-42).

Central nervous system

Spinal cord

The spinal cord is a slender cylinder about as big around as the little finger. Figure 8-43 shows the basic division of the internal structures of the cord. The central butterfly-shaped area is the *gray matter*; it is filled with interneurons, the cell bodies and dendrites of efferent neurons, the entering fibers of afferent neurons, and glial cells. The cell bodies of neurons in the gray matter are often clustered together with other neurons having similar functions into groups called *nuclei*. The gray region is surrounded by *white matter*, which consists largely of bundles of myelinated nerve fibers (*tracts* and *pathways*) running longitudinally through the cord, some descending to convey information from the brain to the spinal cord (or from upper to lower levels of the cord), others ascending to transmit in the opposite direction. The

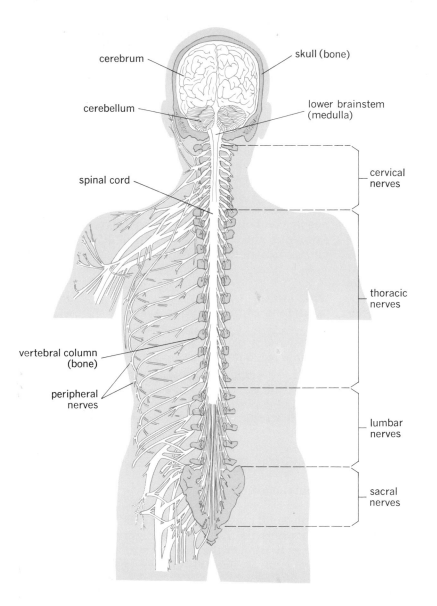

cerebrum

skull (bone)

cerebellum

lower brainstem
(medulla)

cervical
nerves

spinal cord

thoracic
nerves

vertebral column
(bone)

peripheral
nerves

lumbar
nerves

sacral
nerves

Figure 8-42. Nervous system viewed from behind. *(Adapted from Woodburne.)*

fiber bundles are so organized that a given tract or pathway generally contains fibers transmitting one type of information. For example, the fibers which transmit information from light-touch receptors in the skin travel together in one pathway. Some of these spinal-cord pathways are illustrated in Fig. 8-44. The tracts and even the nuclei can be visualized as columns extending the length of the spinal cord.

Groups of afferent fibers enter the spinal cord on the dorsal side (the side toward the back of the body); these groups form the *dorsal roots* and contain the *dorsal root ganglia* (the cell bodies of the afferent neurons). Efferent fibers leave the spinal cord on the opposite side via the *ventral roots.* Shortly after leaving the cord, the ventral and dorsal roots from the same level combine to form a pair of *spinal nerves,* one on each side of the spinal cord. (The spinal nerves are discussed with the peripheral nervous system.)

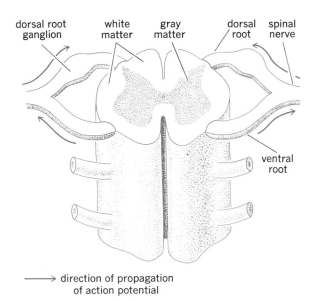

Figure 8-43. Section of the spinal cord (ventral view, i.e., from the front).

Brain

The brain is composed of six subdivisions: the *cerebrum, diencephalon, midbrain, pons, medulla,* and *cerebellum* (Fig. 8-45 and Table 8-6). The midbrain, pons, and medulla together form the *brainstem;* the cerebrum and diencephalon together constitute the *forebrain*.

Brainstem. The brainstem is literally the stalk of the brain, through which pass all the nerve fibers relaying signals of afferent input and efferent output between the spinal cord and higher brain centers. In addition, the brainstem contains the cell bodies of neurons whose axons go out to the periphery to innervate the muscles and glands of the head. It gives rise to 10 of the 12 pairs of *cranial nerves.* The brainstem also receives many afferent fibers from the head and visceral cavities. Running through the entire brainstem is a core of tissue called the *reticular formation* which is composed of a diffuse collection of small, many-branched neurons. The neurons of the reticular formation receive and integrate information from many afferent pathways as well as from many other regions of the brain. Some reticular-formation neurons are clustered together, forming certain of the brainstem nuclei and "centers." We have already discussed the neurons which dictate the stereotyped pattern of swallowing; they make up the swallowing center, which is located in the reticular formation, and there are cardiovascular, respiratory, and vomiting centers, too.

The output of the reticular formation can be divided functionally into descending and ascending systems. The descending components influence both somatic and autonomic efferent neurons and frequently afferent neurons, as well; the ascending components affect such things as

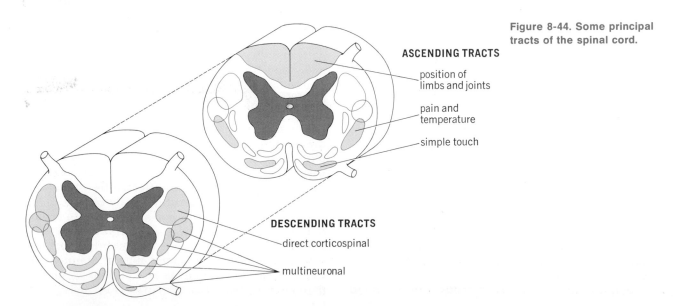

Figure 8-44. Some principal tracts of the spinal cord.

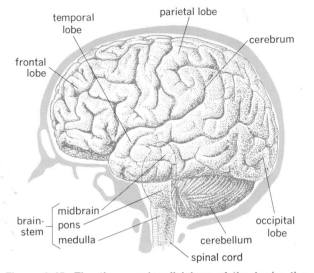

Figure 8-45. The three major divisions of the brain: the brainstem, cerebellum, and forebrain. The outer layer of the forebrain (the cerebrum) consists of four lobes, as shown.

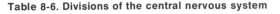

Table 8-6. Divisions of the central nervous system

Brain	Cerebrum; cerebral hemispheres (includes cerebral cortex and basal ganglia) Diencephalon (includes thalamus and hypothalamus)	Forebrain
	Midbrain Pons Medulla Cerebellum	Brainstem
Spinal cord		

wakefulness and the direction of attention to specific events.

Cerebellum. The cerebellum is chiefly involved with skeletal muscle functions; it helps to maintain balance and provide smooth, directed movements (Chap. 19). The location of the cerebellum relative to the brainstem and forebrain can be seen in Fig. 8-45.

Forebrain. The large part of the brain remaining when the brainstem and cerebellum have been excluded is the forebrain. It consists of a central core, the *diencephalon,* and right and left *cerebral hemispheres* (the *cerebrum*). The hemispheres are connected to each other by fiber bundles known as *commissures,* the *corpus callosum* (Fig. 8-46) being the largest. Areas within a single hemisphere are connected to each other by association fibers.

The outer portion of the cerebral hemispheres, the *cerebral cortex,* is a cellular shell about 3 mm thick covering the entire surface of the cerebrum. It forms the outer rim of the brain cross section in Fig. 8-46. Popular opinion calls the cortex the "site of the mind and the intellect." Scientific opinion considers it to be an integrating area necessary for the bringing together of basic afferent information into complex perceptual images and

ultimate refinement of control over all efferent systems. The cortex is divided into several parts, or *lobes* (frontal, parietal, occipital, and temporal [Fig. 8-45]), the functions of which will be discussed in later chapters. The cortex is an area of gray matter, so called because of the predominance of cell bodies. In other parts of the forebrain, nerve-fiber tracts predominate, their whitish myelin coating distinguishing them as white matter.

The *subcortical nuclei* form areas of gray matter lying, as the name suggests, under the surface of the cortex (Fig. 8-46) and contribute to the coordination of muscle movements. The *thalamus,* part of the diencephalon (Fig. 8-46), is a way station and important integrating center for all sensory input (except smell) on its way to the cortex. It also contains a central core, which is part of the reticular system. The *hypothalamus,* which lies below the thalamus, is a tiny region whose volume is about 5 to 6 cm³; it is responsible for the integration of many basic behavioral patterns which involve correlation of neural and endocrine function (Fig. 8-46). Indeed, the hypothalamus appears to be the single most important control area for regulation of the internal environment. It is also one of the brain areas associated with emotions; stimulation of some hypothalamic areas leads to behavior interpreted as rewarding or pleasurable, and stimulation of other areas is associated with unpleasant feelings. Neurons of the hypothalamus are also affected by a variety of hormones and other circulating chemicals.

The *limbic system* is not a single brain region but is an interconnected group of brain structures within the cerebrum, including portions of the frontal-lobe cortex, temporal lobe, thalamus, and hypothalamus, as well as the circuitous neuron pathways connecting all parts (Fig. 8-47). The

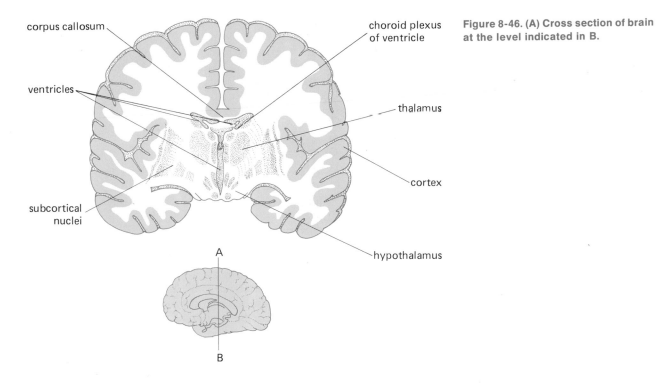

corpus callosum

choroid plexus
of ventricle

ventricles

thalamus

cortex

subcortical
nuclei

hypothalamus

A

B

**Figure 8-46. (A) Cross section of brain
at the level indicated in B.**

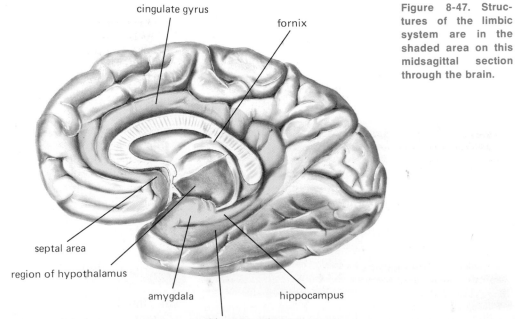

cingulate gyrus

fornix

septal area

region of hypothalamus

amygdala

hippocampus

parahippocampal gyrus

**Figure 8-47. Struc-
tures of the limbic
system are in the
shaded area on this
midsagittal section
through the brain.**

Table 8-7. Divisions of the peripheral nervous system

I Afferent
II Efferent
  a Somatic
  b Autonomic
    1 Sympathetic
    2 Parasympathetic

limbic system is concerned with emotional behavior and learning. Besides being connected with each other, the parts of the limbic system have connections with many other parts of the central nervous system. For example, it is likely that information from all the different afferent modalities can influence activity within the limbic system, whereas activity of the limbic system can result in a wide variety of automatic responses and body movements. This should not be surprising since emotional feelings are accompanied by automatic responses such as sweating, blushing, and heart-rate changes, and by somatic responses such as laughing and sobbing.

Peripheral nervous system

Nerve processes (axons) in the peripheral nervous system are grouped into bundles called *nerves*, whereas the individual processes are *nerve fibers.* (There are no "nerves" in the central nervous system; recall that the bundles there are called tracts or pathways.) The peripheral nervous system consists of 43 pairs of nerves; 12 pairs connect with the brain and are called the *cranial nerves,* and 31 pairs connect with the spinal cord as the *spinal nerves.*

Each nerve fiber is surrounded by a type of glial cell (called a Schwann cell in the peripheral nervous system and an oligodendrocyte in the central nervous system). Some of the axons are wrapped in layers of Schwann-cell (or oligodendroglial) membrane, and these tightly wrapped membranes form the myelin sheath. Other fibers are unmyelinated and are simply tucked into invaginations of the glial cells. The glial cells forming the myelin sheath are like elongated beads strung on a slender thread (the axon), the narrow spaces between consecutive beads representing the nodes of Ranvier.

The individual nerve fibers in a nerve may be processes of either afferent or efferent neurons; accordingly they may be classified as belonging to the *afferent* or *efferent division* of the peripheral nervous system (Table 8-7). All the spinal nerves and some of the cranial nerves contain processes of both afferent and efferent neurons.

Peripheral nervous system: Afferent division

Afferent neurons convey information from receptors in the periphery to the central nervous system. Regardless of whether the initial stimulus is a prick of the skin, stretch of skeletal muscle, distention of the intestine, or a loud sound, the action potentials relaying information to the central nervous system travel along neurons which are structurally very similar. The neurons described in the receptor section of this chapter are typical afferent neurons, and so are the neurons which constitute the afferent pathways of the flexion and swallowing reflexes.

The cell bodies of afferent neurons are in structures called *ganglia* which are outside but close to the brain or spinal cord. From the region of the cell body, one long process extends away from the ganglion to innervate the receptors; commonly, the process branches several times as it nears its destination, each branch containing or innervating one receptor. A second process passes from the cell body into the central nervous system where it branches; these branches terminate in synaptic junctions onto other neurons.

Afferent neurons are sometimes called *primary afferents* or *first-order neurons* because they are the first cells entering the central nervous system in the synaptically linked chains of neurons which handle incoming information. They are also frequently called sensory neurons, but we hesitate to use this term because it implies that the information transmitted by these neurons is destined to reach consciousness, and this is not always true. For example, we have no conscious awareness of our blood pressure even though we have receptors sensitive to this variable.

Peripheral nervous system: Efferent division

The efferent division is more complicated than the afferent, being subdivided into a *somatic nervous*

Table 8-8. Differences between somatic and autonomic nervous systems

Somatic nervous system

1 Fibers do not synapse once they have left the central nervous system.
2 Innervates skeletal muscle.
3 Always leads to excitation of the muscle.

Autonomic nervous system

1 Fibers synapse once in ganglia after they have left the central nervous system.
2 Innervates smooth or cardiac muscle or gland cells.
3 Can lead to excitation or inhibition of the effector cells.

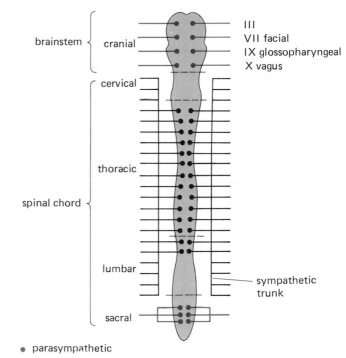

• parasympathetic
• sympathetic

Figure 8-48. Origins of the sympathetic and parasympathetic divisions of the autonomic nervous system.

system and an *autonomic nervous system.* These terms are somewhat unfortunate because they conjure up additional "nervous systems" distinct from the central and peripheral nervous systems. Keep in mind that the terms refer simply to the efferent divisions of the peripheral nervous system. Although separating these two divisions is justified by many anatomical and physiological differences, the simplest distinction between the two is that the somatic nervous system innervates skeletal muscle and the autonomic nervous system innervates smooth and cardiac muscle and glands. Other differences are listed in Table 8-8.

Somatic nervous system. The somatic division of the peripheral nervous system is made up of all the fibers going from the central nervous system to skeletal muscle cells. The cell bodies of these neurons are located in groups within the brain or spinal cord; their large-diameter, myelinated axons leave the central nervous system and pass directly, i.e., without any synapses, to skeletal muscle cells. The transmitter substance released by these neurons is *acetylcholine.* Because activity of somatic efferent neurons causes contraction of the innervated skeletal muscle cells, these neurons are often called *motor neurons.* Motor neurons can be activated by local reflex mechanisms, as in the flexion reflex, or they can be activated by pathways which descend from higher brain centers, but in either case, their excitation always leads to *contraction* of skeletal muscle cells; there are no *inhibitory* somatic motor neurons.

Autonomic nervous system. Fibers of the autonomic division of the peripheral nervous system innervate cardiac muscle, smooth muscle, and glands. Anatomic and physiologic differences within the autonomic nervous system are the basis for its further subdivision into *sympathetic* and *parasympathetic* components. The cell bodies of the first neurons in the two subdivisions are located in different areas of the central nervous system, and their fibers leave at different levels, the sympathetic from the thoracic and lumbar regions of the spinal cord, and the parasympathetic from brain and the sacral portion of the spinal cord (Fig. 8-48). Thus, the sympathetic division is also called the *thoracolumbar division,* and the parasympathetic the *craniosacral division.*

The fibers of the autonomic nervous system synapse once after they have left the central nervous system and before they arrive at the effector cells (Fig. 8-49). These synapses outside the central nervous system occur in cell clusters called ganglia. The two divisions of the autonomic nervous system differ with respect to the locations

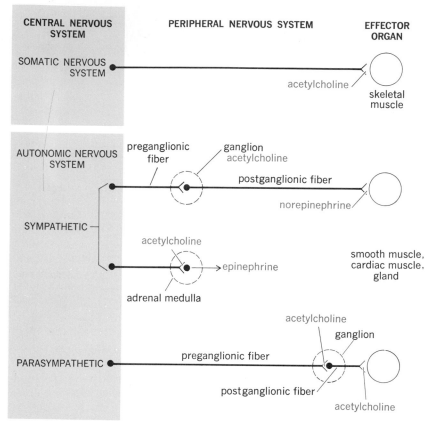

Figure 8-49. Efferent divisions of the peripheral nervous system.

of their ganglia. Many of the sympathetic ganglia lie close to the spinal cord, forming the paired chains of ganglia known as the *sympathetic trunks,* and others lie halfway between the spinal cord and innervated organ. In contrast, the parasympathetic ganglia lie within the walls of the effector organ. The fibers passing between the central nervous system and the sympathetic or parasympathetic ganglia are the *preganglionic* autonomic fibers; those passing between the ganglia and the effector organ are the *postganglionic* fibers.

In both sympathetic and parasympathetic divisions, the chemical transmitter for the ganglionic synapse between pre- and postganglionic fibers is *acetylcholine.* The chemical transmitter in the parasympathetic division between the postganglionic fiber and the effector cell is also acetylcholine. (Fibers that release acetylcholine are called *cholinergic* fibers.) The transmitter in the sympathetic division between the post-ganglionic fiber

and the effector cell is *norepinephrine* (Fig. 8-49).[1] Because early experiments on the function of the sympathetic division were done using epinephrine rather than the closely related norepinephrine, and because the epinephrine was called by its British name, adrenaline, fibers which release norepinephrine came to be called *adrenergic* fibers.

Many drugs stimulate or inhibit various components of the autonomic nervous system. An important site of action for certain of them is the receptor for the neurotransmitters. Indeed, study of these drugs has revealed that there are several different types of receptors for each transmitter. For example, acetylcholine receptors on the postsynaptic neurons of all autonomic ganglia also respond to the drug nicotine and are therefore *nicotinic receptors.* In contrast, the acetylcholine

---

[1] There are a few exceptions to this statement, and they will be identified where appropriate.

receptors on smooth muscle and gland cells are stimulated not by nicotine but by the mushroom poison, muscarine; they are called *muscarinic receptors*. Similarly, there are two classes of *adrenergic receptors* (i.e., receptors for epinephrine and norepinephrine) also distinguished by the specific drugs which stimulate or block them. They are called *alpha-adrenergic* and *beta-adrenergic receptors*.

One "ganglion" in the sympathetic division never developed long postganglionic fibers; instead, upon activation of its preganglionic nerves, the cells of this "ganglion" discharge their transmitters into the blood stream. This "ganglion," called the *adrenal medulla,* is therefore an endocrine gland (Chap. 9). It releases a mixture of about 80 percent epinephrine and 20 percent norepinephrine. These substances, in this case more properly called hormones rather than neurotransmitters, are transported via the blood to receptor sites on effector cells sensitive to them. The receptor sites may be the same ones that sit beneath the terminals of sympathetic postganglionic neurons and are normally activated by the transmitter delivered directly to them, or they may be other, noninnervated sites which are activated only by the circulating epinephrine or norepinephrine.

The heart and many glands and smooth muscles are innervated by both sympathetic and parasympathetic nerve fibers, i.e., they receive *dual innervation*. (Do not confuse this with reciprocal innervation.) Whatever one division does to the effector organ, the other division frequently (but not always) does just the opposite (Table 8-9). For example, action potentials arriving over the sympathetic nerves to the heart increase the heart rate, whereas action potentials arriving over the parasympathetic fibers decrease it. In the intestine, activation of the sympathetic fibers reduces contraction of the smooth muscle in the intestinal wall, whereas the parasympathetics increase contraction. Dual innervation with fibers of opposite action provides for a very fine degree of control over the effector organ, for it is like equipping a car with both an accelerator and a brake. With only an accelerator, one could slow the car simply by decreasing the pressure on the accelerator, but the combined effects of releasing the accelerator and applying the brake provide faster and more accurate control. To prevent the sympathetic and

parasympathetic divisions' opposing effects from conflicting with each other, the two divisions are usually activated reciprocally; i.e., as the activity of one division is enhanced, the activity of the other is depressed. Skeletal muscle cells, in contrast, do not receive dual innervation; they are activated by somatic motor neurons, which are only excitatory.

Because glands, smooth muscles, and the heart participate as effectors in almost all bodily functions, it follows that the autonomic nervous system has an extremely widespread and important role in the homeostatic control of the internal environment. It exerts a wide array of effects (Table 8-9) which can be very difficult to remember. Some common denominators are that, in general, the sympathetic division helps the body to cope with challenges from the outside environment, and the parasympathetic seems to be more responsible for internal housekeeping, such as digestion, defecation, and urination. The sympathetic division is utilized in situations involving stress or strong emotions such as fear or rage, whereas the parasympathetic division is most active during recovery or at rest. The sympathetic division provides the responses to a situation leading to "fight or flight." For example, the sympathetic division increases blood flow to exercising muscles and sustains blood pressure in case of severe blood loss; it decreases activity of the gastrointestinal tract, increases the metabolic production of energy, and increases sweating, changes which provide energy utilization most appropriate to the emergency. These and many other functions of the sympathetic (and parasympathetic) division will be discussed in relevant places throughout the book.

Autonomic responses usually occur without conscious control or awareness as though they were indeed autonomous (in fact, the autonomic nervous system has been called the involuntary nervous system). However, it is wrong to assume that this need always be the case; it has been shown that discrete visceral and glandular responses can be learned. For example, to avoid an electric shock, a rat can learn to selectively increase or decrease its heart rate and a rabbit can learn to constrict the vessels in one ear while dilating those in the other. The implications of such voluntary control of autonomic functions in human medicine are enormous.

These experiments are also important in showing

Table 8-9. Some effects of autonomic nervous system activity*

Effectors	Sympathetic nervous system	Parasympathetic nervous system
Eye		
Muscles of the iris	Contracts radial muscle (widens pupil)	Contracts sphincter muscle (makes pupil smaller)
Ciliary muscle	Relaxation (tightens suspensory ligaments thus flattening lens for far vision)	Contraction (relaxes ligament allowing lens to become more convex for near vision)
Heart		
S-A node	Increases heart rate	Decreases heart rate
Atria	Increases contractility	Decreases contractility
A-V node	Increases conduction velocity	Decreases conduction velocity
Ventricles	Increases contractility	———
Arterioles		
Coronary	Constriction	Dilation
Skin and mucous membrane	Constriction	———
Skeletal muscle	Constriction or dilation	———
Abdominal viscera and kidneys	Constriction	———
Salivary glands	Constriction	———
Penis or clitoris	Constriction	Dilation (causes erection)
Veins	Constriction	———
Lung		
Bronchial muscle	Relaxation	Contraction
Bronchial glands	Inhibits secretion	Stimulates secretion
Salivary glands	Stimulates secretion	Stimulates secretion

that small segments of the autonomic response can be regulated independently; thus, overall autonomic responses, made up of many small components, are quite variable. Rather than being gross, undiscerning discharges, they are finely tailored to the specific demands of any given situation.

Blood supply, blood-brain barrier phenomena, and cerebrospinal fluid

Glucose is the only substrate that can usually be metabolized sufficiently rapidly by the brain to supply its energy requirements, and most of the energy from glucose is transferred to high-energy ATP molecules during its oxidative breakdown. The glucogen stores of brain are negligible; thus the brain is completely dependent upon a continuous blood supply of glucose and oxygen. Although the adult brain is only 2 percent of the body weight, it receives 15 percent of the total blood supply at rest to support the high oxygen utilization. If the oxygen supply is cut off for 4 to 5 min or if the glucose supply is cut off for 10 to 15 min, brain damage will occur. In fact, the most common

cause of brain damage is stoppage of the blood supply (a *stroke*). The cells in the region deprived of nutrients cease to function and die.

The central nervous system receives a rich blood supply, but the exchange of substances between the blood and neurons in the central nervous system is handled differently from the somewhat unrestricted movement of substances from capillaries in other organs. A complex group of *blood-brain barrier* mechanisms closely controls both the kinds of substances which enter the extracellular space of brain and the rate at which they enter. The barrier probably comprises both anatomical structures and physiological transport systems which handle different classes of substances in different ways. The blood-brain barrier mechanisms precisely regulate the chemical composition of the extracellular space of the brain and prevent harmful substances from reaching neural tissue. However, certain areas of the brain lack a blood-brain barrier and can be readily influenced by blood-borne substances.

In addition to its blood supply, the central nervous system is perfused by a second fluid, the *cerebrospinal fluid*. This clear fluid fills the four cavities (*ventricles*) (Fig. 8-50) within the brain

Table 8-9. (*continued*)

Effectors	Sympathetic nervous system	Parasympathetic nervous system
Stomach		
Motility	Decreases	Increases
Sphincters	Contraction	Relaxation
Secretion	Possibly inhibition	Stimulation
Intestine		
Motility	Decreases	Increases
Sphincters	Contraction	Relaxation
Secretion	Possibly inhibition	Stimulation
Gallbladder and ducts	Relaxation	Contraction
Liver	Glycogenolysis, gluconeogenesis	——
Pancreas		
Exocrine glands	Decreased secretion	Stimulates secretion
Endocrine glands (islets)	Inhibits insulin secretion, stimulates glucagon secretion	Stimulates insulin secretion
Fat cells	Stimulates lipid breakdown	——
Urinary bladder	Relaxation	Contraction
Uterus	Pregnant: contraction Nonpregnant: relaxation	Variable
Reproductive tract (male)	Ejaculation	——
Skin		
Muscles causing hair to stand erect	Contraction	——
Sweat glands	Stimulates secretion	Stimulates secretion
Lacrimal glands	——	Stimulates secretion

* Table adapted from Louis S. Goodman and Alfred Gilman, "The Pharmacological Basis of Therapeutics," 5th ed., Macmillan, New York, 1975.

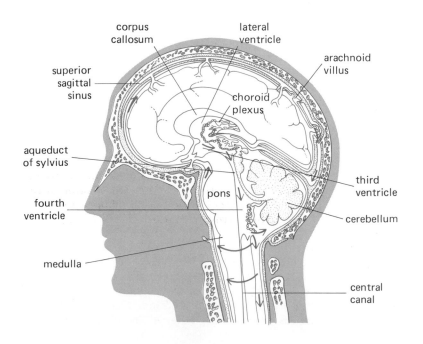

Figure 8-50. The ventricular system of the brain and the distribution of cerebrospinal fluid. (The second lateral ventricle is not shown.) Cerebrospinal fluid is formed in the ventricles (much of it by the choroid plexuses), passes to the subarachnoid space outside the brain, and then through the arachnoid villi which are valvelike structures, into large veins of the head. (*Adapted from Kuffler and Nicholls.*)

and surrounds the outer surfaces of the brain and spinal cord so that the central nervous system literally floats in a cushion of cerebrospinal fluid. Since the brain and spinal cord are soft and delicate tissue about the consistency of jelly, they are thus protected from sudden and jarring movements. Cerebrospinal fluid is formed within the ventricles by highly vascular tissues, the *choroid plexes*. A blood-brain barrier is present here, too, between the capillaries of the choroid plexes and the cerebrospinal fluid which flows next to the neural tissue. Consistent with the barrier mechanisms, cerebrospinal fluid is a selective secretion, not a simple filtrate of plasma. For example, protein, potassium, and calcium concentrations are lower in cerebrospinal fluid than in plasma, whereas the sodium and chloride are higher. The mechanisms responsible for this selective formation of spinal fluid are not known.

As the cerebrospinal fluid flows from its origin at the choroid plexes within the ventricles, substances diffuse between it and the extracellular space of the neural tissue since the walls of the ventricles are permeable to most substances. This exchange across the ventricular walls allows nutrients to enter brain tissue from cerebrospinal fluid and end products of brain metabolism to leave. However, the major sites of nutrient and end-product exchange are the cerebral capillaries. The cerebrospinal fluid moves from its origin, through the interconnected ventricular system to the brainstem, where it passes through small holes from the ventricle out to the surface of the brain and spinal cord. Aided by circulatory, respiratory, and postural pressure changes, the cerebrospinal fluid finally flows to the top of the outer surface of the brain, where it enters large veins through one-way valves. If the path of flow is obstructed at any point between its site of formation and its final reabsorption into the vascular system, cerebrospinal fluid builds up, causing hydrocephalus, or "water on the brain." The resulting pressure in the ventricles may be high enough to damage the brain and cause mental retardation in severe, untreated cases.

9
HORMONAL CONTROL MECHANISMS

Hormone–target-cell specificity
General factors which determine the plasma concentrations
    of hormones
  Rate of secretion
  Rates of inactivation and excretion
  Transport in the blood
  Activation of hormones after secretion
  Summary example: thyroxine
Mechanisms of hormone action
  Cell responses to hormones
  Hormone receptors
  Events elicited by hormone-receptor binding
    Steroid hormones
  Hormone interactions on target cells

  Pharmacological effects
Control of hormone secretion
  The anterior pituitary hormones
  Control of anterior pituitary hormone secretion
  Hypothalamus–posterior-pituitary function
  Epinephrine and the adrenal medulla
  Role of the autonomic nervous system in hormone
    secretion
  Direct control of hormone secretion by plasma ions or
    nutrients
  Summary
Multiple hormone secretion
  Steroid synthesis

Although the term "endocrine" broadly denotes any ductless gland, we shall use it in its more usual restricted sense to include only those glands whose secretory products are hormones; thus, for example, the liver, which primarily secretes non-hormonal materials (glucose and other metabolites) into the blood, is generally excluded from this category. A *hormone* is defined as a chemical substance synthesized by an endocrine gland and secreted into the blood, which carries it to other sites in the body where its actions are exerted. In terms of chemical structure, hormones generally fall into four categories: steroids, amino acid derivatives, peptides, and proteins.

Figure 9-1 illustrates the locations of the major endocrine glands. Clearly, the endocrine system differs from most of the other organ systems of the body in that the various glands are not in anatomic continuity with each other; however, they do form a system in the functional sense. It should also be noted that some of the glands form completely distinct organs (the pituitary, for example)

whereas others are found within larger organs having nonendocrine functions as well (the gonads and pancreas, for example).

The endocrine system constitutes the second great communications system of the body, the hormones serving as blood-borne messengers which control and integrate many bodily functions: reproduction (Chap. 16), organic metabolism and energy balance (Chap. 15), and mineral metabolism (Chap. 13).

In recent years, it has become clear that the nervous and endocrine systems actually function as a single interrelated system. The central nervous system, particularly the hypothalamus, plays a crucial role in controlling hormone secretion, and conversely, hormones markedly alter neural function and strongly influence many types of behavior. These interrelationships form the area of study known as *neuroendocrinology.*

Table 9-1 summarizes the physiology of the major hormones. As emphasized above, the hormones function as components of the body's con-

Table 9-1. Summary of the major hormones

Gland	Hormone	Major function is control of:
Hypothalamus	Releasing hormones	Secretion by the anterior pituitary
	Oxytocin	(See posterior pituitary)
	Antidiuretic hormone	(See posterior pituitary)
Anterior pituitary	Growth hormone (somatotropin, GH, STH)*	Growth; organic metabolism
	Thyroid-stimulating hormone (TSH, thyrotropin)	Thyroid gland
	Adrenocorticotropic hormone (ACTH, corticotropin)	Adrenal cortex
	Prolactin	Breasts (milk synthesis)
	Gonadotropic hormones:	Gonads (gamete production and sex hormone synthesis)
	Follicle-stimulating hormone (FSH)	
	Luteinizing hormone (LH)	
Posterior pituitary†	Oxytocin	Milk secretion; uterine motility
	Antidiuretic hormone (ADH, vaso-pressin)	Water excretion
Adrenal cortex	Cortisol	Organic metabolism; response to stresses
	Androgens	Growth and, in women, sexual activity
	Aldosterone	Sodium and potassium excretion
Adrenal medulla	Epinephrine	Organic metabolism; cardiovas-cular function; response to stresses
	Norepinephrine	
Thyroid	Thyroxine (T-4)	Energy metabolism; growth
	Triiodothyronine (T-3)	
	Calcitonin	Plasma calcium
Parathyroids	Parathyroid hormone (parathormone, PTH, PH)	Plasma calcium and phosphate
Gonads		
Female: ovaries	Estrogens	Reproductive system; growth and development; breasts
	Progesterone	
Male: testes	Testosterone	Reproductive system; growth and development
Pancreas	Insulin	Organic metabolism; plasma glucose
	Glucagon	
	Somatostatin	
Kidneys	Renin	Adrenal cortex; blood pressure
	Erythropoietin (ESF)	Erythrocyte production
	1,25-Dihydroxy vitamin D_3	Calcium balance
Gastrointestinal tract	Gastrin	Gastrointestinal tract; liver; pancreas; gallbladder
	Secretin	
	Cholecystokinin	
	Gastric inhibitory peptide	
	Somatostatin	
Thymus	Thymus hormone (thymosin)	Lymphocyte development
Pineal	Melatonin	?Sexual maturity

* The names and abbreviations in parentheses are synonyms.

† The posterior pituitary stores and secretes these hormones; they are synthesized in the hypothalamus.

NOTE: This table is by no means comprehensive, mainly because we have left out those substances suspected of being hormones but not conclusively established to have a function in people. Perhaps the most interesting of these omissions are the endorphins, whose roles as neurotransmitters are described in Chap. 8 but which are also secreted by the pituitary. Also not shown are the hormones secreted by the placenta (see Chap. 16).

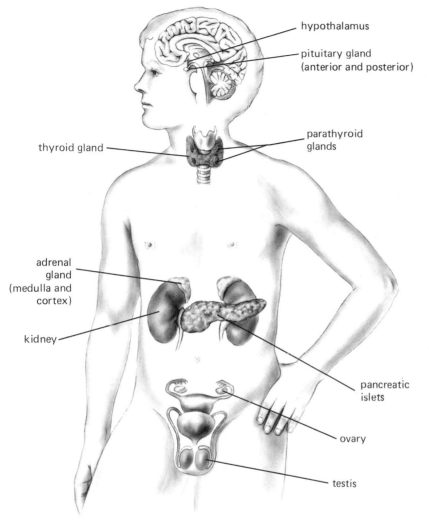

hypothalamus

pituitary gland
(anterior and posterior)

thyroid gland

parathyroid
glands

adrenal
gland
(medulla and
cortex)

kidney

pancreatic
islets

ovary

testis

Figure 9-1. Location of some endocrine glands. Note that the parathyroid glands actually lie on the posterior surface of the thyroid. Also note that this hypothetical bisexual figure has both ovaries and testes; this does not occur in normal human beings. Not shown in the figure are the pineal, thymus, and placenta.

trol systems; accordingly, we have chosen to describe their detailed physiology in later chapters in the contexts of the control systems in which they participate. The aim of this chapter is to provide the foundation for these later descriptions by presenting: (1) the general characteristics and principles which apply to almost all hormones; (2) the types of direct input which act upon endocrine-secreting cells to cause production and release of the hormones. This chapter contains many illustrative examples, but most of these scattered bits of information are given here only to explain general principles. Factual information concern-

ing each hormone will be covered systematically in subsequent chapters.

Hormone–target-cell specificity

Hormones travel in the blood and are therefore able to reach virtually all tissues. This is obviously very different from the efferent nervous system, which can send messages selectively to specific organs. Yet, the body's response to hormones is not all-inclusive but highly specific, in some cases involving only one organ or group of cells. In other

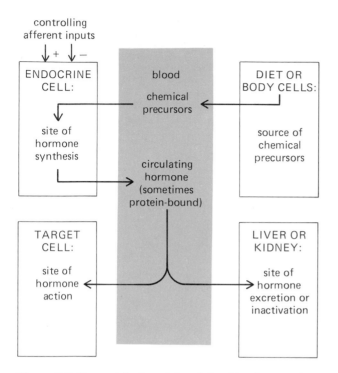

Figure 9-2. General factors determining blood concentrations of hormones.

respond depending upon the presence of specific receptors for those hormones. As described in Chap. 7, the receptor is the site in the target cell of the first interaction of the hormone with the cell. Specialization of target-cell receptors thus explains the specificity of action of hormones; e.g., thyroid-stimulating hormone is produced by the anterior pituitary and affects significantly only the thyroid gland and no other tissue because only thyroid-gland cells possess receptors for it.

General factors which determine the plasma concentrations of hormones[1]

Rate of secretion

With few exceptions, hormones are not secreted at constant rates. Indeed, as emphasized previously, it is essential that any regulatory system be capable of altering its output, and this is the case for the hormonal components of the body's regulatory systems, just as it is true for the neural components. Most, if not all, hormones are released in short bursts with little or no release occurring between bursts; accordingly, the blood concentrations of hormones may fluctuate rapidly over brief time periods (Fig. 9-3). When an endocrine gland is stimulated by appropriate inputs, the bursts of hormone release occur more frequently so that the average plasma concentration of the hormone increases; in contrast, in the absence of stimula-

words, despite the ubiquitous distribution of a hormone via the blood, only certain cells are capable of responding to the hormone; they are known as *target cells*. Cells have become differentiated so as to respond in a highly characteristic manner only to certain hormones, this ability to

[1] See Fig. 9-2.

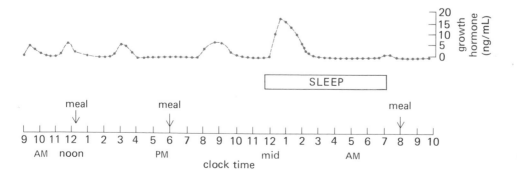

Figure 9-3. Plasma growth hormone concentrations in a 23-year-old male throughout 24 h. Blood was sampled with an indwelling catheter so that sleep would not be disturbed. [*Redrawn from Sassin et al.* (1972).]

tion or in the presence of active inhibitory inputs, the bursts decrease in frequency or stop altogether and the plasma concentration of the hormone decreases. This pattern is quite analogous to the phenomena of facilitation and inhibition manifested by neural control mechanisms.

Many hormones manifest 24-h cyclical variations in their secretory rates (Fig. 9-3). The circadian patterns for no two hormones are identical; some are clearly linked to sleep, during which increased secretion may occur. For example, Fig. 9-3 illustrates how growth hormone secretion is markedly increased during the early period of sleep. The mechanisms underlying these cycles are ultimately traceable to cyclical variations in the activity of the central nervous system, but just how these 24-h "programs" are generated and what their adaptive value is remain unknown.

Rates of inactivation and excretion

The concentration of a hormone in the plasma depends not only upon the rate of secretion but also upon the rate of removal from the blood by either excretion or metabolic transformation. Sometimes the hormone is inactivated by the cells upon which it acts, but for most hormones, the pathway of removal from the blood is the liver or the kidneys. Accordingly, patients with kidney or liver disease may suffer from excess of certain hormones solely as a result of reduced hormone loss. It should be noted, however, that the negative-feedback systems which control plasma hormonal concentrations are usually so effective that should decreased inactivation or excretion occur as a primary event, the ensuing elevation of plasma hormone concentration would usually induce, reflexly, an inhibition of hormone secretion, thus bringing plasma concentration of the hormone back toward normal.

Because, for many hormones, the rate of urinary excretion of the hormone or its metabolites is directly proportional to the rate of glandular secretion, it is often used as an indicator of secretory rate.

Transport in the blood

Most of the nonprotein hormone molecules which circulate in the blood are bound to various plasma proteins; the free moiety may be quite small and is in equilibrium with the bound fraction:

"Free" hormone + protein $\rightleftharpoons$ hormone-protein

It is important to realize that only the free hormone can exert effects on the target cells.

Activation of hormones after secretion

Earlier in this section we discussed metabolic transformation of the hormone in terms of its inactivation. However, certain hormones are relatively inactive in the forms released from their endocrine glands and must undergo further molecular alteration before they are able to exert their hormonal effects; i.e., the metabolism of a hormone may yield molecules which are *more* active, rather than less, compared to the original hormone. An example of this phenomenon is provided by testosterone, which is converted to dihydrotestosterone by certain of its target cells; the latter steroid then elicits the overall response of the target cell.

Particularly interesting (and confusing) is the possibility that different metabolites of the same hormone may be the mediators of different effects of that hormone. For example, several different polypeptide fragments of the protein hormone, growth hormone, circulate simultaneously in the plasma, and it has been extremely difficult to ascertain whether the intact growth-hormone molecule acts on target cells or whether the fragments are the active agents, each perhaps exerting its own unique effect.

Wherever such activation through molecular transformation is required, deficiency of the enzymes which mediate the activation may constitute an important source of malfunction in disease. For example, a deficiency in the relevant target cells of the enzyme which converts testosterone to dihydrotestosterone would result in the failure of testosterone to act on those target cells; the end result for those cells would be the same as if no testosterone were being secreted.

Summary example: thyroxine

In subsequent sections, we shall devote considerable attention to the afferent inputs which control the secretion of various hormones but very little to the other factors described above, which also help determine the concentration of cir-

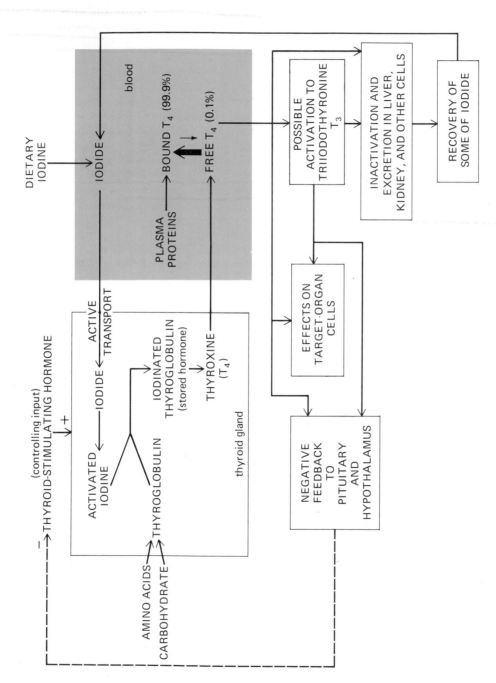

Figure 9-4. Summary of thyroxine pathways. Besides iodine, the diet must also supply the amino acids which are used for the synthesis of tyrosine, thyroglobulin, and plasma protein. Thyroid-stimulating hormone is produced by the anterior pituitary. Iodine is converted to iodide in the process of absorption by the gastrointestinal tract. The sites of conversion of T₄ to T₃ are not known for certain, but may be the target cells themselves. Some T₃ is also synthesized and secreted by the thyroid gland.

culating hormone. Therefore in order to illustrate the general principles, we here describe these factors in some detail as they apply to one hormone, thyroxine (Fig. 9-4).

The thyroid, located in the neck, secretes several hormones; secreted in largest amount is *thyroxine*, an iodine-containing amino acid. Thyroid physiology received its greatest stimulus when it was discovered that the enlarged thyroids (goiters) then common among inland populations were completely preventable by the administration of small quantities of iodine, as little as 4 g/year. The other ingredient for thyroxine synthesis is the amino acid *tyrosine*, which can be produced from a wide variety of substances within the body and therefore offers no supply problem.

Much of the ingested iodine absorbed by the gastrointestinal tract (which converts it to iodide, the ionized form of iodine) is removed from the blood by the thyroid cells, which manifest a remarkably powerful active-transport mechanism for iodide. Once in the gland, the iodide is converted to an active form of iodine, which is then attached to certain tyrosyl residues of a large glycoprotein known as *thyroglobulin.* Some of the doubly-iodinated tyrosyl residues then become linked together to form thyroxine residues of this protein. The normal gland may store several weeks' supply of thyroxine in this bound form. Hormone release into the blood occurs by enzymatic splitting of the thyroxine from the thyroglobulin and the entry of this freed thyroxine into the blood. Finally, the overall process is controlled by a pituitary *thyroid-stimulating hormone*, which stimulates certain key rate-limiting steps and thereby alters the rate of thyroxine synthesis and secretion. A variety of defects—dietary, hereditary, or disease-induced— may decrease the amount of thyroxine released into the blood. One such defect results from dietary iodine deficiency. However, this deficiency need not lead to permanent reduction of thyroxine secretion because the thyroid gland enlarges (goiter) so as to provide greater utilization of whatever iodine is available. This response is mediated by thyroid-stimulating hormone, which induces thyroid enlargement whenever the blood concentration of thyroxine decreases, regardless of cause (the mechanism is described in a subsequent section).

Once in the blood, most of the thyroxine becomes bound to certain plasma proteins and circulates in this form. This protein-bound thyroxine is in equilibrium with a much smaller amount of free thyroxine, the latter being the effective hormone. Thyroxine may be removed from the blood and excreted by the liver and kidneys or inactivated by catabolism in many tissues. These processes may assume considerable importance in certain disease states, particularly those involving the kidneys or liver.

Finally, it is not yet clear as to whether thyroxine, itself, acts upon its target cells or whether it must first be metabolized by the cells to a different molecule (probably triiodothyronine, T_3) which then serves as the active form of the hormone; increasing evidence favors the latter possibility.

Mechanisms of hormone action

Cell responses to hormones

The end result of a hormone's stimulation of its target cells is an alteration in the rate at which the target cells perform a specific activity: muscle cells increase or decrease contraction; epithelial cells (and other cell types as well) alter their rates of water or solute transport; gland cells secrete more or less of their secretory products. The direct cause of many of these observed changes is a hormone-induced change in the activity of a crucial enzyme in the cell. What is meant by a "crucial enzyme" in this context? The reader should review the section on "regulation of multi-enzyme pathways" in Chap. 5. Recall that most metabolic reactions are truly reversible whereas others proceed generally in one direction under the influence of one set of enzymes and in the reverse direction under the influence of a second set of enzymes. The enzymes which catalyze these "one-way" reactions are the primary ones regulated by various hormones. Let us consider, as an example, the relationship between glucose and glycogen in the liver:

$$\text{Glucose} \underset{\text{enzyme B}}{\overset{\text{enzyme A}}{\rightleftharpoons}} \text{glycogen}$$

Although there are actually multiple steps in both pathways, each catalyzed by a different enzyme, two enzymes (A + B) are particularly critical since they catalyze the major irreversible reactions in opposing directions. The hormone insulin increases the activity of enzyme A and reduces that of enzyme B; the result is increased formation of glycogen from glucose. In contrast, the hormone epinephrine increases the activity of enzyme B and decreases that of enzyme A; the result is facilitation of the catabolism of glycogen to glucose.

How is the activity of a particular type of enzyme increased? One way is for the cell to produce more of the enzyme; this may well be the major biochemical action of most steroid hormones (although it is by no means limited to steroid hormones). There are, however, other ways by which enzyme activity may be altered without a change in the total number of enzyme molecules in the cells, i.e., with no change in enzyme synthesis. Many enzymes exist within a cell in both active and inactive forms; thus, the number of active enzymes can be increased by converting some of the inactive molecules into active ones by the mechanisms described in Chap. 4, and many hormones influence this conversion. Certain hormones induce both the synthesis of new enzyme molecules and an increased activity of the enzyme molecules already present in the cell. The advantages of this dual effect are considerable: Induction of new enzyme synthesis requires hours to days, whereas the activation of molecules already present can occur within minutes. Thus, the hormone simultaneously exerts a rapid effect and sets into motion a long-term adaptation.

As widespread as such hormone-induced enzyme changes are, it is not always possible to explain the cell's response in these terms. For example, the facilitation of glucose uptake induced by insulin in its target cells does not seem to be due to an enzyme change but rather to an alteration in the carrier process itself. This is true for several other hormone-induced changes in membrane transport. Changes in muscle tone are another type of hormone-induced response not always explainable in terms of enzyme changes.

Hormone receptors

We have thus far been describing only the end results of a hormone's stimulation of its target cells. However, this response is only the final event in a sequence triggered when the hormone interacts with the target cell. In all cases, this initial interaction occurs between hormone and target-cell receptors, i.e., molecules of the target cells which have a specific capacity to bind the hormone. As emphasized in earlier chapters, this "recognition" (binding) is made possible by the configuration of portions of the receptor molecules such that a "match" exists between them and the specific hormone. It is the presence of the hormone-receptor combination which initiates the chain of intracellular biochemical events leading ultimately to the cell's overall response.

Where in the target cells are the hormone receptors located? The receptors for steroid hormones (and possibly several nonsteriod hormones as well) are soluble proteins within the cytoplasm of the target cells; steriods are quite lipid-soluble and readily cross the plasma membrane to enter the cytoplasm and combine with their specific receptors. In contrast, the receptors for nonsteroid hormones (proteins, peptides, and amines) are proteins located in the plasma membranes of the target cells.

The physiology of hormone receptors is the most rapidly advancing area in endocrinology, and we have come to realize that changes in receptors may be important factors in altering hormone function. For example, certain men have a genetic defect manifested by the absence of receptors for dihydrotestosterone, the form of testosterone active in many target cells; these cells are, therefore, unable to bind dihydrotestosterone and the result is lack of development of male characteristics just as if the hormone were not being produced (note that this defect is quite distinct from the inability to synthesize dihydrotestosterone, described in an earlier section).

Receptor modulation is not limited to disease states but appears to be a normal component of hormonal control systems. It explains why certain hormones, when present chronically in large amounts, frequently manifest a reduction in their effectiveness. For example, many obese, but otherwise normal, people have high concentrations of insulin in their plasma (for reasons to be described in Chap. 15) but do not manifest any signs of insulin excess; this is because of a loss of receptors on insulin's target cells. One factor which causes this change is insulin itself! Thus, a persistently elevated plasma insulin concentration induces (by unknown biochemical events) a loss of insulin receptors, and this prevents the target cells from overreacting to the high hormone concentration. This clearly constitutes an important negative-feedback control over insulin's actions.

Similar controls seem to be exerted by many other hormones over their specific receptors.

Hormones need not cause an alteration of their own receptors only; they may also affect receptors for other hormones. If one hormone induces a loss of a second hormone's receptors, the result will be a reduction of the latter hormone's effectiveness. An example of this is progesterone's inhibition of estrogen's actions on uterine smooth muscle, mediated by a progesterone-induced loss of estrogen receptors. On the other hand, a hormone may induce an increase in the number or affinity of receptors for a second hormone; in this case the effectiveness of the second hormone will be increased. An example of this is thyroxine's ability to increase the number of receptors for epinephrine in the heart; this accounts for the fact that patients with excessive thyroxine (hyperthyroidism) are extremely sensitive to epinephrine.

Finally, the ability of one hormone to influence the receptors of a second may underlie certain situations in which physiological responses require actions of hormones in a sequence. For example, stimulation of release of an ovum (egg) from the ovary requires the action of two different hormones secreted by the anterior pituitary, and the first hormone paves the way for the second by increasing the number of the latter's receptors in the relevant cells of the ovary.

Although we have spoken of "loss" or "increase" in receptors, these are only operational terms, the biochemical meaning of which is still not settled. We do not know whether "lost" receptors are actually degraded or, alternatively, whether their molecular configurations are so altered as to render them incapable of combining with hormone. Similarly, it is not clear whether hormones ever induce synthesis of additional receptors or, alternatively, simply unmask those already present.

Events elicited by hormone-receptor binding

In a sense, we began this section on the mechanisms of hormone action at the end of the story, i.e., with the end result or overall response of the target cell to the hormone. We then jumped to the beginning of the story, the hormone receptors. Now we must fill in the middle, namely the events which lead from the hormone-receptor combination to the overall response; much of this still remains quite uncertain.

As described above, the receptors for most nonsteroid hormones are located on the outer surface of the plasma membrane. For at least 12 of these hormones, the event triggered by hormone-receptor combination is activation of membrane-bound adenylate cyclase with subsequent generation of cyclic AMP inside the cell. Thus, these hormones all utilize the cAMP-protein-kinase second-messenger system described in Chap. 7, the overall response representing the final event in a cascade of biochemical reactions. However, it is clear that all the actions of all nonsteroid hormones are not mediated by activation of this system; in at least one case (insulin), inhibition of adenylate cyclase (with subsequent reduction of cellular cAMP) may be important; in other cases, cGMP (Chap. 7) may be involved; while in still others neither cAMP nor cGMP is adequate to explain the response, the explanation of which remains unknown.

Until very recently it was believed that polypeptide hormones could not gain entry to the interior of cells. Now it is clear that many can, in fact, do so, apparently by endocytosis. However, what they do after they have gained entry remains unknown. One hypothesis is that they (or the peptide fragments resulting from their intracellular degradation) combine with intracellular receptors and exert long-term effects upon the cell.

Steroid hormones. As mentioned above, the common denominator of the effects of the steroid hormones is an increased synthesis of proteins (enzymes, structural proteins, etc.) by their specific target cells. This increased protein synthesis is the result of stimulation of mRNA synthesis by the hormone-receptor complex. Recall that steroid hormones enter the cytoplasm of their target cells and combine with receptors (Fig. 9-5). This binding activates the receptor so that it moves into the cell's nucleus carrying with it the bound hormone. There the bound hormone combines with a DNA-associated protein specific for it (an "acceptor"). This interaction causes the original receptor molecule to split, one fragment remaining bound to the acceptor protein while the other interacts with an adjacent segment of DNA. It is this last reaction which triggers transcription of the DNA segment, i.e., synthesis of messenger RNA, which can then enter the cytoplasm and serve as a template for synthesis of a specific protein. Note that this entire sequence depends upon several "recognitions"—

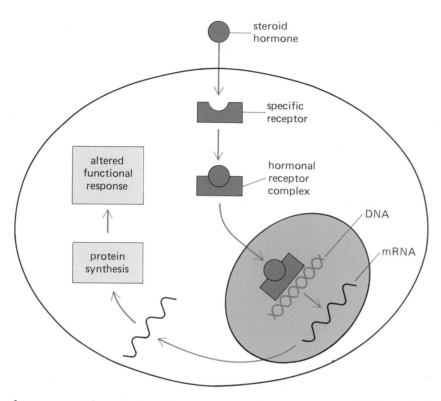

Figure 9-5. Steroid hormones enter the target cell and bind to a specific cytoplasmic receptor. The hormone-receptor complex is then transported into the nucleus, where it binds to specific sites on the DNA molecule and, activates transcription of new RNA, which mediates the response characteristic of the cell by directing protein synthesis. *(Adapted from McEwen)*

hormone with cytoplasmic receptor and receptor with nuclear acceptor. The latter is essential for proper positioning of the receptor-hormone complex along the DNA so that the appropriate gene is transcribed.

Hormone interactions on target cells

Cells are constantly exposed to the simultaneous effects of many hormones. This allows for complex hormone-hormone interactions on the target cells,

Figure 9-6. Ability of thyroxine to permit epinephrine-induced liberation of fatty acids from adipose tissue cells.

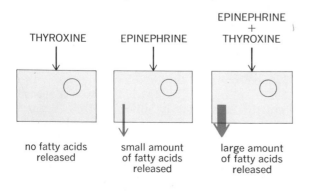

including inhibition, synergism, and the important phenomenon known as *permissiveness*. In general terms, frequently hormone A must be present for the full exertion of hormone B's effect. In essence, A is "permitting" B to exert its action. For example (Fig. 9-6), the hormone epinephrine causes marked release of fatty acids from adipose tissue only in the presence of thyroid hormone. Many of the defects seen when an endocrine gland is removed or ceases to function because of disease actually result from loss of the permissive powers of the hormone secreted by that gland. It is likely that, as described above, many of these interactions may be the result of the influence of one hormone on the number or affinity of the other hormone's receptors.

Pharmacological effects

Administration of very large quantities of a hormone may have results which are never seen in a normal person, although these so-called *pharmacological effects* sometimes occur in endocrine diseases when excessive amounts of hormone are secreted. These effects are of great importance in medicine since hormones in pharmacological

Table 9-2. Summary of the control of hormone secretion*

In response to neural, hormonal, or metabolic inputs, hypothalamic neurons themselves release:

Oxytocin
Antidiuretic hormone } (From the posterior pituitary)
Hypothalamic releasing hormones

Hypothalamic releasing hormones directly control the release from the anterior pituitary of:

Growth hormone
Thyroid-stimulating hormone (TSH)
Adrenocorticotropic hormone (ACTH)
Gonadotropic hormones (FSH and LH)
Prolactin

Anterior pituitary hormones directly control the release of:

Thyroid hormone
Cortisol (from adrenal cortex)
Gonadal hormones
  (female: estrogen and progesterone)
  (male: testosterone)

Autonomic neurons directly control the release of:

Epinephrine and norepinephrine (from adrenal medulla)
Renin (from kidney)
Insulin and glucagon (from pancreas)
Gastrointestinal hormones
?Others

Plasma concentrations of ions or nutrients directly control the release of:

Parathyroid hormone
Insulin and glucagon (from pancreas)
Aldosterone (from adrenal cortex)
Calcitonin (from thyroid glands)

* This table does not necessarily list all the controls of each hormone.

doses are used as therapeutic agents. Perhaps the most famous example is that of the adrenal hormone cortisol, which is highly useful in suppressing allergic and inflammatory reactions. Mental changes, including outright psychosis, may also be induced by large quantities of cortisol and are frequently a striking symptom of patients suffering from hyperactive adrenal glands.

Control of hormone secretion

The immediate inputs which control the secretion of hormones by the endocrine glands fall into one of five categories, and an appreciation of them should greatly facilitate the understanding of each specific hormone as it is discussed in subsequent chapters. Table 9-2 summarizes the categories to be discussed; it must be reemphasized that all the material contained in this table will be presented again in later sections or chapters

where relevant. It should also be noted that the table does not constitute a complete inventory of all hormones or of the immediate inputs for any given hormone but includes only those which help to illustrate the common patterns.

It is evident from Table 9-2 that the *pituitary gland* is of major importance in hormone secretion. This gland lies in a pocket of the bone just below the hypothalamus (Fig. 9-7), to which it is

Figure 9-7. Relationship of the pituitary to the brain and hypothalamus. *(Adapted from Guillemin and Burgus.)*

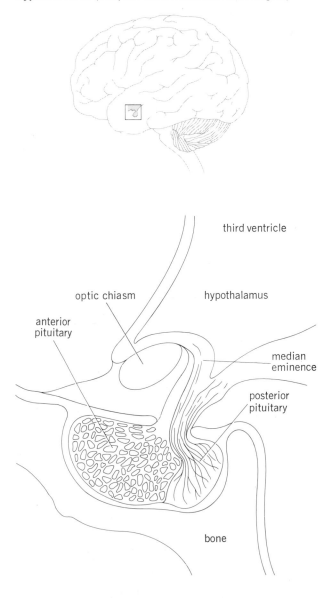

third ventricle

optic chiasm

hypothalamus

anterior pituitary

median eminence

posterior pituitary

bone

connected by a stalk containing neurons and small blood vessels. It is composed of three lobes—*anterior*, *intermediate*, and *posterior*—each of which is a more or less distinct gland. In human beings, the intermediate lobe is rudimentary and contains two substances called melanocyte-stimulating hormones (MSH), which are known to cause skin darkening in lower vertebrates; the function of these substances in human beings is unknown.[1]

The anterior pituitary hormones

The anterior pituitary produces at least six different polypeptide hormones (Table 9-1). Each of four of the hormones is secreted by a distinct cell type; the remaining two (the gonadotropic hormones) may be secreted by a common cell type, although this remains unsettled. In any case, secretion of each of the six hormones may occur more or less independently of the others; i.e., the anterior pituitary comprises, in effect, six endocrine glands anatomically associated in a single structure.

The major function of two of the anterior pituitary hormones is to stimulate the secretion of other

[1] It has also been found recently that both the intermediate and anterior pituitary contain large amounts of peptides known as endorphins. The possible hormonal role of these substances, which are also found in the central nervous system where they function as transmitters (Chap. 8), is unknown.

hormones: (1) *Thyroid-stimulating hormone* (thyrotropin, TSH) induces secretion of thyroid hormones (thyroxine and triiodothyronine) from the thyroid. (2) *Adrenocorticotropic hormone* (corticotropin, ACTH), meaning "hormone which stimulates the adrenal cortex," is responsible for stimulating the secretion of cortisol. Thus, the important target organs for TSH and ACTH are the thyroid and adrenal cortex, respectively (whether these hormones have other functions elsewhere in the body remains controversial).

Two other anterior pituitary hormones, *follicle-stimulating hormone* (FSH) and *luteinizing hormone* (LH), primarily control the secretion of sex hormones (estrogen, progesterone, and testosterone) by the gonads. These *gonadotropins* differ from TSH and ACTH in that, besides controlling the secretion of other hormones, they have a second major role, regulating the growth and development of the reproductive cells to produce the sperm and ova. The gonads are the sole target organs for the anterior pituitary gonadotropins.

It should now be clear why the anterior pituitary is frequently called the master gland; it secretes six hormones itself and controls the secretion of three or four (depending upon the person's sex) other hormones.

The two remaining anterior pituitary hormones have not been shown to exert major control over the secretion of other hormones. *Prolactin's* major target organs are the breasts, and *growth hormone*

Figure 9-8. Target organs and functions of the six anterior pituitary hormones.

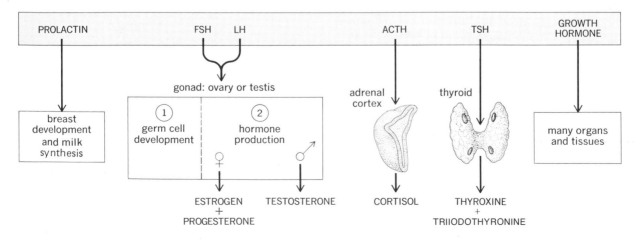

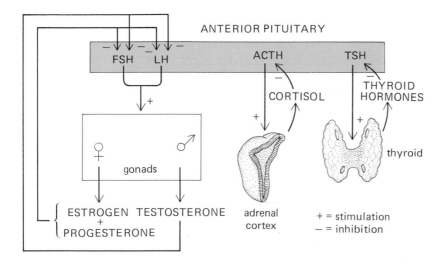

Figure 9-9. Negative feedback of target-organ hormones on their respective anterior pituitary tropic hormones. The gonadal hormones actually exert complex effects on the pituitary as described in Chap. 16.

exerts multiple metabolic effects upon many organs and tissues. The target organs and functions of the anterior pituitary hormones are summarized in Fig. 9-8. It should be noted that the tropic hormones of the anterior pituitary control not only the synthesis and secretion of their target-gland hormones, but the growth and development of the target glands themselves.

Control of anterior pituitary hormone secretion

What direct inputs control secretion of the anterior pituitary hormones? One important type of input for the tropic hormones is the target-gland hormone itself, a beautiful example of negative feedback (Chap. 7). Thus, ACTH stimulates cortisol secretion and increases the blood concentration of cortisol, which acts upon the anterior pituitary to inhibit ACTH release. In this manner, any increase in ACTH secretion is partially prevented by the resultant increase in cortisol secretion, as illustrated in Fig. 9-9. This same pattern of negative feedback is exerted by the thyroid hormones and the sex hormones on their respective pituitary tropic hormones (the sex hormone effects are actually more complex than those shown in the figure, as will be discussed in Chap. 16). It is evident that such a system is highly effective in damping hormonal responses, i.e., limiting the extremes of hormone secretory rates; it also serves to maintain the plasma concentration of the hormone relatively constant whenever a primary change occurs in the catabolism of the hormone.

However, if this negative-feedback relationship were the sole source of anterior pituitary control, there would be no way of altering anterior pituitary output; some unchanging equilibrium blood concentrations of pituitary and target-gland

Figure 9-10. Hypothalamus–anterior-pituitary vascular connections. The hypothalamic neurons, which secrete releasing hormones, end on the capillary loops of the primary capillary plexus of the portal system carrying blood from the hypothalamus to the anterior pituitary.

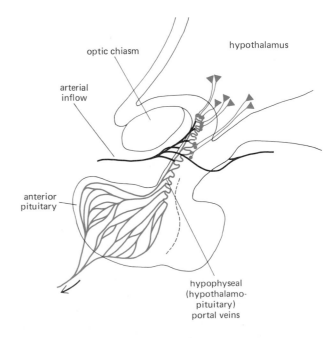

Table 9-3. Major hypothalamic releasing hormones and their effects on the anterior pituitary

Hypothalamic Releasing Hormone	Effect on Anterior Pituitary
Corticotropin releasing hormone (CRH)	Stimulates secretion of ACTH (corticotropin)
Thyrotropin releasing hormone (TRH)	Stimulates secretion of TSH (thyrotropin)
Growth hormone releasing hormone (GRH)	Stimulates secretion of GH
Somatostatin (also known as growth-hormone release inhibiting hormone [GIH]	Inhibits secretion of growth hormone
Gonadotropin releasing hormone (GnRH)*	Stimulates secretion of LH and FSH
Prolactin releasing hormone (PRH)	Stimulates secretion of prolactin
Prolactin release inhibiting hormone (PIH)	Inhibits secrection of prolactin

* It is still unsettled as to whether there is one releasing hormone for both FSH and LH or whether others exist, i.e., an LH-RH and an FSH-RH.

hormone would always be maintained. Obviously, there must be some other type of input to the anterior pituitary. In reality, this other input is the major controller of anterior pituitary function.

Figure 9-11. Control of anterior pituitary secretion by a hypothalamic releasing hormone.

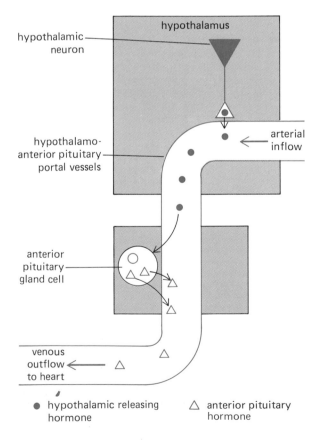

hypothalamic neuron

hypothalamus

arterial inflow

hypothalamo-anterior pituitary portal vessels

anterior pituitary gland cell

venous outflow to heart

● hypothalamic releasing hormone

△ anterior pituitary hormone

The major inputs controlling release of anterior pituitary hormones are a group of so-called releasing hormones produced in the hypothalamus. An appreciation of the anatomical relationships between the hypothalamus and anterior pituitary is essential for understanding this process. Although the anterior pituitary lies just below the hypothalamus, there are no important neural connections between the two, but there is an unusual capillary-to-capillary connection (Fig. 9-10).

Certain of the arteries supplying the hypothalamus end in its base (the *median eminence*) as intricate capillary tufts which recombine into the *hypothalamo-pituitary portal vessels* (the term "portal" denotes vessels which connect distinct capillary beds). These pass down the stalk connecting the hypothalamus and pituitary and enter the anterior pituitary where they break up into a second capillary bed, the anterior pituitary capillaries, which provide most of the blood supply to the gland. Thus, these portal vessels offer a local route for flow of capillary blood from hypothalamus to anterior pituitary. The axons of neurons which originate in diverse areas of the hypothalamus terminate in the median eminence around the capillary tufts, i.e., the origins of the portal vessels. These neurons secrete into the capillaries substances which are carried by the portal vessels to the anterior pituitary where they act upon the various pituitary cells to control their secretions of hormones (Fig. 9-11).

We are dealing with multiple discrete hypothalamic substances, each secreted by a unique group of hypothalamic neurons and influencing the release of one or, in some cases, two of the six anterior pituitary hormones (Table 9-3 and Fig.

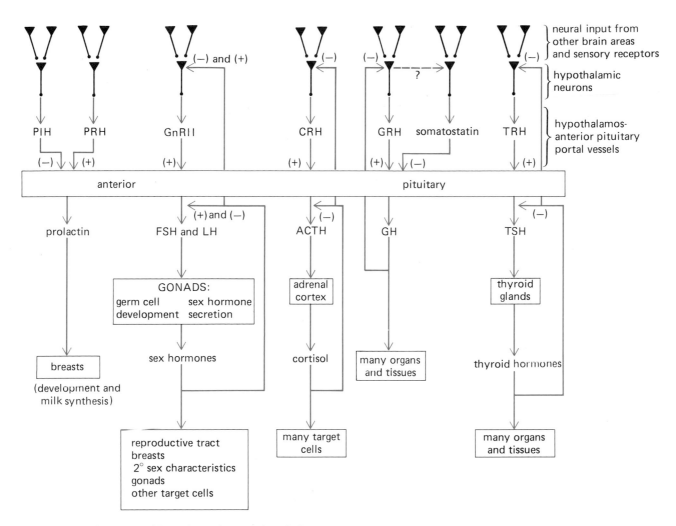

Figure 9-12. Summary of hypothalamic–anterior-pituitary–target-organ relationships. The neural input to the hypothalamus from other brain areas and sensory receptors is both inhibitory and excitatory. It is not known whether somatostatin release is influenced in a feedback manner by GH (the ? in the diagram). The interactions between gonadal hormones and the hypothalamus–anterior pituitary are complex and will be described in Chap. 16. This model will certainly become more complicated as research progresses. For example, current evidence suggests that prolactin may exert a negative-feedback control over its own secretion via the hypothalamus and that this hormone may also be influenced by TRH as well as by PIH and PRH.

9-12). Because most of these substances stimulate release of their relevant hormones, they were originally termed "hypothalamic releasing factors"; however, they fulfill the criteria for the definition of a hormone, and we shall term them "hormone" rather than "factor."[1] Each of the

"factor." The reason is that many of the hypothalamic releasing substances have not yet been isolated in pure form or shown to fulfill the criteria required for establishing them as physiological regulators. In other words, the evidence for their existence and roles remains, to some extent, indirect. Accordingly, many endocrinologists believe that the term "factor" should be used for the incompletely documented substances and "hormone" be reserved for those which have passed all the required tests. To avoid confusion, we have not followed this practice.

[1] There remains considerable disagreement among endocrinologists concerning the use of "hormone" and

hypothalamic releasing hormones is named according to the anterior-pituitary hormone whose secretion it controls; for example, ACTH (also known as corticotropin) secretion is stimulated by corticotropin releasing hormone (CRH). As shown in Table 9-3 and Fig. 9-12, at least two of the hypothalamic hormones inhibit, rather than stimulate, release of anterior pituitary hormones. One of these inhibits the secretion of prolactin and is termed prolactin release inhibiting hormone (PIH). The other inhibits secretion of growth hormone and is most commonly called somatostatin.

This system would be much easier to research if there were always a strict one-to-one correspondence of hypothalamic releasing hormone and anterior pituitary hormone, but this is not always the case, as is clear from Table 9-3 and Fig. 9-12. First, at least one of the hypothalamic hormones influences the secretion of more than one anterior-pituitary hormone; thus, gonadotropin releasing hormone (Gn RH) controls the secretion of both LH and FSH. Secondly, dual systems of hypothalamic hormones, one stimulatory and the other inhibitory, operate on several anterior-pituitary hormones; prolactin secretion is inhibited by hypothalamic PIH, but is stimulated by a second hypothalamic hormone, PRH; the secretion of growth hormone is also controlled by opposing stimulatory and inhibitory hypothalamic releasing hormones. Clearly, in such dual-control cases, the overall anterior-pituitary response depends upon the relative amounts of the two opposing hormones released by the hypothalamic neurons. This is simply one more example of how the output of an endocrine gland cell may reflect the integration of multiple inputs to it.

Before leaving this subject it should be noted that only a few of the hypothalamic releasing hormones have thus far been completely isolated, identified, and synthesized—those which have are peptides. These same peptides have been found in locations other than the hypothalamus, particularly in other areas of the brain and in the gastrointestinal tract. For example, somatostatin (the hypothalamic hormone which inhibits secretion of growth hormone) is found in the stomach and pancreas, in cells adjacent to other endocrine cells; this has raised the possibility that somatostatin is synthesized in these areas and serves locally to modulate the secretion of hormones and other substances from adjacent endocrine and exocrine cells. Of particular interest is the finding of these peptides in areas of brain other than hypothalamus; this has raised the possibility that they may serve as neurotransmitters or neuromodulators in the central nervous sytem.

We stated earlier that the nervous and endocrine systems actually function as a single interrelated system; the anterior pituitary may be the master gland, but its function is primarily controlled by the hypothalamus via the releasing hormones. It now appears that some diseases characterized by inadequate secretion of one or more pituitary hormones really are due to hypothalamic malfunction rather than primary pituitary disease.

Our analysis has pushed the critical question one step further: The hypothalamic releasing hormones control anterior pituitary function, but what controls secretion of the releasing hormones? The answer is neural and hormonal input to the hypothalamic neurons which secrete the releasing hormones. The hypothalamus receives neural input, both facilitory and inhibitory, from virtually all areas of the body; the specific type of input which controls the secretion rate of the individual releasing hormones will be described in future chapters when we discuss the relevant anterior pituitary or target-gland hormone. It suffices for now to point out that endocrine disorders may be generated, via alteration of hypothalamic activity, by all manner of neural activity, such as stress, anxiety, etc. An example is sterility caused by severe emotional upsets.

Hormonal influences upon the hypothalamus are also important. Some of the negative-feedback effect of thyroxine, cortisol, and sex hormones upon pituitary tropic-hormone secretion is actually mediated via the hypothalamus, i.e., by inhibition of releasing-hormone secretion. For example, cortisol acts not only directly upon the anterior pituitary to inhibit ACTH secretion but also upon the hypothalamus to inhibit CRH secretion, an event which also reduces ACTH secretion. Presently, there is still controversy over the quantitative importance of these two negative-feedback sites for the various hormones, but the generalization that both sites are involved, albeit to varying extents for each hormone, seems warranted. In addition, it is quite likely that growth hormone and prolactin, the two anterior pituitary hormones which have no target-organ hormones, exert negative-feedback controls, via the hypothalamus, over their own secretion. In this regard, recent evidence indicates that a significant portion of the blood flow *from* the pituitary actually goes back toward the central nervous system before entering the systemic circulation. Therefore, it is quite possible that pituitary hormones may reach the brain in high concentrations to influence the secretion of hypothalamic hormones, behavior, etc. Our description of the interrelationships of the hypothalamus, anterior pituitary, and target glands is now complete (Fig. 9-12).

Hypothalamus–posterior-pituitary function

The *posterior pituitary* lies just behind the anterior pituitary in the same bony pocket at the base of the hypothalamus, but its structure is totally different from its neighbor's. The posterior pituitary is actually an outgrowth of the hypothalamus and is true neural tissue. Two well-defined clusters of hypothalamic neurons send out nerve fibers which pass by way of the connecting stalk to end within the posterior pituitary in close proximity to capillaries (Fig. 9-13). The two hormones, *oxytocin* and *antidiuretic hormone* (ADH, vasopressin), released from the posterior pituitary are actually synthesized in the hypothalamic cells; enclosed in small vesicles, they move slowly down the cytoplasm of the neuron axons to accumulate at the nerve endings. Release into the capillaries occurs in response to generation of an action potential within the axon. Thus, these hypothalamic neurons secrete hormones in a manner quite analogous to that described previously for hypothalamic releasing hormones, the essential difference being that the releasing hormones are secreted into capillaries which empty directly into the anterior pituitary whereas the posterior pituitary capillaries drain primarily into the general body circulation. It is evident, therefore, that the term posterior pituitary hormones is somewhat of a misnomer since the hormones are actually synthesized in the hypothalamus, of which the posterior pituitary is merely an extension. We must therefore add two more names to our growing list of hormones under direct or indirect control of the hypothalamus.

Epinephrine and the adrenal medulla

Each of the paired adrenal glands constitutes two distinct endocrine glands, an inner *adrenal me-*

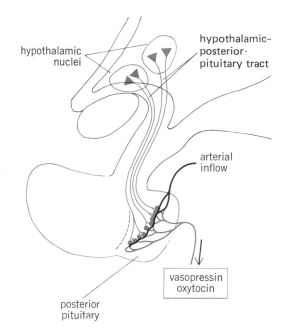

Figure 9-13. Relationship between the hypothalamus and posterior pituitary. *(Adapted from Guillemin and Burgus.)*

dulla and its surrounding *adrenal cortex*. As described in Chap. 8, the adrenal medulla is really an overgrown sympathetic ganglion whose cell bodies do not send out nerve fibers but release their active substances directly into the blood, thereby fulfilling the criteria for an endocrine gland. In man, the hormone released by the medulla is for the most part *epinephrine* (a smaller amount of norepinephrine is also secreted). In controlling epinephrine secretion, the adrenal medulla behaves just like any sympathetic ganglion and is dependent upon stimulation by sympathetic preganglionic fibers. Destruction of these incoming nerves causes marked reduction of epinephrine release and failure to increase secretion in response to the usual physiological stimuli.

The adrenal medulla is best viewed as a general reinforcer of sympathetic activity. Its secretion of epinephrine into the blood serves to increase the overall sympathetic functions of the body. We shall discuss in later chapters the specific reflexes which cause enhanced sympathetic activity and elicit epinephrine secretion; suffice it to say that these reflexes are under the strong control of higher brain centers, including the hypothalamus.

Role of the autonomic nervous system in hormone secretion

We have seen that one hormone—epinephrine—is actually a product of the sympathetic nervous system. This is not the only influence the autonomic nervous system has on the endocrine system. Certain of the endocrine glands receive a rich supply of sympathetic and parasympathetic neurons whose activity influences their rates of hormone secretion. As shown in Table 9-2, examples are the secretion of renin by the kidneys, insulin and glucagon by the pancreas, and the gastrointestinal hormones. Others will almost certainly be discovered. In all cases studied, excepting the adrenal medullary hormones, the autonomic input does not serve as the sole regulator of the hormone's secretion but serves rather as only one of multiple inputs.

Direct control of hormone secretion by plasma ions or nutrients

The endocrine cells of the pancreas and parathyroid glands respond directly to the glucose and calcium concentrations, respectively, of the blood supplying them.[1] As will be described in later chapters, *insulin* and *glucagon* participate in regulation of blood glucose concentration, whereas the major function of *parathyroid hormone* is regulation of blood calcium concentration. The control mechanisms are therefore quite appropriate and remarkably simple; the hormone-secreting cells are themselves sensitive to the plasma substance their hormones regulate. It is likely that such regulatory systems represent the earliest type of control mechanism evolved by complex organisms.

Summary

Table 9-1 should now appear less formidable. The control mechanisms for many of the hormones involve the direct or indirect participation of the hypothalamus and pituitary. The anterior pituitary hormones are controlled primarily by releasing hormones secreted into the hypothalamopituitary portal vessels by neurons in the hypothalamus. In turn, the four anterior pituitary tropic hormones

[1] Other important inputs which control insulin and glucagon secretion will be described in Chap. 15.

control hormone secretion by their target-organ glands; the thyroid, adrenal cortex, and gonads. These glands exert a negative-feedback control over their own secretion via the effects of their hormones on both the hypothalamus and anterior pituitary. The hypothalamus itself also produces two hormones, oxytocin and ADH, which are released from nerve endings in the posterior pituitary. And finally, the hypothalamus exerts important control over the autonomic nervous system, including the adrenal medulla (the hypothalamus is not, of course, the only brain area which controls the output of the autonomic nervous system). Thus, this small area of brain, which weighs 4 g in the adult human being, acts as a compact integrating center, receiving messages, both neural and hormonal, from all areas of the body and sending out efferent messages via both the nerves and hormones. As we shall see, it also regulates body temperature, food intake, water balance, and a host of other autonomic, endocrine, and behavioral activities. Despite the central role of the hypothalamopituitary system, there are important hormones whose secretion is controlled, at least in part, by completely distinct mechanisms. These include aldosterone from the adrenal cortex, insulin and glucagon from the pancreas, parathyroid hormone, several kidney hormones, and a group of gastrointestinal hormones.

With this basic foundation laid, future chapters will describe the relevant functions of each major hormone and the specific environmental changes and afferent inputs which induce reflex alteration of their secretion rates.

Multiple hormone secretion

The phenomenon of multiple hormone secretion by a single gland is clearly evident in Table 9-1. In certain glands, such as the pancreas, the hormones are secreted by completely distinct cells; in other glands, it is likely that, although some separation of function exists, a single cell may secrete more than one hormone. Even in such cases, it is essential to realize that each hormone has its own unique control mechanism; i.e., there is no massive undifferentiated release of the multiple hormones.

A frequent source of confusion concerning the endocrine system concerns not multiple hormone

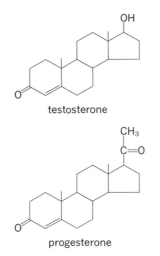

Figure 9-14. Structures of testosterone and progesterone.

secretion by a single gland but the mixed general functions exhibited by the gonads, pancreas, kidneys, and gastrointestinal tract. All these organs contain endocrine gland cells but also perform other completely distinct nonendocrine functions. For example, most of the pancreas is concerned with the production of digestive enzymes; these are produced by exocrine glands; i.e. the secretory products are transported not into the blood but into ducts leading, in this case, into the intestinal tract. The endocrine function of the pancreas is performed by completely distinct nests of endocrine cells scattered throughout the pancreas. This pattern is true for all the organs of mixed function; the endocrine function is always subserved by gland cells distinct from the other cells which constitute the organ.

Steroid synthesis

The adrenal cortex and gonads offer a particularly interesting problem of multiple hormone secretion. The hormones secreted by these organs all have a common ringlike type of lipid structure known as a steroid. Subtle changes in the steroid molecule produce great alterations of physiological activity (compare, for example, the structures of the male sex hormone, testosterone, and one of the female sex hormones, progesterone, in Fig. 9-14).

The biochemical pathways for steroid-hormone synthesis are complex, but Fig. 9-15, which is a highly simplified flow sheet in that it leaves out many intermediate steps, should serve to illustrate several important generalizations. First, note that the pathways are overlapping in that the synthesis of one hormone may require the synthesis of another earlier in the pathway. For example, the synthesis of aldosterone, one of the major hormones of the adrenal cortex, proceeds via progesterone, but note that progesterone may also go to testosterone (and via this latter hormone, on to estrogen) or to cortisol, the second major hormone of the adrenal cortex. Moreover, progesterone is not only a precursor for other hormones, but is itself an important hormone secreted by specific cells of the ovary.

Why is it that progesterone goes exclusively to aldosterone in a particular group of adrenal-cortical cells, mainly to cortisol in a second group of adrenal-cortical cells, and almost entirely to testosterone in the male gonads (the testes)? The answer is that all these biochemical steps are mediated by specific enzymes and the different steroid-synthesizing cells possess different quantities of the key enzymes. Thus, aldosterone is synthesized only by certain adrenal-cortical cells because only they possess the enzymes required for its formation; at the same time, these cells lack the enzymes needed to synthesize cortisol, enzymes which *are* possessed by the second group of adrenal-cortical cells. Similarly, the testicular cells which synthesize testosterone lack the enzymes in both the aldosterone and cortical pathways but possess high concentrations of those enzymes in the testosterone pathway. As might be predicted, the ovarian cells which synthesize estrogen show an enzyme profile similar to this but, in addition, have high concentrations of the enzyme required to transform testosterone into estrogen; accordingly they convert the former into the latter as fast as it is produced and so secrete little, if any, testosterone. The separation of enzyme activities is, however, not total, which explains why the normal testes do actually secrete small amounts of estrogens whereas the ovaries secrete small amounts of testosterone; there are, thus, no uniquely male or female hormones although, of course, the relative concentrations of the "male" and "female" sex hormones are quite different in the two sexes.

Another generalization explainable in these terms is that, in certain disease states, the steroid-producing glands may secrete large amounts of a

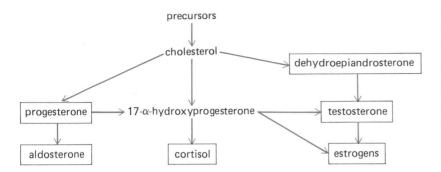

Figure 9-15. A simplified flow sheet for steroid-hormone synthesis; many intermediate steps have been left out. Aldosterone, cortisol, and dehydroepiandrosterone (an androgen) are the major hormones secreted by the adrenal cortex. Testosterone is the major hormone of the testes, and estrogens and progesterone of the ovaries (and placenta).

steroid not normally secreted by them. For example, an enzyme defect in the cortisol pathway may prevent the formation of cortisol by the adrenal cortex and result in the shunting of precursors into the androgen pathway instead; the result in a woman would be masculinization (androgens are hormones that produce maleness; testosterone is only one of several androgens). To take a quite different example, the genes which code for the enzymes leading from dehydroepiandrosterone to testosterone are normally repressed in adrenal-cortical cells, but may be derepressed in diseased cells, leading to secretion of large amounts of testosterone, a hormone normally secreted in only very small amounts by the adrenal cortex.

Finally, it should be noted that this picture has been made even more complex by the discovery that, following their secretion into the blood by the adrenal cortex or gonads, steroid hormones may undergo further interconversion in the blood or other organs. For example, testosterone may be converted to estrogen, and this constitutes another source of estrogen in normal men.

10
MUSCLE

Structure of skeletal muscle fibers
Molecular mechanisms of contraction
   Sliding-filament model
   Regulator proteins and calcium
   Excitation-contraction coupling
      Sarcoplasmic reticulum • Membrane excitation:
      The neuromuscular junction
Mechanics of muscle contraction
   Single twitch
   Summation of contractions
   Length-tension relationship
Control of whole-muscle tension
   Lever action of muscles

Muscle energy metabolism
Muscle fiber differentiation and growth
Types of skeletal muscle fibers
Smooth muscle
   Smooth muscle structure
   Excitation-contraction coupling
      Spontaneous electrical activity • Nerves and
      hormones • Local chemical factors
   Classification of smooth muscle
      Single-unit smooth muscle • Multiunit smooth
      muscles
Summary

Movement, the result of forces generated by the interaction of certain proteins, fueled by chemical energy, is a characteristic of all living cells. It occurs in the form of movements within cells, e.g., during cell division as chromosomes move to opposite sides of the cell and as the cell constricts to form two daughter cells. It may result in propulsion of the entire cell through its environment, e.g., the migration of white blood cells through tissues and of sperm cells on their way to fertilize an egg. Still other movements involve hairlike processes, known as *cilia*, which extend from the surfaces of specialized epithelial cells such as those lining the airways and the oviducts; by moving back and forth, these cilia propel material along the surface of their cells and through the tubes lined by them.

The force-generating apparatus that produces all these movements must have been present at a very early stage of biological evolution. Muscle cells, with their specialized ability to generate force and motion, merely utilize an extension and modification of this basic apparatus common to all cells (just as nerve cells manifest specializations of the permeability characteristics of all cell membranes). In multicellular organisms, control of the forces generated by muscle cells contributes not only to regulation of the internal environment but to the movement of the organism in its external environment.

Three types of muscle cells can be identified on the basis of structure and contractile properties: (1) *skeletal muscle;* (2) *smooth muscle;* (3) *cardiac muscle.* Most skeletal muscle, as the name implies, is attached to bones, and its contraction is responsible for the movements of parts of the skeleton. The contraction of skeletal muscle is controlled by the somatic nervous system and is under voluntary control. The movements produced by skeletal muscle are primarily involved with interactions between the body and the external environment.

Smooth muscle surrounds such hollow organs and tubes as the stomach, intestinal tract, urinary bladder, uterus, blood vessels, and the air passages

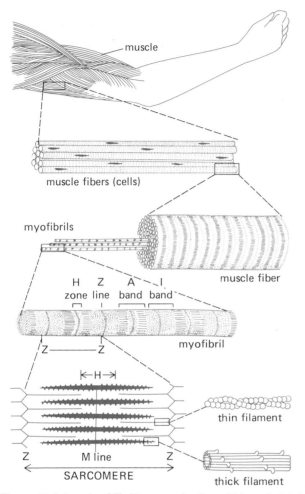

Figure 10-1. Levels of fibrillar organization within a skeletal muscle. *(Redrawn from Bloom and Fawcett.)*

to the lungs. It is also found as single cells distributed throughout organs (for example, the spleen) and in small groups of cells attached to the hairs in the skin or the iris of the eye. The contraction of this smooth muscle may either propel the luminal contents out of or through the hollow organs, or it may regulate the flow of the contents through tubes by changing their diameters without itself initiating the propulsion. Smooth muscle contraction is controlled by factors intrinsic to the muscle itself, by the autonomic nervous system, and by hormones; therefore, it is not normally under direct conscious control.

The third type of muscle, cardiac muscle, is the muscle of the heart, and its contraction propels blood through the circulatory system. Like smooth muscle, it is regulated by intrinsic factors and by the autonomic nervous system and hormones.

Although there are significant differences in the structure, contractile properties, and control mechanisms of the three types of muscle, the force generation will be described first, followed in all of them. The properties of skeletal muscle and the molecular events associated with its force generation will be described first, followed by smooth muscle with emphasis on the ways in which the latter differs from skeletal muscle. Cardiac muscle, which combines some of the properties of both skeletal and smooth muscle, will be described in the next chapter in the context of its role in the circulation.

Structure of skeletal muscle fibers

A single muscle cell is known as a *muscle fiber*. Each skeletal muscle fiber is a cylinder, with a diameter of 10 to 100 μm and a length which may extend up to 300,000 μm (1 ft). The term *muscle* refers to a number of muscle fibers bound together by connective tissue (Fig. 10-1). Thus, the relation between a single muscle fiber (cell) and a muscle is analogous to that between a single nerve fiber (axon) and a nerve composed of many axons.

Figure 10-2 shows a section through a skeletal muscle as seen with a light microscope. The most striking feature is the series of transverse light and dark bands forming a regular pattern along each fiber. Both skeletal and cardiac muscle fibers have this characteristic banding and are known as *striated muscles*; smooth muscle cells show no banding pattern. Although the pattern appears to be continuous across the entire cytoplasm of a single fiber, the bands are actually confined to a number of independent cylindrical elements known as *myofibrils* (Fig. 10-1). Each myofibril is about 1 to 2 μm in diameter and continues through the length of the muscle fiber. Myofibrils occupy about 80 percent of the fiber volume and vary in number from several hundred to several thousand per single fiber, depending on the fiber's diameter (keep in mind that "muscle cell" and "muscle fiber" are synonymous).

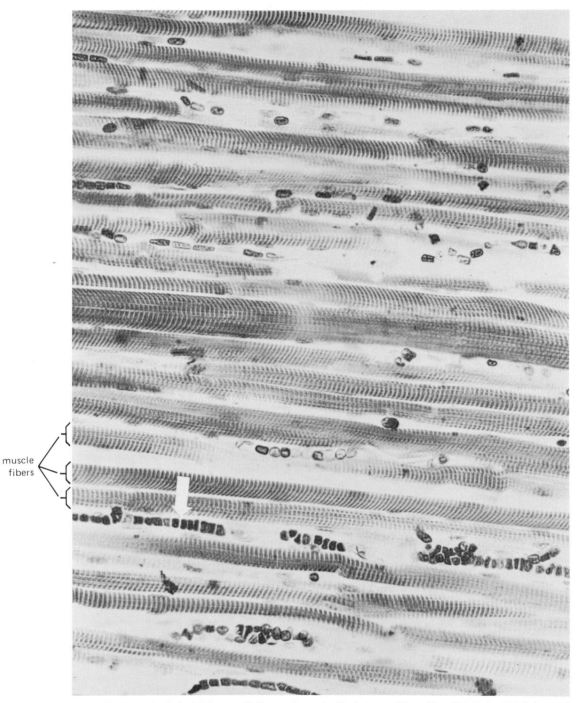

muscle
fibers

Figure 10-2. Photomicrograph of skeletal muscle fibers. Arrow indicates a capillary blood vessel containing red blood cells. *(From Edward K. Keith and Michael H. Ross, "Atlas of Descriptive Histology," Harper & Row, Publishers, Incorporated, New York, 1968.)*

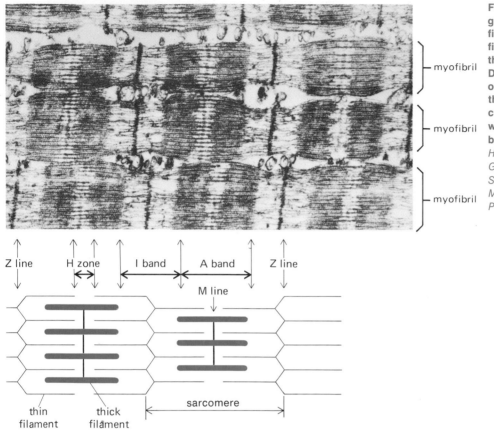

Figure 10-3. Electron micrograph showing three myofibrils in a single muscle fiber. Each myofibril runs the entire length of the fiber. Diagrammed below is the organization of thick and thin filaments in the sarcomeres of a myofibril which gives rise to the banding pattern. [*From H. E. Huxley and J. Hanson, in G. H. Bourne (ed.), "The Structure and Function of Muscle," vol. 1 Academic Press, Inc., New York, 1960.*]

Viewed with the electron microscope, the structures responsible for the banding pattern become evident. The myofibrils consist of smaller *filaments* (Fig. 10-3), which are arranged in a repeating pattern along the length of the fibril. One unit of this repeating pattern is known as a *sarcomere* (little muscle), which is the functional unit of the contractile system in striated muscles. Each sarcomere contains two types of filaments: thick filaments composed of the contractile protein *myosin*, and thin filaments containing the contractile protein *actin*. The thick filaments, 12–18 nm in diameter, are located in the central region of the sarcomere where their orderly parallel arrangement gives rise to the dark bands, known as *A bands*. The thin filaments, 5–8 nm in diameter, are attached at either end of a sarcomere to a structure known as the *Z line*. Two successive Z lines define the limits of one sarcomere. The Z lines are short fibrous structures which interconnect the thin filaments from two adjoining sarcomeres and thus provide an anchoring point for the thin filaments, which extend from the Z lines toward the center of the sarcomere where they overlap with the thick filaments.

Between the ends of the dark A bands of two adjacent sarcomeres is the *I band* (Fig. 10-3), forming the lighter region of the striated pattern (note that unlike the A band, the I band includes segments of two sarcomeres). This band contains those portions of the thin filaments which do not overlap the thick filaments; it is bisected by the Z line. One additional band, the *H zone*, appears as a thin, lighter band in the center of the A band; it corresponds to the space between the ends of the thin filaments; thus, only thick filaments are found in the H zone. Finally, a thin dark band can be seen in the center of the H zone; this is known as the *M line* and is produced by linkages between the thick filaments. The M line, by cross-linking the thick filaments, keeps all these in a single sarcomere in parallel alignment. Thus, neither the thick nor the thin filaments are free floating; each is linked either to Z lines, in the case of the thin filaments, or to M lines, in the case of the thick filaments.

A cross section through several myofibrils in the region of their A bands, where both thick and thin filaments overlap (Fig. 10-4), shows their regular, almost crystalline, arrangement. Each thick filament is surrounded by a hexagonal arrangement of six thin filaments, and each thin filament is surrounded by a triangular arrangement of three thick. Altogether there are twice as many thin as thick filaments in the region of overlap.

At higher magnification, in the region of overlap (Fig. 10-5), the space between adjacent thick and thin filaments is seen to be bridged by projections at intervals along the filaments. These projections, or *cross bridges*, are portions of myosin molecules which extend from the surface of the thick filaments. The cross bridges are arranged in a helix around the thick filament so that those extending from a single thick filament can make contact with each of the six surrounding thin filaments (Fig. 10-4).

Molecular mechanisms of contraction

Sliding-filament model

When a skeletal muscle fiber shortens, the width of the A band in a sarcomere remains constant (Fig. 10-6). Since this width corresponds to the length of the thick filament, the filament must not have changed length during shortening. In contrast, the width of the I band and H zone decreases during shortening as a result of the movement of the thin filaments past the thick filaments. Thus, each H zone decreases as the ends of the thin filaments from opposing ends of its sarcomere approach each other, and it disappears when these thin filaments meet in the center of the A band (further shortening produces a new band in the center of the A band, one which increases in width as the two sets of thin filaments overlap each other more and more). The I band also decreases in width for the same reason, i.e., as more and more of the thin filament length comes to overlap the thick filaments. Thus, the lengths of neither the thick filaments nor the thin filaments change during shortening; rather the two sets of filaments

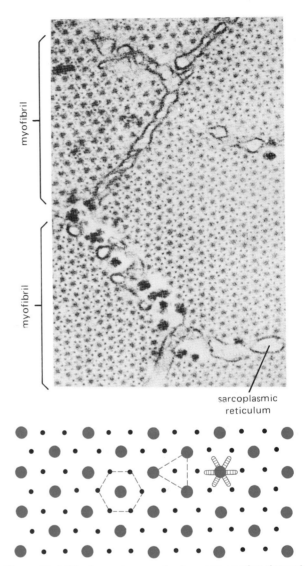

Figure 10-4. Electron micrograph of a cross section through six myofibrils in a single skeletal muscle fiber. Diagrammed below are the hexagonal arrangements of the thick and thin filaments and the bridges extending from a thick filament to each of the surrounding thin filaments. [*From H. E. Huxley,* J. Mol. Biol., *37:507-520 (1968).*]

merely slide past each other. These observations lead to the *sliding-filament model* of muscle contraction.

What produces the movement of these filaments? The answer is the cross bridges which extend from the surface of the thick filaments and

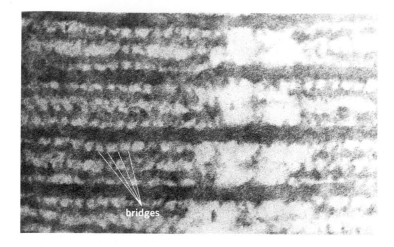

bridges

Figure 10-5. High-magnification electron micrograph in the region of the A band in a myofibril. The H zone can be seen at the righ. Bridges between the thick and thin filaments can be seen at regular intervals along the filaments. [From H. E. Huxley and J. Hanson, in G. H. Bourne (ed.), "The Structure and Function of Muscle," vol. 1, Academic Press, Inc., New York,' 1960.]

make contact with the adjacent thin filaments. When a cross bridge is activated, it moves in an arc parallel to the long axis of the thick filament, much like an oar on a boat. If, at this time, the cross bridge is attached to a thin filament, this swiveling motion slides the thin filament toward the center of the A band, thereby producing shortening of the sarcomere. One stroke of a cross bridge produces only a small displacement of a thin filament relative to a thick, but the cross bridges undergo many cycles during a single contraction, each cycle requiring attachment of the bridge to the thin

Figure 10-6. Changes in filament alignment and banding pattern in a myofibril during shortening.

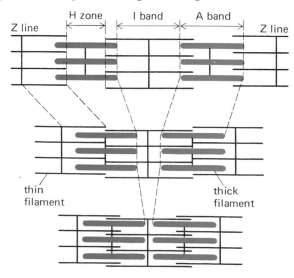

Z line   H zone   I band   A band   Z line

thin
filament

thick
filament

filament, angular movement of the bridge while attached, detachment from the thin filament, re-attachment at a new location, and repetition of the cycle. Each cross bridge undergoes its own independent cycle of movement, so that at any one instant during contraction only about 50 percent of the bridges are attached to the thin filaments, while the others are at intermediate stages of the cycle.

Let us look more closely at the filaments and their cross bridges. Molecules of actin are arranged in two chains helically intertwined to form the thin filaments (Fig. 10-7). Myosin is a much larger molecule with a globular end attached to a long tail. Approximately 200 myosin molecules comprise a single thick filament; the tails of the molecules lie along the axis of the filament and the globular heads extend out to the sides, forming the cross bridges. Each globular head contains a binding site able to bind to a complementary site on an actin molecule. The myosin molecules in the two halves of each thick filament are oriented in opposite directions, such that all their tail ends are directed toward the center of the fiber (Fig. 10-8). Because of this arrangement, the power strokes of the cross bridges in the two ends of each

Figure 10-7. Two helical chains of globular actin molecules form the primary structure of the thin filaments.

thick filament are directed toward the center, thereby moving the thin filaments at both ends of the sarcomere toward the center of the sarcomere during shortening.

In addition to the binding site for actin, the globular end of myosin contains a separate active enzymatic site that catalyzes the breakdown of ATP to ADP and inorganic phosphate, releasing the chemical energy stored in ATP. (This myosin ATPase site requires magnesium as a cofactor to bind ATP to the active site.) The actual splitting of ATP occurs on the myosin molecule before it attaches to actin, but the ADP and inorganic phosphate generated remain bound to the active site on myosin. The chemical energy released at the time of ATP splitting is transferred to myosin (M), producing a high-energy form of myosin (M*)

$$M \cdot ATP \longrightarrow M^* \cdot ADP \cdot P_i$$
(ATP breakdown)

The subsequent binding of this high-energy form of myosin to actin (A) via a cross bridge triggers the discharge of the energy stored in myosin, with the resultant production of the force causing movement of the cross bridge (the ADP and P_i are released from myosin at this time):

$$A + M^* \cdot ADP \cdot P_i \rightarrow A \cdot M^* \cdot ADP \cdot P_i \rightarrow A \cdot M + ADP + P_i$$
(actin binding)　　　　(energy release and bridge movement)

This sequence of energy storage and release by myosin is analogous to a mousetrap: energy is stored in the trap when the spring is cocked (ATP splitting) and released when the trap is sprung (binding to actin). This is, of course, only an analogy, and the actual structural changes in myosin which accompany energy storage and release are unknown.

During contraction, as we have seen, each cross bridge undergoes many cycles of attachment, movement, and dissociation from the thin filaments, each cycle being accompanied by the splitting of one molecule of ATP. However, the myosin cross bridge binds very firmly to actin and this linkage must be broken at the end of each bridge cycle; the binding of a new molecule of ATP to myosin is responsible for breaking this link:

$$A \cdot M + ATP \longrightarrow A + M \cdot ATP$$
(A·M dissociation)

Thus, upon binding (but not splitting) a molecule of ATP, myosin dissociates from actin. The free

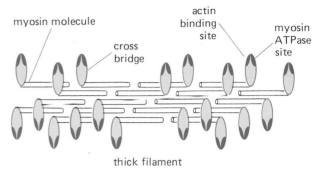

Figure 10-8. Orientation of myosin molecules in thick filaments, with the globular heads of the myosin molecules forming the cross bridges.

myosin bridge then splits its bound ATP, thereby reforming the high energy state of myosin, which can now reattach to a new site on the actin filament, and so on. Thus, ATP performs two distinct roles in the cross-bridge cycle: (1) The energy released from the splitting of ATP provides the energy for cross-bridge movement; (2) the binding (not splitting) of ATP to myosin breaks the link between actin and myosin at the end of a cross-bridge cycle, allowing it to be repeated. Figure 10-9 provides a summary of these chemical and mechanical changes that occur during one cross-bridge cycle.

The importance of ATP in dissociating actin and myosin at the end of a bridge cycle is illustrated by the phenomenon of *rigor mortis* (death rigor), in which the muscles become very stiff shortly after death. Rigor mortis begins 3 to 4 hours after death and is complete after about 12 hours, but then slowly disappears over the next 48 to 60 hours. Rigor mortis results directly from the loss of ATP in the dead muscle cells. In the absence of ATP the myosin cross bridges are able to bind to actin but the bond between them is not broken. The thick and thin filaments become cross-linked to each other, producing the rigid condition of the dead muscle. In contrast, in the living muscle at rest, the myosin bridges are not bound to actin and the filaments readily slide past each other when the muscle is passively stretched.

Regulator proteins and calcium

Since a muscle fiber contains all the ingredients necessary for cross-bridge activity—actin, myosin, ATP, and magnesium ions—the question arises:

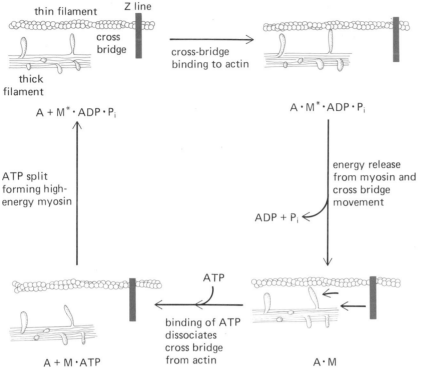

thin filament

Z line

cross bridge

thick filament

$A + M^* \cdot ADP \cdot P_i$

cross-bridge binding to actin

$A \cdot M^* \cdot ADP \cdot P_i$

ATP split forming high-energy myosin

energy release from myosin and cross bridge movement

$ADP + P_i$

ATP

binding of ATP dissociates cross bridge from actin

$A + M \cdot ATP$

$A \cdot M$

Figure 10-9. Chemical and mechanical changes during the stages of a single cross-bridge cycle. Start reading figure at the relaxed fiber in the upper left.

Why are muscles not in a continuous state of contraction? The reason is that in a resting, relaxed muscle fiber, the cross bridges are unable to bind to actin and initiate the cross-bridge cycle that leads to contraction. This inhibition of cross-bridge binding is due to two proteins, *troponin* and *tropomyosin,* which are bound to the thin filaments (Fig. 10-10). Tropomyosin is a rod-shaped molecule with a length equal to approximately seven actin molecules. Tropomyosin molecules are arranged end-to-end along the chains of actin, so that they partially cover the myosin binding sites on the actin molecules, thereby preventing them from binding to the myosin cross bridges. Each tropomyosin molecule is held in this blocking position by a molecule of troponin, itself bound to both tropomyosin and actin.

Having accounted for the mechanism that prevents cross-bridge activity and thus keeps a muscle fiber turned "off," we can now ask what turns the muscle fiber "on," i.e., allows cross-bridge activity to proceed? In order for the cross bridges to bind to actin, the tropomyosin molecules must be moved away from their blocking position. This occurs when calcium binds to a specific site on troponin. The binding produces a change in the shape of troponin such that it pulls the tropomyosin bound to it to one side, uncovering the cross-bridge binding sites on actin (Fig. 10-10). Conversely, removal of calcium from troponin reverses the process and tropomyosin moves back into its blocking position so that cross-bridge activity ceases. Note that the activating effect of calcium on contraction is not directly on either actin or myosin, but is mediated through the two regulator proteins, troponin and tropomyosin.

In summary, the availability of calcium ions to the troponin binding sites determines whether a muscle fiber is turned on or off. We now describe how this availability is coupled to the electrical events occurring in the muscle plasma membrane.

Excitation-contraction coupling

Excitation-contraction coupling in skeletal muscle refers to the process by which an action potential in the plasma membrane of the muscle fiber triggers off the sequence of events leading to

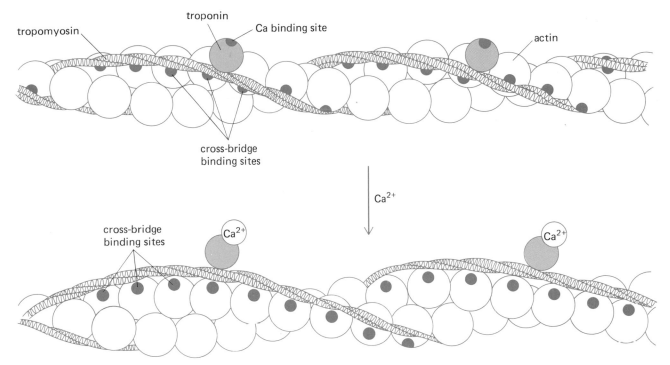

Figure 10-10. In the absence of calcium, tropomyosin blocks the cross-bridge binding sites on actin. Binding of calcium to troponin moves the tropomyosin to one side, exposing the binding sites and allowing cross bridges to bind to the thin filament.

cross-bridge activity and contraction (via increased availability of calcium). This plasma membrane is an excitable membrane capable of generating and propagating action potentials by mechanisms similar to those decribed for nerve cells (Chap. 8). An action potential in a skeletal muscle fiber lasts 1 to 2 ms and is completed before any signs of mechanical activity begin (Fig. 10-11). The period between the initiation of the action potential and the beginning of mechanical activity lasts several milliseconds and is known as the *latent period;* it is during this period that the events of excitation-contraction coupling are occurring. Once begun, the mechanical activity following a single action potential may last 100 ms or more. Note that the mechanical activity far outlasts the duration of the action potential; thus, the electrical activity in the plasma membrane does not directly act upon the contractile proteins but initiates a process which continues to activate the contractile apparatus long after the electrical activity in the membrane has ceased. This process is an increase in the availability of calcium, and

the source of this calcium is the sarcoplasmic reticulum within the fibers. In a resting muscle fiber the concentration of free calcium in the cytosol surrounding the thick and thin filaments is very low, calcium is not bound to troponin, and thus cross-bridge activity is blocked by tropomyosin. Following an action potential there is a rapid increase in the calcium concentration surrounding the contractile filaments, calcium binds to troponin, removing the blocking effect of tropomyosin and thereby initiating contraction.

Sarcoplasmic reticulum. The sarcoplasmic reticulum in muscle is homologous to the endoplasmic reticulum found in most cells. The sarcoplasmic reticulum forms a sleeve-like structure around each of the myofibrils (Fig. 10-12). One segment of the sarcoplasmic reticulum surrounds the region of the A band, while an identical but separate segment surrounds the I-band region. Each of these segments of the sarcoplasmic reticulum possesses at its ends two enlarged sac-like regions, the *lateral sacs,* which are connected to each other

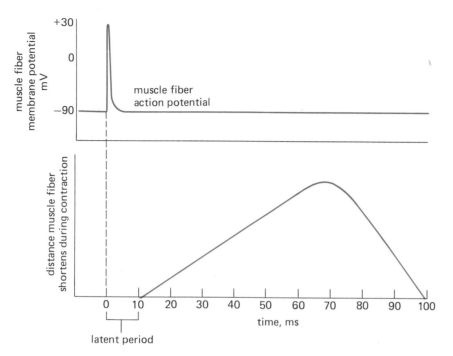

Figure 10-11. Time relations between
a skeletal muscle fiber action potential and mechanical contraction.

by a series of smaller tubular elements. The lateral sacs store the calcium to be released following membrane excitation. A separate, continuous tubular structure, the *transverse tubule (t tubule)* crosses the fiber at the level of each A-I junction, passing between the adjacent segments of the sarcoplasmic reticulum surrounding each myofibril and eventually joining the plasma membrane of the muscle fiber. The lumen of the t tubule is therefore continuous with the extracellular medium surrounding the muscle fiber.

The membrane of the t tubule, like the plasma membrane, is able to propagate action potentials. Thus, an action potential, once initiated in the plasma membrane, is rapidly conducted over the surface of the muscle fiber and into the interior of the fiber along the t tubules. As the action potential in the t tubule passes the lateral sacs of the sarcoplasmic reticulum, they release calcium. Since the reticulum surrounds the myofibrils, the released calcium has only a short distance to diffuse to reach troponin where its binding initiates contraction.

One of the major unsolved problems in the process of excitation-contraction coupling is the mechanism by which an action potential in the membrane of the t tubule triggers the release of calcium from the lateral sacs. There is some evidence to suggest that calcium ions may themselves be the trigger; the action potential increases the permeability of the t tubule membrane to calcium (which is present in the lumen of the tubule at a concentration equal to that in the extracellular medium, a concentration much higher than that of the muscle cytosol), and calcium may diffuse into the fiber across the t tubular membrane to stimulate the membranes of the lateral sacs to release some of their stored calcium.

How is contraction, once initiated by the release of calcium from the sarcoplasmic reticulum, turned off? As mentioned above, the contraction is turned off by removing the calcium from troponin, and this is achieved by once again lowering the calcium concentration in the region of the troponin binding sites. The membrane of the sarcoplasmic reticulum contains a carrier-mediated active transport system that pumps calcium ions from the cytosol into the lumen of the reticulum. Calcium is rapidly released from the reticulum upon arrival of an action potential but the pumping of the released calcium back into the reticulum requires a much longer time. Therefore, the contractile activity continues for some time after the action potential.

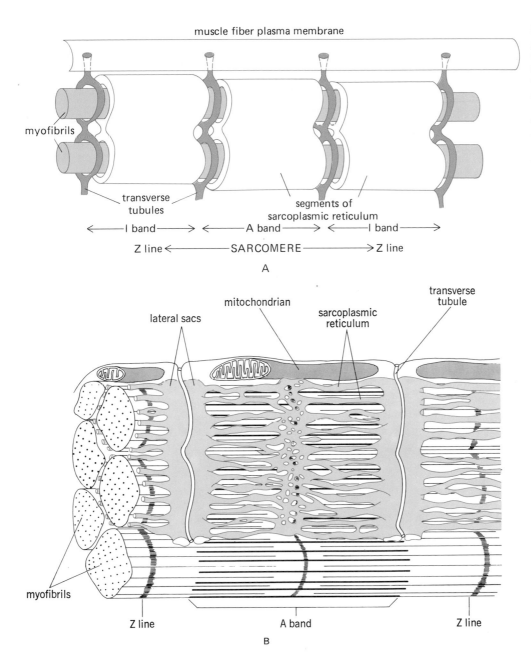

muscle fiber plasma membrane

myofibrils

transverse
tubules

segments of
sarcoplasmic reticulum

←———— I band ————→ ←———— A band ————→ ←———— I band ————→

Z line ←——————— SARCOMERE ———————→ Z line

A

lateral sacs

mitochondrian

sarcoplasmic
reticulum

transverse
tubule

myofibrils

Z line

A band

Z line

B

Figure 10-12. (A) Dia-grammatic representa-tion of the geometrical relationships between the membranes of the sarcoplasmic reticu-lum, the transverse tubules, and the myo-fibrils. (B) Three-di-mensional view of transverse tubules and sarcoplasmic reticulum in human skeletal mus-cle.

To reiterate, just as contraction results from the release of calcium ions stored in the sarcoplasmic reticulum, so relaxation occurs as calcium is pumped back into the reticulum (Fig. 10-13). ATP is required to provide the energy for the calcium pump, and this is the third major role of ATP in the mechanics of muscle contraction, the others being provision of the energy for cross-bridge movement, and inducing the dissociation of actin and myosin at the end of each cross-bridge cycle.

Membrane excitation: The neuromuscular junc-tion. We have just seen that an action potential in

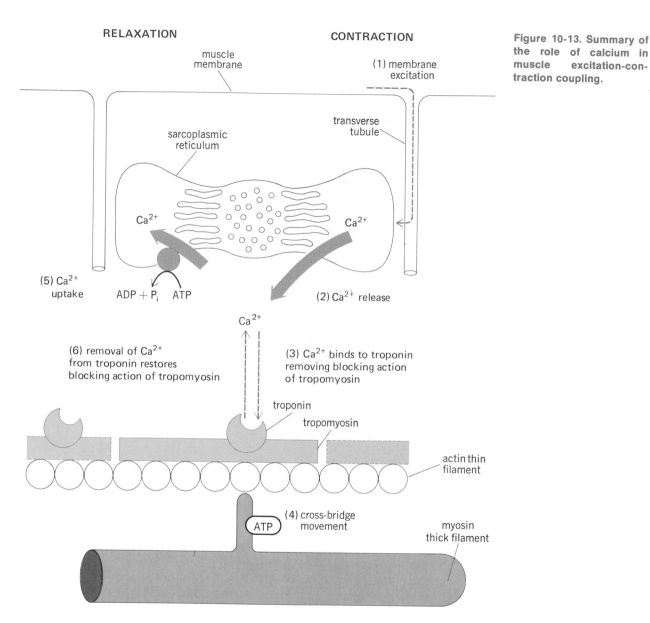

RELAXATION

CONTRACTION

muscle
membrane

(1) membrane
excitation

transverse
tubule

sarcoplasmic
reticulum

Ca^{2+}

Ca^{2+}

(5) Ca^{2+}
uptake

ADP + P$_i$ ATP

(2) Ca^{2+} release

Ca^{2+}

(6) removal of Ca^{2+}
from troponin restores
blocking action of tropomyosin

(3) Ca^{2+} binds to troponin
removing blocking action
of tropomyosin

troponin

tropomyosin

actin thin
filament

(4) cross-bridge
movement

ATP

myosin
thick filament

Figure 10-13. Summary of the role of calcium in muscle excitation-contraction coupling.

the muscle fiber membrane is the signal that leads to mechanical activity by triggering the release of calcium. What are the mechanisms that lead to the initiation of action potentials in muscle-fiber membranes? There are three answers to this question, depending on the type of muscle that is being considered: (1) stimulation by a nerve fiber; (2) stimulation by hormones and local chemical agents; (3) spontaneous electrical activity within the membrane itself. Stimulation by nerve fibers is the only mechanism by which skeletal muscles are normally excited, whereas all three mechanisms are involved in initiating excitation in smooth and cardiac muscle.

The nerve cells whose axons innervate skeletal muscle fibers are known as *motor neurons* (somatic efferent) and their cell bodies are located in the brainstem or spinal cord. The axons of these motor neurons are myelinated and are the largest-diameter axons in the body. They are therefore able to propagate action potentials at high velocities, rapidly initiating muscle activity.

As the motor axon approaches the muscle, it divides into many branches, each of which forms a single junction with a muscle fiber (Fig. 10-14). Thus, each motor neuron is connected through its branching axon to several muscle fibers (although each motor neuron innervates many muscle fibers, each muscle fiber is innervated by only a single motor neuron). The motor neuron plus the muscle fibers it innervates is a *motor unit*. When the motor neuron of a motor unit fires an action potential, all the muscle fibers in that motor unit are activated.

As a branch of the motor axon approaches the muscle surface, it loses its myelin sheath and divides into a fine terminal arborization which lies in grooves on the muscle fiber surface. The region of the muscle membrane which lies directly under the terminal portion of the axon has special properties and is known as the *motor end plate*. The entire junction, including the axon terminal and motor end plate, is known as a *neuromuscular junction*.

The terminal ends of the motor axon contain membrane-bound vesicles resembling the synaptic vesicles found at synaptic junctions. The vesicles contain the chemical transmitter *acetylcholine* (abbreviated ACh). When an action potential in the motor axon arrives at the neuromuscular junction, it depolarizes the nerve membrane increasing its permeability to calcium, which diffuses into the terminal; this calcium triggers a fusion of the transmitter vesicles with the nerve membrane, allowing them to release their contents of acetylcholine into the extracellular cleft separating the nerve and muscle membranes.

The acetylcholine diffuses across this cleft and combines with receptor sites on the motor-end-plate membrane. The binding of ACh to these sites increases the permeability of the membrane to sodium and potassium ions, producing a depolarization of the motor end plate, known as the *end-plate potential* (EPP). The mechanism responsible for the EPP in muscle is similar to that of the EPSP (excitatory postsynaptic potential) produced at synaptic junctions, but the magnitude of a muscle EPP is much larger than a single EPSP, because much larger amounts of transmitter agent are released. The magnitude of a single muscle EPP is sufficiently large to exceed

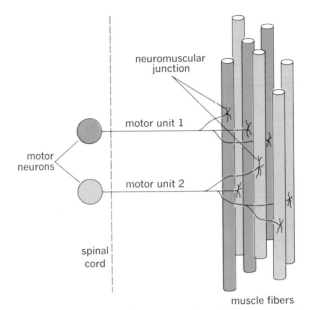

Figure 10-14. Muscle fibers associated with two motor neurons, forming two motor units within a muscle.

the threshold potential of the muscle membrane adjacent to the end-plate region, and local currents flow between these two regions, initiating an action potential in the muscle fiber's plasma membrane. This is then propagated over the surface of the membrane by the same mechanism described for the propagation of action potentials along axon membranes.

Because each EPP produces an action potential in a muscle fiber, every action potential in a motor neuron produces an action potential in the muscle fibers of its motor unit. Thus, there is a one-to-one transmission of activity from the motor neuron to the muscle fiber. This is quite different from the pattern at synaptic junctions in which multiple EPSPs must occur (temporal and spatial summation) in order to reach threshold and elicit an action potential in the postsynaptic membrane. A second difference between synaptic and neuromuscular junctions should be noted. At some synaptic junctions it is possible to produce an inhibitory postsynaptic potential (IPSP) which hyperpolarizes the postsynaptic membrane and decreases the probability of firing an action potential, but such inhibitory potentials are not found in human skeletal muscle; all neuromuscular junctions are excitatory. Thus, the only way to reduce the electrical

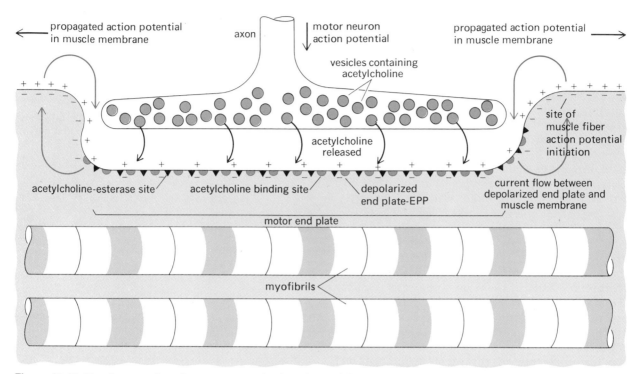

Figure 10-15. Events occurring at a neuromuscular junction which lead to an action potential in the muscle membrane.

activity in the muscle membrane is to inhibit the initiation of action potentials in that muscle fiber's motor neuron.

Motor-end-plate membranes contain, in addition to the receptor sites for acetylcholine, the enzyme acetylcholine-esterase, which breaks down ACh (just as at ACh-mediated synapses in the nervous system). ACh bound to the receptor sites is in equilibrium with free ACh in the cleft between the nerve and muscle membranes; therefore as the concentration of free ACh falls (because of its breakdown by acetylcholine-esterase) bound ACh is released from the receptor sites and is subsequently broken down. When the receptor sites no longer contain bound ACh, the end-plate permeability to sodium and potassium ions decreases to normal and the depolarized end plate returns to the resting potential. The events occurring at the neuromuscular junction are summarized in Fig. 10-15.

There are many ways in which events at the neuromuscular junction can be modified by disease or drugs. For example, the deadly South American Indian arrowhead poison, curare, is strongly bound to the acetylcholine receptor site, but it does not change membrane permeability, nor is it destroyed by acetylcholine-esterase. When a receptor site is occupied by curare, acetylcholine cannot bind to the receptor; therefore, although the motor nerves still conduct normal action potentials and release acetylcholine, there is no resulting EPP or contraction. Since the skeletal muscles responsible for breathing movements depend upon neuromuscular transmission to initiate their contraction, death comes from asphyxiation. Curare and similar drugs are used in small amounts to prevent muscular contractions during certain types of surgical procedures.

Neuromuscular transmission can also be blocked by inhibition of acetylcholine-esterase. Some organophosphates, which are the main ingredients in certain pesticides and nerve gases (the latter developed for biological warfare), inhibit this enzyme; acetylcholine is not destroyed and its prolonged action maintains depolarization of the motor end plate. The failure of repolarization prevents new action potentials from being initiated; thus the muscle does not contract in response to further nerve stimulation, and the result is skeletal muscle paralysis and death from asphyxiation.

A third group of substances, such as botulinus toxin, produced by the bacterium *Clostridium botulinum*, blocks the release of acetylcholine from the nerve terminals, and thus prevents excitation of the muscle membrane. Botulinus toxin is responsible for one type of food poisoning and is one of the most deadly poisons known. Less than 0.0001 mg is sufficient to kill a human being, and 500 g could kill the entire human population.

One form of neuromuscular disease, known as *myasthenia gravis*, which is associated with skeletal muscle weakness and fatigue, is due to decreased numbers of ACh receptor sites at the motor end plates. The release of ACh from the nerve terminals is normal but the magnitude of the muscle EPP is markedly reduced because of the decreased number of receptor sites. As described in Chap. 17, the destruction of the ACh receptors in this disease is brought about by the body's own defense mechanisms gone awry.

The sequence of events leading to contraction of a skeletal muscle fiber is summarized in Table 10-1.

Mechanics of muscle contraction

Contraction refers to the active process of generating a force within a muscle. This force, generated by the sliding filaments, is exerted parallel to the muscle fiber. The force exerted by a contracting muscle on an object is known as the muscle *tension*, and the force exerted on the muscle by the object is the *load*. Thus, muscle tension and load are opposing forces. To move a load the muscle tension must be greater than the load.

When a muscle shortens and moves a load, the muscle contraction is said to be *isotonic* (constant tension) since the load on the muscle remains constant and equal to the muscle tension throughout most of the period of shortening. In contrast, when a muscle develops tension but does not shorten, the contraction is said to be *isometric* (constant length). Such contractions occur when the muscle supports a load in a fixed position or attempts to move a load that is greater than the tension developed by the muscle.

The electrical and chemical events occurring in the muscle fibers are the same in both isotonic and

Table 10-1. Sequence of events between a motor neuron action potential and contraction of a skeletal muscle fiber

1 An action potential is initiated and propagated in the axon of a motor neuron as a result of synaptic events on the neuron's cell body and dendrites within the central nervous system
2 The action potential in the axon causes the release of acetylcholine from the axon terminals at the neuromuscular junction.
3 Acetylcholine is bound to receptor sites on the motor-end-plate membrane.
4 Bound acetylcholine increases the permeability of the motor end plate to sodium and potassium ions, producing an end-plate potential (EPP).
5 The EPP depolarizes the muscle membrane by local current flow to its threshold potential, generating an action potential which is then propagated over the surface of the muscle membrane.
6 The action potential is propagated from the surface into the muscle fiber along the transverse tubules.
7 Depolarization of transverse tubules leads to the release of calcium ions from the lateral sacs of the sarcoplasmic reticulum surrounding the myofibrils.
8 These calcium ions bind to troponin located on the thin filaments, causing tropomyosin to move away from its blocking position covering the cross-bridge binding sites on actin.
9 The energized myosin cross bridges on the thick filaments bind to actin: $A + M^* \cdot ADP \cdot P_i \rightarrow A \cdot M^* \cdot ADP \cdot P_i$
10 This binding triggers the release of energy stored in myosin, producing an angular movement of the cross bridge: $A \cdot M^* \cdot ADP \cdot P_i \rightarrow A \cdot M + ADP \times P_i$
11 ATP binds to myosin, breaking the linkage between actin and myosin, thereby allowing the cross bridge to dissociate from actin: $A \cdot M + ATP \rightarrow A + M \cdot ATP$
12 The ATP bound to myosin is split, transferring energy to the myosin cross bridge and readying it for another cycle: $M \cdot ATP \rightarrow M^* \cdot ADP \cdot P_i$
13 The cross bridges repeat the cycle (10 to 13), leading to the movement of the thin filaments past the thick filaments. These cycles of cross-bridge movement continue as long as calcium remains bound to troponin.
14 The concentration of calcium ions around the myofibrils decreases as calcium is actively transported into the sarcoplasmic reticulum by a membrane pump that uses energy derived from the splitting of ATP.
15 Removal of calcium ions from troponin restores the blocking action of tropomyosin, the cross-bridge cycle ceases, and the fiber relaxes.

isometric contractions, i.e., the cross bridges are activated and exert force on the thin filaments. In an isotonic contraction the thin filaments move

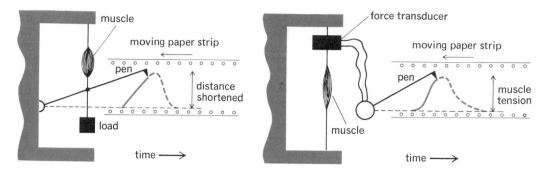

ISOTONIC CONTRACTION   ISOMETRIC CONTRACTION

Figure 10-16. Methods of recording isotonic and isometric muscle contractions.

past the thick filaments, causing the muscle to shorten, whereas in isometric contractions, the cross bridges still exert a force on the thin filaments but there is no overall muscle shortening.

Figure 10-16 illustrates the general method of

Figure 10-17. Isometric and isotonic skeletal muscle twitches following a single action potential.

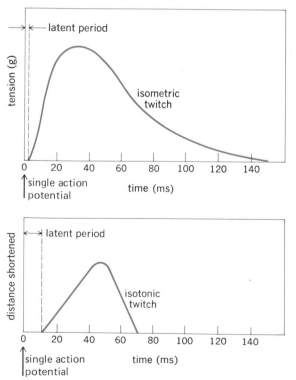

recording isotonic and isometric contractions. During an isotonic contraction the distance the muscle shortens as a function of time is recorded, whereas during an isometric contraction, the force (tension) generated by the muscle as a function of time is measured. Shortening distance can be directly measured by a pen that is attached to a muscle (or to a muscle lever) and leaves a trace on a moving strip of paper as the muscle shortens and relaxes. To measure an isometric contraction the muscle is attached at one end to a rigid support and at the other to a force transducer, which controls the movement of a recording pen in proportion to the force exerted.

Single twitch

The mechanical response of a muscle to a *single* action potential is known as a *twitch*. Figure 10-17 shows the main features of an isometric and an isotonic twitch. In an isometric twitch there is an interval of a few milliseconds, the *latent period*, before the tension begins to increase. It is during this latent period that the processes associated with excitation-contraction coupling are occurring. The time interval from the stimulus to the peak tension is the *contraction time*. Not all skeletal muscles contract at the same rate. Some "fast" fibers have contraction times as short as 10 ms, whereas slower fibers may take 100 ms or longer. The time from peak tension until the tension has decreased to zero is known as the *relaxation time*.

Comparing an isometric twitch with an isotonic twitch in the same muscle, one can see from Fig. 10-17 that the duration of the isotonic twitch is

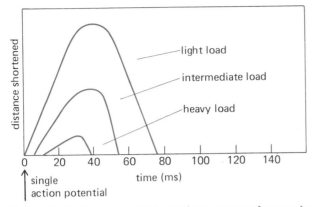

Figure 10-18. Change in the isotonic response of a muscle fiber with different loads.

considerably shorter, whereas the latent period is considerably longer. Once shortening begins, it proceeds at a constant velocity over about 70 percent of the total distance shortened, as can be seen from the straight-line relation between distance shortened and time. The slope of this line indicates the velocity of shortening. Both the velocity of shortening and the duration of an isotonic twitch depend upon the magnitude of the load being lifted (Fig. 10-18). At heavier loads the latent period lasts longer but the velocity of shortening, the duration of the twitch, and the distance shortened all decrease. During the latent period of an isotonic contraction, the cross bridges begin to develop force but actual shortening does not begin

Figure 10-19. Shortening velocity as a function of load.

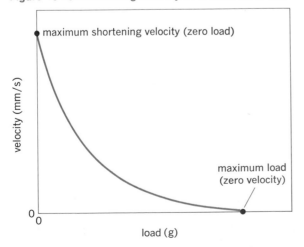

until the muscle tension becomes equal to the load; therefore, the heavier the load, the longer the latent period. As the load on a muscle is increased, eventually a load will be reached that the muscle is unable to lift, the velocity of shortening will be zero, and the contraction becomes isometric. The maximum velocity of shortening occurs when there is no load on the muscle (Fig. 10-19).

Summation of contractions

Since a muscle action potential lasts 1 to 2 ms, whereas the mechanical response to it (single twitch) may last for a hundred milliseconds, it is possible for a second action potential to be initiated during the period of mechanical activity. Figure 10-20 illustrates the isometric contractions of a muscle in response to three successive stimuli. In Fig. 10-20A the isometric twitch following the

Figure 10-20. Summation of isometric contractions produced by shortening the time between stimulus S₂ and S₃.

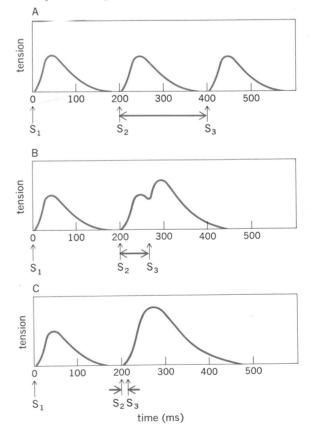

first stimulus S_1 lasts 150 ms. The second stimulus S_2, applied to the muscle 200 ms after S_1 when the muscle has completely relaxed, causes a second identical twitch. In Fig. 10-20B the interval between S_1 and S_2 remains 200 ms, but a third stimulus is applied 60 ms after S_2, when the mechanical response resulting from S_2 is beginning to decrease. Stimulus S_3 induces a contractile response whose peak tension is greater than that produced by S_2. In Fig. 10-20C the interval between S_2 and S_3 is further reduced to 10 ms and the resulting peak tension is even greater; the mechanical response to S_3 is a continuation of the mechanical response already induced by S_2.

The increase in the mechanical response of a muscle to action potentials occurring in rapid succession is known as *summation*. The greater the frequency of stimulation, the greater is the intensity of the mechanical response (summation) until a frequency is reached beyond which the response no longer increases (Fig. 10-21). This is the greatest tension the muscle can develop and is generally about three to four times greater than the isometric twitch tension produced by a single stimulus. The maximal response to high-frequency stimulation is known as a *tetanus*. A muscle contracting isotonically can also undergo summation and tetanus, repetitive stimulation leading in this case to greater shortening.

Just as the contraction time of different muscle fibers varies considerably, so does the stimulus frequency that will produce a tetanus. Frequencies of about 30 per second may produce a tetanus in

"slow" fibers, whereas frequencies of 100 per second or more are necessary in very rapidly contracting fibers.

How long a muscle fiber can be maintained in a tetanically contracted state depends upon the ability of the muscle's metabolism to supply ATP to the contractile proteins. If the ATP concentration of the muscle declines, the force of contraction lessens and eventually falls to zero. This drop in tension with prolonged stimulation is known as *muscle fatigue*. The time of fatigue onset during repetitive stimulation varies considerably for different types of muscle fibers; some of the factors contributing to fatigue will be discussed later in the chapter.

What is the mechanism responsible for the increase in the mechanical response of a muscle fiber to repetitive stimulation? The explanation involves the passive elastic properties of the muscle which cause a dissociation between the *internal* tension generated by the myofibrils and the *external* manifestation of this activity—shortening or tension development by the muscle fiber. Let us first look once more at the internal events. In skeletal muscle fibers, the amount of calcium released from the sarcoplasmic reticulum by a single action potential is sufficient to produce nearly complete saturation of all troponin sites so that all of the cross bridges are turned "on" and the filaments exert their maximal force (this is probably not the case in cardiac and smooth muscle, as we shall see). This internal tension generated by the filaments as a result of cross-bridge cycling

Figure 10-21. Isometric contractions produced by multiple stimuli of (B) 10 stimuli per second and (C) 100 stimuli per second as compared with a single twitch (A).

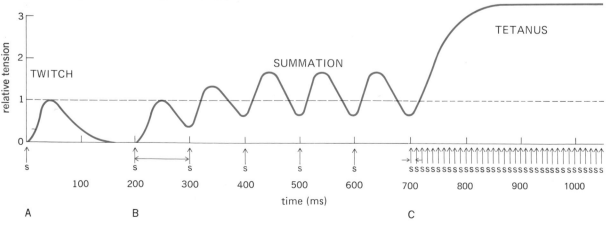

at any instant is known as the *active state* of the muscle fiber. The active state declines as calcium is pumped back into the sarcoplasmic reticulum (i.e., the active state parallels calcium availability). As can be seen from Fig. 10-22, the time course of the active state is markedly different from that of the external tension being exerted on a load during an isometric twitch. What causes this dissociation?

Tension is transmitted from the cross bridges through the thick and thin filaments, across the Z lines, and eventually through the extracellular connective tissue and tendons to the load on the muscle fiber. All these structures have a certain amount of elasticity and are collectively known as the *series elastic component* (SEC). The series elastic component has the properties of a spring placed between the force-generating cross bridges and the external load. The force generated by the cross bridges stretches this spring (the series elastic component), which in turn transmits its tension to the external load. The tension in the series elastic component is therefore the external tension, and it depends on the extent to which the SEC has been stretched by cross-bridge activity; the greater the stretch, the greater the tension in the SEC and the greater the tension transmitted to the load on the muscle fiber. Here is the crucial point: It takes time to stretch the series elastic component. In a single twitch, while the SEC is being stretched by the forces generated by the cross bridges, these very forces are declining (Fig. 10-22) as calcium is pumped back into the sarcoplasmic reticulum. Therefore, the peak tension developed by the SEC (the external tension) during the single twitch is not as great as the peak force generated by the cross bridges.

In contrast, when a muscle is stimulated repetitively, the cross bridges remain active for a longer period of time (i.e., the active state is prolonged) because the repetitive release of calcium from the sarcoplasmic reticulum maintains the required free calcium concentration. Therefore, because more time is available to stretch the SEC, it is stretched farther, and the tension transmitted to the load is greater, i.e., summation occurs. In a tetanus, the continuous maintenance of the active state at its maximal level allows sufficient time to fully stretch the SEC so that the external tension becomes equal to the force exerted by the cross bridges. To summarize, the increase in external

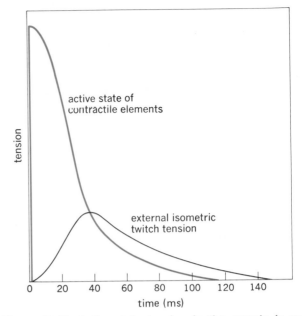

Figure 10-22. Active-state tension in the muscle in response to a single action potential as a function of time.

tension (summation and tetanus) which accompanies repeated stimulation is the result of an increase in the length of time the cross bridges remain active, thereby allowing more time to stretch the SEC and increase the amount of tension transmitted to the load on the muscle.

Length-tension relationship

One of the classic observations in muscle physiology is the relationship between a muscle's length and the tension it develops. A muscle fiber can be passively stretched to various lengths, and the magnitude of the isometric tetanus tension (maximal tension) during stimulation measured at each length (Fig. 10-23). The length at which the muscle develops the greatest tension is termed l_0. When the muscle is set at a length 60 percent of l_0, it develops no tension when stimulated. As the length of the muscle is increased, the developed tension rises to a maximum at l_0 and further lengthening of the muscle causes a drop in tension. When the muscle is stretched to 175 percent of l_0 or beyond, it again develops no tension.

This relationship can be explained in terms of the sliding-filament model. Stretching a muscle

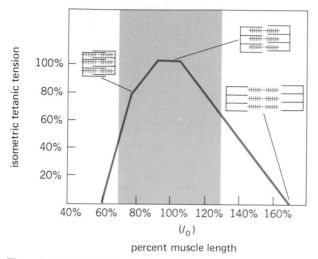

Figure 10-23. Variation in isometric tetanus tension with muscle fiber length. The shaded band represents the range of length changes (from 70 to 130 percent) that can occur in the body while the muscles are attached to bones. *(Adapted from Gordon, Huxley, and Julian.)*

changes the amount of overlap between the thick and thin filaments in the myofibrils. Stretching a muscle to 175 percent of l_0 pulls the thick and thin filaments so far apart that there is no overlap between the two, and since there is no overlap there can be no bridge interaction and no tension developed. A larger and larger region of the thick filaments overlaps with thin filaments as the muscle length is decreased from 175 percent l_0 to l_0; the tension developed increases in proportion to the increased number of interacting cross bridges in the overlap region. At l_0 there is a maximal overlap of thick and thin filaments, and tension is maximal. At lengths less than l_0 two mechanical factors lead to decreasing tension: (1) The thin filaments in the two halves of the sarcomere begin to overlap each other, interfering with cross-bridge binding and decreasing the total number of active cross bridges; (2) the thick filaments become compressed against the two Z lines and resist further shortening of the sarcomeres. In addition to these mechanical factors, at muscle lengths below about 80 percent of l_0 the process of excitation-contraction coupling becomes defective for some reason and all the cross bridges are not activated following stimulation.

In the body, where muscles are attached to bones, the relaxed length of the muscle is very

nearly l_0 and thus is optimal for force generation. The total range of length changes that a skeletal muscle can undergo while still attached to bone is limited to about a 30 percent increase or decrease of its resting length and is often much less. Therefore, the length-tension relation contributes to some extent to variations in skeletal muscle force within the body, but as we shall see, it plays an even more important role in determining the force generated by cardiac muscle. The length of cardiac muscle fibers is not limited by their attachment to bones and thus they can undergo much larger changes in length, leading to corresponding changes in their capacity to develop tension.

Control of whole-muscle tension

Thus far we have described the structure and contractile properties of individual skeletal muscle fibers, these fibers arranged in bundles and sheets to form muscles. The human body contains over 600 different skeletal muscles, which taken all together comprise the largest tissue in the body, accounting for 40 to 45 percent of the total body weight. Some muscles are very small, consisting of only a few hundred fibers, while larger muscles may contain several hundred thousand.

Surrounding the individual fibers in a muscle is a network of collagen fibers and connective tissue through which pass blood vessels and nerves. Collagen is a fibrous protein which has great strength but no active contractile properties. Each end of a muscle is usually attached to bone by bundles of collagen fibers known as *tendons.* The collagen fibers in the connective tissue network surrounding the fibers and in the tendons act as a structural framework which transmits the contractile force of the muscle fibers to the bones.

In some muscles, the individual fibers extend the entire length of the muscle, but in most, the fibers are shorter. When short fibers are parallel to the axis of a muscle, their ends are anchored to the connective tissue network within the muscle. In other cases, the fibers are aligned diagonally to the long axis of the muscle with their ends attached to connective tissues in the center and along the sides of the muscle. The transmission of force from muscle to bone is like a number of people pulling on a rope, each one corresponding to a single muscle fiber and the rope to the connective tissue and tendons.

Some tendons are very long, and the site of attachment of the tendon to bone is far removed from the muscle. For example, some of the muscles which move the fingers are in the forearm, as one can observe by wiggling one's fingers and feeling the movement of the muscles in the lower arm. These muscles are connected to the fingers by long tendons. If the muscles which move the fingers were located in the fingers themselves, we would have very fat fingers.

The total tension a muscle can develop depends upon two factors (Table 10-2): (1) the number of muscle fibers in the muscle that are contracting at any given time; (2) the amount of tension developed by each contracting fiber. Let us look at these, in turn.

The number of muscle fibers contracting at any time depends upon both the number of motor neurons activated and the number of muscle fibers associated with each of them (recall that each motor neuron innervates several muscle fibers, forming a motor unit [Fig. 10-14]). In other words, one way to alter the total tension a muscle develops is to vary the number of motor units that are activated. This is, of course, determined by the activity of the synaptic inputs to the motor neurons. A single motor neuron may receive as many as 15,000 synaptic endings, which converge from many different sources, the balance between excitatory and inhibitory synaptic input determining whether a given motor neuron will fire or not fire an action potential. An increase in the number of motor neurons that are actively discharging action potentials to a given muscle is known as *recruitment.*

The number of muscle fibers associated with a single motor unit varies considerably in different types of muscles. In muscles, such as those in the hand and eye, which are able to produce very delicate movements, the size of the individual motor units is small (e.g., in an eye muscle, one motor neuron innervates only about 13 muscle fibers). In contrast, in the more coarsely controlled muscles of the back and legs, each motor unit contains hundreds of muscle fibers (e.g., a single motor unit in the large calf muscle of the leg contains about 1,700 muscle fibers). If a muscle is composed of small motor units, the total tension produced by the muscle can be increased in small steps by the recruitment of additional motor units, whereas if the motor units are large, big jumps in tension

Table 10-2. Factors determining total muscle tension

I Number of muscle fibers contracting
  a. Recruitment of motor units
  b. Number of muscle fibers per motor unit
  c. Asynchronous activity of motor units
II Tension produced by each contracting muscle fiber
  a. Frequency of action potentials in motor neuron (summation and tetanus)
  b. Muscle-fiber length (degree of overlap of thick and thin filaments)
  c. Duration of activity (fatigue)
  d. Characteristics of different types of muscle fibers

occur as each additional motor unit is recruited. Thus, finer control of muscle tension is achieved in muscles with small motor units.

The motor neurons to a given muscle fire asynchronously. Thus, some motor units may be active while other motor units are momentarily inactive. In muscles which are active for long periods of time, such as the postural muscles which support the weight of the body, this asynchronous activity helps prevent the fatigue that might otherwise result from prolonged continuous activity. Asynchronous activity is able to maintain a nearly constant tension in a muscle, as illustrated in Fig. 10-24.

Now for the second factor determining total-muscle tension. Not only can the number of active motor units be varied, but the tension developed by the individual motor units can be altered. The mechanisms which do so are those discussed previously—summation and tetanus, the length-tension relationship, and fatigue. As regards the first of these mechanisms, in general when a motor neuron is stimulated, it discharges a burst of action potentials rather than a single one. Thus the contractions of motor units are brief summations or more prolonged tetanic contractions, rather than single twitches.

Lever action of muscles

A contracting muscle exerts a force on bones through its connecting tendons. When the force is great enough, the bone moves as the muscle shortens. A contracting muscle exerts only a pulling force, so that as the muscle shortens, the bones to which it is attached are pulled toward each other. *Flexion* of a limb is its bending at a joint, and *extension* is straightening. These motions require

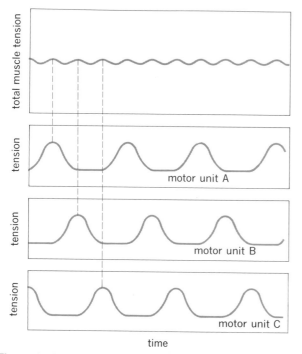

Figure 10-24. Asynchronous motor-unit activity can maintain a nearly constant tension in the total muscle.

The arrangement of the muscles, bones, and joints in the body forms lever systems. The basic principle of a lever is illustrated by the flexion of the arm by the biceps muscle (Fig. 10-27), which exerts an upward pulling force on the forearm about 5 cm away from the elbow. In this example, a 10-kg weight held in the hand exerts a downward force of 10 kg about 35 cm from the elbow. A law of physics tells us that the forearm is in mechanical equilibrium when the product of the downward force (10 kg) and its distance from the elbow (35 cm) is equal to the product of the upward force (X) exerted by the muscle and its distance from the elbow (5 cm), i.e., $10 \times 35 = 5X$; thus $X = 70$ kg. In other words, this system is working at a mechanical disadvantage since the force exerted by the muscle is considerably greater than the load it is supporting (in individuals with very powerful muscles, the large forces produced under conditions of maximum exertion sometimes tear the tendon away from the muscle or bone, or in rare cases break the bone). However, the me-

at least two separate muscles, one to cause flexion and the other extension. From Fig. 10-25 it can be seen how contraction of the biceps causes flexion of the elbow and how contraction of the triceps causes its extension (note that both muscles exert only a pulling force upon the forearm when they contract, and that these forces cause flexion or extension because of the arrangement of the muscles on the bones).

Groups of muscles which produce oppositely directed movements of a limb are known as *antagonists*. Sets of antagonistic muscles are required not only for flexion-extension, but for side-to-side movements or rotation of a limb. The contraction of some muscles leads to two types of limb movement; for example, contraction of the gastrocnemius muscle in the calf causes both extension of the foot and flexion of the leg at the knee, as in walking (Fig. 10-26). Contraction of the gastrocnemius at the same time as that of the quadriceps femoris (which causes extension of the lower leg) prevents the knee joint from bending, leaving only the ankle joint capable of moving; the foot is extended, and the body rises on tiptoe.

Figure 10-25. Antagonistic muscle for flexion and extension of the forearm.

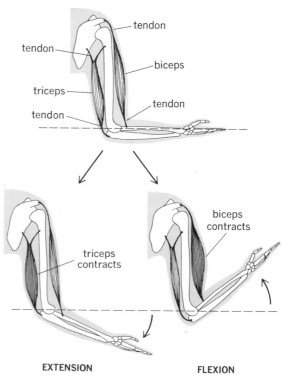

EXTENSION   FLEXION

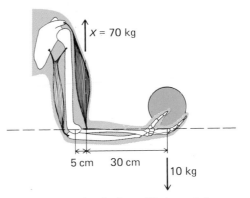

$X = 70$ kg

5 cm   30 cm

10 kg

Figure 10-27. (A) Mechanical equilibrium of forces acting on the forearm while supporting a 10-kg load.

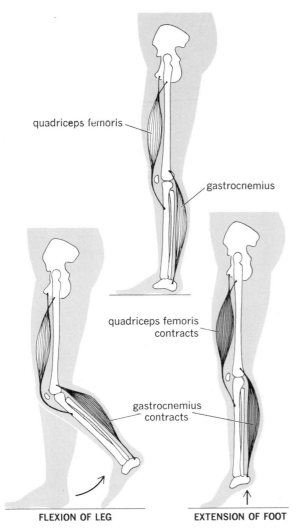

quadriceps femoris

gastrocnemius

quadriceps femoris
contracts

gastrocnemius
contracts

FLEXION OF LEG   **EXTENSION OF FOOT**

Figure 10-26. Flexion of the leg or extension of the foot follows contraction of the gastrocnemius muscle, depending on the activity of the quadriceps femoris muscle.

throw a baseball at 160 km/h even though his muscles shorten at only a fraction of this velocity.

Muscle energy metabolism

As we have seen, ATP performs three major functions in muscle contraction and relaxation: (1) The

Figure 10-28. Small movements of the biceps muscle are amplified by the lever system of the arm, producing large movements of the hand.

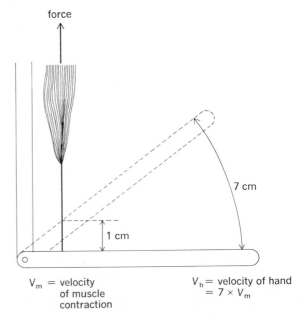

force

7 cm

1 cm

V_m = velocity
of muscle
contraction

V_h = velocity of hand
$= 7 \times V_m$

chanical disadvantage under which most muscles operate is offset by increased maneuverability. In Fig. 10-28, when the biceps shortens 1 cm, the hand moves through a distance of 7 cm. Since the muscle shortens 1 cm in the same amount of time that the hand moves 7 cm, the *velocity* at which the hand moves is seven times faster than the rate of muscle shortening. The lever system amplifies the movements of the muscle so that short, relatively slow movements of the muscle produce faster movements of the hand. Thus, a pitcher can

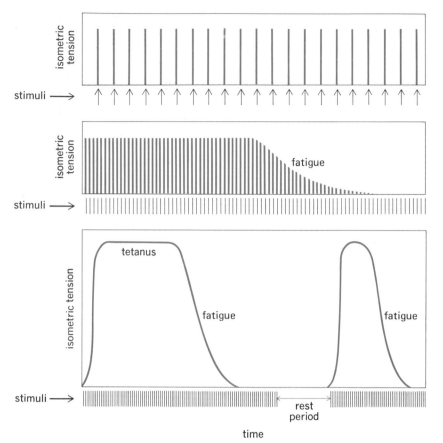

stimuli ⟶

stimuli ⟶

stimuli ⟶

time

Figure 10-29. Muscle fatigue resulting from prolonged stimulation and recovery of ability to contract after a period of rest.

energy released from ATP splitting is directly coupled to the movement of the cross bridges; (2) ATP binding to myosin is necessary to break the link between the cross bridges and actin, allowing the cross bridges to operate cyclically; (3) energy released from ATP splitting is utilized by the sarcoplasmic reticulum to re-accumulate calcium ions, producing relaxation.

When a muscle receives an adequate supply of oxygen and nutrients which can be used to produce ATP, it can continue to give a series of twitch responses to low-frequency stimulation for very long periods of time. In other words, under these conditions the muscle is able to synthesize ATP at a rate sufficient to keep up with the rate of ATP breakdown. However, if the rate of stimulation is increased, at some point the contractile responses soon begin to grow weaker (in some types of skeletal muscle) and eventually fall to zero as the rate of ATP breakdown exceeds the rate at which ATP can be formed by metabolism (Fig.

10-29). When stimulation is rapid enough to produce tetanic contractions, fatigue occurs even sooner. When the stimulation is stopped and a period of rest is allowed before resuming stimulation, the muscle recovers its ability to contract, because its ATP concentration is restored by metabolism, from the very low concentration typical of fatigued muscle. (In contrast to *metabolic fatigue* of muscle, *psychological fatigue* may cause an individual to stop exercising even though the muscles are not depleted of ATP and are still able to contract; an athlete's performance depends not only on the physical state of the muscles but also upon the "will to win.")

To reiterate, if a muscle had to rely solely on the ATP in its fibers at the start of contraction, it would be completely fatigued within a few twitches; therefore, if a muscle is to maintain its contractile activity, molecules of ATP must be synthesized as rapidly as they are broken down. There are three sources for this ATP (Fig. 10-30):

(1) *creatine phosphate*; (2) oxidative phosphorylation in the mitochondria; (3) substrate phosphorylation during glycolysis.

Creatine phosphate (CP) provides the most rapid means of forming ATP in the muscle cell. This molecule contains energy and phosphate, both of which can be transferred to a molecule of ADP to form ATP and creatine (C):

$$CP + ADP \rightleftharpoons C + ATP$$

Energy is stored as creatine phosphate in resting muscle by the reversal of this reaction. The high levels of ATP in a resting muscle favor, by mass action, the formation of creatine phosphate, and during periods of rest the muscle fibers build up a concentration of creatine phosphate approximately five times that of ATP. When the ATP level begins to fall at the beginning of contraction, mass action favors the formation of ATP from creatine phosphate. This is so efficient that the actual concentration of ATP in the cell changes very little at the start of contraction while the concentration of creatine phosphate falls rapidly.

If contractile activity is to be continued, the muscle must be able to derive ATP from sources other than the limited creatine phosphate stores. At moderate levels of muscle activity (moderate rates of ATP breakdown) most of this ATP can be formed by the process of oxidative phosphorylation, using fatty acids as the predominant source of nutrient and to a lesser extent carbohydrates (see Chap. 15). However, during very intense exercise, when the breakdown of ATP is very rapid, a number of factors begin to limit the cell's ability to replace ATP by oxidative phosphorylation: (1) the delivery of oxygen to the muscle; (2) the availability of nutrients; (3) the rates at which the enzymes in the metabolic pathways can process these nutrients. Any of these may become rate-limiting for oxidative phosphorylation under appropriate conditions, and the rate at which oxidative phosphorylation can produce ATP may therefore become inadequate to keep pace with the rapid rate of ATP breakdown.

Accordingly, when the level of exercise exceeds about 50 percent of maximum (50 percent of the

Figure 10-30. Biochemical pathways producing ATP utilized during muscle contraction.

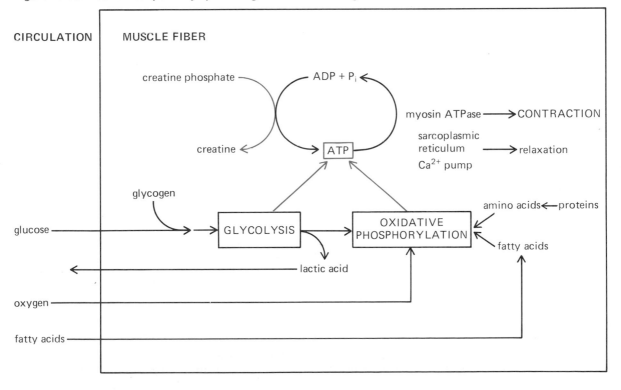

maximal rate of ATP breakdown), glycolysis begins to contribute an increasingly significant fraction of the total ATP produced by the muscle. The glycolytic pathway, although producing only small quantities of ATP from each molecule of glucose metabolized, can operate at a high rate. Thus, while oxidative phosphorylation can produce 36 molecules of ATP from 1 glucose molecule, 64 molecules of ATP may be formed by glycolysis in the same period of time through the breakdown of 32 molecules of glucose to lactic acid (this lactic acid diffuses out of the muscle into the blood). Not only can glycolysis produce ATP rapidly, but it can proceed in the absence of oxygen.

Although glycolysis can produce ATP very rapidly, it has the disadvantage of requiring very large quantities of glucose to produce relatively small amounts of ATP, as the example above illustrates. The ability of muscle to store glucose in the form of glycogen provides a certain degree of independence from externally supplied glucose, and during intense exercise, the glycogen content of the muscle falls, the rate of fall depending upon the intensity of the exercise. The onset of fatigue in these types of exercise correlates closely with the depletion of muscle glycogen stores and thus with the ability of the fibers to replace ATP rapidly by glycolysis. Finally, in very intense exercise (necessarily of short duration) ATP may be broken down faster than even glycolysis can replace it and fatigue occurs rapidly even though glycogen is still present in the muscle.

In summary (Fig. 10-30), creatine phosphate provides a very rapid mechanism for replacing ATP at the onset of contraction until the metabolic pathways can adjust to the increased demand for ATP production. During mild exercise, the metabolism of fatty acid (and, to a lesser extent, glucose) by oxidative phosphorylation provides most of the additional ATP needed. As the intensity of exercise increases, more and more of the ATP is supplied by glycolysis of internal glycogen stores. When the glycogen stores are depleted, the muscle becomes fatigued because it is unable to supply ATP as rapidly as it is broken down. At very intense levels of exercise, fatigue may occur before the glycogen supplies have been depleted, because the very high rate of ATP breakdown is faster than new ATP can be supplied by glycolysis.

Following an intense period of exercise, creatine phosphate levels have decreased, and much of the muscle glycogen may have been converted to lactic acid. To return the cell to its original state, the glycogen stores must be replaced and creatine phosphate resynthesized; both processes require energy. Therefore, to provide the energy necessary for these synthetic processes, a muscle continues to consume oxygen at a high rate for some time after it has ceased to contract (as evidenced by the fact that one continues to breathe deeply and rapidly for a period of time immediately following intense exercise). The longer and more intense the previous exercise, the longer it takes to restore the muscle to its original state.

The overall efficiency with which muscle converts chemical energy to work is only about 20 percent, the remaining 80 percent of the energy released during muscle contraction appearing as heat. Therefore, the more intense the exercise, the greater the amount of heat produced. This increased heat production may place a severe stress upon the ability to maintain a constant body temperature, especially on a hot day. On the other hand, the process of shivering (small rapidly oscillating contractions of skeletal muscles) reflects the use of this same source of heat energy to maintain body temperature in a cold environment. The various mechanisms regulating body temperature will be discussed in Chap. 15.

Muscle fiber differentiation and growth

Each muscle fiber is formed during embryological development by the fusion of a number of small, mononucleated myoblasts to form a single long, cylindrical, multinucleated fiber. It is only after the fusion of the myoblasts that the muscle fibers begin to form actin and myosin filaments and become capable of contracting. Once the myoblasts have fused, the resulting muscle fibers no longer have the capacity for cell division but do retain the ability to grow in length and diameter. This stage of muscle differentiation is completed around the time of birth. Therefore, in an adult, if skeletal muscle fibers are destroyed, they cannot be replaced by division of other existing, differentiated fibers. There may occur some formation of new fibers from undifferentiated cells in the adult, but the major compensation for a loss of muscle tissue occurs through the increased growth in the size of the remaining differentiated fibers.

Following the fusion of the myoblasts in the embryo into muscle fibers, motor neurons send axon processes into the muscle, forming neuromuscular junctions and bringing the muscle under the control of the central nervous system. From this point on there is a very critical dependence of the muscle fiber on its motor neuron, not only to provide a means of initiating contraction but also for its continued survival and development. If the nerve fibers to a muscle are severed or otherwise destroyed, the denervated muscle fibers become progressively smaller, and their content of actin and myosin decreases. This is known as *denervation atrophy* (a muscle can also atrophy with its nerve supply intact if it is not used for a long period of time, as when a broken arm or leg is immobilized in a cast, and this is known as *disuse atrophy*). Denervation atrophy can be prevented by initiating action potentials in the muscle membrane using direct electrical stimulation, a fact which demonstrates that electrical activity in the muscle membrane is the key requirement for the maintenance of the functional state of the muscle. In contrast to the decrease in muscle mass that results from a lack of regular neural stimulation, increased amounts of neural activity, such as accompany repeated exercise, may produce a considerable increase in the size of muscle fibers (*hypertrophy*) as well as other changes in their chemical composition. Action potentials in nerve fibers appear to release chemical substances which influence the biochemical activities of the muscle fiber, but the identity of these tropic agents is unknown.

Types of skeletal muscle fibers

Skeletal muscle fibers differ in their speed of contraction and capacity to split ATP. This is because all myosin molecules are not identical, some having a higher ATPase activity than others. The speed with which a fiber contracts depends upon the rate at which its myosin splits ATP, because the latter determines the rate of cross-bridge cycling. Thus, myosin isolated from muscle fibers with a fast contraction speed is found to split ATP at a very high rate, whereas myosin from slowly contracting fibers has a low ATPase activity.

A second major difference between skeletal muscle fibers is the type of enzymatic machinery available for synthesizing ATP. Some fibers contain numerous mitochondria and thus have a high capacity for oxidative phosphorylation; the activity of the glycolytic enzymes in these fibers is relatively low. Therefore, most of the ATP produced by such fibers is dependent upon a supply of oxygen, and they are surrounded by numerous capillaries. These high-oxidative fibers also contain a protein known as *myoglobin* which is similar to hemoglobin, found in red blood cells. Myoglobin binds oxygen and increases the rate of oxygen diffusion into the muscle cell, as well as providing a small store of oxygen within the fiber. Fibers containing large amounts of myoglobin have a dark red color which distinguishes them from the paler fibers which lack appreciable amounts of myoglobin.

In contrast to these high-oxidative fibers, other types have few mitochondria but a very high capacity for glycolysis and a large store of glycogen. These fibers are specialized for the production of ATP by glycolysis in the absence of oxygen. Corresponding to their low requirement for oxygen, they are surrounded by relatively few capillaries and contain little myoglobin.

Three types of skeletal muscle fibers can be distinguished based on differences in myosin ATPase activity and ATP-synthesizing characteristics (Fig. 10-31):

1 *"Slow twitch–resistant to fatigue."* These fibers combine low myosin-ATPase activity with high oxidative capacity. They manifest a low speed of contraction (because of the low ATPase activity), and their ATP production by oxidative phosphorylation can readily keep pace with ATP breakdown. Moreover, they are well supplied with oxygen and nutrients by their surrounding capillary network. The net effect is that such fibers do not undergo fatigue.

2 *"Fast twitch–resistant to fatigue."* These combine high myosin-ATPase activity with high oxidative capacity and a copious blood supply; they manifest a rapid speed of contraction and will undergo fatigue only if maintained in a contracted state for long periods of time.

3 *"Fast twitch–fatigable."* This type of fiber combines high myosin-ATPase activity with low oxidative capacity, a high glycolytic capacity, and a poor blood supply. Their speed of contraction is fast but they rapidly fatigue as they exhaust their internal supply of glycogen and their poor blood supply fails to provide them with the

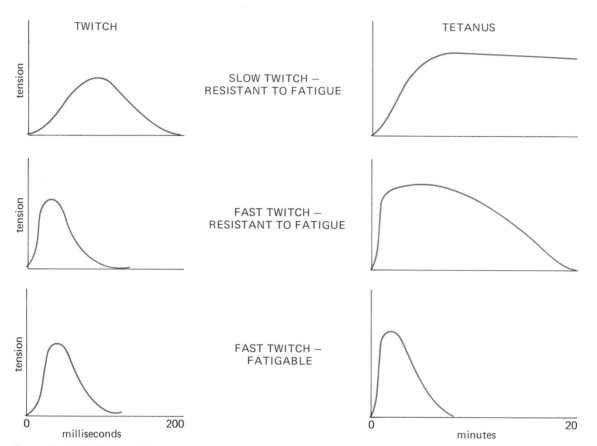

Figure 10-31. Twitch and fatigue characteristics of the three types of skeletal muscle fibers.

additional glucose required to maintain contraction.

It must be recognized that although the speed of contraction of each fiber type depends upon the type of myosin-ATPase it contains, the tension developed by a single cross bridge does not and is the same in all three types of fibers. Therefore, the total tension a given fiber develops depends on the total number of cross bridges that are active in parallel; this reflects both the degree of overlap between the thick and thin filaments (length-tension relationship) and the cross-sectional area of the fiber (i.e., the number of myofibrils in parallel). "Fast twitch–fatigable" fibers tend to have larger fiber diameters than either of the other two fiber types and therefore develop greater maximal tensions.

In summary, three variables determine the char-

acteristics of any given fiber: The myosin-ATPase activity determines the speed of contraction; the type of metabolism, in relation to the myosin-ATPase activity, determines the fibers' susceptibility to fatigue; the diameter of the fiber determines the maximal amount of tension that can be developed at any given fiber length. The characteristics of the three types of skeletal muscle fibers are summarized in Table 10-3.

Some muscles may contain predominantly one type of fiber, but most muscles have a mixture of three types interspersed with each other (Fig. 10-32) (however, all the muscle fibers of a given motor unit are of the same type). Moreover, they contain different proportions of the three types of fibers, so that the contractile properties of whole muscles, compared with single muscle fibers, show a range of contraction speeds and fatigability.

The significance of having distinct muscle-fiber

Table 10-3. Properties of three types of skeletal muscle fibers

	Slow twitch–resistant to fatigue	Fast twitch–resistant to fatigue	Fast twitch– fatigable
Speed of contraction	Slow	Fast	Fast
Myosin-ATPase activity	Low	High	High
Primary source of ATP production	Oxidative phosphorylation	Oxidative phosphorylation	Anaerobic glycolysis
Glycolytic enzyme activity	Low	Intermediate	High
Number of mitochondria	Many	Many	Few
Capillaries	Many	Many	Few
Myoglobin content	High	High	Low
Glycogen content	Low	Intermediate	High
Fiber diameter	Small	Intermediate	Large
Rate of fatigue	Slow	Intermediate	Fast

types is, in part, that skeletal muscles are called upon to perform various functions in different locations. The muscles which support the weight of the body (the postural muscles of the back and legs) must be able to maintain their activity for long periods of time without fatigue; the muscles in the arms may be called upon to produce large amounts of tension rapidly, as in the lifting of heavy objects; the leg muscles perform the movements of walking and running in addition to supporting the body's posture. Moreover, the activities that a given skeletal muscle may be called upon to perform also vary.

Let us look at how these three types of fibers

Figure 10-32. Cross sections of a skeletal muscle, showing individual muscle fibers which have been stained according to their chemical composition. (A) The capillaries surrounding the muscle fibers have been stained. Note the large number of capillaries surrounding the smaller-diameter, high-oxidative fibers. (B) Darkly stained fibers reveal the presence of high concentrations of oxidative enzymes in the high-oxidative, smaller-diameter fibers. *(Courtesy of John A. Faulkner.)*

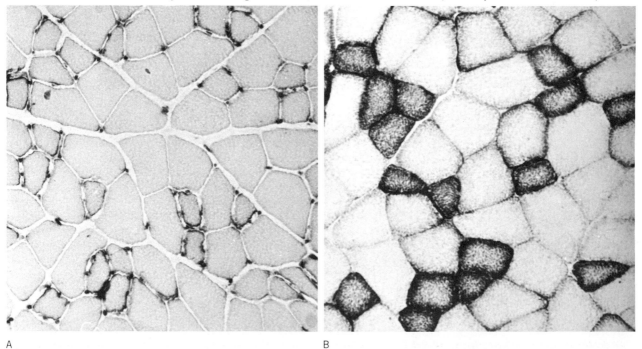

A   B

interact in a single muscle during different types of exercise. In exercise of short duration, the fibers involved depend upon the strength the muscle is called upon to generate: When only weak contractions are required, just the "slow twitch–resistant to fatigue" fibers are recruited. Stronger contractions result from the additional recruitment of the "fast twitch–resistant to fatigue" fibers and eventually the "fast twitch–fatigable" fibers. In contrast, during endurance exercise of long duration and low intensity, it is mainly the first two types of fibers that are recruited.

The types of motor units recruited during a contraction is of course dependent on the pattern of neural activity to the muscle. In addition, this neural activity can gradually produce changes in the muscle: (1) transformation of one fiber type into another; (2) changes in the size of the muscle fibers. For example, endurance types of exercise (such as running and swimming) lead to a transformation of "fast twitch–fatigable" fibers into "fast twitch–resistant to fatigue" fibers, as manifested by an increase in the number of mitochondria and in the number of capillaries surrounding the fibers (these changes are accompanied by only small increases in the diameter of the fibers and their strength). This transformation markedly increases the capacity for long-duration activity without fatigue.

In contrast, high-intensity, short-duration exercise (such as weight lifting) produces quite a different pattern of change in the muscle, namely hypertrophy of the "fast twitch–fatigable" fibers, i.e., increased synthesis of actin and myosin filaments and a large increase in the fiber diameter; the result is a corresponding increase in the muscle's strength of contraction. The extreme result of this type of exercise is the bulging muscles of a professional weight lifter.

Because different types of exercise produce quite different chemical changes in skeletal muscle, an individual performing regular exercises to improve muscle performance must be careful to choose a type of exercise that is compatible with the type of activity he or she ultimately wishes to perform. Thus lifting weights will not improve the endurance of a long-distance runner, and jogging will not produce the increased strength desired by a weight lifter. As we shall see in later chapters, endurance exercise produces changes not only in the skeletal muscles but also in the respiratory and

circulatory systems, changes which improve the delivery of oxygen and nutrients to the muscle fibers.

It should be reemphasized that all of these changes in muscle fiber metabolism with different types of exercise occur without a change in the total number of fibers. The molecular mechanisms by which changes in the pattern of neural activity influence metabolism and myosin composition are not known.

Smooth muscle

Smooth muscle, like skeletal muscle, uses cross-bridge movements between actin and myosin filaments to generate force, and calcium ions to control cross-bridge activity. However, the structural organization of the contractile filaments, the process of excitation-contraction coupling, and the time-course of contraction are quite different in the two types of muscle. Furthermore, there is considerable diversity in the properties of smooth muscle in different organs, especially with respect to the mechanism of excitation-contraction coupling. Nevertheless, two anatomical characteristics are common to all smooth muscles: They lack the cross-striated banding pattern found in skeletal and cardiac fibers (thus the name "smooth muscle"); the nerves to them are derived from the autonomic division of the nervous system rather than the somatic division (thus, smooth muscle is not normally under direct voluntary control).

Smooth muscle structure

Unlike skeletal muscle fibers, the precursor cells of smooth muscle fibers do not fuse during embryological development. Rather, each differentiated smooth muscle fiber is spindle-shaped and contains a single nucleus in its central portion. The diameter of these small fibers ranges from 2 to 10 μm, compared to a range of 10 to 100 μm for skeletal muscle fibers. In most hollow organs, the smooth muscle fibers are arranged in bundles organized into two layers—an outer longitudinal layer and an inner circular layer. In blood vessels, bundles of muscle fibers are arranged in a circular or helical fashion around the vessel wall.

The cytoplasm of smooth muscle fibers is filled with filaments oriented approximately parallel to the long axis of the fiber (Fig. 10-33). Three types

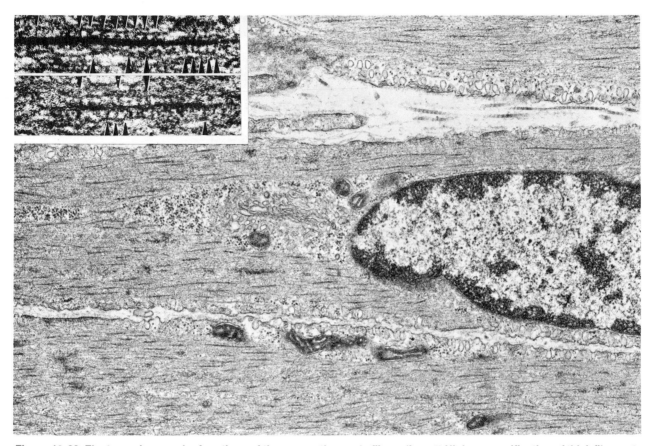

Figure 10-33. Electron micrograph of portions of three smooth muscle fibers. (Insert) Higher magnification of thick filaments with projections (arrows) suggestive of cross bridges connecting adjacent thin filaments. [*From A. P. Somlyo, C. E. Devine, Avril V. Somlyo, and R. V. Rice, Phil. Trans. R. Soc. Lond. B., **265**:223–229 (1973).*]

of filaments are present: thick myosin-containing filaments (which are longer than those of skeletal muscle); thin actin-containing filaments; intermediate-sized filaments that are not found in striated muscles (the latter filaments do not seem to play a role in the active generation of force and probably function as an elastic framework or cytoskeleton to maintain the shape of the cell). There are no Z or M lines to cross-link filaments of similar type, nor are there sarcomeres or any other regular alignment of the filaments; this accounts for the absence of a banding pattern in smooth muscle fibers. However, the actin filaments are anchored, either to the plasma membrane or to cytoplasmic structures (known as *dense bodies*).

Although, as stated above, smooth muscle contains both actin and myosin proteins, the relative amounts of these proteins differ from those found in skeletal muscle. The amount of myosin is only about one-third that in skeletal muscle, whereas the actin content can be as much as twice that of skeletal muscle. These differences are reflected in the relative numbers of thick and thin filaments in the two types of muscle; in the region of thick and thin filament overlap, there are two thin filaments for every thick filament in striated muscle, whereas in smooth muscle there are 10 to 15 thin filaments for each thick. The thick filaments contain myosin cross bridges, although they do not appear to have the opposed orientation in the two halves of the filament present in skeletal muscle.

The tension developed by smooth muscle fibers varies with muscle length in a manner qualitatively similar to skeletal muscle. However, the

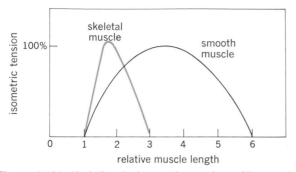

Figure 10-34. Variation in isometric tension with muscle length for skeletal and smooth muscle.

range of muscle lengths at which smooth muscle is able to develop tension is considerably greater (Fig. 10-34). The longer thick filaments, the lack of Z lines to limit shortening, and the lack of a regular alignment of overlapping filaments, all help to account for this wider range of tension development. This property is highly adaptive, since smooth muscle surrounds hollow organs which undergo change in volumes and thus in the lengths of the smooth muscle fibers in their walls. Even with relatively large increases in volume (as during the accumulation of large amounts of urine in the bladder), the smooth muscle fibers in the wall retain some ability to develop tension, whereas such distortion would have stretched skeletal muscle fibers beyond the length of thick and thin filament overlap. The maximal tension per unit of cross-sectional area developed by smooth muscles at their optimum length is comparable to that developed by skeletal muscle, despite the former's lower myosin content.

The presence of myosin cross bridges, overlapping thick and thin filaments, and a length-tension relationship, all suggest that smooth muscle contraction occurs by a sliding-filament mechanism similar to that in skeletal muscle.

Excitation-contraction coupling

Changes in the calcium concentration around the filaments controls the contractile activity of smooth muscle fibers, as it does in skeletal muscle. However, there are significant differences between the two types of muscle as regards the way in which calcium exerts its effects on cross-bridge activity and the mechanisms by which its intracellular concentration is controlled. Smooth

muscle thin filaments contain tropomyosin, as in skeletal muscle, but the site of calcium binding is on the myosin-containing thick filaments, rather than on the thin filaments. Whether calcium binds directly to myosin or to a troponin-like molecule, itself bound to myosin, is still unclear, as is the mechanism by which calcium binding allows cross-bridge activity to proceed. Once activated by calcium, the rate at which ATP is split is much slower than in skeletal muscle, and this accounts for the relatively slow speed of smooth muscle contraction. A single twitch may last several seconds in smooth muscle but only a fraction of a second in skeletal muscle.

There are two sources of the increased free intracellular calcium which triggers contraction in smooth muscle: Some is released from the sarcoplasmic reticulum and some enters the cell across the plasma membrane. Different types of smooth muscles rely more on one source of calcium than the other, but most make use of both sources to some extent. Relaxation is brought about by the removal of calcium from the cytosol through the action of calcium pumps either in the membrane of the sarcoplasmic reticulum or in the plasma membrane.

In smooth muscle, the volume of the fiber occupied by sarcoplasmic reticulum is only a fraction of that in skeletal muscle. Furthermore, there are no transverse tubules connected to the plasma membrane, although regions of the sarcoplasmic reticulum are located near the plasma membrane, forming junctions similar to those between a tubule and the lateral sacs in skeletal muscle. Thus the elaborate skeletal muscle mechanism for rapidly activating and relaxing the contractile apparatus is not present in smooth muscle, but it is not really required since smooth muscle has such a slow rate of contraction anyway. Furthermore, the small diameter of smooth muscle fibers, which are little larger than individual myofibrils within a skeletal muscle fiber, allows fairly rapid diffusion of calcium to the filaments.

In a sense we have approached the question of excitation-contraction coupling backward by first describing the "coupling" (changes in free cytosolic calcium), but now we must ask what constitutes the "excitation" which elicits the changes in calcium.

Many smooth muscles undergo a continuous, low-level contractile activity, known as *tone*, in the absence of any known external stimuli. This tone can be altered and, in addition, phasic contractions can be superimposed upon it in response to the variety of inputs summarized in Table 10-4. (This multiplicity of controls is in contrast to skeletal muscle, in which contractile activity is normally completely dependent on a single input—the somatic neurons to the muscle.) To reiterate, all of these inputs have the ultimate effect of altering either the release of calcium from the sarcoplasmic reticulum or the permeability of the plasma membrane to calcium; they do this either by changing the electrical potential across the plasma membrane or by actions which are independent of changes in membrane potential (Fig. 10-35). Let us examine these in turn.

What is the relationship between membrane potential and free intracellular calcium in smooth muscle? A decrease in membrane potential (depolarization) induced by the input increases the membrane permeability to calcium. Since calcium is present in higher concentration outside the muscle fiber than inside (and since the inside is

Table 10-4. Inputs influencing smooth muscle contractile activity

1 Spontaneous electrical activity in the smooth muscle plasma membrane
2 Neurotransmitters released by autonomic neurons
3 Hormones
4 Locally induced changes in the chemical composition of the extracellular fluid surrounding the smooth muscle fibers (paracrines, acidity, oxygen, osmolarity, ion concentrations, etc.)
5 Rapid stretching of the smooth muscle

negatively charged), any increase in the membrane permeability to calcium results in an increased flux of calcium into the cell. The greater the depolarization, the greater the influx of calcium (this is analogous to the voltage-dependent increase in membrane permeability to sodium that occurs in other excitable membranes, as described in Chap. 8) and the greater the contractile response. The amount of calcium released by the sarcoplasmic reticulum also increases with the degree of plasma-membrane depolarization but

Figure 10-35. Electrical and chemical pathways leading to an increase in intracellular free calcium ion concentration in smooth muscle. The dashed lines denote second-messenger pathways. Not shown are the pathways by which locally induced changes in extracellular composition influence the muscle.

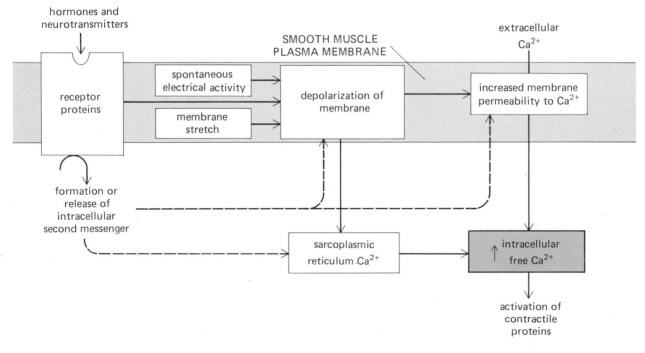

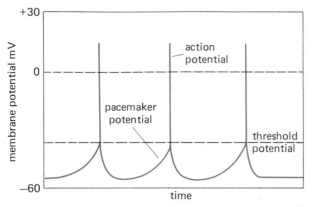

Figure 10-36. Generation of action potentials in a smooth muscle fiber resulting from spontaneous depolarizations of the membrane potential (pacemaker potentials).

the mechanism is not known. Note that we have said nothing thus far of action potentials. In smooth muscle, unlike skeletal muscle, small changes in membrane potential, which do not lead to action potentials, still produce changes in intracellular free calcium and corresponding changes in contractile activity; indeed some smooth muscles never manifest action potentials. On the other hand, if action potentials are elicited by an input, there occurs a much greater increase in calcium and contraction. (Interestingly, calcium ions, rather than sodium ions, are the major carrier of the positive charge into smooth muscle cells during the rising phase of the action potential.)

How is calcium influx or release from the sarcoplasmic reticulum altered by inputs which do not cause changes in membrane potential? In these cases the chemical messenger or an intracellular second messenger, generated by its binding to the smooth muscle plasma membrane, acts on the sarcoplasmic reticulum or on specific calcium channels in the plasma membrane.

We now turn to a description of the specific inputs listed in Table 10-4.

Spontaneous electrical activity. Some types of smooth muscle fibers generate action potentials spontaneously in the absence of any neural or hormonal input. The membrane in fibers exhibiting such spontaneous activity is unstable and does not maintain a constant potential, but gradually depolarizes until it reaches the threshold potential at

which point an action potential is initiated. Following repolarization, the membrane again begins to depolarize (Fig. 10-36), so that a sequence of action potentials occurs, the frequency being determined by the rate at which the membrane depolarizes and the nearness of the average membrane potential to threshold. The spontaneous depolarization to threshold is known as the *pacemaker potential*. (As will be described later, certain cardiac muscle fibers and some neurons in the central nervous system also have pacemaker potentials and can spontaneously generate action potentials.)

In addition to these pacemaker potentials, some types of smooth muscle have even slower oscillations of membrane potential, with cycle times of seconds or even minutes. These *slow wave potentials* arise from spontaneous cyclical changes in the rates at which ions are actively transported across the plasma membrane (this is an example of an electrogenic pump, Chap. 8). Bursts of action

Figure 10-37. Slow wave oscillations in membrane potential trigger bursts of action potentials. The mechanical activity of the smooth muscle correlates with the frequency of the slow wave potential changes and the magnitude of the mechanical response to the frequency of action potentials in each burst.

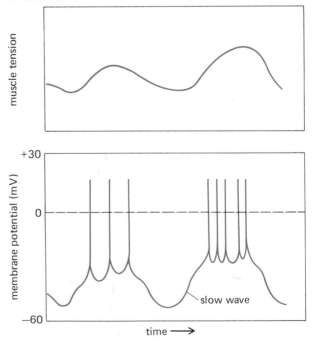

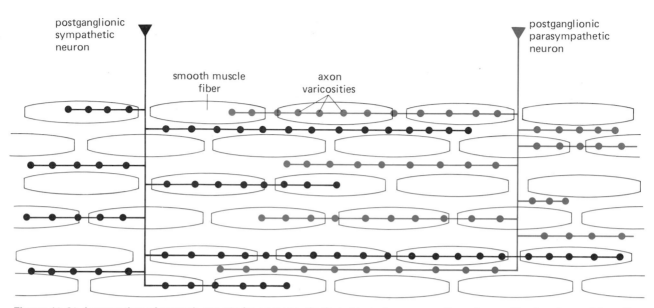

Figure 10-38. Innervation of smooth muscle by postganglionic autonomic neurons. Neurotransmitter is released from the varicosities along the branched axons adjacent to or overlapping smooth muscle fibers. (Pattern of innervation is typical of the multiunit type of smooth muscle.)

potentials are set off by pacemaker potentials at the peaks of the slow wave potential, i.e., at the membrane potential closest to threshold (Fig. 10-37). It should be noted that pacemaker potentials and slow wave potentials do not occur in all types of smooth muscle cells, nor are slow wave potentials present in all cells which show spontaneous pacemaker activity.

What is the significance of these bursts of action potentials? A twitch produced by a single action potential in smooth muscle is very small compared to a skeletal muscle twitch. Therefore, in order for significant changes in tension to occur, bursts of action potentials leading to summation of mechanical activity and to tetanus are required.

Nerves and hormones. Some smooth muscles are sensitive to a variety of hormones and are innervated by both sympathetic and parasympathetic fibers; others are innervated by fibers from only one autonomic division, and still others receive no innervation at all. Unlike skeletal muscle, smooth muscle does not have a single neuromuscular junction for each fiber or a specialized motor-end-plate region. Rather, as the axon of a postganglionic neuron enters the region of smooth muscle tissue, it divides into numerous branches, each

branch containing a series of swollen regions known as varicosities. Each varicosity contains numerous vesicles filled with neurotransmitter, which is released as an action potential conducted along the axon passes the varicosity. Several varicosities from a single axon may be located along a single muscle fiber, and a single muscle fiber may be located near varicosities belonging to postganglionic fibers of both sympathetic and parasympathetic neurons (Fig. 10-38). The concentration of released neurotransmitter at the surface of the muscle fiber's membrane is dependent on the distance between the varicosity and the muscle surface, since the neurotransmitter becomes diluted as it diffuses away from its site of release. Therefore, the magnitude of the smooth muscle response to nerve stimulation may vary considerably in different organs.

The neurotransmitters (released from nearby varicosities) or hormones (reaching the muscle cells via the blood) bind to specific receptor sites on the smooth muscle membranes, and this binding triggers the events leading to changes in free intracellular calcium. Both pathways leading to changes in calcium may be involved, i.e., changes in the smooth muscle membrane potential and potential-independent alterations of the specific

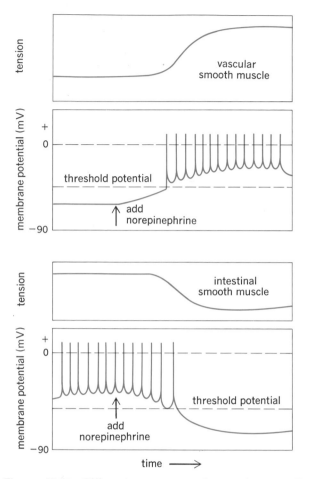

Figure 10-39. Different responses of vascular smooth muscle and intestinal smooth muscle to norepinephrine released from a sympathetic nerve ending.

membrane calcium channels or sarcoplasmic reticulum.

When a particular hormone or neurotransmitter exerts its effect by depolarizing the muscle membrane, it may produce a range of responses, depending on its concentration at the membrane surface and the type of muscle fiber involved. Small depolarizations produced by a low concentration of the neurotransmitter may not initiate action potentials and may increase the tonic tension of the fiber without producing phasic contractions. If the smooth muscle fiber is spontaneously active, such a depolarization may bring the membrane potential close enough to threshold so to increase the frequency of action potentials,

thereby initiating or increasing the intensity of phasic contractions. Finally, if the depolarization is great enough, it may initiate action potentials and phasic contractions in a fiber that does not have spontaneous electrical activity.

Whereas some hormones and neurotransmitters depolarize smooth muscle membranes, others produce a hyperpolarization which leads to inhibition of contractile activity. Thus, in contrast to skeletal muscle, which receives only excitatory input from its motor neurons, smooth muscle can be either excited or inhibited by neural activity (and by circulating hormones). Moreover, a given neurotransmitter or hormone may produce opposite effects in different smooth muscle tissues. For example, norepinephrine, the neurotransmitter released from postganglionic sympathetic neurons, depolarizes certain types of vascular smooth muscle, thereby initiating action potentials and contraction of the muscle; whereas in intestinal smooth muscle, it inhibits spontaneous activity and produces a relaxation of the muscle (Fig. 10-39). Thus, the nature of the response depends on events initiated by the binding of the chemical agent to receptor sites in the membrane and is not an inherent property of the agent, itself.

A spectrum of changes in contraction similar to that just described can also be produced when, as is often the case, neurotransmitters or hormones act via mechanisms independent of changes in the membrane potential.

Local chemical factors. Local factors, including paracrines, acidity, oxygen concentration, osmolarity, and the ion composition of the extracellular fluid may produce changes in the smooth muscle membrane potential and/or alter the permeability of the membrane to calcium and the release of calcium from the sarcoplasmic reticulum by mechanisms independent of potential changes. This responsiveness to local stimuli provides a direct means for altering smooth muscle tension in accord with changes in the metabolic activity of the surrounding tissues. Such responses play an important role in the local regulation of the internal environment within a tissue undergoing moment-to-moment fluctuations in activity. In contrast, neural and hormonal inputs to smooth muscle generally regulate contractile activity so as to control the internal environment of the body as a whole.

Classification of smooth muscle

The great diversity of smooth muscle characteristics has made it difficult to classify the different types found in various organs. One of the most general classifications divides smooth muscles into two groups based on the excitability characteristics of the muscle membranes and on the conduction of electrical activity from fiber to fiber within the muscle tissue: (1) *single-unit smooth muscles*, whose membranes are capable of propagating action potentials from fiber to fiber and may manifest spontaneous action potentials; (2) *multiunit smooth muscles* which exhibit little, if any, propagation of electrical activity from fiber to fiber and whose contractile activity is closely coupled to the neural activity to it.

Single-unit smooth muscle. The smooth muscles of the intestinal tract, the uterus, and small-diameter blood vessels are examples of single-unit smooth muscles. All the many muscle fibers making up a single-unit smooth muscle undergo synchronous activity, both electrical and mechanical; i.e., the whole muscle responds to stimulation as a single unit. Propagation of action potentials from fiber to fiber is responsible for this phenomenon. Each muscle fiber is linked to adjacent fibers by gap junctions (see Chap. 6), through which small ions, such as potassium and chloride, can move, carrying electric current; therefore, action potentials occurring in one cell are propagated by local current flow through the gap junctions into adjacent cells. Certain of the fibers in a single-unit smooth muscle are pacemaker cells in that they spontaneously generate the action potentials which are then propagated into the other fibers, which do not have pacemaker potentials (Fig. 10-40).

The contractile activity of single-unit smooth muscles can be altered by nerves and hormones. The extent to which these muscles are innervated varies considerably in different organs, and some are not innervated at all. The nerve terminals are often restricted to the pacemaker regions of the muscle tissue (Fig. 10-41), but any neural alteration of pacemaker cells of course alters the activity of the entire muscle.

One additional property characteristic of single-unit smooth muscle is the contractile response that is induced by a rapid stretching of the muscle.

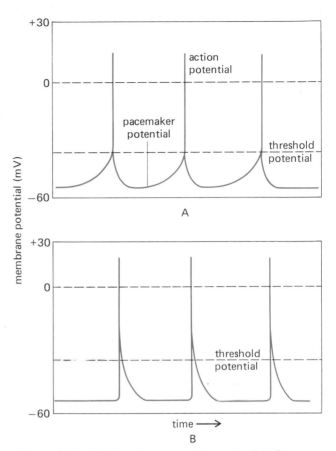

Figure 10-40. (A) Spontaneous action potentials in pacemaker cells. (B) Action potentials in nonpacemaker cells connected to above pacemaker cell by gap junctions.

This may result from a membrane depolarization caused by the stretch.

Multiunit smooth muscles. The smooth muscle in the large airways to the lungs, in large arteries, and those attached to the hairs in the skin are examples of multiunit smooth muscles. These muscles are richly innervated by branches of the autonomic nervous system, and like the motor units in skeletal muscle, a number of smooth muscle fibers are innervated by a single nerve fiber (of course, skeletal muscle is innervated by somatic neurons, not autonomic neurons). However, unlike skeletal muscle, there may be considerable overlapping innervation; i.e., a single smooth muscle fiber may be innervated by more than one neuron (Fig. 10-38). Again, like skeletal muscle, the neurally influenced contractile re-

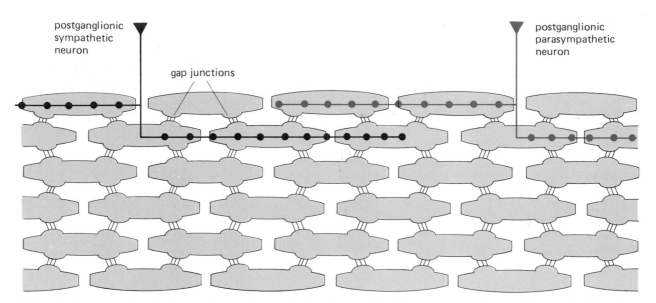

Figure 10-41. Innervation of single-unit smooth muscle is often restricted to only a few fibers in the muscle. Electrical activity is conducted from fiber to fiber throughout the muscle by way of the gap junctions between the fibers.

Table 10-5. Characteristics of muscle fibers

| Characteristic | Skeletal muscle | Smooth muscle | | Cardiac muscle |
		Single-unit	Multiunit	
Thick and thin filaments	yes	yes	yes	yes
Sarcomeres—banding pattern	yes	no	no	yes
Transverse tubules	yes	no	no	yes
Sarcoplasmic reticulum (SR)	+++	+	+	++
Source of activating calcium	SR	SR & extra-cellular	SR & extra-cellular	SR & extra-cellular
Site of calcium regulation	troponin (thin filaments)	myosin (thick filaments)	myosin (thick filaments)	troponin (thin filaments)
Speed of contraction	fast-slow	very slow	very slow	slow
Inherent tone (low levels of maintained tension)	no	yes	no	no
Spontaneous production of action potentials by pacemakers	no	yes	no	yes
Gap junctions between fibers	no	yes	few	yes
Effect of nerve stimulation	excitation	excitation or inhibition	excitation or inhibition	excitation or inhibition
Physiological effects of hormones on excitability and contraction	no	yes	yes	yes
Quick stretch of fiber produces contraction	no	yes	no	yes
Slow wave potentials may be present	no	yes	no	no

sponse of the whole muscle depends on the number of motor units that are activated and on the frequency of nerve stimulation to each of them (summation). Although stimulation of the nerve fibers to the muscle leads to a depolarization and a contractile response, action potentials may not occur in certain types of multiunit smooth muscles. Finally, circulating hormones can also initiate contraction of the muscle.

Although gap junctions have been found in multiunit smooth muscle, they are much less frequent than in single-unit muscles so that cell-to-cell conduction is quite limited. Another difference from single-unit smooth muscle is that multiunit smooth muscle does not contract when rapidly stretched.

Summary

Table 10-5 provides a comparison of some of the properties of the different types of muscle. Cardiac muscle has been included for completeness, although its properties have not been discussed in this chapter (see Chap. 11). In brief, cardiac muscle has a filament organization very similar to that of skeletal muscle, with myofibrils, sarcomeres, transverse tubules, and a well-developed sarcoplasmic reticulum. Like single-unit smooth muscle, some cardiac muscle fibers act as pacemakers to spontaneously generate action potentials which are then propagated throughout the heart by way of gap junctions between the cardiac fibers.

PART THREE

COORDINATED BODY FUNCTIONS

11
CIRCULATION

Overall design of the circulation

Section A: Blood

Plasma
Cellular elements of the blood
    Erythrocytes
       Iron • Vitamin B$_{12}$ and folic acid • Regulation of erythrocyte production • Anemia
    Leukocytes
    Platelets

Section B: The heart

Anatomy
Heart-beat coordination
    Origin of the heart beat
    Sequence of excitation
    Refractory period of the heart
    The electrocardiogram
    Excitation-contraction coupling
Mechanical events of the cardiac cycle
    Mid-to-late diastole (ventricular filling)
    Systole
    Early diastole (isovolumetric ventricular relaxation)
    Pulmonary circulation pressures
    Heart sounds
The cardiac output
    Control of heart rate
    Control of stroke volume
       Relationship between end-diastolic volume and stroke volume • The sympathetic nerves
    Summary of cardiac output control
    Cardiac energetics

Section C: The vascular system

Basic principles of pressure, flow, and resistance
    Determinants of resistance
       Nature of the fluid • Geometry of the tube
Arteries
    Arterial blood pressure
    Measurement of arterial pressure

Arterioles
    Local controls (autoregulation)
       Active hyperemia • "Pressure" autoregulation • Response to injury
    Extrinsic (reflex) controls
       Sympathetic nerves • Parasympathetic nerves • Hormones
Capillaries
    Anatomy of the capillary network
    Resistance of the capillaries
    Velocity of capillary blood flow
    Diffusion across the capillary wall: Exchanges of nutrients and metabolic end products
    Bulk flow across the capillary wall: Distribution of the extracellular fluid
Veins
    Determinants of venous pressure
    Effects of venous constriction on resistance to flow
    The venous valves
Lymphatics
    Functions of the lymphatic system
       Return of excess filtered fluid • Return of protein to the blood • Specific transport functions • Lymph nodes
    Mechanism of lymph flow

Section D: Integration of cardiovascular function: Regulation of systemic arterial pressure

Arterial pressure, cardiac output, and arteriolar resistance
Cardiovascular control centers in the brain
Receptors and afferent pathways
    Arterial baroreceptors
    Other baroreceptors
    Chemoreceptors
    Conclusion

Section E: Cardiovascular patterns in health and disease

Hemorrhage, hypotension, and shock
    Shock
The upright posture

Exercise
Hypertension
Congestive heart failure
"Heart attacks" and atherosclerosis

Section F: Hemostasis: The prevention of blood loss

Hemostatic events prior to clot formation
  Initial constriction of the injured vessel

Sticking of the endothelial surfaces
  Formation of a platelet plug
  Humoral facilitation of vasoconstriction
Blood coagulation: Clot formation
Clot retraction
The anticlotting system
Excessive clotting: Intravascular thrombosis

Overall design of the circulation

Beyond a distance of a few cell diameters, diffusion is not sufficiently rapid to meet the metabolic requirements of cells. If multicellular organisms were to evolve to a size larger than a microscopic cluster of cells, some mechanism other than simple diffusion would be needed to rapidly transport molecules over the long distances between the cells and the body's surface and between the various specialized tissues and organs. This problem was solved in the animal kingdom by the evolution of the circulation, which comprises the *blood,* the set of tubes, *blood vessels,* through which the blood flows, and a pump, the *heart,* which produces this flow; the heart and blood vessels together are termed the *cardiovascular system.*

Rapid *bulk flow* of blood through all parts of the body via the blood vessels is produced by pressures created by the pumping action of the heart. The extraordinary degree of branching of these vessels assures that all cells of the body are within a few cell diameters of at least one of the smallest branches, the *capillaries. Diffusion* across the walls of the capillaries and through the interstitial fluid permits the exchange of nutrients and metabolic end products between the capillary blood and the cells near the capillary. Thus, the circulation utilizes bulk flow to solve the problem of delivering blood to the various organs and tissues, but uses diffusion for the actual exchanges between the blood and the cells of these organs and tissues (Fig. 11-1).

Physiology as an experimental science began in 1628, when William Harvey demonstrated that the cardiovascular system forms a circle, so that blood is continuously being pumped out of the heart through one set of vessels and returned to the heart via a different set. There are actually two circuits (Fig. 11-2), both originating and terminating in the heart, which is divided longitudinally into two functional halves. Blood is pumped via one circuit (the *pulmonary circulation*) from the right half of the heart through the lungs and back to the left half of the heart. It is pumped via the second circuit (the *systemic circulation*) from the left half of the heart through all the tissues of the body, except, of course, the lungs, and back to the right half of the heart. In both circuits, the vessels carrying blood away from the heart are called *arteries,* and the vessels carrying blood from the lungs and tissues back to the heart are called *veins.* In the systemic circuit, blood leaves the left half of the heart via a single large artery, the *aorta.* From the aorta, branching arteries conduct blood to the various organs and tissues. These arteries divide in a highly characteristic manner into progressively smaller branches, much of the branching occurring within the specific organ or tissue sup-

Figure 11-1. Bulk-flow brings blood to the capillaries and diffusion links plasma and the interstitial fluid.

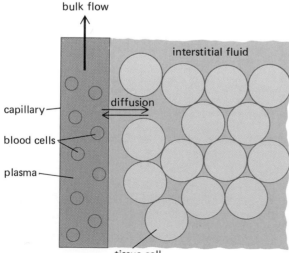

bulk flow

interstitial fluid

capillary

diffusion

blood cells

plasma

tissue cell

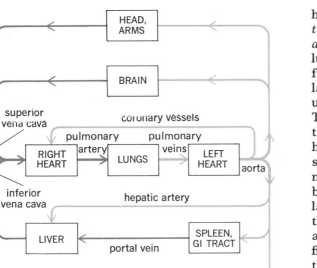

Figure 11-2. Diagrammatic representation of the cardiovascular system in the adult human being. Lighter shading indicates blood with high oxygen content.

plied. The smallest arteries branch into *arterioles*, which differ structurally and functionally from the arteries. Ultimately the arterioles branch into a huge number of very small, thin vessels, termed *capillaries*. The capillaries unite to form larger vessels *(venules)* which, in turn, unite to form fewer and still larger vessels, the veins. The veins from different organs and tissues unite to form two large veins, the *inferior vena cava* (from the lower portion of the body) and the *superior vena cava* (from the upper half of the body). By these two veins blood is returned to the right half of the heart. The entire systemic circuit can be visualized, therefore, as two trees, one arterial and the other venous, having the same origin (the heart) and being connected by fine twigs (capillaries) which unite the smallest branches of each tree (Fig. 11-3). As we shall see, the blood vessels do not merely constitute inert plumbing; each type has a characteristic structure and function.

The pulmonary circulation is composed of a similar circuit. Blood leaves the right half of the heart via a single large artery, the *pulmonary trunk*, which divides into the two *pulmonary arteries*, one supplying each lung. Within the lungs, the arteries continue to branch, ultimately forming arterioles, which then divide into capillaries. These capillaries unite to form small venules, which unite to form larger and larger veins. The blood leaves the lungs via the largest of these, the *pulmonary veins*, which empty into the left half of the heart. The blood flowing through the systemic veins, right half of the heart, and pulmonary arteries has a low oxygen content. As this blood flows through the lung capillaries, it picks up large quantities of oxygen; therefore, the blood in the pulmonary veins, left heart, and systemic arteries has a high oxygen content. As this blood flows through the capillaries of tissues and organs throughout the body, much of this oxygen leaves the blood, resulting in the low oxygen content of systemic venous blood.

In a normal person, blood can pass from the systemic veins to the systemic arteries only by first being pumped through the pulmonary circuit, thus oxygenating all the blood returning from the body tissues before it is pumped back to them. Normally the total volumes of blood pumped through the pulmonary and systemic circuits during a given period of time are equal. In other words, the right heart pumps the same amount of blood as the left heart. Only under unusual circumstances do these volumes differ from each other, and then only transiently.

It should be evident that all the blood pumped by the right heart flows through the lungs; in contrast, only a fraction of the total left ventricular output flows through any single organ or tissue. In

Figure 11-3. Systemic circulation as two trees connected by capillaries. As indicated by the color change, oxygen leaves the blood during passage through the capillaries. *(Adapted from Rushmer.)*

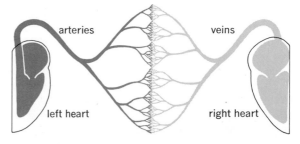

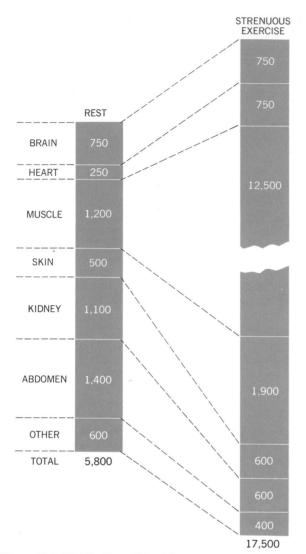

Figure 11-4. Distribution of blood flow to the various organs and tissues of the body at rest and during strenuous exercise. The numbers show blood flow in milliliters per minute. *(Adapted from Chapman and Mitchell.)*

ing normal person, the amount of blood pumped simultaneously by each half of the heart is approximately 5 L/min. During heavy work or exercise in a well-trained person, the volume may increase as much as sevenfold to 35 L/min. Each organ or tissue obviously receives only a fraction of the total left-ventricular output; the typical distribution for a normal adult at rest and during exercise is given in Fig. 11-4. Perhaps the most striking aspect of brain blood flow is its relative constancy. It really requires little, if any, additional energy to think; i.e., whether we are staring blankly into space or contemplating the theory of relativity, the total energy consumption of the brain remains virtually unchanged. In contrast, the energy consumption of muscular tissues of the body (heart, skeletal muscle, uterus, etc.) varies directly with the degree of muscle activity. Thus, there is a large increase in blood flow to the exercising skeletal muscle and heart. Skin blood flow also increases, kidney flow decreases, and brain flow is unchanged. Obviously, in exercise the total amount of blood pumped by the heart has increased; but, more important for our present purposes, the distribution of flow has greatly changed. Blood is distributed to the various organs and tissues according to their metabolic activity at any given moment.

This completes our brief survey of the overall design of the circulation, and we now turn to a systematic description of its components and their control. In so doing, we might very easily lose sight of the forest for the trees if we don't persistently ask of each section: "How does this component of the circulation contribute either to an adequate flow of blood to the various organs or to an adequate exchange of materials across their capillaries?"

Section A
Blood

Blood is composed of specialized cellular elements and a liquid, *plasma*, in which they are suspended. The cells are the *red blood cells*, or *erythrocytes*; the *white blood cells*, or *leukocytes*; and the *platelets* (which are really cell fragments). Ordinarily, the constant motion of the blood keeps the cells well dispersed throughout the plasma, but if a sample of blood is allowed to stand (clotting pre-

other words, the systemic circulation comprises numerous different pathways "in parallel." They all originate as large arteries branching off from the aorta. One significant deviation from this pattern is the blood supply to the liver, much of which is not arterial but venous blood, which has just left the spleen and gastrointestinal tract.

The dominant feature of the cardiovascular system is the pumping of blood by the heart. In a rest-

vented), the cells slowly sink to the bottom. This process can be speeded up by centrifuging. By this means, the percentage of total blood volume which is cells, known as the *hematocrit*, can be determined. The normal hematocrit is approximately 45 percent (since the vast majority of all blood cells are erythrocytes, the total cell volume is approximately equal to the erythrocyte volume). The total blood volume of an average person is approximately 8 percent of total body weight. Accordingly, for a 70-kg person

$$\text{Total blood weight} = 0.08 \times 70 \text{ kg} = 5.6 \text{ kg}$$

One kilogram of blood occupies approximately one liter; therefore,

$$\text{Total blood volume} = 5.6 \text{ L}$$

The hematocrit is 45 percent; therefore,

$$\text{Total cell volume} = 0.45 \times 5.6 \text{ L} = 2.5 \text{ L}$$
$$\text{Plasma volume} = 5.6 - 2.5 \text{ L} = 3.1 \text{ L}$$

Plasma

Plasma is an extremely complex liquid. It consists of a large number of organic and inorganic substances dissolved in water (Table 11-1). The most abundant solutes by weight are the proteins, which together compose approximately 7 percent of the total plasma weight. The *plasma proteins* vary greatly in their structure and function, but they can be classified, according to certain physical and chemical reactions, into three broad groups: the *albumins, globulins,* and *fibrinogen*. The albumins are more abundant than the globulins and usually are of smaller molecular weight. The plasma proteins, with notable exceptions, are synthesized by the liver, the major exception being the group known as *gamma globulins*, which are formed in the lymph nodes and other lymphoid tissues (Chap. 17). The plasma proteins serve a host of important functions, which will be described in relevant chapters, but it must be emphasized that normally they are *not* taken up by cells and utilized as metabolic fuel. Accordingly, they must be viewed quite differently from most other organic constituents of plasma, such as glucose, which use the plasma as a vehicle for transport but function in cells. The plasma proteins function in the plasma itself or, under certain circumstances, in the interstitial fluid. Finally, plasma is distin-

guished from *serum*, which is plasma from which fibrinogen has been removed as a result of clotting.

In addition to the organic solutes—proteins, nutrients, and metabolic end products—plasma contains a large variety of mineral electrolytes, the concentrations of which are also shown in Table 11-1. Comparison of the concentrations in millimoles per liter for these electrolytes and protein may cause puzzlement in view of the previous statement that protein is the most abundant plasma solute by *weight*. Remember, however, that molarity is a measure not of the weight but of the *number* of molecules or ions per unit volume. Protein molecules are so large in comparison with sodium ions that a very small number of them greatly outweighs a much larger number of sodium ions.

Cellular elements of the blood

Erythrocytes

Each milliliter of blood contains approximately 5 billion erythrocytes. Since there is approximately 5000 mL of blood in the average person, the total number of erythrocytes in the human body is about 25 trillion. The shape of these cells is that of a biconcave disk, i.e., a disk thicker at the edge than in the middle, like a doughnut with a center depression on each side instead of a hole (Fig. 11-5). This shape and their small size (7 μm in diameter) have adaptive value in that oxygen and carbon dioxide, carriage of which is the major function of the erythrocytes, can rapidly diffuse throughout the entire cell interior. The plasma membranes of erythrocytes contain specific proteins which differ from person to person, and these confer upon the blood its so-called "type" as will be described in Chap. 17.

The outstanding physiological characteristic of erythrocytes is the presence of the iron-containing protein *hemoglobin*, which binds oxygen and constitutes approximately one-third of the total cell weight. Each hemoglobin molecule is made up of four subunits, each consisting of a *heme* portion conjugated to a polypeptide; the four polypeptides are collectively called *globin*. It is the heme portion of the molecule (Fig. 11-6) which contains the iron (Fe^{2+}), and it is to this iron that oxygen binds. The heme portion of hemoglobin is identical to the

Table 11-1. Constituents of arterial plasma

Constituent	Amount/concentration	Major functions
Water	93% of plasma weight	Medium for carrying all other constituents
Electrolytes (inorganic)	Total < 1% of plasma weight	Keep H_2O in extra-cellular compartment. Act as buffers. Function in membrane excitability.
Na^+	142 meq/L (142 mM)	
K^+	4 meq/L (4 mM)	
Ca^{2+}	5 meq/L (2.5 mM)	
Mg^{2+}	3 meq/L (1.5 mM)	
Cl^-	103 meq/L (103 mM)	
HCO_3^-	27 meq/L (27 mM)	
Phosphate (mostly HPO_4^{2-})	2 meq/L (1 mM)	
SO_4^{2-}	1 meq/L (0.5 mM)	
Proteins	7.3 g/100 mL (2.5 mM)	Provide colloid osmotic pressure of plasma. Act as buffers. Bind other plasma constituents (lipids, hormones, vitamins, metals, etc.). Clotting factors. Enzymes, enzyme precursors. Antibodies (immune globulins). Hormones
Albumins	4.5 g/100 mL	
Globulins	2.5 g/100 mL	
Fibrinogen	0.3 g/100 mL	
Gases		
CO_2	2 mL/100 mL plasma	
O_2	0.2 mL/100 mL	
N_2	0.9 mL/100 mL	
Nutrients		
Glucose and other carbohydrates	100 mg/100 mL (5.6 mM)	
Total amino acids	40 mg/100 mL (2 mM)	
Total lipids	500 mg/100 mL (7.5 mM)	
Cholesterol	150–250 mg/100 mL (4–7 mM)	
Individual vitamins	0.0001–2.5 mg/100 mL	
Individual trace elements	0.001–0.3 mg/100 mL	
Waste products		
Urea	34 mg/100 mL (5.7 mM)	
Creatinine	1 mg/100 mL (0.09 mM)	
Uric acid	5 mg/100 mL (0.3 mM)	
Bilirubin	0.2–1.2 mg/100 mL (0.003–0.018 mM)	
Individual hormones	0.000001–0.05 mg/100 mL	

heme found in the cytochromes (Chap. 5) and myoglobin (Chap. 12).

The amino acid sequences in the polypeptide chains are, of course, genetically determined, and mutant genes cause the production of abnormal hemoglobins, some of which may cause serious malfunction. For example, *sickle cell anemia* is caused by the presence of a particular form of abnormal hemoglobin molecules which, at the low oxygen concentrations existing in many capillaries, interact with each other to form fiberlike structures; these distort the erythrocyte membrane and cause the cell to form sickle shapes or other bizarre forms. This results both in the blockage of the capillaries, with consequent tissue damage, and in the destruction of the deformed erythrocytes, with consequent anemia. All this because of a change in one of 287 amino acids in the globin molecule!

There are other abnormal hemoglobins which cause changes in the structure of the erythrocyte, making it more fragile, and still others which alter the ability of the hemoglobin to bind oxygen; the latter lead not to erythrocyte destruction but to impairment of oxygen carriage and delivery to the tissues.

Erythrocytes are incomplete cells in that they lack nuclei and the metabolic machinery to synthesize new proteins. Thus, they can neither reproduce themselves nor maintain their normal structure indefinitely. As essential enzymes within them deteriorate and are not replaced, the cells age and ultimately die. Fortunately, their oxygen-carrying ability is not significantly diminished during the aging period. The average life span of an erythrocyte is approximately 120 days, which

Figure 11-5. Scanning electron micrograph of red blood cells. *(From R. G. Kessel and C. Y. Shih, "Scanning Electron Microsopy in Biology," Springer-Verlag, New York, 1974, p. 265.)*

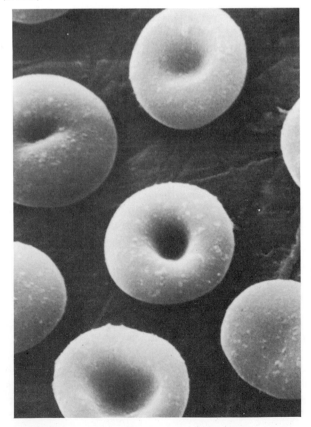

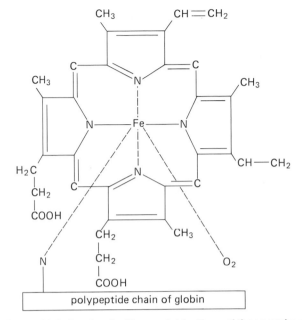

polypeptide chain of globin

Figure 11-6. A subunit of hemoglobin. Four of these make a single hemoglobin molecule. *(Adapted from Ganong.)*

means that almost 1 percent of the total erythrocytes in the body are destroyed every day. Destruction of erythrocytes is accomplished by a group of cells, the *macrophages*, found in liver, spleen, bone marrow, and lymph nodes, usually lining the blood vessels or lying close to them. These cells, the physiology of which is described in detail in Chap. 17, ingest and destroy the erythrocytes by breaking down their large complex molecules with lysosomal enzymes.

In the process, hemoglobin molecules are split to yield heme and the polypeptide chains of globin. Iron is then removed from the heme, enters the blood, and is transported to the bone marrow (and other cells) to be reused in the synthesis of new heme groups. The remainder of the heme is broken down in several steps to the yellow substance, bilirubin, which is released into the blood, picked up by liver cells, and secreted into the bile; its further metabolism is described in Chap. 14.

Obviously, in a normal person, a quantity of erythrocytes equal to that destroyed must be simultaneously synthesized and released into the circulatory system. The site of erythrocyte production is the soft, highly cellular interior of bones called *bone*

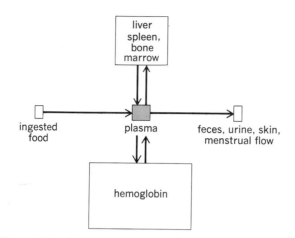

Figure 11-7. Summary of iron balance. The sizes of the boxes represent the quantity of iron involved. The magnitude of exchange in a particular direction varies according to conditions.

marrow. In the adult, erythrocyte formation occurs only in the bones of the chest, the base of the skull, and the upper arms and legs. The erythrocytes are originally descended from large bone-marrow cells which contain no hemoglobin but do have nuclei and are therefore capable of cell division. After several cell divisions, cells emerge which are identifiable as immature erythrocytes because they contain hemoglobin. As maturation continues, these cells accumulate increased amounts of hemoglobin and their nuclei become progressively smaller until they ultimately disappear completely. The mature erythrocytes leave the bone marrow and enter the general circulation, in which they will travel for some 120 days.

This growth process requires a number of different raw materials. The formation of the erythrocyte itself requires the usual nutrients and structural materials: amino acids, lipids, and carbohydrates. In addition, certain growth factors, including vitamin B_{12} and folic acid, are essential for normal erythrocyte formation. Finally, the production of an erythrocyte requires the materials which go into the making of hemoglobin: iron, amino acids, and the organic molecules which are incorporated into the protein portion of hemoglobin. A lack of any of these growth factors or raw materials results in the failure of normal *erythropoiesis* (erythrocyte formation). The substances which are most commonly lacking are iron, folic acid, and vitamin B_{12}.

Iron. Iron is obviously an essential component of the hemoglobin molecule since it is to this element that the oxygen binds. The balance of iron and its distribution within the body are shown schematically in Fig. 11-7. About 70 percent of the total body iron is in hemoglobin, and the remainder is stored primarily in the liver, spleen, and bone marrow. As erythrocytes are destroyed, most of the iron released from hemoglobin is returned to these depots, from which it can again be released for hemoglobin synthesis. Small amounts of iron, however, are lost each day via the urine, feces, sweat, and cells sloughed from the skin. In addition, women lose a significant quantity of iron via menstrual blood. In order to remain in iron balance, the amount of this metal lost from the body must be replaced by ingestion of iron-containing foods. An upset of this balance results either in iron deficiency and inadequate hemoglobin production or in an excess of iron in the body and serious toxic effects.

The homeostatic control of iron balance resides primarily in the intestinal epithelium, which actively absorbs the iron from ingested food. Only a small fraction of ingested iron is absorbed, and, what is more important, this fraction is increased or decreased depending upon the state of body iron balance. These fluctuations appear to be mediated by changes in the iron content of the intestinal epithelium itself, as described in Chap. 14. Iron absorption also depends on the type of food in which it is contained and which is ingested along with it. This is because the iron becomes bound to negatively charged ions or complexes, which can either retard or enhance its transport. For example, iron in liver is much more absorbable than is iron in egg yolk, since the latter contains phosphates which bind the iron to form an insoluble complex.

The problem of iron forming insoluble precipitates has been circumvented in the body by the evolution of iron-binding proteins which act both as carrier molecules for transferring iron in the plasma (and between cells) and as the storage molecules for iron in the various storage depots; these iron-protein complexes are known as *transferrin* and *ferritin*, respectively.

Vitamin B_{12} and folic acid. Normal erythrocyte formation requires extremely small quantities (one-millionth of a gram per day) of a cobalt-containing molecule, vitamin B_{12}, which, by mechanisms still unknown, permits final maturation of the erythrocyte. This substance is not synthesized within the body, which therefore depends upon its dietary intake. Absorption of vitamin B_{12} from the gastrointestinal tract into the blood is described in Chap. 14. Dietary deficiency of vitamin B_{12} or impaired intestinal absorption of it (see Chap. 14) leads to the failure of erythrocyte proliferation and maturation known as *pernicious anemia*. Since vitamin B_{12} is also needed for normal functioning of the nervous system, specifically for myelin formation, a variety of neurological symptoms may accompany the anemia.

Folic acid is a vitamin essential for the formation of DNA because it is involved in the synthesis of purines and thymine. Accordingly, its deficiency results in failure of normal erythrocyte multiplication and maturation, and in poor growth and development of many other tissues, as well.

Regulation of erythrocyte production. In a normal person, the total volume of circulating erythrocytes remains remarkably constant. Such constancy is required both for delivery of oxygen to the tissues and for maintenance of the blood pressure. From the preceding paragraphs it should be evident that a constant number of circulating erythrocytes can be maintained only by balancing erythrocyte pro-duction with the sum of destruction and loss. Such balance is achieved by controlling the rate of erythrocyte production. During periods of severe erythrocyte destruction or hemorrhage, the normal rate of erythrocyte production can be increased more than sixfold.

It is easiest to discuss the control of erythrocyte production by first naming the mechanisms *not* involved. In the previous section, we listed a group of nutrient substances, such as iron and vitamin B_{12}, which must be present for normal erythrocyte production. However, none of these substances actually *regulates* the rate of erythrocyte production.

The direct control of erythropoiesis is exerted by a hormone called *erythropoietin*. The cells producing erythropoietin are located in the kidneys, although other sites in the body may also contribute to its production. Normally, a small quantity of erythropoietin is circulating, which stimulates the bone marrow to produce erythrocytes at a certain basal rate. An increase in circulating erythropoietin further stimulates the bone marrow and increases erythropoiesis, whereas a decrease in circulating erythropoietin below the normal level results in a decrease of erythropoiesis.

The common denominator of changes which increase circulating erythropoietin is precisely

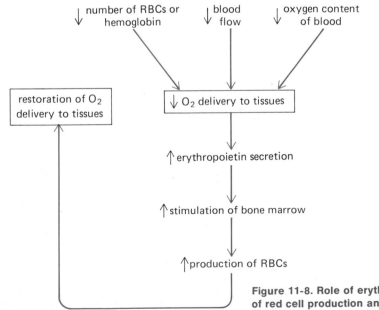

Figure 11-8. Role of erythropoietin in regulation of red cell production and oxygen delivery.

what one would logically expect—a decreased oxygen delivery to the kidneys (and other erythropoietin-producing tissues). As will be described in Chap. 12, this can result either from a decrease in the number of erythrocytes or their hemoglobin content, decreased blood flow, or decreased oxygen delivery from the lungs into the blood. As a result, erythropoietin secretion is increased, erythrocyte production is increased, the oxygen-carrying capacity of the blood is increased, and oxygen delivery to the tissues is returned toward normal (Fig. 11-8). Unfortunately, how the erythropoietin-producing cells detect changes in oxygen delivery has not been discovered. It is interesting that testosterone in physiological amounts stimulates the release of erythropoietin, and this may account for the fact that hemoglobin concentration in men is 1 to 2 percent higher than the concentration in women.

Anemia. Anemia is a reduced total blood hemoglobin. It may be due to a decrease in the total number of erythrocytes (each having a normal quantity of hemoglobin) or to a diminished concentration of hemoglobin per erythrocyte. Thus, anemia has a wide variety of causes, including dietary deficiencies (of iron, vitamin B_{12}, or folic acid), bone-marrow failure (due to toxic drugs and cancer, for example), excessive blood loss from the body, excessive destruction of erythrocytes (as in sickle cell anemia), and inadequate secretion of erythropoietin.

Leukocytes

If one takes a drop of blood, adds appropriate dyes, and examines it under a microscope, the various cell types can be seen (Fig. 11-9). Over 99 percent of all the cells are erythrocytes (Table 11-2). The

Table 11-2. Numbers and distribution of erythrocytes and white cells in normal human blood

Total erythrocytes = 5,000,000,000 cells per milliliter of blood
Total white cells = 7,000,000 cells per milliliter of blood
Percent of total white cells:
  Polymorphonuclear granulocytes:
    Neutrophils 50–70
    Eosinophils 1–4
    Basophils 0.1
  Mononuclear cells
    Monocytes 2–8
    Lymphocytes 20–40

remaining cells, the *leukocytes* (white blood cells), all have as their major function defense against foreign cells, such as bacteria, and other foreign matter. Their roles are described in detail in Chap. 17 and we introduce them here to complete our survey of blood. Leukocytes are classified according to their structure and affinity for various dyes. The name *polymorphonuclear granulocytes* refers to the three types of leukocytes with lobulated nuclei and abundant cytoplasmic granules. The granules of one group show no dye preference, and the cells are therefore called *neutrophils.* The granules of the second group take up the red dye eosin, thus giving the cells their name *eosinophils.* Cells of the third group have an affinity for a basic dye and are called *basophils.* All three types of granulocytes are produced in the bone marrow and released into the circulation. It should be understood that, unlike the erythrocytes, the major functions of granulocytic leukocytes are exerted not within the blood vessels but in the interstitial fluid; i.e., granulocytes utilize the circulatory system only as the route for reaching a damaged or invaded area. Once there, they leave the blood vessels to enter the tissue and perform their functions.

The primary function of the neutrophil is *phagocytosis,* the ingestion and digestion of particulate material. It has a survival time in the circulation of only a few hours and is incapable of division, so that its supply must be continuously replenished by the bone marrow (it is estimated that 125 billion neutrophils are produced each day). The eosinophil may also be involved in phagocytosis, but its precise function is unknown. The basophil, in contrast, is not a phagocytic cell; rather it contains powerful chemicals, such as histamine, which it releases locally, a response which, as we shall see, may importantly contribute to tissue damage and allergy. The basophil is virtually identical to the group of cells known as *mast cells,* which are found in connective tissue throughout the body and do not circulate.

A fourth type of leukocyte quite different in appearance from the three granulocytic types is the *monocyte.* These cells, which are also produced by the bone marrow, are somewhat larger than the granulocytes, with a single oval or horseshoe-shaped nucleus and relatively few cytoplasmic granules. Monocytes are continually leaving the circulation and being transformed into *macrophages,* a second type of phagocytic cell (neutro-

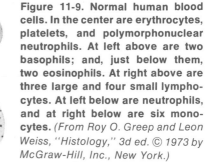

Figure 11-9. Normal human blood cells. In the center are erythrocytes, platelets, and polymorphonuclear neutrophils. At left above are two basophils; and, just below them, two eosinophils. At right above are three large and four small lymphocytes. At left below are neutrophils, and at right below are six monocytes. *(From Roy O. Greep and Leon Weiss, "Histology," 3d ed. © 1973 by McGraw-Hill, Inc., New York.)*

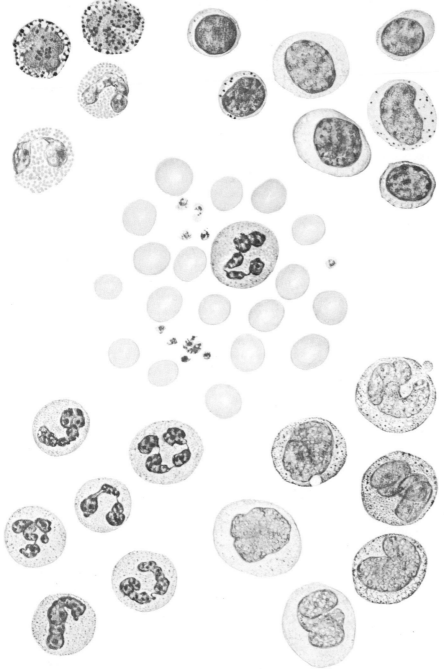

phils being the other). Thus, the function of circulating monocytes is to provide to the tissues a source of phagocytic macrophages.

The final class of leukocyte is the *lymphocyte*. Its outstanding structural features are a relatively large single nucleus and scanty surrounding cytoplasm. The circulating pool of lymphocytes continually travels from the blood across capillary

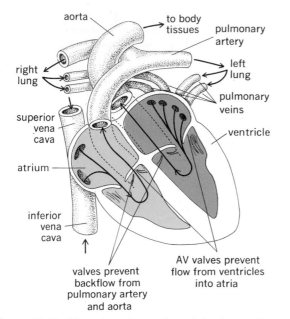

aorta

to body
tissues

pulmonary
artery

right
lung

left
lung

pulmonary
veins

superior
vena
cava

ventricle

atrium

inferior
vena
cava

valves prevent
backflow from
pulmonary artery
and aorta

AV valves prevent
flow from ventricles
into atria

Figure 11-10. Diagrammatic section of the heart. The arrows indicate the direction of blood flow. The intracardiac pulmonary artery and aorta have been cut away; their locations are indicated by the dashed lines. *(Adapted from McNaught and Callender.)*

walls, into lymphocytes, then to lymph nodes, and, finally, back to the blood. However, circulating lymphocytes constitute only a very small fraction of the total body lymphocytes, most of which are found at any instant in the lymphoid tissues. The life history of lymphocytes and the interactions between them, the bone marrow, and the various lymphoid organs are quite complex and will be described in Chap. 17. Suffice it here to point out that the lymphocytes (and an antibody-secreting daughter-cell line, the *plasma cells*) are responsible for so-called "specific defenses" against foreign invaders.

Platelets

In human beings, the circulating platelets (Fig. 11-9) are colorless corpuscles much smaller than erythrocytes and containing numerous granules; there are approximately 250 million per milliliter of blood (compare this with the number of white blood cells given in Table 11-2). The platelets, which are actually cell fragments and lack nuclei, originate from certain large cells (*megakaryocytes*)

found in bone marrow; they are portions of cytoplasm of these cells which are pinched off and enter the circulation. They play crucial roles in blood clotting, as described in a subsequent section.

**Section B
The heart**

Anatomy

The heart is a muscular organ located in the chest (*thoracic*) cavity and covered by a fibrous sac, the *pericardium*. Its walls are composed primarily of cardiac muscle (*myocardium*). The inner surface of the myocardium, i.e., the surface in contact with the blood within the heart chambers, is lined by a thin layer of epithelial cells (*endothelium*).

The human heart is divided longitudinally into right and left halves (Fig. 11-10), each consisting of two chambers, an *atrium* and a *ventricle*. The cavities of the atrium and ventricle on each side of the heart communicate with each other, but the right chambers do not communicate directly with those on the left. Thus, right and left atria and right and left ventricles are distinct.

Perhaps the easiest way to picture the architecture of the heart is to begin with its fibrous skeleton, which comprises four interconnected rings of dense connective tissue (Fig. 11-11), and which separate atria from ventricles. To the tops of these rings are anchored the muscle masses of the atria, pulmonary trunk, and aorta. To the bottom are attached the muscle masses of the ventricles. The connective-tissue rings form the openings between the atria and ventricles and between the ventricles and great arteries. To these rings are attached four sets of valves (Fig. 11-11).

Between the cavities of the atrium and ventricle in each half of the heart are the *atrioventricular valves (AV valves)*, which permit blood to flow from atrium to ventricle but not from ventricle to atrium (Fig. 11-12). The right and left AV valves are called, respectively, the *tricuspid* and *mitral valves*. When the blood is moving from atrium to ventricle, the valves lie open against the ventricular wall, but when the ventricles contract, the valves are brought together by the increasing pressure of the ventricular blood and the atrioventricular opening is closed. Blood is therefore forced into the pulmonary arteries (from the right ventricle)

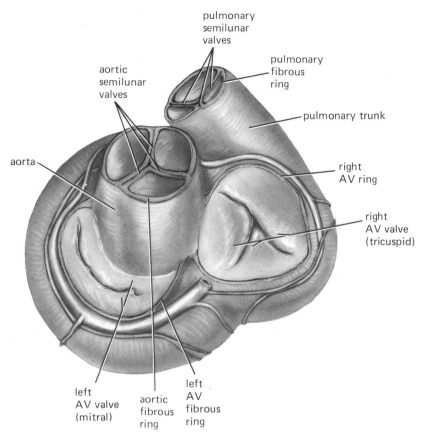

pulmonary
semilunar
valves

aortic
semilunar
valves

pulmonary
fibrous
ring

pulmonary trunk

aorta

right
AV ring

right
AV valve
(tricuspid)

left
AV valve
(mitral)

aortic
fibrous
ring

left
AV
fibrous
ring

Figure 11-11. Four connective-tissue rings compose the fibrous skeleton of the heart. To these are attached the valves, the trunks of the aorta and the pulmonary artery, and the muscle masses of the cardiac chambers.

and into the aorta (from the left ventricle) instead of back into the atria. To prevent the valves themselves from being forced upward into the atrium, they are fastened by fibrous strands to muscular projections of the ventricular walls. These muscular projections do *not* open or close the valves: They act only to limit the valves' movements and prevent them from being everted.

The openings of the ventricles into the pulmonary trunk and aorta are also guarded by valves, the *pulmonary* and *aortic valves*, respectively, which permit blood to flow into these arteries but close immediately preventing reflux of blood in the opposite direction. There are no true valves at the entrances of the venae cavae and pulmonary veins into the right and left atrium, respectively.

We can now list the structures through which blood flows in passing from the systemic veins to the systemic arteries: superior or inferior venae cavae; right atrium; tricuspid valve; right ventricle; pulmonary valve; pulmonary trunk and arteries, arterioles, capillaries, venules, veins; left atrium; mitral valve; left ventricle; aortic valve; aorta. The driving force for this flow of blood, as we shall see, comes solely from the active contraction of the cardiac muscle. The valves play no part at all in initiating flow and only prevent the blood from flowing in the opposite direction.

The blood within the heart chambers does not exchange nutrients and metabolic end products with the cells constituting the heart walls. The heart, like all other organs, receives its blood supply via arterial branches *(coronary arteries)* which arise from the aorta.

The walls of the atria and ventricles are composed of layers of cardiac muscle which are tightly bound together and completely encircle the blood-filled chambers. Thus, when the walls of a chamber contract, they come together like a squeezing fist, thereby exerting pressure on the blood they

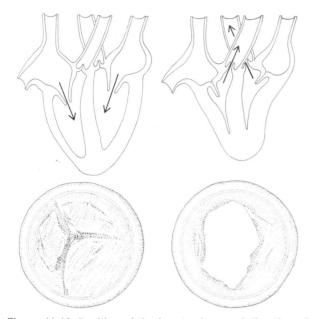

Figure 11-12. Position of the heart valves and direction of blood flows during ventricular relaxation *(left)* **and ventricular contraction** *(right) (Adapted from Carlson, Johnson, and Cavert.)*

The heart receives a rich supply of sympathetic and parasympathetic nerve fibers, the latter contained in the vagus nerves. The postganglionic sympathetic fibers terminate upon the cells of the specialized conducting system of the heart as well as on the ordinary myocardial cells of the atria and ventricles. The parasympathetic neurons also innervate the conducting system and the atrial myocardial cells but not the ventricular myocardium. The sympathetic postganglionic fibers release norepinephrine, the parasympathetics, acetylcholine. The receptors, for norepinephrine are β-adrenergic and those for acetylcholine are muscarinic.

Heart-beat coordination

Contraction of cardiac muscle, like other types, is triggered by depolarization of the muscle membrane. In dealing with the mechanisms by which membrane excitation is initiated and spread through the heart we may wonder what would happen if all the many muscle fibers in the heart were to contract in a random manner. One result would

enclose. Cardiac muscle cells combine certain of the properties of smooth and skeletal muscle. The individual cell is striated (Fig. 11-13) due to the repeating sarcomeres containing both the thick myosin and thin actin filaments, as described for skeletal muscle. Cardiac cells are considerably shorter than the long, cylindrical skeletal fibers and have several branching processes. Adjacent cells are joined end to end at structures known as intercalated disks, within which are two types of membrane junctions: (1) desmosomes, which hold the cells together and to which the myofibrils are attached; (2) gap junctions which allow action potentials to be transmitted from one cardiac cell to another, in a manner similar to that in smooth muscle.

Besides the usual type of cardiac muscle shown in Fig. 11-13, certain areas of the heart contain specialized muscle fibers which have a different appearance and are essential for normal excitation of the heart. They constitute a network known as the *conducting system* of the heart and also are in contact with fibers of the usual cardiac muscle via gap junctions which permit passage of action potentials from one cell to another.

Figure 11-13. Electron micrograph of cardiac muscle. Note the striations similar to those of skeletal muscle. The dark bands indicate interdigitating areas called intercalated disks. *(Courtesy of D. W. Fawcett.)*

intercalated
disc

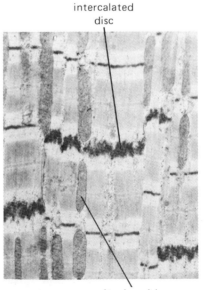

mitochondrion

be lack of coordination between pumping by each corresponding atrium and ventricle, but this defect is dwarfed by the more serious lack of muscle co-ordination within the ventricles. The blood would be sloshed back and forth within the ventricular cavities instead of being ejected into the aorta and pulmonary trunk. In other words, the complex muscle masses which form the ventricular pumps must contract more or less simultaneously for efficient pumping.

Such coordination is made possible by two factors, already mentioned: (1) The gap junctions allow spread of an action potential from one fiber to the next so that excitation in one muscle fiber spreads throughout the heart; (2) the specialized conducting system within the heart facilitates the rapid and coordinated spread of excitation. Where and how does the action potential first arise, and what is the path and sequence of excitation?

Origin of the heart beat

Certain cardiac muscle cells, like certain forms of smooth muscle, are autorhythmic; i.e., they are capable of spontaneous, rhythmical self-excitation. When the individual cells of a salamander embryo heart are separated and placed in salt solution, the individual cells are seen beating spontaneously. But they are beating at different rates. Figure 11-14 shows recordings of membrane potentials from two such cells, the most important feature of which is the gradual depolarization causing the membrane potential to reach threshold, at which point an action potential occurs. Following the action potential, the membrane potential returns to the initial resting value, and the gradual depolarization begins again. It should be evident that the slope of this depolarization, i.e., the rate of membrane potential change per unit time, determines how quickly threshold is reached and the next action potential elicited. Accordingly, cell A has a faster rate of firing than cell B. This capacity for autonomous depolarization toward threshold makes the rhythmical self-excitation of the muscle cells possible. It is due to a decreasing membrane permeability to potassium, but just how this change is generated "spontaneously" remains obscure.

In the course of the salamander experiment above, many individual cells form gap junctions. When such a gap junction is formed between two

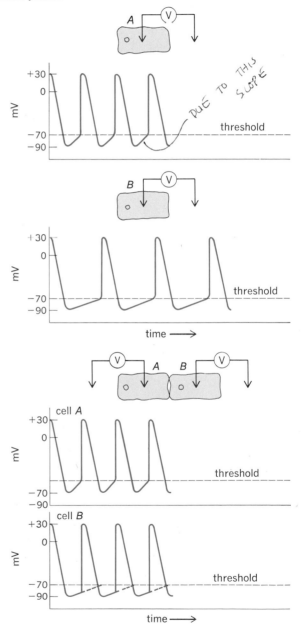

Figure 11-14. Transmembrane potential recordings from cardiac muscle cells grown in tissue culture. The dashed lines in the bottom recording of cell B indicate the course depolarization would have followed if the cells had not been joined.

cells previously contracting autonomously at different rates, both the joined cells contract at the faster rate (Fig. 11-14). In other words, the faster cell sets the pace, causing the initially slow cell to contract at the faster rate. The mechanism is straightforward: The action potential generated by the faster cell causes depolarization, via the gap junction, of the second cell's membrane to threshold, at which point an action potential occurs in this second cell. The important generalization emerges that, because of the gap junctions between cardiac muscle cells, all the cells are excited at the rate set by the cell with the fastest autonomous rhythm.

Precisely the same explanation holds for the origination of the heart beat in the intact heart; several areas in the conducting system of the adult mammalian heart demonstrate these same characteristics of autorhythmicity and pacemaking, the one with the fastest inherent rhythm being a small mass of specialized myocardial cells embedded in the right atrial wall near the entrance of the superior vena cava (Fig. 11-15). Called the *sinoatrial* (SA) *node,* it is the normal pacemaker for the entire heart. Figure 11-16 is an intracellular recording from an SA node cell; note the slow depolarization toward threshold which initiates the action potential, and compare this SA nodal action potential to that of unspecialized nonautorhythmical atrial cells, which fall to show the pace-

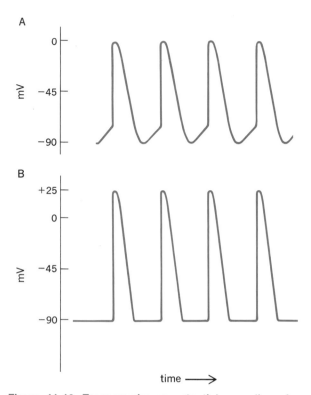

Figure 11-16. Transmembrane potential recordings from (A) SA-node cell and (B) atrial muscle fiber.

maker potential. If the activity of the SA node is depressed or conduction from it blocked, another portion of the conducting system may take over as pacemaker. In contrast, ordinary atrial and ventricular muscle fibers, which constitute 99 percent of the total cardiac muscle, are not normally capable of generating pacemaker potentials and do so only under abnormal conditions.

Sequence of excitation

The cells of the SA node make contact with the surrounding atrial myocardial fibers (Figs. 11-15 and 11-17). From the SA node, the wave of excitation spreads throughout the right and left atria along ordinary atrial myocardial cells (as well as several specialized bundles of fibers), passing from cell to cell by way of the gap junctions. How does the excitation spread to the ventricles? At the base of the right atrium very near the wall between the ventricles (*interventricular septum*), the wave of excitation encounters a second small mass of

Figure 11-15. Conducting system of the heart. The interatrial and SA-node–AV-node conducting bundles are not shown in the figure (*Adapted from Rushmer.*)

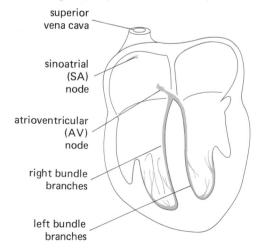

specialized cells, the *atrioventricular* (AV) *node.* This node and the bundle of fibers leaving it constitute the only conducting link between the atria and ventricles, all other areas being separated by nonconducting connective tissue. This anatomical pattern ensures that excitation will travel from atria to ventricles only through the AV node, but it also means that malfunction of the AV node may completely dissociate atrial and ventricular contraction. The AV node manifests one particularly important characteristic: The propagation of action potentials through the node is delayed for approximately 0.1 s (mainly because of the small diameters of these cells), allowing the atria to contract and empty their contents into the ventricles before ventricular contraction. The wave of excitation travels between the SA and AV nodes, in large part, along ordinary atrial myocardial fibers. However, there may also be some contribution by several more or less specialized fiber bundles which conduct directly between these two nodes.

After leaving the AV node, the impulse travels rapidly along specialized conducting fibers which run down the interventricular septum as bundles which then spread throughout much of the right and left ventricular myocardium. Finally, these fibers make contact with unspecialized myocardial cells through which the impulse spreads from cell to cell in the remaining myocardium. The rapid conduction along these fibers and their highly diffuse distribution cause depolarization of all right and left ventricular cells more or less simultaneously and ensure a single coordinated contraction.

Refractory period of the heart

The pumping of blood requires alternate periods of contraction and relaxation. Imagine the result of a prolonged tetanic contraction of cardiac muscle like that described for skeletal muscle. Obviously, pumping would cease and death would ensue. In reality such contractions never occur in the heart because of the long *refractory period* of cardiac muscle. Recall that in any excitable membrane an action potential is accompanied by a period during which the membrane is completely insensitive to a stimulus regardless of intensity. Following this absolute refractory period comes a second period during which the membrane can be depolarized again but only by a more intense stimulus. In skeletal muscle, the absolute refractory periods are very short (1 to 2 ms) compared with the duration of contraction (20 to 100 ms), and a second contraction can be elicited before the first is over. In contrast, the absolute refractory period of cardiac muscle lasts almost as long as the contraction (250 ms), and the muscle cannot be excited in time to produce summation (Fig. 11-18).

It is evident from Fig. 11-18 that the long refractory period of ventricular muscle fibers is associated with the long plateau in the action potential manifested by these fibers, a plateau not seen in the action potentials of skeletal muscle or

Figure 11-17. Sequence of cardiac excitation. Atrial excitation is complete before ventricular excitation begins because of the delay at the AV node. *(Adapted from Rushmer.)*

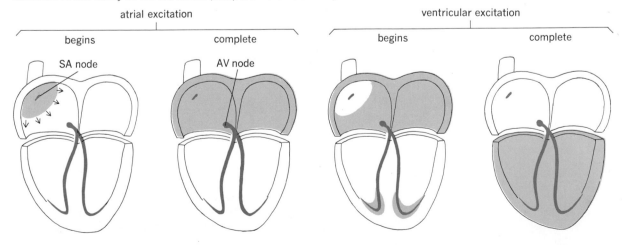

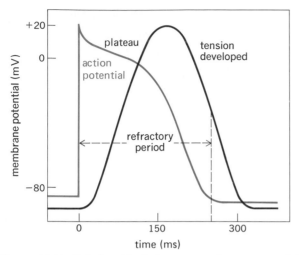

Figure 11-18. Relationship between membrane potential changes and contraction in a single ventricular muscle cell. The refractory period lasts almost as long as the contraction.

even other cardiac fibers (see Fig. 11-16, for example). The initial spike of the action potential is, as everywhere else, due to an explosive increase in membrane permeability to sodium; the plateau reflects a combination of (relative to the resting state) decreased permeability to potassium and increased permeability to both sodium and calcium; the return of the membrane potential to the resting state reflects the return of these three permeabilities to their resting values.

A common situation explainable in terms of the

refractory period is shown in Fig. 11-19. In many people, drinking several cups of coffee causes increased excitability of areas other than SA node due to the action of caffeine. When such an area of the atria or ventricles actually initiates an action potential, it is known as an *ectopic focus.* When one of these areas fires just after completion of a normal contraction but before the next SA nodal impulse, a premature wave of excitation occurs. As a result, the next normal SA nodal impulse fires during the refractory period of the premature beat and is not propagated since the myocardial cells are not excitable (the SA node still fires because it has a shorter refractory period). The second SA nodal impulse after the premature contraction is propagated normally. The net result is an unusually long delay between beats. The contraction after the delay is unusually strong, and the person is aware of his heart pounding. If the ectopic focus continued to discharge at a rate higher than that of the SA node, it might then capture the role of pacemaker and drive the heart at rates as high as 200 to 300 beats per minute, compared with a normal rate of 70.

Many similar examples are important clinically and help to clarify the normal physiological process. For example, disease may damage cardiac tissue, hampering conduction through the AV node, so that only a fraction of the atrial impulses are transmitted into the ventricles; thus, the atria may have a rate of 80 beats per minute and the ventricles only 60. If there is complete block at the AV node, none of the atrial impulses get through

Figure 11-19. Effect of an ectopic discharge on ventricular contraction. The arrows indicate the times at which the SA node or ectopic focus fires. The premature beat induced by the ectopic discharge makes the ventricular muscle refractory at the time of the next SA-node impulse. The failure of this impulse to induce a contraction results in a longer period than normal before the next beat.

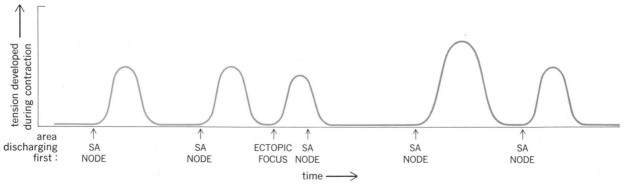

and the AV node itself begins to initiate excitation at its own spontaneous rate; i.e., the AV node (or a portion of the conducting system just below) is capable of pacemaker action when released from capture by impulses from the SA node. This rate is quite slow, generally 25 to 40 beats per minute, and completely out of synchrony with the atrial contractions, which continue at the normal higher rate. Under such conditions, the atria are totally ineffective as pumps since they are usually contracting against closed AV valves, but atrial pumping, as we shall see, is relatively unimportant for cardiac functioning (except during relatively strenuous exercise). Some patients have transient recurrent episodes of complete AV block signaled by fainting spells (due to decreased brain blood flow). These spells result because the ventricles do not begin their own impulse generation immediately and their pumping ceases temporarily.

Partial AV block is not always caused by disease but may represent a normal life-saving adaptation. Imagine a patient with an ectopic area driving his atria at 300 beats per minute. Ventricular rates this high are very inefficient because there is inadequate time to fill the ventricles between contractions. Fortunately, the long refractory period of the AV node may prevent passage of a significant fraction of the impulses and the ventricles beat at a slower rate.

Another group of abnormalities apparently is characterized by a prolonged or unusual conduction route so that the impulse constantly meets an area which is no longer refractory and keeps traveling around the heart in a so-called "circus" movement. This may lead to continuous, completely disorganized contractions (*fibrillation*), which can cause death if they occur in the ventricles. For example, ventricular fibrillation is the immediate cause of death in many instances of heart attack.

The electrocardiogram

The *electrocardiogram* (ECG)[1] is primarily a tool for evaluating the electric events within the heart. The action potentials of cardiac muscle can be viewed as batteries which cause current flow

[1] Traditionally, electrocardiogram has been abbreviated EKG, the K being derived from the Greek word for heart. ECG is now the preferred term.

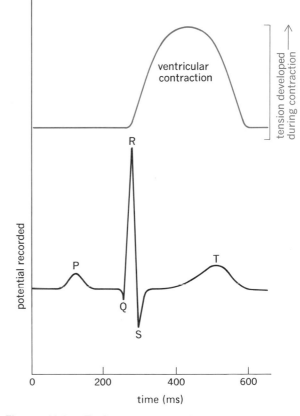

Figure 11-20. Typical electrocardiogram. P: atrial depolarization; QRS: ventricular depolarization; T: ventricular repolarization.

throughout the body fluids. These currents produce voltage differences at the body surface which can be detected by attaching small metal plates at different places on the body. Figure 11-20 illustrates a typical normal ECG recorded as the potential difference between the right and left wrists. The first wave P represents atrial depolarization. The second complex QRS, occurring approximately 0.1 to 0.2 s later, represents ventricular depolarization. The final wave T represents ventricular repolarization. No manifestation of atrial repolarization is evident because it occurs during ventricular depolarization and is masked by the QRS complex. Figure 11-21A gives one example of the clinical usefulness of the ECG. This patient is suffering from partial AV nodal block so that only one-half the atrial impulses are being transmitted. Note that every second P wave is not followed by a

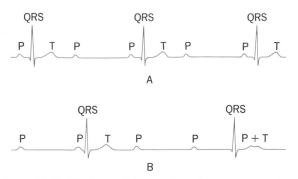

Figure 11-21. Electrocardiograms from two persons suffering from atrioventricular block. (A) Partial block; one-half of the atrial impulses are transmitted to the ventricles. (B) Complete block; there is absolutely no synchrony between atrial and ventricular electric activities.

QRS and T. Because many myocardial defects alter normal impulse propagation, and thereby the shapes of the waves, the ECG is a powerful tool for diagnosing heart disease.

Excitation-contraction coupling

The depolarization ("excitation") of cardiac muscle cells triggers their contraction. As described in Chap. 10, the mechanism coupling excitation and contraction is an increase in the cytosolic concentration of calcium; this calcium combines with the regulator protein, troponin, with the resulting removal of tropomyosin's inhibition of cross-bridge formation between actin and myosin. There are two sources of the increased cytosolic calcium during the action potential: diffusion of calcium from interstitial fluid across the plasma membrane into the cytosol during the plateau of the action potential; release of calcium from the sarcoplasmic reticulum. During repolarization, cytosolic calcium is restored to its original extremely low value by active transport of the calcium back across the plasma membrane or into the sarcoplasmic reticulum, and the muscle relaxes. As we shall see, changes in cytosolic calcium account for many situations in which the strength of cardiac muscle contraction is altered.

Mechanical events of the cardiac cycle

Fluid always flows from a region of higher pressure to one of lower pressure. (This important con-

cept will be developed formally in a later section; at present, we need deal with it only as an intuitively obvious phenomenon.) The sole function of the heart is to generate the pressures which produce blood flow, the heart valves serving to direct the flow. The orderly process of depolarization described in the previous section triggers contraction of the atria followed rapidly by ventricular contraction. The volume and pressure changes occurring during this cardiac cycle are summarized in Fig. 11-22. The reader should follow our analysis of the heart picture and the pressure profile in each phase carefully since an understanding of the cardiac cycle is essential.

The cardiac cycle is divided into two major phases: the period of ventricular contraction (*systole*), followed by the period of ventricular relaxation (*diastole*). We start our analysis at the far left of Fig. 11-22 with the events of mid-to-late diastole, considering first only the left heart; events on the right side have an identical sequence, but the pressure changes are smaller.

Mid-to-late diastole (ventricular filling)

The left atrium and ventricle are both relaxed; left atrial pressure is very slightly higher than left ventricular (because blood is entering the atrium from the pulmonary veins); therefore, the AV valve is open, and blood is passing from atrium to ventricle. This is an important point: The ventricle receives blood from the atrium throughout most of diastole, not just when the atrium contracts. Indeed, at rest, approximately 80 percent of ventricular filling occurs before atrial contraction.[1] Note that the aortic valve (between aorta and left ventricle) is closed because the aortic pressure is higher than the ventricular pressure. The aortic pressure is slowly falling because blood is moving out of the arteries and through the vascular tree; in contrast, ventricular pressure is rising slightly because blood is entering from the atrium, thereby expanding the ventricular volume. At the very end of diastole, the SA node discharges, the atrium de-

---

[1] It is for this reason that the conduction defects discussed above, which eliminate the atria as efficient pumps, do not seriously impair cardiac function, at least at rest. Many persons lead relatively normal lives for many years despite atrial fibrillation. Thus, in many respects, the atrium may be conveniently viewed as merely a continuation of the large veins.

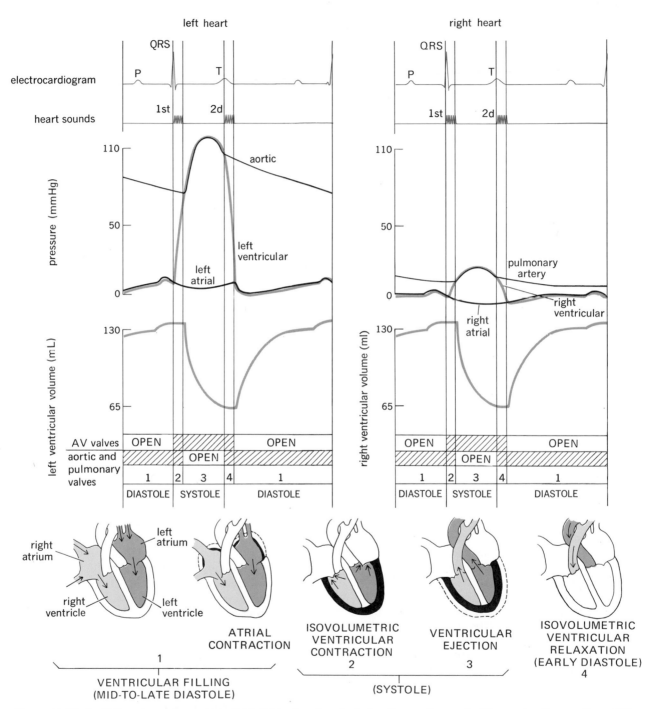

Figure 11-22. *(Left)* **Summary of events in the left heart and aorta during the cardiac cycle. The contracting portions of the heart are shown in black.** *(Right)* **Summary of events in the right heart and pulmonary arteries during the cardiac cycle.** *(Adapted from Ganong.)*

polarizes (as shown by the P wave of the ECG), the atrium contracts (note the small rise in atrial and ventricular pressure), and a small volume of blood is added to the ventricle. The amount of blood in the ventricle just prior to systole is called the *end-diastolic volume.*

Systole

The wave of depolarization passes through the ventricle (QRS complex) and triggers ventricular contraction. As the ventricle contracts, ventricular pressure rises steeply. Almost immediately, this pressure exceeds the atrial pressure and closes the AV valve, thus preventing backflow into the atrium. Since for a brief period the aortic pressure still exceeds the ventricular, the aortic valve remains closed and the ventricle does not empty despite contraction. This early phase of systole is called *isovolumetric ventricular contraction* because ventricular volume is constant; i.e., the lengths of the muscle fibers remain approximately constant as in an isometric skeletal-muscle contraction. This brief phase ends when ventricular pressure exceeds aortic, the aortic valve opens, and *ventricular ejection* occurs. The ventricular-volume curve shows that ejection is rapid at first and then tapers off. *The ventricle does not empty completely;* the amount remaining after ejection is called the *end-systolic volume.* As blood flows into the aorta, the aortic pressure rises with ventricular pressure. Atrial pressure also rises slowly throughout the entire period of ventricular ejection because of continued flow of blood from the veins. Note that peak aortic pressure is reached before the end of ventricular ejection; i.e., the pressure actually is beginning to fall during the last part of systole despite continued ventricular ejection. This phenomenon is explained by the fact that the rate of blood ejection during this last part of systole is quite small (as shown by the ventricular volume curve) and is less than the rate at which blood is leaving the aorta (and other large arteries) via the arterioles; accordingly the volume and, therefore, the pressure within the aorta begin to decrease.

Early diastole (isovolumetric ventricular relaxation)

When contraction stops, the ventricular muscle relaxes rapidly owing to release of tension created during contraction. Ventricular pressure therefore falls almost immediately below aortic pressure, and the aortic valve closes. However, ventricular pressure still exceeds atrial so that the AV valve remains closed. This phase of early diastole, obviously the mirror image of early systole, is called *isovolumetric ventricular relaxation.* It ends as ventricular pressure falls below atrial, the AV valves open, and ventricular filling begins. Filling occurs rapidly at first and then slows down as atrial pressure decreases. The fact that ventricular filling is almost complete during early diastole is of the greatest importance; it ensures that filling is not seriously impaired during periods of rapid heart rate, e.g., exercise, emotional stress, fever, despite a marked reduction in the duration of diastole. However, when rates of approximately 200 beats per minute or more are reached, filling time does become inadequate so that the volume of blood pumped during each beat is decreased. Significantly, the AV node in normal adults does not conduct at rates greater than 200 to 250 beats per minute.

Pulmonary circulation pressures

Figure 11-22 also summarizes the simultaneously occurring events in the right heart and pulmonary arteries, the sequence being virtually identical to those just described for the left heart. There is one striking quantitative difference: The ventricular and arterial pressures are considerably lower during systole. The pulmonary circulation is a low-pressure system (for reasons to be described in a later section). This difference is clearly reflected in the ventricular architecture, the right ventricular wall being much thinner than the left. Note, however, that despite the lower pressure, the right ventricle ejects the same amount of blood as the left.

Heart sounds

Two heart sounds are normally heard through a stethoscope placed on the chest wall (see Fig. 11-22). The first sound, a low-pitched *lub*, is associated with closure of the AV valves at the onset of systole; the second, a high-pitched *dub*, is associated with closure of the pulmonary and aortic valves at the onset of diastole. These sounds, which result from vibrations caused by valvular closure, are perfectly normal, but heart murmurs are frequently (although not always) a sign of heart disease. When blood flows smoothly in a streamline manner, i.e., layers of fluid sliding evenly over one another, it makes no sound, but turbulent flow produced by unusually high velocities makes a noise. This noise is heard as a murmur or sloshing sound. Turbulence can be produced by blood flowing rapidly in the usual direction through an abnormally narrowed valve (stenosis), backward through a damaged leaky valve (regurgitation), or between the two atria or two ventricles via a small hole in the septum. The exact timing and location of the murmur provide the physician with a powerful diagnostic clue. For example, a murmur heard throughout systole suggests a narrowed pulmonary or aortic valve or a hole in the interventricular septum.

The cardiac output

The volume of blood pumped *by each ventricle* per minute is called the *cardiac output,* usually expressed as liters per minute. It must be remembered that the cardiac output is the amount of blood pumped by *each* ventricle, *not* the total amount pumped by both ventricles. The cardiac output is determined by multiplying the *heart rate* and the volume of blood ejected by each ventricle during each beat *(stroke volume):*

cardiac output = heart rate × stroke volume
   L/min   beats/min   L/beat

For example, if each ventricle has a rate of 72 beats per minute and ejects 70 mL with each beat, what is the cardiac output?

CO = 72 beats/min × 0.07 L/beat = 5.0 L/min

These values are approximately normal for a resting adult. During periods of exercise, the cardiac output may reach 30 to 35 L/min. Obviously, heart rate or stroke volume or both must have increased. Physical exercise is but one of many situations in which various tissues and organs require a greater flow of blood; e.g., flow through skin vessels increases when heat loss is required, and flow through intestinal vessels increases during digestion. Some of the increased flow can be obtained merely by decreasing blood flow to some other organ, i.e., by redistributing

the cardiac output, but most of the supply must come from a greater total cardiac output. The following description of the factors which alter the two determinants of cardiac output, heart rate and stroke volume, applies in all respects to both the right and left heart since stroke volume and heart rate are the same for both the right and left ventricles.

Control of heart rate

The rhythmic discharge of the SA node occurs spontaneously in the complete absence of any nervous or hormonal influences. However, it is under the constant influence of both nerves and hormones. As mentioned earlier a large number of parasympathetic and sympathetic fibers end on the SA node as well as on other areas of the conducting system. As shown in Fig. 11-23, stimulation of the parasympathetic (vagus) nerves (or local application of acetylcholine) causes slowing of the heart and, if strong enough, may stop the heart completely for some time. The effects of the sympathetic nerves are just the reverse; their stimulation (or local application of norepinephrine) increases the heart rate, whereas cutting the sympathetics slows the heart. Finally, cutting the parasympathetics causes the heart rate to increase. The experiments in which the nerves are sectioned demonstrate that both the sympathetic and parasympathetic nerves normally are discharging at some finite rate. In the resting state, the parasympathetic influence is dominant since simultaneous removal of all nerves causes the heart rate to increase to approximately 100 beats per minute. This is the inherent autonomous discharge rate of the SA node.

Figure 11-24 illustrates the nature of the sympathetic and parasympathetic influence on SA node function. Sympathetic stimulation increases the slope of the pacemaker potential, which causes the cell to reach threshold more rapidly and is responsible for the rate increase. Stimulation of the parasympathetics has the opposite effect; the slope of the pacemaker potential decreases, threshold is reached more slowly, and heart rate decreases. How do these autonomic mediators alter the slope of the pacemaker potential? Acetylcholine does so by increasing the permeability of the membrane to potassium; the story for norepinephrine is not yet clear, but the most likely explanation is that it does

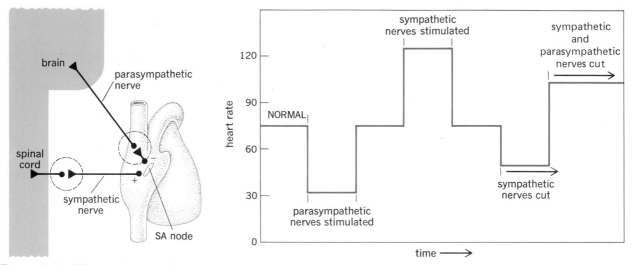

Figure 11-23. Effects of autonomic nerves on heart rate. Sympathetic nerves stimulate the SA node whereas parasympathetics inhibit it. The parasympathetic fibers to the heart are contained in the vagus nerves. The values shown
during nerve stimulation do not represent the maximal changes obtainable.

just the opposite, namely decreases the membrane permeability to potassium.

Factors other than the cardiac nerves also can alter heart rate. Epinephrine, the hormone liberated from the adrenal medulla, speeds the heart; this is not surprising since epinephrine is a blood-borne sympathetic mediator similar in structure to norepinephrine (Chaps. 8 and 9).

The heart rate is also sensitive to many other factors, including temperature, plasma electrolyte concentrations, and hormones other than epinephrine. However, these are generally of lesser importance, and the heart rate is primarily regulated very precisely by balancing the slowing effects of parasympathetic discharge against the accelerating effects of sympathetic discharge, both operating on the SA node (Fig. 11-25).

Control of stroke volume

The second variable which determines cardiac output is *stroke volume*, the volume of blood ejected during each ventricular contraction. It is important to recognize that the ventricles never completely empty themselves of blood during contraction. Therefore, a more forceful contraction, i.e., a greater myocardial-fiber shortening, can produce an increase in stroke volume. Changes in the force of contraction can be produced by a variety of factors, but two are dominant under most physiological conditions: (1) changes in the degree of stretching of the ventricular muscle secondary to changes in ventricular volume; (2) alterations in the magnitude of sympathetic nervous system input to the ventricles.

Relationship between end-diastolic volume and stroke volume. It is possible to study the completely intrinsic adaptability of the heart by means of the so-called heart-lung preparation (Fig. 11-26). Tubes are placed in the heart and vessels of an anesthetized animal so that blood flows from the

Figure 11-24. Effects of sympathetic and parasympathetic nerve stimulation on the slope of the pacemaker potential of an SA-node cell. *(Adapted from Hoffman and Cranefield.)*

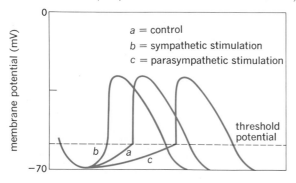

a = control
b = sympathetic stimulation
c = parasympathetic stimulation

very first part of the aorta (just above the exit of the coronary arteries) into a blood-filled reservoir and from there into the right atrium. The blood then is pumped by the right heart as usual via the lungs into the left heart from which it is returned to the reservoir. The net effect is to nourish the heart and lungs and to abolish all nervous and hormonal input to the heart. A key feature of this preparation is that the pressure causing blood flow into the heart ("venous pressure") can be altered simply by raising or lowering the reservoir. This is analogous to altering the venous and right atrial pressures and causes changes in the quantity of blood entering the right ventricle during diastole. Thus when the reservoir is raised, ventricular filling increases, thereby increasing end-diastolic volume. The more-distended ventricle responds with a more forceful contraction. The net result is a new steady state, in which the ventricle is distended and diastolic filling and stroke volume are both increased, but equal.

The mechanism underlying this completely intrinsic adaptation is that cardiac muscle, like other muscle, increases its strength of contraction when

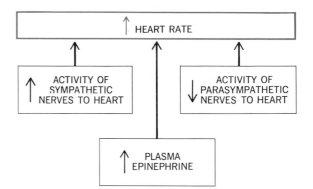

Figure 11-25. Summary of major factors which influence heart rate. All effects are exerted upon the SA node. The figure, as drawn, shows how heart rate is increased; conversely, heart rate is slowed when sympathetic activity and epinephrine are decreased and when parasympathetic activity is increased.

it is stretched. Thus, in the experiment above, the increased diastolic volume stretches the ventricular muscle fibers and causes them to contract more forcefully. This relationship was expounded

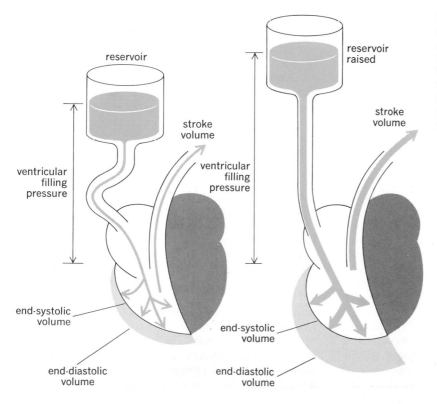

Figure 11-26. Experiment for demonstrating the relationship between end-diastolic volume and stroke volume (Starling's law of the heart). By raising the reservoir, the pressure causing ventricular filling is increased. The increased filling distends the ventricle, which responds with an increased strength of contraction.

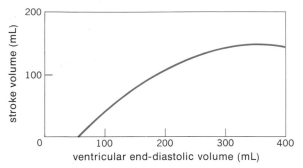

Figure 11-27. Relationship between ventricular end-diastolic volume and stroke volume (Starling's law of the heart). The data were obtained by progressively increasing ventricular filling pressure, as in Fig. 11-26. The horizontal axis could have been labelled "sarcomere length," and the vertical "contractile force," i.e., this is a length-tension curve.

by the British physiologist Starling, who observed that there was a direct proportion between the diastolic volume of the heart, i.e., the length of its muscle fibers, and the force of contraction of the following systole. It is now referred to as *Starling's law of the heart*. A typical response curve obtained by progressively increasing end-diastolic volume is shown in Fig. 11-27. Note that marked overstretching causes the force of contraction, and thereby the stroke volume, to fall off. Thus, heart muscle manifests a length-tension relationship very similar to that described earlier for skeletal muscle and explainable in terms of the sliding-filament mechanism of muscle contraction, i.e., in terms of the number of active cross-bridges between actin and myosin. However, unlike skeletal muscle, cardiac muscle length in the resting state is less than that which yields maximal tension during contraction so that an increase in length produces an increase in contractile tension.

This intrinsic relationship between end-diastolic volume and stroke volume, originally demonstrated in the heart-lung preparation, applies equally to the intact human being. End-diastolic volume, therefore, becomes a crucial determinant of cardiac output.

What then are the factors which determine end-diastolic volume, i.e., the degree of ventricular distension just before systole? The simplest way to approach this question is to view the ventricle as an elastic chamber, much like a balloon. A balloon enlarges when one blows into it because the internal pressure acting upon the wall becomes greater than the external pressure. The more air blown in, the higher the internal pressure becomes. The degree of distension therefore depends upon the pressure difference across the wall and the distensibility of the wall. This is precisely the situation for the ventricle. We shall ignore the problem of ventricular distensibility (which changes little under physiological conditions) and concentrate only on the *transmural*, or "across-the-wall," difference in pressure. Where are the pressures acting across the ventricular wall at the end of diastole? The internal pressure, of course, is the fluid pressure exerted by the blood against the walls; the external pressure surrounding the heart is the pressure within the chest (thoracic) cavity, the intrathoracic pressure.

We can now answer our original question. End-diastolic ventricular volume and distension can be increased either by increasing the intraventricular blood pressure or by decreasing the intrathoracic pressure, or both. The latter occurs during inspiration (see Chap. 12) and accounts, in part, for the increased stroke volume which usually occurs during inspiration. However, it is primarily by changes in the end-diastolic intraventricular pressure that end-diastolic volume is controlled. As can be seen from Fig. 11-22, the pressure within the ventricles during diastole is only slightly less than the atrial pressure, since the valves are open and the chambers are connected by the wide AV orifice through which blood is flowing into the ventricle. Accordingly, the diastolic ventricular blood pressure is determined by the atrial pressure. As we shall see, the atrial pressure, in turn, is determined by the pressure in the veins emptying into the atria.

The significance of this mechanism should now be apparent; an increased flow of blood from the veins into the heart automatically forces an equivalent increase in cardiac output by distending the ventricle and increasing stroke volume, just like the heart-lung apparatus when we increased "venous return" by elevating the reservoir. This is probably the single most important mechanism for maintaining equality of right and left output. Should the right heart, for example, suddenly begin to pump more blood tan the left, the increased blood flow to the left ventricle would automatically produce an equivalent increase in left ventricular output and blood would not be allowed to accumulate in the lungs. Another example of Starling's law has already been described above, namely, the pounding that occurs after a premature contraction. Recall that an unusually long period elapses between the premature contraction and the next contraction; the period for diastolic filling is increased, end-diastolic volume increases, and the force of contraction is increased (Fig. 11-19). It is this strong contraction, which may actually lift the heart against the chest wall, that the person is aware of.

The sympathetic nerves. Sympathetic nerves are distributed not only to the SA node and conducting system but to all myocardial cells. The effect of the sympathetic mediator, norepinephrine, is to increase ventricular (and atrial) strength of contraction at any given initial end-diastolic volume; i.e., unlike the relationship described by Starling's law, the increased force of contraction secondary to sympathetic-nerve stimulation is independent of a change in end-diastolic ventricular muscle-fiber length. It is termed increased *contractility.* Note that "contractility" and "force of contraction" are *not* synonyms; a change in force of contraction due to increased end-diastolic volume (Starling's law) does not reflect increased contractility, since the latter term is specifically defined as an increased force of contraction *not* due to altered end-diastolic volume. Circulating epinephrine also produces myocardial changes similar to those induced by the sympathetic nerves to the heart.

Not only does enhanced sympathetic nerve activity to the myocardium cause the contraction to be more powerful, it causes both the contraction and relaxation of the ventricles to occur more quickly. These latter effects are quite important since, as described earlier, increased sympathetic activity to the heart also increases heart rate. As heart rate increases, the time available for diastolic filling decreases, but the more rapid contraction and relaxation induced simultaneously by the sympathetic neurons partially compensate for this problem by permitting a larger fraction of the cardiac cycle to be available for filling. Moreover, because the ventricles relax so rapidly after a contraction, the intraventicular pressure falls rapidly, thereby creating an enhanced pressure gradient for flow of blood into the ventricles. The ability of these effects to maintain diastolic filling

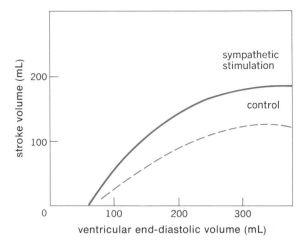

Figure 11-28. Effects on stroke volume of stimulating the sympathetic nerves to the heart.

is, of course, not unlimited, and diastolic filling is significantly reduced at very high heart rates. The significance of this interplay between diastolic filling time, heart rate, and contractility will be analyzed further in the later section on exercise.

Norepinephrine and epinephrine increase contractility by inducing an increased movement of calcium into the cytosol during excitation. They speed relaxation by enhancing the rate at which this calcium is transported out of the cytosol. In contrast to the sympathetic nerves, the parasympathetic nerves to the heart have almost no effect on ventricular contractility.

The interrelationship between Starling's law and the cardiac nerves as measured in a heart-lung preparation is illustrated in Fig. 11-28. The dashed line is the same as the line shown in Fig. 11-27 and was obtained by slowly raising ventricular pressure while measuring end-diastolic volume and stroke volume; the solid line was obtained similarly for the same heart but during sympathetic-nerve stimulation. Starling's law still applies, but during nerve stimulation the stroke volume is greater at any given end-diastolic volume. In other words, the increased contractility leads to *a more complete ejection* of the end-diastolic ventricular volume.

In summary (Fig. 11-29) stroke volume is controlled both by an intrinsic cardiac mechanism dependent only upon changes in end-diastolic volume and by an extrinsic mechanism mediated by the cardiac sympathetic nerves (and circu-

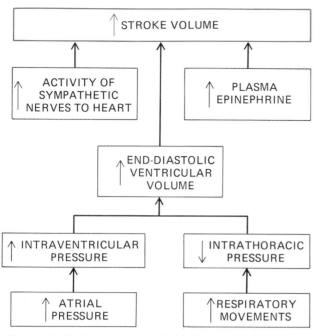

Figure 11-29. Major factors which influence stroke volume. The figure as drawn shows how stroke volume is increased; a reversal of all arrows in the boxes would illustrate how stroke volume is decreased. Refer to the text for details.

lating epinephrine), which causes increased ventricular contractility. The contribution of each mechanism in specific physiological situations and the reflexes controlling the nerves are described in a later section.

Summary of cardiac output control

A summary of the major factors which determine cardiac output is presented in Fig. 11-30, which combines the information of Fig. 11-25 (factors influencing heart rate) and Fig. 11-29 (factors influencing stroke volume).

Cardiac energetics

During contraction, the heart, like other kinds of muscle, converts chemical energy stored in ATP into mechanical work and heat. The total energy utilized per unit time is determined by a large number of factors, but the most important are the arterial pressure against which the ventricles must pump and the heart rate (i.e., the number of times

per minute this pressure must be developed). This means that the energy expended by the heart may vary markedly in different situations having the same cardiac output. For example, if the heart rate is doubled and the stroke volume halved, cardiac output does not change but the energy required increases markedly. To take another example, the hearts of persons with hypertension (high arterial blood pressure) consume more energy in pumping blood even though their cardiac output is normal. From these facts it should be clear why people with impaired coronary blood flow (i.e., decreased supply of oxygen and nutrients) are subjected to the greatest danger when their heart rate and arterial blood pressure increase, as during emotional excitement.

Section C
The vascular system

As described at the beginning of this chapter, blood leaving the ventricles passes in turn through arteries, arterioles, capillaries, venules, and veins to return to the heart. To comprehend the functional characteristics of these various blood-vessel types, one must be familiar with the basic physical principles underlying the *bulk flow* of fluids.

Basic principles of pressure, flow, and resistance

Fluid flows through a tube in response to a difference in pressure, ΔP, between the two ends of the tube. The total volume flowing per unit time is directly proportional to the pressure difference. It is not the absolute pressure in the tube which determines flow but the difference in pressure between the two ends (Fig. 11-31). This direct proportionality of flow F and pressure difference ΔP can be written

$$F = k \, \Delta P$$

where k is the proportionality constant describing how much flow occurs for a given pressure difference. Knowing only the pressure difference between two ends of a tube is not enough to determine how much fluid will flow; one must also know the numerical value of k. This constant is simply a measure of the ease with which fluid will flow

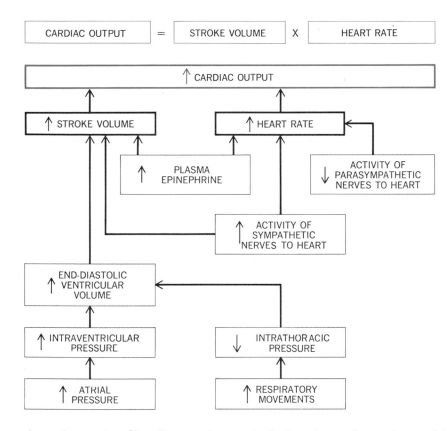

Figure 11-30. Major factors determining cardiac output (an amalgamation of Figs. 11-25 and 11-29).

through a tube. Usually we do not deal directly with k but with the reciprocal of k, known as *resistance R:*

$$R = \frac{1}{k}$$

Thus, the larger k is, the smaller R is. R answers the question: "How difficult is it for fluid to flow through a tube at any given pressure?" Its name, resistance, is therefore quite appropriate. Substituting $1/R$ for k, the basic equation becomes

$$F = \frac{\Delta P}{R}$$

In words, flow is directly proportional to the pressure difference and inversely proportional to the resistance.

Determinants of resistance

Resistance is essentially a measure of friction since it is basically the friction between tube wall and fluid and between the molecules of the fluid

themselves which opposes flow. Resistance depends upon the nature of the fluid and the geometry of the tube.

Nature of the fluid. Syrup flows less readily than water because there is greater friction between the

Figure 11-31. Flow between two points within a tube is proportional to the pressure *difference* between the points.

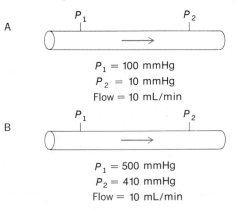

A

$P_1 = 100$ mmHg
$P_2 = 10$ mmHg
Flow = 10 mL/min

B

$P_1 = 500$ mmHg
$P_2 = 410$ mmHg
Flow = 10 mL/min

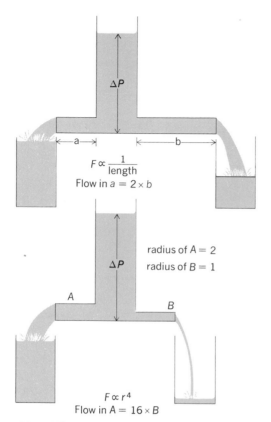

Figure 11-32. Effects of tube length and radius on resistance to flow.

molecules of syrup as they slide past each other. This property of fluids is called *viscosity*; the more friction, the higher the viscosity. In abnormal states, changes in blood viscosity can have important effects on the resistance to flow, but under most physiological conditions, the viscosity of blood is relatively constant.

Geometry of the tube. Both the length and radius of a tube affect its resistance by determining the surface area of the tube in contact with the fluid. Resistance to flow is directly proportional to the length of the tube, but since the lengths of the blood vessels remain constant in the body, length is not a factor in the *control* of resistance. The resistance increases markedly as tube radius decreases; the exact relationship is given by the following formula, which states that the resistance is inversely proportional to the fourth power of the radius (the radius multiplied by itself four times):

$$R \propto \frac{1}{r^4}$$

The extraordinary dependence of resistance upon radius can be appreciated by the fact (shown in Fig. 11-32) that doubling the radius increases the flow sixteenfold. As we shall see, the radius (at constant pressure gradient) of blood vessels can be changed significantly and constitutes the most important factor in the control of resistance to blood flow.

Arteries

The aorta and other arteries have thick walls containing large quantities of elastic tissue. Although they also have smooth muscle, there is relatively little fluctuation in the state of their muscle activity, and the arteries can be viewed most conveniently as simple elastic tubes. Because the arteries have large radii, they serve as low-resistance pipes conducting blood to the various organs. Their second major function, related to their elasticity, is to act as a pressure reservoir for maintaining blood flow through the tissues.

Arterial blood pressure

Recall once more the factors determining the pressure within an elastic container, e.g., a balloon filled with water. The pressure inside the balloon depends upon the volume of water within it and the distensibility of the balloon, i.e., how easily its walls can be stretched. If the walls are very stretchable, large quantities of water can go in with only a small rise in pressure; conversely, a small quantity of water causes a large pressure rise in a balloon with low distensibility.

These principles can now be applied to an analysis of arterial function. The contraction of the ventricles ejects blood into the pulmonary and systemic arteries during systole. If a precisely equal quantity of blood were to flow simultaneously out of the arteries via arterioles, the total volume of blood in the arteries would remain constant and arterial pressure would not change. Such, however, is not the case. As shown in Fig 11-33, a volume of blood equal to only about one-third the stroke volume leaves the arteries during systole. The excess volume distends the arteries by raising the

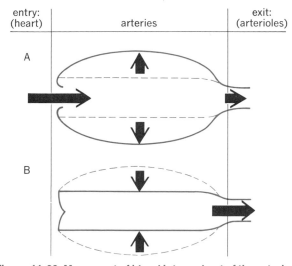

entry:
(heart)   arteries   exit:
                                              (arterioles)

A

B

Figure 11-33. Movement of blood into and out of the arteries during the cardiac cycle. The lengths of the arrows denote relative quantities. During systole (A) less blood leaves the arteries than enters and the arterial walls stretch. During diastole (B) the walls recoil passively, driving blood out of the arteries.

arterial pressure. When ventricular contraction ends, the stretched arterial walls recoil passively (like a stretched rubber band upon release) and the arterial pressure continues to drive blood through the arterioles. As blood leaves the arteries, the arterial volume and, therefore, the pressure slowly falls, but the next ventricular contraction occurs while there is still adequate blood in the arteries to stretch them partially, so that the arterial pressure does not fall to zero. In this manner, the arterial pressure provides the immediate driving force for tissue blood flow. The aortic pressure pattern shown in Fig. 11-34 is typical of the pressure changes which occur in all the large systemic arteries. The maximum pressure is reached during peak ventricular ejection and is called *systolic pressure.* The minimum pressure obviously occurs just before ventricular contraction and is called *diastolic pressure.* They are generally recorded as systolic/diastolic, that is, 125/75 mmHg in our example. The pulse, which can be felt in an artery, is due to the difference between systolic and diastolic pressure. This difference ($125 - 75 = 50$) is called the *pulse pressure.* The factors which alter pulse pressure are the following: (1) An increased stroke volume tends to elevate systolic

pressure because of greater arterial stretching by the additional blood. (2) Decreased arterial distensibility, as in atherosclerosis ("hardening of the arteries"), may cause a marked increase in systolic pressure because the wall is stiffer; i.e., any given volume of blood produces a greater pressure rise.

It is evident from the figure that arterial pressure (and, therefore, flow) is constantly changing throughout the cardiac cycle and the average pressure *(mean pressure)* throughout the cycle is not merely the value halfway between systolic and diastolic pressure, because diastole usually lasts longer than systole. The true mean arterial pressure can be obtained only by complex methods, but for most purposes it is approximately equal to the diastolic pressure plus one-third of the pulse pressure. Thus, in our example,

$$\text{Mean pressure} = 75 + (\tfrac{1}{3} \times 50) = 92 \text{ mmHg}$$

The mean arterial pressure is the most important of the pressures described because it is the *average* pressure driving blood into the tissues throughout the cardiac cycle. In other words, if the pulsatile pressure changes were eliminated and the pressure throughout the cardiac cycle were always equal to the mean pressure, the total flow would be unchanged. It is a closely regulated quantity; indeed, the reflexes which accomplish this regulation constitute the basic cardiovascular control mechanisms and will be described in detail in a later section.

One last point: We can refer to "arterial" pressure without specifying to which artery we are referring because the aorta and other arteries have such large diameters that they offer only negligible

Figure 11-34. Typical aortic pressure fluctuations during the cardiac cycle.

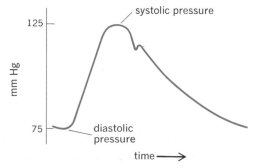

125

mm Hg

75

systolic pressure

diastolic pressure

time ———→

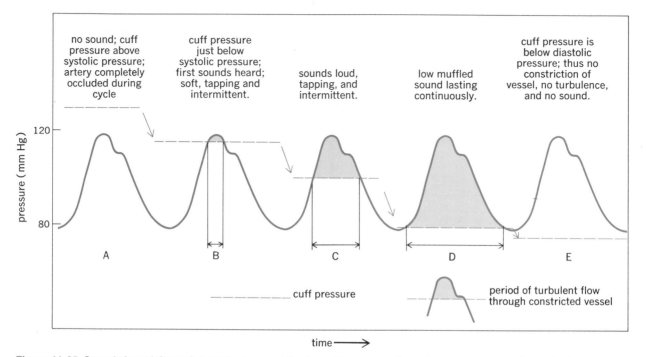

Figure 11-35. Sounds heard through a stethoscope while the cuff pressure of a sphygmomanometer is gradually lowered. Systolic pressure is recorded at B and diastolic pressure at the point of sound disappearance.

resistance to flow and the pressures are therefore similar everywhere in the arterial tree.

Measurement of arterial pressure

Both systolic and diastolic blood pressure are readily measured in human beings with the use of a sphygmomanometer (Fig. 11-35). A hollow cuff is wrapped around the arm and inflated with air to a pressure greater than systolic blood pressure (Fig. 11-35A). The high pressure in the cuff is transmitted through the tissues of the arm and completely collapses the arteries under the cuff, thereby preventing blood flow to the lower arm. The air in the cuff is now slowly released, causing the pressure in the cuff and arm to drop. When cuff pressure has fallen to a point just below the systolic pressure (Fig. 11-35B), the arterial blood pressure at the peak of systole is greater than the cuff pressure, causing the artery to expand and allow blood flow for this brief time. During this interval, the blood flow through the partially occluded artery occurs at a very high velocity because of the small opening for blood passage and

the large pressure gradient. The high-velocity blood flow produces turbulence and vibration, which can be heard through a stethoscope placed over the artery just below the cuff. The pressure measured on the manometer attached to the cuff at which sounds are first heard as the cuff pressure is lowered is identified as the systolic blood pressure. These first sounds are soft tapping sounds, corresponding to the peak systolic pressure reached during ejection of blood from the heart. As the pressure in the cuff is lowered further, the time of blood flow through the artery during each cycle becomes longer (Fig. 11-35C). The tapping sound becomes louder as the pressure is lowered. When the cuff pressure reaches the diastolic blood pressure, the sounds become dull and muffled, as the artery remains open throughout the cycle, allowing continuous turbulent flow (Fig. 11-35D). Just below diastolic pressure all sound stops as flow is now continuous and nonturbulent through the completely open artery. Thus, systolic pressure is measured as the cuff pressure at which sounds first appear and diastolic pressure as the cuff pressure at which sounds disappear.

Arterioles

The arterioles are primarily responsible for blood-flow distribution. Figure 11-36 illustrates the major principles of blood-flow distribution in terms of a simple model, a fluid-filled tank with a series of compressible outflow tubes. What determines the rate of flow through each exit tube? As always,

$$\text{Flow} = \frac{\Delta P}{R}$$

Since the driving pressure is identical for each tube, differences in flow are completely determined by differences in the resistance to flow offered by each tube. The lengths of the tubes are approximately the same, and the viscosity of the fluid is a constant; therefore, differences in resistance offered by the tubes are due solely to differences in their radii. Obviously, the widest tube has the greatest flow. If we equip each outflow tube with an adjustable cuff, we can obtain any combination of *relative* flows we wish.

This analysis can now be applied to the cardiovascular system. The tank is analogous to the arteries, which serve as a pressure reservoir, the major arteries themselves being so large that they contribute little resistance to flow. The smaller arteries begin to offer some resistance, but it is relatively slight. Therefore, all the arteries of the body can be considered a single pressure reservoir. The arteries branch within each organ into the next series of smaller vessels, the arterioles, which are now narrow enough to offer considerable resistance. The arterioles are the major site of resistance in the vascular tree and are therefore analogous to the outflow tubes in the model. Their radii are subject to precise physiological control; their walls consist of relatively little elastic tissue

but much smooth muscle, which can relax or constrict, thereby changing the radius of the inside (lumen) of the arteriole. The pattern of blood-flow distribution depends primarily upon the degree of arteriolar smooth muscle constriction within each organ and tissue.

The smooth muscles surrounding arterioles possess a large degree of inherent myogenic activity, i.e., "spontaneous" contraction. This basal tone is responsible for a large portion of the basal resistance offered by the arterioles. However, what is more important is that a variety of physiological factors play upon these smooth muscles to either increase or decrease their degree of contraction, i.e., decrease or increase the vessels' resistance. The mechanisms controlling these changes in arteriolar resistance fall into two general categories: (1) local controls; (2) extrinsic (reflex) controls. The local controls serve to couple blood flow through an organ with the metabolic requirements of that organ, whereas the external controls are efferent pathways of reflexes which serve to integrate the overall activity of the cardiovascular system.

Local controls (autoregulation)

The blood flow through any organ is given by the following equation, assuming for simplicity that the venous pressure, the other end of the pressure gradient, is zero:

$$F_{\text{organ}} = \frac{\text{mean arterial pressure}}{R_{\text{organ}}}$$

Since the arterial pressure, i.e., the driving force for flow through each organ, is identical throughout the body, differences in flows between organs

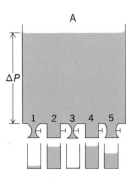
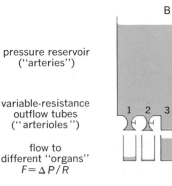

A

B

pressure reservoir
("arteries")

variable-resistance
outflow tubes
("arterioles")

flow to
different "organs"
$F = \Delta P / R$

ΔP

1 2 3 4 5

1 2 3 4 5

Figure 11-36. Physical model of the relationship between arterial pressure, arteriolar radius in different organs, and blood-flow distribution. Blood has been shifted from organ 2 to organ 3 (in going from A to B) by constricting the "arterioles" of 2 and dilating those of 3.

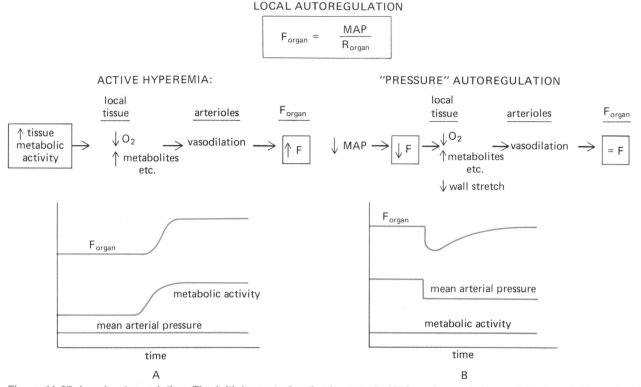

Figure 11-37. Local autoregulation. The initial event of active hyperemia (A) is a change in the metabolic activity, that of "pressure" autoregulation (B) a change in the blood pressure within the vessel. The chemical mediators may be similar in the two situations.

depend entirely on the relative resistances (R_{organ}) offered by the arterioles of each organ. The term "local autoregulation" denotes the fact that individual organs, to greater or lesser extent, possess completely inherent mechanisms for altering their own arteriolar resistances, thereby conferring upon them the capacity to self-regulate ("autoregulate") their blood flows.

Active hyperemia. Certain organs and tissues, particularly the heart, skeletal muscle, and other muscular organs, manifest an increased blood flow (*hyperemia*) any time their metabolic activity is increased (Fig. 11-37A). For example, the blood flow to exercising skeletal muscle increases in direct proportion to the increased activity of the muscle. This phenomenon, known as *active* (or *exercise*) *hyperemia*, is the direct result of arteriolar dilation within the more active organ. This vasodilation does not depend upon the presence of

nerves or hormones but is a locally mediated response. The adaptive value of the phenomenon should be readily apparent; an increased rate of activity in an organ automatically produces an increased blood flow to that organ by relaxing its arterioles.

The arterioles dilate because their smooth muscle relaxes, i.e., loses a fraction of its basal contractile tone. The factors acting upon smooth muscle in active hyperemia to cause it to relax are local chemical changes resulting from the increased metabolic cellular activity, but the relative contributions of the various factors implicated seem to vary, depending upon the organs involved and the duration of the increased activity. At present, therefore, we can only name but not quantify some of the factors which appear to be involved: decreased *oxygen concentration*, increased concentrations of *carbon dioxide, hydrogen ion*, and *metabolic intermediates*, such as *adenosine;* increased concentration of *potassium* (perhaps as a result of enhanced movement out of skeletal muscle cells during the more frequent action potentials); increased *osmolarity* (i.e., decreased water concentration) (as a result of the increased breakdown of high-molecular-weight substances); increased concentrations of *prostaglandins.* Changes in all these variables—decreased oxygen, increased carbon dioxide, hydrogen ion, potassium, osmolarity, adenosine, and prostaglandins—have been shown to cause arteriolar dilation under controlled experimental conditions, and they all may contribute to the active-hyperemia response. It must be emphasized that all these chemical changes act locally upon the arteriolar smooth muscle, causing it to relax (dilate), no nerves or hormones being involved.

It should not be too surprising that the phenomenon of active hypermia is most highly developed in heart and skeletal muscle, which show the widest range of normal metabolic activities of any organs or tissues in the body. It is highly efficient, therefore, that their supply of blood be primarily determined locally by their rates of activity. The gastrointestinal tract also manifests a great capacity for active hyperemia in keeping with its relatively wide range of metabolic activity.

"Pressure" autoregulation. The discussion of local metabolic controls has thus far emphasized increased metabolic activity of the tissue or organ as the initial event leading to vasodilation. However, similar changes in the local concentrations of the implicated metabolic factors may occur under a very different set of conditions, namely when a tissue or organ suffers a reduction in its blood supply (*ischemia*) due to a reduction in the blood pressure. The supply of oxygen to the tissue is diminished and the local oxygen concentration decreased. Simultaneously, the concentrations of carbon dioxide, hydrogen ion, and metabolites all increase because they are not removed by the blood as fast as they are produced, and prostaglandin synthesis is increased. In other words, the local metabolic changes which occur during both increased metabolic activity and ischemia are similar (although not identical) because both situations reflect an imbalance of blood supply and level of cellular metabolic activity. The adaptive value of vasodilation in response to ischemia, say, as a result of partial occlusion of the artery supplying the tissue, is that it automatically tends to maintain the blood supply to the tissue (Fig. 11-37B).

Pressure autoregulation is not limited to circumstances in which arterial pressure goes down. The opposite events occur when, for various reasons, arterial pressure increases; the arterioles constrict in response to the increased arterial pressure thereby maintaining local flow relatively constant in the face of the increased pressure.

Although our description has emphasized the likely role of *metabolic* factors in the local resistance changes induced by changes in arterial pressure, it should be noted that other mechanisms may also participate; for example, arteriolar smooth muscle may respond directly to an increased stretch (secondary to the increased arterial pressure) by increasing its basal myogenic tone. Finally, it must be emphasized that various organs differ markedly in their autoregulatory capacities.

Response to injury. Tissue injury causes local release from cells or generation from plasma precursors of a variety of chemical substances which make arteriolar smooth muscle relax and are the causes of vasodilation in an injured area. This phenomenon, a part of the general process known as inflammation, will be described in detail in Chap. 17. In contrast, injury to blood vessels per se causes vasoconstriction, as described at the end of this chapter.

Extrinsic (reflex) controls

Sympathetic nerves. Most arterioles in the body receive a rich supply of sympathetic postganglionic nerve fibers. These neurons (with one major exception) release norepinephrine, which combines with α-adrenergic receptors on the vascular smooth muscle to cause vasoconstriction. If almost all the nerves to arterioles are constrictor in action, how can reflex arteriolar dilation be achieved? Since the sympathetic nerves are seldom completely quiescent but discharge at some finite rate, which varies from organ to organ, the nerves always cause some degree of tonic constriction; from this basal position, further constriction is produced by increased sympathetic activity, whereas dilation can be achieved by decreasing the rate of sympathetic activity below the basal level. The skin offers an excellent example of these processes (Fig. 11-

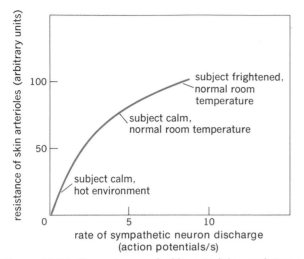

Figure 11-38. Dependence of skin arteriolar resistance upon activity of sympathetic nerves to the arterioles.

38). Skin arterioles of a normal unexcited person at room temperature are already under the influence of a high rate of sympathetic discharge; an appropriate stimulus (fear, loss of blood, etc.) causes reflex enhancement of this activity; the arterioles constrict further, and the skin pales. In contrast, an increased body temperature reflexly inhibits the sympathetic nerves to the skin, the arterioles dilate, and the skin flushes. This generalization cannot be stressed too strongly: Control of the sympathetic constrictor nerves to arteriolar smooth muscle can accomplish either dilation or constriction.

In contrast to the processes of active hyperemia and pressure autoregulation, the primary functions of these nerves are concerned *not* with the coordination of *local* metabolic needs and blood flow but with reflexes that help maintain an adequate blood supply at all times to vital organs such as the brain and heart. The common denominator of these reflexes is the regulation of arterial blood pressure, to be described in detail in a subsequent section.

There is, as we said, one exception to the generalization that sympathetic nerves to arterioles release norepinephrine. A group of sympathetic (*not* parasympathetic) nerves to the arterioles in skeletal muscle instead releases acetylcholine, which causes arteriolar dilation and increased blood flow. It still must be emphasized that *most* sympathetic nerves to skeletal muscle arterioles release norepinephrine; thus skeletal muscle arterioles receive a

dual set of sympathetic nerves. The only known function of the vasodilator fibers is in the response to exercise or stress and will be described in a later section; the vasoconstrictor fibers mediate all other situations involving neural control of skeletal muscle arterioles.

Parasympathetic nerves. With but few major exceptions (notably the blood vessels supplying the external genitals), there is no significant parasympathetic innervation of arterioles. It is true that stimulation of the parasympathetic nerves to certain glands is associated with an increased blood flow, but this is secondary to the increased metabolic activity induced in the gland by the nerves, with resultant local active hyperemia.

Hormones. Several hormones cause constriction or dilation of arteriolar smooth muscle. One of these is epinephrine, the hormone released from the adrenal medulla. In most vascular beds, epinephrine, like the sympathetic nerves, causes vasoconstriction; surprisingly, in other vascular beds, epinephrine may induce vasodilation. However, it is likely that the effects of circulating epinephrine on arterioles are quantitatively of little significance when compared with those exerted by norepinephrine released from sympathetic-nerve endings.

Another hormone which exerts a potent vasoconstrictor action on most arterioles is angiotensin; this peptide is generated in the blood as the result of the action of an enzyme, *renin*, on a plasma protein, *angiotensinogen.* The control and multiple functions of the so-called renin-angiotensin system will be described in Chap. 13; suffice it here to emphasize that an increase in plasma angiotensin will produce a widespread increase in the degree of arteriolar vasoconstriction.

Figure 11-39 summarizes the factors which determine arteriolar radius. Of course, the local metabolic factors and extrinsic reflex controls do not exist in isolation from one another but manifest important interactions. For example, let us suppose that the arterioles in the kidneys are receiving a large vasoconstrictor input via the sympathetic nerves to them. The resulting vasoconstriction (induced by norepinephrine released from the nerves) will reduce blood flow to the kidneys, but this will, in turn, generate within the kidneys changes in the concentrations of the various vasodilator meta-

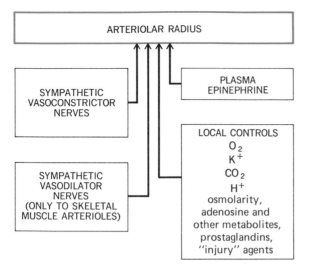

Figure 11-39. Major factors affecting arteriolar radius.

bolic factors such that the latter act upon the arterioles to induce vasodilation, i.e., to offset much of the reflex vasoconstriction mediated by norepinephrine. The result is that the arterioles do not constrict nearly as much as if the neural input were the only factor operating, and the kidneys are protected against an excessive and potentially damaging reduction in their blood flow. Locally generated prostaglandins, in particular, seem to play important roles in buffering such reflexly induced alterations of blood flow; however, the role of prostaglandins in control of local blood flow is extremely complex since there is a large number of different prostaglandins, some of which are vasodilators and others vasoconstrictors.

Let us take the blood vessels supplying the heart (the *coronary vessels*) as another example. The coronary vessels were once thought not to be innervated by sympathetic fibers because when all the sympathetic nerves to the heart are experimentally stimulated, vasoconstriction is not usually seen; rather, intense vasodilation of the coronary vessels occurs with consequent increase in coronary blood flow. What causes the vasodilation? Recall that the sympathetic nerves to the heart induce an increase in heart rate and myocardial contractility; these are associated with a marked increase in myocardial oxygen consumption, which, in turn, causes active hyperemia, i.e., vasodilation secondary to local metabolic factors. Thus, the increased coronary blood flow associated with in-

creased sympathetic nerve activity is an *indirect* consequence of increased myocardial work. Only recently have we recognized that the coronary vessels are, indeed, innervated by vasoconstrictor sympathetic fibers but that the vasodilation induced by the local metabolic factors when sympathetic nerves to the heart are stimulated is so intense that it masks the simultaneously occurring, neurally mediated vasoconstrictor input. In other words, the overall response of the vessels is the algebraic sum of opposing inputs. Under abnormal conditions (for example, in a patient with abnormally narrowed coronary vessels having reduced ability to dilate in response to metabolic factors), the direct neural vasoconstriction may prevent an adequate increase in coronary blood flow during exercise or stress, thereby precipitating damage to the cardiac muscle.

Capillaries

At any given moment, approximately 5 percent of the total circulating blood is in the capillaries. Yet it is this 5 percent which is performing the ultimate function of the entire system, namely, the exchange of nutrients and metabolic end products. All other segments of the vascular tree subserve the overall aim of getting adequate blood flow through the capillaries. The capillaries permeate every tissue of the body; no cell is more than 0.01 cm from a capillary. Therefore, diffusion distances are very small, and exchange is highly efficient. There are thousands of kilometers of capillaries in an adult person, each individual capillary being only about 1 mm long.

Capillaries throughout the body vary somewhat in structure, but the typical capillary (Fig. 11-40) is a thin-walled tube of endothelial cells one layer

Figure 11-40. Typical capillary.

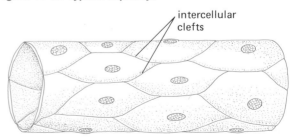

intercellular clefts

thick without elastic tissue, connective tissue, or smooth muscle to impede transfer of water and solutes. The flat cells which constitute the endothelial lining interlock like pieces of a jigsaw puzzle.

Anatomy of the capillary network

Figure 11-41 illustrates diagrammatically the general anatomy of the small vessels which constitute the so-called microcirculation. Blood enters the capillary network from the arterioles. Most tissues appear to have two distinct types of capillaries: "true" capillaries and thoroughfare channels. The thoroughfare channels connect arterioles and venules directly. From these channels exit and re-

Figure 11-41. Diagram of microcirculation. Note the thinning of the smooth muscle coat in the thoroughfare channels and its complete absence in the true capillaries. The black lines on the surface of the vessels are nerve fibers leading to smooth muscle cells. *(Adapted from Zweifach.)*

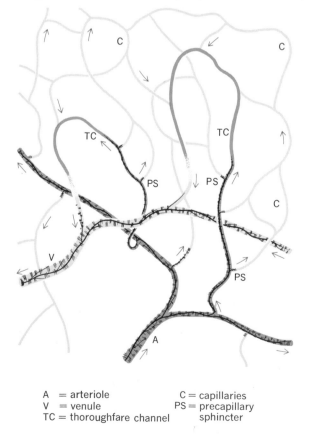

A   = arteriole
V   = venule
TC = thoroughfare channel
C = capillaries
PS = precapillary
  sphincter

enter the network of true capillaries across which materials actually exchange. The site at which a true capillary exits from a thoroughfare channel is protected by a ring of smooth muscle, the *precapillary sphincter*, which continually opens and closes (due to local autoregulation) so that flow through any given capillary is usually intermittent. Generally, the more active the tissue, the more precapillary sphincters are open at any moment. The sphincters are best visualized as functioning in concert with arteriolar smooth muscle to regulate not only the total flow of blood through the tissue capillaries but the number of functioning capillaries as well.

Resistance of the capillaries

Since a capillary is very narrow, it offers a considerable resistance to flow, but for two reasons, the resistance is not of critical importance for cardiovascular function: (1) Despite the fact that the capillaries are actually narrower than the arterioles, the huge total number of capillaries provides such a great cross-sectional area for flow that the *total* resistance of *all* the capillaries is considerably less than that of the arterioles. (2) Because capillaries have no smooth muscle, their radius (and, therefore, their resistance) is not normally subject to *active* control and simply reflects the volume of blood delivered to them via the arterioles (and the volume leaving via the venules).

Velocity of capillary blood flow

Figure 11-42 illustrates a simple mechanical model of a series of 1-cm-diameter balls being pushed down a single tube which branches into narrower tubes. Although each tributary tube has a smaller cross section than the wide tube, the sum of the tributary cross sections is much greater than the area of the wide tube. Let us assume that in the wide tube each ball moves 3 cm/min. If the balls are 1 cm in diameter and they move two abreast, six balls leave the wide tube per minute and enter the narrow tubes. Obviously, then, six balls must be leaving the narrow tubes per minute. At what speed does each ball move in the small tubes? The answer is 1 cm/min. This example illustrates the following important generalization: When a continuous stream moves through consecutive sets of tubes, the velocity of flow decreases as the sum of the cross-sectional areas of the tubes increases. This is precisely the case in the cardiovascular system (see Fig. 11-43); the blood velocity is very great in the aorta, progressively slows in the arteries and arterioles, and then markedly slows as it passes through the huge cross-sectional area of the capillaries (600 times the cross-sectional area of the aorta). The speed then progressively increases in the venules and veins because the cross-sectional area decreases. The adaptive significance of this phenomenon is very great; blood flows through the capillaries so slowly that there is adequate time for exchange of nutrients and metabolic end products between the blood and tissues.

Diffusion across the capillary wall: Exchanges of nutrients and metabolic end products

There is no active transport of solute across the capillary wall, and materials cross by diffusion. As described in the next section, there is some movement of fluid by bulk flow, but it is of negligible importance for the exchange of nutrients and metabolic end products. The factors determining diffusion rates were described in Chap. 6. Because fat-soluble substances penetrate cell membranes easily, they pass directly through the endothelial capillary cells. In contrast, many ions and molecules are poorly soluble in fat and must pass through the water-filled clefts between adjacent endothelial cells.

Because these clefts (or pores) constitute less than 1 percent of the total area of the capillary

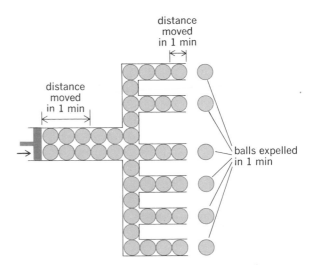

Figure 11-42. Relationship between cross-sectional area and velocity of flow. The total cross-sectional area of the small tubes is three times greater than that of the large tube. Accordingly, velocity of flow is one-third as great in the small tubes.

endothelium, more than 100 times more area is available for diffusion of lipid-soluble substances than for water-soluble substances. Moreover, the size of the clefts varies from tissue to tissue and accounts for the great differences in the "leakiness" of capillaries in different tissues and organs. At one extreme are the extremely "tight" capillaries of the brain, whereas at the other end of the spectrum are liver capillaries which have such large clefts that even protein molecules can readily pass; the latter is adaptive in that among the major

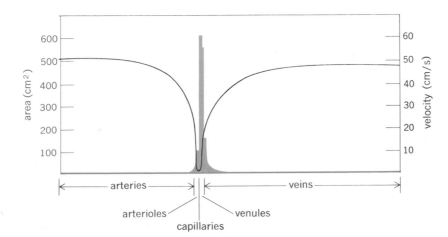

Figure 11-43. Relation between cross-sectional area and velocity of flow in the systemic circulation. The values are those for a 30-lb dog. Velocity of blood flow through the capillaries is about 0.07 cm/s. *(Adapted from Rushmer.)*

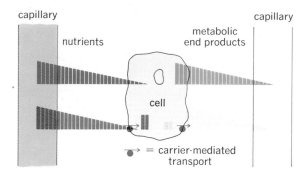

Figure 11-44. Movements of nutrients and metabolic end products between plasma and tissue cells. The substance may enter or leave the cell by diffusion or carrier-mediated transport, but it moves across the capillary wall only by diffusion dependent upon capillary–interstitial-fluid concentration gradients. (Two capillaries are shown for convenience; actually, of course, nutrients and end products leave and enter the same capillary.)

functions of the liver are the synthesis of plasma proteins and metabolism of protein-bound substances. Most capillaries between these extremes allow the passage of small amounts of protein, mainly by endocytosis and exocytosis (see Chap. 6).

What is the sequence of events involved in capillary-cell transfers (Fig. 11-44)? Tissue cells do not exchange material *directly* with blood; the interstitial fluid always acts as middleman. Thus, nutrients diffuse across the capillary wall into the interstitial fluid, from which they gain entry to cells. Conversely, metabolic end products move first across cell membranes into interstitial fluid, from which they diffuse into the plasma. Thus, two membrane-transfer processes must always be considered, that across the capillary wall and that across the tissue cell membrane. The cell-membrane step may be by simple diffusion or by carrier-mediated transport, but, as described above, the transcapillary movement is always by diffusion. Since to achieve *net* movement of any substance by diffusion, a concentration gradient is required, net transcapillary diffusion of nutrients and metabolic end products proceeds in the direction of these diffusion gradients for these substances between the blood and interstitial fluid.

How do these diffusion gradients arise? Let us take two examples, those of glucose and carbon dioxide transcapillary movement in muscle. Glucose is continuously consumed after being transported from interstitial fluid into the muscle cells by carrier-mediated transport mechanisms; this removal from interstitial fluid lowers the interstitial-fluid glucose concentration below that of plasma and creates the gradient for glucose diffusion out of the capillary. Carbon dioxide is continuously produced by muscle cells, thereby creating an increased intracellular carbon dioxide concentration, which causes diffusion of carbon dioxide into the interstitial fluid; in turn, this causes the interstitial carbon dioxide concentration to be greater than that of plasma and produces carbon dioxide diffusion into the capillary. We chose these particular examples to emphasize the fact that net movement of a substance between interstitial fluid and cells is the event which establishes the transcapillary plasma-interstitial diffusion gradients but that in some cases the substance moves across the cell membrane by diffusion and in others by carrier-mediated transport.

When a tissue increases its rate of metabolism, it must obviously obtain more nutrients from the blood and eliminate more metabolic end products. One important mechanism for achieving this is the increase in diffusion gradients between plasma and cell which occurs for two reasons (Fig. 11-45): (1) Increased cellular consumption of oxygen and nutrients lowers their intracellular concentrations while increased production of carbon dioxide and other end products raises their cell concentrations; (2) simultaneously, blood flow to the area increases, due to the action of local metabolic factors on arterioles, and this additional blood flow prevents the plasma concentrations of oxygen and nutrients from falling rapidly and those of carbon dioxide and other end products from rising rapidly as diffusion proceeds. The vasodilation caused by the local metabolic factors facilitates diffusion by another mechanism as well — it dilates precapillary sphincters, and the resulting increase in the number of open capillaries increases the total surface area available for diffusion and decreases the average distance between cells and capillaries.

Bulk flow across the capillary wall: Distribution of the extracellular fluid

Since the capillary wall is highly permeable to water and to almost all the solutes of the plasma with the exception of the plasma proteins, it behaves like a porous filter through which protein-free plasma (ultrafiltrate) moves by bulk flow

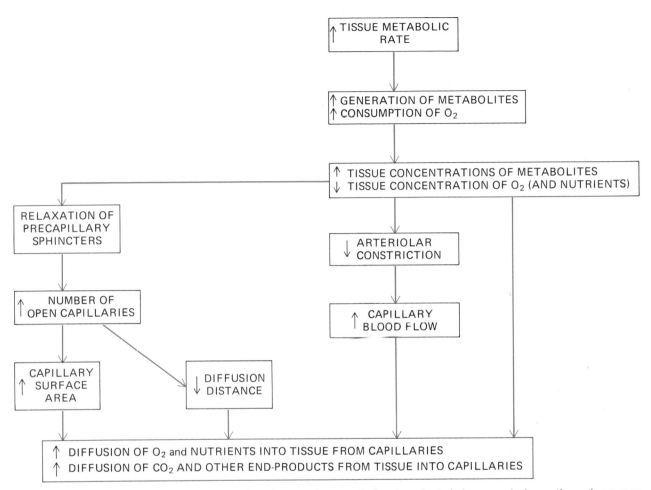

Figure 11-45. Mechanisms by which supply of nutrients and removal of end products is increased when a tissue increases its activity.

through the clefts between the cells under the influence of a hydrostatic pressure gradient. The magnitude of the bulk flow is directly proportional to the hydrostatic pressure difference between the inside and outside of the capillary, i.e., between the capillary blood pressure and the interstitial-fluid pressure. Normally, the former is much larger than the latter, so that a considerable hydrostatic pressure gradient exists to drive the filtration of protein-free plasma out of the capillaries into the interstitial fluid. Why then does all the plasma not filter out into the interstitial space? The explanation was first elucidated by Starling (the same scientist who expounded the law of the heart which bears his name) and depends upon the principles of osmosis.

In Chap. 6 we described how a net movement of water occurs across a semipermeable membrane from a solution of high water concentration to a solution of low water concentration. Recall that the concentration of water depends upon the concentration of solute molecules or ions dissolved in the water. When two solutions A and B, which are separated by a membrane, have identical concentrations of all solutes, the water concentrations are identical and no net water movement occurs. When, however, a quantity of a nonpermeating substance is added to solution A, the water concentration of A is reduced below that of solution B and a net movement of water will occur by osmosis from B into A. This osmotic flow of water "drags" along with it any dissolved solutes to which the membrane is highly permeable. Thus, a difference in water concentration can result in the movement of a solution containing both water and permeating solute in a manner similar to the bulk flow produced by a hydrostatic pressure difference. Units of pressure (millimeters of mercury) are used to express the difference in water concentration resulting from the presence of the nonpenetrating solute.

This analysis can now be applied to capillary fluid movements. The plasma within the capillary and the interstitial fluid outside it contain large quantities of low-molecular-weight solutes (crystalloids), e.g., sodium, chloride, or glucose. Since the capillary lining is highly permeable to all these crystalloids, they all have almost identical concentrations in the two solutions. There are small concentration differences occurring for substances consumed or produced by the cells, but these tend to cancel each other, and, accordingly, no significant water-concentration difference is caused by the presence of the crystalloids. In contrast, the plasma proteins cross capillary walls only very slightly and therefore have a very low interstitial-fluid concentration. This difference in protein concentration between plasma and interstitial fluid means that the water concentration of the plasma is lower than that of interstitial fluid, inducing an osmotic flow of water from the interstitial compartment into the capillary. Along with the water are carried all the different types of crystalloids dissolved in the interstitial fluid. Thus, osmotic flow of fluid, like bulk flow, does not alter the concentrations of the low-molecular-weight substances of plasma or interstitial fluid.

In summary, two opposing forces act to move fluid across the capillary: (1) The hydrostatic pressure difference between capillary blood pressure and interstitial-fluid pressure favors the filtration of a protein-free plasma out of the capillary; (2) the water-concentration difference between plasma and interstitial fluid, which results from the protein-concentration differences, favors the osmotic movement of interstitial fluid into the capillary. Accordingly, the movements of fluid depend directly upon four variables: the capillary hydrostatic pressure, interstitial hydrostatic pressure, plasma protein concentration, and interstitial-fluid protein concentration.

We may now consider quantitatively how these variables act to move fluid across the capillary wall (Fig. 11-46). Much of the arterial blood pressure has already been dissipated as the blood flows through the arterioles, so that pressure at the beginning of the capillary is 35 mmHg. Since the capillary also offers resistance to flow, the pressure continuously decreases to 15 mmHg at the end of the capillary. The interstitial pressure is essentially zero.[1] The difference in protein concentration between plasma and interstitial fluid causes a difference in water concentration (plasma water concentration less than interstitial-fluid water concentration), which induces an osmotic flow of fluid into the capillary equivalent to that produced by a hydrostatic pressure difference of 25 mmHg. It is evident that in the first portion of the capillary the hydrostatic pressure difference is greater than the osmotic forces and a net movement of fluid out of the capillary occurs; in the last portion of the capillary, however, a net force causes fluid movement into the capillary (termed *absorption*). The net result is that the early and late capillary events tend to cancel each other out, and there is little overall net loss or gain of fluid (Fig. 11-46A). In a normal person there is a small net filtration; as we shall see, this is returned to the blood by lymphatics.

The analysis of capillary fluid dynamics in terms of different events occurring at the arterial and venous ends of the capillary is oversimplified. It is very likely that many capillaries manifest only net filtration or net absorption along their entire lengths because the arterioles supplying them are either so dilated or so constricted as to yield a capillary hydrostatic pressure above or below 25 mmHg along the entire length of the capillary. This does not alter the basic concept that, taken as a unit, a capillary bed manifests net absorption or filtration, depending upon the average levels of hydrostatic pressures within the individual capillaries constituting the bed.

Figure 11-46B and C illustrates the effects on this equilibrium of changing capillary pressure. In B, the arterioles in the organ have been dilated, and the capillary pressure therefore increases since less of the arterial pressure is dissipated in the passage through the arterioles. Outward filtration

---

[1] The exact value of interstitial hydrostatic pressure remains highly controversial. Many physiologists believe that it is not zero but is actually subatmospheric. The outcome of this controversy will not, however, alter the basic concepts being presented here.

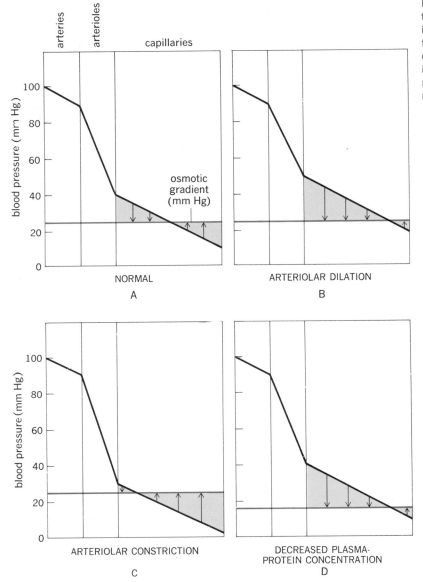

Figure 11-46. Relevant filtration-absorption forces acting across the capillary wall in several situations. Arrows down indicate filtration out of the capillary. Arrows up indicate fluid movement from interstitial fluid into capillary. The shaded areas denote relative magnitudes of the fluid movements.

now predominates, and some of the plasma enters the interstitial fluid. In contrast, marked arteriolar constriction (Fig. 9-46C) produces decreased capillary pressure and net movement of interstitial fluid into the vascular compartment. Figure 9-46D shows how net absorption or filtration can be produced in the absence of capillary pressure changes whenever plasma protein concentration is altered. Thus, in liver disease, protein synthesis decreases,

plasma protein concentration is reduced, plasma water concentration is increased, net filtration occurs, and fluid accumulates in the interstitial space *(edema)*.

The transcapillary protein-concentration difference can also be decreased by a quite different event, namely, the leakage of protein across the capillary wall into the interstitium whenever the capillary lining is damaged. This eliminates the protein-concentration difference, and interstitial fluid accumulation (edema) occurs as a result of the unchecked hydrostatic pressure difference still acting across the capillary. The fluid accumulation in a blister is an excellent example.

The major function of this capillary filtration-absorption equilibrium should now be evident: It determines the distribution of the extracellular-fluid volume between the vascular and interstitial compartments. Obviously, the ability of the heart to pump blood depends upon the presence of an adequate volume of blood within the system. Recall that the interstitial-fluid volume is three to four times larger than the plasma volume; therefore, the interstitial fluid serves as a reservoir which can supply additional fluid to the circulatory system or draw off excess. The important role this equilibrium plays in the physiological response to many situations, such as hemorrhage, will be described in a later section.

It should be stressed again that capillary filtration and absorption do not alter *concentrations* of any substance (other than protein) since movement is by bulk flow; i.e., everything in the plasma (except protein) or the interstitial fluid moves together. The reason this process of filtration plays no significant role in the exchange of nutrients and metabolic end products between capillary and tissues is that the total quantity of a substance (such as glucose or oxygen) moving into or out of the capillary during bulk flow is extremely small in comparison with the quantities moving by diffusion. For example, during a single day approximately 20,000 g of glucose crosses the capillary into the interstitial fluid by diffusion[1] but only 20 g enters by bulk flow.

Veins

Most of the pressure imparted to the blood by the heart is dissipated as blood flows through the arterioles and capillaries, so that pressure in the small venules is only approximately 15 mmHg and only a small pressure remains to drive blood back to the heart. One of the major functions of the veins is to act as low-resistance conduits for blood flow from

[1] Of course, only a small fraction of this glucose is utilized by the cells, the remainder moving back into the blood, again almost entirely by diffusion.

the tissues back to the heart. This function is performed so efficiently that the total pressure drop from venule to right atrium is only about 10 mmHg, the right atrial pressure being 0 to 5 mmHg. The resistance is low because the veins have a large diameter.

In addition to acting as low-resistance conduits, the veins perform a second important function: Their total *capacity* is adjusted in the face of variations in blood volume so as to maintain venous pressure and, thereby, venous return to the heart. The veins are the last set of tubes through which the blood must flow on its trip to the heart, and the force immediately driving this venous return is the pressure gradient between the veins and atria. In turn, the rate of venous return, i.e., inflow to the atria, is one of the most important determinants of atrial pressure. In a previous section on the control of cardiac output, we emphasized that the atrial pressure is the major determinant of ventricular end-diastolic volume and thereby of intrinsic control of stroke volume. Combining these two statements, we now see that venous pressure is a crucial determinant of stroke volume via the intermediation of atrial pressure and ventricular end-diastolic volume.

Determinants of venous pressure

The factors determining pressure in any elastic tube, as we know, are the volume of fluid within it and the distensibility of its wall. Accordingly, total blood volume is one important determinant of venous pressure. The veins differ from the arteries in that their walls are thinner and much more distensible and can accommodate large volumes of blood with a relatively small increase of internal pressure. This is illustrated by comparing Fig. 11-47 with Fig. 11-48; approximately 60 percent of the total blood volume is present in the systemic veins at any given moment, but the venous pressure averages less than 10 mmHg. In contrast, the systemic arteries contain less than 15 percent of the blood at a pressure of approximately 100 mmHg. This pressure-volume relationship of the veins allows them to act as a reservoir for blood.

The walls of the veins contain smooth muscle richly innervated by sympathetic vasoconstrictor nerves, stimulation of which causes venous constriction, thereby increasing the stiffness of the wall, i.e., making it less distensible, and raising the

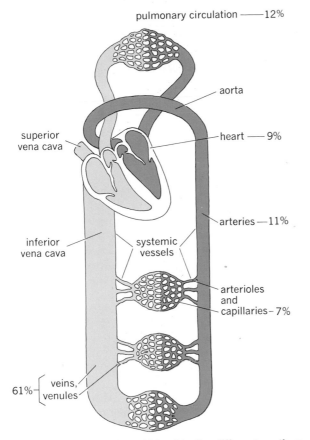

pulmonary circulation —— 12%

aorta

heart —— 9%

superior vena cava

arteries —— 11%

inferior vena cava

systemic vessels

arterioles and capillaries — 7%

veins, venules

61%

Figure 11-47. Distribution of blood in the different portions of the cardiovascular system. Compare this distribution of blood volumes with the pressures for the relevant areas shown in Fig. 11-48. *(Adapted from Guyton.)*

Figure 11-48. Summary of pressures in the vascular system.

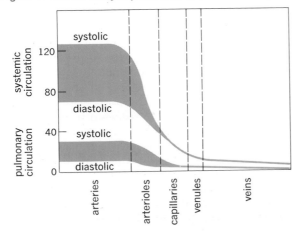

systolic

systemic circulation

120

80

diastolic

40

pulmonary circulation

systolic

diastolic

0

arteries   arterioles   capillaries   venules   veins

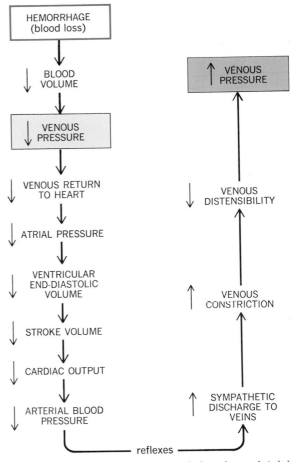

HEMORRHAGE (blood loss)

↓ BLOOD VOLUME

↓ VENOUS PRESSURE

↓ VENOUS RETURN TO HEART

↓ ATRIAL PRESSURE

↓ VENTRICULAR END-DIASTOLIC VOLUME

↓ STROKE VOLUME

↓ CARDIAC OUTPUT

↓ ARTERIAL BLOOD PRESSURE

reflexes

↑ VENOUS PRESSURE

↓ VENOUS DISTENSIBILITY

↑ VENOUS CONSTRICTION

↑ SYMPATHETIC DISCHARGE TO VEINS

Figure 11-49. Role of venoconstriction in maintaining venous pressure during blood loss. The increased venous smooth muscle constriction returns the decreased venous pressure toward, but not to, normal. The reflexes involved in this response are described later in the chapter.

pressure of the blood within the veins. Increased venous pressure drives more blood out of the veins into the right heart. Thus, venous constriction exerts precisely the same effect on venous return as giving a transfusion.

The great importance of this effect can be visualized by the example in Fig. 11-49. A large decrease in total blood volume initially reduces the pressures everywhere in the circulatory system, including the veins; venous return to the heart decreases, and cardiac output decreases. However, reflexes to be described cause increased sympa-

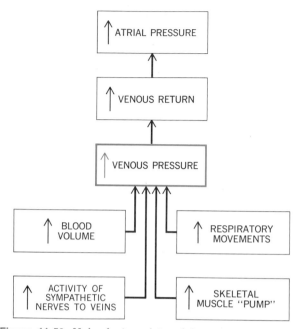

Figure 11-50. Major factors determining venous pressure and thereby atrial pressure. The figure as drawn shows how venous and atrial pressures are increased; reversing the arrows in the boxes indicates how these pressures can be reduced.

thetic discharge to the venous smooth muscle, which contracts, thereby returning venous pressure toward normal, restoring venous return and cardiac output.

Two other mechanisms can decrease venous capacity, increase venous pressure, and facilitate venous return; these are the skeletal muscle "pump" and the effects of respiration upon thoracic and abdominal veins (respiratory "pump"). During skeletal muscle contraction, the veins running through the muscle are partially compressed, thereby reducing their diameter and decreasing venous capacity. As will be described in Chap. 12, during inspiration, the diaphragm descends, pushes on the abdominal contents, and increases abdominal pressure. The large veins which pass through the abdomen are partially compressed by this increased pressure; this facilitates movement of blood, but only toward the heart because venous valves in the legs prevent backflow. Simultaneously, the pressure in the chest (thorax) decreases, and this decrease is transmitted passively to the intrathoracic veins and

right atria. The net effect is to increase the pressure gradient between the right atrium and veins outside the thorax; accordingly, venous return to the heart is enhanced during inspiration.

In summary (Fig. 11-50), the effects of venous smooth muscle contraction, the skeletal muscle pump, and the respiratory pump are to facilitate return of blood to the heart. The net result is that atrial pressure and, thereby, cardiac output are determined in large part by these factors.

This completes our description of the pressure changes throughout the vascular tree. The normal pressure profiles for the systemic and pulmonary circulation are given in Fig. 11-48. Note that the pulmonary pressures are considerably smaller than the systemic pressures for reasons shortly to be described. Note also that the resistance offered by the arterioles effectively damps the pulse; by doing so, the arterioles convert the pulsatile arterial flow into a continuous capillary flow.

Effects of venous constriction on resistance to flow

We have seen that decreasing the diameter of the veins increases venous pressure, which increases venous return to the heart. However, this decreased diameter also increases resistance to flow, a phenomenon which would retard venous return if the effect of venous constriction upon resistance were not so slight as to be negligible. The veins have such large diameters that a slight decrease in size (which has great effects on venous capacity) produces little increase in resistance. This is just the opposite of the arterioles, which are so narrow that they contain little blood at any moment (Fig. 11-47), further decrease having little effect on blood displacement but even a slight decrease in diameter producing a marked increase in resistance to flow. Flow back to the heart, therefore, tends to be impaired by arteriolar constriction and enhanced by venous constriction. It should be stressed, however, that abnormally great increases in venous resistance, say from an internal blood clot or a tumor compressing from the outside, may markedly impair blood flow. Under such conditions, blood accumulates behind the lesion, pressures in the small veins and capillaries drained by the occluded vein increase, capillary filtration increases, and the tissue becomes edematous, i.e., swollen with excess interstitial fluid.

The venous valves

The veins in the limbs have valves which close so as to allow flow only toward the heart (Fig. 11-51). Why are these valves necessary if the pressure gradient created by cardiac contraction always moves blood toward the heart? We have seen how two other forces, the muscle pump and inspiratory movements, facilitate flow of venous blood. When these forces squeeze the veins, blood would be forced in both directions if the valves were not there to prevent backward flow. As we shall see, valves also play a role in counteracting the effects of upright posture.

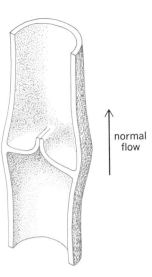

Figure 11-51. Normal venous valve. Any tendency toward retrograde flow would immediately push the valve leaflets together.

normal flow

Lymphatics

The lymphatics are not part of the circulatory system per se but constitute a one-way route for the movement of interstitial fluid to blood. The lymphatic system in human beings constitutes an extensive network of thin vessels resembling the veins. It arises as a group of blind-end lymph capillaries which are present in almost all organs of the body. These capillaries are permeable to virtually all interstitial-fluid constituents, including protein. The lymph then moves through the vessels, which converge to form larger and larger vessels. Ultimately, the two largest of these lymphatics drain into veins in the lower neck (valves here permit only one-way flow). Thus, the lymphatics carry fluid from the interstitial fluid into the blood.

Functions of the lymphatic system

Return of excess filtered fluid. In a normal person, the fluid filtered out of the capillaries each day exceeds that reabsorbed by approximately 3 L. This excess is returned to the blood via lymphatics. Partly for this reason, obstruction of lymphatics leads to increased interstitial fluid, i.e., *edema.*

Return of protein to the blood. Most capillaries in the body have a slight permeability to protein, and accordingly, there is a small steady loss of protein from the blood into the interstitial fluid. This protein is returned to the circulatory system via the lymphatics. The breakdown of this cycle is the most important cause of the marked edema seen in patients with lymphatic malfunction. Because protein (in small amounts) is normally lost from the capillaries, failure of the lymphatics to remove it allows the interstitial protein concentration to increase. This reduces or eliminates the protein-concentration difference and thus the water-concentration difference across the capillary wall and permits the net movement of increased quantities of fluid out of the capillary into the interstitial space.

Specific transport functions. In addition to these nonspecific transport functions, the lymphatics also provide the pathway by which certain specific substances reach the blood. The most important is fat absorbed from the gastrointestinal tract. Certain high-molecular-weight hormones also may reach the blood via the lymphatics.

Lymph nodes. Besides its transport functions, the lymphatic system plays a critical role in the body's defenses against disease. This function, which is mediated by the lymph nodes located along the larger lymphatic vessels, is described in Chap. 17.

Mechanism of lymph flow

How does the lymph move with no heart to push it? The best explanation at present is that lymph flow depends primarily upon forces external to the vessels, e.g., the pumping action of the skeletal muscles through which the lymphatics flow and the effects of respiration on thoracic-cage pressures. Since the lymphatics have valves similar to those in veins, external pressures cause only flow toward the points of entry of the lymphatics into the circulatory system.

Section D
Integration of cardiovascular function: Regulation of systemic arterial pressure

In Chap. 7 we described the fundamental ingredients of all reflex control systems: (1) the internal-environmental variable being regulated, i.e., maintained relatively constant, and the receptors sensitive to it; (2) afferent pathways passing information from the receptors to (3) a control center, which integrates the different afferent inputs; (4) efferent pathways controlling activity of (5) effector organs, whose output raises or lowers the level of the regulated variable. The control and integration of cardiovascular function will be described in these terms. *The major variable being regulated is the systemic arterial blood pressure.* The central role of arterial pressure and the adaptive value of keeping it relatively constant should be apparent from the ensuing discussion.

Arterial pressure, cardiac output, and arteriolar resistance

Adequate blood flow through the vital organs (brain and heart) must be maintained at all times; the brain, for example, suffers irreversible damage within 3 min of ischemia. In contrast, many areas of the body, e.g., the gastrointestinal tract, the kidneys, skeletal muscle, and skin, can withstand moderate reductions of blood flow for longer periods of time or even severe reductions if only for a few minutes. The mean arterial blood pressure is the driving force for blood flow through all the organs. The distribution of flow, i.e., the actual flow through the various organs at any given arterial pressure, depends primarily upon the radii of the arterioles in each vascular bed. A critical relationship has not been emphasized before, although it is implicit in the basic pressure-flow equation; these two factors, arterial pressure and arteriolar resistance, are *not* independent variables; *arteriolar resistance is one of the major determinants of arterial pressure.* This can be illustrated by the simple mechanical model shown in Fig. 11-52. A pump pushes fluid into a cylinder at the rate of 1 L/min; at steady state, fluid leaves the cylinder via the outflow tubes at 1 L/min, and the height of the

Figure 11-52. Model to illustrate the dependency of arterial blood pressure upon arteriolar resistance, showing the effects of dilating one arteriolar bed upon arterial pressure and organ blood flow if no compensatory adjustments occur. The middle panel is a transient state before the new equilibrium occurs. In one respect, the illustration of the model is misleading in that the arterial reservoir is shown containing very large quantities of blood. In fact, as we have seen, the volume of blood in the arteries is quite small.

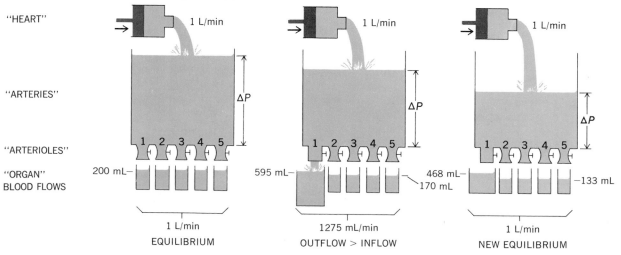

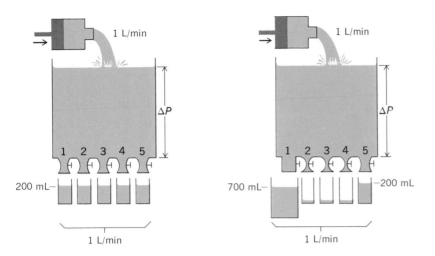

1 L/min

ΔP

1 2 3 4 5

200 mL—

1 L/min

1 L/min

ΔP

1 2 3 4 5

700 mL—   —200 mL

1 L/min

Figure 11-53. Compensation for dilation in one bed by constriction in others. When outflow tube 1 is opened, outflow tubes 2 to 4 are simultaneously tightened so that the *total* outflow resistance remains constant, total rate of runoff remains constant, and reservoir pressure remains constant.

fluid column, which is the driving pressure for outflow, remains stable. Assuming that the radii of the adjustable outflow tubes are all equal so that the flows through them are equal, we disturb the steady state by loosening the cuff on the outflow tube 1, thereby increasing its radius, reducing its resistance, and increasing its flow. The total outflow for the system is now greater than 1 L/min, more fluid leaves the reservoir than enters via the pump, and the height of the fluid column begins to decrease. In other words, a change in outflow resistance must produce changes in the pressure of the reservoir (unless some compensatory mechanism is brought into play). As the pressure falls, the rate of outflow via all tubes decreases. Ultimately, in our example, a new steady state is reached when the reservoir pressure is low enough to cause only 1 L/min outflow despite the decreased resistance of tube 1.

This analysis can be applied to the cardiovascular system by equating the pump with the heart, the reservoir with the arteries, and the outflow tubes with various arteriolar beds. An analogy to opening outflow tube 1 is exercise; during exercise, the skeletal muscle arterioles dilate, primarily because of active hyperemia, thereby decreasing resistance. If the cardiac output and the arteriolar diameters of all other vascular beds remain unchanged, the increased runoff through the skeletal muscle arterioles causes a decrease in arterial pressure. This, in turn, decreases flow through all other organs of the body. Indeed, even the exercising muscles themselves suffer a lessening of flow (below that seen immediately after they dilated)

as arterial pressure falls. Thus, the only way to guarantee the essential flow to the vital organs and the additional flow to the exercising muscle is to prevent the arterial pressure from falling.

This can be accomplished by changing cardiac output or the radii of the other arteriolar vascular beds or both. How these factors contribute can be visualized by returning to Fig. 11-52. When outflow tube 1 is loosened, how can one prevent a drop in the height of the reservoir fluid column, i.e., the driving pressure? Figure 11-53 demonstrates the first major possibility, simultaneously tightening one or more of outflow tubes 2 to 5. This partially compensates for the decreased resistance of tube 1, and the total outflow resistance of all tubes can be shifted back toward normal. Therefore, the total outflow remains near 1 L/min; of course, the distribution of flow is such that flow in tube 1 is increased and all the others are decreased. If, for some reason, tube 5 is declared a vital pathway requiring constant flow, the adjustments can always be made in tubes 1 to 4.

Applied to the body, this process is obviously analogous to control of total vascular resistance. When the skeletal muscle arterioles dilate during exercise, the *total* resistance of *all* vascular beds can still be maintained if arterioles constrict in other organs, such as the kidneys and gastrointestinal tract, which can readily suffer moderate flow reductions for at least short periods of time. In contrast, the brain arterioles remain unchanged, thereby assuring constant brain blood supply.

This type of resistance juggling, however, can compensate only within limits. Obviously if tube 1

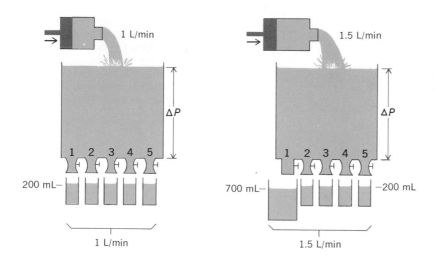

Figure 11-54. Compensation for dilation by increasing pump output. When outflow tube 1 is dilated, the total resistance decreases and total rate of runoff increases. The pump output is increased by precisely the same account, so that reservoir pressure remains constant.

opens very wide, even total closure of the other tubes cannot compensate completely. Moreover, if the closure is prolonged, absence of flow will cause severe tissue damage. There must therefore be a second compensatory mechanism: increasing the inflow by increasing the activity of the pump (Fig. 11-54). When tube 1 widens and total outflow increases, the reservoir column can be completely maintained by simultaneously increasing the inflow from the pump. Thus, at the new equilibrium, the total outflow and inflow are still equal, the reservoir pressure is unchanged, outflow through tubes 2 to 5 is unaltered, and the entire increase in outflow occurs through tube 1. Applied to the body, it should be evident that, when the blood vessels dilate, arterial pressure can be maintained constant by stimulating the heart to increase cardiac output. Thus, the regulation of arterial pressure not only assures blood supply to the vital organs but provides a means for coordinating cardiac output with total tissue requirements.

In summary, the regulation of arterial blood pressure is accomplished by control of both cardiac output and arteriolar resistance. Figure 11-55 shows both mechanisms in operation simultaneously.

It should now be possible to formalize these qualitative relationships for the entire cardiovascular system, using our basic pressure-flow equation. Flow through any tube is directly proportional to the pressure gradient between the ends of the tube and inversely proportional to the resistance:

$$\text{Flow} = \frac{\Delta P}{R}$$

Rearranging terms algebraically

$$\Delta P = \text{flow} \times R$$

This is simply another way of looking at the same equation, a way which clearly shows the dependence of pressure upon flow and resistance, which we have just described, using our models. Because the vascular tree is a continuous closed series of tubes, this equation holds for the entire system, i.e., from the very first portion of the aorta to the last portion of the vena cava just at the entrance to the heart. Therefore

Flow = cardiac output
ΔP = mean aortic pressure − late vena cava pressure
R = total resistance

where total resistance means the sum of the resistances of all the vessels in the systemic vascular tree usually termed *total peripheral resistance.*

Since the late vena cava pressure is very close to 0 mmHg, the formula

ΔP = mean aortic pressure − late vena cava pressure

becomes

ΔP = mean aortic pressure − 0 or
ΔP = mean aortic pressure

Moreover, since the mean pressure is essentially

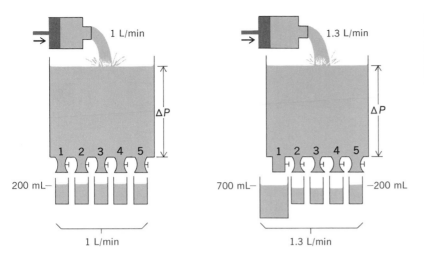

Figure 11-55. Compensation for dilation in one vascular bed by a combination of increases in pump output and constriction in other vascular beds, a combination of the compensatory adjustments in Figs. 11-53 and 11-54.

the same in the aorta and all large arteries, the pressure term in the equation becomes

$$\Delta P = \text{mean arterial pressure}$$

The pressure-flow equation for the entire vascular tree now becomes

Mean arterial pressure
  = cardiac output × total peripheral resistance

Recall that the arteries and veins are so large that they contribute very little to the total peripheral resistance. The major sites of resistance are the arterioles. The capillaries also offer significant resistance, but they have no muscle and their diameter reflects primarily the diameter of the arterioles supplying them. For these reasons, it is convenient to consider changes in arteriolar radius as virtually the only determinant of variations in total peripheral resistance. Thus, the equation formally and quantitatively states the basic relationships described earlier, namely, that arterial blood pressure can be increased, either by increasing cardiac output or total peripheral resistance.

This equation is the fundamental equation of cardiovascular physiology. Given any two of the variables, the third can be calculated. For example, we can now explain why pulmonary arterial pressure is much lower than the systemic arterial pressure (Fig. 11-22). The blood flow per minute, i.e., cardiac output through the pulmonary and systemic arteries, is, of course, the same; therefore, the pressures can differ only if the resistances differ. Thus, the pulmonary vessels must be wider and offer much less resistance to flow than the systemic arterioles. In other words, the total pulmonary vascular resistance is lower than the total systemic peripheral resistance. This permits the pulmonary circulation to function as a low-pressure system.

Figure 11-56 presents the grand scheme of effector mechanisms and efferent pathways which regulate systemic arterial pressure.[1] None of this information is new, all of it having been presented in previous figures. The reader can now appreciate how the function of the heart and various vascular segments are coordinated to achieve this. A change in any single variable shown in the figure will, all others remaining constant, produce a change in mean arterial pressure by altering either cardiac output or total peripheral resistance. Conversely, any such deviation in mean

---

[1] Any model of a system as complex as the circulatory system must, of necessity, be an oversimplification. One basic deficiency in our model is that its chain of causal links is entirely unidirectional (from bottom to top in Fig. 11-56) whereas, in fact, there are also important causal interactions in the reverse direction. For example, as we have shown, a change in cardiac output influences arterial pressure, but it is also true that a change in arterial pressure has an important effect on cardiac output. To take another example, atrial pressure is shown influencing cardiac output, but it is also true that cardiac output influences atrial pressure. A reader interested in delving further into these complex interactions and feedback loops should consult the more advanced works listed at the back of the book.

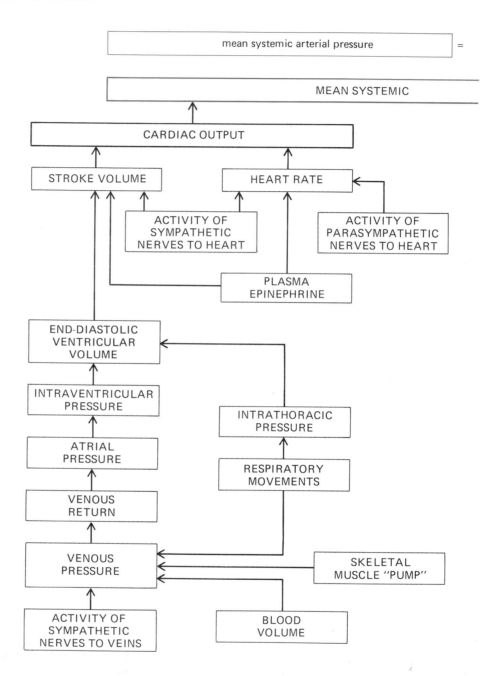

arterial pressure can be minimized by the reflex alteration of some other variable. It should be evident from the figure that the reflex control of cardiac output and peripheral resistance involves primarily: (1) sympathetic nerves to heart, arterioles, and veins; (2) parasympathetic nerves to the heart. In one sense, we have approached arterial pressure regulation backwards, in that the past sections have described the effector sites (heart, arterioles, veins) and motor pathways (autonomic nervous system). Now we must complete the reflexes by describing the monitoring systems (receptors and afferent pathways to the brain) and the control centers in the brain.

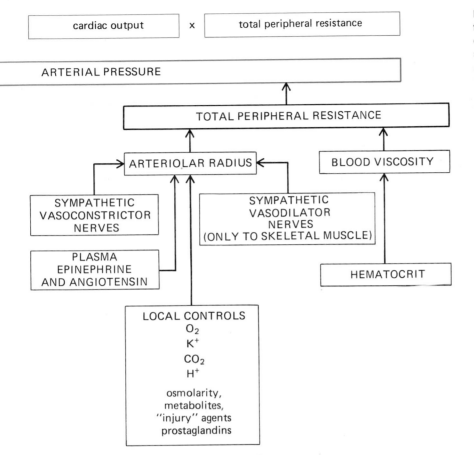

Figure 11-56. Summary of effector mechanisms and efferent pathways which regulate systemic arterial pressure, an amalgamation of Figs. 11-30, 11-39, and 11-50, with the addition of the effect of hematocrit on resistance.

Before doing so, however, we must emphasize that, despite the preeminent role of the autonomic nervous system in the reflexes controlling arterial pressure, it certainly is not the only factor, as shown in Fig. 11-56. Blood volume, in particular, is a crucial determinant, and the regulation of blood volume is, therefore, a critical component of blood-pressure control. Short-term regulation of blood volume is achieved mainly by alteration of the so-called Starling forces which determine the distribution of extracellular fluid between the vascular and interstitial compartments; these were described earlier in the section dealing with capillaries, and an example of their potent role will be given in the section on hemorrhage appearing later in this chapter. Long-term control is achieved principally by homeostatic regulation of red-cell volume (through the erythropoietin system described earlier in this chapter) and plasma volume (through the control of salt and water excretion by the kidneys, as will be described in Chap. 15).

Cardiovascular control centers in the brain

The primary cardiovascular control center is in the medulla, the first segment of brain above the spinal cord. The nerve fibers from this center synapse with the autonomic neurons and exert dominant influence over them. The medullary cardiovascular center is absolutely essential for blood-pressure regulation. The relevant medullary neurons are sometimes divided into cardiac and vasomotor centers, which are then further subdivided and

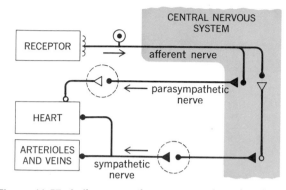

Figure 11-57. A diagrammatic representation of reciprocal innervation in the control of the cardiovascular system. Afferent input, which stimulates the parasympathetic nerves to the heart, simultaneously inhibits the sympathetic nerves to the heart, arterioles, and veins.

classified, but because these areas actually constitute diffuse networks of highly interconnected neurons, we prefer to call the entire area the *cardiovascular center*.

The synaptic distribution of the medullary axons and the input to their cell bodies are such that when the parasympathetic nerves to the heart are stimulated, the sympathetic nerves to the heart, as well as to the arterioles and the veins, are usually inhibited (Fig. 11-57). Conversely, parasympathetic inhibition and sympathetic stimulation are usually elicited simultaneously. This pattern is important because there is always some continuous discharge of the autonomic nerves. Therefore, the heart can be slowed by two simultaneous events: inhibition of the sympathetic activity to the SA node and enhancement of the parasympathetic activity to the SA node. The converse is also true for accelerating the heart. In contrast, only sympathetic fibers significantly innervate the ventricular muscle itself and the arteriolar and venous smooth muscle. However, the muscle activity can still be decreased below normal by inhibiting the basal sympathetic activity.

Other areas of the brain, particularly in the cerebral cortex and hypothalamus, have an important influence on blood pressure, but there is good reason to believe that they exert their effects mainly via the medullary centers; i.e., nerve im-

pulses from them descend to the medulla and through synaptic connections alter the discharge of the primary medullary neurons. It is through these pathways that factors such as hunger, pain, anger, body temperature, and many others can alter blood pressure. There is at least one major exception: The sympathetic vasodilator fibers to skeletal muscle arterioles are apparently not controlled by the medullary centers but are under the direct influence of neuronal pathways originating in the cerebral cortex and hypothalamus. These pathways and the sympathetic vasodilators are activated only during exercise and stress and play no role in any of the many other cardiovascular responses.

Receptors and afferent pathways

We have now to discuss the last (really the first) link in arterial pressure regulation, namely, the receptors and afferent pathways bringing information into the medullary centers. The most important of these are the *arterial baroreceptors*.

Arterial baroreceptors

It is only logical that the reflexes which homeostatically regulate arterial pressure originate primarily with arterial receptors which are pressure-sensitive. High in the neck each of the major arteries (carotids) supplying the brain divides into two smaller arteries. At this bifurcation, the wall of the artery is thinner than usual and contains a large number of branching, vine-like nerve endings (Fig. 11-58). This small portion of the artery is called the *carotid sinus*, and its nerve endings are apparently highly sensitive to stretch or distortion; since the degree of wall stretching is directly related to the pressure within the artery, the carotid sinus actually serves as a pressure receptor *(baroreceptor)*. The nerve endings come together to form afferent neurons which travel to the medulla, where they eventually synapse upon the neurons of the cardiovascular center. An area functionally similar to the carotid sinuses found in the *arch of the aorta* constitutes a second important arterial baroreceptor.

Action potentials recorded in single afferent fibers from the carotid sinus demonstrate the pattern of response by these receptors. When the arterial pressure within the carotid sinus is artifi-

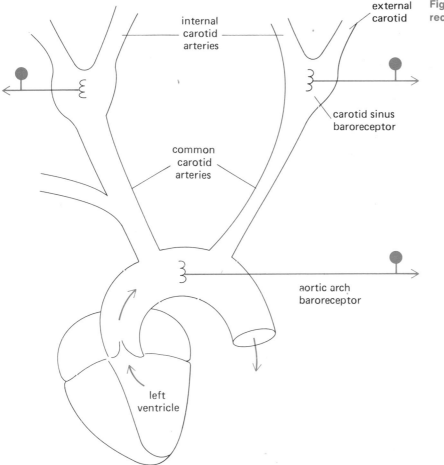

internal
carotid
arteries

external
carotid

Figure 11-58. Location of arterial baroreceptors.

carotid sinus
baroreceptor

common
carotid
arteries

aortic arch
baroreceptor

left
ventricle

cially controlled at a steady nonpulsatile pressure of 100 mmHg, there is a tonic rate of discharge by the nerve. This rate of firing can be increased by raising the arterial pressure; it can be decreased by lowering the pressure. The receptors show no fatigue or adaptation. Figure 11-59 illustrates the same type of experiment, except that pulsatile perfusion is used. The arterial baroreceptors are responsive not only to the mean arterial pressure but to the pulse pressure as well, a responsiveness that adds a further degree of sensitivity to blood-pressure regulation since small changes in certain important factors (such as blood volume) cause changes in pulse pressure before they become serious enough to affect mean pressure.

Our description of the major blood-pressure-regulating reflex is now complete (Fig. 11-60); an increase in arterial pressure increases the rate of discharge of the carotid sinus and aortic arch baroreceptors; these impulses travel up the afferent nerves to the medulla and, via appropriate synaptic connections with the neurons of the medullary cardiovascular centers, induce (1) slowing of the heart because of decreased sympathetic discharge and increased parasympathetic discharge, (2) decreased myocardial contractility because of decreased sympathetic activity, (3) arteriolar dilation because of decreased sympathetic discharge to arteriolar smooth muscle, and (4) venous dilation because of decreased sympathetic discharge to venous smooth muscle. The net result is a decreased cardiac output (decreased heart rate and stroke volume), decreased peripheral resistance, and return of blood pressure toward normal.

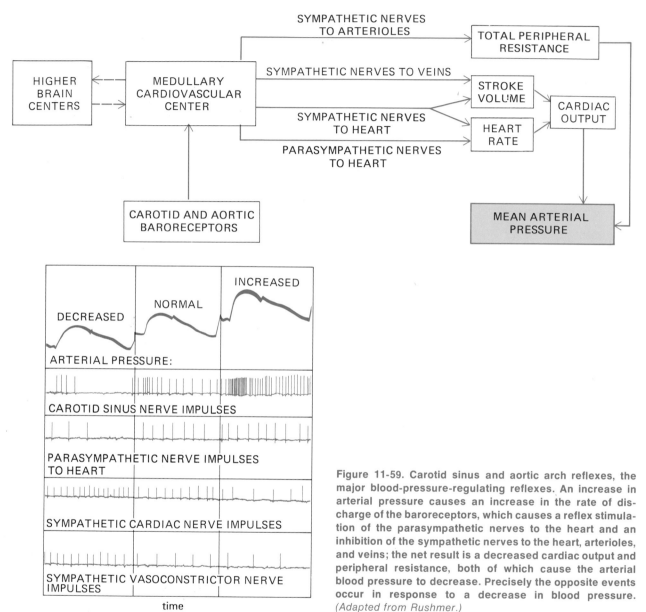

Figure 11-59. Carotid sinus and aortic arch reflexes, the major blood-pressure-regulating reflexes. An increase in arterial pressure causes an increase in the rate of discharge of the baroreceptors, which causes a reflex stimulation of the parasympathetic nerves to the heart and an inhibition of the sympathetic nerves to the heart, arterioles, and veins; the net result is a decreased cardiac output and peripheral resistance, both of which cause the arterial blood pressure to decrease. Precisely the opposite events occur in response to a decrease in blood pressure. *(Adapted from Rushmer.)*

Other baroreceptors

Other portions of the vascular tree contain nerve endings sensitive to stretch, namely, other large arteries, the large veins, the pulmonary vessels, and the cardiac walls themselves. Most of them seem to function like the carotid sinus and aortic arch in that they show increased rates of discharge with increasing pressure. By means of these recep-

tors, the medulla is kept constantly informed about the venous, atrial, and ventricular pressures, and a further degree of sensitivity is gained. Thus, a slight decrease in atrial pressure begins to facilitate the sympathetic nervous system even before the change becomes sufficient to lower cardiac output and arterial pressure far enough to be detected by the arterial baroreceptors. As we shall see in Chap. 13, the atrial baroreceptors are particularly

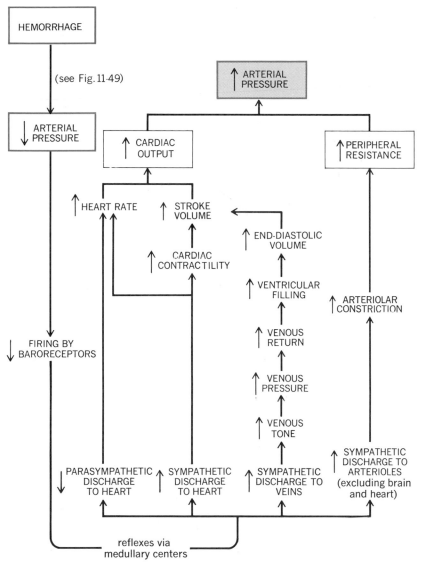

Figure 11-60. Reflex mechanisms by which lower arterial pressure following blood loss is brought back toward normal. The compensatory mechanisms do not restore arterial pressure completely to normal.

important for the control of body sodium and water, and, therefore, blood volume.

Chemoreceptors

The aortic and carotid arteries contain specialized structures sensitive primarily to the concentrations of arterial oxygen but also to those of carbon dioxide and hydrogen ion. Since these receptors are far more important for the control of respiration, they are described in Chap. 12, but they also send information to the medullary cardiovascular

centers, the result being that blood pressure tends to be reflexly increased by decreased arterial oxygen. Changes in carbon dioxide and hydrogen-ion concentrations also alter blood pressure reflexly, but the effects are small and the pathways quite complex.

Conclusion

The medullary cardiovascular control centers are true integrating centers, receiving a wide variety of information from baroreceptors, chemoreceptors, peripheral sensory receptors of all kinds (pain, cold, etc.), and many higher brain centers, particularly the hypothalamus. Therefore, it is not surprising that at every moment arterial pressure reflects the resultant response to all these inputs. Sudden anger increases the pressure; fright may actually cause hypotension (low blood pressure) severe enough to cause fainting, but this complexity should not obscure the important generalization that the primary regulation of arterial pressure is exerted by the baroreceptors, particularly those in the carotid sinus and aortic arch. Other inputs may alter the pressure somewhat from minute to minute, but the mean arterial pressure in a normal person is maintained by the baroreceptors within quite narrow limits.

Section E
Cardiovascular patterns in health and disease

In order to demonstrate how the cardiovascular components we have been discussing are integrated, we now examine the responses of the entire systemic circulation to a variety of normal and diseased states. Most of the necessary facts and concepts are already familiar to the reader.

Table 11-3. Cardiovascular effects of hemorrhage

	Prehemorrhage	Posthemorrhage	
		Immediate	5 min
Arterial pressure, mmHg	125/75	80/55	115/75
Left atrial pressure, mmHg	4	2	2.5
End-diastolic volume, mL	150	75	90
Stroke volume, mL	75	40	53
Heart rate, beats/min	70	70	91
Cardiac output, mL/min	5250	2800	4775
Kidney blood flow, mL/min	1300	1000	850
Brain blood flow, mL/min	1300	1000	1275

Hemorrhage, hypotension, and shock

The decrease in blood volume caused by bleeding produces a drop in blood pressure (*hypotension*) by the sequence of events previously shown in Fig. 11-49. The most serious consequences of the lowered blood pressure are the reduced blood flow to the brain and cardiac muscle. Compensatory mechanisms restoring arterial pressure toward normal are summarized in Fig. 11-60; their effects can best be appreciated from the data of Table 11-3. Kidney flow is even lower 5 min after the hemorrhage, despite the improved arterial pressure, but we recall that one of the important compensatory mechanisms is increased arteriolar constriction in many organs; thus, kidney blood flow is reduced in order to maintain arterial blood pressure and thereby brain and heart blood flow.

A second important compensatory mechanism involving capillary fluid exchange results from both the decrease in blood pressure and the increase in arteriolar constriction, both of which decrease capillary hydrostatic pressure, thereby favoring absorption of interstitial fluid (Fig. 11-61). Thus, the initial event—blood loss and decreased blood volume—is in large part compensated for by the movement of interstitial fluid into the vascular system. Indeed, as shown in Table 11-4, 12 to 24 h after a moderate hemorrhage, the blood volume may be restored virtually to normal.

The entire compensation is due to expansion of the plasma volume; replacement of the lost erythrocytes requires many days. Note that at 18 h, much of the plasma albumin lost in the hemorrhage has been replaced. This phenomenon is of great importance for expansion of plasma volume, as can be seen by considering the capillary filtration-absorption equilibrium (Fig. 11-46). As capillary hydrostatic pressure decreases as a result of the hemorrhage, interstitial fluid enters the plasma; this fluid, however, contains virtually no protein, so that its entrance dilutes the plasma proteins and increases the plasma water concentration. The resulting reduction of the water-concentration difference between the capillaries and the interstitial fluid hinders further fluid reabsorption and would prevent the full compensatory expansion if it were not that replacement of the plasma protein minimizes this fall in protein concentration, and movement of interstitial fluid into the plasma can continue. Synthesis of albumin contributes to

Table 11-4. Fluid shifts after hemorrhage

	Normal	Immediately after hemorrhage	18 h after hemorrhage
Total blood volume, mL	5000	4000 (↓20%)	4900
Erythrocyte volume, mL	2300	1840 (↓20%)	1840
Plasma volume, mL	2700	2160 (↓20%)	3060
Plasma albumin mass g	135	108 (↓20%)	125

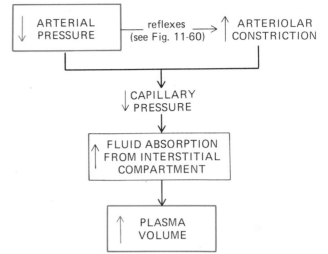

Figure 11-61. Mechanisms compensating for blood loss by movement of interstitial fluid into the capillaries. This response is diagramed in Fig. 11-46C.

this replacement to only a minor degree during the first 24 h; most of the replacement albumin which appears in the plasma during this early stage is interstitial protein carried by the lymphatics to the vascular compartment as a result of increased lymph flow.[1]

We must emphasize that this capillary mechanism has only *redistributed* the extracellular fluid; ultimate replacement of the plasma lost from the body involves the control of fluid ingestion and kidney function, both described in Chap. 13. Similarly, replacement of the lost red cells requires stimulation of erythropoiesis (by erythropoietin). Both these replacement processes require days to weeks in contrast to the rapidly occurring reflex compensations described in Fig. 11-61.

Loss from the body of large quantities of extracellular fluid rather than whole blood can also cause hypotension. This may occur via the skin, as in severe sweating or burns, via the gastrointestinal tract, as in diarrhea or vomiting, or via unusually large urinary losses. Regardless of the route, the loss decreases circulating blood volume and produces symptoms and compensatory phenomena similar to those seen in hemorrhage.

Hypotension may be caused by events other than blood or fluid loss. One such form is due to strong emotion and can result in fainting. Somehow the higher brain centers involved with emotions act upon the medullary cardiovascular centers to inhibit sympathetic activity and enhance parasympathetic activity resulting in decreased arterial pressure and brain blood flow. Fortunately, this whole process is usually transient, with no after-

effects, although a weak heart may suffer damage during the period of reduced blood flow to the cardiac muscle.

Other important causes of hypotension seem to have a common denominator in the liberation within the body of chemicals which relax arteriolar smooth muscle. There the cause of hypotension is clearly excessive arteriolar dilation and reduction of peripheral resistance, an important example being the hypotension which occurs during severe allergic responses.

It may be of interest to point out the physiological reasons for not treating patients with hypotension in ways commonly favored by the uninformed, namely, administering alcohol and covering the person with mounds of blankets. Both alcohol and excessive body heat, by actions on the central nervous system, cause profound dilation of skin arterioles, thus lowering peripheral resistance and decreasing arterial blood pressure still further. As shown below, the worst thing is to try to get the person to stand up.

Shock

The compensatory mechanisms described above for hemorrhage are highly efficient so that losses of as much as 1 to 1.5 L of blood (approximately 20 percent of total blood volume) can be sustained

[1] The reasons for this increased lymph flow are presently unclear.

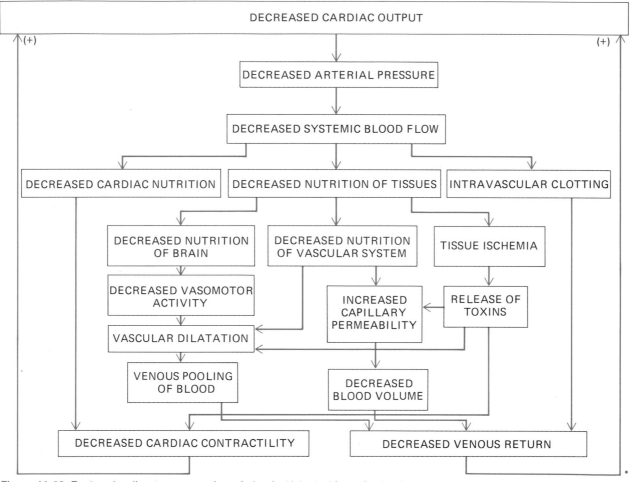

Figure 11-62. Factors leading to progression of shock. *(Adapted from Guyton.)*

with only slight reductions of mean arterial pressure or cardiac output. In contrast, when much greater losses occur or when compensatory adjustments are inadequate, the decrease in cardiac output may be large enough to seriously impair blood supply to the various tissues and organs. This is known as *circulatory shock.* The cardiovascular system, itself, suffers damage if shock is prolonged, and as it deteriorates, cardiac output declines even more and shock becomes progressively worse and ultimately irreversible — the person dies even after blood transfusions and other appropriate therapy. The factors that lead to progressive and irreversible shock are summarized in Fig. 11-62. Note that the chains of events are all positive feedbacks; i.e., they bring about further decreases in cardiac out-

put which then cause even greater deterioration of the cardiovascular system. Clearly, there must be a critical cardiac output above which the negative-feedback compensatory mechanisms lead to recovery, and below which the positive-feedback factors elicit a vicious cycle of deterioration. Damage to the heart by its low blood flow and by toxins released from other damaged tissues is one common denominator of these positive-feedback events; the other is progressive decrease in venous return to the heart because of clotting, blood pooling, and leakage of plasma out of capillaries.

Just as hemorrhage is not the only cause of hypotension and decreased cardiac output, so it is not the only cause of shock. Low volume due to loss of fluid other than blood, excessive release of vasodilators as in allergy and infection, loss of neural tone to the cardiovascular system, trauma—all these can lead to severe reductions of cardiac output and the positive-feedback cycles culminating in irreversible shock.

The upright posture

The simple act of getting out of bed and standing up is equivalent to a mild hemorrhage, because the changes in the circulatory system in going from a lying, horizontal position to a standing, vertical position result in a decrease in the effective circulating blood volume. The decrease results from the action of gravity upon the long continuous columns of blood in the vessels between the heart and the feet. All the pressures we have given in previous sections of this chapter were for the horizontal position (Fig. 11-63A), in which all blood vessels are at approximately the same level as the heart, and the weight of the blood produces negligible pressure. In the vertical position (Fig. 11-63C), the intravascular pressure everywhere becomes equal to the usual pressure resulting from cardiac contraction plus an additional pressure equal to the weight of a column of blood from the heart to the point of measurement (80 mmHg; Fig. 11-63B). In a foot capillary, for example, the pressure increases from 25 to 105 mmHG (80 + 25).

The veins are highly distensible structures, as we have seen. The increased hydrostatic pressure in the veins of the legs which occurs upon standing pushes outward upon the vein walls, causing marked distension with resultant pooling of blood; i.e., much of the blood emerging from the capillaries simply remains in the expanding veins rather than returning to the heart. Simultaneously, the marked increase in capillary pressure caused by the gravitational force produces increased filtra-

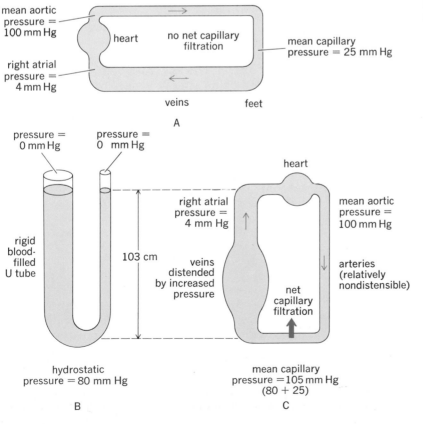

Figure 11-63. Effects of standing upon blood pressure, blood flow, and capillary filtration.

tion of fluid out of the capillaries into the interstitial space. Most of us have experienced swollen feet after a day's standing. The combined effects of venous pooling and increased capillary filtration are a significant reduction in the effective circulating blood volume in a manner very similar to a mild hemorrhage. The ensuing decrease in arterial pressure causes reflex compensatory adjustments similar to those shown in Fig. 11-60 for hemorrhage.

Perhaps the most effective compensation is contraction of skeletal muscles of the leg which produces intermittent, complete emptying of veins within the upper leg, so that uninterrupted columns of venous blood from the heart to the feet no longer exist. The result is a decrease in venous distension and pooling and a marked reduction in capillary hydrostatic pressure and fluid filtration out of the capillaries (Fig. 11-64). An example of the importance of this compensation (really its absence) is when soldiers faint after standing very still, i.e., with minimal contraction of the abdominal and leg muscles, for long periods of time. (Here the fainting may be considered adaptive in that venous and capillary pressure changes induced by gravity are eliminated once the person is prone, the pooled venous blood is mobilized, and

Figure 11-64. Role of contraction of the leg skeletal muscles in reducing capillary pressure and filtration in the upright position. The skeletal muscle contraction compresses the veins, causing intermittent complete emptying so that the column of blood is interrupted.

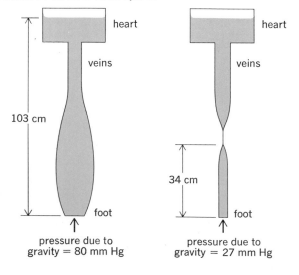

the previously filtered fluid is reabsorbed into the capillaries; thus, the wrong thing to do to anyone in a faint is to hold him upright.) Varicose veins (irregularly swollen superficial veins of the lower extremities) are also explainable in terms of hydrostatic pressures and the venous valves. They are the result of abnormal structural properties of the wall such that they are overdistensible. Therefore, they become so distended at normal hydrostatic pressures that the valves become incompetent and there is a backflow of blood from the deep veins through connections to the superficial veins. As described in a subsequent section, a major complication of venous stasis in varicose veins is the possibility of clot formation.

Thus far, we have described only the effects of normal gravitational forces, but modern airplanes and space vehicles have created the further problem of unusually large gravitational forces. During any form of marked change in acceleration—a flyer pulling out of a dive or an astronaut blasting off—the body is subjected to large gravitational forces which cause venous pooling and increased capillary filtration. If the forces are great enough, cardiac output may be so compromised as to produce fainting or blackout. At the other extreme, the phenomenon of weightlessness is encountered in space travel. The major problem encountered so far by astronauts after prolonged periods of weightlessness is a tendency toward hypotension upon returning to earth and standing to leave the spaceship. This is not yet clearly understood but seems to reflect a loss of the usual compensatory reflexes, as though these reflexes, used so frequently each day beginning with first getting out of bed in the morning, are temporarily lost during prolonged disuse. This is probably analogous to the dizziness or actual fainting commonly encountered when a long-bedridden patient first tries to arise.

Exercise

In order to maintain muscle activity during exercise, a large increase in blood flow is required to provide the oxygen and nutrients consumed and to carry away the carbon dioxide and heat produced. Thus, cardiac output may increase from a resting value of 5 L/min to the maximal values of 35 L/min obtained by trained athletes. The increased skeletal muscle blood flow results from marked dilation of the skeletal muscle arterioles mediated by local factors associated with active hyperemia. In addition, in a person just about to begin exercising, the skeletal muscle flow actually increases before the onset of muscular activity and therefore of active hyperemia; this anticipatory response providing a rapid initial supply of blood to the muscle is mediated by the sympathetic vasodilator fibers. It must be stressed, however, that once exercise has begun, these sympathetic vasodilator nerves are of little importance and active hyperemia plays the primary role in producing vasodilation.

The cardiovascular response to exercise has already been shown in our series of models (Figs. 11-52 to 11-55). The decrease in peripheral resistance resulting from dilation of skeletal muscle arterioles is partially offset by constriction of arterioles in other organs, particularly the gastrointestinal tract and kidneys. However, the "resistance juggling" is quite incapable of compensating for the huge dilation of the muscle arterioles, and the net result is a marked decrease in total peripheral resistance.

The cardiac output increase during exercise is associated with greater sympathetic activity and less parasympathetic activity to the heart. Thus, heart rate and stroke volume both rise, causing an increased cardiac output. The heart-rate changes are usually much greater than stroke-volume changes. Note (Fig. 11-65) that, in our example, the increased stroke volume occurs without change in end-diastolic ventricular volume; accordingly, the former is ascribable completely to the increased contractility induced by the cardiac sympathetic nerves. The stability of end-diastolic volume in the face of reduced time for ventricular filling (increased heart rate) is, in part, attributable to the fact, described earlier, that the sympathetic nerves increase the speed of relaxation.

However, it would be incorrect to leave the impression that enhanced sympathetic activity to the heart completely accounts for the elevated cardiac output which occurs in exercise, for such is not the case. The fact is that cardiac output could not be increased to high levels unless the venous return to the heart were not simultaneously facilitated to the same degree, for otherwise end-diastolic volume would fall and stroke volume would decrease (because of Starling's law). Therefore, factors promoting venous return during exercise are extremely important. They are (1) the marked activ-

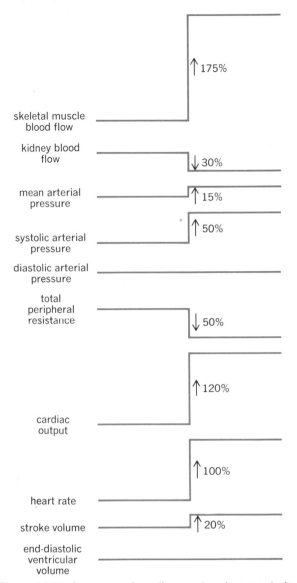

Figure 11-65. Summary of cardiovascular changes during mild exercise. (See also Fig. 11-4.)

ity of the skeletal muscle pump, (2) the increased respiratory movements, (3) the sympathetically mediated increase in venous tone, and (4) the ease with which blood flows from arteries to veins through the dilated skeletal muscle arterioles. These factors may be so powerful that venous return is enhanced enough to cause an increase in end-diastolic ventricular volume; under such conditions, stroke volume (and, thereby, cardiac output) is further enhanced.

Such an effect has been strikingly demonstrated in experiments on greyhounds whose hearts were completely denervated, yet who were able, after recovery, to run a race almost as fast as before. How is this possible? Apparently, the heart deprived of its nerves follows the relationship expressed in Starling's law more closely. As we have seen, various factors facilitate venous return during exercise; normally, increased venous return does not cause an increased end-diastolic pressure because the sympathetically driven heart pumps the blood as fast as it arrives. After cardiac denervation, however, the sympathetically mediated heart-rate and contractility increases do not occur as exercise begins and the increased venous return causes an immediate large increase in end-diastolic volume. This, by Starling's law, causes a marked increase in stroke volume and cardiac output. Thus, the heart now uses primarily a stroke-volume change rather than a heart-rate increase to raise cardiac output. The net overall result is, however, the same. Surprisingly, during very severe exercise, the denervated heart actually does manifest increased heart rate and contractility; this appears to be mediated by the circulating epinephrine released during severe exercise and to which the denervated heart has become "supersensitive." This problem is far from academic since we can now transplant the heart from a recently deceased person to a person dying of heart disease. Such a heart, of course, lacks nerves, and the entire question of cardiac control in response to exercise, posture, etc., is of the utmost importance.

What happens to arterial blood pressure during exercise? As always, the mean arterial pressure depends only upon the cardiac output and peripheral resistance. During most forms of exercise, the cardiac output tends to increase somewhat more than the peripheral resistance decreases so that mean arterial pressure usually increases slightly. However, the pulse pressure may show a marked increase because of greater systolic pressure and a relatively constant diastolic pressure. The former is due primarily to the faster ejection. These changes are all shown schematically in Fig. 11-65.

It is evident from this description that the sympathetic nerves (and inhibition of the parasympathetics) play an important role in the cardiovascular response to exercise. However, a problem arises when we try to understand the mechanisms which control the autonomic nervous system during exercise. The pulsatile (and mean) arterial pressure tends to be elevated, which should cause the arterial baroreceptors to signal the medullary centers to decrease cardiac output and dilate arterioles of the abdominal organs. Obviously, then, the arterial baroreceptors not only cannot be the origin of the cardiovascular changes in exercise but actually oppose these changes. Similarly, other possible inputs, such as oxygen and carbon dioxide, can be eliminated since they show little, if any, change (Chap. 12). We shall meet this same problem again when we describe (Chap. 12) increased respiration during exercise. A major clue is the finding that electric stimulation of certain hypothalamic areas in resting unanesthetized dogs produces all the cardiovascular changes usually observed during exercise. On the basis of these and other experiments, the best present working hypothesis is that a center in the hypothalamus acts, via descending pathways, upon the medullary centers (and directly upon the sympathetic vasodilators to skeletal muscle arterioles) to produce the changes in autonomic function so characteristic of exercise. What is the input to the hypothalamus? We do not know for certain, but it is probably information coming from the same motor areas of the cerebral cortex which are responsible for the skeletal muscle contraction. It is likely that as these fibers descend from the cerebral cortex to the spinal-cord motor neurons, they give off branches to the relevant hypothalamic centers. This system would nicely coordinate the skeletal muscle contraction with the blood supply needed to support it. This combination of events is also triggered not only during exercise but during times of excitement, stress, and anger, and, in such cases, is viewed as a preparation for exercise ("fight or flight").

Hypertension

Hypertension (high blood pressure) is defined as a chronically increased arterial pressure. In general, the dividing line between normal pressure and hypertension is taken to be 140/90 mmHg, although systolic pressures above 140 are frequently not associated with ill effects; the diastolic pressure is usually the most important index of hypertension. Hypertension is one of the major causes of illness and death in America today. It is estimated that approximately 6 million people suffer from this disease, the lethal end result of which may be heart attack or failure (see below), brain stroke (occlusion or rupture of a cerebral blood vessel), or kidney damage, all caused by prolonged hypertension and its attendant strain on the various organs.

Theoretically, hypertension could result from an increase in cardiac output or peripheral resistance or both. In fact, at least in well-established hypertension, the major abnormality is increased peripheral resistance due to abnormally reduced arteriolar diameter. What causes arteriolar narrowing? In most cases, we do not know, and hypertension of unknown cause is called *essential hypertension.* In a small fraction of cases, the cause of hypertension is known: (1) Certain tumors of the adrenal medulla secrete excessive amounts of epinephrine; (2) certain tumors of the adrenal cortex secrete excessive amounts of hormones which lead to hypertension by as yet unknown mechanisms; (3) many diseases which damage the kidneys or decrease their blood supply are associated with hypertension, but despite intensive efforts by many investigators, the actual factor(s) mediating *renal hypertension* remains unknown; increased release of renin with subsequent increased generation of angiotensin almost certainly plays a role, but it is unlikely that this is the only causal factor.

Besides the three known causes of hypertension described above and several others, we are left with the vast majority of patients in the essential, or unknown, category. Many hypotheses have been proposed for increased arteriolar constriction but none proved. At present, much evidence seems to point to excessive sodium ingestion or retention within the body as a common denominator of renal, adrenal, and essential hypertension, but how the sodium is involved in the increased arteriolar constriction remains unknown. In any case, sodium restriction has become a major form of therapy in hypertension. Virtually all other forms of therapy involve drugs which act upon some aspect of autonomic function to produce arteriolar dilation. This does not mean that excessive sympathetic tone was the original cause of the hypertension (it probably is, in certain people) but only that, whatever the cause, anything which dilates the arterioles reduces the blood pressure.

The perceptive reader might also wonder why the arterial baroreceptors do not, by way of the reflexes they initiate, return the blood pressure to normal. The reason seems to be that, in chronic hypertension, the baroreceptors are "reset" at a higher level; i.e., they regulate blood pressure but at a greater pressure.

Congestive heart failure

The heart may become weakened for many reasons; regardless of cause, however, the failing heart induces a similar procession of signs and symptoms grouped under the category of *congestive heart failure.* Patients with early mild heart disease may show at rest no significant abnormalities because of the great safety factor, or reserve, in cardiac function. However, the ability to perform exercise is impaired, as evidenced by shortness of breath and early fatigue. Ultimately, the cardiac reserve becomes inadequate to supply normal amounts of blood even at rest, and the patient becomes bedridden. Finally, the cardiac output may become too low to support life.

The basic defect in heart failure is a decreased contractility of the heart, but the molecular mechanism is unknown. As shown in Fig. 11-66, the failing heart shifts downward to a lower Starling

Figure 11-66. Relationship between end-diastolic ventricular volume and stroke volume in normal and failing hearts. The normal curves are those shown previously in Figs. 11-27 and 11-28. The failing heart can still eject an adequate stroke volume if the sympathetic activity to it is increased or if the end-diastolic volume increases, i.e., if the ventricle becomes more distended.

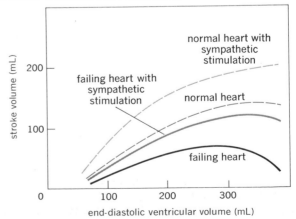

curve. How can this be compensated for? Increased sympathetic stimulation would help to increase contractility, and this does occur. However, an even more striking compensation is increased ventricular end-diastolic volume. The failing heart is generally engorged with blood, as are the veins and capillaries, the major cause being an increase (sometimes massive) in plasma volume. The sequence of events is as follows: decreased cardiac output causes a decrease in mean and pulsatile arterial pressure; this triggers reflexes (to be described in Chap. 13) which induce the kidneys to reduce their excretion of sodium and water; the retained fluid then causes expansion of the extracellular volume, increasing venous pressure, venous return, and end-diastolic ventricular volume and thus tending to restore stroke volume toward normal.

Another result of elevated venous and capillary pressure is increased filtration out of the capillaries, with resulting edema. This accumulation of tissue fluid may be the chief feature of ventricular failure; the legs and feet are usually most prominently involved (because of the additional effects of gravity), but the same engorgement is occurring in other organs and may cause severe malfunction. The most serious result occurs when the left ventricle fails; in this case, the excess fluid accumulates in the lung air sacs (*pulmonary edema*) because of increased pulmonary capillary pressure, and the patient may actually drown in his own fluid. This situation usually worsens at night; during the day, because of the patient's upright posture, fluid accumulates in the legs, but it is slowly absorbed when he lies down at night, the plasma volume expands, and an attack of pulmonary edema is precipitated.

Thus, what began as a useful compensation becomes potentially lethal because the tension-length relationship for muscle holds only up to a point, beyond which further stretching of the muscle may actually cause decreased strength of contraction. Thus, expansion of plasma volume may so increase end-diastolic volume as to decrease contractility (Fig. 11-66) and produce a rapidly progressing downhill course.

The treatment for congestive heart failure is easily understood in these terms: The precipitating cause should be corrected if possible; contractility can be increased by a drug known as digitalis; excess fluid should be eliminated by the use of drugs which increase excretion of sodium and water by the kidneys; the patient's activity level should be adjusted so as to reduce the cardiac output required to fulfill the body's metabolic needs.

"Heart attacks" and atherosclerosis

We have seen that the myocardium does not extract oxygen and nutrients from the blood within the atria and ventricles but depends upon its own blood supply via the coronary vessels. The coronary arteries exit from the aorta just above the aortic valves and lead to a branching network of small arteries, arterioles, capillaries, venules, and veins similar to those in all other organs. The rate of blood flow depends primarily upon the arterial diastolic blood pressure (the coronary vessels are completely occluded during systole by the constriction of the surrounding myocardial cells) and the resistance offered by the coronary vessels. The degree of arteriolar constriction, or dilation, is normally determined mainly by local metabolic control mechanisms (with some degree of neural control). Insufficient coronary blood flow leads to myocardial damage and, if severe enough, to death of the myocardium (*infarction*), a so-called *heart attack.* This may occur as a result of decreased arterial pressure but is more commonly due to increased vessel resistance following coronary atherosclerosis.

Atherosclerosis is a disease characterized by a thickening of the arterial wall with abnormal smooth muscle cells and deposits of cholesterol and other substances. The mechanisms which initiate this thickening are not clear, but it is known that cigarette smoking, a high plasma cholesterol, hypertension, diabetes, and a variety of other factors are associated with this disease. The suspected relationship between atherosclerosis and blood concentrations of cholesterol has probably received the most widespread attention, and many studies have documented that high blood concentrations of this lipid increase the rate and the severity of the atherosclerotic process. Cholesterol metabolism is described in Chap. 15.

The incidence of coronary atherosclerosis in the United States is extraordinarily great; it is estimated to cause 500,000 deaths per year. The mechanism by which atherosclerosis reduces coronary blood flow is quite simple; the extra muscle cells and various deposits narrow the vessels and increase resistance to flow. This is usually progressive, leading often ultimately to complete occlusion. Acute coronary occlusion may occur because of sudden formation of a clot on the roughened vessel surface or breaking off of a deposit, which then lodges downstream, completely blocking a smaller vessel. If, on the other hand, the atherosclerotic process causes only gradual occlusion, the heart may remain uninjured because of the development, over time, of new accessory vessels supplying the same area of myocardium. It should be stressed that before complete occlusion many patients experience recurrent transient episodes of inadequate coronary blood flow, usually during exertion or emotional tension. The pain associated with this is termed *angina pectoris.*

The cause of death from coronary occlusion and myocardial infarction may be either severe hypotension resulting from weakened contractility or disordered cardiac rhythm resulting from damage to the cardiac conducting system, but in addition the severe hypotension which may be associated with a heart attack is frequently due to reflex inhibition of the sympathetic nervous system and enhancement of the parasympathetics. The origin of these totally inappropriate and frequently lethal reflexes is not known. Finally, should the patient survive an acute coronary occlusion, the heart may be left permanently weakened, and a slowly progressing heart failure may ensue. On the other hand, many people lead quite active and normal lives for many years after a heart attack.

We do not wish to leave the impression that atherosclerosis attacks only the coronary vessels, for such is not the case. Most arteries of the body are subject to this same occluding process. For example, cerebral occlusions *(strokes)* are extremely common in the aged and constitute an important cause of sickness and death (200,000 per year). Wherever the atherosclerosis becomes severe, the resulting symptoms always reflect the decrease in blood flow to the specific area.

Section F
Hemostasis: The prevention of blood loss

All animals with a vascular system must be able to minimize blood loss consequent to vessel dam-

age. In human beings blood coagulation is only one of several important mechanisms for hemostasis. The mechanism which predominates varies, depending upon the kind and number of vessels damaged and the location of the injury.

The basic prerequisites for bleeding are (1) loss of vessel continuity or marked increase in permeability so that cells and fluid can leak out and (2) a pressure inside the vessel greater than outside, since hemorrhage occurs by bulk flow. Accordingly, bleeding ceases if at least one of two requirements is met: The pressure difference favoring blood loss is eliminated, or the damaged portion of the vessel is sealed. All hemostatic processes accomplish one of these two requirements. We shall discuss the probable sequence of events in damage to small vessels —arterioles, capillaries, venules—because they are the most common source of bleeding in everyday life and because the hemostatic mechanisms are most effective in dealing with such injuries. In contrast, the bleeding from a severed artery of medium or large size is not usually controllable by the body and requires radical aids such as application of pressure and ligatures. Venous bleeding is less dangerous because of the vein's low hydrostatic pressure; indeed, the drop in hydrostatic pressure induced by simple elevation of the bleeding part may stop the hemorrhage. In addition, if the venous bleeding is into the tissues, the accumulation of blood *(hematoma)* may increase interstitial pressure enough to eliminate the pressure gradient required for continued blood loss.

The hemostatic events in small vessels are summarized in Table 11-5. The first four events usually occur within seconds, whereas the formation of a clot (step 5) takes several minutes; however, the different events actually do not occur in neat orderly sequence but overlap in time and are closely interrelated functionally. The platelets play a critical role in the last four hemostatic components of our synopsis.

Table 11-5. Synopsis of hemostatic events in small blood vessels

1 Initial constriction of the damaged vessel
2 Sticking together of injured endothelium
3 Clumping of platelets to form a plug
4 Facilitation of the initial vasoconstriction
5 Blood coagulation, i.e., formation of a fibrin clot
6 Retraction of the clot

Hemostatic events prior
to clot formation

Initial constriction of the injured vessel

When a blood vessel is severed or injured, its immediate response is to constrict. This response, which slows the flow of blood in the affected area, occurs in all vessels, even in the capillaries, which have no muscular layers. This constriction, which may be intense enough to close the injured vessel completely, is soon enhanced, as we shall see, by chemicals released from platelets, but the mechanism of the initial constriction is not yet understood.

Sticking of the endothelial surfaces

This event occurs as a direct result of the initial constriction which presses the opposed endothelial surfaces of the vessel together. This contact induces a stickiness capable of keeping them "glued" together against high pressures even after the active vasoconstriction has begun to wane. The injury itself also alters the surface properties of the vessel and facilitates the adhesiveness of the damaged membranes. This entire process is probably of great importance only in the very smallest vessels of the microcirculation.

Formation of a platelet plug

The involvement of platelets in hemostatic events requires their adhesion to a surface. Although platelets have a propensity for adhering to many foreign or rough surfaces, they do not adhere to the normal endothelial cells lining the blood vessels. However, injury to a vessel disrupts the endothelium and exposes the underlying connective tissue with its collagen molecules, fibrous proteins which constitute a major component of the interstitial matrix. Platelets adhere strongly to collagen, and this attachment somehow triggers the release of the platelets' granules containing potent chemical agents, including adenosine diphosphate (ADP). This ADP then causes the surface of the adhered platelets to become extremely sticky so that new platelets adhere to the old ones and an aggregate or plug of platelets is rapidly built up by this self-perpetuating (positive-feedback) process. The role of prostaglandins (released by the plate-

lets or generated locally) in platelet aggregation is also being studied extensively at present, but no clear picture has yet emerged. The platelet plug may occlude small vessels so as to slow or even stop bleeding from them, but it is relatively unstable by itself and cannot withstand for long the high pressures of the larger vessels. However, regardless of its role as a blood stauncher, platelet adhesion is one of the crucial events in hemostasis because, as we shall see below, the release of the platelet chemicals triggered by adhesion also induces vasoconstriction and blood coagulation.

Humoral facilitation of vasoconstriction

As we have seen, the initial vascular constriction following vessel injury is a direct response of the damaged vessel itself. The maintenance of vasoconstriction in small vessels, however, is due to the local release from the aggregated platelets of powerful vasoconstrictor chemicals (including serotonin). This secondary, prolonged vasoconstriction does not occur unless there have been platelet adhesion and granule discharge. The first four events in hemostasis are summarized in Fig. 11-67.

Blood coagulation:
Clot formation

Despite the participation of the four mechanisms just described, blood coagulation is the dominant hemostatic defense in people, as attested by the fact that, with few exceptions, clotting defects are the cause of abnormal bleeding. Pure vascular defects, interfering with the preclot hemostatic mechanisms, do occur but are much less frequently the cause of abnormal bleeding.

The event transforming blood into a solid gel is the conversion of the plasma protein *fibrinogen* to *fibrin*. Fibrinogen is a soluble, large, rod-shaped protein (molecular weight approximately 340,000) produced by the liver and always present in the plasma of normal persons. Its conversion to fibrin is catalyzed by the enzyme *thrombin*:

$$\text{Fibrinogen} \xrightarrow{\text{thrombin}} \text{fibrin}$$

In this reaction, several small negatively charged polypeptides are split from fibrinogen, conferring upon the remaining large molecule the ability to bind to other molecules of fibrin. They join each other end-to-end and side-to-side to form the

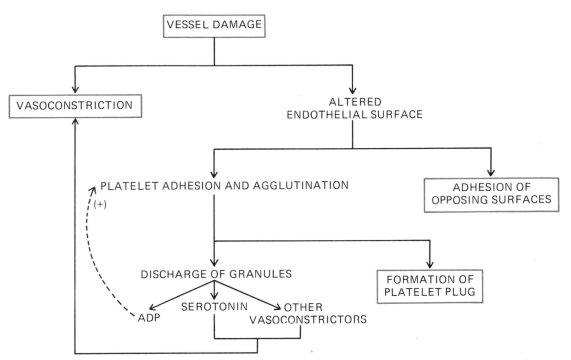

Figure 11-67. Summary of hemostatic mechanisms not dependent upon blood coagulation. The dashed line indicates the positive-feedback effect of ADP on platelet adhesion and agglutination. Prostaglandins, not shown in the figure, may either inhibit or facilitate platelet clumping.

polymer known as fibrin (Fig. 11-68). This polymerization causes the fluid portion of the blood to gel, rather like a gelatin dessert. In addition, cellular elements of the blood become entangled in the meshwork and contribute to its strength. It must be emphasized that the clot is basically due to fibrin and can occur in the absence of blood cells (except the platelets).

Since fibrinogen is always present in the blood, the enzyme thrombin must normally be absent and its formation triggered by vessel damage. The generation of thrombin follows the same general principle as that of fibrin, in that an inactive precursor, *prothrombin*, is produced by the liver and is normally present in the blood; prothrombin is converted to thrombin during clot formation by a specific enzyme which splits off an inactive peptide from the prothrombin molecule:

$$\text{Prothrombin} \xrightarrow{\text{enzyme}} \text{thrombin}$$
$$\text{Fibrinogen} \xrightarrow{} \text{fiibrin}$$

We have now only pushed the essential question one step further back: Where does the enzyme come from which catalyzes the conversion of the prothrombin to thrombin? The answer is that this enzyme is always present in the plasma in an inactive form and is converted to its active form by yet another enzyme which itself was activated by another enzyme, which. . . . Thus, we are dealing with a *cascade* of plasma proteins, each normally an inactive proteolytic enzyme until activated by the previous one in the sequence (Fig. 11-69). Ultimately, the final enzyme in the sequence is activated and in turn catalyzes the conversion of prothrombin to thrombin. We must point out that the designations A and B in our figure are arbitrarily chosen to demonstrate the general principle. The first factor is actually called *factor XII* or *Hageman factor* (after the patient in whom it was first discovered) and, as will be described in subsequent sections and Chap. 17, it has several other important functions in addition to initiating clotting.

The adaptive value of this type of cascade sys-

Figure 11-68. Scanning electron micrograph of fibrin. *(From R. G. Kessel and C. Y. Shih, "Scanning Electron Microscopy in Biology," Springer-Verlag, New York, 1974, p. 265.)*

tem lies in the amplification gained at each step, i.e., in the manyfold increase in the number of active molecules produced (recall from Chap. 9 that an analogous cascade of enzyme activations was described for the cyclic AMP-protein kinase system). However, there is at least one disadvantage in such a cascade in that a single defect, either hereditary or disease-induced, anywhere in

the system can block the entire cascade, thereby interfering with clot formation.

In addition to these plasma proteins, calcium is required as cofactor for several enzymatic steps. However, calcium deficiency is never a cause of clotting defects in people since only very small concentrations are required. Finally, a phospholipid substance exposed on the surface of platelets during their aggregation is required as cofactor for several of the steps in the catalytic sequence; it seems to provide a surface for interaction of certain of the plasma clotting factors.

Figure 11-69 reveals that we have still not answered the basic question of what *initiates* clotting. What activates the first enzyme (Hageman factor or factor A in the figure) in the catalytic sequence? The answer is contact of this protein with a damaged vessel surface, most likely with the negatively charged collagen fibers underlying the damaged endothelium (as was true for platelet aggregation). At last we have reached the connecting link between vessel injury and initiation of clotting. Thus, for two reasons, the critical event initiating clot formation is contact of the blood with a damaged surface: (1) It activates the first factor in the activation sequence; (2) it causes platelet adhesion and exposure of a phospholipid cofactor. One final detail is that thrombin markedly

Figure 11-69. Cascade theory of blood clotting. Each substance left of an arrow is normally present in plasma but requires activation by the action of the previous substance in the sequence. The name of factor A is Hageman factor. The term "factor" signifies "enzyme."

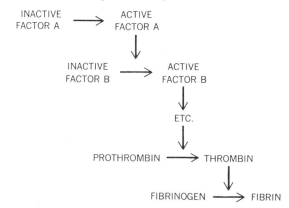

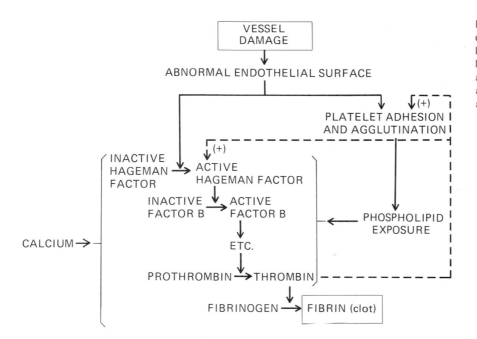

Figure 11-70. Summary of blood-clotting mechanism. The dashed line indicates the positive-feed-back effect of thrombin on platelet adhesion (recall that ADP exerts a similar positive feedback) and activation of Hageman factor.

enhances both the aggregation of platelets and the activation of Hageman factor; thus, once thrombin formation has begun, the overall reaction progresses explosively owing to the positive-feedback effects of thrombin. Our growing figure can now be completed (Fig. 11-70).

This entire process occurs only locally at the site of vessel damage. Each active component is formed, functions, and then is rapidly inactivated by enzymes in the plasma or local tissue, without spilling over into the rest of the circulation. Otherwise, because of the chain-reaction nature of the response, the appearance of exposed platelet phospholipid and any single activated factor in the overall circulation would induce massive widespread clotting throughout the body.

The reason should also be clear why blood coagulates when it is taken from the body and put in a glass tube. This has nothing whatever to do with exposure to air, as popularly supposed, but happens because the glass surface induces the same activation of Hageman factor and aggregation of platelets as does a damaged vessel surface. A silicone coating markedly delays clotting by reducing the activating effects of the glass surface.

The liver plays several important indirect roles in the overall functioning of the clotting mechanism (Fig. 11-71). First, it is the site of production for many of the plasma factors, including prothrombin and fibrinogen. Second, the bile salts produced by the liver are required for normal gastrointestinal absorption of the fat-soluble *vitamin K*, which is an essential cofactor in the normal hepatic synthesis of prothrombin and several other plasma factors; when vitamin K is not present in adequate amounts, these clotting factors are still produced, but they lack the critical binding sites which would permit them to interact with platelet phospholipid and become activated. It should be clear, therefore, why patients with liver disease or

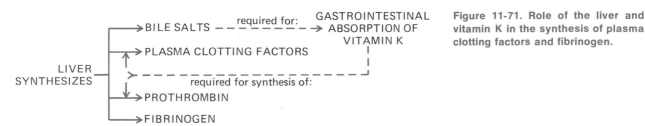

Figure 11-71. Role of the liver and vitamin K in the synthesis of plasma clotting factors and fibrinogen.

defective gastrointestinal fat absorption frequently have serious bleeding problems.

Finally, a word must be said about the contribution of a tissue (rather than blood) factor to clotting. If one extracts almost any of the body's tissues and injects the extract into unclotted normal blood in a siliconized tube, clotting occurs within seconds. The explanation is that the tissues contain a lipoprotein substance known as *tissue thromboplastin* which can substitute for platelet phospholipid as well as several of the plasma factors; thus an abnormal surface is no longer required to initiate clotting. This is known as the *extrinsic clotting pathway* to distinguish it from the *intrinsic pathway* described above. Its quantitative contribution to normal intravascular clotting is unclear but it may play an essential role in the response to many bacterial infections by initiating interstitial fibrin clots which may block further spread of the bacteria.

This completes our discussion of the events leading to clot formation. The initiating event is the presence of a damaged vessel surface, which induces platelet adhesion and activation of the first plasma factor in the catalytic sequence leading to generation of thrombin. Calcium and a platelet phospholipid are required for the overall reaction. As the final step, thrombin enzymatically splits off several small polypeptides from fibrinogen, leading to clot formation by generating polymerizing fibrin strands. A defect or lack of any of these components may induce inadequate clotting and prolonged bleeding.

Clot retraction

When blood is carefully collected and placed in a glass test tube, clotting usually occurs within 5 to 8 min, the entire volume of blood appearing as a coagulated gel. However, during the next 30 min, a striking transformation occurs; the clot literally retracts, squeezing out the fluid which constituted a large fraction of the gel. The end result is a small, hard clot at the bottom of the tube with a large volume of serum floating on top (plasma without fibrinogen is called serum). The fibrin meshwork with its entangled cells has thus become denser and stronger. This same process in the body is known as *clot retraction.* Besides increasing the strength of the clot, it has the advantage of pulling the vessel walls adhering to the clot closer together.

Clot retraction is due to the platelets. As fibrin strands form around them during clotting, the agglutinated mass of platelets sends out adhering pseudopods along them. The pseudopods then contract, pulling the fibrin fibrils together and squeezing out the serum. This contraction is produced by actomyosinlike contractile proteins in the platelets. The various roles of platelets in hemostasis are summarized in Fig. 11-72.

The anticlotting system

Clots dissolve or fail to form in several normal circumstances. Usually this is due to a proteolytic enzyme called *plasmin*, which is able to decompose fibrin, thereby dissolving a clot. The physiology of plasmin bears some striking similarities to that of the coagulation factors in that it circulates in blood in an inactive form (plasminogen) which is enzymatically converted to active plasmin by the action of activated Hageman factor as well as by other substances found in tissue fluid (Fig. 11-73). It may seem paradoxical that the same substance, Hageman factor, triggers off simultaneously the clotting and anticlotting systems. Yet, in fact, this makes

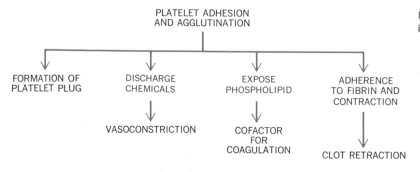

Figure 11-72. Summary of platelet functions in hemostasis.

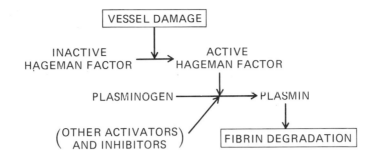

Figure 11-73. Summary of the plasmin anticlotting system.

sense since the generated plasmin becomes trapped within the newly formed clot and very slowly dissolves it, thereby contributing to tissue repair at a time when the danger of hemorrhage is past.

The anticlotting system no doubt has other functions as well. It may be that small amounts of fibrin are constantly being laid down throughout the vascular tree and that plasmin acts on this fibrin to prevent clotting. The lung tissue, for example, contains a substance which activates plasmin; this probably explains the lung's ability to dissolve the fibrin clumps which its capillaries filter from the blood. Moreover, the uterine wall is extremely rich in a similar activator, and thus normal menstrual blood generally does not clot.

A second naturally occurring anticoagulant is *heparin*. This substance, found in various cells of the body, acts by interfering with the activation of several of the clotting factors and with the ability of thrombin to split fibrinogen. Despite its presence in the body and the fact that it is the most powerful anticoagulant known, it is not clear whether heparin plays a normal physiological role in clot prevention. On the other hand, heparin is widely used as an anticoagulant drug in medicine.

Excessive clotting: Intravascular thrombosis

Formation of a clot in a bleeding vessel is obviously a homeostatic physiological response, but the formation of clots within intact vessels is pathological. It may occur in the veins, the microcirculation, or the arteries. Coronary arterial occlusion secondary to thrombosis (thrombus means clot) is one of the major killers in the United States today. The sequence of events leading to thrombosis is the subject of intensive study, and numerous theories

have been proposed. A brief synopsis of some of them is in order because of the great importance of the subject and its illustration of the basic physiological processes.

One of the dominant theories today postulates that the clotting mechanism in persons prone to thrombosis is hyperactive, as manifested by the reduced time it takes for withdrawn blood to clot in a test tube. Perhaps one of the plasma factors is present in excessive amounts, or a normally occurring anticoagulant is deficient. This theory emphasizes that the blood itself is the cause of excessive clotting. However, there seems little question that hypercoagulability is not always essential for thrombosis, since hemophiliacs have been known to suffer from coronary thrombosis.

A second category of theories puts the blame on the blood vessels. Since initiation of blood clotting is primarily dependent upon the state of the blood vessel lining, even minor transient alterations in the endothelial surface could trigger the cascade sequence leading to clot formation. These vessel-oriented theories can explain many of the situations associated with an increased probability of

Figure 11-74. Damaged-vessel theory of abnormal intravascular clotting.

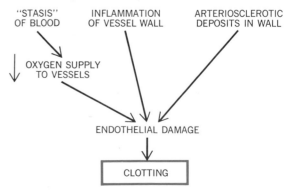

vascular thrombosis (Fig. 11-74): (1) Stasis, i.e., decreased movement, of blood in veins, which occurs during quiet standing, valve malfunction, or cardiac insufficiency, may induce damage in the vein wall as a result of oxygen lack. (2) Inflammation of veins and other vessels caused by bacteria, allergic reactions, or toxic substances may cause vessel damage. (3) Accumulation of lipids and smooth muscle cells in arterial walls (atherosclerosis) causes marked thickening and irregularity of the arterial lining.

Thus, three of the major conditions predisposing to clot formation are consistent with the damaged-lining theory. It should be noted that this concept and the hypercoagulability theory are not mutually exclusive. Both probably are valid, depending upon the circumstances.

Regardless of the initiating event, there is no question that a clot, no matter how small, provides a suitable surface upon which more clot can form. Thus the thrombus grows and may eventually occlude the entire vessel, thereby leading to damage of the tissue supplied or drained by the vessel. A second important factor in vessel closure during clot growth is the release of vasoconstrictors from freshly adhered platelets. Finally, the chances are greatly increased of clot fragments breaking off and being carried to the lungs (if from a vein) or other organs (if from an artery).

The prevention of new clot growth with its associated consequences is the major reason for giving patients anticoagulant drugs. Anticoagulants now in use include heparin and a group of drugs which interfere with the normal synthesis of several of the plasma clotting factors, including prothrombin, by blocking the action of vitamin K. Indeed, one of the important observations leading to the discovery of vitamin K was that animals on a diet of spoiled sweet clover hay, which contains these anticoagulants, manifested serious bleeding tendencies. Of course, a patient receiving any anticoagulant is prone to bleeding.

12
RESPIRATION

Organization of the respiratory system
  Conducting portion of the respiratory system
  Site of gas exchange in the lungs: The alveoli
  Relation of the lungs to the thoracic cage
Inventory of steps involved in respiration
Exchange of air between atmosphere and alveoli: Ventilation
  The concept of intrapleural pressure
  Inspiration
  Expiration
  Quantitative relationship between atmosphere-intraalveolar pressure gradients and air flow: Airway resistance
  Pulmonary vessels and their control
  Lung-volume changes during breathing
  Air distribution within the lungs
    Anatomic dead space • *Alveolar dead space*
  The work of breathing
Exchange and transport of gases in the body
  Basic properties of gases
  Behavior of gases in liquids
  Pressure gradients of oxygen and carbon dioxide within the body
  Transport of oxygen in the blood: The role of hemoglobin
    Effect of P_{O_2} on hemoglobin saturation • *Effect of acidity on hemoglobin saturation* • *Effect of tem-*

perature on hemoglobin saturation • *Effect of DPG on hemoglobin saturation*
Transport of carbon dioxide in the blood
Transport of hydrogen ions between tissues and lungs
Control of respiration
  Neural generation of rhythmic breathing
  Control of ventilatory volume
    Control of ventilation by oxygen • *Control of ventilation by carbon dioxide* • *Control of ventilation by changes in arterial hydrogen-ion concentration not due to altered carbon dioxide*
  Control of ventilation during exercise
    Decreased P_{CO_2} as the stimulus • *Increased P_{CO_2} as the stimulus* • *Increased hydrogen ion as the stimulus*
  Control of respiration by other factors
    Temperature • *Epinephrine* • *Reflexes from joints and muscles* • *Baroreceptor reflexes* • *Protective reflexes* • *Pain* • *Emotion* • *Voluntary control of breathing*
Maximal oxygen uptake: A synthesis of cardiovascular-respiratory interactions
Hypoxia
  Response to high altitude
Metabolic functions of the lung

Most cells in the human body obtain the bulk of their energy from chemical reactions involving oxygen. In addition, cells must be able to eliminate carbon dioxide, the major end product of these reactions. A unicellular organism can exchange oxygen and carbon dioxide directly with the external environment, but this is obviously impossible for most cells of a complex organism like the human body, since only a small fraction of the cells (skin, gastrointestinal lining, respiratory lining) is in direct contact with the external environment. In

order to survive, large animals had to develop specialized systems for the supply of oxygen and elimination of carbon dioxide. These systems are not the same in all complex animals since evolution often follows several pathways simultaneously. The organs of gas exchange with the external environment in fish are gills; those in humans are *lungs*. Specialized blood components have also evolved which permit the transportation of large quantities of oxygen and carbon dioxide between the lungs and cells.

At rest, the body's cells consume approximately 200 mL of oxygen per minute. Under conditions of high oxygen requirement, e.g., exercise, the rate of oxygen consumption may increase as much as thirtyfold. Equivalent amounts of carbon dioxide are simultaneously eliminated. It is obvious, therefore, that mechanisms must exist which coordinate breathing with metabolic demands. We shall see in Chap. 13 that the control of breathing also plays an important role in the regulation of the acidity of the extracellular fluid.

Before describing the basic processes of oxygen supply, carbon dioxide elimination, and breathing control, we must first define the terms *respiration* and *respiratory system*. Respiration has two quite different meanings: (1) the metabolic reaction of oxygen with carbohydrate and other organic molecules and (2) the exchange of gas between the cells of an organism and the external environment. The various steps of the second process form the subject matter of this chapter; the first process was described in Chap. 5. The term *respiratory system* refers only to those structures which are involved in the exchange of gases between the blood and external environment; it does not include the transportation of gases in the blood or gas exchange between blood and the tissues. Admittedly, this definition is arbitrary since it includes only half of the processes involved in respiration, but it has become firmly established by long usage. The respiratory system comprises the lungs, the series of airways leading to the lungs, and the chest structures responsible for movement of air in and out of the lungs.

Organization of the respiratory system

In order for air to reach the lungs, it must first pass through a series of air passages connecting the lungs to the nose and mouth (Fig. 12-1). There are two lungs, the right and left, each divided into several lobes. Together with the heart, great vessels, esophagus, thymus, and certain nerves, the lungs completely fill the *chest (thoracic) cavity.* The lungs are not simply hollow balloons but have a highly organized structure consisting of air-containing tubes, blood vessels, and elastic connective tissue. The airways within the lungs (Fig. 12-2) are actually the continuation of those which con-

nect the lungs to the nose and mouth. Together they are termed the *conducting portion* of the respiratory system and constitute a series of highly branched hollow tubes becoming smaller in diameter and more numerous at each branching, much like arteries and arterioles. The smallest of these tubes end in tiny blind sacs, the *alveoli*, which number approximately 300 million and are the actual sites of gas exchange within the lungs. All portions of these air passageways and alveoli receive a rich supply of blood by way of blood vessels, which constitute a large portion of the total lung substance (Fig. 12-2).

Conducting portion of the respiratory system

Air can enter the respiratory passages either by nose or mouth, although the nose is the normal route. It then passes into the *pharynx* (throat), a passage common to the routes followed by air and food. The pharynx branches into two tubes, one (the *esophagus*) through which food passes to the stomach and one through which air passes to the lungs. The first portion of the air passage, called the *larynx*, houses the *vocal cords.* It is protected against the entry of food by closure of the vocal cords across the opening of the larynx. The larynx opens into a lung tube (the *trachea*), which, in turn, branches into the two *bronchi*, one of which enters each lung. Within the lungs, these major bronchi branch many times into progressively smaller bronchi, thence into bronchioles, and finally into the alveolar portions of the lungs.

The conducting system of tubes serves several important functions:

1 The epithelial linings contain hairlike projections, called *cilia*, which constantly beat toward the pharynx (Fig. 12-3). These cilia line the respiratory airways to the end of the bronchioles; in the same regions are epithelial glands which secrete *mucus*. Particulate matter, such as dust contained in the inspired air, sticks to the mucus, which is constantly moved by the cilia to the pharynx, and then is swallowed. Besides keeping the lungs clean, this mucus escalator is important in the body's total defenses against bacterial infection, since many bacteria enter the body on dust particles. A major cause of lung infection is probably reduction of ciliary

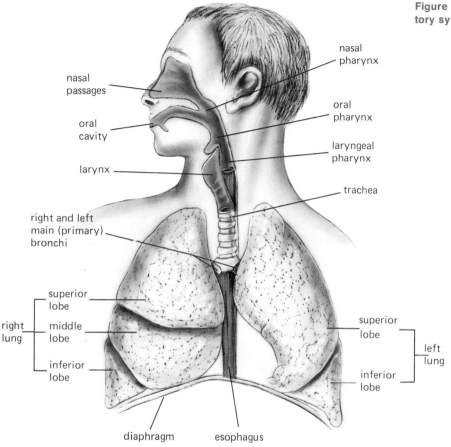

Figure 12-1. Organization of the respiratory system.

nasal pharynx

nasal passages

oral pharynx

oral cavity

laryngeal pharynx

larynx

trachea

right and left main (primary) bronchi

superior lobe

superior lobe

right lung

middle lobe

left lung

inferior lobe

inferior lobe

diaphragm   esophagus

activity by noxious agents; a single cigarette can cause the cilia to be nonmotile for several hours. This, coupled with the stimulation of mucus secretion induced by these same agents, may result in partial or complete airway obstruction by the stationary mucus. (A smoker's early-morning cough is the attempt to clear this obstructive mucus from the airways.) A second protective mechanism is provided by the phagocytic cells, which are present in the respiratory-tract lining in great numbers. These cells, which engulf dust, bacteria, and debris, are also injured by cigarette smoke and other air pollutants.

2 As air flows through the respiratory passages, it is warmed and moistened by contact with the epithelial lining.

3 The vocal cords, two strong bands of elastic tissue, lie stretched across the lumen of the larynx. The movement of air past them causes them to vibrate, providing the tones of phonation.

4 The walls of the respiratory airways contain richly innervated smooth muscle sensitive to certain circulating hormones, e.g., epinephrine. Contraction or relaxation of this muscle alters resistance to air flow.

Site of gas exchange in the lungs: The alveoli

The alveoli are tiny cup-shaped hollow sacs whose open ends are continuous with the lumens of the smallest bronchioles and the alveolar ducts (Fig. 12-2). The alveoli (Fig. 12-4) are lined by a continuous single thin layer of epithelial cells resting on a thin basement membrane, in turn resting on a very loose mesh of connective tissue elements which constitutes the interstitial space of the alveolar

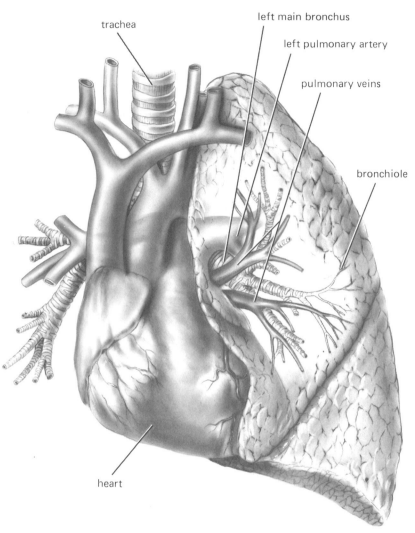

trachea

left main bronchus

left pulmonary artery

pulmonary veins

bronchiole

heart

Figure 12-2. Relationships between respiratory airways and blood vessels.

walls. Most of the alveolar wall is occupied by capillaries, the endothelial lining of which is separated from the alveolar epithelial lining only by the very thin interstitial space. Indeed, the interstitial space may be absent altogether; i.e., the basement membranes of the epithelium and endothelium may actually fuse. Thus the blood within a capillary is separated from the air within an alveolus only by an extremely thin barrier (0.2 μm compared with 7 μm, which is the diameter of an average red blood cell). The total area of alveoli in contact with capillaries is 70 m² in human beings (the size of a badminton court). This immense area combined with the thin barrier permits the rapid exchange of large quantities of oxygen and carbon dioxide.

In addition to its thin cells, the alveolar epithelium also contains smaller numbers of thicker specialized cells (*Type II cells*) which produce surfactant, to be discussed below. The interstitial space contains phagocytic cells (macrophages) and other connective-tissue cells which function in the lung's defense mechanisms. Finally, there are pores in the alveolar membranes which permit

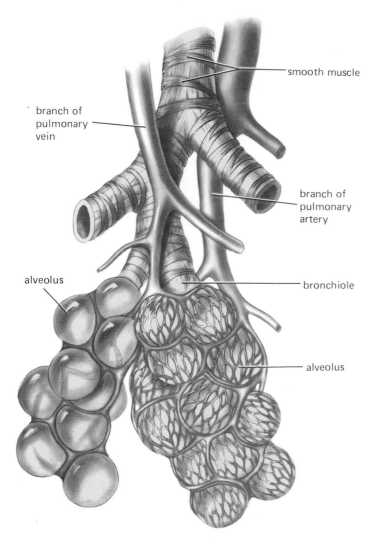

smooth muscle

branch of
pulmonary
vein

branch of
pulmonary
artery

bronchiole

alveolus

alveolus

some flow of air between alveoli. This "collateral ventilation" can be very important when the duct leading to an alveolus is occluded by disease, since some air can still enter this alveolus by way of pores between it and adjacent alveoli.

Relation of the lungs to the thoracic cage

To understand how breathing occurs we must know something about the tissues which make up the chest wall. The thoracic cage is a closed compartment. It is bounded at the neck by muscles and connective tissue, and it is completely separated from the abdomen by a large dome-shaped sheet of

Figure 12-3. Epithelial lining of the airways. The arrows indicate the upward direction in which the cilia move the overriding layer of mucus, to which foreign particles are stuck.

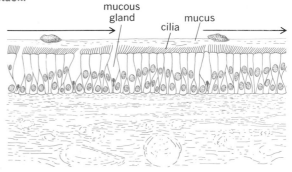

mucous
gland

cilia

mucus

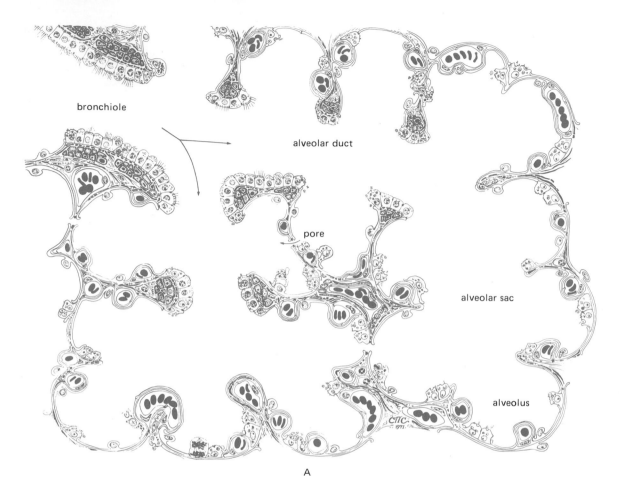

bronchiole

alveolar duct

pore

alveolar sac

alveolus

A

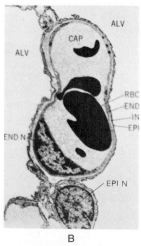

ALV

CAP

ALV

RBC

END

IN

EPI

END N

EPI N

B

Figure 12-4. (A) Diagram of the respiratory subdivisions in the lung. *(From R. O. Creep and L. Weiss, "Histology," 3d ed., McGraw-Hill Book Company, New York, 1973.)* **(B) Higher magnification of a portion of an alveolar wall, showing a single capillary (CAP) surrounded by an alveolar epithelial cell. The nucleus (END N) and cytoplasm (END) of a capillary endothelial cell are visible, as are the nucleus (EPI N) and cytoplasm of an alveolar epithelial cell. Note that the blood in the capillary is separated from air in the alveoli (ALV) only by the thin layer consisting of endothelium, interstitial fluid (IN), and epithelium.** *[Courtesy of E. R. Weibel, from* Physiol. Rev., **53:**424 *(1973)].*

skeletal muscle, the diaphragm. The outer walls of the thoracic cage are formed by the breastbone *(sternum),* 12 pairs of *ribs,* and the muscles which lie between the ribs (the *intercostal muscles).*

These walls also contain large amounts of elastic connective tissue.

Firmly attached to the entire interior of the thoracic cage is a thin sheet of cells, the *pleura,* which folds back upon itself to form two completely enclosed sacs within the thoracic cage, one on each side of the midline. The relationship between the lungs and pleura can be visualized by imagining what happens when one punches a fluid-filled balloon (Fig. 12-5): The arm represents the major bronchus leading to the lung, the fist is the lung, and the balloon is the pleural sac. The outer portion of the fist becomes coated by one surface of the balloon. In addition, the balloon is pushed back upon itself so that its surfaces lie close together. This is precisely the relation between the lung and pleura except that the pleural surface coating the lung is firmly attached to the lung surface. This layer of pleura and the outer layer which lines the interior thoracic wall are so close to each other that they are virtually in contact, being separated only by a very thin layer of *intrapleural fluid.*

Inventory of steps involved in respiration

1 **Exchange of air between the atmosphere (external environment) and alveoli.** This process includes the movement of air in and out of the lungs and the distribution of air within the lungs. Not only must a large volume of new air be delivered constantly to the alveoli but it must be distributed proportionately to the millions of alveoli within each lung. This entire process is called *ventilation* and occurs by bulk flow.
2 **Exchange of oxygen and carbon dioxide between alveolar air and lung capillaries by diffusion.** The volume and distribution of the pulmonary (lung) blood flow are extremely important for normal functioning of this process.
3 **Transportation of oxygen and carbon dioxide by the blood.** This includes the flow of blood and the forms of the gases within the blood.
4 **Exchange of oxygen and carbon dioxide between the blood and tissues of the body by diffusion as blood flows through tissue capillaries.**

Exchange of air between atmosphere and alveoli: Ventilation

Like blood, air moves by bulk flow from a high pressure to a low pressure. We have seen (Chaps. 6 and 11) that bulk flow can be described by the equation

$$F = k(P_1 - P_2)$$

That is, flow is proportional to the pressure difference between two points, k being the proportionality constant. For air flow into or out of the

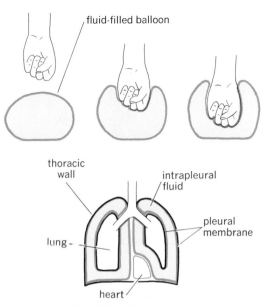

Figure 12-5. Relationship of lungs, pleura, and thoracic cage, analogous to pushing one's fist into a fluid-filled balloon. Note that there is no communication between the right and left intrapleural fluids. The volume of intrapleural fluid is greatly exaggerated; normally it consists of an extremely thin layer of fluid between the pleural membrane lining the inner surface of the thoracic cage and the pleural membrane lining the surface of the lungs.

lungs, the two relevant pressures are the *atmospheric pressure* and the pressure in the alveoli (*alveolar pressure*):

$$F = K(P_{atm} - P_{alv})$$

The atmospheric pressure is 760 mmHg at sea level and is obviously not subject to control, short of putting a person in a space suit or diving bell. Since atmospheric pressure remains relatively constant, if air is to be moved in and out of the lungs, the alveolar pressure must be made alternately less than and greater than atmospheric pressure (Fig. 12-6).

The concept of intrapleural pressure

When the chest is opened during surgery, care being taken to cut only the thoracic wall but not the lung, the lung on the side of the incision collapses immediately (Fig. 12-7). Normally the highly elastic lungs are stretched within the intact chest, and the force responsible is eliminated

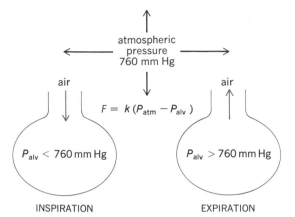

Figure 12-6. Relationships required for breathing. When the alveolar pressure P_{alv} is less than atmospheric pressure, air enters the lungs. Flow F is directly proportional to the pressure difference.

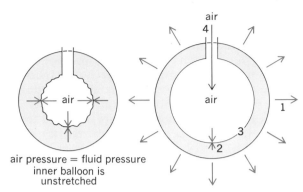

Figure 12-8. How fluid between two balloons causes the inner balloon to expand whenever the outer one does. 1 Outer balloon is expanded by an outside force. 2 Expansion of the fluid space causes the fluid pressure to decrease. 3 Fluid pressure is less than internal air pressure; therefore, the wall of the inner balloon is pushed out. 4 As the inner balloon expands, its internal air pressure decreases and air moves in from the atmosphere.

when the chest is opened. This force is the pressure difference across the lung wall.

An analogy (Fig. 12-8) may help illustrate the mechanism. Imagine two balloons of slightly different size, one inside the other. The space between the balloons is completely filled with water. Since the inner, smaller balloon is open at the top so that there is free communication between its interior and the atmospheric air surrounding the larger balloon, it contains air at atmospheric pressure. The forces acting upon the wall of the inner balloon are the inner air pressure and the water pressure surrounding it. Initially, as shown in the left half of Fig. 12-8, these pressures are approximately equal, and there is no tension in the wall of the inner balloon. When we enlarge the outer balloon by pulling on it in all directions, the inner balloon expands by an almost equal amount and

Figure 12-7. Lung collapse caused by a stab wound piercing the thoracic cage. Note that the air in the pleural space did *not* come from the lungs since the lung wall is still intact.

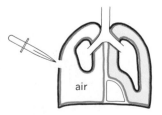

its walls become highly stretched and taut. This occurs because the fluid pressure of the water surrounding the inner balloon becomes less than the air pressure inside it. Water is highly *indistensible*; i.e., any attempt to expand or compress a completely water-filled space causes a marked decrease or increase, respectively, of the fluid pressure within the space (it is much more difficult to compress a water-filled balloon than an air-filled balloon of the same size). Thus the pull on the external balloon produces a drop in the fluid pressure surrounding the inner balloon, which now becomes less than the air pressure within the inner balloon; this difference in pressure across the wall pushes out the wall of the inner balloon. As the inner wall moves outward, the internal air pressure falls slightly, but atmospheric air immediately enters through the opening so that atmospheric pressure is restored. Thus, the fluid pressure remains lower than the internal air pressure and the inner balloon expands until the force of its elastic recoil becomes great enough to balance this distending pressure difference.

We can apply this analogy to the lungs (air-filled inner balloon), thoracic cage (outer balloon), and intrapleural fluid (the water between the balloons). At rest, i.e., even when no respiratory-muscle contraction is occurring, elastic forces in the tissues of the thoracic cage tend to pull it away from the outer surface of the lungs. This drops the *intrapleural fluid pressure* below that of the alveolar air pressure, a pressure difference that forces the lungs to distend virtually the same degree as the thoracic cage, and their elastic walls become stretched. The tendency for the lungs to recoil as a result of this stretch is balanced by the difference between the alveolar air pressure and the intrapleural fluid pressure.

Why the lung collapses when the chest wall is opened should now be apparent. The low intrapleural pressure is significantly less than the pressure of the atmospheric air outside the body; i.e., it is subatmospheric. When the chest wall is pierced, atmospheric air rushes into the intrapleural space, the pressure difference across the lung wall is eliminated, and the stretched lung collapses. Air in the intrapleural space is known as a *pneumothorax*.

The subatmospheric pressure of the intrapleural fluid is normally maintained throughout life (only during a forced expiration does intrapleural pressure exceed atmospheric pressure). Regardless of whether the person is inspiring, expiring, or not breathing at all, the intrapleural pressure is always lower than the air pressure within the lungs, and the lungs are considerably stretched. However, the gradient between intrapleural and alveolar pressures does vary during breathing and directly causes the changes in lung size which occur during inspiration and expiration. The magnitude of the change in lung volume (ΔV) caused by a given change in the pressure difference across the lung wall (ΔP) is defined as the *lung compliance*. The higher this ratio ($\Delta V/\Delta P$), the more compliant (i.e., the more stretchable) is the lung. In the disease *emphysema*, the internal structures of the lung become more stretchable, i.e., their compliance increases; this results in a marked increase in resting lung size since the resting alveolar-intrapleural pressure difference distends the lung much more than usual. This phenomenon has important deleterious consequences for gas exchange, as will be described later.

Since intrapleural pressure is transmitted throughout the intrathoracic fluid surrounding not only the lungs but the heart and other intrathoracic structures as well, it is frequently termed the *intrathoracic pressure.* Recall that (Chap. 11) subatmospheric intrathoracic pressure was mentioned in describing the forces which determine end-diastolic ventricular volume.

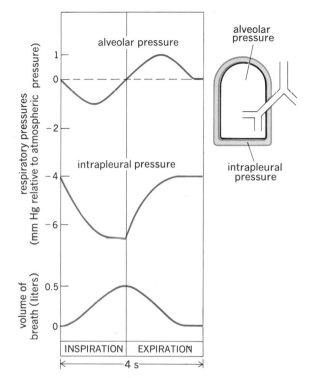

Figure 12-9. Summary of alveolar and intrapleural pressure changes and air flow during inspiration and expiration of 500 mL of air. Note that, on the pressure scale at the left, normal atmospheric pressure (760 mmHg) has a scale value of zero.

Inspiration

The left half of Fig. 12-9 summarizes events which occur during inspiration. Just before the inspiration begins, i.e., at the conclusion of the previous expiration, the respiratory muscles are relaxed and no air is flowing. The intrapleural pressure is subatmospheric (for reasons described above). The alveolar pressure, i.e., the air pressure within the alveoli, is exactly atmospheric because the alveoli are in free communication with the atmosphere via the airways. Inspiration is initiated by the contraction of the diaphragm and intercostal muscles (Fig. 12-10). When the diaphragm contracts, its dome moves downward into the abdomen, thus enlarging the volume of the thoracic cage. Simultaneously, the inspiratory intercostal muscles, which insert on the ribs, contract, leading to an upward and outward movement of the ribs and a further increase in thoracic cage size. As the

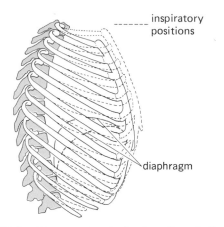

Figure 12-10. Movements of chest wall and diaphragm during breathing. The contracting intercostal muscles move the ribs upward and outward during inspiration while the contracting diaphragm moves downward. *(Adapted from McNaught and Callender.)*

thoracic cage begins to move away from the lung surface, the intrapleural fluid pressure abruptly decreases, i.e., becomes even more subatmospheric. This increases the difference between the alveolar and intrapleural pressures, and the lung wall is pushed out. Thus, when the inspiratory muscles increase the thoracic dimensions, the lungs are also forced to enlarge because of the changes in intrapleural pressure. This further stretching of the lung causes an increase in the volumes of all the air-containing passages and alveoli within the lung. As the alveoli enlarge, the air pressure within them drops to less than atmospheric, and this difference in pressure causes bulk flow of air from the atmosphere through the airways into the alveoli until their pressure again equals atmospheric.

Expiration

The expansion of thorax and lungs produced during inspiration by active muscular contraction stretches both lung and thoracic wall elastic tissue. When inspiratory-muscle contraction ceases and these muscles relax, the stretched tissues recoil to their original length since there is no force left to maintain the stretch. An obvious analogy is the snap of a stretched rubber band when it is released. The tissue recoil causes a rapid and complete reversal of the inspiratory process, as

shown in the right side of Fig. 12-9. The thorax and lungs spring back to their original sizes, alveolar air becomes temporarily compressed so that its pressure exceeds atmospheric, and air flows from the alveoli through the airways out into the atmosphere. Expiration at rest is thus completely passive, depending only upon the cessation of inspiratory-muscle activity and the relaxation of these muscles. Under certain conditions (during heavy exercise, for example) expiration is facilitated by the contraction of another group of intercostal muscles, which actively decrease thoracic dimensions. Contraction of the abdominal muscles also helps by increasing intraabdominal pressure and forcing the diaphragm into the thorax.

It should be noted that the analysis of Fig. 12-9 treats the lungs as a single alveolus. The fact is that there are significant regional differences in both alveolar and intrapleural pressures throughout the lungs and thoracic cavity. These differences are due, in part, to the effects of gravity and to local differences in the elasticity of the chest structures. They are of great importance in determining the pattern of ventilation.

Quantitative relationship between atmosphere-alveolar pressure gradients and air flow: Airway resistance

What is the quantitative relationship between the atmosphere-alveolar pressure gradients and the volume of air flow? It is expressed by precisely the same equation given for the circulatory system:

$$Flow = pressure\ gradient/resistance$$

The volume of air which flows in or out of the alveoli per unit time is directly proportional to the pressure gradient between the alveoli and atmosphere and inversely proportional to the *resistance* to flow offered by the airways. Normally the magnitude of this pressure gradient is increased by increasing the strength of contraction of the inspiratory muscles. This causes, in order, a more rapid expansion of the thoracic cage, a greater pressure gradient across the lung wall, increased expansion of the lungs, and increased flow of air into the lungs (Fig. 12-11).

What factors determine airway resistance? Resistance is (1) directly proportional to the magnitude of the interactions between the flowing gas molecules, (2) directly proportional to the length of the airway, and (3) inversely proportional to the fourth power of the airway radius. These factors, of course, are counterparts of the factors determining resistance in the circulatory system; and in the respiratory tree, just as in the circulatory tree, resistance is largely controlled by the radius of the airways.

The airway diameters are normally so large that they offer little total resistance to air flow, and interaction between gas molecules is also usually negligible, as is the contribution of airway length. Therefore, the total resistance remains so small that minute pressure gradients suffice to produce large volumes of air flow. As we have seen (Fig. 12-9), the average pressure gradient during a normal breath at rest is less than 1 mmHg; yet, approximately 500 mL of air is moved by this tiny gradient.

The diameter of the airways becomes critically important in certain conditions such as asthma, which is characterized both by severe airway smooth muscle constriction and by plugging of the airways by mucus. Airway resistance to air flow may become great enough to prevent air flow completely, regardless of the atmosphere-alveolar pressure gradient.

Airway size and resistance may be altered by physical, nervous, or chemical factors (Table 12-1). The most important normal physical factor is simply expansion of the lungs; during inspiration, airway resistance decreases because as the lungs expand, the airways within the lung are also widened. Conversely, during expiration, airway resistance increases. For this reason persons with abnormal airway resistance, as in asthma, have much less difficulty inhaling than exhaling, with the result that air may be trapped in the lungs and the lung volume progressively increased.

Nervous regulation of airway size is mediated by the autonomic nervous system, the sympathetic

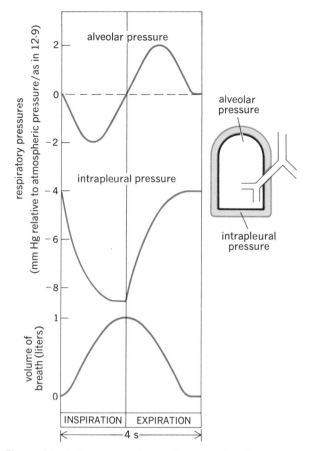

Figure 12-11. Summary of alveolar and intrapleural pressure changes and air flow during inspiration of 1000 mL of air. Compare these values with those given in Fig. 12-9.

neurons causing relaxation of the airway smooth muscle (decreased resistance) and the parasympathetics causing smooth muscle contraction (increased resistance). These reflexes are important in causing airway constriction upon inhalation of chemical irritants but their precise contribution to control of airway resistance under normal conditions is unclear.

As would be expected from knowledge of the effects of the sympathetic nerves on airway resistance, circulating epinephrine also causes airway dilation. This is a major reason for administering epinephrine or epinephrine-like drugs to patients suffering from airway constriction, as in an asthmatic attack. In contrast, histamine causes airway constriction (and increased mucus secretion as well) and may be the cause of the airway

Table 12-1. Major factors which control airway and pulmonary vascular resistance

	Airways	Arterioles
Constricted by:	Histamine	$\downarrow O_2$
	Parasympathetic nerves	$\uparrow H^+$
	$\downarrow CO_2$	
Dilated by:	Epinephrine	$\uparrow O_2$
	Sympathetic nerves	$\downarrow H^+$
	$\uparrow CO_2$	

constriction observed in allergic attacks. This explains the use of antihistamines to relieve the respiratory symptoms of allergies. Finally, the effects of the prostaglandins on pulmonary airways in asthma and other diseases may be of particular importance since the lungs take up, metabolize, and release various members of the prostaglandin family, some of which are airway constrictors, some dilators.

So far we have been discussing *total* airway resistance, but discrete *local* changes in airway resistance are important in promoting efficient gas exchange. The airway smooth muscle is highly responsive to carbon dioxide, high carbon dioxide producing bronchodilation and low carbon dioxide bronchoconstriction. These effects are exerted locally and are independent of nerves or hormones. What is the significance of this sensitivity? The lungs are composed of approximately 300 million discrete alveoli, each receiving carbon dioxide from the pulmonary capillary blood. To be most efficient, the right proportion of alveolar air and capillary blood should be available to each alveolus, a pattern local changes in bronchiolar tone help maintain. For example, if an alveolus is receiving too much air for its blood supply, the concentration of carbon dioxide in it and its surrounding tissue will be low; the airway supplying the alveolus is exposed to this low tissue carbon dioxide concentration and constricts; by this completely local mechanism, the ventilation and blood supply of the alveolus are matched (Fig. 12-12).

Pulmonary vessels and their control

The previous chapter described in detail how the systemic arterioles control the distribution of blood to the various organs and tissues. The pulmonary vessels perform an analogous function in controlling the distribution of blood to different alveolar capillaries, thus providing a second mechanism for matching air flow and blood flow. A decreased oxygen concentration causes pulmonary vessel constriction, whereas an increased oxygen concentration causes vasodilation; this may represent a direct effect of oxygen on pulmonary vascular smooth muscle or it may be mediated by local release of a chemical mediator such as prostaglandin. In addition, the pulmonary vessels are also sensitive to their local hydrogen-ion concentration; an increased hydrogen-ion concentration causes constriction whereas a decreased concentration causes vasodilation. (As will be emphasized subsequently, the hydrogen-ion concentration reflects, in large part, the partial pressure of carbon dioxide.) It should be noted that these purely local effects of oxygen and hydrogen ion on pulmonary vessels are precisely the opposite of those exerted by them on systemic arterioles (Chap. 11).

How does this provide matching of air and blood supplies? Recall the example above in which an alveolus has a large ventilation and small blood flow. The area around it will therefore have a high oxygen concentration and a low hydrogen-ion concentration. This causes vasodilation of the blood

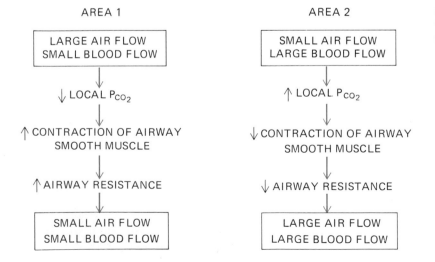

Figure 12-12. Local matching of ventilation and blood flow through effects of P_{CO_2} on airway resistance.

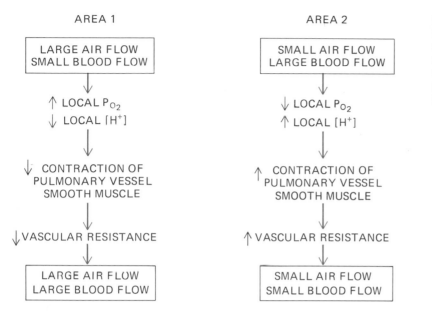

AREA 1

```
┌─────────────────────┐
│ LARGE AIR FLOW      │
│ SMALL BLOOD FLOW    │
└─────────────────────┘
           ↓
    ↑ LOCAL P_{O_2}
    ↓ LOCAL [H⁺]
           ↓
  ↓ CONTRACTION OF
  PULMONARY VESSEL
   SMOOTH MUSCLE
           ↓
 ↓ VASCULAR RESISTANCE
           ↓
┌─────────────────────┐
│ LARGE AIR FLOW      │
│ LARGE BLOOD FLOW    │
└─────────────────────┘
```

AREA 2

```
┌─────────────────────┐
│ SMALL AIR FLOW      │
│ LARGE BLOOD FLOW    │
└─────────────────────┘
           ↓
    ↓ LOCAL P_{O_2}
    ↑ LOCAL [H⁺]
           ↓
  ↑ CONTRACTION OF
  PULMONARY VESSEL
   SMOOTH MUSCLE
           ↓
 ↑ VASCULAR RESISTANCE
           ↓
┌─────────────────────┐
│ SMALL AIR FLOW      │
│ SMALL BLOOD FLOW    │
└─────────────────────┘
```

Figure 12-13. Local matching of ventilation and blood flow through effects of P_{O_2} and [H+] on pulmonary vascular resistance. Events of this figure and those of Fig. 12-12 occur simultaneously; i.e., local mechanisms control the distribution of both ventilation and blood flow.

vessel supplying it and an increased blood supply to match the high air supply (Fig. 12-13).

Before leaving the subject of the pulmonary circulation, we wish to reemphasize the fact, described in Chap. 11, that the pulmonary circulation is a low-resistance, low-pressure circuit. The normal pulmonary capillary pressure which is the major force favoring movement of fluid out of the pulmonary capillaries into the interstitium is only 15 mmHg. It is below the major force favoring absorption, namely, the plasma colloid osmotic pressure of 25 mmHg. Accordingly, the alveoli normally do not accumulate fluid, a feature essential for normal gas exchange. Should pulmonary capillary pressure increase above the colloid osmotic pressure, as occurs when the left ventricle "fails," fluid *(pulmonary edema)* accumulates in the interstitial space of the alveoli (and even in the alveoli themselves); this impairs gas exchange between alveolus and capillary.

Lung-volume changes during breathing

The volume of air entering or leaving the lungs during a single breath is called the *tidal volume* (Fig. 12-14). For a breath under resting conditions, this is approximately 500 mL. We are all aware that the resting thoracic excursion is small compared with a maximal breathing effort. The vol-

ume of air which can be inspired over and above the resting tidal volume is called the *inspiratory reserve* and amounts to 2500 to 3500 mL of air. At the end of a normal expiration, the lungs still contain a large volume of air, part of which can be exhaled by active contraction of the expiratory muscles; it is called the *expiratory reserve* and measures approximately 1000 mL of air. Even after a maximal expiration, some air (approximately 1000 mL) still remains in the lungs and is termed the *residual volume.*

What, then, is the maximum amount of air which can be moved in and out during a single breath? It is the sum of the normal tidal, inspiratory-reserve, and expiratory-reserve volumes. This total volume is called the *vital capacity.* During heavy work or exercise, a person uses part of both the inspiratory and expiratory reserves (particularly the former) but rarely uses more than 50 percent of his total vital capacity, because deeper breaths than this require exhausting activity of the inspiratory and expiratory muscles. The greater depth of breathing during exercise greatly increases pulmonary ventilation, and still larger increases are produced by increasing the rate of breathing as well.

The *total pulmonary ventilation* per minute is determined by the tidal volume times the respiratory rate (expressed as breaths per minute). For example, at rest, a normal person moves approximately 500 mL of air in and out of the lungs with

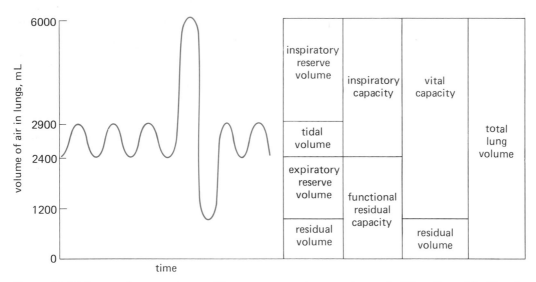

Figure 12-14. Lung volumes and capacities as measured on a spirograph. When the subject inspires, the pen moves up; with expiration, it moves down.

each breath, and takes 10 breaths each minute. The total pulmonary ventilation is therefore 500 mL × 10 = 5000 mL of air per minute. However, as we shall see, not all this air is available for exchange with the blood.

Air distribution within the lungs

Anatomic dead space. The respiratory tract, as we have seen, is composed of conducting airways and the alveoli. Within the lungs, exchanges of gases with the blood occur only in the alveoli and not in the conducting airways, the total volume of which is approximately 150 mL. Picture, then, what occurs during expiration: 500 mL of air is forced out of the alveoli and through the airways. Approximately 350 mL of this air is exhaled at the nose or mouth, but approximately 150 mL still remains in the airways at the end of expiration. During the next inspiration, 500 mL of air flows into the alveoli, but the first 150 mL of entering air is not atmospheric but the 150 mL of alveolar air left behind. Thus, only 350 mL of new atmospheric air enters the alveoli during the inspiration. At the end of inspiration, 150 mL of fresh air also fills the conducting airways, but no gas exchange with the blood can occur there. At the next expiration, this fresh air will be washed out and again replaced by old alveolar air, thus completing the cycle. The end result is that 150 mL of the atmospheric air enter-

ing the respiratory system during each inspiration never reaches the alveoli but is merely moved in and out of the airways. Because these airways do not permit gas exchange with the blood, the space within them is termed the *anatomic dead space.*

The volume of fresh atmospheric air entering the alveoli during each inspiration, then, equals the total tidal volume minus the volume of air in the anatomic dead space. Thus, for a normal breath

$$\begin{aligned}
\text{Tidal volume} &= 500 \text{ mL} \\
\text{Anatomic dead space} &= \underline{150 \text{ mL}} \\
\text{Fresh air entering alveoli} &= 350 \text{ mL}
\end{aligned}$$

This total is called the *alveolar ventilation.* The term is somewhat confusing, because it seems to indicate that only 350 mL air enters and leaves the alveoli with each breath. This is not true—the total is 500 mL of air, but only 350 mL is fresh air.

What is the significance of the anatomic dead space and alveolar ventilation? Total pulmonary ventilation is equal to the tidal volume of each breath multiplied by the number of breaths per minute. But since only that portion of inspired atmospheric air which enters the alveoli, i.e., the alveolar ventilation, is useful for gas exchange with the blood, the magnitude of the alveolar ventilation is of much greater significance than the magnitude of the total pulmonary ventilation, as can be demonstrated readily by the data in Table 12-2.

In this experiment subject A breathes rapidly and shallowly, B normally, and C slowly and deeply. Each subject has exactly the same total pulmonary ventilation, i.e., each is moving the same amount of air in and out of the lungs each minute. Yet, when we subtract the anatomic-dead-space ventilation from the total pulmonary ventilation, we find marked differences in alveolar ventilation. Subject A has no alveolar ventilation (and would become unconscious in several minutes), whereas C has a considerably greater alveolar ventilation than B, who is breathing normally. The important deduction to be drawn from this example is that increased depth of breathing is far more effective in elevating alveolar ventilation than an equivalent increase of breathing rate. Conversely, a decrease in depth can lead to a critical reduction of alveolar ventilation. This is because a fraction of *each* tidal volume represents anatomic-dead-space ventilation. If the tidal volume decreases, this fraction increases until, as in subject A, it may represent the entire tidal volume. On the other hand, any increase in tidal volume goes entirely toward increasing alveolar ventilation. Alveolar ventilation per minute is:

Alveolar ventilation (mL/min)
  = frequency (breaths/min)
    × [tidal volume − dead space] (mL/breath)

These concepts have important physiological implications. Most situations, such as exercise, which necessitate an increased oxygen supply (and carbon dioxide elimination) reflexly call forth a relatively greater increase in breathing depth than rate. Indeed, well-trained athletes can perform moderate exercise with very little increase, if any, in respiratory rate. The mechanisms by which rate and depth of respiration are controlled will be described in a later section of this chapter.

Alveolar dead space. Some inspired air is not useful for gas exchange with the blood even though it reaches the alveoli because some alveoli, for various reasons, receive too little blood supply for their size. Air which enters these alveoli during inspiration cannot exchange gases efficiently because there is insufficient blood. The inspired air in this *alveolar dead space* must be distinguished from that in the anatomic dead space. It is quite small in normal persons but may reach lethal proportions in several kinds of lung disease. As we have seen, it is minimized by the local mechanisms which match air and blood flows.

The work of breathing

During inspiration, active muscular contraction provides the energy required to expand the thorax and lungs. What determines how much work these muscles must perform in order to provide a given amount of ventilation? First, there is simply the stretchability of the thorax and lungs. To expand these structures they must be stretched. The easier they stretch, i.e., the more compliant the lung, the less energy is required for a given amount of expansion. Much of the work of breathing goes into stretching the elastic tissue of the lung, but an even larger fraction goes into stretching a different kind of "tissue"—water itself! The air within each alveolus is separated from the alveolar membranes by an extremely thin layer of fluid; in a sense, therefore, the alveoli may be viewed as air-filled bubbles lined with water. At an air-water interface, the attractive forces between water molecules cause them to squeeze in upon the air within the bubble (Fig. 12-15). This force, known as *surface tension*, makes the water lining very like highly stretched rubber which constantly tries to shorten and resists further stretching. Thus inspiration requires considerable energy to expand the lungs because of the difficulty of dis-

Table 12.2. Effect of breathing patterns on alveolar ventilation

Subject	Tidal volume, mL/breath	× Frequency, breaths/min	= Total pulmonary ventilation, mL/min	Dead space ventilation, mL/min	Alveolar ventilation, mL/min
A	150	40	6000	150 × 40 = 6000	0
B	500	12	6000	150 × 12 = 1800	4200
C	1000	6	6000	150 × 6 = 900	5100

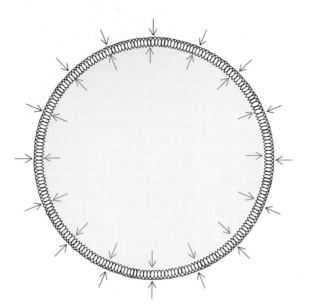

Figure 12-15. Forces acting on the surface of a bubble. The springs and dark arrows represent the surface tension resulting from the cohesive forces of water molecules at the air-water interface. This tension is opposed by the air pressure within the bubble (colored arrows).

tending these alveolar bubbles. Indeed, the surface tension of pure water is so great that lung expansion would require exhausting muscular effort and the lungs would tend to collapse. It is extremely important, therefore, that the Type II alveolar cells produce a phospholipoprotein complex, known as *pulmonary surfactant,* which intersperses with the water molecules on the alveolar surface and markedly reduces their cohesive force, thereby lowering the surface tension. A striking example of what occurs when insufficient surfactant is present is provided by the disease known as "respiratory-distress syndrome of the newborn," which frequently afflicts premature infants in whom the surfactant-synthesizing cells are too immature to function adequately. The infant is able to inspire only by the most strenuous efforts which may ultimately cause complete exhaustion, inability to breathe, lung collapse, and death. Normal maturation of the surfactant-synthesizing apparatus is facilitated by the hormone cortisol, the secretion of which is increased late in pregnancy. Accordingly, it has been found that administration of cortisol to the pregnant woman may provide an important means of combating this

disease in premature infants. Unfortunately, postnatal administration of cortisol to the infant is ineffective.

The second factor determining the degree of muscular work required for a certain amount of ventilation is the magnitude of the airway resistance. When airway resistance is increased by smooth muscle contraction or by secretions, the usual pressure gradient does not suffice for adequate air inflow and a deeper breath is required to create a larger pressure gradient.

One might imagine from this discussion (and from observing an athlete exercising hard) that the work of breathing uses up a major portion of the energy spent by the body. Not so; in a normal person, even during heavy exercise, the energy needed for breathing is only about 3 percent of the total expenditure. It is only in disease, when the work of breathing is markedly increased by structural changes in the lung or thorax, by loss of surfactant, or by an increased airway resistance, that breathing itself becomes an exhausting form of exercise.

Exchange and transport of gases in the body

We have completed our discussion of alveolar ventilation, but this is only the first step in the total respiratory process. Oxygen must move across the alveolar membranes into the pulmonary capillaries, be transported by the blood to the tissues, leave the tissue capillaries, and finally cross cell membranes to gain entry into cells. Carbon dioxide must follow a similar path in reverse (Fig. 12-16). At rest, during each minute, body cells consume approximately 200 mL of oxygen and produce approximately the same amount of carbon dioxide. The relative amounts depend primarily upon what nutrients are being used for energy; e.g., when glucose is utilized, one molecule of carbon dioxide is produced for every molecule of oxygen consumed:

$$C_6H_{12}O_6 + 6O_2 \longrightarrow 6CO_2 + 6H_2O + energy$$

The ratio $(CO_2$ produced$)/(O_2$ consumed$)$ is known as the *respiratory quotient* (RQ) and accordingly is 1 for glucose. When fat is utilized, only 7 molecules of carbon dioxide are produced for every 10 molecules of oxygen consumed, and RQ = 0.7. For simplicity, Fig. 12-16 assumes that the carbon

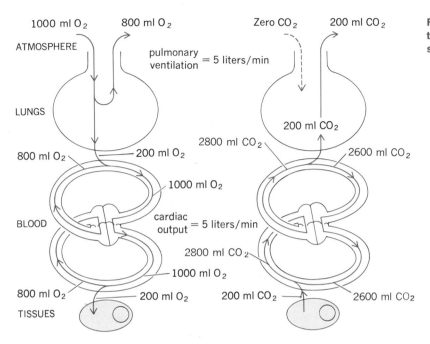

dioxide and oxygen amounts are equal and the total volumes of air inspired and expired therefore identical.

At rest, the total pulmonary ventilation equals 5 L of air per minute. Since only 20 percent of atmospheric air is oxygen (most of the remainder is nitrogen), the total oxygen input is $20\% \times 5$ L $= 1$ L of O_2 per minute. Of this inspired oxygen, 200 mL crosses the alveoli into the pulmonary capillaries, and the remaining 800 mL is exhaled. This 200 mL of oxygen is carried by 5 L of blood, which is the pulmonary blood flow (cardiac output) per minute. Note, however, that blood entering the lungs already contains large quantities of oxygen, to which this 200 mL is added. This blood is then pumped by the left ventricle through the tissue capillaries of the body, and 200 mL of oxygen leaves the blood to be taken up and utilized by cells. Because only a fraction of the total blood oxygen actually leaves the blood, some oxygen remains in the blood when it returns to the heart and lungs. It is obvious but important that the quantities of oxygen added to the blood in the lungs and removed in the tissues are identical. As shown by Fig. 12-16, the story reads in reverse for carbon dioxide. As we shall see, most of the blood carbon dioxide is actually in the form HCO_3^-, but we have shown it as CO_2 for simplicity.

The pumping of blood by the heart obviously propels oxygen and carbon dioxide between the lungs and tissues by bulk flow, but what forces induce the net movement of these molecules across the alveolar, capillary, and cell membranes? The answer is diffusion; there is no active membrane transport for oxygen or carbon dioxide. As described in Chap. 5, diffusion can effect the net transport of a substance only when a concentration gradient exists for it. Understanding the mechanisms involved depends upon familiarity with some basic chemical and physical properties of gases, to which we now turn.

Basic properties of gases

A gas consists of individual molecules constantly moving at great speeds. Since rapidly moving molecules bombard the walls of any vessel containing them, they therefore exert a *pressure* against the walls. The magnitude of the pressure is increased by anything which increases the bombardment. The pressure a gas exerts is proportional to (1) the temperature, because heat increases the speed at which molecules move, and (2) the concentration of the gas, i.e., the number of molecules per unit volume. In other words, when a certain number of molecules are compressed into a smaller volume, there are more collisions with the walls. The pressure of a gas is therefore a measure of the concentration and speed of its molecules.

Of great importance is the relationship between different gases, i.e., different kinds of molecules, such as oxygen and nitrogen, in the same container. In a mixture of gases, the pressure exerted by each gas is independent of the pressure exerted by the others because gas molecules are normally so far apart that they do not interfere with each other. Since each gas behaves as though the other gas were not present, the total pressure of a mixture of gases is simply the sum of the individual pressures. These individual pressures, termed *partial pressures*, are denoted by a P in front of the symbol for the gas, the partial pressure of oxygen thus being represented by Po_2. Thus, the partial pressure of a gas is a measure of the concentration of that gas in a mixture of gases. Net diffusion of a gas will occur from a region where its partial pressure (concentration) is high to a region where it is low. Gas partial pressures are usually expressed in millimeters of mercury, the same units used for the expression of hydrostatic pressure.

Behavior of gases in liquids

Several factors determine the uptake of gases by liquids and the behavior of gases dissolved in liquids. When a free gas comes into contact with a liquid, the number of gas molecules which dissolve in the liquid is directly proportional to the pressure of the gas. This phenomenon is clear from the basic definition of pressure. Suppose, for example, that oxygen is placed in a closed container half full of water. Oxygen molecules constantly bombard the surface of the water, some entering the water and dissolving. Since the number of molecules striking the surface is directly proportional to the pressure of the oxygen gas Po_2, the number of molecules entering the water is also directly proportional to Po_2. How many entering molecules actually stay in the water? Since the dissolved oxygen molecules are also constantly moving, some of them strike the water surface from below and escape into the free oxygen above. The rate of escape from the water and the rate of entry into the water are equal when the rates of bombardment are equal. At this point we say that

the partial pressure of the oxygen in the liquid is equal to its partial pressure in the gas phase. Thus, we come back to our earlier statement: The number of gas molecules which will dissolve in a liquid is directly proportional to the partial pressure of the gas. When the partial pressure in the gas phase is higher than the partial pressure in a liquid, there will be a net diffusion into the liquid. Conversely, if a liquid containing a dissolved gas at high partial pressure is exposed to that same free gas at lower partial pressure, gas molecules will diffuse from the liquid into the gas phase until the partial pressures in the two phases become equal. These are precisely the phenomena occurring between alveolar air and pulmonary capillary blood.

It should also be apparent that dissolved gas molecules diffuse *within* the liquid from a region of higher partial pressure to a region of lower partial pressure, an effect which underlies the exchange of gases between cells, tissue fluid, and capillary blood throughout the body.

This discussion has been in terms of proportionalities rather than absolute amounts. The number of gas molecules which will dissolve in a liquid is *proportional* to the partial pressure, but the *absolute* number also depends upon the *solubility* of the gas in the liquid. Thus, if a liquid is exposed to two different gases at the same partial pressures, the numbers of molecules of each gas which are dissolved at equilibrium are not identical but reflect the solubilities of the two gases. Nevertheless, doubling the partial pressures doubles the number of gas molecules dissolved.

Pressure gradients of oxygen and carbon dioxide within the body

With these basic gas properties as foundation, we can discuss the diffusion of oxygen and carbon dioxide across alveolar, capillary, and cell membranes. The pressures of these gases in atmospheric air and in various sites of the body are given in Fig. 12-17 for a resting person at sea level. The rest of this section is devoted to an elaboration of this figure.

Atmospheric air consists primarily of nitrogen and oxygen with very small quantities of water vapor, carbon dioxide, and inert gases such as argon. The sum of the partial pressures of all these gases is termed *atmospheric pressure* or *barometric pressure*. It varies in different parts of the world as a result of differences in altitude, but at sea level it is 760 mmHg. Since air is 20 percent oxygen, the P_{O_2} of inspired air is $20\% \times 760 = 152$ mmHg at sea level.

The first question suggested by Fig. 12-17 is why the partial pressures of the constituents of expired air are not identical to those of alveolar air. Recall that approximately 150 mL of the inspired atmospheric air during each breath never gets down into the alveoli but remains in the airways (dead space). This air does not exchange carbon dioxide or oxygen with blood and is expired along with alveolar air during the subsequent expiration. Therefore, the P_{O_2} and P_{CO_2} of the total expired air are higher and lower, respectively, than those of alveolar air.

The next question concerns the alveolar gas pressures themselves. One might logically reason that the alveolar gas pressures must vary considerably during the respiratory cycle, since new atmospheric air enters only during inspiration. In fact, however, the variations in alveolar P_{O_2} and P_{CO_2} during the cycle are so small as to be negligible because, as explained in the section on lung volumes, a large volume of gas is always left in the lungs after expiration. This remaining alveolar gas contains large quantities of oxygen and carbon dioxide, and when the relatively small volume of new air enters, it mixes with the alveolar air already present, lowering its P_{CO_2} and raising its P_{O_2}, but only by a small amount. For this reason, the alveolar-gas partial pressures remain *relatively* constant throughout the respiratory cycle, and we may use the single alveolar pressures shown in Fig. 12-17 in our subsequent analysis of alveolar-capillary exchange, ignoring the minor fluctuations.

Our next question concerns the exchange of gases between alveoli and pulmonary capillary blood. The blood which enters the pulmonary capillaries is, of course, systemic venous blood pumped to the lungs via the pulmonary arteries. Having come from the tissues, it has a high P_{CO_2} (46 mmHg) and a low P_{O_2} (40 mmHg). As it flows through the pulmonary capillaries, it is separated from the alveolar air only by an extremely thin layer of tissue. The differences in the partial pressures of oxygen and carbon dioxide on the two sides of this alveolar-capillary membrane result in the net diffusion of oxygen into the blood and of carbon dioxide into the alveoli. As this diffusion occurs, the capillary blood P_{O_2} rises above its original value

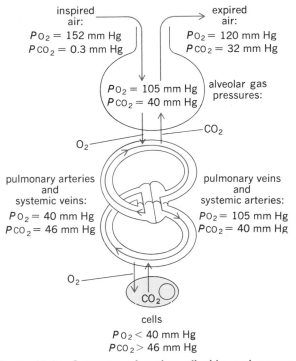

Figure 12-17. Summary of carbon dioxide and oxygen pressures in the inspired and expired air and various places within the body.

and the P_{CO_2} falls. The net diffusion of these gases ceases when the alveolar and capillary partial pressures become equal. In a normal person, the rates at which oxygen and carbon dioxide diffuse are so rapid and the blood flow through the capillaries so slow that complete equilibrium is usually reached (at rest, a red blood cell takes 0.75 s to pass through the pulmonary capillaries). Thus, the blood that leaves the lungs to return to the heart has essentially the same P_{O_2} and P_{CO_2} as alveolar air.

The diffusion of gases between alveoli and capillaries may be impaired in a number of ways, resulting in failure of gas equilibration between alveolar air and blood. The disease emphysema, which is intimately related to cigarette smoking, is characterized by the breakdown of the alveolar walls with the formation of fewer but larger alveoli. The result is a reduction in the total area available for diffusion. In a different kind of defect, caused by membrane thickening or by pulmonary edema, the area available for diffusion is normal but the molecules must travel a greater distance. Finally, without becoming thicker the alveolar walls may become denser and less permeable, as for example when beryllium is inhaled and deposited on the walls.

As the arterial blood enters capillaries throughout the body, it is separated from the interstitial fluid only by the thin, highly permeable capillary membrane. The interstitial fluid, in turn, is separated from intracellular fluid by cell membranes which are also quite permeable to oxygen and carbon dioxide. Metabolic reactions occurring within these cells are constantly consuming oxygen and producing carbon dioxide. Therefore, as shown in Fig. 12-17, intracellular P_{O_2} is lower and P_{CO_2} higher than in blood. As a result, a net diffusion of oxygen occurs from blood to cells, and a net diffusion of carbon dioxide from cells to blood. In this manner, as blood flows through capillaries, its P_{O_2} decreases and its P_{CO_2} increases. This accounts for the venous blood values shown in Fig. 12-17. Venous blood returns to the right ventricle and is pumped to the lungs, where the entire process begins again.

In summary, the consumption of oxygen in the cells and the supply of new oxygen to the alveoli create P_{O_2} gradients which produce net diffusion of oxygen from alveoli to blood in the lungs and from blood to cells in the rest of the body. Conversely, the production of carbon dioxide by cells and its elimination from the alveoli via expiration create P_{CO_2} gradients which produce net diffusion of carbon dioxide from blood to alveoli in the lungs and from cells to blood in the rest of the body.

Transport of oxygen in the blood: The role of hemoglobin

Table 12-3 summarizes the oxygen content of arterial blood. Each liter of arterial blood contains the same number of oxygen molecules as 200 mL of pure gaseous oxygen. Oxygen is present in two

Table 12-3. Oxygen content of arterial blood

1 L arterial blood contains:

	3 mL O_2 physically dissolved
	197 mL O_2 chemically bound to hemoglobin
Total	200 mL O_2

Cardiac output = 5 L/min
O_2 carried to tissues/min = 5 × 200 = 1000 mL

forms: (1) physically dissolved in the blood water and (2) reversibly combined with *hemoglobin* molecules. The amount of oxygen which can be dissolved in blood is directly proportional to the P_{O_2} of the blood, but because oxygen is relatively insoluble in water, only 3 mL of oxygen can be dissolved in 1 L of blood at the normal alveolar and arterial P_{O_2} of 100 mmHg. In contrast, in 1 L of blood, 197 mL of oxygen, more than 98 percent of the total, is carried within the red blood cells bound to hemoglobin.

Recall that the shape of the erythrocyte is a biconcave disk, i.e., thicker at the edge than in the middle, like a doughnut with a center depression on each side instead of a hole. Its shape and small dimensions have adaptive value in that oxygen and carbon dioxide can rapidly diffuse throughout the entire cell interior.

The outstanding physiological characteristic of erythrocytes is the presence of the iron-containing protein hemoglobin, constituting approximately one-third of the total cell weight. Hemoglobin is an oligomeric protein composed of four polypeptide chains, each of which contains a single atom of iron. It is with the iron that the oxygen combines; thus each hemoglobin molecule can combine with four molecules of oxygen. However, the equation for the reaction of oxygen and hemoglobin is usually written in terms of a single polypeptide chain:

$$O_2 + Hb \rightleftharpoons HbO_2 \qquad (12\text{-}1)$$

Hemoglobin combined with oxygen (HbO_2) is called *oxyhemoglobin;* not combined (Hb), it is called *reduced hemoglobin* or deoxyhemoglobin. Because the number of sites on the hemoglobin molecule which bind oxygen is limited, there is a maximum number of oxygen molecules which can be combined with the hemoglobin molecule. When all the iron atoms in hemoglobin are combined with O_2, the hemoglobin solution is said to be *fully saturated.* The *percentage saturation* of hemoglobin is a measure of the fraction of the hemoglobin molecules combined with oxygen.

What factors determine the extent to which oxygen will combine with hemoglobin? By far the most important is the P_{O_2} of the blood, i.e., the concentration of physically dissolved oxygen. The blood hydrogen-ion concentration and temperature also play significant roles as do certain chemicals produced by the red blood cells themselves.

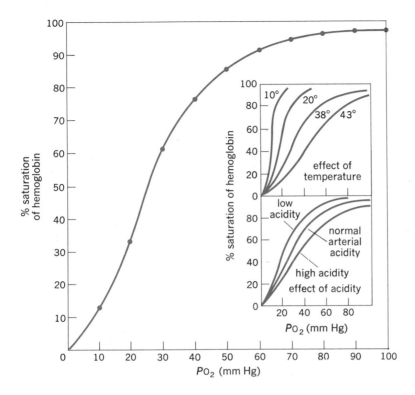

Figure 12-18. Hemoglobin-oxygen dissociation curve, achieved by equilibrating samples of whole blood with gaseous oxygen at 10 different values of P_{O_2}. The large curve applies to blood at 38°C and the normal arterial hydrogen-ion concentration (acidity). The inset curves illustrate the effects of altering temperature and acidity on the relationship between P_{O_2} and hemoglobin saturation with oxygen. (Adapted from Comroe.)

Effect of P_{O_2} on hemoglobin saturation. From inspection of Eq. 12-1 and the law of mass action, it is obvious that raising the P_{O_2} of the blood should increase the combination of oxygen with hemoglobin. The experimentally determined quantitative relationship between these variables is shown in Fig. 12-18. When a sample of blood is exposed to a large volume of oxygen, a net diffusion of oxygen occurs from the gas into the blood until the blood and gas partial pressures become equal. If the volume of gas is extremely large compared with the volume of blood, the final equilibrium P_{O_2} is very close to that of the gas. Using such a procedure, therefore, we can achieve any blood P_{O_2} we wish. By this means, we produce in 10 different samples of blood 10 different oxygen partial pressures ranging from 10 to 100 mmHg and analyze the effect of P_{O_2} on hemoglobin saturation by measuring the fraction of hemoglobin combined with oxygen in each case. Data from such an experiment are plotted in Fig. 12-18, which is called an *oxygen-hemoglobin dissociation*, (on *saturation) curve.* It is an S-shaped curve with a steep slope between 10 and 60 mmHg P_{O_2} and a flat portion between 70 and 100 mmHg P_{O_2}. In other words, the extent to which hemoglobin combines with oxygen increases very rapidly from 10 to 60 mmHg so that, at a P_{O_2} of 60 mmHg, 90 percent of the total hemoglobin is combined with oxygen. From this point on, a further increase in P_{O_2} produces only a small increase in oxygen uptake. The adaptive importance of this plateau at higher P_{O_2} values is very great for the following reason. Many situations (severe exercise, high altitudes, cardiac or pulmonary disease) are characterized by a moderate reduction of alveolar and arterial P_{O_2}. Even if the P_{O_2} fell from the normal value of 100 to 60 mmHg, the total quantity of oxygen carried by hemoglobin would decrease by only 10 percent, since hemoglobin saturation is still close to 90 percent at a P_{O_2} of 60 mmHg. The plateau therefore provides an excellent safety factor in the supply of oxygen to the tissues.

We now retrace our steps and reconsider the movement of oxygen across the various membranes, this time including hemoglobin in our analysis. It is essential to recognize that the oxygen which is bound to hemoglobin does not contribute directly to the P_{O_2} of the blood. Only gas molecules which are *free in solution,* i.e., dis-

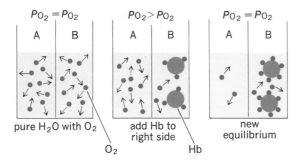

Figure 12-19. Effects of adding hemoglobin on the distribution of oxygen between two compartments separated by a semipermeable membrane. At the new equilibrium, the P_{O_2} values are equal to each other but lower than before the hemoglobin was added; however, the total oxygen, i.e., both dissolved and chemically combined with hemoglobin, is much higher on the right side of the membrane. *(Adapted from Comroe.)*

solved, do so. Therefore, the diffusion of oxygen is governed only by that portion which is dissolved, a fact which permitted us to ignore hemoglobin in discussing transmembrane pressure gradients. However, the presence of hemoglobin plays a critical role in determining the *total amount* of oxygen which will diffuse, as illustrated by a simple example (Fig. 12-19). Two solutions separated by a semipermeable membrane contain equal quanti-

ties of oxygen, the gas pressures are equal, and no net diffusion occurs. Addition of hemoglobin to compartment B destroys this equilibrium because much of the oxygen combines with hemoglobin. Despite the fact that the *total quantity of oxygen* in compartment B is still the same, the number of molecules *dissolved* has decreased; therefore, the P_{O_2} of compartment B is less than that of A, and net diffusion of oxygen occurs from A to B. At the new equilibrium, the oxygen pressures are once again equal, but almost all the oxygen is in compartment B and is combined with hemoglobin.

Let us now apply this analysis to the lung and tissue capillaries (Fig. 12-20). The plasma and erythrocytes entering the lungs have a P_{O_2} of 40 mmHg, and the hemoglobin is 75 percent saturated. Oxygen diffuses from the alveoli because of its higher P_{O_2} (100 mmHg) into the plasma; this increases plasma P_{O_2} and induces diffusion of oxygen into the erythrocytes, elevating erythrocyte P_{O_2} and causing increased combination of oxygen and hemoglobin. Thus, the vast preponderance of the oxygen diffusing into the blood from the alveoli does not remain dissolved but combines with hemoglobin. In this manner, the blood P_{O_2} remains less than that of the alveolar P_{O_2} until hemoglobin is virtually completely saturated, and the diffusion gradient favoring oxygen movement into the blood

Figure 12-20. Oxygen movements in the lungs and tissues. Movement of air into the alveoli is by bulk flow; all movements across membranes are by passive diffusion.

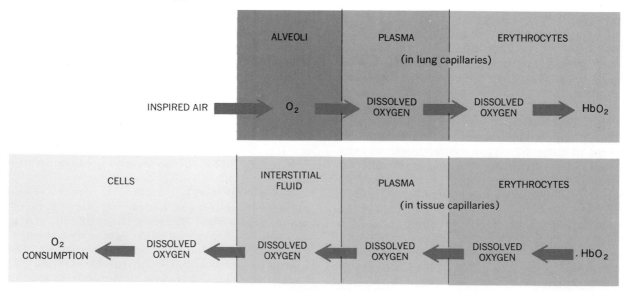

is maintained despite the very large transfer of oxygen. In the tissue capillaries, the procedure is reversed: As the blood enters the capillaries, plasma P_{O_2} is greater than interstitial fluid P_{O_2} and net oxygen diffusion occurs across the capillary membrane; plasma P_{O_2} is now lower than erythrocyte P_{O_2}, and oxygen diffuses out of the erythrocyte into the plasma. The lowering of erythrocyte P_{O_2} causes the dissociation of HbO_2, thereby liberating oxygen; simultaneously, the oxygen which diffused into the interstitial fluid is moving into cells along the concentration gradient generated by cell utilization of oxygen. The net result is a transfer of large quantities of oxygen from HbO_2 into cells purely by passive diffusion.

The fact that hemoglobin is still 75 percent saturated at the end of tissue capillaries under resting conditions underlies an important automatic mechanism by which cells can obtain more oxygen whenever they increase their activity. An exercising muscle consumes more oxygen, thereby lowering its intracellular P_{O_2}; this increases the overall blood-to-cell P_{O_2} gradient and the diffusion of oxygen out of the blood; in turn, the resulting reduction of erythrocyte P_{O_2} causes additional dissociation of hemoglobin and oxygen. An exercising muscle can thus extract virtually all the oxygen from its blood supply. This process is so effective because it takes place in the steep portion of the hemoglobin dissociation curve. Of course, an increased blood flow to the muscle also contributes greatly to the increased oxygen supply.

Effect of acidity on hemoglobin saturation. The large oxygen-hemoglobin dissociation curve illustrated in Fig. 12-18 is for blood having a specific hydrogen-ion concentration equal to that found in arterial blood. When the same experiments are performed at a different level of acidity, the curve changes significantly. A group of such curves is shown in the inset of Fig. 12-18 for different hydrogen-ion concentrations. It is evident that, regardless of the existing acidity, the percentage saturation of hemoglobin is still determined by the P_{O_2}. However, a change in acidity causes highly significant shifts in the entire curve. An increased hydrogen-ion concentration moves the curve downward and to the right, which means that, at any given P_{O_2}, hemoglobin has less affinity for oxygen when the acidity is high. For reasons to be described later, the hydrogen-ion concentration in the tissue capillaries is greater than in arterial blood. Blood flowing through tissue capillaries becomes exposed to this elevated hydrogen-ion concentration and therefore loses even more oxygen than if the decreased P_{O_2} had been the only factor involved. Conversely, the hydrogen-ion concentration is lower in the lung capillaries than in the systemic venous blood, so that hemoglobin picks up more oxygen in the lungs than if only the P_{O_2} were involved. Finally, the more active a tissue is, the greater its hydrogen-ion concentration; accordingly, hemoglobin releases even more oxygen during passage through these tissue capillaries, thereby providing the more active cells with additional oxygen. The hydrogen ion exerts this effect on the affinity of hemoglobin for oxygen by combining with the hemoglobin and altering its molecular structure. The importance of this combination for the regulation of extracellular acidity is described in Chap. 13.

Effect of temperature on hemoglobin saturation. The effect of temperature on the oxygen-hemoglobin dissociation curve (inset of Fig. 12-18) resembles that of an increase in acidity. The implication is similar: Actively metabolizing tissue, e.g., exercising muscle, has an elevated temperature, which facilitates the release of oxygen from hemoglobin as blood flows through the muscle capillaries.

Effect of DPG on hemoglobin saturation. Red cells contain large quantities of the substance 2,3-diphosphoglycerate (DPG) which is present in only trace amounts in other mammalian cells. DPG, which is produced by the red cells during glycolysis, binds reversibly with hemoglobin, causing it to change its conformation and release oxygen. Therefore, the effect of increased DPG is to shift the curve downward and to the right (just as does an increased temperature or hydrogen-ion concentration). The net result is that whenever DPG is increased there is enhanced unloading of oxygen as blood flows through the tissues. Such an increase is triggered by a variety of conditions associated with decreased oxygen supply to the tissues and helps to maintain oxygen delivery.

In summary, we have seen that oxygen is transported in the blood primarily in combination with hemoglobin. The extent to which hemoglobin binds oxygen is dependent upon the P_{O_2}, hydrogen-ion concentration, temperature, and DPG. These factors cause the release of large quantities of oxygen from hemoglobin during blood flow through tissue capillaries and virtually complete conversion of reduced hemoglobin to oxyhemoglobin during blood flow through lung capillaries. An active tissue increases its extraction of oxygen from the blood because of its lower P_{O_2} and higher hydrogen-ion concentration and temperature.

Transport of carbon dioxide in the blood

The quantity of carbon dioxide which is dissolved in blood at physiological carbon dioxide partial pressures is quite small, certainly much smaller than the large volume of carbon dioxide which must be constantly transported from the tissues to the lungs.

Carbon dioxide can undergo the reaction

$$CO_2 + H_2O \rightleftharpoons H_2CO_3$$
Carbonic acid

which goes quite slowly unless it is catalyzed by the enzyme *carbonic anhydrase*. The quantities of both dissolved carbon dioxide and carbonic acid are directly proportional to the P_{CO_2} of the solution. The actual amount of carbonic acid in blood is small because most of the carbonic acid ionizes

according to the equation

$$H_2CO_3 \rightleftharpoons HCO_3^- + H^+$$
Bicarbonate

This reaction proceeds very rapidly and requires no enzyme. Combining these two equations, we find

$$CO_2 + H_2O \xrightleftharpoons{\text{carbonic anhydrase}}$$
$$H_2CO_3 \rightleftharpoons HCO_3^- + H^+ \quad (12\text{-}2)$$

Thus, the addition of carbon dioxide to a liquid results ultimately in bicarbonate and hydrogen ions.

Carbon dioxide can also react directly with proteins, particularly hemoglobin, to form *carbamino* compounds.

$$CO_2 + Hb \rightleftharpoons HbCO_2 \quad (12\text{-}3)$$

When arterial blood flows through tissue capillaries, oxyhemoglobin gives up oxygen to the tissues and carbon dioxide diffuses from the tissues into the blood, where the following processes occur (Fig. 12-21):

1 A small fraction (8 percent) of the carbon dioxide remains physically dissolved in the plasma and red blood cells.

2 The largest fraction (81 percent) of the carbon dioxide undergoes the reactions described in

Figure 12-21. Summary of carbon dioxide movements and reactions as blood flows through tissue capillaries. All movements across membranes are by passive diffusion. Note that most of the CO_2 ultimately is converted to HCO_3^-; this occurs almost entirely in the erythrocytes (because the carbonic anhydrase is located there), but most of the HCO_3^- then diffuses out of the erythrocytes into the plasma.

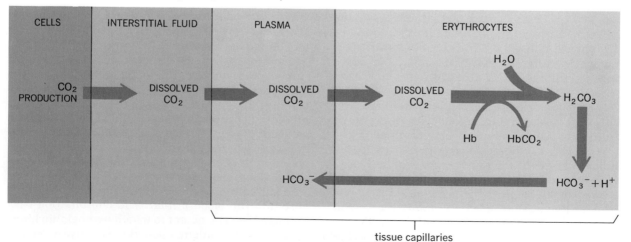

tissue capillaries

Eq. 12-2 and is converted into bicarbonate and hydrogen ions. This occurs primarily in the red blood cells because they contain large quantities of the enzyme carbonic anhydrase but the plasma does not. However, most of the bicarbonate diffuses out of the red cells and is carried in the plasma; bicarbonate, in contrast to carbon dioxide, is extremely soluble. The reaction described in Eq. 12-2 explains why tissue capillary hydrogen-ion concentration is higher than that of the arterial blood and increases as metabolic activity increases. The fate of these hydrogen ions will be discussed in the next section.

3 The remaining fraction (11 percent) of the carbon dioxide reacts directly with hemoglobin to form $HbCO_2$, as in Eq. 12-3.

Since these are all reversible reactions, i.e., they can proceed in either direction, depending upon the prevailing conditions, why do they all proceed primarily to the right, toward generation of HCO_3^- and $HbCO_2$, as blood flows through the tissues? Once again, the answer is provided by the law of mass action: It is the increase in carbon dioxide concentration which drives these reactions to the right as blood flows through the tissues.

Obviously, a sudden lowering of blood P_{CO_2} has just the opposite effect. HCO_3^- and H^+ combine to give H_2CO_3, which generates carbon dioxide and water. Similarly, $HbCO_2$ generates hemoglobin and free carbon dioxide. This is precisely what happens as venous blood flows through the lung capillaries (Fig. 12-22). Because the blood P_{CO_2} is higher than

alveolar, a net diffusion of carbon dioxide from blood into alveoli occurs. This loss of carbon dioxide from the blood lowers the blood P_{CO_2} and drives these chemical reactions to the left, thus generating more dissolved carbon dioxide. Normally, as fast as this carbon dioxide is generated from HCO_3^- and H^+ and from $HbCO_2$, it diffuses into the alveoli. In this manner, all the carbon dioxide delivered into the blood in the tissues now is delivered into the alveoli, from which it is expired and eliminated from the body.

Transport of hydrogen ions between tissues and lungs

In the previous section, we pointed out [Eq. (12-2)] that hydrogen ions and bicarbonate ions are generated by the dissociation of H_2CO_3 after its formation by the hydration of CO_2. As blood flows through the lungs, the reaction is reversed, and H^+ and HCO_3^- recombine. However, most of these hydrogen ions do not remain free in the blood during transit from tissues to lungs (if they did, a toxic level of free hydrogen ion would result), but rather become bound to plasma proteins and hemoglobin. This latter protein is quantitatively far more important for the following reason: Reduced hemoglobin has a much greater affinity for hydrogen ion than oxyhemoglobin does. As blood flows through the tissues, a fraction of oxyhemoglobin loses its oxygen and is transformed into reduced hemoglobin. Simultaneously, a large quantity of carbon dioxide enters the blood and undergoes (primarily

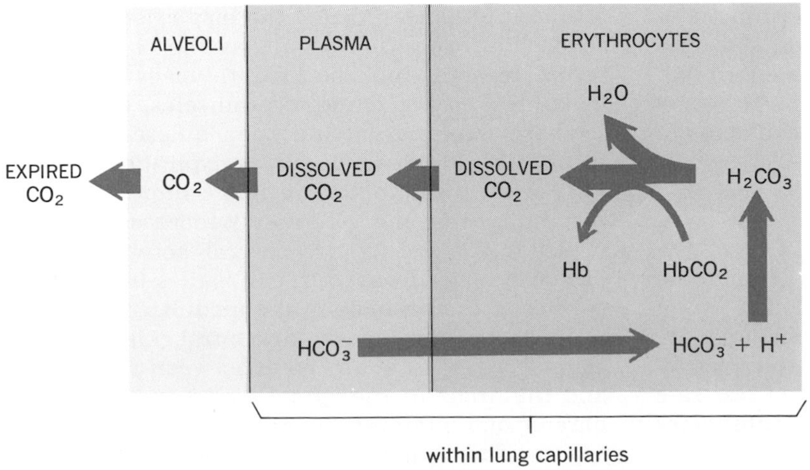

Figure 12-22. Summary of carbon dioxide movements and reactions as blood flows through the lung capillaries. All movements across membranes are by passive diffusion. The plasma-erythrocyte phenomena are simply the reverse of those occurring during blood flow through the tissue capillaries, as shown in Fig. 12-21. The breakdown on H_2CO_3 and CO_2 is catalyzed by carbonic anhydrase.

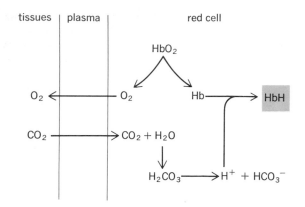

Figure 12-23. Buffering of hydrogen ions by hemoglobin as blood flows through tissue capillaries.

in the red blood cells) the reactions which ultimately generate HCO_3^- and H^+. Because reduced hemoglobin has a strong affinity for hydrogen ion, most of these hydrogen ions become bound to hemoglobin (Fig. 12-23). In this manner only a small number of hydrogen ions remain free, and the acidity of venous blood is only slightly greater than that of arterial blood. As the venous blood passes through the lungs, all these reactions are reversed. Hemoglobin becomes saturated with oxygen, and its ability to bind hydrogen ions decreases. The hydrogen ions are released whereupon they react with HCO_3^- to give CO_2, which diffuses into the alveoli and is expired.

In a previous section, we described how the hydrogen-ion concentration of the blood is an important determinant of the ability of hemoglobin to bind oxygen. Now we have shown how the presence of oxygen is an important determinant of hemoglobin's ability to bind hydrogen ion. The adaptive value of these phenomena is enormous. Their combination in one molecule marks hemoglobin as a remarkable evolutionary development.

Control of respiration

In dealing with the mechanisms by which the basic respiratory processes are controlled, we shall be concerned primarily with two questions: By what mechanisms are rhythmical breathing movements generated? What factors control the rate and depth of breathing, i.e., the total ventilatory volume?

Neural generation of rhythmic breathing

Like cardiac muscles, the inspiratory muscles normally contract rhythmically; however, the origins of these contractions are quite different. Cardiac muscle has automaticity; i.e., it is capable of self-excitation; the nerves to the heart merely alter this basic inherent rate and are not actually required for cardiac contraction. On the other hand, the diaphragm and intercostal muscles consist of skeletal muscle, which cannot contract unless stimulated by nerves. Thus, breathing depends entirely upon cyclical respiratory muscle excitation by the phrenic nerves (to the diaphragm) and the intercostal nerves (to the intercostal muscles). These nerves originate in the spinal cord at the levels of the neck and thorax. Destruction of them or the spinal cord areas from which they originate (as in poliomyelitis, for example) results in paralysis of the respiratory muscles and death, unless some form of artificial respiration can be rapidly instituted.

At the end of expiration, when the chest is at rest, a few impulses are still passing down these nerves. Like other skeletal muscles, therefore, the respiratory muscles have a certain degree of resting tonus. This muscular contraction is too slight to move the chest but plays a role in maintaining normal posture. Inspiration is initiated by an increased rate of firing of these inspiratory motor units. As more and more new motor units are recruited thoracic expansion increases. In addition, the firing frequency of the individual units increases. By these two measures, the force of inspiration increases as it proceeds. Then almost all these units stop firing, the inspiratory muscles relax, and expiration occurs as the elastic lungs recoil. In addition, when expiration is facilitated by contraction of expiratory muscles, the nerves to these muscles, having been quiescent during inspiration, begin firing during expiration.

By what mechanism are nerve impulses to the respiratory muscles alternately increased and decreased? Control of this neural activity resides primarily in neurons with cell bodies in the lower portion of the brainstem, the medulla, which also contains the cardiovascular control centers. If the spinal cord is cut at any point between the medulla and the areas of the spinal cord from which the phrenic and intercostal nerves originate, breathing ceases. This experiment demonstrates that these efferent nerves are controlled by synaptic connections with neurons which descend, in the spinal cord, from the medulla.

By means of tiny electrodes placed in various parts of the medulla to record electric activity, neurons have been found which discharge in synchrony with inspiration, and cease discharging during expiration. These neurons are called *inspiratory neurons*. They provide, via either direct or interneuronal connections, the rhythmic input to the motor neurons innervating the inspiratory muscles.

What are the factors which determine the discharge pattern of the medullary inspiratory neurons? There are really two distinct components of this problem: (1) What factors are essential for the *generation* of rhythmical bursts of firing? (2) What factors modify the timing of the bursts and their intensity? This section deals only with the first question, the subsequent section with the second.

It is quite likely, although not definitely proven, that the medullary inspiratory neurons have inherent automaticity and rhythmicity, i.e., the capacity for cyclical self-excitation. This alone could account for alternating cycles of inspiration and expiration. However, synaptic inputs to them from other neurons also play important roles. There are at least three sources of such input: (1) other neurons in the medulla; (2) neurons with cell bodies in the pons, the area of brainstem just above the medulla; (3) stretch receptors in the lung airways.

(1) The medullary inspiratory neurons give off, in addition to their descending fibers, processes which synapse upon and stimulate adjacent medullary neurons. These neurons, in turn, send processes back to the inspiratory neurons and inhibit the latter. Thus the firing of the medullary inspiratory neurons generates, via this loop, an inhibitory input to itself, thereby helping to terminate its own firing.

(2) The medullary inspiratory neurons receive a rich synaptic input, either directly or via the inhibitory neurons adjacent to them, from neurons in various areas in the pons. The precise contribution of these areas to the generation of the normal respiratory rhythm is still the subject of debate.

(3) Pulmonary stretch receptors. These receptors lie within the airway smooth muscle and are activated by lung inflation. The afferent nerve fibers from them travel in the vagus nerves to the medulla and inhibit the inspiratory neurons either directly or by way of the adjacent inhibitory neurons. Thus feedback from the lungs themselves helps to terminate inspiration. However, the threshold of these receptors in human beings is very high, as evidenced by the fact that blockade of the vagus nerves produces no effect on the ventilatory pattern unless the tidal volume is 1.5 to 2 times normal. Accordingly, this pulmonary stretch receptor reflex plays a role in setting respiratory rhythm only under conditions of very large tidal volumes, as in exercise.

It must be emphasized that cessation of inspiratory neuronal activity is all that is required for expiration to occur, since expiration is normally a passive process dependent only upon relaxation of the inspiratory muscles. In contrast, during active expiration, cessation of firing of the medullary inspiratory neurons is accompanied by the activation of other descending pathways which synapse upon and stimulate the motor neurons to those muscles whose contractions produce active expiration. Just which medullary neurons and interactions are responsible for controlling active expiration is not known.

Control of ventilatory volume

In the previous section, we were concerned with the mechanisms that generate rhythmical breathing. It is obvious that the actual respiratory rate is not fixed but can be altered over a wide range. Similarly, the depth and force of breathing movements can also be altered. As we have seen, these two factors, rate and depth, determine the alveolar ventilatory volume. Generally, rate and depth change in the same direction, although there may be important quantitative differences. For simplicity, we shall describe the control of total ventilation without attempting to discuss whether rate or depth makes the greatest contribution to the change.

Depth of respiration depends upon the number of motor units firing and their frequency of discharge, whereas respiratory rate depends upon the length of time elapsing between the bursts of motor unit activity (Fig. 12-24). As described above, the respiratory motor units are directly controlled by descending pathways from the medullary respiratory neurons. The efferent pathways for control of ventilation are therefore clear-cut,

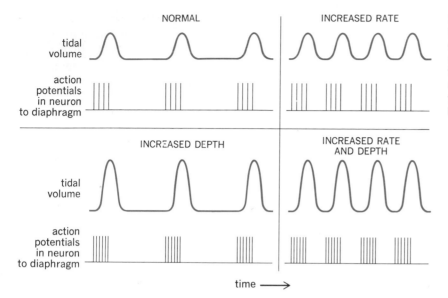

Figure 12-24. Recordings of depths of respiration (tidal volume) and frequency of action potentials in motor neurons to the diaphragm. Note that the rate of discharge during the inspiratory "burst" determines the tidal volume, whereas the time between bursts determines the respiratory rate.

and the critical question becomes: What is the nature of the afferent input to these centers? In other words, what variables does the control of ventilatory volume regulate?

This may seem a ridiculously complex way of phrasing a question when the answer seems so intuitively obvious. After all, respiration ought to supply oxygen as fast as it is consumed and ought to excrete carbon dioxide as fast as it is produced. *But,* how do the respiratory neurons "know" what the body's oxygen requirements are? The logical way to approach this question is to ask what detectable changes would result from imbalance of metabolism and ventilation. Certainly the most obvious candidates are the blood P_{O_2} and P_{CO_2}. Inadequate ventilation would lower the P_{O_2} because consumption would get ahead of supply and would elevate the P_{CO_2} because production would exceed elimination. Less obvious, perhaps, is the fact that arterial hydrogen-ion concentration also is exquisitely sensitive to changes in ventilation. Recall the equilibrium between H^+, HCO_3^-, and CO_2:

$$CO_2 + H_2O \rightleftharpoons H_2CO_3 \rightleftharpoons HCO_3^- + H^+$$

As described by the law of mass action any increase in carbon dioxide concentration drives this reaction to the right, thereby liberating additional hydrogen ions. *Any increase or decrease in blood P_{CO_2} is accompanied by changes in blood hydrogen-ion concentration.*

However, these statements of fact in no way prove that any of these blood concentrations actually are involved in respiratory regulation. The next step is to ask whether there are, indeed, receptors which can detect the levels of these variables in plasma and transmit the information to the medullary respiratory neurons. If so, where are these receptors located and what contribution do they make to overall control of ventilation? We shall describe the answers to these questions first for oxygen and then for carbon dioxide and hydrogen ion.

Control of ventilation by oxygen. The rationale of the experiments to be described is quite simple: Alteration of inspired P_{O_2} produces changes in blood P_{O_2}; if blood P_{O_2} is important in controlling ventilation, we should observe definite changes in ventilation. If a normal person takes a single breath of 100 percent oxygen (without being aware of it), after a latent period of about 8 s, a transient reduction in ventilation occurs of approximately 10 to 20 percent. Such studies have demonstrated that the normal sea-level blood P_{O_2} of 100 mmHg exerts a tonic stimulatory effect adequate to account for approximately 20 percent of total ventilation. Thus, the increase in blood P_{O_2} produced by the 100 percent oxygen removed this tonic stimulation and reduced ventilation. Conversely, a reduction in blood P_{O_2} below normal stimulates ventilation.

The receptors stimulated by low P_{O_2} are located

at the bifurcation of the common carotid arteries and in the arch of the aorta, quite close to, but distinct from, the baroreceptors described in Chap. 11 (Fig. 12-25). Known as the *carotid and aortic bodies*, they are composed of epithelial-like cells and neuron terminals in intimate contact with the arterial blood; of the two groups, the carotid bodies are far more important. Afferent nerve fibers arising from these terminals pass to the medulla where they synapse ultimately with the respiratory neurons. A low P_{O_2} increases the rate at which the receptors discharge, resulting in an increased number of action potentials traveling up the afferent nerve fibers and a stimulation of the medullary inspiratory neurons. If an animal is exposed to a low P_{O_2} after the afferent nerves from the carotid and aortic bodies have been cut, the usual rapid increase in alveolar ventilation is not observed, demonstrating that the immediate stimulatory effects of a low P_{O_2} are completely mediated via these pathways. (Indeed, in the absence of the carotid bodies, low P_{O_2} actually reduces ventilation through a direct depressive effect on the medullary respiratory neurons.)

What is the precise stimulus to these chemoreceptors? It is most likely their own internal P_{O_2}, that is, the concentration of dissolved oxygen within them. Because their blood supply is extremely large, relative to their size, the chemoreceptors receive so much oxygen that their oxygen requirements are met without an appreciable fall in P_{O_2}; i.e., their internal P_{O_2} is virtually identical to the arterial P_{O_2}. Accordingly, any time arterial P_{O_2} is reduced by lung disease, hypoventilation, or high altitude, the internal chemoreceptor P_{O_2} will change; similarly, the chemoreceptors will be stimulated and initiate a compensatory increase in ventilation. Because the chemoreceptors respond to P_{O_2}, not to total blood oxygen content, they will not be stimulated in situations in which there are modest reductions in total oxygen content (e.g., moderate anemia) but no change in arterial P_{O_2}. However, the close correspondence between arterial P_{O_2} and internal-chemoreceptor P_{O_2} is eliminated whenever there is a very marked reduction in the total amount of oxygen reaching the chemoreceptors, as in severe anemia or severe hypotension (which results in diminished blood supply to the receptors). In these situations, arterial P_{O_2} is still normal but chemoreceptor P_{O_2} falls because the total supply of

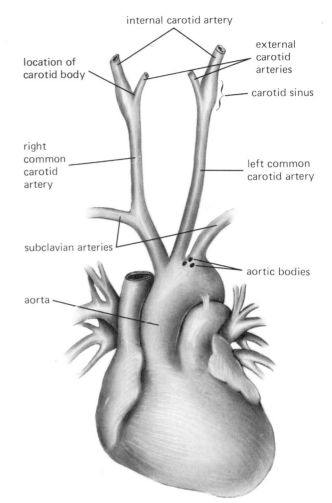

Figure 12-25. Location of the carotid and aortic bodies. Both right and left common carotid bifurcations contain a carotid sinus and a carotid body.

oxygen to its cells is not adequate to maintain chemoreceptor P_{O_2} at the original arterial value. Accordingly, the chemoreceptors are stimulated and elicit increased ventilation; however, it must be reemphasized that even in these cases, the receptors are responding directly to their own reduction in internal P_{O_2}, not to the reduction in total oxygen content of the blood. This same analysis holds true when total oxygen delivery is reduced by carbon monoxide, a gas which reacts with the same iron-binding sites on the hemoglobin molecule as does oxygen, and has such a high affinity for these sites that even small amounts reduce the amount of oxygen combined with hemoglobin.

Since carbon monoxide does not affect the amount of oxygen which can dissolve in blood, the arterial P_{O_2} is unaltered and no increase in peripheral chemoreceptor output occurs (until the carbon monoxide poisoning becomes very severe).

Control of ventilation by carbon dioxide. The important role that CO_2 plays in the control of ventilation can be demonstrated by a few relatively simple experiments:

1 Figure 12-26 illustrates the effects on respiratory volume of increasing the P_{CO_2} of inspired air. Normally, atmospheric air contains virtually no carbon dioxide. In the experiment illustrated, the subject breathed from bags of air containing variable quantities of carbon dioxide. The presence of carbon dioxide in the inspired air caused an elevation of alveolar P_{CO_2} and thereby an elevation of arterial P_{CO_2} as well. This increased P_{CO_2} markedly stimulated ventilation, an increase of 5 mmHg in alveolar P_{CO_2}, causing a 100 percent increase in ventilation.
2 A subject is asked to breathe as rapidly and deeply as he can. When the period of voluntary hyperventilation is over, the subject is told

Figure 12-26. Effects on respiration of increasing alveolar P_{CO_2} by adding carbon dioxide to inspired air.

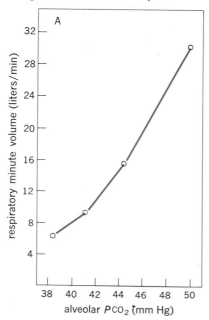

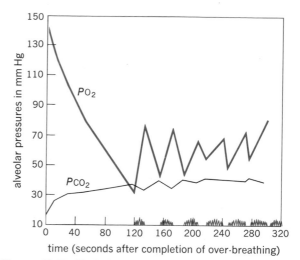

Figure 12-27. Variations in alveolar gas pressures after voluntary overbreathing for 2 min. The actual breathing movements are shown by the jagged lines at the bottom of the figure. *(Adapted from Douglas and Haldane.)*

merely to breathe naturally. All subjects manifest markedly reduced breathing for the next few minutes, and a few stop breathing completely (apnea), often for 1 to 2 min (Fig. 12-27). Ventilation is reduced because during the period of hyperventilation carbon dioxide was blown off faster than it was produced; accordingly, at the moment when the subject ceased to hyperventilate, his blood P_{CO_2} was lower than normal, and ventilation was reduced until blood P_{CO_2} returned toward normal as a result of accumulation of metabolically produced carbon dioxide. Note in Fig. 10-27 that the subject began breathing again, at least intermittently, before the P_{CO_2} was completely back to normal; this is due to stimulation of the carotid and aortic bodies by the extremely low P_{O_2}.

3 That carbon dioxide is a more potent controller of ventilation than is oxygen can be shown by an experiment in which healthy subjects breathed low P_{O_2} gas mixtures for 8 min (Fig. 12-28). Note that despite the fact that reduced P_{O_2} is known to increase stimulatory input from the peripheral chemoreceptors to the medullary inspiratory neurons, no significant increase in respiration was observed until the oxygen content of the inspired air was reduced by 40 percent (the arterial P_{O_2} at this point was 40 mmHg!). Only when the oxygen lack was greater

than this did all subjects manifest a sustained large increase in ventilation. What apparently happened to the subjects was that, although even mild reductions of P_{O_2} actually raised the ventilation at first (not shown in the figure), this increase lowered their blood P_{CO_2} very slightly, which reduced ventilation and thereby counteracted the stimulatory effects of the reduced P_{O_2} (Fig. 12-29). When this type of experiment is repeated with the P_{CO_2} held constant, all subjects manifest a considerably greater increase in ventilation when exposed to low P_{O_2} gas mixtures. Thus, people are indeed sensitive to a reduction of blood P_{O_2} but the respiratory stimulation may be overridden by other simultaneously occurring changes to which they are even more sensitive, particularly changes in P_{CO_2}.

4 Why can a person hold her breath for only a relatively short time? The lack of ventilation causes an accumulation of carbon dioxide and increased blood P_{CO_2}; the ability of this increased P_{CO_2} to stimulate the respiratory neurons is so powerful that it overcomes the voluntary inhibition of respiration. Unfortunately, underwater swimmers have misguidedly made use of these facts. They voluntarily hyperventilate

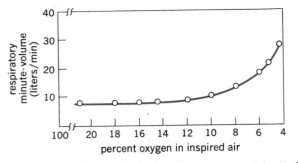

Figure 12-28. Effects of altering the oxygen content of inspired air on ventilation in normal man. The points are the average values for all subjects. *(Adapted from Comroe.)*

for several minutes before submerging and are therefore able to hold their breath for long periods of time. This is a very dangerous procedure, particularly during exercise, when oxygen consumption is high; the low P_{O_2} may cause unconsciousness and drowning.

The experiments we have discussed strongly support the hypothesis that P_{CO_2} is the major determinant of respiratory neuron activity. An increase in P_{CO_2} stimulates ventilation and thereby promotes the excretion of additional carbon

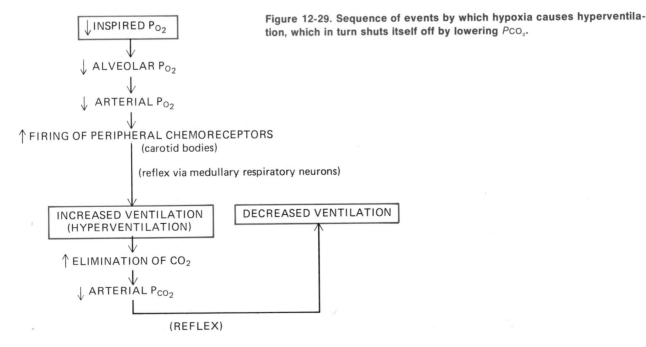

Figure 12-29. Sequence of events by which hypoxia causes hyperventilation, which in turn shuts itself off by lowering P_{CO_2}.

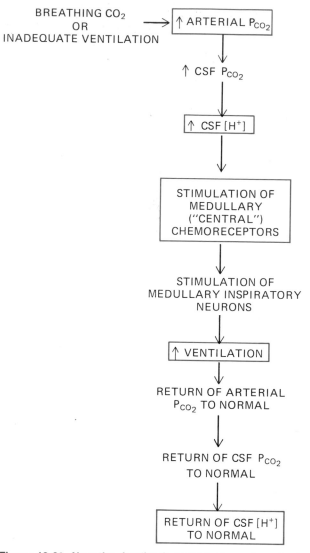

Figure 12-30. Negative-feedback control of cerebrospinal fluid [H+] via regulation of ventilation.

dioxide. Conversely, a decrease in P_{CO_2} below normal removes some of the stimulus for ventilation, thereby reducing ventilation and allowing metabolically produced carbon dioxide to accumulate and return the P_{CO_2} to normal. In this manner, the arterial P_{CO_2} is stabilized at the normal value of 40 mmHg.

What is the mechanism and afferent pathway by which the P_{CO_2} controls ventilation? The evidence, to date, indicates that the effects of carbon dioxide on ventilation are due not to carbon dioxide

itself but to the associated changes in hydrogen-ion concentration. For example, the stimulant effects of breathing mixtures containing large amounts of carbon dioxide are probably due not to the effects of molecular carbon dioxide but to those of the increased hydrogen-ion concentration resulting from the chemical reactions described above. This is why we stressed the relationship between P_{CO_2} and hydrogen-ion concentration.

There are two groups of H^+ receptors which participate in this reflex response. One is the same peripheral chemoreceptors — the carotid bodies — which initiated the hyperventilatory response to low P_{O_2}. Thus, these receptors are stimulated by both low P_{O_2} and high hydrogen concentration. However, whereas they are the only important receptors in the low-oxygen reflex, they play only a minor role in the CO_2-dependent H^+ reflex. The major receptors for this latter reflex are the so-called "central" chemoreceptors, located in the medulla of the brain; they monitor the H^+ concentration of the brain's extracellular fluid, the cerebrospinal fluid. The sequence of events is illustrated in Fig. 12-30: CO_2 diffuses rapidly across the membranes separating arterial blood and brain tissue so that any increase in arterial P_{CO_2} causes a rapid identical increase in cerebrospinal fluid P_{CO_2}; this increased P_{CO_2}, by mass action, causes the generation of an increased number of hydrogen ions, i.e., an increased H^+ concentration; the latter stimulates the central chemoreceptors which, via synaptic connections, stimulate the medullary respiratory neurons to increase ventilation; the end result is a return of arterial and cerebrospinal-fluid P_{CO_2} and hydrogen-ion concentration to normal.

We are left with the rather startling conclusion that the control of breathing (at least during rest) is aimed primarily at the regulation of *brain hydrogen-ion concentration!* However, this actually makes perfectly good sense, in terms of survival of the organism. As emphasized above, the relationship between carbon dioxide and hydrogen ion ensures that the regulation of hydrogen-ion concentration also produces relative constancy of P_{CO_2} as well. To continue the chain, the close relationship between oxygen consumption and carbon dioxide production ensures that ventilation adequate to maintain P_{CO_2} constant by excreting carbon dioxide as fast as it is produced also suffices to supply adequate oxygen (except in certain kinds of lung disease). Moreover, the shape of the hemoglobin dissociation curve minimizes any minor deficiency of oxygen supply, and any major oxygen deficit induces reflex respiratory stimulation via the carotid and aortic bodies. In contrast, brain function is extremely sensitive to changes in hydrogen-ion concentration so that even small increases or decreases in brain hydrogen-ion concentration could induce serious malfunction.

Throughout this section we have described the *stimulatory* effects of carbon dioxide on ventilation. It should also be noted that very high levels of carbon dioxide *depress* the entire central nervous system, including the respiratory neurons, and may therefore be lethal. Closed environments, such as submarines and space capsules, must be designed so that carbon dioxide is removed as well as oxygen supplied.

Finally, it should be noted that the effects of increased P_{CO_2} (via changes in H$^+$), on the one hand, and decreased P_{O_2}, on the other, not only exist as independent inputs to the medulla but manifest synergistic interactions as well; ventilatory response to combined hypoxia and increased P_{CO_2} is considerably greater than the sum of the responses to each alone.

Control of ventilation by changes in arterial hydrogen-ion concentration not due to altered carbon dioxide. In the previous section we described how changes in hydrogen-ion concentration secondary to altered P_{CO_2} influence ventilation. However there are many situations in which a change in hydrogen-ion concentration occurs due to some cause other than a primary change in P_{CO_2}. In contrast to their relative lack of importance in mediating CO$_2$-induced H$^+$ changes, the peripheral chemoreceptors play the major role in stimulating ventilation whenever arterial H$^+$ is changed by any means other than by elevated P_{CO_2}. For example, increased addition of lactic acid to the blood causes hyperventilation (Fig. 12-31) almost entirely by way of the peripheral chemoreceptors. The central chemoreceptors do not respond in this case because hydrogen ions penetrate the blood-brain barrier so poorly that cerebrospinal-fluid H$^+$ concentration is not increased appreciably by the administration of the lactic acid; this is in contrast to the ease with which changes in arterial P_{CO_2} produce equivalent changes in cerebrospinal-fluid H$^+$ concentration.

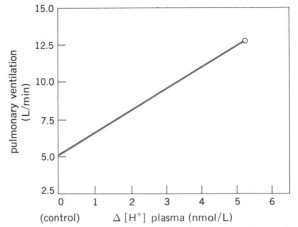

Figure 12-31. Change in ventilation in response to an elevation of plasma hydrogen-ion concentration, produced by the administration of lactic acid. *(Adapted from Lambertsen.)*

The converse of the above situation is also true, i.e., when arterial hydrogen-ion concentration is lowered by any means other than by a reduction of P_{CO_2} (for example, by loss of H$^+$ from the stomach in vomiting), ventilation is inhibited because of decreased peripheral chemoreceptor output. The adaptive value such reflexes have in regulating arterial hydrogen-ion concentration is described in Chap. 13.

Control of ventilation during exercise

During heavy exercise, the alveolar ventilation may increase ten- to twentyfold to supply the additional oxygen needed and excrete the excess carbon dioxide produced. On the basis of our three variables—P_{O_2}, P_{CO_2}, and hydrogen-ion concentration—it might seem easy to explain the mechanism which induces this increased ventilation. Unhappily, such is not the case.

Decreased P_{CO_2} as the stimulus? It would seem logical that, as the exercising muscles consume more oxygen, plasma P_{O_2} would decrease and stimulate respiration. But, in fact, arterial P_{O_2} is *not* significantly reduced during exercise (Fig. 12-32). The alveolar ventilation increases in exact proportion to the oxygen consumption; therefore, P_{O_2} remains constant. Indeed, in exhausting exercise, the alveolar ventilation may actually increase

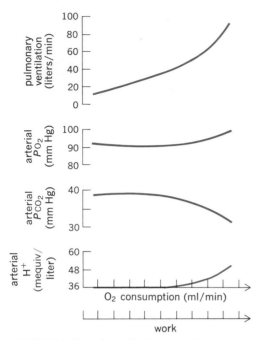

Figure 12-32. Relation of ventilation, arterial gas pressures, and hydrogen-ion concentration to the magnitude of muscular exercise. *(Adapted from Comroe.)*

relatively more than oxygen consumption, resulting in an *increased* P_{O_2}.

Increased P_{CO_2} as the stimulus? This is virtually the same story. Despite the marked increase in carbon dioxide production, the precisely equivalent increase in alveolar ventilation excretes the carbon dioxide as rapidly as it is produced, and arterial P_{CO_2} remains constant[1] (Fig. 12-32). Indeed, for the same reasons given above for oxygen increasing, the arterial P_{CO_2} may actually decrease during exhausting exercise.

Increased hydrogen ion as the stimulus? Since the arterial P_{CO_2} does not change (or decreases) during exercise, there is no accumulation of excess hydrogen ion as a result of carbon dioxide accumulation. Although there *is* an increase in arterial hydrogen-ion concentration for quite a different reason,

[1] For this reason the increased ventilation of exercise is not termed hyperventilation; this latter term refers specifically to increased ventilation adequate to lower P_{CO_2}.

namely, generation and release into the blood of lactic acid and other acids during exercise, the changes in hydrogen-ion concentration are not nearly great enough, particularly in only moderate exercise, to account for the increased ventilation.

We are left with the fact that, despite intensive study for more than 100 years by many of the greatest respiratory physiologists, we do not know what input stimulates ventilation during exercise. Our big three—P_{O_2}, P_{CO_2}, and hydrogen ion—appear presently to be inadequate, but many physiologists still believe that they will ultimately be shown to be the critical inputs. They reason that the fact that P_{CO_2} remains constant during moderate exercise is very strong evidence that ventilation is actually controlled by P_{CO_2}. In other words, if P_{CO_2} were not the major controller, how else could it remain unchanged in the face of the marked increase in its production? They reason that there may be a change in sensitivity of the chemoreceptors to hydrogen ion so that changes undetectable by experimental methods might be responsible for the stimulation of ventilation.

Moreover, the problem is even more complicated; as shown in Fig. 12-33, there is an abrupt increase (within seconds) in ventilation at the onset of exercise and an equally abrupt decrease at the end. Clearly, these changes occur too rapidly to be explained by alteration of chemical constituents of the blood.

In Chap. 11, we described how the cardiovascular responses to exercise were probably mediated by input to the hypothalamus from branches of the neuron pathways descending from the cerebral cortex to the motor nerves. It is likely that

Figure 12-33. Ventilation changes in time during exercise. Note the abrupt increase 1 at the onset of exercise and the equally abrupt but larger decrease 2 at the end of exercise.

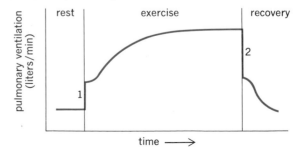

these fibers send branches into the medullary respiratory neurons as well. However, the magnitude of this contribution to stimulation of ventilation is not known.

Control of respiration by other factors

Temperature. An increase in body temperature frequently occurs as a result of increased physical activity and contributes to stimulation of alveolar ventilation. This facilitation of the respiratory centers is probably due both to a direct physical effect of increased temperature upon the respiratory neurons and to stimulation via pathways from thermoreceptors of the hypothalamus.

Epinephrine. Injection of epinephrine into a person produces stimulation of respiration by an action on the brain. Epinephrine secretion is uniformly increased during heavy exercise and probably contributes to the increased ventilation.

Reflexes from joints and muscles. Many receptors in joints and muscles can be stimulated by the physical movements which accompany muscle contraction. It is quite likely that afferent pathways from these receptors play a significant role in stimulating respiration during exercise. Thus, the mechanical events of exercise help to coordinate alveolar ventilation with the metabolic requirements of the tissues. This input is usually considered to be a stimulus not only during steady-state exercise but to be the cause of the abrupt increase in ventilation at the onset of exercise; however, recent data suggest that the latter actually represents a conditioned (i.e., learned) response.

Baroreceptor reflexes. We have already discussed the carotid sinus and aortic arch reflexes as the primary controllers of cardiovascular function. These reflexes can also alter alveolar ventilation, but their effect in people is usually minor.

Protective reflexes. A group of responses protect the respiratory tract against irritant materials, most familiar being the *cough* and the *sneeze.* These reflexes originate in receptors which line the respiratory tract. When they are excited, the result is stimulation of the medullary respiratory neurons in such a manner as to produce a deep inspiration and a violent expiration. In this man-

ner, particles can be literally exploded out of the respiratory tract. The cough reflex is inhibited by alcohol, which may contribute to the susceptibility of alcoholics to choking and pneumonia. Another example of a protective reflex is the immediate cessation of respiration which is frequently triggered when noxious agents are inhaled.

Pain. Painful stimuli anywhere in the body can produce reflex stimulation of the respiratory neurons.

Emotion. Emotional states are often accompanied by marked respiratory neuron stimulation, as evidenced by the rapid breathing rate which characterizes fright and many similar emotions. In addition, the movement of air in or out of the lungs is an absolute requirement for such involuntary expressions of emotion as laughing and crying. In such situations, the respiratory neurons must be primarily controlled by descending pathways from higher brain centers.

Voluntary control of breathing. Although we have discussed in detail the involuntary nature of most respiratory reflexes, it is quite obvious that we retain considerable voluntary control of respiratory movements. This is accomplished by descending pathways from the cerebral cortex to the motor neurons of the respiratory muscles, bypassing the medullary respiratory neurons. As we have seen, this voluntary control of respiration cannot be maintained when the involuntary stimuli, such as an elevated P_{CO_2} or hydrogen-ion concentration, become intense. Besides the obvious forms of voluntary control, e.g., breath holding, respiration must also be controlled during the production of complex actions such as speaking and singing.

Maximal oxygen uptake: A synthesis of cardiovascular-respiratory interactions

During exercise the supply of additional oxygen to the exercising muscles and the elimination of carbon dioxide depend upon precise integration of cardiovascular and respiratory functions, and this section reviews and amplifies several of these critical interactions.

When a person is subjected to progressively increasing work loads, there is a linear increase in

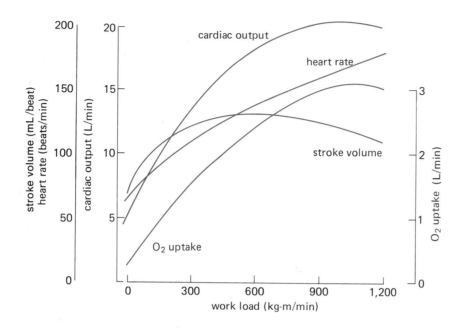

Figure 12-34. Changes in heart rate, stroke volume, cardiac output, and oxygen uptake with increasing work loads (measured as kg-m/min). Note that cardiac output plateaus because stroke volume decreases at maximal levels of oxygen uptake.

the amount of oxygen taken up by the muscle cells until a point is reached beyond which a further increase in work load does not result in further oxygen uptake (Fig. 12-34). This is defined as the maximal oxygen uptake. Work loads greater than this can be achieved only for very short periods of time since the muscles are operating by anaerobic glycolysis to a large extent, as manifested by a large increase in lactic acid production. What sets this upper limit for oxygen uptake and therefore work? In a normal person it is usually *not* set by any "saturation" of intracellular oxidative reactions, i.e., by an inability of the muscle cells to use more oxygen. Rather it is set by the maximal ability of the cardiovascular-respiratory systems to provide oxygen.

There are two means by which the muscle cells obtain more oxygen during exercise: (1) increased blood flow to the muscle; (2) extraction of more oxygen from each liter of blood. We have already described the mechanisms responsible for these changes. Local chemical changes resulting from increased metabolic activity cause vasodilation of the arterioles, thereby lowering local vascular resistance and increasing blood flow; the degree to which blood flow can be increased depends ultimately upon the ability to increase cardiac output. Simultaneously, as the contracting muscles use more oxygen, the local tissue P_{O_2} decreases toward

zero, thereby causing increased release of oxygen from hemoglobin. Recall that below 60 mmHg the hemoglobin dissociation curve is very steep so that a large additional amount of oxygen is released for each mmHg decrease in P_{O_2} below the value seen at rest. Moreover, the more work done by the muscle, the greater is its local temperature and hydrogen-ion concentration, both of which shift the dissociation curve downward and to the right, with the resultant release of even more oxygen. Red cell DPG also increases during exercise and contributes to the enhanced release of oxygen. The result of all these factors is that extraction of oxygen by the exercising muscles increases proportionately to work load until virtually all the oxygen is extracted at high work loads.

This analysis reveals that maximal oxygen uptake depends upon the product of maximal cardiac output and arteriovenous oxygen difference:

Maximal O_2 uptake =
maximal cardiac output $\times$ (A-V)O_2 difference

As we shall see, in a normal person at sea level, the increased ventilation triggered by exercise is quite capable of maintaining virtually complete saturation of hemoglobin during even the most severe exercise; accordingly, maximal oxygen uptake is not limited by the respiratory system. Thus, we are left with the fact that cardiac output is the rate-limiting variable in endurance-type exercise, and measurement of maximal oxygen uptake provides a sensitive index to the person's cardiovascular function.

Finally, we ask: What limits cardiac output during exercise? The answer is the interaction between heart rate and stroke volume (Fig. 12-21). Stroke volume increases with work load, although not to the same degree as does heart rate, but then decreases from maximal values at work loads beyond maximal oxygen uptake (Fig. 12-21). The major factors responsible for this decrease are the very rapid heart rate (which decreases diastolic filling time) and failure of the peripheral factors favoring venous return (muscle pump, respiratory pump, venoconstriction, arteriolar dilation; see Chap. 10) to elevate venous pressure high enough to maintain adequate ventricular filling during the very short time available. The net result is a decrease in end-diastolic volume which causes, by Starling's law, a decrease in stroke volume.

A person's maximal oxygen uptake is not fixed at any given value but can be altered (in either direction) by the habitual level of physical activity. For example, prolonged bed rest may decrease it 25 percent whereas intense long-term physical training may increase it a similar amount. It must be emphasized that, to be effective in raising maximal oxygen uptake, the training must be of an endurance type, i.e., involve large muscle groups for an extended period of time. We are still uncertain of the relative combinations of intensity and duration which are most effective but, as an example, jogging 10 to 15 min three times weekly at 5 to 8 mi/h definitely produces some increase in most people. Training induces an increased stroke volume and decreased heart rate at any given work load and an increased cardiac output at maximal work loads; the latter is due to an increased maximal stroke volume since maximal heart rate is not altered by training. How much of this increased maximal stroke volume is due to changes in the heart itself, as opposed to changes in skeletal muscle vasculature (which would facilitate venous return), remains unclear. Although still controversial, present evidence suggests that a training program adequate to elevate maximal oxygen uptake may offer some protection against the occurrence of heart attacks.

Thus far, our analysis has been in terms of normal persons. It must be emphasized that disease processes may reduce maximal oxygen uptake by interfering with any step in the transfer of oxygen from the atmosphere to the cells: lung disease, which prevents normal alveolar ventilation or alveolar-capillary diffusion and thereby reduces arterial Po_2 during exercise; reduced cardiac output secondary to heart disease; anemia; high altitude; failure of venous or arteriolar function. In other words, cardiac output is the limiting variable in normal individuals, but some other factor may become rate-limiting in the presence of disease.

At the other end of the spectrum are the champion marathon runners. They have maximum oxygen uptakes of 5 to 6 L/min compared with an untrained person's 3 L/min. Although training can increase the latter's value by approximately 20 percent, it certainly cannot produce the champion's values. Some of this difference may be ascribed to genetic variation, but a certain fraction is likely to be due to the pattern of the individual during early life—another example of the concept of "critical periods" described in Chap. 7.

Hypoxia

Hypoxia is defined as a deficiency of oxygen at the tissue level. There are many potential causes of hypoxia, but they can be classed in four general categories: (1) *hypoxic hypoxia*, in which the arterial Po_2 is reduced; (2) *anemic hypoxia*, in which the arterial Po_2 is normal but the total oxygen content of the blood is reduced because of inadequate numbers of red cells, deficient or abnormal hemoglobin, or competition for the hemoglobin molecule by carbon monoxide; (3) *ischemic hypoxia*, in which the basic defect is too little blood flow to the tissues; and (4) *histotoxic hypoxia*, in which the quantity of oxygen reaching the tissue is normal but the cell is unable to utilize the oxygen because a toxic agent (cyanide, for example) has interfered with its metabolic machinery.

Clearly, anemic hypoxia is mainly the result of abnormal blood formation or destruction, ischemic hypoxia is caused by failure of the cardiovascular system, and histotoxic hypoxia represents malfunction of the tissue cells, themselves. In contrast to these three categories, hypoxic hypoxia is due either to high-altitude exposure or to malfunction of the respiratory system. The causes are many,

Table 12-4. Some causes of hypoxic hypoxia

1 Decreased P_{O_2} in inspired air (high altitude)*
2 Hypoventilation
   Increased airway resistance (foreign body, asthma)
   Paralysis of respiratory muscles (poliomyelitis)
   Skeletal deformities
   Inhibition of medullary respiratory centers (morphine)
   Decreased "stretchability" of the lungs and thorax (deficient surfactant, thickened lung or chest tissues)
   Lung collapse (pneumothorax)
3 Deficient alveolar-capillary diffusion
   Decreased area for diffusion (pneumonia)
   Thickening of alveolar-capillary membranes (beryliosis)
4 Abnormal matching of ventilation and blood flow (emphysema)

* The phrases in parentheses are specific examples in each category.

and a list of certain of the major ones provides an excellent review of the interplay between the various components of the respiratory system described in this chapter (Table 12-4). We shall describe the response to high altitude as a representative example of the range of reflex adjustments which can be brought into play to offset the reduced arterial P_{O_2} characteristic of hypoxic hypoxia, regardless of cause.

Response to high altitude

Barometric pressure progressively decreases as altitude increases. Thus, at the top of Mt. Everest (approximately 9000 m), the atmospheric pressure is 245 mmHg. The air is still 20 percent oxygen which means that the P_{O_2} is 49 mmHg. Obviously, the alveolar and arterial P_{O_2} must decrease as one ascends unless pure oxygen is breathed. The effects of oxygen lack vary with individuals, but most persons who ascend rapidly to altitudes above 10,000 ft experience some degree of mountain sickness, consisting of breathlessness, palpitations, headache, nausea, fatigue, and impairment of a host of mental processes (vision, judgment, etc.). Over the course of several days, these symptoms diminish and ultimately disappear, although maximal physical capacity remains reduced. This process, known as acclimatization, is achieved by the compensatory mechanisms listed below. The highest villages permanently inhabited by people are in the Andes at 18,000

ft. These villagers work quite normally, and apparently the only major precaution they take is that the women come down to lower altitudes during late pregnancy in order to protect the fetus.

In response to the hypoxia, oxygen supply to the tissues is maintained in three ways: (1) by increased alveolar ventilation; (2) by increased erythrocyte and hemoglobin content of blood; (3) by a shift in the hemoglobin dissociation curve to the right. (This shift occurs despite the fact that, as described below, the blood hydrogen-ion concentration is decreased at high altitude, an event which would, itself, cause a left shift of the curve; the explanation is that the hypoxia and low hydrogen-ion concentration stimulate the production of 2,3-DPG by the red cells and this substance causes the right shift.) These three responses have the net effect of supplying normal amounts of oxygen to the tissues with minimal deviation of tissue P_{O_2} from normal.

The first acute response to high altitude is increased ventilation, mediated via the peripheral chemoreceptors, mainly the carotid bodies. As we have seen, the arterial P_{O_2} must generally fall by approximately 50 percent in order to elicit increased ventilation. The arterial P_{O_2} does not reach this low level until the altitude is 10,000 ft or greater. However, as described earlier in the chapter, the increased ventilation produced by low arterial P_{O_2} induces decreases in blood and cerebrospinal-fluid P_{CO_2} and hydrogen-ion concentration, which act as a brake on ventilation. However, over several days, mechanisms are brought into play which return cerebrospinal-fluid hydrogen-ion concentration back toward normal despite reduced P_{CO_2}. The net result is that the hypoxic drive is less offset by changes in CO_2 and H^+ and ventilation progressively increases. Other factors, not yet identified, must also contribute to the stimulation of ventilation.

Over time, erythrocyte and hemoglobin synthesis are stimulated by the hormone *erythropoietin* (see Chap. 11), and the person's total circulating red cell mass increases considerably. Another slowly developing compensation, the mechanism of which is unknown, is the growth of many new capillaries. This has the important effect of decreasing the distance which oxygen must diffuse from blood to cell. Tissue mitochondria and muscle myoglobin are also increased.

Metabolic functions of the lungs

The lungs have a variety of functions in addition to their roles in gas exchange and regulation of H^+ concentration. Most notable are their contributions to altering the concentrations in arterial blood of a large number of biologically active substances. As described in Chap. 8, hormones are added to the blood by the various endocrine glands of the body, but in addition many substances (neurotransmitters and paracrines, for example) which are released locally into the interstitium of their respective organs may diffuse into capillaries and thus make their way into the systemic venous system. The lungs may partially or completely remove these latter substances from the blood and thereby prevent them from reaching other locations in the body via the arteries (recall that the entire cardiac output passes through the lungs). The cells which perform this function are the endothelial cells lining the pulmonary capillaries. An important example of this function is the removal of certain prostaglandins from the blood during passage through the lungs.

In contrast, the lung may add new substances to the blood. These substances generally play local roles within the lungs, but if produced in large enough quantity, they may diffuse into the pulmonary capillaries, and be carried to the rest of the body. For example, inflammatory responses (Chap. 17) in the lung may, via excessive release of potent chemicals such as histamine, lead to profound alterations of systemic blood pressure or flow. Moreover, in at least one case, the lungs contribute a hormone (angiotensin II) to the blood not by secreting it themselves but by enzymatically activating a hormone precursor during passage of the blood through the pulmonary capillaries (see the renin-angiotensin system in Chap. 13).

Finally, the lungs also act as a "sieve" which traps and dissolves small blood clots generated in the systemic circulation, thereby preventing them from reaching the systemic arterial blood where they could occlude blood vessels in other organs.

13

REGULATION OF WATER AND ELECTROLYTE BALANCE

Section A: Basic principles of renal physiology

Structure of the kidney and urinary system
Basic renal processes
  Glomerular filtration
    Rate of glomerular filtration
  Tubular reabsorption
    Types of reabsorption
  Tubular secretion
Micturition (urination)

Section B: Regulation of sodium and water balance

Basic renal processes for sodium, chloride, and water
Control of sodium excretion: Regulation of extracellular
    volume
  Control of GFR
  Control of tubular sodium reabsorption
    Aldosterone and the renin-angiotensin system ●
    Factors other than aldosterone
  Conclusion
  ADH secretion and extracellular volume
Renal regulation of extracellular osmolarity
  Urine concentration: The countercurrent system
  Osmoreceptor control of ADH secretion
Thirst and salt appetite

Section C: Regulation of potassium, calcium, and hydrogen-ion concentrations

Potassium regulation
  Renal regulation of potassium
Calcium regulation
  Effector sites for calcium homeostasis
    Bone ● *Gastrointestinal tract* ● *Kidney*
  Parathormone
  Vitamin D
  Calcitonin
Hydrogen-ion regulation
  Basic definitions
  Strong and weak acids
  Buffer action and buffers
  Generation of hydrogen ions in the body
  Buffering of hydrogen ions in the body
    Bicarbonate-CO_2
  Renal regulation of extracellular hydrogen-ion concentration
  Classification of disordered hydrogen-ion concentration
Kidney disease
  The artificial kidney

One of the major themes of this book is the regulation of the internal environment. A cell's function depends not only upon receiving a continuous supply of organic nutrients and eliminating its metabolic end products but also upon the existence of stable physicochemical conditions in the extracellular fluid bathing it. Among the most important substances contributing to these conditions are water, sodium, potassium, calcium, and hydrogen ion. This chapter is devoted to a discussion of the mechanisms by which the total amounts of these substances in the body and their concentrations in the extracellular fluid are maintained relatively constant, mainly as a result of the kidney's activities.

A substance appears in the body either as a result of ingestion or as a product of metabolism. Conversely, a substance can be excreted from the body or consumed in a metabolic reaction. Therefore, if the quantity of any substance in the body is to be maintained at a constant level over a period of time, the total amounts ingested and produced must equal the total amounts excreted and consumed. This is a general statement of the *balance concept* described in Chap. 7. For water and hydrogen ion, all four possible pathways apply. However, for the mineral electrolytes balance is simpler since they are neither synthesized nor consumed by cells, and their total body balance thus reflects only ingestion versus excretion.

As an example, let us describe the balance for total body water (Table 13-1). It should be recognized that these are average values, which are subject to considerable variation. The two sources of body water are metabolically produced water, resulting largely from the oxidation of organic nutrients, and ingestion of water in liquids and so-called solid food (a rare steak is approximately 70 percent water). There are four sites from which water is lost to the external environment: skin, lungs, gastrointestinal tract, and urinary tract. (Menstrual flow constitutes a fifth source of water loss.) The loss of water by evaporation from the cells of the skin and the lining of respiratory passageways is a continuous process, often referred to as *insensible loss* because the person is unaware of its occurrence. Additional water can be made available for evaporation from the skin by the production of sweat.

Under normal conditions, as can be seen from Table 13-1, water loss exactly equals water gain, and no net change of body water occurs. This is obviously no accident but the result of precise regulatory mechanisms. The question then is: Which processes involved in water balance are controlled to make the gains and losses balance? The answer, as we shall see, is voluntary intake *(thirst)* and urinary loss. This does not mean that none of the other processes is controlled but that their control is not primarily oriented toward water balance. Catabolism of organic nutrients, the source of the water of oxidation, is controlled by mechanisms directed toward regulation of energy balance. Sweat production is controlled by mechanisms directed toward temperature regulation. Insensible loss (in people) is truly uncontrolled, and normal gastrointestinal loss of water (in feces) is generally quite small (but can be severe in vomiting or diarrhea).

The mechanism of thirst is certainly of great importance, since body deficits of water, regardless of cause, can be made up only by ingestion of water, but it is also true that our fluid intake is

Table 13-1. Normal routes of water gain and loss in adults

	Milliliters per day
Intake:	
Drunk	1200
In food	1000
Metabolically produced	350
Total	2550
Output:	
Insensible loss (skin and lungs)	900
Sweat	50
In feces	100
Urine	1500
Total	2550

often influenced more by habit and sociological factors than by the need to regulate body water. The control of urinary water loss is the major mechanism by which body water is regulated.

By similar analyses, we find that the body balances of most of the ions in the extracellular fluid are regulated primarily by the kidneys. To appreciate the importance of these kidney regulations one need only make a partial list of the more important simple inorganic substances which constitute the internal environment and which are regulated by the kidney: water, sodium, potassium, chloride, calcium, magnesium, sulfate, phosphate, and hydrogen ion. It is worth repeating that normal biological processes depend on the constancy of this internal environment, the implication being that the amounts of these substances must be held within very narrow limits, regardless of large variations in intake and abnormal losses resulting from disease (hemorrhage, diarrhea, vomiting, etc.). Indeed, the extraordinary number of substances which the kidney regulates and the precision with which these processes normally occur accounted for the kidney's being the last stronghold of the nineteenth-century vitalists, who simply would not believe that the laws of physics and chemistry could fully explain renal function. By what mechanism does urine flow rapidly increase when a person ingests several glasses of liquid? How is it that the patient on an extremely low salt intake and the heavy salt eater both urinate precisely the amounts of salt required to maintain their sodium balance? What mechanisms decrease the urinary calcium excretion of children deprived of milk?

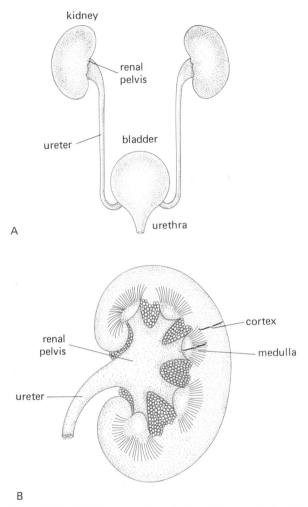

A

B

Figure 13-1. (A) Urinary system. The urine, formed by the kidneys, collects in the renal pelvis and then flows through the ureters into the bladder, from which it is eliminated via the urethra. (B) Section of a human kidney. Half the kidney has been sliced away. Note that the structure shows regional difference. The outer portion, which has a granular appearance, contains all the glomeruli. The collecting ducts form a large portion of the inner kidney, giving it a striped pyramidlike appearance, and drain into the renal pelvis.

This regulatory role is obviously quite different from the popular conception of the kidneys as glorified garbage disposal units which rid the body of assorted wastes and "poisons." It is true that the chemical reactions which occur in cells result ultimately in end products collectively called waste products; e.g., the catabolism of protein produces

approximately 30 g of urea per day. Other waste substances produced in relatively large quantities are uric acid (from nucleic acids), creatinine (from muscle creatine), and the end products of hemoglobin breakdown. There are many others, not all of which have been completely identified. Most of these substances are eliminated from the body as rapidly as they are produced, primarily by way of the kidneys. Some of these waste products are harmless, but the accumulation of certain of them during periods of renal malfunction accounts for certain of the disordered body functions in severe kidney disease. However, many of the problems which occur in renal disease are due simply to disordered water and electrolyte metabolism.

The kidneys have another excretory function which is presently assuming increased importance, namely, the elimination from the body of foreign chemicals (drugs, pesticides, food additives, etc.). A final kidney function is to act as endocrine glands secreting at least three substances which are components of hormonal systems: erythropoietin (Chap. 11), renin, and the active form of vitamin D (the last two to be discussed later in this chapter).

Section A
Basic principles of renal physiology

Structure of the kidney and urinary system

The kidneys are paired organs which lie in the back abdominal wall, one on each side of the vertebral column (Fig. 13-1). Each kidney is composed of approximately 1 million tiny units, all similar in structure and function, and bound together by small amounts of connective tissue containing blood vessels, nerves, and lymphatics. One such unit, or *nephron*, is shown in Fig. 13-2. Each nephron consists of a *vascular component* (the *glomerulus*) and a tubule. Throughout its course, the tubule is composed of a single layer of epithelial cells which differ in structure and function along its length. It originates as a sac, known as *Bowman's capsule*, which is lined with epithelial cells. On one side, Bowman's capsule is intimately associated with the glomerulus; on the other it opens into the first portion of the tubule, which is highly coiled and is known as the *proximal convoluted tubule*. The next portion of the tubule is a sharp hairpinlike loop, called *Henle's loop*. The tubule once more becomes coiled (the *distal convoluted tubule*) and finally runs a straight course as the *collecting duct*. From Bowman's capsule to the beginning of the collecting duct, each of the million tubules is completely separate from its neighbors. The tiny collecting ducts form separate tubules then join to form larger ducts, which in turn form even larger ducts, which finally empty into a large central cavity, the *renal pelvis*, at the base of each kidney (Fig. 13-1). The renal pelvis is continuous with the *ureter*, which empties into the *urinary bladder*, where urine is temporarily stored

and from which it is eliminated during urination. The urine is not altered after it leaves the collecting ducts. From the renal pelvis on, the remainder of the urinary system simply serves as plumbing.

To return to the other component of the nephron: What is the origin and nature of the glomerulus? Blood enters the kidney via the renal artery, which then divides into progressively smaller branches. Each of the smallest arteries gives off, at right angles to itself, a series of arterioles, *afferent arterioles* (Fig. 13-3), each of which leads to a compact tuft of capillaries. This tuft of capillaries is the glomerulus, which protrudes into the side of Bow-

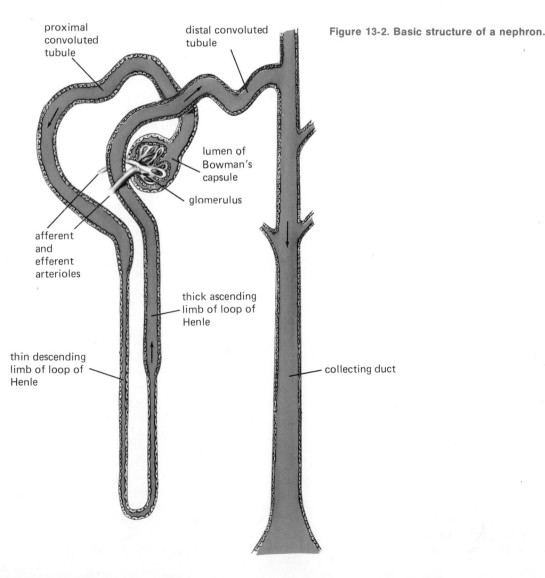

Figure 13-2. Basic structure of a nephron.

proximal convoluted tubule

distal convoluted tubule

lumen of Bowman's capsule

glomerulus

afferent and efferent arterioles

thick ascending limb of loop of Henle

thin descending limb of loop of Henle

collecting duct

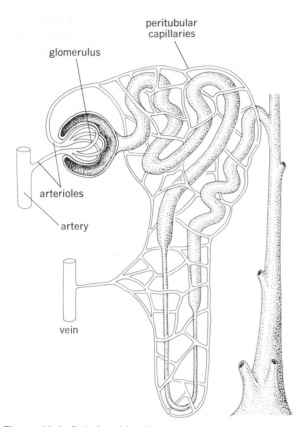

peritubular
capillaries

glomerulus

arterioles

artery

vein

Figure 13-3. Relationships between vascular and tubular components of the nephron. *(Adapted from Smith.)*

Basic renal processes

Urine formation begins with the filtration of essentially protein-free plasma through the glomerular capillaries into Bowman's capsule. This *glomerular filtrate* contains all low-molecular-weight substances in virtually the same concentrations as plasma. The final urine which enters the renal pelvis is quite different from the glomerular filtrate because, as the filtered fluid flows from Bowman's capsule through the remaining portions of the tubule, its composition is altered. This change occurs by two general processes, *tubular reabsorption* and *tubular secretion.* The tubule is at all points intimately associated with the peritubular capillaries, a relationship that permits transfer of materials between peritubular plasma and the inside of the tubule *(tubular lumen)*. When the direction of transfer is from tubular lumen to peritubular capillary plasma, the process is called tubular reabsorption. Movement in the opposite direction, i.e., from peritubular plasma to tubular lumen, is called tubular secretion. (This term must not be confused with *excretion;* to say that a substance has been excreted is to say only that it

man's capsule and is completely covered by the epithelial lining of the capsule. The functional significance of this anatomical arrangement is that blood in the glomerulus is separated from the lumen of Bowman's capsule by only a thin layer of tissue composed of: (1) the single-celled capillary lining; (2) a layer of basement membrane; (3) the single-celled lining of Bowman's capsule. *Filtration* of fluid occurs from the capillaries through this thin barrier into Bowman's capsule.

The glomerular capillaries, instead of combining to form veins, recombine to form another set of arterioles, the *efferent arterioles*, through which blood leaves the glomerulus. Each efferent arteriole soon divides into a second set of capillaries (Fig. 13-3), the *peritubular capillaries*, which are profusely distributed to, and intimately associated with, all the remaining portions of the tubule. They eventually join to form veins by which blood ultimately leaves the kidney.

Figure 13-4. The three basic components of renal function.

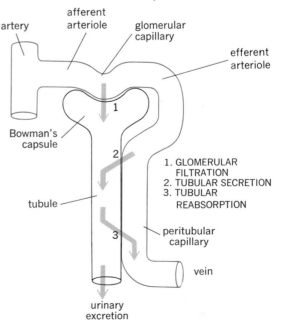

artery

afferent
arteriole

glomerular
capillary

efferent
arteriole

Bowman's
capsule

tubule

1

2

3

1. GLOMERULAR
   FILTRATION
2. TUBULAR SECRETION
3. TUBULAR
   REABSORPTION

peritubular
capillary

vein

urinary
excretion

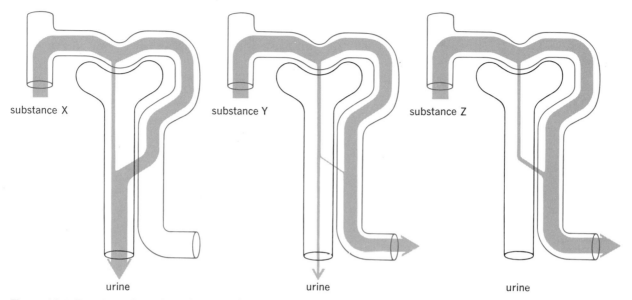

Figure 13-5. Renal manipulation of three substances X, Y, and Z. X is filtered and secreted but not reabsorbed. Y is filtered, and a fraction is then reabsorbed. Z is filtered and is completely reabsorbed.

appears in the final urine.) These relationships are illustrated in Fig. 13-4.

The most common relationships between these basic renal processes—glomerular filtration, tubular reabsorption, and tubular secretion—are shown in Fig. 13-5. Plasma containing substances X, Y, and Z enters the glomerular capillaries. A certain quantity of protein-free plasma containing these substances is filtered into Bowman's capsule, enters the proximal tubule, and begins its flow through the rest of the tubule. The remainder of the plasma, also containing X, Y, and Z, leaves the glomerular capillaries via the efferent arteriole and enters the peritubular capillaries. The cells composing the tubular epithelium can actively transport X (not Y or Z) from the peritubular plasma into the tubular lumen, but not in the opposite direction. By this combination of filtration and tubular secretion all the plasma which originally entered the renal artery is cleared of substance X, which leaves the body via the urine, thus reducing the amount of X remaining in the body.

If the tubule were incapable of reabsorption, the Y and Z originally filtered at the glomerulus would also leave the body via the urine, but the tubule can transport Y and Z from the tubular lumen back into the peritubular plasma. The amount of reabsorption of Y is small, so that most of the filtered material does escape from the body, but for Z the reabsorptive mechanism is so powerful that virtually all the filtered material is transported back into the plasma, which flows through the renal vein back into the vena cava. Therefore no Z is lost from the body. Hence the processes of filtration and reabsorption have canceled each other out, and the net result is as though Z had never entered the kidney at all.

Each substance in plasma is handled in a characteristic manner by the nephron, i.e., by a particular combination of filtration, reabsorption, and secretion. The critical point is that the rates at which the relevant processes proceed for many of these substances are subject to physiological control. What is the effect, for example, if the Y filtration rate is increased or its reabsorption rate decreased? Either change means that more Y is lost from the body via the urine. By triggering such changes in filtration or reabsorption whenever the plasma concentration of Y rises above normal, homeostatic mechanisms regulate plasma Y.

Glomerular filtration

As described in Chap. 11, capillaries are freely permeable to water and small solutes. They are relatively impermeable to large molecules (col-

Table 13-2. Forces involved in glomerular filtration

Forces	Millimeters of mercury
Favoring filtration:	
Glomerular capillary blood pressure	50
Opposing filtration:	
Fluid pressure in Bowman's capsule	10
Osmotic gradient	30
(water concentration difference	
due to protein)	
Net filtration pressure	10

loids), the most important of which are plasma proteins. The barrier offered by the glomerulus and adherent epithelium of Bowman's capsule behaves qualitatively like any capillary; accordingly, the glomerular filtrate, i.e., the fluid within Bowman's capsule, is essentially protein-free and contains all low-molecular-weight substances in virtually the same concentrations as the plasma.

These facts were demonstrated by the theoretically simple but technically complicated method of micropuncture. The nephrons of amphibians are relatively large and are not packed together into an enclosed organ like the human kidney. Tiny pointed tubes (micropipets) were inserted into Bowman's capsule and extremely small quantities of fluid (approximately 1/100 of an average-size drop) were withdrawn and analyzed. Micropuncture has also been used to withdraw fluid from almost all portions of the tubule. Its application to mammals as well as amphibia has provided many of the fundamental facts of renal physiology.

The hydrostatic pressure of the blood within the glomerular capillaries is less than the mean blood pressure in the large arteries of the body (approximately 100 mmHg), since pressure is dissipated as the blood passes through the arterioles connecting the renal artery branches to the glomeruli; the glomerular capillary pressure is usually about 50 mmHg. This is about half of mean arterial pressure and is considerably higher than in other capillaries of the body because the afferent arterioles leading to the glomeruli are wider than most arterioles and therefore offer less resistance to flow. The capillary hydrostatic pressure favoring filtration is not completely unopposed. There is, of course, fluid within Bowman's capsule which results in a capsular hydrostatic pressure of 10 mmHg resisting further filtration into the capsule.

A second opposing force results from the presence of protein in the plasma and its absence in Bowman's capsule. As in other capillaries, this unequal distribution of protein causes the water concentration of the plasma to be less than that of the fluid in Bowman's capsule. Again, as in other capillaries, the water-concentration difference is due completely to the plasma protein since all the low-molecular-weight solutes have virtually identical concentrations in the plasma and Bowman's capsule. The difference in water concentration induces an osmotic flow of fluid (water plus all the low-molecular-weight solutes) from Bowman's capsule into the capillary, a flow that opposes filtration. Its magnitude is equivalent to the bulk flow produced by a hydrostatic pressure difference of 30 mmHg.

As can be seen from Table 13-2, the net filtration pressure is approximately 10 mmHg. This pressure initiates urine formation by forcing an essentially protein-free filtrate of plasma through glomerular pores into Bowman's capsule and thence down the tubule into the renal pelvis. The glomerular membranes serve only as a filtration barrier and play no active, i.e., energy-requiring, role. The energy producing glomerular filtration is the energy transmitted to the blood (as hydrostatic pressure) when the heart contracts.

Before leaving this topic, we must point out the reason for use of the term "essentially protein-free" in describing the glomerular filtrate. In reality there is a very small amount of protein in the filtrate since the glomerular membranes are not perfect sieves for protein. Normally, less than 1 percent of serum albumin and no globulin are filtered; whatever protein is filtered is normally completely reabsorbed so that no protein appears in the final urine. However, in diseased kidneys, the glomerular membranes may become much more leaky to protein so that large quantities are filtered and some of this protein appears in the urine.

Rate of glomerular filtration. In a 70-kg person, the average volume of fluid filtered from the plasma into Bowman's capsule is 180 L/day (approximately 45 gal)! The implications of this remarkable fact are extremely important. When we recall that the average total volume of plasma is approximately 3 L, it follows that the entire plasma volume is filtered by the kidneys some 60 times a day. It is, in part, this ability to process such huge volumes of plasma that enables the kidneys to excrete large quantities of waste products and to regulate the constituents of the internal environment so precisely. The second implication concerns the magnitude of the reabsorptive process. The average person excretes between 1 and 2 L of urine per day. Since 180 L of fluid is filtered, approximately 99 percent of the filtered water must have been reabsorbed into the peritubular capillaries, the remaining 1 percent escaping from the body as urinary water. One must not gain the impression from this description that the rate of glomerular filtration is fixed, for as we shall see, it is subject to physiological control.

How can the rate of glomerular filtration be measured? To answer this question, let us use substance W (Fig. 13-6). Over a 24-h period (or any other convenient period of time) the subject's urine is collected. The amount of W in this volume of urine is measured and found to be 720 mg. How did this amount of W get into the urine? From other experiments it was determined that W is filtered at the glomerulus and is not secreted. Therefore, W can get into the urine only by filtration. Moreover, it also was previously determined that W is not reabsorbed by the tubules. Thus, all the filtered W appears in the final urine. Therefore, in order to excrete 720 mg of W in 24 h, the subject must have filtered 720 mg of W during the same 24-h period. The question now becomes: How much glomerular filtrate contains 720 mg of W? Since W is freely filterable, it must have the same concentration in the glomerular filtrate as in plasma. Several samples of the person's blood are obtained during the same 24-h period, and the plasma W concentration is found to be 4 mg/L of plasma. Since 1 L of plasma contains 4 mg of W, then 720/4, or 180, L contain 720 mg of W. In other words, 180 L of plasma having a W concentration of 4 mg/L must have been filtered during the 24-h period in order to account for the appearance of 720 mg of W in the final urine. The validity of this analysis depends upon the fact that W is freely filterable at the glomerulus and is neither reabsorbed nor secreted by the tubules. A polysaccharide called *inulin* (not insulin) completely fits this description and is used for just such determinations in people and experimental animals.

This type of so-called "clearance" analysis (measurement of the volume of plasma cleared of a particular substance) has proved invaluable for

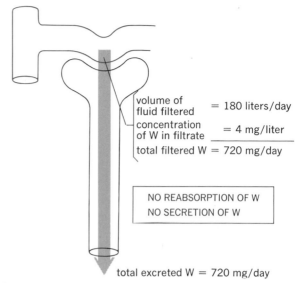

volume of fluid filtered = 180 liters/day

concentration of W in filtrate = 4 mg/liter

total filtered W = 720 mg/day

NO REABSORPTION OF W
NO SECRETION OF W

total excreted W = 720 mg/day

Figure 13-6. Measurement of glomerular filtration. W is filtered but is neither reabsorbed nor secreted.

gaining information about tubular reabsorption and secretion. For example, suppose we were interested in learning whether there is tubular reabsorption of phosphate. Using inulin, we would determine the glomerular filtration rate (GFR). (This must be repeated for every experiment because the GFR is not fixed but varies significantly.) The GFR in this particular experiment is found to be 165 L/day. The plasma concentration of phosphate is 1 mmol/L. Since phosphate is freely filterable at the glomerulus, 165 L/day × 1 mmol/L = 165 mmol/day is filtered. Finally, the amount of phosphate in the urine during this same 24-h period is found to be 40 mmol. Therefore, 165 − 40 mmol = 125 mmol of phosphate must have been reabsorbed by the tubules per 24 h. The generalization emerging from this example is that whenever the quantity of a substance excreted in the urine is *less* than the amount filtered during the same period of time, tubular reabsorption must have occurred. Conversely, if the amount excreted in the urine is *greater* than the amount filtered during the same period of time, tubular secretion must have occurred.

To complete this discussion of glomerular filtration we must consider the magnitude of the *total renal blood flow*. We have seen that none of the red blood cells and only a portion of the plasma which enters the glomerular capillaries are filtered into

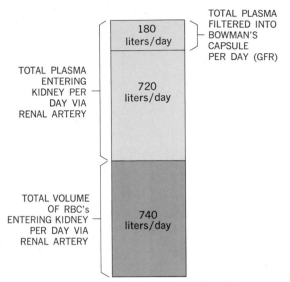

TOTAL PLASMA FILTERED INTO BOWMAN'S CAPSULE PER DAY (GFR)

180 liters/day

TOTAL PLASMA ENTERING KIDNEY PER DAY VIA RENAL ARTERY

720 liters/day

TOTAL VOLUME OF RBC's ENTERING KIDNEY PER DAY VIA RENAL ARTERY

740 liters/day

total renal blood flow = 1,640 liters/day

Figure 13-7. Magnitude of glomerular filtration rate, total renal plasma flow, and total renal blood flow. Only 20 percent of the plasma entering the kidneys is filtered from the glomerulus into Bowman's capsule. The remaining 80 percent flows through the glomerulus into the efferent arteriole and thence into the peritubular capillaries.

Table 13-3. Average values for several components handled by filtration and reabsorption.

Substance	Amount filtered per day	Amount excreted per day	Percent reabsorbed
Water, liters	180	1.8	99.0
Sodium, g	630	3.2	99.5
Glucose, g	180	0	100
Urea, g	54	30	44

from Table 13-3, which summarizes data for a few plasma components which are handled by filtration and reabsorption.

These are typical values for a normal person on an average diet. There are at least three important conclusions to be drawn from this table: (1) The quantities of material entering the nephrons via the glomerular filtrate are enormous, generally larger than their total body stores; e.g., if reabsorption of water ceased but filtration continued, the total plasma water would be urinated within 30 min. (2) The quantities of waste products, such as urea, which are excreted in the urine are generally sizable fractions of the filtered amounts; thus, in mammals, coupling a large glomerular filtration rate with a limited urea reabsorptive capacity permits rapid excretion of the large quantities of this substance produced constantly as a result of protein breakdown. (3) In contrast to urea and other waste products, the amounts of most useful plasma components, e.g., water, electrolytes, and glucose, which are excreted in the urine represent much smaller fractions of the filtered amounts because of reabsorption. For this reason one often hears the generalization that the kidney performs its regulatory function by *completely* reabsorbing all these biologically important materials and thereby preventing their loss from the body. This is a misleading half-truth, the refutation of which serves as an excellent opportunity for reviewing the essential features of renal function and regulatory processes in general.

Let us begin by pointing out the part of the generalization that is true. Certain substances, notably glucose, are not normally excreted in the urine because the amounts filtered are completely reabsorbed by the tubules. But does such a system permit the kidneys to *regulate* the plasma concentration of glucose, i.e., set it at some specific concentration? The following example will point out why

Bowman's capsule, the remainder passing via the efferent arterioles into the peritubular capillaries. Normally, the glomerular filtrate constitutes approximately one-fifth of the total plasma entering the kidney. Thus the total renal plasma flow is equal to 5 × 180 = 900 L/day. Since plasma constitutes approximately 55 percent of whole blood, the total renal blood flow, i.e., erythrocytes plus plasma, must be approximately 1640 L/day (1.1 L/min). Thus, the kidneys receive one-fifth to one-fourth of the total cardiac output (5 L/min) although their combined weight is less than 1 percent of the total body weight! These relationships are illustrated in Fig. 13-7.

Tubular reabsorption

Many filterable plasma components are either completely absent from the urine or present in smaller quantities than were originally filtered at the glomerulus. This fact alone is sufficient to prove that these substances undergo tubular reabsorption. An idea of the magnitude and importance of these reabsorptive mechanisms can be gained

the answer is *no*. Suppose the plasma glucose concentration is 100 mg/100 mL of plasma. Since reabsorption of this carbohydrate is complete, no glucose is lost from the body via the urine, and the plasma concentration remains at 100 mg/100 mL. If, instead of 100 mg/100 mL, we set our hypothetical plasma glucose concentration at 60 mg/100 mL, the analysis does not change; no glucose is lost in the urine, and the plasma glucose stays at 60 mg/100 mL. Obviously the kidney is merely maintaining whatever plasma glucose concentration happens to exist and is not involved in the regulatory mechanisms by which the original "setting" of the plasma glucose was accomplished. It is not the kidney but primarily the liver and the endocrine system which set and regulate the plasma glucose concentration. For comparison, consider what happens when a person drinks a lot of water: Within 1 to 2 h all the excess has been excreted in the urine, chiefly, as we shall see, as the result of decreased renal tubular reabsorption of water. In this example the kidney is the effector organ of a reflex which maintains plasma water concentration within very narrow limits. The critical point is that for many plasma components, particularly certain of the inorganic ions and water, the kidney does not always completely reabsorb the total amounts filtered. The rates at which these substances are reabsorbed (and therefore the rates at which they are excreted) are constantly subject to physiological control.

Types of reabsorption. A bewildering variety of ions and molecules is found in the plasma. With the exception of the proteins (and a few ions tightly bound to protein), these materials are all present in the glomerular filtrate, and most are reabsorbed, to varying extents. It is essential to realize that tubular reabsorption is a different process than glomerular filtration. The latter occurs completely by bulk flow in which water and all low-molecular-weight solutes move together. In contrast, tubular reabsorption of various substances is by more or less discrete mediated-transport mechanisms and by diffusion. The phrase "more or less" in the above sentence denotes several facts: First, the mediated-transport systems for reabsorption of different ions and molecules are frequently coupled, just as they are in other cell membranes (Chap. 6); second, in many cases a single reabsorptive system transports several different components if they are similar in structure (for example, many of the simple carbohydrates are reabsorbed by the same system).

As described in Chap. 6, transport processes can be categorized broadly as *active* or *passive*, and there are many examples of each in the kidney. The process is passive if no cellular energy is involved in the transport of the substance, i.e., if the substance moves downhill by simple or facilitated diffusion as a result of an electric or chemical concentration gradient. Active reabsorption, on the other hand, can produce net movement of the substance uphill against its electrochemical gradient and therefore requires energy expenditure by the transporting cells.

Transport of any substance across the renal tubule involves a sequence of steps. In Chap. 6 when we described the basic characteristics of transport processes, we were dealing only with transport across a single membrane, i.e., from the outside of the cell to the inside or vice versa. However, to cross the renal tubule, a substance must traverse not just one but a sequence of membranes. For example, to be reabsorbed, a sodium ion must gain entry to the tubular cell by crossing the cell membrane lining the lumen. It must then move through the cell's cytoplasm and cross the cell membrane separating the cytoplasm from the interstitial fluid. Finally, it must cross the capillary membranes to enter the plasma. The entire process is known as *transepithelial transport* and occurs not only in the kidney but in the gastrointestinal tract and other epithelial linings of the body.

In transepithelial transport the overall process is called *active* if one or more of the individual steps in the sequence is active. Sodium ions, for example, diffuse across the first tubular cell membrane and the cytoplasm of the cell. They are then actively transported out of the cell into the interstitial fluid; water moves from luminal to interstitial fluid by osmosis and the entire fluid then gains entry into the capillary by the bulk-flow process typical of all capillaries. Thus, three of the four steps are passive, but the crucial step is mediated by an active carrier process, and the overall process of sodium reabsorption is therefore said to be active. In this chapter we shall ignore the fact that multiple membranes lie between the lumens of the tubule and capillary and treat the tubular epithelium as a single membrane separating tubular fluid and plasma.

Most, if not all, of the active reabsorptive systems

in the renal tubule have a limit to the amounts of material they can transport per unit time, because the membrane carrier responsible for the transport becomes saturated. The classic example is the tubular transport process for glucose. As we know, normal persons do not excrete glucose in their urine because all filtered glucose is reabsorbed; but it is possible to produce urinary excretion of glucose in a completely normal person merely by administering large quantities of glucose directly into a vein. Recall that the filtered quantity of any freely filterable plasma component such as glucose is equal to its plasma concentration multiplied by the glomerular filtration rate. If we assume that, as we give our subject intravenous glucose, the glomerular filtration remains constant, the filtered load of glucose will be directly proportional to the plasma glucose concentration (Fig. 13-8). We shall find that even after the plasma glucose concentration has doubled, the urine will still be glucose-free, indicating that the *maximal tubular capacity* (T_m) for reabsorbing glucose has not yet been reached. But as the plasma glucose and the filtered load continue to rise, glucose finally appears in the urine. From this point on any further increase in plasma glucose is accompanied by an increase in excreted glucose, because the T_m has now been exceeded. The tubules are now reabsorbing all the glucose they can, and any amount filtered in excess of this quantity cannot be reabsorbed and appears in the urine. This is precisely what occurs in the

patient with diabetes mellitus. In this disease the hormonal control of plasma glucose is defective and it may rise to extremely high values. The filtered load of glucose becomes great enough to exceed the T_m, and glucose appears in the urine. There is nothing wrong with the tubular transport mechanism for glucose, it is simply unable to reabsorb the huge filtered load.

In normal persons the plasma glucose never becomes high enough to cause urinary excretion of glucose because the reabsorptive capacity for glucose is much greater than necessary for normal filtered loads. However, for certain other substances the reabsorptive T_m is very close to the normal filtered load; thus small increases in plasma concentrations of such substances will result in increased excretion.

Just as glucose provides an excellent example of an actively transported solute, urea provides an example of passive transport. Since urea is filtered at the glomerulus, its concentration in the very first portion of the tubule is identical to its concentration in peritubular capillary plasma. Then, as the fluid flows along the tubule water reabsorption occurs, increasing the concentration of any intratubular solute not being reabsorbed at the same rate as the water, with the result that intratubular urea concentration becomes greater than the peritubular-plasma urea concentration. Accordingly, urea is able to diffuse passively down this concentration gradient from tubular lumen to peritubular capillary. Urea reabsorption is thus a passive process and completely dependent upon the reabsorption of water (itself passive, as we shall see), which establishes the diffusion gradient. In people urea reabsorption varies between 40 and 60 percent of the filtered urea, the lower figure holding when water reabsorption is low and the higher when it is high.

Passive reabsorption is also of considerable importance for many foreign chemicals. The renal tubular epithelium acts in many respects as a lipid barrier; accordingly, lipid-soluble substances, like urea, can penetrate it fairly readily. Recall that one of the major determinants of lipid solubility is the polarity of a molecule—the less polar, the more lipid-soluble. Many drugs and environmental pollutants are nonpolar and, therefore, highly lipid-soluble. This makes their excretion from the body via the urine quite difficult since they are filtered at the glomerulus and then reabsorbed, like urea,

Figure 13-8. Saturation of the glucose transport system. Glucose is administered intravenously to a person so that plasma glucose and, thereby, filtered glucose are increased. The curves have been idealized, for the sake of clarity.

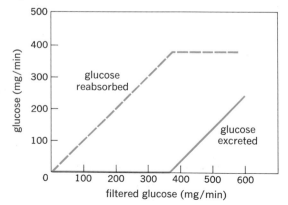

as water reabsorption causes their intratubular concentrations to increase. But evolution has provided a way out of this dilemma: The liver transforms most of these substances to progressively more polar metabolites which, because of their reduced lipid solubility, are poorly reabsorbed by the tubules and can therefore be excreted.

Tubular secretion

Tubular secretory processes, which transport substances from peritubular capillaries to tubular lumen, i.e., in the direction opposite to tubular reabsorption, constitute a second pathway into the tubule, the first being glomerular filtration. Like tubular reabsorption processes, secretory transport may be either active or passive. Of the large number of different substances transported into the tubules by tubular secretion, only a few are normally found in the body, the most important being hydrogen ion and potassium. We shall see that most of the excreted hydrogen ion and potassium enters the tubules by secretion rather than filtration. Thus, renal regulation of these two important substances is accomplished primarily by mechanisms which control the rates of their tubular secretion. The kidney is also able to secrete a large number of foreign chemicals, thereby facilitating their excretion from the body; penicillin is an example.

In the remainder of this chapter we shall see how the kidney functions in a variety of homeostatic processes and how renal function is coordinated with that of other organs. Before turning to the individual variables being regulated, we complete our basic story by describing the mechanisms for eliminating urine from the body.

Micturition (urination)

From the kidneys, urine flows to the bladder through the ureters (Fig. 13-1), propelled by peristalic contractions of the smooth muscle which makes up the ureteral wall. The bladder is a chamber, with walls of thick layers of smooth muscle. Upon contraction, the walls squeeze inward, increasing the pressure of the urine in the bladder. The muscle layer at the base of the bladder constitutes the first portion of the urethra. Although it is sometimes called the internal urethral sphincter,

it is not a distinct muscle but the last portion of the bladder; however, when the bladder is relaxed, this smooth muscle functions as a closed sphincter. As the bladder contracts, this sphincter is pulled open by changes in bladder shape during contraction. In other words, no special mechanism is required for its relaxation.

Another muscle important in the process of *micturition* (elimination of urine from the bladder) is a circular layer of skeletal muscle which surrounds the urethra farther down from the base of the bladder. When contracted, this *external urethral sphincter* can hold the urethra closed against strong bladder contractions.

Micturition is basically a local spinal reflex which is influenced by higher brain centers. The bladder muscle receives a rich supply of parasympathetic neurons, the stimulation of which causes bladder contraction (Fig. 13-9). The external urethral

Figure 13-9. Basic structure and nerve supply of the bladder and external urethral sphincter. Stimulation of the mechanoreceptors in the bladder wall causes reflex stimulation of the parasympathetic nerves to the bladder and inhibition of the motor nerves to the external sphincter. The result is bladder contraction and sphincter relaxation. Higher-center input allows voluntary initiation or delay of micturition.

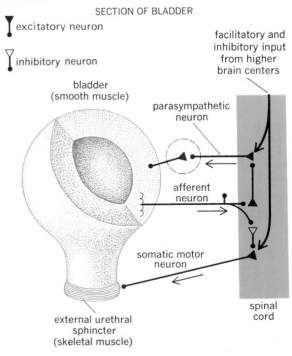

SECTION OF BLADDER

excitatory neuron

inhibitory neuron

bladder (smooth muscle)

facilitatory and inhibitory input from higher brain centers

parasympathetic neuron

afferent neuron

somatic motor neuron

spinal cord

external urethral sphincter (skeletal muscle)

Table 13-4. Normal routes of sodium chloride intake and loss

	Grams per day
Intake:	
Food	10.5
Output:	
Sweat	0.25
Feces	0.25
Urine	10.0
Total output	10.5

sphincter is innervated by somatic motor neurons just like any other skeletal muscle. The bladder wall contains stretch receptors whose afferent neurons enter the spinal cord and eventually synapse with these parasympathetic and somatic motor neurons, stimulating the former and inhibiting the latter. Via descending pathways, higher brain centers also synaptically facilitate and inhibit these motor pathways.

The following sequence of events leads to bladder emptying in an infant, in whom higher centers have only minor influence. When the bladder contains only small amounts of urine, its internal pressure is low, there is little stimulation of the bladder stretch receptors, the parasympathetics are relatively quiescent, and the somatic neurons are discharging at a moderately rapid rate. As the bladder fills with urine, it becomes distended, and the stretch receptors are gradually stimulated until their output becomes great enough to contract the bladder while simultaneously relaxing the external sphincter. Thus, the entire process is quite analogous to any other spinal reflex.

This process describes micturition adequately in the infant, but it is obvious that adults have the capacity either to delay micturition or to induce it voluntarily. In an adult, the volume of urine in the bladder required to initiate the spinal reflex for bladder contraction is approximately 300 mL. Delay is accomplished via descending pathways from the cerebral cortex which inhibit the bladder parasympathetics and stimulate the motor neurons to the external sphincter, thereby overriding the opposing synaptic input from the bladder stretch receptors. Voluntary initiation of micturition is just the opposite: Descending pathways from the cerebral cortex inhibit the motor neurons to the sphincter, thereby summating with the afferent input from the stretch receptors and initiating micturi-

tion. It is by learning to control these pathways that a child achieves the ability to control the timing of micturition.

Section B
Regulation of sodium and water balance

Table 13-1 is a typical balance sheet for water; Table 13-4, for sodium chloride. As with water, the excretion of sodium via the skin and gastrointestinal tract is normally quite small but may increase markedly during severe sweating, vomiting, or diarrhea. Hemorrhage, of course, can result in loss of large quantities of both sodium and water.

Control of the renal excretion of sodium and water constitutes the most important mechanism for the regulation of body sodium and water. The excretory rates of these substances can be varied over an extremely wide range; e.g., a gross consumer of salt may ingest 20 to 25 g of sodium chloride per day whereas a patient on a low-salt diet may ingest only 50 mg. The normal kidney can readily alter its excretion of salt over this range. Similarly, urinary water excretion can vary physiologically from approximately 400 mL/day to 25 L/day, depending upon whether one is lost in the desert or participating in a beer-drinking contest.

Basic renal processes for sodium, chloride, and water

Sodium, chloride, and water do not undergo tubular secretion; each is freely filterable at the glomerulus and approximately 99 percent is reabsorbed as it passes down the tubules (Table 13-3). Indeed, the vast majority of all renal energy production is used to accomplish this enormous reabsorptive task. The tubular mechanisms for reabsorption of these substances can be summarized by two generalizations:

1 The reabsorption of either sodium and/or chloride is an active process; i.e., it is carrier-mediated, requires an energy supply, and can occur against an electrochemical gradient. In most segments of the tubule, sodium is the actively transported ion with the reabsorption of chloride coupled to it by several mechanisms; in at least

one tubule segment, the situation is reversed; i.e., chloride is the actively transported ion, sodium movement being coupled to it. The end result is the same, and for simplicity we shall usually refer simply to "sodium reabsorption," implicitly including chloride.

2 The reabsorption of water is by diffusion (osmosis) and depends upon the reabsorption of sodium chloride.

How does the active reabsorption of sodium lead to the passive reabsorption of water? Since the concentrations of all low-molecular-weight solutes are virtually identical in plasma and Bowman's capsular fluid, no significant transtubular concentration gradients exist for sodium, chloride, or water in the very first portion of the proximal tubule. As the fluid flows down the tubule, sodium is actively reabsorbed into the peritubular capillaries. What happens to the water concentration of the tubular fluid as a result of this reabsorption? Obviously the removal of solute lowers osmolarity, i.e., raises water concentration, below that of plasma. Thus, a water-concentration gradient is created between tubular lumen and peritubular plasma which constitutes a driving force for water reabsorption via osmosis. If the water permeability of the tubular epithelium is very high, water molecules are reabsorbed passively almost as rapidly as the actively transported sodium ions, so that the tubular fluid is only slightly more dilute than plasma. In this manner almost all the filtered sodium and water could theoretically be reabsorbed and the final urine would still have approximately the same osmolarity as plasma. However, this reabsorption of water can occur only if the tubular epithelium is highly permeable to water. No matter how great the water-concentration gradient, water cannot move if the epithelium is impermeable to it.

The permeability of the last portions of the tubules (the distal tubules and collecting ducts) to water is subject to physiological control. The major determinant of this permeability is a hormone known as *antidiuretic hormone* (ADH) or *vasopressin*. In the absence of ADH (Fig. 13-10) the water permeability of the distal tubule and collecting duct is very low sodium reabsorption proceeds normally because ADH has little or no effect on sodium reabsorption, but water is unable to follow and thus remains in the tubule to be excreted as a large volume of urine. On the other hand, in the

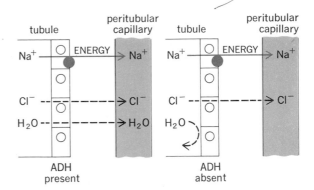

Figure 13-10. Effect of ADH on water reabsorption. ADH causes the tubule to be highly permeable to water so that water movement can accompany sodium and chloride reabsorption. Without ADH, the tubule becomes quite impermeable to water, and water reabsorption is impeded. Note that ADH does not alter reabsorption of sodium or chloride ions.

presence of ADH, the water permeability of these last nephron segments is very great, water reabsorption is able to keep up with sodium reabsorption, and the final urine volume is small.

ADH is an octapeptide (a molecule consisting of eight amino acids) closely related in structure to another hormone, oxytocin, the function of which is discussed in Chap. 16. ADH exerts its action by stimulating production of cyclic AMP in the tubular cells; but precisely how this results in increased permeability to water remains unclear. It is interesting that ADH exerts this same effect on other animal tissues, such as frog skin and toad bladder, which actually serve as useful tools for studying its cellular mechanism of action. How ADH secretion is regulated will be discussed in subsequent sections.

Several crucial aspects of sodium and water reabsorption should be emphasized:

1 The tubular response to ADH is not all or none, like an action potential, but shows graded increases as the concentration of ADH is elevated, thus permitting fine adjustments of water permeability and excretion.

2 Excretion of large quantities of sodium *always* results in the excretion of large quantities of water. This follows from the passive nature of water reabsorption, since water can be reabsorbed only if sodium is reabsorbed first. As we shall see, this relationship has considerable im-

portance for the regulation of extracellular volume.

3 In contrast, large quantities of water can be excreted even though the urine is virtually free of sodium. This process we shall find critical for the renal regulation of extracellular osmolarity.

Control of sodium excretion: Regulation of extracellular volume

The renal compensation for increased body sodium is excretion of the excess sodium. Conversely, a deficit in body sodium is prevented by reducing urinary sodium to an absolute minimum, thus re-

Figure 13-11. Sodium excretion is increased by increasing the GFR (B), by decreasing reabsorption (C), or by a combination of both (D). The arrows indicate relative magnitudes of filtration, reabsorption, and excretion.

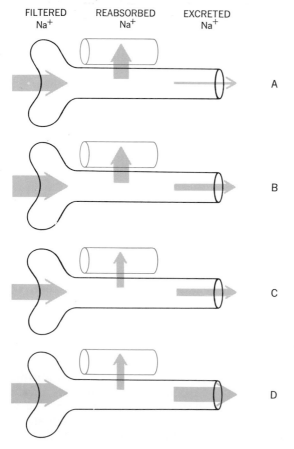

taining within the body the amount which would otherwise have been lost via the urine.

Since sodium is freely filterable at the glomerulus and actively reabsorbed but not secreted by the tubules, the amount of sodium excreted in the final urine represents the resultant of two processes, glomerular filtration and tubular reabsorption.

$$\text{Sodium excretion} =$$
$$\text{sodium filtered} - \text{sodium reabsorbed}$$

It is possible, therefore, to adjust sodium excretion by controlling one or both of these two variables (Fig. 13-11). For example, what happens if the quantity of filtered sodium increases (as a result of a higher GFR) but the rate of reabsorption remains constant? Clearly, sodium excretion increases. The same final result could be achieved by lowering sodium reabsorption while holding the GFR constant. Finally, sodium excretion could be raised greatly by elevating the GFR and simultaneously reducing reabsorption. Conversely, sodium excretion could be decreased below normal levels by lowering the GFR or raising sodium reabsorption or both. Control of GFR and sodium reabsorption is in fact the mechanism by which renal regulation of sodium balance is accomplished.

The reflex pathways by which changes in total body sodium balance lead to changes in GFR and sodium reabsorption include: (1) "volume" receptors (the reasons for the use of quotation marks will soon be apparent) and the afferent pathways leading from them to the central nervous system and endocrine glands; (2) efferent neural and hormonal pathways to the kidneys; (3) renal effector sites: the renal arterioles and tubules.

That cardiovascular volume receptors are appropriate receptors for regulating total body sodium is shown by the following analysis. What happens when a person ingests a liter of isotonic sodium chloride, i.e., a solution of salt with exactly the same osmolarity as the body fluids? It is absorbed from the gastrointestinal tract, all the salt and water remain in the extracellular fluid (plasma and interstitial fluid), and none enters the cells. Because of the active sodium "pumps" in cell membranes, sodium is effectively barred from the cells and therefore remains in the extracellular fluid. The water, too, remains since only the volume and not the osmolarity of the extracellular

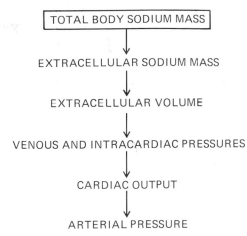

Figure 13-12. Dependence of cardiovascular parameters on total body sodium mass. The magnitude of each variable in the flow sheet is determined, in large part, by the magnitude of the one above it. Therefore, receptors which detect changes in pressure or distension supply the information required for regulation of body sodium.

compartment has been changed; i.e., no osmotic gradient exists to drive the ingested water into cells. This example is one of many which could be used to illustrate the important generalization that total extracellular fluid volume depends primarily upon the mass of extracellular sodium, which in turn correlates directly with total body sodium. However, rather than extracellular volume, per se, the variables monitored in sodium-regulating reflexes are those that are altered as a result of changes in this volume, specifically intravascular and intracardiac pressure (Fig. 13-12). For example, a decrease in plasma volume tends to lower the hydrostatic pressures in the veins, cardiac chambers, and arteries. These changes are detected by receptors (for example, the carotid sinus baroreceptor) in these structures and initiate the reflexes leading to renal sodium retention which helps restore the plasma volume toward normal.

Control of GFR

Fig. 13-13 summarizes the reflex pathway by which negative sodium balance (as, for example, in diarrhea) causes a decrease in GFR. Note that this is simply the basic baroreceptor reflex described in Chap. 11, where it was pointed out that a decrease in cardiovascular pressures causes reflex vasoconstriction in many areas of the body.

Conversely, an increased GFR can result from increased plasma volume and contribute to the increased renal sodium loss which returns extracellular volume to normal.

Figure 13-13. Pathway by which the GFR is decreased when plasma volume decreases. The baroreceptors which initiate the sympathetic reflex are probably located in large veins and the walls of the heart, as well as in the carotid sinuses and aortic arch.

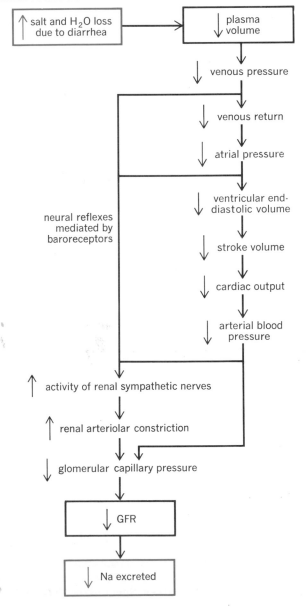

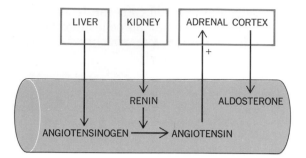

Figure 13-14. Summary of the renin-angiotensin-aldosterone system. The plus sign denotes the stimulatory effect of angiotensin on aldosterone secretion. "Angiotensin" denotes a family of peptides (see text).

Control of tubular sodium reabsorption

So far as long-term regulation of sodium excretion is concerned, the control of tubular sodium reabsorption is probably more important than that of GFR. One major controller of the rate of tubular sodium reabsorption is aldosterone.

Aldosterone and the renin-angiotensin system. An early clue to the control of sodium reabsorption was the observation that patients whose adrenal glands are diseased or missing excrete large quantities of sodium in the urine due to decreased tubular reabsorption; indeed, if untreated, they may die because of low blood pressure resulting from depletion of plasma volume. The adrenal cortex produces a hormone, called *aldosterone*, which stimulates sodium reabsorption, specifically by the distal tubules and collecting ducts. In the complete absence of this hormone, the patient may excrete 25 g of salt per day, whereas excretion may be virtually zero when aldosterone is present in large quantities. In a normal person, the amounts of aldosterone produced and salt excreted lie somewhere between these extremes, varying with the amount of salt ingested. (It is interesting that aldosterone also stimulates sodium transport by other epithelia in the body, namely, by sweat and salivary glands and the intestine; the net effect is the same as that exerted on the renal tubules—a movement of sodium out of the luminal fluid into the blood.)

Aldosterone secretion (and thereby tubular sodium reabsorption) is controlled by reflexes involving the kidneys themselves. Specialized cells lining the arterioles in the kidneys synthesize and secrete into the blood an enzyme known as *renin* which splits off a small polypeptide, *angiotensin* I, from a large plasma protein, *angiotensinogen* (Fig. 13-14). Angiotensin I then undergoes further enzymatically mediated cleavage to form a series of other angiotensins (II, III, etc.), but we shall refer collectively to these molecules simply as angiotensin. Angiotensin is a profound stimulator of aldosterone secretion and constitutes the major input to the adrenal gland controlling the production and release of this hormone.

Angiotensinogen is synthesized by the liver and is always present in the blood; therefore, the rate-limiting factor in angiotensin formation is the concentration of plasma renin, which, in turn, depends upon the rate of renin secretion by the kidneys. The critical question now becomes: "What controls the rate of renin secretion?" There are multiple inputs to the renin-secreting cells and it is not yet possible to assign quantitative roles to each of them. It is likely that the renal sympathetic nerves and decreased renal arterial pressure constitute important inputs in normal persons; this makes excellent sense, teleologically, since a reduction in body sodium and extracellular volume lowers blood pressure and triggers off increased sympathetic discharge to the kidneys (as shown in Fig. 13-13), thereby setting off the hormonal chain of events which restores sodium balance and extracellular volume to normal (Fig. 13-15).

Clearly, by helping to regulate sodium balance and thereby plasma volume, the renin-angiotensin system contributes to the control of arterial blood pressure. However, this is not the only way in which it influences arterial pressure; recall from Chap. 11 that angiotensin is a potent constrictor of arterioles and this effect on peripheral resistance increases arterial pressure. Abnormal increases in the activity of the renin-angiotensin system contribute to the development of hypertension (high blood pressure) via both the sodium-retaining (plasma-volume increasing) and arteriole-constricting effects.

Factors other than aldosterone. There are a large number of factors other than aldosterone which are capable of altering tubular sodium reabsorption, and renal physiologists are presently attempting to quantitate their relative contributions to the control of sodium excretion in both health and disease. This task is far from complete and we present here only a few examples: (1) In the previous sections we pointed out how the sympathetic neurons to the kidney participate in the control of both the renal arterioles (thereby influencing GFR) and the renin-secreting cells (thereby ultimately influencing aldosterone secretion); they also seem to act directly upon the tubules to enhance sodium reabsorption; (2) there is much indirect evidence suggesting the existence of one or more hormones, as yet unidentified, which inhibit the reabsorption of sodium; if true, this would mean that the tubules are probably under the continuous influence of opposing hormonal inputs—aldosterone enhancing reabsorptive rate and the hypothesized other hormone reducing it.

Conclusion

The control of sodium excretion depends upon the control of two variables of renal function, the GFR and sodium reabsorption. The latter is controlled, at least in part, by the renin-angiotensin-aldosterone hormone system and, in part, by other less well-defined factors. The reflexes which control both GFR and sodium reabsorption are essentially blood-pressure-regulating reflexes since they are probably most frequently initiated by changes in arterial or venous pressure (or cardiac output). This is only fitting since cardiovascular function depends upon an adequate plasma volume, which, as a component of the extracellular fluid volume, normally reflects the mass of sodium in the body. In normal persons, these regulatory mechanisms are so precise that sodium balance does not vary by more than 2 percent despite marked changes in dietary intake or in losses due to sweating, vomiting, or diarrhea. In several types of diseases, however, sodium balance becomes deranged by the failure of the kidneys to excrete sodium normally. Sodium excretion may fall virtually to zero despite continued sodium ingestion, and the patient may retain huge quantities of sodium and water, leading to abnormal expansion of the extracellular fluid and edema. The most important example of this phenomenon is congestive heart failure (Chap. 11). A patient with a failing heart manifests a decreased GFR and increased aldosterone secretion rate, both of which contribute to the virtual absence of sodium from the urine. In addition to aldosterone, the other less well-defined factors also

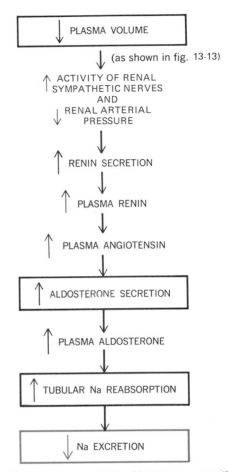

Figure 13-15. Pathway by which aldosterone secretion is increased when plasma volume is decreased. (The pathways by which the GFR is reduced are shown in Fig. 13-13.)

cause decreased sodium excretion by enhancing tubular sodium reabsorption. The net result is expansion of plasma volume, increased capillary pressure, and filtration of fluid into the interstitial space, i.e., edema. Why are these sodium-retaining reflexes all stimulated despite the fact that the expanded extracellular volume should call forth sodium-losing responses (increased GFR and decreased aldosterone)? We do not know, although it seems fairly certain that the lower cardiac output as a result of cardiac failure is somehow responsible. In addition to treating the basic heart disease, physicians use *diuretics*, or drugs which inhibit tubular sodium or chloride reabsorption, thereby leading to greater sodium and water excretion.

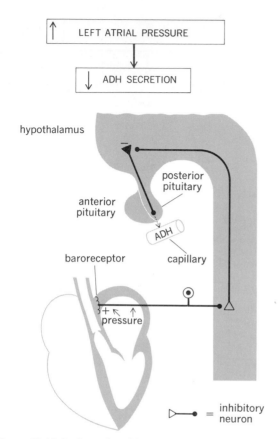

Figure 13-16. Pathway by which ADH secretion is decreased when plasma volume is increased. The greater plasma volume raises left-atrial pressure, which stimulates the atrial baroreceptors and inhibits ADH secretion.

ADH secretion and extracellular volume

Although we have spoken of extracellular-volume regulation only in terms of the control of sodium excretion, it is clear that to be effective in altering extracellular volume, the changes in sodium excretion must be accompanied by equivalent changes in water excretion. We have already pointed out that the ability of water to be reabsorbed when sodium is reabsorbed depends upon ADH. Accordingly, a decreased extracellular volume must reflexly call forth increased ADH secretion as well as increased aldosterone secretion. What is the nature of this reflex? As described in Chap. 9, ADH is produced by a discrete group of hypothalamic neurons whose axons terminate in the posterior pituitary, from which ADH is released into the

blood. These hypothalamic cells receive input from several vascular baroreceptors, particularly a group located in the left atrium (Fig. 13-16). The baroreceptors are stimulated by increased atrial blood pressure, and the impulses resulting from this stimulation are transmitted via afferent neurons and ascending pathways to the hypothalamus, where they *inhibit* the ADH-producing cells. Conversely, decreased atrial pressure causes less firing by the baroreceptors and stimulation of ADH synthesis and release (Fig. 13-17). The adaptive value

Figure 13-17. Pathway by which ADH secretion is increased when plasma volume decreases. This figure is merely the converse of Fig. 13-16.

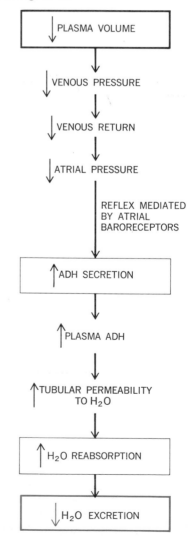

of this baroreceptor reflex, one more in our expanding list, should require no comment.

Renal regulation of extracellular osmolarity

We turn now to the renal compensation for pure water losses or gains, e.g., a person drinking 2 L of water, where no change in total salt content of the body occurs, only total water. The most efficient compensatory mechanism is for the kidneys to excrete the excess water without altering their usual excretion of salt, and this is precisely what they do. ADH secretion is reflexively inhibited, as will be described below, tubular water permeability of the collecting ducts becomes very low, sodium reabsorption proceeds normally but water is not reabsorbed, and a large volume of extremely dilute urine is excreted. In this manner, the excess pure water is eliminated.

Thus it is easy to see how the kidneys produce a final urine having the same osmolarity as that of plasma or one having a lower osmolarity than plasma (hypoosmotic urine), the latter occurring whenever water reabsorption lags behind solute reabsorption, i.e., when plasma ADH is reduced. Clearly the formation of a hypoosmotic urine is a good compensation for an excess of water in the body.

At the other end of the spectrum, the final urine may indeed reach a concentration considerably greater than that of the plasma; moreover, this concentrated urine is produced without violating the generalization that water reabsorption is always passive. But how can the kidneys produce a hyperosmotic urine, i.e., a urine having an osmolarity greater than that of plasma? For this to occur, does not water reabsorption have to "get ahead" of solute reabsorption? How can this happen if water reabsorption is always secondary to solute (particularly sodium chloride) reabsorption? To this question we now turn.

Urine concentration: The countercurrent system

The ability of the kidneys to produce concentrated urine is not merely an academic problem. It is a major determinant of one's ability to survive without water. The human kidney can produce a maximal urinary concentration of 1400 millios-

mol/L. The urea, sulfate, phosphate, and other waste products (plus the smaller number of nonwaste ions) which are excreted each day amount to approximately 600 milliosmol; therefore, the water required for their excretion constitutes an obligatory water loss and equals

$$\frac{600 \text{ milliosmol/day}}{1400 \text{ milliosmol/L}} = 0.444 \text{ L/day}$$

As long as the kidneys are functioning, excretion of this volume of urine will occur, despite the absence of water intake. In a sense, a person lacking access to water may literally urinate to death (due to fluid depletion). If the body could produce a urine with an osmolarity of 6000 milliosmol/L, then only 100 mL of obligatory water need be lost each day and survival time would be greatly expanded. A desert rodent, the kangaroo rat, does just that; this animal never drinks water, the water produced in its body by oxidation of foodstuffs being ample for maintenance of water balance.

The kidneys produce concentrated urine by a complex interaction of events involving the so-called *countercurrent multiplier system* residing in the loop of Henle. Recall that the loop of Henle, which is interposed between the proximal and distal convoluted tubules, is a hairpin loop extending into the inner portions of the kidney (the renal medulla). The fluid flows in opposite directions in the two limbs of the loop, thus the name countercurrent. Let us list the critical characteristics of this loop.

1 The ascending limb of the loop of Henle (i.e., the limb leading to the distal tubule) *actively* transports sodium chloride out of the tubular lumen into the surrounding interstitium. (In the ascending limb of the loop of Henle chloride is the actively transported ion with sodium following passively; for simplicity, we shall refer to the process as "sodium chloride transport.") It is fairly impermeable to water, so that water cannot follow the sodium chloride.
2 The descending limb of the loop of Henle (i.e., the limb into which drains fluid from the proximal tubule) does *not actively* transport sodium chloride; it is the only tubular segment that does not. Moreover, it is highly permeable to water, but relatively impermeable to sodium chloride.

Given these characteristics, imagine the loop of Henle filled with a stationary column of fluid supplied by the proximal tubule. At first, the concentration everywhere would be 300 milliosmol/L, since fluid leaving the proximal tubule is isosmotic to plasma, equivalent amounts of sodium chloride and water having been reabsorbed by the proximal tubule.

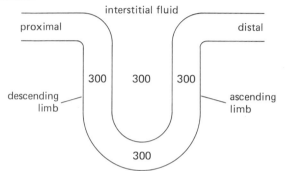

Now let the active pump in the ascending limb transport sodium chloride into the interstitium until a limiting gradient (say 200 milliosmol/L) is established between ascending-limb fluid and interstitium.

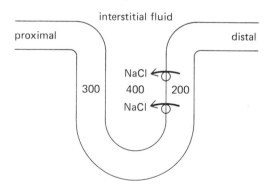

A limiting gradient is reached because the ascending limb is *not* completely impermeable to sodium and chloride; accordingly, passive backflux of ions into the lumen counterbalances active outflux and a steady-state limiting gradient is established.

Given the relatively high permeability of the descending limb to water, what net flux now occurs between interstitium and descending limb? There is a net diffusion of water out of the descending limb into the interstitium until the osmolarities are equal. The interstitial osmolarity is maintained at 400 milliosmol/L during this equilibration be-

cause of continued active sodium chloride transport out of the ascending limb.

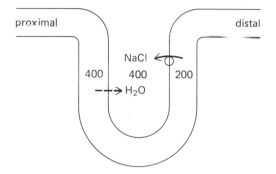

Note that as a result of this purely passive water movement across the descending limb, the osmolarities of the descending limb and interstitium become equal and both are higher than that of the ascending limb.

So far we have held the fluid stationary in the loop, but of course it is actually continuously flowing. Let us look at what occurs under conditions of flow (Fig. 13-18), simplifying the analysis by assuming that flow through the loop, on the one hand, and ion and water movements, on the other, occur in discontinuous out-of-phase steps. During the stationary phase as described above, sodium chloride is actively transported out of the ascending limb to establish a gradient of 200 milliosmol, and water diffuses out of the descending limb until descending limb and interstitium have the same osmolarity. During the "flow" phase, fluid leaves the loop via the distal tubule and new fluid enters the loop from the proximal tubule.

Note that the fluid is progressively concentrated as it flows down the descending limb and then is progressively diluted as it flows up the ascending limb. Whereas only a 200-milliosmol/L gradient is maintained across the ascending limb at any given *horizontal level* in the medulla, there exists a much larger osmotic gradient from the top of the medulla to the bottom (312 milliosmol/L versus 700 milliosmol/L). In other words, the 200-milliosmol/L gradient established by active sodium chloride transport has been *multiplied* because of the *countercurrent* flow within the loop (i.e., flow in opposing directions through the two limbs of the loop). It should be emphasized that the active sodium chloride transport mechanism in the ascending limb is the essential component of the entire system; without it, the countercurrent flow

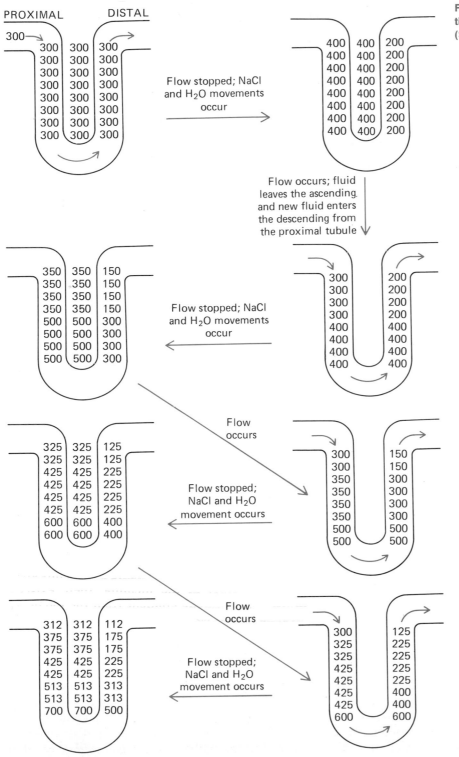

Figure 13-18. Countercurrent multiplier system in the loop of Henle. (See text for explanation.)

PROXIMAL   DISTAL

Flow stopped; NaCl and H₂O movements occur

Flow occurs; fluid leaves the ascending, and new fluid enters the descending from the proximal tubule

Flow stopped; NaCl and H₂O movements occur

Flow occurs

Flow stopped; NaCl and H₂O movement occurs

Flow occurs

Flow stopped; NaCl and H₂O movement occurs

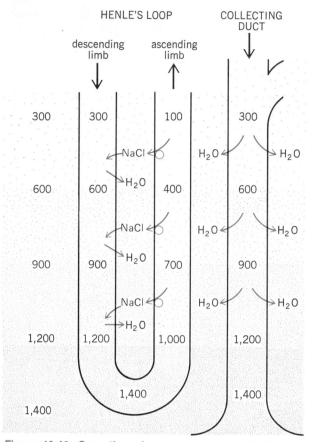

HENLE'S LOOP

descending limb   ascending limb

COLLECTING DUCT

Figure 13-19. Operation of countercurrent multiplier system in the formation of hypertonic urine. *(Adapted from R. F. Pitts.)*

would have no effect whatever on concentrations.

But what has this system really accomplished? Certainly it concentrates the loop fluid, but then it immediately redilutes it so that the fluid entering the distal tubule is actually more dilute than the plasma. Where is the final urine concentrated and how? The site of final concentration is the collecting ducts which pass through the renal medulla parallel to the loops of Henle, and are bathed by the interstitial fluid of the medulla. In the presence of maximal levels of ADH, fluid in the late distal tubules reequilibrates with peritubular plasma and is isosmotic to plasma (that is, 300 milliosmol/L) when it enters the collecting ducts. As this fluid then flows through the collecting ducts it equilibrates with the ever-increasing osmolarity of the interstitial fluid. Thus, the real function of the loop countercurrent multiplier

system is to concentrate the *medullary interstitium*. Under the influence of ADH, the collecting ducts are highly permeable to water, which diffuses out of the collecting ducts into the interstitium as a result of the osmotic gradient (Fig. 13-19). The net result is that the fluid at the end of the collecting duct has equilibrated with the interstitial fluid at the tip of the medulla. By this means, the final highly concentrated urine contains relatively less of the filtered water than solute, which is precisely the same as adding pure water to the extracellular fluid, and thereby compensating for a pure water deficit.

In contrast, in the presence of low plasma-ADH concentration, the collecting ducts become relatively impermeable to water and the interstitial osmotic gradient is ineffective in inducing water movement out of the collecting ducts; therefore a large volume of dilute urine is excreted, thereby compensating for a pure water excess.

Osmoreceptor control of ADH secretion

To reiterate, pure water deficits or gains are compensated for by partially dissociating water excretion from that of salt through changes in ADH secretion. What variable controls ADH under such conditions? The answer is changes in extracellular osmolarity. The adaptive rationale should be obvious, since osmolarity is the variable most affected by pure water gains or deficits. Osmoreceptors are located in the hypothalamus, and information from them is transmitted by neurons to the hypothalamic cells which secrete ADH. Via these connections an increase in osmolarity increases the rate of ADH secretion, and, conversely, decreased osmolarity inhibits ADH secretion (Fig. 13-20).

We have now described two different afferent pathways controlling the ADH-secreting hypothalamic cells, one from baroreceptors and one from osmoreceptors. These hypothalamic cells are therefore true integrating centers whose rate of activity is determined by the total synaptic input. Thus, a simultaneous increase in extracellular volume (detected by the baroreceptors) and decrease in extracellular osmolarity (detected by the osmoreceptors) cause maximal inhibition of ADH secretion; conversely, the opposite changes produce maximal stimulation. But what happens in the following situation? A person suffering from severe diarrhea loses 3 L of salt and water during the same time that he drinks 2 L of pure water. His total extracellular volume is decreased, but his osmolarity is also decreased. As a result, the ADH-producing cells receive opposing input from the baroreceptors and osmoreceptors. Which predominates depends completely upon the strength of the two inputs.

To add to the complexity, these cells receive synaptic input from many other brain areas, so that ADH secretion (and therefore urine flow) can be altered by pain, fear, and a variety of other factors. However, these effects are usually short-lived and should not obscure the generalization that ADH secretion is determined primarily by the states of extracellular volume and osmolarity. Alcohol is a powerful inhibitor of ADH release and probably accounts for much of the large urine flow accompanying the ingestion of alcohol.

The disease *diabetes insipidus*, which is distinct from diabetes mellitus, i.e., sugar diabetes, illustrates what happens when the ADH system is disrupted. This disease is characterized by the constant excretion of a large volume of highly dilute urine (as much as 25 L/day). In most cases, the flow can be restored to normal by the administration of ADH. These patients apparently have lost the ability to produce ADH, usually as a result of damage to the hypothalamus. Thus, renal tubular permeability to water is low and unchanging regardless of extracellular osmolarity or volume. The very thought of having to urinate (and therefore to drink) 25 L of water per day underscores the importance of ADH in the control of renal function and body water balance.

Figure 13-21 shows all these factors which control renal sodium and water excretion in operation in response to severe sweating, as in exercise; the renal retention of fluid helps to compensate for the water and salt lost in the sweat.

Thirst and salt appetite

Now we must turn to the other component of the balance, control of intake. It should be evident that large deficits of salt and water can be only partly compensated by renal conservation and that ingestion is the ultimate compensatory mechanism. The subjective feeling of thirst, which drives one to obtain and ingest water, is stimulated both by a lower extracellular volume and a higher

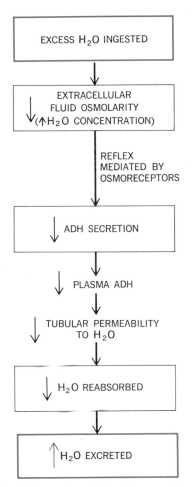

Figure 13-20. Pathway by which ADH secretion is lowered and water excretion raised when excess water is ingested.

plasma osmolarity, the adaptive significance of both being self-evident. Note that these are precisely the same changes which stimulate ADH production. Indeed, the centers which mediate thirst are also located in the hypothalamus very close to those areas which produce ADH. Damage to these centers abolishes thirst completely. Conversely electric stimulation of them may induce profound and prolonged drinking. These water-intake centers are very close to, but distinct from, the food-intake centers to be described in a later chapter. Because of the similarities between the stimuli for ADH secretion and thirst, it is tempting to speculate that the receptors (osmoreceptors and atrial baroreceptors) which initiate the ADH-controlling reflexes are identical to those for thirst.

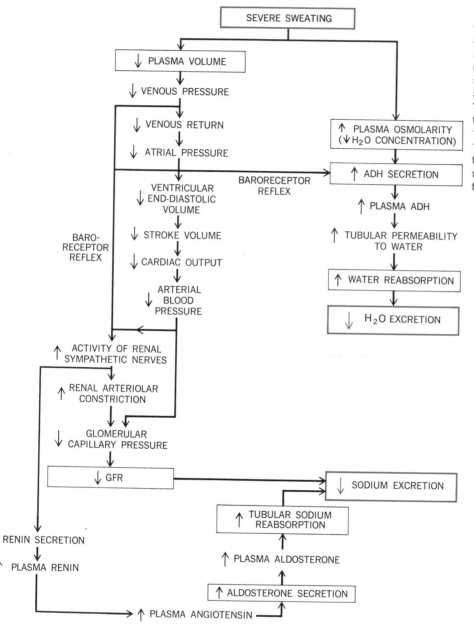

Figure 13-21. Pathways by which sodium and water excretion are decreased in response to severe sweating. (Sweat is a hypotonic salt solution, approximately half as concentrated as plasma.) This figure is an amalgamation of Figs. 13-13, 13-15, 13-17, and the converse of Fig. 13-20. Non-aldosterone factors controlling sodium reabsorption are not shown in the figure.

Much evidence indicates that this is, indeed, the case.

Of considerable interest is the fact that angiotensin stimulates thirst by a direct effect on the brain. Thus, the renin-angiotensin system is not only an important regulator of sodium balance but

of water balance as well and constitutes one of the pathways by which thirst is stimulated when extracellular volume is decreased.

There are also other pathways controlling thirst. For example, dryness of the mouth and throat causes profound thirst, which is relieved by merely moistening them. It is fascinating that animals such as the camel (and people, to a lesser extent) which have been markedly dehydrated will rapidly drink just enough water to replace their previous losses and then stop; what is amazing is that when they stop, the water has not yet had time to be absorbed from the gastrointestinal tract into the blood. Some kind of "metering" of the water intake by the gastrointestinal tract has occurred, but its nature remains a mystery.

This last phenomenon may be only one example of the larger area concerning "learned," "feed-forward," or "anticipatory" control of thirst. As described above, there exist osmoreceptor and "volume receptor" reflexes influencing water intake; yet a person may normally regulate water intake with only minimal use of these reflexes. Much of our drinking is done in association with eating and is therefore called *prandial drinking.* The quantity of fluid drunk with each meal is, in large part, a learned response determined by past experience. Thus, as described in Chap. 7, such "anticipatory" responses help to *prevent* dehydration; in contrast to these "feedforward" phenomena, the osmoreceptor and volume-receptor reflexes go into operation only *after* a deficit (i.e., internal changes) has already occurred.

The analog of thirst for sodium, salt appetite, is also an extremely important component of sodium homeostasis in most mammals. It is clear that salt appetite is innate and consists of two components: "hedonistic" appetite and "regulatory" appetite; i.e., animals "like" salt and eat it whenever they can, regardless of whether they are salt-deficient, and, in addition, their drive to obtain salt is markedly increased in the presence of deficiency. The significance of these animal studies for people is unclear. Salt craving seems to occur in human beings who are severely salt-depleted but the contribution of regulatory salt appetite to everyday sodium homeostasis in normal persons is probably slight. On the other hand, people seem to have a strong hedonistic appetite for salt as manifested by almost universally large intakes of sodium whenever it is cheap and readily available. Thus, the average American intake of salt is 10 to 15 g/day despite the fact that human beings can survive quite normally on less than 0.5 g/day. Present evidence strongly suggests that a large salt intake may be an important contributor to the pathogenesis of hypertension.

Section C
Regulation of potassium, calcium, and hydrogen-ion concentrations

Potassium regulation

The potassium concentration of extracellular fluid is a closely regulated quantity. The importance of maintaining this concentration in the internal environment stems primarily from the role of potassium in the excitability of nerve and muscle. Recall that the resting-membrane potentials of these tissues are directly related to the relative intracellular and extracellular potassium concentrations. Raising the external potassium concentration depolarizes membranes, thus increasing cell excitability. Conversely, lowering the external potassium hyperpolarizes cell membranes and reduces their excitability.

Since most of the body's potassium is found within cells, primarily as a result of active ion-transport systems located in cell membranes (Chap. 6), even a slight alteration of the rates of ion transport across cell membranes can produce a large change in the amount of extracellular potassium. Unfortunately, little is known about the physiological control of these transport mechanisms, and our understanding of the regulation of extracellular potassium concentration will remain incomplete until further data are obtained on this critical subject.

The normal person remains in total body potassium balance (as is true for sodium balance) by daily excreting an amount of potassium in the urine equal to the amount ingested minus the small amounts normally eliminated in the feces and sweat. Again the control of renal function is the major mechanism by which body potassium is regulated.

Renal regulation of potassium

Potassium is freely filterable at the glomerulus. The amounts of potassium excreted in the urine are generally a small fraction (10 to 15 percent) of the quantity filtered, thus establishing the existence of tubular potassium reabsorption. However, under certain conditions, the excreted quantity may actually exceed that filtered, thus establishing the existence of tubular potassium secretion. The subject is therefore complicated

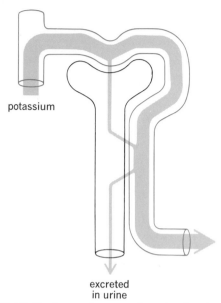

potassium

excreted
in urine

Figure 13-22. Basic renal processing of potassium. Since most of the filtered potassium is reabsorbed, potassium excreted in the urine results mainly from tubular secretion.

by the fact that potassium can be both reabsorbed and secreted by the tubule. Present evidence suggests, however, that normally most of the filtered potassium is reabsorbed (by active transport) regardless of changes in body potassium balance. In other words, the reabsorption of potassium does not seem to be controlled so as to achieve potassium homeostasis. The important result of this phenomenon is that changes in potassium *excretion* are due to changes in potassium *secretion* (Fig. 13-22). Thus, during potassium depletion when the homeostatic response is to reduce potassium excretion to a minimal level, there is no significant potassium secretion, and only the small amount of potassium escaping reabsorption is excreted. In all other situations, to this same small amount of unreabsorbed potassium is added a variable amount of secreted potassium. Thus, in describing the homeostatic control of potassium excretion we may ignore changes in GFR or potassium reabsorption and focus only on the factors which alter the rate of tubular potassium secretion.

One of the most important of these factors is the potassium concentration of the renal tubular cells themselves. When a high-potassium diet is ingested, potassium concentration in most of the

body's cells increases, including the renal tubular cells. This higher concentration facilitates potassium secretion into the lumen and raises potassium excretion. Conversely, a low-potassium diet or a negative potassium balance, e.g., from diarrhea, lowers renal tubular cell potassium concentration; this reduces potassium entry into the lumen and decreases potassium excretion, thereby helping to reestablish potassium balance.

A second important factor controlling potassium secretion is the hormone aldosterone, which besides assisting tubular sodium reabsorption enhances tubular potassium secretion. The reflex by which changes in extracellular volume control aldosterone production is completely different from the reflex initiated by an excess or deficit of potassium. The former constitutes a complex pathway, involving renin and angiotensin; the latter, however, seems to be much simpler (Fig. 13-23): The aldosterone-secreting cells of the adrenal cortex are apparently themselves sensitive to the potassium concentration of the extracellular fluid bathing them. For example, an increased intake of potassium leads to an increased extra-

Figure 13-23. Pathway by which an increased potassium intake induces greater potassium excretion mediated by aldosterone.

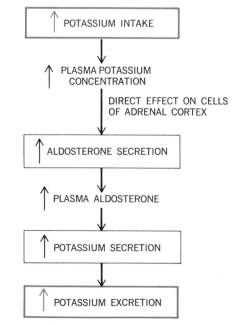

cellular potassium concentration, which in turn directly stimulates aldosterone production by the adrenal cortex. This extra aldosterone circulates to the kidney, where it increases tubular potassium secretion and thereby eliminates the excess potassium from the body. Conversely, a lowered extracellular potassium concentration would decrease aldosterone production and thereby inhibit tubular potassium secretion. Less potassium than usual would be excreted in the urine, thus helping to restore the normal extracellular potassium concentration. Again, complete compensation depends upon the ingestion of additional potassium. The control and renal tubular effects of aldosterone are summarized in Fig. 13-24.

Calcium regulation

Extracellular calcium concentration is also normally held relatively constant, the requirement for precise regulation stemming primarily from the profound effects of calcium on neuromuscular excitability. A low calcium concentration increases the excitability of nerve and muscle cell membranes so that patients with diseases in which low calcium occurs suffer from *hypocalcemic tetany*, characterized by skeletal muscle spasms. Hypercalcemia is also dangerous in that it causes cardiac arrhythmias as well as depressed neuromuscular excitability.

Effector sites for calcium homeostasis

At least three effector sites are involved in the regulation of extracellular calcium concentration: bone, the kidney, and the gastrointestinal tract.

Bone. Approximately 99 percent of total body calcium is contained in bone, which consists primarily of a framework of organic molecules upon which calcium phosphate crystals are deposited. Contrary to popular opinion, bone is not an absolutely fixed, unchanging tissue but is constantly being remolded and, what is more important, is available for either the withdrawal or deposit of calcium from extracellular fluid.

Gastrointestinal tract. Under normal conditions considerable amounts of ingested calcium are not absorbed from the intestine, but are eliminated in

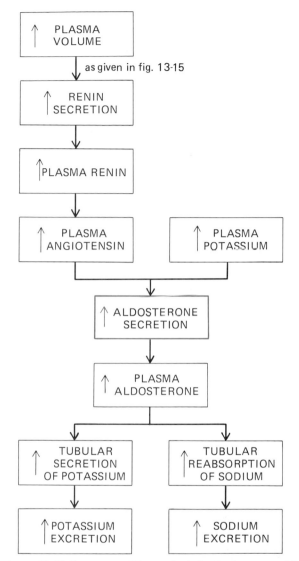

Figure 13-24. Summary of the control of aldosterone and its functions.

the feces. Accordingly, control of the active transport system which moves this ion from gut lumen to blood can result in large increases or decreases in the rate of absorption.

Kidney. The kidney handles calcium by filtration and reabsorption. In addition, as we shall see, the renal handling of phosphate also plays an important role in the regulation of extracellular calcium.

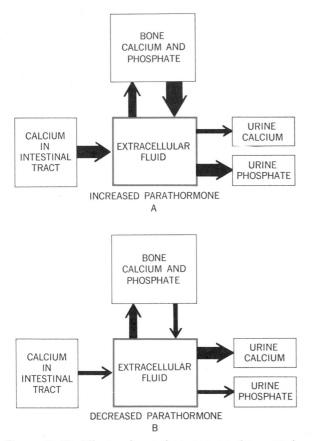

Figure 13-25. Effects of parathormone on the gastrointestinal tract, kidneys, and bone, the arrows signifying relative magnitudes. Note that when parathormone is decreased there is net movement of calcium and phosphate into bone, urine calcium is raised, and gastrointestinal absorption of calcium is reduced. The action on the intestinal tract is actually mediated by vitamin D, which parathormone causes to be activated.

Parathormone

All three of the effector sites described above are subject, directly or indirectly, to control by a protein hormone, called *parathormone* (also called parathyroid hormone), produced by the parathyroid glands. Parathormone production is controlled by the calcium concentration of the extracellular fluid bathing the cells of these glands. Lower calcium concentration stimulates parathormone production and release, and a higher concentration does just the opposite. It should be emphasized that extracellular calcium concentration acts directly upon the parathyroids (just as was true of the relation between extracellular potassium and aldosterone production) without any intermediary hormones or nerves.

Parathormone exerts at least four distinct effects on the sites described earlier (Fig. 13-25):

1 It increases the movement of calcium (and phosphate) from bone into extracellular fluid, making available this immense store of calcium for the regulation of extracellular calcium concentration; how it does this is still controversial.
2 It increases gastrointestinal absorption of calcium. This is not a direct action of parathormone on the intestine but is the indirect result of the hormone's effect on vitamin D, as discussed below.
3 It increases renal tubular calcium reabsorption, thus decreasing urinary calcium excretion.
4 It reduces the renal tubular reabsorption of phosphate, thus raising its urinary excretion and lowering extracellular phosphate concentration.

The adaptive value of the first three should be obvious: They all result in a higher extracellular calcium concentration, thus compensating for the lower concentration which originally stimulated parathormone production. Conversely, an increase in extracellular calcium concentration inhibits normal parathormone production, thereby producing increased urinary and fecal calcium loss and net movement of calcium from extracellular fluid into bone (Fig. 13-25).

The adaptive value of the fourth effect requires further explanation. Because of the solubility characteristics of undissociated calcium phosphate, the extracellular concentrations of ionic calcium and phosphate bear the following relationship to each other: The product of their concentrations, i.e., calcium times phosphate, is approximately a constant. In other words, if the extracellular concentration of phosphate increases, it forces the deposition of some extracellular calcium in bone, lowering the calcium concentration and keeping the calcium phosphate product a constant. The converse is also true. Imagine now what happens when parathormone releases calcium from bone. Both calcium and phosphate are released into the extracellular fluid. If the phosphate concentration is allowed to increase, further movement of calcium from bone is retarded; but in addition to this effect on bone, we have seen that parathormone also decreases tubular reabsorption of phosphate, thus permitting the excess phosphate to be eliminated in the urine. Indeed, extracellular phosphate may actually be reduced by this mechanism, which would allow even more calcium to be mobilized from bone.

Parathormone has other functions, notably its role in milk production, but the four effects discussed above constitute the major mechanisms by which it integrates various organs and tissues in the regulation of extracellular calcium concentration.

Vitamin D

Vitamin D plays an important role in calcium metabolism, as attested by the fact that its deficiency results in poorly calcified bones. Vitamin D really should be called a hormone since it can be produced by the skin in the presence of sunlight; only when adequate sunlight is not available are people dependent on dietary intake of vitamin D. The major action of vitamin D is to stimulate active calcium absorption by the intestine. Thus, the major event in vitamin D deficiency is decreased gut calcium absorption, resulting in decreased plasma calcium. In children, the newly formed bone protein matrix fails to be calcified normally because of the low plasma calcium, leading to the disease *rickets.*

The molecular form in which vitamin D is ingested or formed in the skin is relatively inactive and must undergo several transformations in the liver and kidneys before it is able to stimulate transport of calcium by the intestine. The molecular alteration which occurs in the kidneys seems to be particularly critical and is, itself, subject to physiological regulation by several inputs. One of these is parathormone; accordingly, a decreased plasma calcium stimulates the secretion of parathormone which, in turn, enhances the activation of vitamin D which, by its actions on the intestine, helps to restore plasma calcium to normal. Thus, the feedback loops for parathormone and vitamin D are closely intertwined.

Calcitonin

Another hormone, known as *calcitonin*, has significant effects on plasma calcium. Calcitonin is secreted by cells in the thyroid gland which surround but are completely distinct from the thyroxine-secreting follicles. Calcitonin lowers plasma calcium primarily by inhibiting bone resorption. Its secretion is controlled directly by the calcium concentration of the plasma supplying the thyroid gland; increased calcium causes increased calcitonin secretion. Thus, this system constitutes a second feedback control over plasma calcium concentration, one that is opposed to the parathormone system. However, its overall contribution to calcium homeostasis is minor compared with that of parathormone.

Hydrogen-ion regulation

Most metabolic reactions are highly sensitive to the hydrogen-ion concentration[1] of the fluid in which they occur. This sensitivity is due primarily to the influence on enzyme function exerted by the hydrogen ion. Accordingly, the hydrogen-ion concentration of the extracellular fluid is closely regulated.

Basic definitions

The hydrogen ion is an atom of hydrogen which has lost its only electron. When dissolved in water, many compounds dissociate reversibly to produce negatively charged ions (anions) and hydrogen ions, e.g.,

$$\text{Lactic acid} \rightleftharpoons \text{H}^+ + \text{lactate}^-$$
$$\underset{\text{Carbonic acid}}{\text{H}_2\text{CO}_3} \rightleftharpoons \text{H}^+ + \underset{\text{Bicarbonate}}{\text{HCO}_3^-}$$

Any compound capable of liberating a hydrogen ion in this manner is called an *acid.* Conversely, any substance which can accept a hydrogen ion is termed a *base.* Thus, in the reactions above, lactate and bicarbonate are bases since they can bind hydrogen ions. The hydrogen-ion concentration often is referred to in terms of *acidity*: the higher the hydrogen-ion concentration, the greater the acidity (it must be understood that the hydrogen-ion con-

---

[1] Hydrogen-ion concentration is frequently expressed in terms of pH, which is defined as the negative logarithm to the base 10 of the hydrogen-ion concentration: $\text{pH} = -\log \text{H}^+$. This can be confusing for several reasons, not the least of which is that pH decreases as H^+ increases. We have chosen not to use pH in this text.

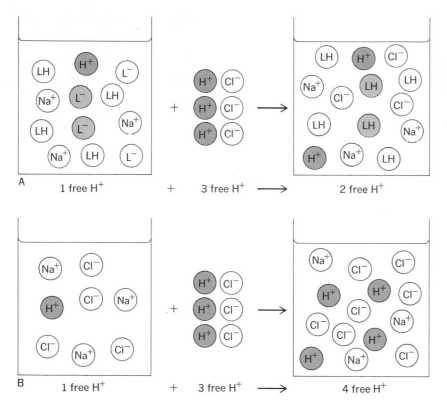

Figure 13-26. (A) Example of a buffer system. LH = lactic acid, L⁻ = lactate. When HCl is added to the beaker, two of the hydrogen ions react with lactate to give lactic acid; therefore, only one of the three added hydrogen ions remains free in solution. Contrast this to B, in which no buffer is present and all three of the added hydrogen ions remain free in solution.

centration of a solution refers only to the hydrogen ions which are *free* in solution).

Strong and weak acids

Strong acids dissociate completely when they dissolve in water. For example, hydrochloric acid added to water gives

$$\text{HCl} \longrightarrow \text{H}^+ + \text{Cl}^-$$
Hydrochloric
acid

Virtually no HCl molecules exist in the solution, only free hydrogen ions and chloride ions. On the other hand, *weak acids* are those which do not dissociate completely when dissolved in water. For example, when dissolved, only a fraction of lactic acid molecules dissociate to form lactate and hydrogen ions, the other molecules remaining intact. This characteristic of weak acids underlies an important chemical and physiological phenomenon, *buffering.*

Buffer action and buffers

Figure 13-26 pictures a solution made by dissolving lactic acid and sodium lactate in water. The sodium lactate dissociates completely into sodium ions and lactate ions, but only a very small fraction of the lactic acid molecules dissociate to form hydrogen ions and lactate ions. Accordingly, the solution has a relatively low concentration of hydrogen ion and relatively high concentrations of undissociated lactic acid molecules, sodium ions, and lactate ions. Lactic acid, hydrogen ion, and lactate are in equilibrium with each other:

$$\text{Lactic acid} \rightleftharpoons \text{lactate}^- + \text{H}^+$$

By the mass-action law, an increase in the concentration of any substances on one side of the arrows forces the reaction in the opposite direction. Conversely, a decrease in the concentration of any substances on one side of the arrows forces the reaction toward that side, i.e., in the direction which generates more of that substance. What happens if we add hydrochloric acid to this solution? Hydrochloric acid, a strong acid, completely dissociates and liberates hydrogen ions. This excess of hydrogen ions drives the lactic acid reaction to the left, and many of the hydrogen ions liberated from the dissociation of hydrochloric acid thus combine with lactate to give undissociated lactic acid. As a result, many of the hydrogen ions generated by the dissociation of hydrochloric acid do not remain free in solution but become incorporated into lactic acid molecules. The final hydrogen-ion concentration is therefore smaller than if the hydrochloric acid had been added to pure water (Fig. 13-26).

Conversely, if instead of adding HCl, we add a chemical which *removes* hydrogen ions, the lactic acid reaction is driven to the right, lactic acid molecules dissociate to generate more hydrogen ions, and the original fall in hydrogen-ion concentration is minimized. This process preventing large changes in hydrogen-ion concentration when hydrogen ion is added or removed from a solution is termed *buffering*, and the chemicals (in this case lactate and lactic acid) are called *buffers* or *buffer systems*.

Generation of hydrogen ions in the body

There are three major sources of hydrogen ion in the body:

1 Phosphorus and sulfur are present in large quantities in many proteins and other biologically important molecules. The catabolism of these molecules releases phosphoric and sulfuric acids into the extracellular fluid. These acids, to a large extent, dissociate into hydrogen ions and anions (phosphate and sulfate). For example,

$$H_2PO_4^- \longrightarrow H^+ + HPO_4^{2-}$$
$$H_2SO_4 \longrightarrow 2\,H^+ + SO_4^{2-}$$

2 Many organic acids, e.g., fatty acids and lactic acid, are produced as end products of metabolic reactions and also liberate hydrogen ion by dissociation.
3 We have already described (Chap. 12) the major source of hydrogen ion, namely, liberation of hydrogen ion from metabolically produced carbon dioxide via the reactions

$$CO_2 + H_2O \longrightarrow H_2CO_3 \longrightarrow H^+ + HCO_3^-$$
$$\text{Bicarbonate}$$

As described in Chap. 12, the lungs normally eliminate carbon dioxide from the body as rapidly as it is produced in the tissues. As blood flows through the lung capillaries and carbon dioxide diffuses into the alveoli, the chemical reactions which originally generated HCO_3^- and H^+ from carbon dioxide and water in the venous blood are reversed:

$$CO_2 + H_2O \longleftarrow H_2CO_3 \longleftarrow H^+ + HCO_3^-$$

As a result, all the hydrogen ion generated from carbonic acid is reincorporated into water molecules. Therefore, there is *normally* no net gain or loss of hydrogen ion in the body from this source, but what happens when lung disease prevents adequate elimination of carbon dioxide? The retention of carbon dioxide means an elevated extracellular P_{CO_2}. Some of the hydrogen ions generated from the reaction of this carbon dioxide with water are also retained and raise the extracellular hydrogen-ion concentration. This hydrogen ion must be eliminated from the body (via the kidneys) in order for hydrogen-ion balance to be maintained.

Buffering of hydrogen ions in the body

Between the generation of hydrogen ions in the body and their excretion (to be described) what happens? The hydrogen-ion concentration of the extracellular fluid is extremely small, approximately 0.00004 mmol/L. What would happen to this concentration if just 2 mmol of hydrogen ion remained free in solution following its dissociation from an acid? Since the total extracellular fluid is approximately 12 L, the hydrogen-ion concentration would increase by 2/12, or approximately 0.167 mmol/L. Since the original concentration was only 0.00004, this would represent more than a 4000-fold increase. Obviously, such a rise does not occur, for we have already observed that the extracellular hydrogen ion is kept remarkably constant. Therefore, of the 2 mmol of hydrogen ion liberated in our example, only an extremely small portion can have remained free in solution. The vast majority has been bound (buffered) by other ions. The kidneys ultimately eliminate excess hydrogen ions, but it is buffering which minimizes hydrogen-ion concentration changes until excretion occurs.

The most important body buffers are bicarbonate —CO_2, large anions such as plasma proteins and intracellular phosphate complexes, and hemoglobin. Recall that only free hydrogen ions contribute to the acidity of a solution. These buffers all act by binding hydrogen ions according to the general reaction

$$\text{Buffer}^- + H^+ \rightleftharpoons \text{H-buffer}$$

It is evident that H-buffer is a weak acid in that it can exist as the undissociated molecule or can dissociate to buffer$^-$ + H$^+$. When hydrogen-ion concentration increases, the reaction is forced to the right and more hydrogen ion is bound. Conversely, when hydrogen-ion concentration decreases, the reaction proceeds to the left and hydrogen ion is released. In this manner, the body buffers stabilize hydrogen-ion concentration against changes in either direction.

Bicarbonate-CO_2. In this section and Chap. 12 we have already described the relationships between HCO_3^-, H^+, and CO_2. Let us once more write the pertinent equations in their true forms as reversible equations:

$$H^+ + HCO_3^- \rightleftharpoons H_2CO_3 \rightleftharpoons H_2O + CO_2$$

The basic mechanism by which this system acts as a buffer should be evident: An increased extracellular hydrogen-ion concentration drives the reaction to the right, H^+ and HCO_3^- combine, hydrogen ion is thereby removed from solution, and the hydrogen-ion concentration returns toward normal. Conversely, a decreased extracellular hydrogen-ion concentration drives the reaction to the left, CO_2 and H_2O combine to generate hydrogen ion, and this additional hydrogen ion returns

hydrogen-ion concentration toward normal. One reason for the importance of this buffer system is that the extracellular bicarbonate concentration is normally quite high and is closely regulated by the kidney. A second reason stems from the relationship between extracellular hydrogen-ion concentration and carbon dioxide elimination from the body.

When additional hydrogen ion is added to the extracellular fluid, i.e., when the H^+ combines with HCO_3^-, the extent to which this reaction can restore hydrogen-ion concentration to normal depends upon precisely how much additional H^+ actually combines with HCO_3^- and is thereby removed from solution. Complete compensation (which never actually occurs) can be obtained only if the HCO_3^-—CO_2 reaction proceeds to the right until all the additional H^+ has combined with HCO_3^-. However, this reaction obviously generates CO_2, which seriously hinders the further buffering ability of HCO_3^- because, as can be seen from the equation, any increase in the concentration of CO_2, by the mass-action law, tends to drive the reaction back to the left, thus preventing further net combination of $H^+ + HCO_3^-$. In reality, however, this expected increase in extracellular CO_2 does not occur. Indeed, during periods of increased body hydrogen-ion production from organic, phosphoric, or sulfuric acids, the extracellular CO_2 is actually *decreased!* What causes this? We have already studied the mechanism, in Chap. 12: A greater extracellular hydrogen-ion concentration stimulates the respiratory centers to increase alveolar ventilation and thereby causes greater elimination of carbon dioxide from the body (Fig. 13-27). Thus, although an increased combination of H^+ and HCO_3^- generates more

Figure 13-27. Effects of excess hydrogen ions on plasma carbon dioxide. The direct effect, by mass action, is to increase the production of carbon dioxide, but the indirect effect is to lower carbon dioxide by reflexly stimulating breathing. Since the latter effect predominates, the net effect is a reduction of plasma carbon dioxide.

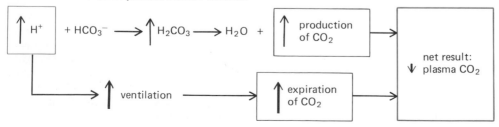

carbon dioxide, respiratory stimulation produced by the higher extracellular hydrogen-ion concentration results in the elimination of carbon dioxide even faster than it is generated. As a result, the extracellular carbon dioxide decreases, and, by the mass-action law, the further combination of H^+ and HCO_3^- is actually facilitated.

A lower extracellular hydrogen-ion concentration resulting from either decreased hydrogen-ion production or increased hydrogen-ion loss from the body is compensated for by just the opposite buffer reactions: (1) The lower hydrogen-ion concentration drives the HCO_3^-—CO_2 reaction to the left, carbon dioxide and water combine to generate H^+ + HCO_3^-, and this additional hydrogen ion returns hydrogen-ion concentration toward normal; (2) the lower hydrogen-ion concentration decreases alveolar ventilation and carbon dioxide elimination by inhibiting the medullary respiratory centers. The elevated extracellular carbon dioxide resulting from this process also serves to drive the HCO_3^-—CO_2 reaction to the left, thus allowing further generation of hydrogen ion.

In Chap. 12 we pointed out that most metabolically produced carbon dioxide is carried from the tissues to the lungs in the form of HCO_3^-. These bicarbonate ions are generated by the hydration of CO_2 to form H_2CO_3 and the dissociation of H_2CO_3 to HCO_3^- and H^+. As blood flows through the lungs, the reaction is reversed, and H^+ and HCO_3^- recombine. However, these hydrogen ions must be buffered while they are in transit from the tissues to the lungs, a function performed primarily by hemoglobin. This phenomenon and the physiological basis for it was described in Chap. 12.

Renal regulation of extracellular hydrogen-ion concentration

It should be emphasized that none of the buffer systems described above eliminates hydrogen ion from the body; instead they combine with the hydrogen ion. Binding thus removes the hydrogen ion from solution, preventing it from contributing to the free hydrogen-ion concentration; the actual elimination of hydrogen ion from the body is normally performed only by the kidneys.

In a normal person the quantity of phosphoric, sulfuric, and organic acids formed depends primarily upon the type and quantity of food ingested. A high-protein diet, for example, results in increased protein breakdown and release of large quantities of sulfuric acid. The average American diet results in the liberation of 40 to 80 mmol of hydrogen ion each day. If hydrogen-ion balance is to be maintained, the same quantity must be eliminated from the body. This loss occurs via the kidneys. In addition, the kidneys must be capable of *altering* their hydrogen-ion excretion in response to changes in body hydrogen-ion production, regardless of whether the source is carbon dioxide (in lung disease) or phosphoric, sulfuric, or organic acids. The kidneys must also be able to compensate for any gastrointestinal loss or gain of hydrogen ion resulting from disease.

Virtually all the hydrogen ion excreted in the urine enters the tubules via tubular secretion, the mechanism and its control being quite complex. Suffice it to say that the presence of excess acid in the body induces an increased urinary hydrogen-ion excretion. Conversely, the kidney responds to a decreased amount of acid in the body by lowering urinary hydrogen-ion excretion. The controlling effect of the acid appears to be primarily exerted directly upon the tubular cells, with no nerve or hormone intermediates. Obviously, such a system is effective in stabilizing hydrogen-ion concentration at the normal value.

The ability of the kidneys to excrete H^+ depends both upon tubular hydrogen-ion secretion and upon the presence of buffers in the urine to combine with the hydrogen ions. The major urinary buffers are HPO_4^{2-} and ammonia NH_3:

$$HPO_4^{2-} + H^+ \longrightarrow H_2PO_4^-$$

$$NH_3 + H^+ \longrightarrow NH_4^+$$

The HPO_4^{2-} in the tubular fluid has been filtered and not reabsorbed. In contrast, the ammonia of the tubular fluid is formed by the tubular cells themselves by the deamination of certain amino acids transported into the renal tubular cells from the peritubular capillary plasma. From the cells the ammonia diffuses into the lumen. The amount which *remains* in the lumen depends upon the hydrogen-ion concentration there, since the tubular cell membrane is quite permeable to ammonia but not to NH_4^+. Accordingly, the more hydrogen ion there is in the lumen, the greater the conversion of NH_3 to NH_4^+, increased secretion of hydrogen ion automatically inducing an increased excretion of NH_4^+. Another important feature of this system is that the rate of ammonia production by the renal tubular cells increases whenever extracellular hydrogen-ion concentration remains elevated for more than 1 to 2 days; this extra ammonia provides the additional buffers required for combination with the increased hydrogen ions secreted.

Classification of disordered hydrogen-ion concentration

Acidosis refers to any situation in which the hydrogen-ion concentration of arterial blood is elevated; *alkalosis* denotes a reduction. It should be evident that acidosis and alkalosis are the results of an imbalance between hydrogen-ion gain and loss, and all such situations fit into two distinct categories: (1) *respiratory acidosis or alkalosis*; (2) *metabolic acidosis or alkalosis*.

As its name implies, the first category results either from failure of the lungs to eliminate CO_2 as fast as it is produced (acidosis) or from elimination of CO_2 faster than it is produced (alkalosis). As described earlier, the imbalance of arterial hydrogen-ion concentrations in such cases is completely explainable in terms of mass action. Thus, the hallmark of a respiratory acidosis is an elevated arterial CO_2 and H^+, that of respiratory alkalosis is a reduction in both values.

The second category—metabolic acidosis or alkalosis—includes all situations other than those in which the primary problem is respiratory. Some common forms of metabolic acidosis are excessive production of lactic acid (during severe exercise or hypoxia) or of ketone acids (in uncontrolled diabetes mellitus or in fasting, Chap. 15). Metabolic acidosis can also result from excessive loss of bicarbonate as in diarrhea. A frequent cause of metabolic alkalosis is persistent vomiting, with its associated loss of hydrogen ions (as HCl) from the stomach.

What is the arterial CO_2 in metabolic, as opposed to respiratory, acidosis or alkalosis? Since, by definition, metabolic acidosis and alkalosis must be due to something other than excess retention or loss of CO_2, one might have predicted that arterial CO_2 would be unchanged, but such is not the case. As described earlier in this chapter (and in Chap. 12) the elevated hydrogen-ion concentration associated with the metabolic acidosis, say due to retention of lactic acid, reflexly stimulates ventilation via the peripheral chemoreceptors and lowers

arterial CO_2; by mass action this helps restore the hydrogen-ion concentration toward normal. Conversely, a person with metabolic alkalosis, say due to vomiting, will reflexly have ventilation inhibited; the result is a rise in arterial CO_2 and, by mass action, an associated elevation of hydrogen-ion concentration. To reiterate, the CO_2 changes in metabolic acidosis and alkalosis are not the *cause* of the acidosis or alkalosis but rather are compensatory reflex responses to primary nonrespiratory abnormalities.

Kidney disease

The term kidney disease is no more specific than "car trouble," since many diseases affect the kidneys. Bacteria, allergies, congenital defects, stones, tumors, and toxic chemicals are some possible sources of kidney damage. Obstruction of the urethra or a ureter may cause injury due to a buildup of pressure and may predispose the kidneys to bacterial infection. Disease can attack the kidney at any age. Experts estimate that there are at present more than 3 million undetected cases of kidney infection in the United States and that 25,000 to 75,000 Americans die of kidney failure each year.

Early symptoms of kidney disease depend greatly upon the type of disease involved and the specific part of the kidney affected. Although many diseases are self-limited and produce no permanent damage, others progress if untreated. The end stage of progressive diseases, regardless of the nature of the damaging agent (bacteria, toxic chemical, etc.), is a shrunken, nonfunctioning kidney. Similarly, the symptoms of profound renal malfunction are independent of the damaging agent and are collectively known as *uremia*, literally "urine in the blood."

The severity of uremia depends upon how well the impaired kidneys are able to preserve the constancy of the internal environment. Assuming that the patient continues to ingest a normal diet containing the usual quantities of nutrients and electrolytes, what problems arise? The key fact to keep in mind is that the kidney destruction has markedly reduced the number of functioning nephrons. Accordingly, the many substances which gain entry to the tubule primarily by filtration are filtered in diminished amounts. Of the substances described in this chapter, this category includes sodium chloride, water, calcium, and a number of waste products. In addition, the excretion of potassium, hydrogen ion, and certain other substances is impaired because there are too few nephrons capable of normal tubular secretion. The buildup of all these substances in the blood causes the symptoms of uremia.

The artificial kidney

The artificial kidney is an apparatus that eliminates the excess ions and wastes which accumulate in the blood when the kidneys fail. Blood is pumped from one of the patient's arteries through tubing which is bathed by a large volume of fluid. The tubing then conducts the blood back into the patient by way of a vein. The tubing is generally made of a cellophane, which is highly permeable to most solutes but relatively impermeable to protein—characteristics quite similar to those of capillaries. The bath fluid, which is constantly replaced, is a salt solution similar in ionic concentrations to normal plasma. The basic principle is simply that of dialysis, or diffusion. Because the

cellophane is permeable to most solutes, as blood flows through the tubing, solute concentrations tend to equilibrate in the blood and bath fluid. Thus, if the plasma potassium concentration of the patient is above normal, potassium diffuses out of the blood into the bath fluid. Similarly, waste products and excesses of other substances diffuse across the cellophane tubing and thus are eliminated from the body.

At first patients were placed on the artificial kidney only when there was reason to believe that their kidney damage was only temporary and that recovery would occur if the patient could be kept alive during temporary renal failure. However, technical improvements have permitted many patients to utilize the artificial kidney several times a week for unlimited periods, and thus patients with permanent kidney failure can be kept alive.

The other major hope for patients with permanent renal failure is kidney transplantation. Although great strides have been made, the major problem remains the frequent rejection of the transplanted kidney by the recipient's body (see Chap. 17).

14

THE DIGESTION AND ABSORPTION OF FOOD

Overview: Role of the various organs of the gastrointestinal system
Digestion and absorption
  Structure of the small intestine
  Carbohydrate
  Protein
  Fat
  Vitamins
  Nucleic acids
  Water and minerals
    Potassium • Calcium • Iron
Regulation of gastrointestinal processes
  Basic principles
    Neural regulation • Hormonal regulation • Phases of gastrointestinal control
  Mouth, pharynx, and esophagus

    Chewing • Saliva • Swallowing
Stomach·
    HCl secretion • Pepsin secretion • Gastric motility
Pancreatic secretions
Bile secretion
Small intestine
    Secretions • Motility
Large intestine
    Motility and defecation
Malfunctioning of the gastrointestinal tract
  Vomiting
  Ulcers
  Gallstones
  Lactose deficiency (milk intolerance)
  Constipation and diarrhea

The gastrointestinal system (Fig. 14-1) includes the gastrointestinal tract (mouth, esophagus, stomach, small and large intestines), salivary glands, and portions of the liver and pancreas; its function is to transfer organic nutrients, salts, and water from the external environment to the internal environment, where they can be distributed to cells by the circulatory system. Most food is taken into the mouth as large particles, consisting of high-molecular-weight substances such as proteins and polysaccharides, which are unable to cross cell membranes. Before these substances can be absorbed, they must be broken down into smaller molecules, such as amino acids and monosaccharides. This breaking-down process, *diges-*

tion, is accomplished mainly by the action of acid and digestive enzymes secreted into the tract or contained in its lining cells (Fig. 14-2). The small molecules resulting from digestion cross the cells of the intestinal tract *(absorption)* entering the blood and lymph. While these processes are taking place, smooth muscle contractions move the luminal contents through the tract.

Contrary to popular belief, the gastrointestinal system is not a major excretory organ for eliminating wastes from the body. It is true that small amounts of certain end products are normally eliminated in the feces, but the elimination of most wastes from the internal environment is performed by the lungs and kidneys. Feces consist primarily of bacteria and ingested material which failed to be digested and absorbed during its passage along the gastrointestinal tract, i.e., material that was never actually part of the internal environment.

The adult gastrointestinal tract is a tube approximately 4.5 m (15 ft) long, running through the body from mouth to anus. The lumen of this tube, like the hole in a doughnut, is continuous with the external environment, which means that its contents are technically outside of the body. This fact is relevant to an understanding of some of the tract's properties. For example, the lower portion of the intestinal tract is inhabited by millions of bacteria, most of which are harmless and even beneficial in this location; however, if the same bacteria enter the body, as may happen, for example, in the case of a ruptured appendix, they are extremely harmful and even lethal.

Throughout its entire length the wall of the gastrointestinal tract has the general structural organization illustrated in Fig. 14-3. The luminal surface of the tube is generally not flat and smooth, but highly convoluted, with many ridges and valleys which greatly increase the total surface area available for absorption. From the stomach on, it is covered by a single layer of epithelial cells, across which absorption takes place. Included in this epithelial layer are cells which secrete *mucus* and other cells which release hormones into the blood. Invaginations of the epithelial layer into the underlying tissue form tubular exocrine glands which secrete mucus, acid, enzymes, water, and ions into the lumen. Just below the epithelial surface is a layer of connective tissue, the lamina propria, through which pass small blood vessels, nerve fibers and lymphatic ducts. The lamina propria is separated from the underlying tissues by a thin layer of smooth muscle, the muscularis mucosa. The combination of these three outermost layers—the epithelium, lamina propria, and muscularis mucosa, is known as the *mucosa*.

Beneath the mucosa is a second connective tissue layer, the *submucosa*, through which pass the larger blood vessels and lymphatics from which branches penetrate into the overlying mucosal layer and the underlying muscular layer. Also located in this layer is a network of nerve cells known as the *submucus plexus*.

Beneath (i.e., surrounding) the submucosa is a layer of smooth muscle tissue, contractions of which provide the forces for moving the gastro-

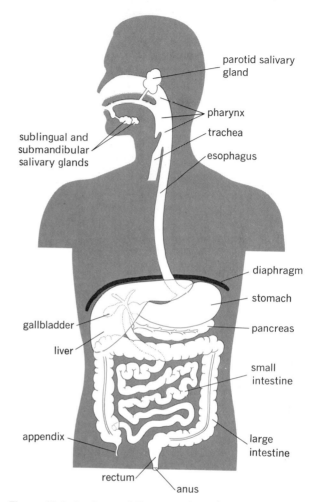

Figure 14-1. Anatomy of the gastrointestinal system.

intestinal contents. It consists of two layers—a relatively thick inner layer whose fibers are oriented in a *circular* pattern around the tube such that contraction produces a narrowing of the lumen, and a thinner, outer layer of *longitudinally* oriented fibers whose contraction shortens the tube. Located between these two layers of smooth muscle is another, more extensive, plexus of nerve cells, the *myenteric plexus*.

Finally, surrounding the outer surface of the tube is a layer of connective tissue, the *serosa*. Thin sheets of connective tissue, the *mesenteries*, which connect the serosa to the abdominal wall, support and anchor various segments of the gastrointestinal tract. At various points along the tube,

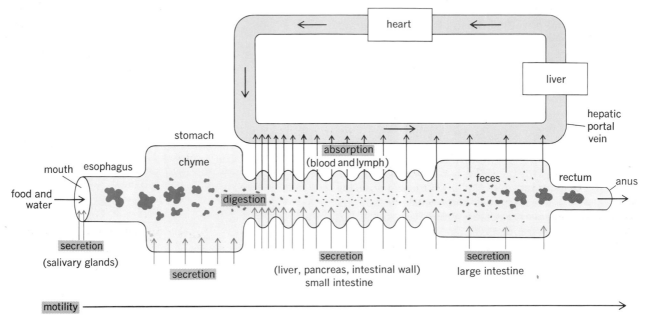

Figure 14-2. Summary of gastrointestinal activity involving motility, secretion, digestion, and absorption.

ducts coming from exocrine organs lying outside the gastrointestinal tract (for example, the liver and pancreas), penetrate the wall of the tract and empty into its lumen.

The activities of the gastrointestinal tract and the other organs of the gastrointestinal system (summarized in Fig. 14-4) are so complex that it is easy to lose sight of the fact that the overall function of this system is to get ingested nutrients digested and absorbed; the many substances secreted, the variety of smooth muscle contractile patterns, and the complex controls over all these processes, have as their basic end the achievement of optimal conditions within the tract for digestion and absorption. Accordingly we begin with a brief overview of the roles of the various organs in these two processes.

Overview: Role of the various organs of the gastrointestinal system

Digestion begins in the mouth with chewing, which breaks up large food particles into smaller particles that can be swallowed without choking. Prolonged chewing of food does not appear to be essential to the digestive process (many animals,

such as the dog and cat, swallow their food almost immediately). Although chewing prolongs the subjective pleasure of taste, it does not appreciably alter the rate at which the food is ultimately digested and absorbed from the small intestine. On the other hand, attempting to swallow a particle of food too large to enter the esophagus may lead to choking, as the particle lodges over the trachea, blocking the entry of air into the lungs. A surprising number of preventable deaths occur each year from choking, the symptoms of which are often confused with the sudden onset of a heart attack, so that no attempt is made to remove the obstruction from the airway (rapid, forceful compression of the thoracic cavity will often generate sufficient pressure within the lungs to dislodge the obstructing particle from the airways).

The salivary glands secrete into the mouth a mucus solution which moistens and lubricates the food particles prior to swallowing. Mucus is secreted by gland cells throughout the gastrointestinal tract. In addition to its lubricating function, mucus (in some regions) may provide a layer at the epithelial surface which protects underlying cells from abrasion and from the enzymes and other chemicals in the lumen. Saliva also contains a carbohydrate-digesting enzyme *amylase*, which begins the breakdown of large polysaccharides into smaller molecular fragments. A third function of saliva is to dissolve some of the molecules in the food particle; only in this dissolved state can they react with the chemoreceptors in the mouth, giving rise to the sensation of taste.

The next segments of the tract, the pharynx and esophagus, contribute nothing to digestion but simply provide the connection between the mouth and the stomach.

The stomach is a hollow bag which initially stores the contents of a meal. The glands lining the wall of the stomach secrete a very strong acid, *hydrochloric acid,* which dissolves the particles of food, forming a solution of molecules known as *chyme.* The high acidity of the chyme alters the ionization of carboxyl and amino groups in protein,

thereby changing protein structure so as to break up connective tissue and cells in the ingested food. Hydrochloric acid thus continues the process begun by chewing, namely reducing large particles of food to smaller particles, functioning as a solvent to yield a solution of individual molecules; however, acid has little ability to break down proteins and polysaccharides all the way to amino acids and glucose and it has virtually no digestive action at all on fats. A second function of gastric acid is to kill most of the bacteria that enter along with food. This process is not 100% effective and some bacteria normally survive to take up residence and multiply in the intestinal tract, particularly the large intestine.

Also secreted by the stomach glands are several protein-digesting enzymes collectively known as

Figure 14-3. Representative longitudinal section of wall of gastrointestinal tract. Not shown are the smaller blood vessels, neural connections between the two plexuses, and neural terminations on muscles and glands.

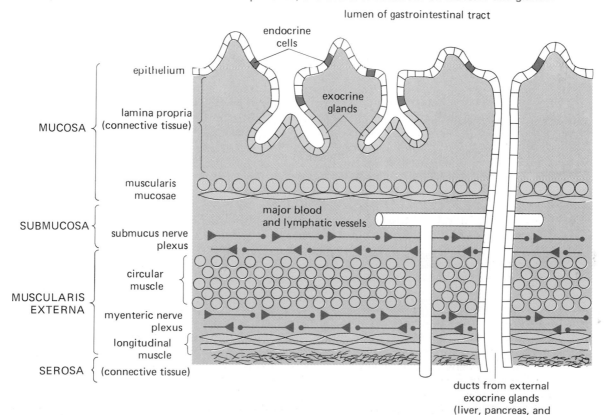

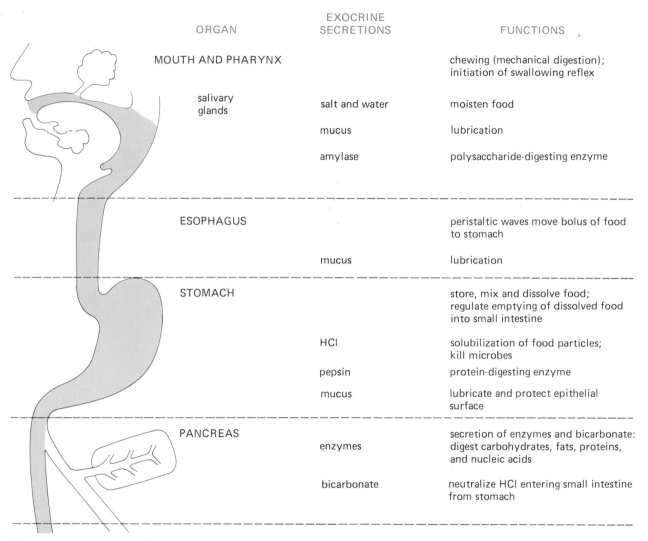

ORGAN	EXOCRINE SECRETIONS	FUNCTIONS
MOUTH AND PHARYNX		chewing (mechanical digestion); initiation of swallowing reflex
salivary glands	salt and water	moisten food
	mucus	lubrication
	amylase	polysaccharide-digesting enzyme
ESOPHAGUS		peristaltic waves move bolus of food to stomach
	mucus	lubrication
STOMACH		store, mix and dissolve food; regulate emptying of dissolved food into small intestine
	HCl	solubilization of food particles; kill microbes
	pepsin	protein-digesting enzyme
	mucus	lubricate and protect epithelial surface
PANCREAS	enzymes	secretion of enzymes and bicarbonate: digest carbohydrates, fats, proteins, and nucleic acids
	bicarbonate	neutralize HCl entering small intestine from stomach

Figure 14-4. Functions of the gastrointestinal organs.

pepsin, which split proteins into small peptide fragments. Although polysaccharides and proteins are partially digested while in the stomach by the actions of salivary amylase and gastric pepsin, the end products are large and electrically charged, so that they cannot diffuse across the epithelium, which also lacks carrier-mediated transport systems for them. The fats in the gastric chyme, which are not soluble in water, tend to aggregate into large droplets, and are not available for absorption. Thus, little absorption of organic nutrients occurs across the wall of the stomach,

and its major functions are to store, dissolve, and partially digest the contents of a meal, and to deliver chyme to the small intestine in amounts optimal for digestion and absorption in that segment.

Most digestion and absorption occurs in the small intestine; large molecules of intact or partially digested carbohydrate, fat, and protein are broken down by enzymes into monosaccharides, fatty acids, and amino acids, all of which are able to cross the membranes of the epithelial cells either by diffusion or mediated transport and be

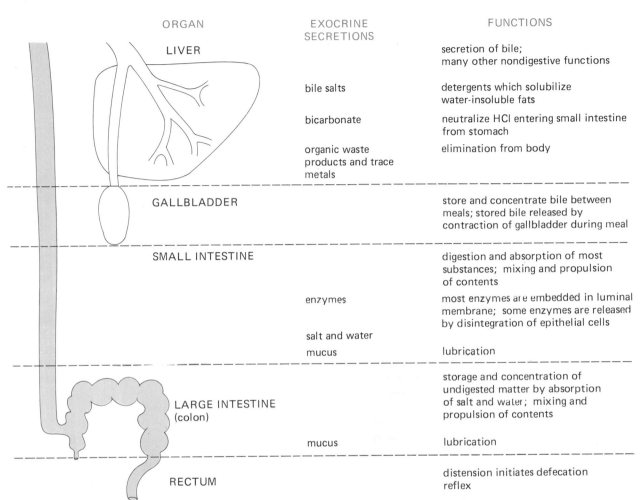

ORGAN	EXOCRINE SECRETIONS	FUNCTIONS
LIVER		secretion of bile; many other nondigestive functions
	bile salts	detergents which solubilize water-insoluble fats
	bicarbonate	neutralize HCl entering small intestine from stomach
	organic waste products and trace metals	elimination from body
GALLBLADDER		store and concentrate bile between meals; stored bile released by contraction of gallbladder during meal
SMALL INTESTINE		digestion and absorption of most substances; mixing and propulsion of contents
	enzymes	most enzymes are embedded in luminal membrane; some enzymes are released by disintegration of epithelial cells
	salt and water	
	mucus	lubrication
LARGE INTESTINE (colon)		storage and concentration of undigested matter by absorption of salt and water; mixing and propulsion of contents
	mucus	lubrication
RECTUM		distension initiates defecation reflex

absorbed into the blood and lymph. Other organic nutrients, as well as salts and water, are also absorbed in this segment. Some of the enzymes which carry out this digestive process are on the luminal membranes of the epithelial cells of the small intestine, but most are secreted by the pancreas into the lumen.

The exocrine portion of the pancreas secretes enzymes specific for each of the major classes of organic molecules. These enzymes enter the small intestine through a duct leading from the pancreas to the first portion of the small intestine, the *duodenum.* The high acidity of the chyme coming from the stomach would inactivate these enzymes if the hydrochloric acid were not neutralized by the sodium bicarbonate also secreted in large amounts by the exocrine pancreas.

Because fat is insoluble in water, the digestion

of fat in the small intestine requires special processes to solubilize these molecules. This is brought about by a group of detergent molecules, known as *bile salts,* which are secreted by the liver into bile ducts which eventually join the pancreatic duct and empty into the duodenum. In addition to bile salts, bile contains sodium bicarbonate which helps to neutralize acid from the stomach. Between meals, secreted bile is stored in a small sac underneath the liver, the *gallbladder,* which concentrates the bile by absorbing salts and water. During a meal, the gallbladder contracts, causing a concentrated solution of bile to be injected into the small intestine.

Monosaccharides and amino acids are absorbed across the wall of the small intestine, mainly by specific carrier-mediated transport processes in the epithelial membranes, while fatty acids enter the

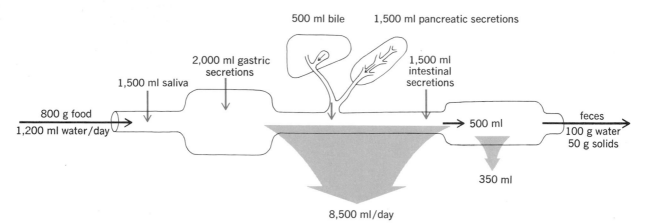

Figure 14-5. Average amounts of food and fluid ingested, secreted, absorbed, and excreted from the gastrointestinal tract daily.

epithelial cells by diffusion. Salts are also actively absorbed, and water follows passively down osmotic gradients. Most digestion and absorption has been completed by the middle portion of the small intestine.

The motility of the small intestine mixes the contents of the lumen and slowly advances the contents toward the large intestine. Only a small volume of salts, water, and undigested material is passed on to the large intestine, which stores the undigested material prior to defecation and concentrates it by absorbing salts and water.

It must be recognized that although the average adult consumes about 800 g of solid food and 1200 mL of water per day, this is only a fraction of the total material entering the gastrointestinal tract. To the approximately 2000 mL of ingested food and drink are added about 7000 mL of fluid from the salivary glands, stomach, pancreas, liver, and intestinal tract (Fig. 14-5). Of this total volume, only about 100 mL is lost in the feces, the rest being absorbed across the walls of the intestine. Almost all of the secreted salts and enzymes (the latter after digestion to amino acids) are also absorbed back into the circulation. Thus, little of the huge volume of secreted fluid is actually lost from the body.

It should be clear from this overview that the small intestine is the major site for digestion and absorption, and we now turn to these processes.

Digestion and absorption

At the beginning of this chapter we described in general terms the structure of the entire gastrointestinal tract; a more detailed description of the small intestine is required to appreciate this segment's dominant role in digestion and absorption.

Structure of the small intestine

The small intestine is a coiled tube approximately 4 cm (1½ in) in diameter, leading from the stomach to the large intestine. This tube, 275 cm (9 ft) in length, is divided into three segments: an initial short 20-cm (8-in) segment, the *duodenum*, followed by the *jejunum* and ending with the longest segment, the *ileum*. Normally, most digestion and absorption has occurred by the middle of the jejunum. Thus, the intestine has a considerable functional reserve, making it almost impossible to exceed its absorptive capacity even when exceedingly large quantities of food are ingested.

The mucosa of the small intestine is highly folded, and the surface of these folds is further convoluted by fingerlike projections known as *villi*. The surface of each villus is covered with a single layer of epithelial cells whose surface membranes form small projections known as *microvilli* (Fig. 14-6). The combination of folded mucosa, villi, and microvilli increases the total surface area of the

Figure 14-6. Microvilli on the surface of intestinal epithelial cells. [*From D. W. Fawcett*, J. Histochem. Cytochem, **13**:*75–91 (1965). Courtesy of Dr. Susumu Ito.*]

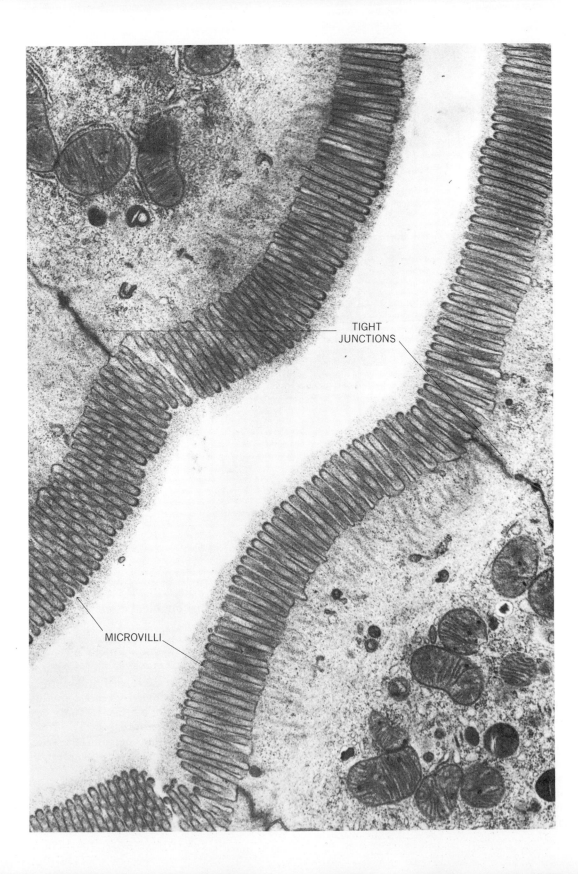

TIGHT
JUNCTIONS

MICROVILLI

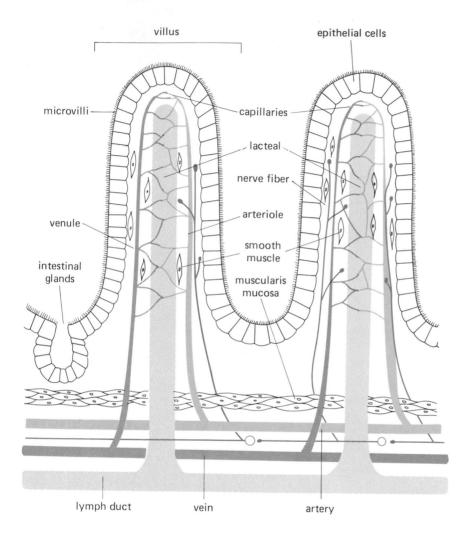

Figure 14-7. Structure of intestinal villi.

small intestine available for absorption about 600-fold over that of a flat-surfaced tube having the same length and diameter. The total surface area of the human small intestine is estimated to be about 300 sq m, or equivalent to the area of a tennis court.

The structure of a villus is illustrated in Fig. 14-7. The center of each villus is occupied by a single blind-ended lymph vessel, a *lacteal,* and by a capillary network which branches from an arteriole and drains into a venule. The blood flow to the small intestine at rest averages about 1 L/min (one-fifth the resting cardiac output), but increases during periods of digestive activity as a result both of local autoregulation (produced by the increased metabolic activity of the intestinal cells) and of

reflexes triggered by mechanical distension of the lumen by chyme.

The villi of the intestine move back and forth independently of each other, propelled by a small number of smooth muscle fibers in each of them. The motion of the villi is increased following a meal, and associated with their increased motion is a greater flow of lymph as the lacteals are "pumped," i.e., alternately compressed. The folds of the intestinal mucosa are not permanent structures but change their pattern with contraction of the underlying muscularis mucosae.

The epithelial cells of the intestinal tract are continually being replaced by new cells arising from the mitotic activity of the cells at the base of the villi. These new cells differentiate and migrate to the top of the villi replacing older cells, which disintegrate and are discharged into the lumen of the intestine. These disintegrating epithelial cells release their intracellular enzymes into the lumen which contribute to the digestive process. The entire epithelium is replaced approximately every five days, and this continuous discharge of cells amounts to about 250 g per day. The surface of the intestinal tract is very sensitive to damage by x-rays, atomic radiation, and anticancer drugs during the period of the mitotic cycle when DNA is being replicated.

The epithelial cell layer forms the barrier separating molecules in the lumen from the interstitial fluid bathing the capillaries and lacteals. Movement across this layer of cells occurs both by diffusion and by a number of specific carrier-mediated transport systems. Different carriers and passive permeability characteristics are present in the plasma membrane exposed to the luminal surface compared to the plasma membrane exposed to the interstitial fluid; this asymmetry contributes to the unidirectional nature of much transport across the epithelium. In particular, an active-transport carrier system at one surface and a leaky membrane (or a facilitated-diffusion carrier system) on the opposite surface permit the movement of various substances across the epithelium against concentration gradients. (We have described similar transport processes in the epithelial cells of the kidney tubules.) The digestive enzymes embedded in the plasma membranes of the epithelial cells at the luminal surface are not only frequently responsible for the final states of enzymatic digestion but are also closely associated with the membrane carrier mechanisms for the digestive end products.

The plasma membranes of adjacent epithelial cells are joined near their luminal surfaces by tight junctions (Fig. 14-6) which reduce the passive diffusion of substances through the extracellular spaces between the cells. The leakiness of these tight junctions varies in different regions of the gastrointestinal tract, the tight junctions in the stomach and large intestine forming much tighter seals than those in the small intestine. Net passive movements of molecules across the epithelium in either direction by way of these extracellular spaces between cells occur whenever there is a concentration gradient, an electrical gradient (in the case of charged particles), or a pressure gradient, as long as the tight junction is permeable to the substance.

Finally, as described in Chap. 11, the venous drainage from the intestines (as well as from the pancreas and portions of the stomach) is unusual in that the blood does not empty directly into the vena cava, but passes first into the liver via the hepatic portal vein; there it flows through a second capillary network before leaving the liver to return to the heart. Thus, material absorbed into the capillaries of the intestine may be processed by the liver before entering the general circulation.

Carbohydrate

The daily intake of carbohydrate varies considerably, ranging from 250 to 800 g/day in a typical American diet. Most of the carbohydrate (Table 14-1) is in the form of the plant polysaccharide, starch, with smaller amounts of the disaccharides sucrose (table sugar) and lactose (milk sugar). In infants, lactose from milk makes up the majority of the carbohydrate in the diet. Only very small amounts of monosaccharides are normally present. The polysaccharide cellulose, which is present in plant cells, cannot be broken down by the enzymes secreted by the gastrointestinal tract, so cellulose passes through the small intestine in undigested form and enters the large intestine, where it is partially digested by bacteria.

As mentioned earlier, starch is partially digested by salivary amylase during passage through the stomach, and this digestion is continued in the small intestine by pancreatic amylase. The products formed by these amylases are the disaccharide maltose and short chains composed of about eight glucose molecules. Enzymes that then split these molecules, as well as the disaccharides sucrose

Table 14-1. Carbohydrates

Class	Examples	Made up of
Polysaccharides	Starch	Glucose
	Cellulose	Glucose
	Glycogen	Glucose
Disaccharides	Sucrose	Glucose-fructose
	Lactose	Glucose-galactose
	Maltose	Glucose-glucose
Monosaccharides	Glucose	
	Fructose	
	Galactose	

and lactose, into monosaccharides are located in the plasma membranes of the epithelial cells.

The monosaccharides, glucose and galactose, liberated from this breakdown of polysaccharides and disaccharides, are actively transported across the intestinal epithelium into the blood. The carriers which transport them into the cells are located very near the membrane disaccharidases, such that the products released from these enzymes immediately combine with the transport carriers. This active-transport system requires sodium ions; the latter combine with the same carrier transporting the sugar, and both are moved into the cell simultaneously.

Protein

An intake of about 50 g of protein each day is required by an adult to supply essential amino acids and replace the amino acid nitrogen converted to urea. A typical American diet contains about 125 g of protein. In addition to the protein in the diet, 10 to 30 g of protein (mostly as enzymes) is secreted into the gastrointestinal tract by the various glands, and about 25 g of protein is derived from the disintegration of epithelial cells. All this protein is normally broken down into amino acids and absorbed by the small intestine.

Proteins are broken down to peptide fragments by pepsin in the stomach and by *trypsin* and *chymotrypsin* secreted by the pancreas. The peptide fragments are further digested to free amino acids by *carboxypeptidase* from the pancreas and *aminopeptidase* located in the intestinal epithelial membranes. (These enzymes' names designate the fact that they split off amino acids from the carboxyl and amino ends of the peptide chains, respectively). The free amino acids are actively transported across the walls of the intestine. Several different carrier systems are available for transporting different classes of amino acids. Some of these require sodium, just as is true for carbohydrate transport. In addition to single amino acids, short chains of two or three amino acids are also actively absorbed.

Small amounts of intact proteins are able to cross the epithelium and gain access to the interstitial fluid without being digested. These proteins are engulfed by the plasma membrane of the epithelial cells (endocytosis), move through the cytoplasm, and are released on the opposite side of the cell by the reverse process, exocytosis. The capacity to absorb intact proteins is greater in a newborn infant than in adults, and antibodies (proteins involved in the immunological defense system of the body, Chap. 17) secreted in the mother's milk may be absorbed by the infant in this manner, providing a short-term passive immunity until the child begins to produce its own antibodies. Moreover, scattered throughout the mucosa of the small intestine are specialized cells of the immune system that respond to foreign proteins that reach them. These cells provide a local defense against certain foreign proteins that may enter the body across this large surface area exposed to the external environment.

Fat

The amount of fat in the diet varies from about 25 to 160 g/day. Most of this fat is in the form of triacylglycerols, and its digestion occurs in the small intestine. The major digestive enzyme in this process is pancreatic *lipase*, which catalyzes the splitting of the bonds linking the fatty acids to the first and third carbon atoms of glycerol, producing free fatty acids and 2-monoglycerides as products. The triacylglycerols entering the small intestine are insoluble in water and are aggregated into large lipid droplets. Since only the lipids at the surface of the droplet are accessible to the water-soluble lipase, digestion in this state would proceed very slowly; furthermore, the products of lipase action, fatty acids and 2-monoglycerides, are themselves insoluble in water. The role of bile salts, supplied by the liver, is to eliminate both these problems: They break up the large lipid droplets into a number of smaller lipid droplets, a fat-solubilizing process known as *emulsification,* and they combine with the fatty acids and 2-monoglycerides produced by the action of lipase to form the water-soluble particles known as *micelles.*

Bile salts are formed from the nonpolar steroid molecule, cholesterol, to which are attached several polar hydroxyl groups and a short carbon chain having a terminal ionized carboxyl group (Fig. 14-8). The several bile salts secreted by the liver differ in the number of their hydroxyl groups and the types of ionized side chain present, but all interact with lipids in a similar way; the nonpolar steroid ring dissolves in the surface of a large nonpolar lipid droplet, leaving the polar hydroxyl

---

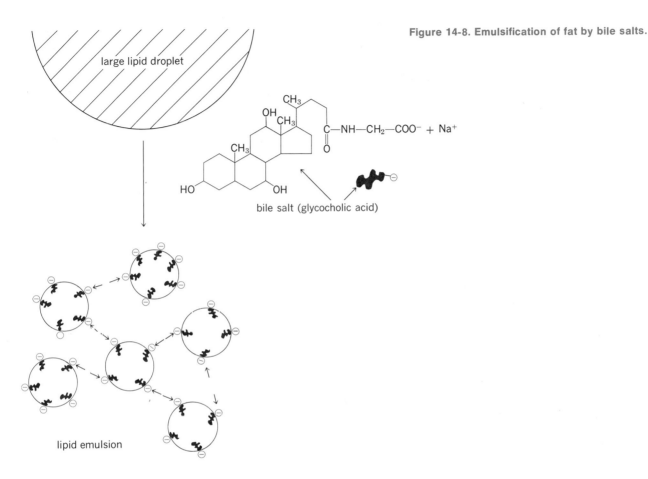

large lipid droplet

bile salt (glycocholic acid)

$C-NH-CH_2-COO^- + Na^+$

lipid emulsion

Figure 14-8. Emulsification of fat by bile salts.

groups and ionized side chain exposed at the surface where they can interact with polar water molecules. As mechanical agitation in the intestine breaks up the large droplets, the resulting smaller droplets become coated in this manner with bile salts. The polar and ionized groups on the bile salts at the surface of these droplets prevent their coalescing back into larger droplets because of the repulsion of the charged surfaces. The resulting suspension of small lipid droplets, each about 1 μm in diameter, is known as an *emulsion* (Fig. 14-8). Because the surface area of lipid accessible to lipase in an emulsion is much greater than in a single large droplet, the rate of lipid digestion is increased.

Although digestion is speeded up by emulsification, absorption of the insoluble products would be very slow if it were not for the second action of bile salts, facilitation of the formation of micelles. Micelles are similar to the droplets in an emulsion,

but they are much smaller. (Whereas a lipid emulsion appears cloudy because of the relatively large size of its emulsion droplets, a solution of micelles is perfectly clear.) Micelles are only 3 to 10 nm in diameter and consist of bile salts, fatty acids, phospholipids, and 2-monoglycerides, all clustered together with the polar ends of each molecule oriented to the micelles' surface where they interact with polar water molecules.

Although free fatty acids and monoglycerides have an extremely low solubility in water, a few individual molecules can exist free in solution, and it is in this form that they are absorbed across the cell membrane by simple diffusion because of their high degree of solubility in the structural lipids of the cell membrane. As the free fatty acids and monoglycerides diffuse into the epithelial cells, they are replaced by others leaving the micelles, because the lipids in the micelles are in equilibrium with those free in solution. In this manner, the luminal concentration of free lipids is maintained and diffusion into cells continues. In the absence of micelles, fat absorption occurs so slowly that it is not completely absorbed and some is passed on to the large intestine and excreted in the feces.

Fatty acids are the primary form of fat entering the epithelial cells, but very little free fatty acid is released from the cell, since during their passage through the epithelial cells, the products of fat digestion are resynthesized into triacylglycerols. Electron micrographs of the intestinal epithelium during fat absorption show the accumulation of fat droplets in the cells' endoplasmic reticulum, where the enzymes involved in triacylglycerol synthesis are located. Release of the lipid droplet is believed to be similar to the release of protein secretory granules: The membrane of the endoplasmic reticulum surrounding the lipid droplet fuses with the cell membrane, releasing the droplet into the extracellular space. These small droplets, 0.1 to 3.5 μm in diameter and known as *chylomicrons*, contain about 90 percent triacylglycerol and small amounts of phospholipid, cholesterol, free fatty acids, and protein.

The released chylomicrons pass into the lacteals (lymph vessels) located in each villus. Entrance of the chylomicrons into the lacteals rather than into the capillaries is explainable by the relative permeabilities to them of the two vessel types. (A basement membrane composed of polysaccharides covers the outer surface of the capillary as in other capillaries of the body, but not the lacteal and may be the barrier keeping the chylomicrons out of the capillaries.) Thus, fat is absorbed into the lymphatic system, which eventually empties into the large veins; in contrast, most other substances absorbed from the intestinal tract enter the capillaries and pass through the liver by way of the hepatic portal vein before reaching the general circulation. In the next chapter we shall discuss how the lipids in the chylomicron are made available to the cells of the body. Figure 14-9 summarizes the pathway taken by fat in moving from the lumen into the lymphatic system.

Vitamins

The fat-soluble vitamins, A, D, E, and K, require no digestion (i.e., molecular alteration) for their absorption. They are released from food by the diges-

tive juices of the stomach and become dissolved in the fat droplets that pass into the small intestine. As the fat is solubilized by the action of bile salts, these vitamins become incorporated into the micelles along with the products of fat digestion; the free vitamin molecules, in equilibrium with the vitamins in the micelles, diffuse across the membranes of the epithelial cells as do the products of fat digestion. Any interference with the secretion of bile or the action of bile salts in the intestine decreases the absorption of fat-soluble vitamins.

Most water-soluble vitamins are absorbed by diffusion or carrier-mediated transport across the wall of the intestinal tract. However, one, vitamin B_{12}, is a very large, charged molecule which is unable to move through plasma membranes. In order to be absorbed, vitamin B_{12} must combine with a special protein, known as *intrinsic factor*. This protein, secreted in the stomach, binds to sites on the luminal surface of epithelial cells in the lower portion of the ileum and this binding initiates membrane endocytosis, which leads to the absorption of the vitamin. Vitamin B_{12} is required for red blood cell formation (see Chap. 11), and when it is deficient a form of anemia known as pernicious anemia develops. This may occur when the stomach has been removed (as a treatment for ulcers or gastric cancer) or because intrinsic factor is not being secreted by the stomach. Since the absorption of vitamin B_{12} occurs specifically in the lower part of the ileum, removal of this segment of the intestine can also result in pernicious anemia.

Nucleic acids

Small amounts of nucleic acids—DNA and RNA—are present in food and are broken down to nucleotides by enzymes secreted by the pancreas. Enzymes in the intestinal epithelial membranes act upon these nucleotides, releasing free bases and monosaccharides, which are then actively transported across the epithelium by specific carrier systems.

Water and minerals

Water is the most abundant substance in the intestinal chyme. Approximately 9000 mL of fluid enter the intestine each day and only 500 mL are passed on to the large intestine, 95 percent of this fluid being absorbed in the small intestine. The

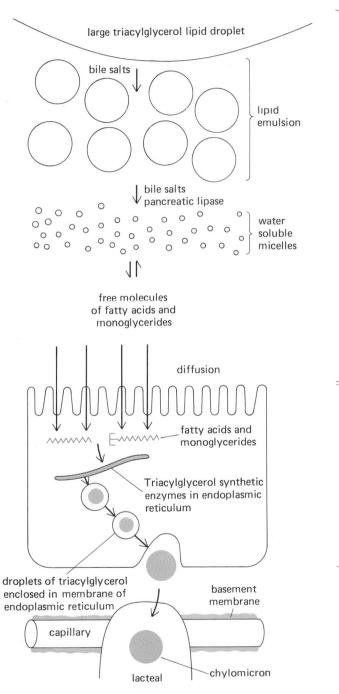

Figure 14-9. Summary of fat absorption across the walls of the small intestine.

large triacylglycerol lipid droplet

bile salts

lipid emulsion

bile salts
pancreatic lipase

water soluble micelles

free molecules of fatty acids and monoglycerides

diffusion

fatty acids and monoglycerides

Triacylglycerol synthetic enzymes in endoplasmic reticulum

droplets of triacylglycerol enclosed in membrane of endoplasmic reticulum

basement membrane

capillary

lacteal

chylomicron

lumen of intestinal tract

epithelial cell

membranes of the epithelial cells, as well as the channels between the cells, are very permeable to water. Therefore, a net diffusion of water (osmosis) occurs across the epithelium whenever a water concentration gradient is established as a result of differences in the total solute concentration (osmolarity) on the two sides. Such differences are established by the active transport of solutes. In other words, active solute transport establishes the osmotic gradients leading to a net movement of water (this mechanism of coupling water movement to the transport of solutes has already been described for the kidney in Chap. 13, and the same basic principles apply to water reabsorption in the intestinal tract). Although, as described above, many organic solutes are actively absorbed across the epithelium, sodium and chloride ions still account for most of the total solute actively transported because they constitute the majority of the total solutes present in the intestinal chyme.

The net absorption of water has an important effect upon the absorption of other substances which cross the epithelium by simple diffusion. As water is absorbed, the volume of the luminal contents decreases, thereby concentrating any solutes not absorbed at the same rate. This rise in concentration secondary to water reabsorption provides the concentration gradient for the net diffusion of these substances across the intestinal wall.

In addition to sodium and chloride, other minerals present in smaller concentrations, such as potassium, magnesium, and calcium, are also absorbed, as are trace elements, such as iron, zinc, and iodide. Consideration of the transport processes associated with all of these is beyond the scope of this book, and we shall briefly consider as examples the absorption of only three of them—potassium, calcium and iron.

Potassium. Potassium moves across the intestinal epithelium by simple diffusion. As water is absorbed, the resulting rise in luminal potassium concentration produces a concentration gradient for potassium absorption. Interference with the intestine's ability to absorb water (as occurs in certain forms of diarrhea) diminishes potassium absorption, and in fact can result in net movement of potassium from the interstitial fluid into the lumen. This secretion of potassium is explained by a small electrical potential across the epithelial cells such that the luminal surface is negatively charged relative to the interstitial surface, and this electrical force pulls potassium into the lumen.

Calcium. We have repeatedly emphasized that for most substances, especially the major organic nutrients, the total amount of the substance delivered to the small intestine gets absorbed, i.e., the amount of nutrient absorbed is not regulated by the nutritional state of the body. The story for calcium is quite different in that the amount absorbed is controlled by reflexes that respond to changes in plasma calcium concentration. As described in Chap. 13, this regulation is mediated by the active form of vitamin D; via this pathway decreases in the plasma concentration of calcium lead to increased active vitamin D and to enhancement of calcium active transport across the intestinal epithelium.

Iron. Ferrous (Fe^{2-}) and, to a lesser extent, ferric (Fe^{3-}) ions are actively transported into intestinal epithelial cells where most are incorporated into *ferritin*, a protein-iron complex which functions as an intracellular store of iron. Some of the absorbed iron is released from the cell and becomes bound to a specific protein in the plasma, *transferrin*, in which form it circulates throughout the body. The absorption of iron, like that of calcium, varies, depending upon the iron content of the body. When body stores are ample the amount of iron bound to ferritin is increased and this somehow reduces further iron absorption. When the body stores of iron drop, as a result of hemorrhage, for example, so does the amount of iron bound to intestinal ferritin and this increases iron absorption. The absorption of iron is typical of that of other trace metals in several respects: Cellular storage proteins and plasma carrier proteins are involved; control of absorption is a major mechanism for the homeostatic control of the body's content of the metal.

Regulation of gastrointestinal processes

Basic principles

A special environment must be created in the lumen of the gastrointestinal tract in order for digestion and absorption to occur. Accordingly, glandular secretions and muscle contractions must be controlled in a manner that will provide this optimal environment. Unlike other control systems, which regulate some variable in the extracellular fluid, the control mechanisms of the gastrointestinal system regulate conditions in the lumen of the tract. In other words, these control mechanisms are not governed to any great extent by the nutritional state of the body but rather by the volume and composition of the luminal contents.

Thus, the control systems must be able to detect the state of the luminal environment, and so it is appropriate that the majority of the receptors for these systems are located in the wall of the gastrointestinal tract, itself. Most gastrointestinal reflexes are initiated by a relatively small number of luminal stimuli: (1) distension of the wall by the luminal contents; (2) osmolarity (total solute concentration) of the chyme; (3) acidity (H⁺ concentration) of the chyme; (4) products formed by the enzymatic digestion of carbohydrates, fats, and proteins—monosaccharides, fatty acids, peptides, and amino acids. Signals initiated by these stimuli acting on mechanoreceptors, osmoreceptors, and chemoreceptors trigger reflexes which influence the effectors—the muscle layers in the wall of the tract and the exocrine glands which secrete substances into its lumen. Thus, both the receptors and effectors of these reflexes are in the gastrointestinal system.

These reflexes, like other negative-feedback systems, prevent large changes in the variables which are the stimuli for the reflexes; in so doing they maintain optimal conditions for digestion and absorption. For example, increased acidity in the duodenum, resulting from acid emptying into it from the stomach, reflexly inhibits the secretion of acid by the stomach, thereby preventing overacidification of the duodenum (the adaptive value of this reflex is that digestive enzymes in the duodenum are inhibited by high acidity). To take another example, distension of the small intestine, due to the presence of material not yet absorbed, reflexly increases its contractile activity, thereby facilitating mixing of the chyme with the intestinal secretions and bringing it into contact with the epithelium; at the same time, the contractions of the stomach are reflexly inhibited, thereby decreasing the rate at which additional material is emptied into the intestine.

Let us now look at the general types of neural and hormonal pathways that mediate these reflexes.

Neural regulation. The gastrointestinal tract has in its walls virtually its own local nervous system, in the form of two major nerve plexuses, the myenteric plexus and the submucous nerve plexus (Fig. 14-3). These two plexuses are found throughout the length of the gastrointestinal tract beginning with the esophagus. They are composed of neurons forming synaptic junctions with other neurons within the plexus or ending near smooth muscles and glands. Moreover, many axons leave the myenteric plexus and synapse with neurons in the submucous plexus, and vice versa, so that neural activity in one plexus influences the activity in the other. The axons in both plexuses branch profusely, and stimulation at one point in the plexus leads to impulses conducted both up and down the tract. Thus, activity initiated in the plexus in the upper part of the small intestine may affect smooth muscle and gland activity in the stomach as well as in the lower part of the intestinal tract. Many of the receptors mentioned in the previous section are, in fact, the dendritic endings of the plexus neurons, so that action potentials arising in them are conducted directly to the plexuses.

The neural connections in the plexuses permit intratract neural reflexes that are independent of the central nervous system. This is not to say that the tract is not subject to neural control via the central nervous system. Nerve fibers from both the sympathetic and parasympathetic branches of the autonomic nervous system enter the intestinal tract and synapse with neurons in the plexuses. Via these pathways, the central nervous system can influence the motor and secretory activity of the gastrointestinal tract. Thus, two types of neural reflex pathways linking a stimulus to a response are found (Fig. 14-10): so-called *short* (intratract) reflexes from receptor through the nerve plexuses to effector cells; *long* reflexes traveling from receptor by way of external nerves to the central nervous system and back to the nerve plexuses and effector cells by way of the autonomic nerve fibers. Some controls are mediated solely by short pathways or long pathways, whereas others use both simultaneously.

Finally, it should be noted that not all reflexes controlling the gastrointestinal system are initiated by signals within the tract; whenever the reflexes are initiated by other receptors, e.g., by the sight of food, the central nervous system must be involved in the response. Complex behavioral influences, e.g., emotions, also operate, in large part, through the central nervous system.

Hormonal regulation. The first hormone ever discovered (in 1902) was a substance extracted from the wall of the small intestine which, when injected into the blood of a dog, caused the pancreas

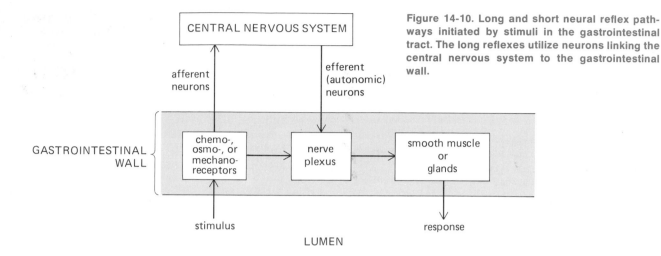

Figure 14-10. Long and short neural reflex pathways initiated by stimuli in the gastrointestinal tract. The long reflexes utilize neurons linking the central nervous system to the gastrointestinal wall.

to secrete a solution containing a high concentration of sodium bicarbonate. This gastrointestinal hormone was named *secretin*, and with its discovery, the science of endocrinology was born. Unlike the endocrine glands of the gonads, adrenals, pituitary, or thyroid, the hormone-secreting cells of the gastrointestinal tract are not clustered into discrete organs, but consist of individual cells scattered throughout the epithelium of the stomach and intestine. One surface of the cell is exposed to the lumen of the gastrointestinal tract; at this surface, various chemical substances in the chyme stimulate the cell, causing hormonal release from the other surface, the basal portion of the cell, from which point the hormone enters capillaries (the circulation provides the route by which the hormone gets from its site of secretion in the tract to its target cells in the gastrointestinal system).

Over the years a large number of extracts from various regions of the gastrointestinal tract have been injected into animals and found to produce changes in gastrointestinal secretory and contractile activity. However, as described in Chap. 9, tissue extracts may contain, in addition to hormones, other chemicals, such as neurotransmitters and paracrine substances, capable of stimulating glandular and muscular activity. Over a dozen different substances extracted from the gastrointestinal tract are currently being investigated as regards their possible hormonal activity, but only four have met all the criteria required for acceptance as a proven hormone. These are *secretin*, *cholecystokinin* (CCK), *gastrin*, and *gastric-in-*

hibitory-peptide (GIP), all of which are small polypeptides. Further investigations are almost certain to add other hormones to this list.

There are three generalizations concerning these gastrointestinal hormones that we wish to emphasize: (1) Each hormone participates in a feedback system for regulating the luminal environment; (2) in doing so, each hormone may affect multiple target organs; (3) each target organ is usually affected by more than one of the hormones. Table 14-2, which summarizes the major characteristics of the four established hormones, not only serves as a reference for future discussions but can be used to illustrate these generalizations.

The first two generalizations can be seen by looking *down* the entire column for any given hormone. Take cholecystokinin for example: The presence of fatty acids in the duodenum triggers its release from the duodenal wall and it, in turn, slows gastric emptying (so as not to overload the small intestine), stimulates release from the pancreas of enzymes (including the enzyme which digests fat), and causes the gallbladder to contract (delivering to the intestine bile salts which are also required for fat digestion and absorption).

The third generalization becomes apparent when one looks *across* columns for any given process. In at least one case—acid secretion by the stomach— opposing responses occur (gastrin stimulating and GIP inhibiting acid secretion); in several others (bicarbonate secretion by the pancreas, for example), both controlling hormones—secretin and CCK—are stimulatory. The term "potentiation" signifies the fact that in the presence of CCK (a very weak stimulator on its own), secretin produces a bicarbonate secretion that is greater than that produced by the sum of the effects of each of the hormones alone.

It should be emphasized that Table 14-2 lists only the major effects produced by physiological concentrations of these hormones. Many additional responses have been observed following the injection of these hormones, often at very high concentrations, into animals and people, and it remains to be determined whether these responses play any role in the normal regulation of the gastrointestinal tract.

Finally, not shown in the table is the fact that gastrin and CCK stimulate epithelial growth in the stomach, intestine, and pancreas. This function is of considerable importance in various disorders of gastrointestinal function.

Phases of gastrointestinal control. The neural and hormonal control of the gastrointestinal system is, in large part, divisible into three phases—*cephalic*,

gastric, and *intestinal* phases—according to the location of the stimuli which initiate a reflex (Fig. 14-11). The cephalic phase is initiated by stimulation of receptors in the head (cephalic)—sight, smell, taste, and chewing, as well as various emotional states. The efferent pathways for the reflex changes elicited by these stimuli involve the parasympathetic fibers in the vagus nerves which pass to almost all regions of the gastrointestinal tract. These fibers activate neurons in the nerve plexuses, which directly stimulate the cells which secrete acid and which also stimulate the release of gastrin, which in turn stimulates acid secretion.

Stimuli applied to the wall of the stomach initiate reflexes which constitute the gastric phase of regulation. These stimuli are distension, acid, and peptides formed during the digestion of protein. The responses to these stimuli are mediated by the nerve plexuses (short reflexes), by long reflex pathways involving the external nerves to the gastrointestinal tract, and by the release of gastrin.

Finally, the intestinal phase is initiated by stimuli

Table 14-2. Properties of gastrointestinal hormones

	Gastrin	Secretin	Cholecystokinin (CCK)	Gastric-inhibitory-peptide (GIP)
Major location of hormone-producing cells	Stomach	Small intestine	Small intestine	Small intestine
Stimuli for hormone release	Peptides in stomach; distension of stomach; nerve stimulation; (increased gastric acid inhibits release)	Acid in small intestine	Amino acids or fatty acids in small intestine	Fatty acids or monosaccharides in small intestine
Major target cell responses:				
Stomach				
Acid secretion	Stimulates			Inhibits
Gastric motility			Inhibits	
Pancreas		Stimulates	Potentiates secretin action	
Bicarbonate secretion				
Enzyme secretion		Potentiates CCK action	Stimulates	
Liver				
Fluid and bicarbonate secretion in the bile		Stimulates	Potentiates secretin action	
Gallbladder contraction			Stimulates	

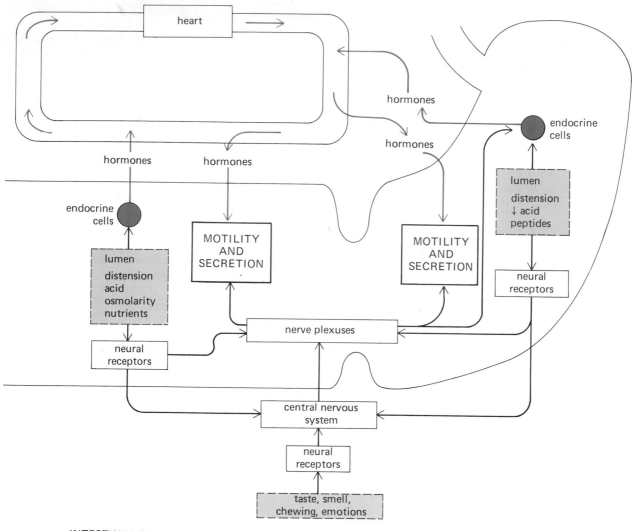

Figure 14-11. Reflex pathways which regulate gastrointestinal tract motility and secretion initiated by cephalic, gastric, and intestinal stimuli. The hormones also influence the secretion by the pancreas and liver.

in the lumen of the intestinal tract—distension, acidity, osmolarity, and the various digestive products of carbohydrates, fats, and proteins. Like the gastric phase, the intestinal phase is mediated by both long and short neural reflexes and by the release of hormones, in this case those from the intestine—secretin, CCK, and GIP.

It must be emphasized that each of these phases is named for the site at which the various stimuli initiate the reflex and not for the sites of effector activity, since each phase is characterized by ef-

ferent output to virtually all organs in the gastrointestinal tract. These phases do not occur in temporal sequence except at the very beginning of a meal. Rather, during ingestion and the much longer absorptive period, reflexes characteristic of all three phases are occurring together.

Keeping in mind the neural and hormonal mechanisms available for regulating gastrointestinal activity, we can now examine the specific contractile and secretory processes that occur in each segment of the gastrointestinal system.

Mouth, pharynx, and esophagus

Chewing. The rhythmic act of chewing is controlled by the somatic nerves to the skeletal muscles of the mouth and jaw. In addition to the voluntary control of these skeletal muscles, rhythmic chewing motions are reflexly activated by the pressure of food against the gums, teeth, hard palate, and tongue. Activation of these pressure receptors leads to inhibition of the muscles holding the jaw closed; the resulting relaxation of the jaw reduces the pressure, allowing a new cycle of contractile activity.

Saliva. Saliva is secreted by three pairs of exocrine glands: the parotid, the submandibular, and the sublingual (Fig. 14-1). Water accounts for 99 percent of the secreted fluid, the remaining 1 percent consisting of various salts and proteins. As described earlier, the major proteins of saliva are the *mucins,* which when mixed with water form the highly viscous solution known as mucus, and the enzyme amylase.

The secretion of saliva is controlled by both the sympathetic and parasympathetic divisions of the autonomic nervous system; however, unlike their antagonistic activity in most organs, both systems stimulate salivary secretion, the parasympathetic branch producing the greatest secretory response. In the awake state a basal rate of about 0.5 mL/min keeps the mouth moist. Food in the mouth increases the rate of salivary secretion; this reflex response is initiated by chemoreceptors and pressure receptors in the walls of the mouth and tongue. Afferent fibers from the receptors enter the medulla in the brainstem, which contains the integrating center controlling the autonomic output to the salivary glands. The most potent stimuli for salivary secretion are acid solutions, e.g., fruit juices and lemons, which can lead to a maximal secretion of 4 mL of saliva per minute. Salivation initiated by the sight, sound, or smell of food is very slight in human beings in contrast to the marked increase produced by these stimuli in dogs.

Swallowing. Swallowing is a complex reflex initiated when the tongue forces a bolus of food into the rear of the mouth, where pressure receptors in the walls of the pharynx are stimulated. They send afferent impulses to the swallowing center in the medulla, which coordinates the sequence of events during the swallowing process via efferent impulses to the muscles in the pharynx, larynx, esophagus, and respiratory muscles.

As the bolus of food moves into the pharynx, the soft palate is elevated and lodges against the back wall of the pharynx, sealing the nasal cavity and preventing food from entering it (Fig. 14-12). The swallowing center inhibits respiration, raises the larynx, and closes the glottis (the opening between the vocal cords), keeping food from getting into the trachea. As the tongue forces the food further back into the pharynx, the bolus tilts the epiglottis backward to cover the closed glottis. This pharyngeal

Figure 14-12. Movement of a bolus of food through the pharynx and upper esophagus during swallowing.

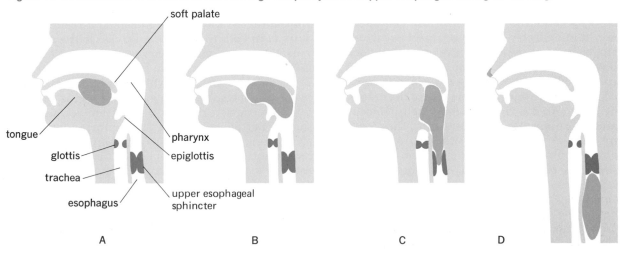

A   B   C   D

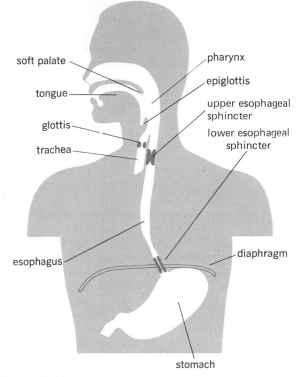

soft palate

tongue

glottis

trachea

pharynx

epiglottis

upper esophageal
sphincter

lower esophageal
sphincter

esophagus

diaphragm

stomach

Figure 14-13. Location of upper and lower esophageal sphincters.

period of swallowing, coordinated by the swallowing center, lasts about 1 s.

The esophagus is a 25-cm-long tube connecting the pharynx to the stomach. It passes through the thoracic cavity, penetrates the diaphragm muscle, which separates the thoracic from the abdominal cavity, and joins the stomach a few centimeters below the diaphragm. Skeletal muscles surround the upper third of the esophagus, smooth muscles the lower two-thirds. As was described in Chap. 12, the pressure in the thoracic cavity is 5 to 10 mmHg less than atmospheric, and this is also the pressure in the lumen of the thoracic esophagus, since the latter is in equilibrium with pressure in the thoracic cavity. In contrast, the pressure in the pharynx, which is continuous with the mouth and nasal passages, is equal to atmospheric pressure and thus is greater than the pressure in the thoracic esophagus. Furthermore, the pressure in the stomach, due to its compression by other abdominal organs, is 5 to 10 mmHg greater than atmospheric. Thus, at both ends of the esophagus there

is a pressure difference tending to move material into it. However, in the absence of swallowing, these pressure differences produce no movement because the upper and lower ends of the esophagus are closed: Skeletal muscles surrounding the esophagus, just below the pharynx, form the *upper esophageal sphincter* which seals the entrance; the smooth muscles in the last 4 cm of the esophagus form the *lower esophageal sphincter*, sealing off the exit (Fig. 14-13).

The esophageal phase of swallowing begins with the relaxation of the upper esophageal sphincter. Immediately after the bolus has passed, the sphincter closes, the glottis opens, and breathing resumes. Once in the esophagus, the bolus is moved toward the stomach by a progressive wave of muscle contractions that proceeds along the esophagus at a steady rate, compressing the lumen and forcing the bolus ahead of it. Such waves of contraction in the muscle layers surrounding a tube are known as *peristaltic waves*. An esophageal peristaltic wave takes about 9 s to reach the stomach. The lower esophageal sphincter is relaxed throughout the period of swallowing, allowing the bolus to enter the stomach. After the bolus has passed, the sphincter closes, resealing the junction between the esophagus and the stomach. Swallowing can occur while a person is upside down since it is not primarily gravity but the peristaltic wave which moves the bolus to the stomach.

Swallowing is an example of a reflex in which multiple responses occur in a regular temporal sequence that is predetermined by the synaptic connections between neurons in the coordinating center. Since both skeletal and smooth muscles are involved, the swallowing center must direct efferent activity in both the somatic nerves (to the skeletal muscle) and autonomic nerves (to the smooth muscle) of the esophagus (the latter nerves being the parasympathetic fibers which reach the plexuses in the esophageal wall by way of the vagus nerves). Simultaneously, afferent fibers from receptors in the wall of the esophagus send information to the swallowing center which can then alter the efferent activity. For example, if a large particle of food does not reach the stomach during the initial peristaltic wave, the distension of the esophagus activates receptors which initiate repeated waves of peristaltic activity (secondary peristalsis) which are not preceded by the pharyngeal phase of swallowing.

The ability of the lower esophageal sphincter to maintain a barrier between the stomach and the esophagus is aided by the fact that its last portion lies below the diaphragm and is subject to the same pressures in the abdominal cavity as is the stomach. Thus, if the pressure of the abdominal cavity is raised, e.g., during cycles of respiration or by contraction of the abdominal muscles, the pressures of both the stomach contents and this terminal segment of the esophagus are raised together and there is no change in the pressure difference between the two. During pregnancy, the growth of the fetus increases the pressure on the abdominal contents and can displace the terminal segment of the esophagus through the diaphragm into the thoracic cavity. The sphincter is therefore no longer assisted by changes in abdominal pressure, and during the last 5 months of pregnancy there is a tendency for the increased pressures in the abdominal cavity to force some of the contents of the stomach up into the esophagus. The hydrochloric acid from the stomach contents irritates the walls of the esophagus, causing contractile spasms of the smooth muscle which is experienced as pain (known generally as *heartburn* because the sensation appears to be located in the region over the heart). Heartburn often subsides in the last weeks of pregnancy as the uterus descends prior to delivery, decreasing the pressure on the abdominal organs. Another potential cause of heartburn is a large meal, which may raise the pressure in the stomach enough to produce reflux of acid into the esophagus.

Stomach

The epithelial layer of cells lining the stomach invaginates into the mucosa forming numerous tubular glands (Fig. 14-14). The cells at the opening of these glands secrete mucus. Deeper within the glands are the *parietal* (wall) cells, also known as *oxyntic* cells, which secrete hydrochloric acid (and intrinsic factor). Yet a third group of cells, *chief cells*, secrete the enzyme precursor *pepsinogen*. Finally, the endocrine cells which secrete the hormone gastrin are scattered throughout the epithelium of the antrum [the lower portion of the stomach which does not secrete acid (Fig. 14-15)]. Thus each of the four major secretions of the stomach—mucus, acid, pepsinogen, and the hormone gastrin—is secreted by a distinct cell type.

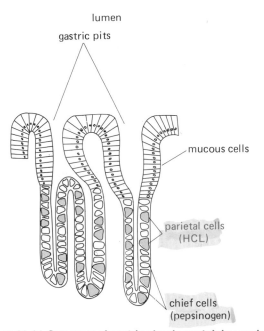

Figure 14-14. Structure of gastric glands containing parietal, chief, and mucous cells.

HCl secretion. The human stomach secretes about 2 L of hydrochloric acid per day. Hydrochloric acid is a strong acid which completely dissociates in water into hydrogen and chloride ions. The concentration of hydrogen ions in the lumen of the stomach may reach 150 mM, three million times greater than the concentration in the blood. The secretion of hydrochloric acid involves the active transport of both hydrogen and chloride ions. The two ions appear to be transported by separate pumps in the luminal membrane of the parietal cell. The enzyme carbonic anhydrase, similar to the enzyme in red blood cells, is present in the gastric mucosa and is believed to be involved in the secretion of acid, probably by way of the general pathway illustrated in Fig. 14-16. Note that for each hydrogen ion secreted into the lumen of the stomach, a bicarbonate ion is released into the blood in exchange for chloride; therefore, the venous blood leaving the stomach is more alkaline than the arterial blood entering it.

During a meal, the rate of hydrochloric acid secretion increases markedly, reaching maximal rates of about 3 mL/min. Table 14-3 summarizes the many factors controlling HCl secretion. During the cephalic phase of the meal, the message

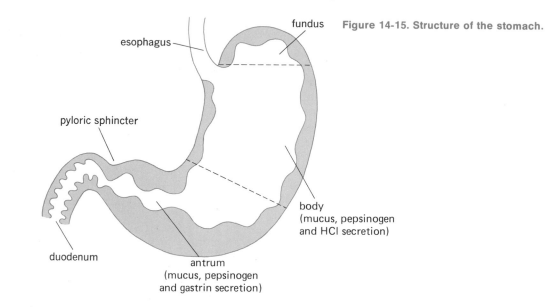

Figure 14-15. Structure of the stomach.

esophagus

fundus

pyloric sphincter

duodenum

antrum
(mucus, pepsinogen
and gastrin secretion)

body
(mucus, pepsinogen
and HCl secretion)

for increased acid secretion reaches the stomach by way of the vagus nerves which go to the nerve plexuses; fibers of these plexuses then directly stimulate both the acid-secreting parietal cells and the cells which release gastrin, which in turn also stimulates the parietal cells.

Once food has reached the stomach the gastric phase of acid secretion is initiated by a variety of intragastric stimuli—distension, peptides, and changes in acidity. These stimuli act directly upon the parietal cells and/or indirectly upon them by triggering short and long neural reflexes and the release of gastrin. The greater the protein content of a meal, the greater the acid secretion via these reflexes. There are two reasons for this. First, peptides formed by the digestive action of pepsin on protein stimulate acid secretion. Second, the presence of protein reduces the acidity of the gastric chyme. The crucial fact to be kept in mind during the following analysis of this second reason is that acid secretion is inhibited by high concentrations of hydrogen ion, because acid inhibits the

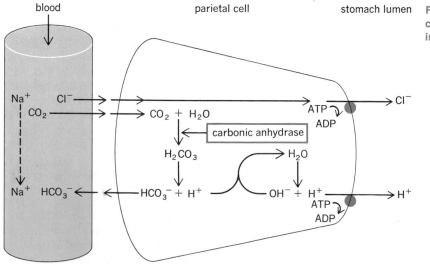

Figure 14-16. Secretion of HCl by parietal cells and the release of bicarbonate ions into the blood.

Table 14-3. Control of HCl secretion during a meal

Stimuli	Pathways to the parietal cells	Result

Cephalic stimuli
  Sight
  Smell
  Taste
  Chewing
	Vagus nerves via
	gastric nerve
	plexuses and gastrin
		↑ HCl secretion

Gastric contents
  Distension
  ↓H$^+$ concentration (protein buffering)
  Peptides
	Long and short neural reflexes;
	gastrin;
	direct action of peptides
		↑ HCl secretion

Intestinal contents
  Distension
  ↑H$^+$ concentration
  Nutrients
  ↑Osmolarity
	* Long and short neural reflexes
	gastric-inhibitory-peptide;
	other hormones (?)
		↓ HCl secretion

*The neural and hormonal pathways triggered during the intestinal phase have as one of their final common pathways a decrease in gastrin secretion.

release of gastrin. Before food enters the stomach, the hydrogen-ion concentration is high, inhibiting gastrin release and, thereby, acid secretion. The statement that hydrogen-ion *concentration* is high before a meal may seem paradoxical since we have pointed out that hydrogen-ion *secretion* is low; the paradox is resolved when one realizes that a small volume of secreted fluid can have a very high concentration of hydrogen ions. The protein in food is an excellent buffer so that as food enters the stomach, the number of free hydrogen ions (i.e., the concentration) drops. This decrease in acidity removes the inhibition of gastrin secretion and, thereby, of acid secretion. The more protein, the greater the buffering of acid, the more gastrin secreted, the more acid secreted—clearly this sequence provides an excellent negative-feedback control which acts to maintain a high luminal acid secretion.

Ingested substances other than protein can stimulate acid secretion when they reach the stomach. For example, caffeine, a substance found in coffee, tea, and cola drinks, causes the release of gastrin and thus increases acid secretion. Alcohol, contrary to popular belief, has little effect on acid secretion in people.

We now come to the intestinal phase of the control of acid secretion, the phase in which events occurring in the early portion of the small intestine reflexly influence acid secretion by the stomach. Because digestive processes in the small intestine are markedly inhibited by acid solutions it is es-

sential that the chyme entering the small intestine from the stomach not contain so much acid that it cannot be rapidly neutralized by the bicarbonate-rich fluids simultaneously being secreted into the intestine by the liver and pancreas. Accordingly it is highly adaptive that excess acidity in the duodenum triggers reflexes which inhibit gastric acid secretion (other reflexes to be described later are elicited which slow gastric emptying and stimulate pancreatic secretion, two other ways of lowering duodenal acidity). Acidity is not the only variable in duodenal chyme capable of triggering reflex inhibition of gastric acid secretion; distension, hypertonic solutions (low water concentration), and solutions containing amino acids, fatty acids, or monosaccharides also do so. Thus, the extent to which acid secretion is inhibited during the intestinal phase by each of the above stimuli varies, depending upon the volume and composition of the meal, but the net result is the same—balancing the secretory activity of the stomach with the digestive and absorptive functions of the small intestine. Long and short neural reflexes and intestinal hormones have been shown to mediate all these inhibitory reflexes, gastric-inhibitory-peptide being the most potent of the intestinal hormones.

Before leaving the control of acid secretion, we should like to point out that many substances and pathways other than those mentioned have been hypothesized to influence it physiologically. Perhaps the most studied is histamine, a paracrine agent found normally in the stomach; it can pro-

LUMEN

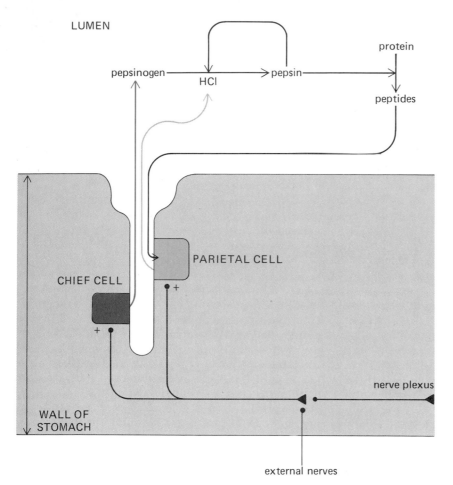

Figure 14-17. Pathways regulating the secretion and activation of pepsinogen in the stomach.

duce a marked increase in parietal-cell acid secretion (indeed, histamine is often administered to measure the acid secreting capacity of the stomach in health and disease), but its physiological role in the control of normal secretion of acid remains unclear.

Pepsin secretion. Pepsin is secreted by the chief cells in an inactive form known as *pepsinogen.* The high acidity in the lumen of the stomach converts pepsinogen into the active enzyme, *pepsin,* by breaking off a small fragment of the molecule, thereby changing the shape of the protein and exposing its active site.

Pepsin catalyzes the splitting of specific peptide bonds in proteins, producing peptide fragments composed of several amino acids. Since high acidity is required for pepsin activation, it is appropri-

ate that, as we have seen, these peptide fragments provide one of the stimuli for acid secretion. Once activated, pepsin acts autocatalytically upon pepsinogen to form more pepsin (Fig. 14-17). The synthesis and secretion of inactive pepsinogen followed by its activation to pepsin provides an example of a process that occurs frequently in enzyme secretion; by synthesizing an inactive form of the enzyme, the cell is protected from internal digestion by the enzyme.

The primary pathway for stimulating pepsinogen secretion is input to the chief cells from the nerve plexuses. In general, pepsinogen secretion parallels the secretion of acid.

Gastric motility. The empty stomach has a volume of about 50 mL and the diameter of its lumen is little larger than that of the small intestine. When a meal is swallowed, the smooth muscle in the body of the stomach relaxes allowing the volume of the lumen to increase with little increase in pressure. This *receptive relaxation* is mediated by the vagus (parasympathetic) nerves and coordinated by the swallowing center. Peristaltic waves proceed along the walls of the stomach at the rate of about 3 per minute. Each wave begins near the entry of the esophagus and produces only a weak ripple as it proceeds over the body of the stomach, one too weak to produce much mixing of the luminal contents with the acid and pepsin secreted by the stomach. As the wave approaches the larger mass of wall muscle surrounding the antrum, it speeds up and produces a more powerful wave of contraction that does mix the contents of the stomach and expels them into the duodenum. The *pyloric sphincter* is a ring of smooth muscle and connective tissue between the terminal antrum and the duodenum. It exerts only a slight pressure at the junction and is open most of the time. When a strong peristaltic wave arrives at the antrum, the pressure of the antral contents is increased, but the antral contraction also closes the pyloric sphincter, with the result that the pressure forces most of the antral contents back into the body of the stomach, and only small amounts pass into the duodenum (Fig. 14-18).

What is responsible for the characteristics of the peristaltic waves? Its rhythmicity (3-per-minute) is due to the fact that pacemaker cells in the longitudinal muscle layer near the esophagus undergo spontaneous depolarization-repolarization cycles at this rate, termed the *basic electrical rhythm.* The depolarizations are propagated through gap junctions along the longitudinal muscle layer of the stomach and induce similar waves in the overlying circular muscle layer (perhaps via gap junctions between the longitudinal and circular layers). However, in the absence of other input, these depolarizations are too small to cause the muscle membranes to reach threshold and fire the action potentials required to elicit peristalic contractions. Excitatory neurotransmitters and hormones act upon the muscle to depolarize the membrane, thereby bringing it closer to threshold, so that action potentials are generated at the peak of the basic electrical rhythm cycle (Fig. 10-37); the number of spikes fired with each wave, in turn, determines the strength of the elicited smooth muscle contraction (see Fig. 10-37). Thus, whereas

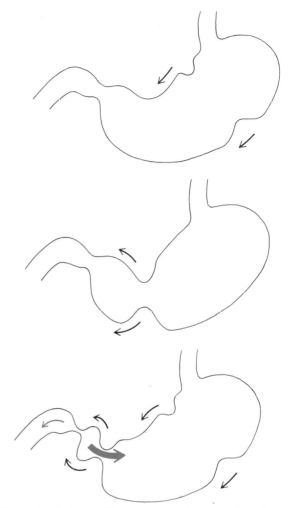

Figure 14-18. Peristaltic waves passing over the stomach empty a small amount of material into the duodenum. Most of the material is forced back into the antrum.

the frequency of contraction is determined by the basic electrical rhythm and remains essentially constant, the force of contraction (and therefore, the amount of gastric emptying per contraction) is determined by neural and hormonal input, i.e., by regulatory reflexes.

There are important gastric and intestinal phases to these reflexes. Distension of the stomach increases the rate of gastric emptying; therefore, the larger a meal the faster the initial rate of emptying, and as the volume of the stomach decreases, so does the rate of emptying. Long and short neural reflexes triggered by mechanoreceptors in the

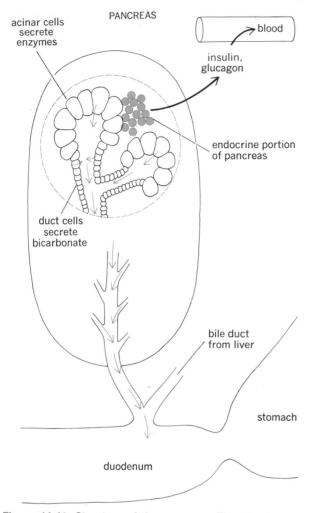

Figure 14-19. Structure of the pancreas. The gland areas are greatly enlarged relative to the entire pancreas and ducts.

stomach mediate this response to distension. In contrast, distension of the duodenum as well as the presence in it of fat, acid, or hypertonic solutions all produce inhibition of gastric emptying mediated by release of intestinal hormones (and to a lesser extent by neural reflexes); fat in the duodenum is the most potent of the chemical stimuli leading to this reflex inhibition of gastric motility.

In addition to the control of gastric emptying by the gastric volume and duodenal contents, motility is influenced by external nerve fibers to the stomach: Inhibition of motility occurs when parasym-pathetic activity is decreased or sympathetic activity increased. Via these pathways, pain and emotions such as sadness, depression, and fear tend to decrease motility, whereas aggression or anger tend to increase it. These relationships are not always predictable, however, and different people show different responses to apparently similar emotional states.

Pancreatic secretions

The pancreas, located below and behind the stomach, is a mixed gland containing both endocrine and exocrine portions. The endocrine cells secrete into the blood the hormones insulin and glucagon, to be discussed in Chap. 15. The exocrine portion of the pancreas secretes two solutions involved in the digestive process, one containing a high concentration of sodium bicarbonate and the other a large number of digestive enzymes. Both these solutions are secreted into the same ducts, which converge into the single pancreatic duct, the latter joining the bile duct from the liver just before entering the duodenum (Fig. 14-19). In the intestinal lumen the bicarbonate neutralizes the hydrochloric acid entering from the stomach, and the enzymes digest the various organic nutrients. The enzyme secretions of the pancreas are released from the acinar cells at the base of the exocrine glands, whereas the bicarbonate solution is secreted by the cells lining the early portions of the ducts leading from the acinar cells.

The secretion of bicarbonate by the pancreas is an active process. The mechanism of bicarbonate secretion is similar to the process of hydrochloric acid secretion by the stomach, the crucial difference being a reverse orientation of the transport systems, such that the bicarbonate ions are released into the lumen rather than the blood. Secretion of an alkaline solution of bicarbonate ions necessitates the formation of an equivalent amount of acid, and the blood leaving the pancreas is therefore more acid than the blood entering it (Fig. 14-20). Accordingly, loss of large quantities of bicarbonate ions from the intestinal tract during periods of prolonged diarrhea leads to acidification of the blood, just as loss of acid from the stomach by vomiting leads to alkalinization of the blood.

The enzymes secreted by the pancreas digest fat, polysaccharides, and proteins to fatty acids, sugars, and amino acids, respectively. A partial list of these enzymes and their activity is given in Table 14-4. Most are secreted in an inactive form, similar to the secretion of pepsinogen by the stomach, and are then activated in the duodenum by another enzyme: *Enterokinase,* which is embedded in the luminal plasma membrane of the intestinal epithelium, is a proteolytic enzyme which splits off a peptide from pancreatic trypsinogen, forming the active enzyme trypsin; trypsin, like enterokinase, is a proteolytic enzyme and once activated, it then proceeds to activate the other enzyme precursors from the pancreas in a similar manner (Fig. 14-21), as well as performing its major role of digesting protein.

Increased pancreatic secretion during a meal is mediated mainly by the intestinal hormones secretin and CCK (Table 14-2). The major effect of secretin is to stimulate bicarbonate secretion, whereas CCK stimulates mainly the secretion of enzymes. (However, as was noted earlier, each of these hormones has a potentiating effect on the activity of the other.) Since bicarbonate neutralizes the acid entering the duodenum from the stomach, it is appropriate that the major stimulus for secretin release from the wall of the duodenum is acid. As the acid is neutralized, the stimulus for

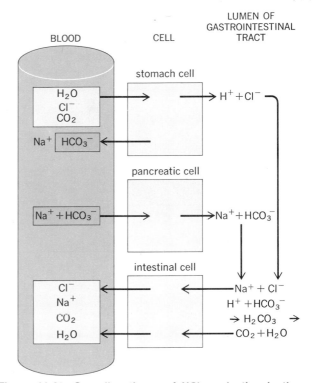

Figure 14-20. Overall pathway of HCl production in the stomach, neutralization by sodium bicarbonate secreted by the pancreas, and reabsorption in the small intestine of the products formed by the neutralization. Note that there is no net gain or loss of any substance.

secretin release is decreased and less bicarbonate is secreted by the pancreas. In analogous fashion, CCK stimulates mainly the secretion of digestive enzymes, including those for fat and protein digestion, and the stimuli for its release are, appropriately, fatty acids and amino acids. Thus, the organic nutrients in the small intestine initiate their own digestion via this reflex. Figure 14-22 summarizes the factors controlling pancreatic secretion.

Bile secretion

Bile is a salt solution containing five primary ingredients: (1) bile salts; (2) cholesterol; (3) lecithin (a phospholipid); (4) bile pigments and certain other end products of organic metabolism; (5) certain trace metals. The first three of these are involved in the solubilization of fat in the small

TABLE 14-4. Pancreatic enzymes

Enzyme	Substrate	Action
Trypsin, chymotrypsin	Proteins	Breaks amino acid bonds in the interior of proteins, forming peptide fragments.
Carboxypeptidase	Proteins	Splits off terminal amino acid from end of protein containing a free carboxyl group.
Lipase	Lipids (triacylglycerols)	Splits off fatty acids from positions 1 and 3 of triacylglycerols, forming free fatty acids and 2-monoglycerides.
Amylase	Polysaccharides	Similar to salivary amylase: splits polysaccharides into a mixture of glucose and maltose.
Ribonuclease, deoxyribonuclease	RNA, DNA	Splits nucleic acids into free mononucleotides.

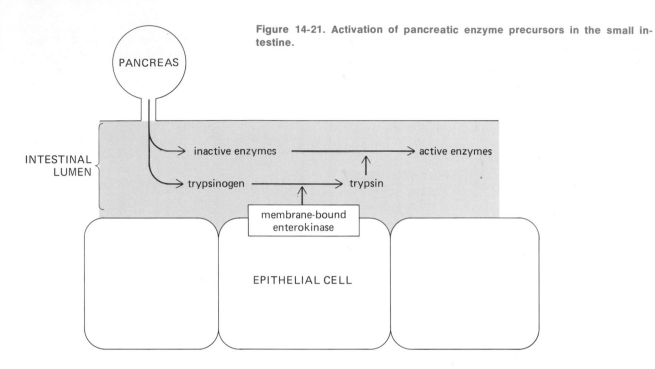

Figure 14-21. Activation of pancreatic enzyme precursors in the small intestine.

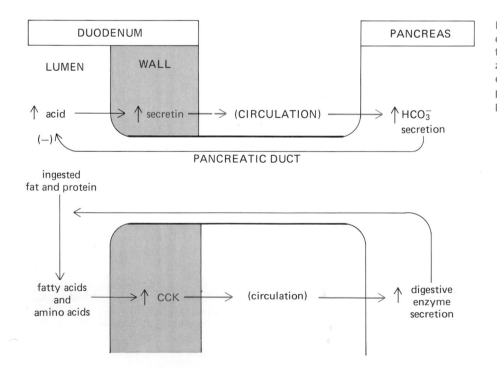

Figure 14-22. Feedback loops controlling pancreatic secretion of bicarbonate and enzymes. Only the major effects of CCK and secretin on the pancreas are shown. Hormonal pathways are colored.

intestine. The fourth and fifth are substances normally excreted from the body in the feces. The major bile pigment is bilirubin, which is a breakdown product of hemoglobin, as described in Chap. 11. The cells of the liver extract bilirubin from the circulation and secrete it into the bile by an active process. Bile pigments are yellow and give bile its golden color. These pigments are modified in the intestinal tract by digestive enzymes to form the pigments which give feces their brown color. In the absence of bile secretion, feces are grayish white. Some of the pigments are reabsorbed during their passage through the intestinal tract and are eventually excreted in the urine, giving urine its yellow color.

Bile is secreted by the liver cells into a number of small ducts, the bile canaliculi (Fig. 14-23), which converge to form the common bile duct which empties into the duodenum at the same site at which the pancreatic duct enters the duodenum. In people, a small sac branches from the bile duct on the underside of the liver, forming the *gallbladder*. The gallbladder can be surgically removed without impairing the secretion of bile or its reaching the intestinal tract; in fact many animals which secrete bile do not have a gallbladder.

From the standpoint of gastrointestinal function, bile salts are the most important components of bile since they are involved in the digestion and absorption of fats. The total amount of bile salt in the body is about 3.6 g; yet during the digestion of a single fatty meal, as much as 4 to 8 g of bile salts may be emptied into the duodenum. This is possible because most of the bile salts entering the intestinal tract are reabsorbed in the lower part of the small intestine and returned to the liver, which actively secretes them once again into the bile. This "recycling" pathway from the intestine to the liver (via the portal vein) and back to the intestine (via the bile duct) is known as the *enterohepatic circulation*.

The volume of bile secreted depends upon the active transport into the bile canaliculi of solutes, which produce an osmotic gradient between the bile and the blood, causing water to follow by diffusion. Inorganic ions (sodium, chloride, and bicarbonate) and bile salts constitute the majority of the solutes producing this water flow. Since bicarbonate ions in the bile help to neutralize acid in the duodenum, just as is true for bicarbonate ions coming from the pancreas, it is appropriate that bile flow, like pancreatic flow, is stimulated by secretin in response to the presence of acid in the duodenum; secretin does so by stimulating ion secretion by the liver.

The secretion of bile salts is stimulated by the bile salts themselves; the greater the plasma concentration of bile salts, the greater is the rate of their secretion (and the greater the volume of bile flow). Thus, between meals, when there is little bile salt in the intestine and, therefore, little absorbed, the plasma concentration of bile salts is

Figure 14-23. Portion of liver showing location of bile ducts with respect to liver cells. *(Adapted from Kappas and Alvares.)*

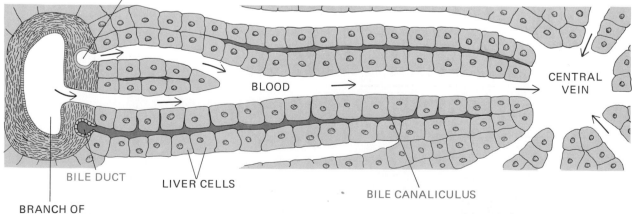

BRANCH OF HEPATIC ARTERY

BLOOD

CENTRAL VEIN

BILE DUCT

LIVER CELLS

BILE CANALICULUS

BRANCH OF PORTAL VEIN

low and little is secreted (i.e., the enterohepatic cycling of bile salts is minimal). During a meal, the higher plasma concentration of absorbed bile salts leads to an increased rate of bile secretion. This, along with the secretin-induced stimulation of ion secretion, increases bile flow. The liver also synthesizes new bile salts to replace the 5 percent of the total bile salt pool that is not absorbed in the small intestine.

Although, as we have seen, the greatest volume of bile is secreted during and just after a meal, some bile is always being secreted by the liver. Surrounding the bile duct at the point where it enters the duodenum is a ring of smooth muscle known as the *sphincter of Oddi.* When this sphincter is closed, the bile secreted by the liver is shunted into the gallbladder. The cells lining the gallbladder

actively transport sodium from the bile back into the plasma. As solute is pumped out of the bile, water follows by osmosis. The net result is a five- to tenfold concentration of the constituents of the bile. Shortly after beginning a meal, the sphincter of Oddi relaxes and the gallbladder contracts, discharging concentrated bile into the duodenum. The signal for gallbladder contraction and sphincter relaxation is the intestinal hormone CCK, appropriately so because the stimulus for this hormone's release is the presence of fatty acids and amino acids in the duodenum. (It is from this ability to cause contraction of the gallbladder that cholecystokinin received its name: *chole*, bile; *kystis*, bladder; *kinin*, to move.) Figure 14-24 summarizes the factors controlling the release of bile.

Figure 14-24. Pathways regulating bile secretion. Start following the loops beginning with ingested fat, ingested protein, and acid. Note the interaction with the pancreatic loop, diagramed in Fig. 14-22. Hormonal pathways are colored. The thick outer lines denote the enterohepatic circulation.

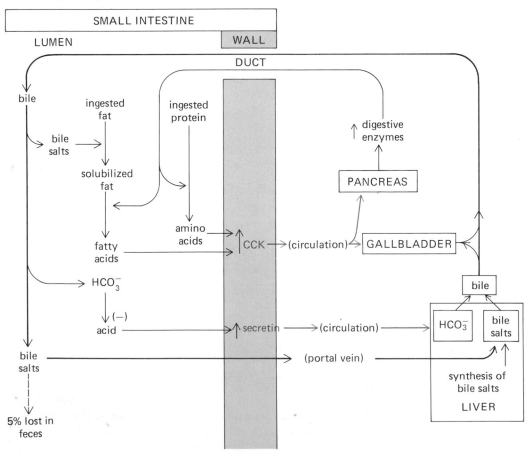

Small intestine

Secretions. In addition to the fluids entering the small intestine from the stomach, liver, and pancreas, 2000 mL of fluid moves across the wall of the small intestine from the blood into the lumen each day. Still larger volumes of fluid are simultaneously moving in the opposite direction—from the lumen into the blood—so that normally there is an overall net reabsorption of fluid from the small intestine. The study of intestinal secretion is difficult, and the factors controlling these secretions are less well understood that those controlling other gastro-intestinal processes.

One reason for water movement into the lumen is that the chyme entering the intestine from the stomach may be hypertonic due to a high concentration of solutes in a particular meal; this causes osmotic diffusion of water from the isotonic interstitial fluids of the wall of the intestine into the hypertonic solution in the lumen. A second factor producing a hypertonic solution in the small intestine is the process of digestion, in which large molecules, such as proteins and polysaccharides, are broken down into many small molecules of amino acids and monosaccharides, thus increasing the solute concentration (osmolarity) of the lumen and inducing a net flow of fluid into the intestinal tract. As we have seen, the osmolarity of the duodenum constitutes one of the stimuli initiating changes in secretion and motility.

In the two cases above, water movement was the result of events in the lumen. In addition, the intestine itself secretes ions (such as sodium, potassium, chloride and bicarbonate) across its epithelial membranes into the lumen, and water moves with these ions. These ion movements involve both carrier-mediated processes and passive membrane permeability, the characteristics of which vary along the length of the intestinal tract (similar complex movements occur across the wall of the large intestine but are smaller than in the small intestine). The factors controlling the overall net movements of these ions under various physiological conditions are poorly understood at the present time.

A third source of material entering the small intestine is the continuous disintegration of the intestinal epithelium. These cells divide, differentiate, function for a few days, and then are discharged from the surface of the intestine. Their breakdown

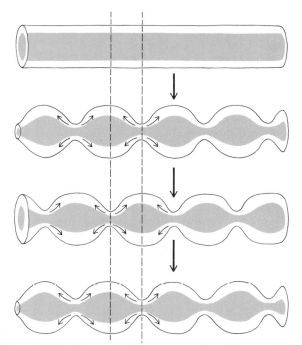

Figure 14-25. Segmentation movements of the small intestine. The small arrows indicate the movements of the luminal contents.

releases enzymes, other organic materials, ions, and water into the lumen of the intestine.

The secretion of mucus throughout the length of the intestine provides yet a fourth source of organic materials and fluid.

Motility. When the motion of the small intestine is observed by x-ray fluoroscopic examination, the contents of the lumen are seen to move back and forth with little apparent net movement toward the large intestine. Thus, in contrast to the waves of peristaltic contraction that sweep over the surface of the stomach, the primary motion of the small intestine is a stationary oscillating contraction and relaxation of rings of smooth muscle (Fig. 14-25). Each contracting segment is about 1 to 4 cm long and the contraction lasts about 5 s. The chyme in the lumen of a contracting segment is forced both up and down the intestine. This rhythmical contraction and relaxation of the intestinal sections, known as *segmentation,* produces a continuous division and subdivision of the intestinal contents which thoroughly mixes the chyme in the lumen and brings it into contact with the intestinal wall.

These segmenting movements of the small intestine are initiated by electric activity generated by pacemaker cells located in the longitudinal smooth muscle. The frequency of segmentation is set by the frequency of the intestinal basic electrical rhythm but unlike the stomach which normally has a single basic electrical rhythm, the intestinal rhythm varies from segment to segment, each successive pacemaker having a slightly lower frequency than the one above. Segmentation in the duodenum occurs at a frequency of about 12 contractions per minute whereas in the terminal portion of the ileum it occurs at a rate of only 9 contractions per minute. Since the frequency of contraction is greater in the upper portion of the small intestine than in the lower, more chyme is forced, on the average, downward than is forced upward.

The magnitude of segmentation can be altered by the nerve plexuses, external nerves (parasympathetic stimulation of the intestine increases contractile activity, and sympathetic stimulation decreases it), and hormones. As is true for the stomach, these reflexes produce changes in the intensity of smooth muscle contraction but do not change the natural frequency of the pacemakers. Following a meal, distension increases the intensity of segmenting contractions (paradoxically, the actual propulsion achieved is decreased, because the contracting segments narrow the lumen of the intestine, producing a greater resistance to flow). These changes in contractile activity are mediated by both short and long reflexes.

After most of a meal has been absorbed, the segmenting type of contractions decline, and instead, peristaltic waves begin to sweep over the intestine, beginning in the duodenum. These waves travel about 60 to 70 cm and then die out; however, the site at which the wave is initiated slowly moves down the small intestine. By the time the waves reach the end of the ileum, new waves are beginning in the duodenum and the process is repeated. These peristaltic waves sweep any material still remaining in the lumen of the small intestine on into the large intestine. Upon the arrival of a new meal in the stomach, they cease and are replaced by the segmenting type of contractile activity. The mechanism controlling the generation and progression of these waves is unknown.

Local distension of the intestine produces yet another characteristic response. The region just above the distended portion contracts and the region just below (toward the large intestine) relaxes. The contracted segment then progresses several centimeters down the intestine, producing peristaltic waves much shorter than those described in the previous paragraph. The coordination of this response is mediated by the nerve plexuses. These short peristaltic waves always proceed in the direction of the large intestine (a property known as the *law of the intestine*).

Several other specific reflexes are worth mentioning: Contractile activity in the ileum increases during periods of gastric emptying, and this is known as the *gastroileal reflex* (conversely, distension of the ileum produces decreased gastric motility, the *ileogastric reflex*); large distensions of the intestine, injury to the intestinal wall, and various bacterial infections in the intestine lead to a complete cessation of motor activity, the *intestino-intestinal reflex.* These reflexes appear to have as their efferent pathways the external nerves.

A person's emotional state can also affect the contractile activity of the intestine and thus the rate of propulsion of chyme and the time available for digestion and absorption. Fear tends to decrease motility whereas hostility increases it, although these responses vary greatly in different individuals. These responses are all mediated by the external nerves.

As much as 500 mL of air may be swallowed along with a meal. Most of this air travels no farther than the esophagus where it is eventually expelled by belching. Some of the air, however, reaches the stomach and is passed on to the intestines, where its percolation through the chyme, as the intestinal contents are mixed, produces gurgling sounds which are often quite loud.

Large intestine

The colon (large intestine), a tube about 6 cm (2.5 in) in diameter, forms the last 120 cm (4 ft) of the gastrointestinal tract. The first portion of the large intestine, the *cecum,* forms a blind-ended pouch from which the *appendix,* a small fingerlike projection, extends; it has no known digestive function. The colon is not coiled like the small intestine but consists of three relatively straight segments—

the ascending, transverse, and descending portions (Fig. 14-1). The terminal portion of the descending colon is S-shaped, forming the sigmoid colon, which empties into a short section, the *rectum*. Although the large intestine has a greater diameter than the small intestine and is about half as long, its epithelial surface area is only about 1/30 that of the small intestine because its mucosa lacks villi and is not convoluted. The large intestine secretes no digestive enzymes and is responsible for the absorption of only about 4 percent of the total intestinal contents per day. Its primary function is to store and concentrate fecal material prior to defecation.

Chyme enters the colon through the ileocecal sphincter separating the ileum from the cecum. This sphincter is normally closed, but after a meal, when the gastroileal reflex increases the contractile activity of the ileum, the sphincter relaxes each time the terminal portion of the ileum contracts, allowing chyme to enter the large intestine. Distension of the colon, on the other hand, produces a reflex contraction of the sphincter, preventing further material from entering.

About 500 mL of chyme enters the colon from the small intestine each day. Most of this material is derived from the secretions of the lower small intestine, since most of the ingested food has been absorbed before reaching the large intestine. The secretions of the colon are scanty and consist mostly of mucus.

The primary absorptive process in the large intestine is the active transport of sodium from the lumen to blood with the accompanying osmotic reabsorption of water. If fecal material remains in the large intestine for a long time, almost all the water is reabsorbed, leaving behind dry fecal pellets. The cells lining the large intestine are unable to actively transport either glucose or amino acids. There is a small net leakage of potassium into the colon, and severe depletion of total body potassium can result from diarrhea when large volumes of fluid are excreted in the feces.

The large intestine also absorbs some of the products (vitamins, for example) synthesized by the bacteria inhabiting this region. Although this source of vitamins generally provides only a small part of the normal daily requirement, it may make a significant contribution when dietary intake of vitamins is low.

Other bacterial products contribute to the production of intestinal gas (*flatus*). This gas is a mixture of nitrogen and carbon dioxide with small amounts of the inflammable gases hydrogen, methane, and hydrogen sulfide. Bacterial fermentation produces gas in the colon at the rate of about 400 to 700 mL/day.

Motility and defecation. Contractions of the circular smooth muscle of the colon produce a segmentation motion considerably slower than that in the small intestine; a contraction may occur only once every 30 min. Because of this slow movement, material entering the colon from the small intestine remains for 18 to 24 h, providing time for bacteria to grow and multiply. Three to four times a day, generally after meals, a marked increase in motility occurs (this usually coincides with the gastroileal reflex, described earlier). This increased motility, in which large segments of the ascending and transverse colon contract simultaneously, is known as a *mass movement*; it propels fecal material one-third to three-fourths of the length of the colon in a few seconds.

The sudden distension of the walls of the rectum produced by the mass movement of fecal material is the normal stimulus for defecation. It initiates the defecation reflex, which is mediated primarily by the nerve plexuses but can be reinforced by external nerves to the terminal end of the large intestine. The reflex response consists of a contraction of the rectum, relaxation of the anal sphincters, and increased peristaltic activity in the sigmoid colon. This activity is sufficient to propel the feces through the anus. Defecation is normally assisted by a deep inspiration followed by closure of the glottis and contraction of the abdominal and chest muscles, causing a marked increase in intraabdominal pressure, which is transmitted to the contents of the large intestine and assists in the elimination of feces. This maneuver also causes a rise in intrathoracic pressure which leads to a sudden rise in blood pressure followed by a fall as the venous return to the heart is decreased (Chap. 11). In elderly people the cardiovascular stress resulting from the strain of defecation may precipitate a stroke or heart attack.

The *internal anal sphincter* is composed of smooth muscle, but the *external anal sphincter* is skeletal muscle under voluntary control. Brain centers may, via descending pathways, override the afferent input from the rectum, thereby keeping the external sphincter closed and allowing a person to delay defecation. The conscious urge to defecate accompanies the initial distension of the rectum. If defecation does not occur, the tension in the walls of the rectum decreases and the urge to defecate, mediated by stretched mechanoreceptors in the wall, subsides until the next mass movement propels more feces into the rectum, increasing its volume and again initiating the defecation reflex.

About 150 g of feces, consisting of about 100 g of water and 50 g of solid material, is eliminated from the body each day. The solid matter is made up mostly of bacteria, undigested cellulose, debris from the turnover of the intestinal epithelium, bile pigments, and small amounts of salts.

Malfunctioning of the gastrointestinal tract

Since the function of the gastrointestinal system is to digest and absorb nutrients along with salts and water, most malfunctions of this organ system affect either the nutritional state of the body or its salt and water content. The effects of gastrointestinal malfunctioning on nutrition should be obvious but the consequences for maintaining body fluid balance may require amplification. We have seen that the gastrointestinal system secretes 7000 mL into the lumen each day, the salt and water of these secretions being derived from plasma whose volume is only 3000 mL. Therefore, any significant decrease in the absorption of fluid by the intestinal tract leads to a decrease in plasma volume. The following provide only a few familiar examples of disordered gastrointestinal function.

Vomiting

Vomiting is the forceful expulsion of the contents of the stomach and upper intestinal tract through the mouth. Like swallowing, vomiting is a complex reflex coordinated by a region of the brainstem medulla, in this case the vomiting center. Neural input to this center from receptors in many different regions of the body can initiate this reflex. For example, excessive distension of the stomach or duodenum, various substances acting upon chemoreceptors in the wall of the small intestine or the brain, increased pressure within the skull, rotating movements of the head (motion sickness), intense

pain or tactile stimuli applied to the back of the throat, can all initiate vomiting. What is the adaptive value of the vomiting reflex? Obviously, the removal of ingested toxic substances before they can be absorbed is of benefit. Moreover, the nausea that usually accompanies vomiting may have the adaptive value of conditioning the individual to avoid the future ingestion of foods containing the same toxic substances. Why other types of stimuli, such as those producing intense pain or motion sickness, have become linked to the vomiting center is less clear.

Vomiting is usually preceded by increased salivation, sweating, faster heart rate, and feelings of nausea—all characteristic of a general discharge of the autonomic nervous system. Vomiting begins with a deep inspiration, closure of the glottis, and elevation of the soft palate. The abdominal and thoracic muscles contract, raising the abdominal pressure, which is transmitted to the contents of the stomach; the lower esophageal sphincter relaxes, and the high abdominal pressure forces the contents of the stomach into the esophagus. This initial sequence of events may occur repeatedly without vomiting and is known as *retching.* Vomiting occurs when the pressure applied to the contents of the stomach becomes great enough to force the contents past the upper esophageal sphincter and into the mouth (and nasal passages). Vomiting is also accompanied by strong contractions of the upper portion of the small intestine which tend to force some of the intestinal contents back into the stomach. Thus some bile may be present in the vomitus.

Excessive vomiting can lead to large losses of fluids and salts which normally would be reabsorbed in the small intestine. This can result in severe dehydration, upset the salt balance of the body, and produce circulatory problems due to a decrease in plasma volume. In particular, the loss of acid results in a lowering of body acidity. If the amount of vomiting is slight, the control mechanisms acting by way of the lungs and kidneys adjust the acidity toward normal levels, by hypoventilation and by excreting bicarbonate in the urine.

Ulcers

Considering the high concentration of acid and pepsin secreted by the stomach, it is natural to wonder why the stomach does not digest itself.

Several factors protect the walls of the stomach (and duodenum). The surface of the mucosa is lined with cells which secrete a slightly alkaline mucus, which forms a layer 1.0 to 1.5 mm thick over the stomach surface. The protein content of mucus and its alkalinity tend to neutralize hydrogen ions in the immediate area of the epithelium, forming a chemical barrier between the highly acid contents of the lumen and the cell surface. In addition, the membranes of cells lining the stomach have a very low permeability to hydrogen ions, preventing their entry into the underlying mucosa. Moreover, as we have seen, the lateral surfaces of the epithelial cells are joined by tight junctions so that there is no extracellular passage between the cells by which material could diffuse from the lumen into the mucosa. Finally, the epithelial cells lining the walls of the stomach are continually replaced every few days by cell division. The mucous layer, the cell membranes, and cell replacement all contribute to maintaining a barrier between the contents of the lumen and the underlying tissues.

Yet, in some people these protective mechanisms are inadequate, and erosions (*ulcers*) of the gastric wall occur. Ulcers may also occur in the lower part of the esophagus as a result of the reflux of acid into the esophagus from the stomach, and in the upper part of the small intestine which is bathed by the acid from the stomach. If severe enough, the ulcer may damage the underlying blood vessels and cause bleeding into the lumen. On occasion, the ulcer may penetrate the entire wall, with the leakage of luminal contents into the abdominal cavity. About 10 percent of the population of the United States are found at autopsy to have ulcers, which are about 10 times more frequent in the walls of the duodenum than in the stomach.

Ulcer formation requires that the mucosal barrier be broken, exposing the underlying tissue to the corrosive action of acid and pepsin, but it is not clear what produces the initial damage to the barrier, and it is likely that many factors are involved; genetic susceptibility, drugs, decreased blood flow, bile salts, excess acid and pepsin are just some of the possibilities.

Regardless of the cause of barrier breakdown, reducing the level of acid and pepsin secretion, as for example by cutting the vagus nerves, promotes healing and prevents reoccurrence. At the other end of the spectrum, the secretion of very large quantities of acid (due to the presence of gastrin-releasing tumors) results in very severe duodenal ulcers. One should not get the impression, however, that a simple correlation exists between acid secretion and ulcer formation; many patients with ulcers have normal or even subnormal rates of acid secretion.

Despite popular notions that ulcers are due to emotional stress, and the existence of a pathway (the vagus nerves) for mediating stress-induced changes in acid secretion, the actual role of stress in initiating ulcer formation remains uncertain.

The pain associated with ulcers probably results from irritation of exposed nerve fibers in the region of the ulcer, as well as contractile spasms of the smooth muscles irritated by acid. The pain from a duodenal ulcer often subsides following a meal, which buffers the acid secreted by the stomach. The pain is most intense during the early hours of the morning, when unbuffered acid is entering the duodenum.

Gallstones

Bile secreted by the liver into the bile duct contains not only bile salts but cholesterol and phospholipids, which are water-insoluble and are maintained in the bile fluid as micelles in association with the bile salts. However, if the proportions of cholesterol, phospholipids, and bile salts do not fall within certain limited ranges, only part of the cholesterol will be solubilized in the micelle and the rest precipitates in the bile duct or gallbladder, forming *gallstones*. The concentration of secreted cholesterol, in proportion to bile salts and phospholipid, is normally very near the point at which cholesterol precipitates; therefore, small increases in the concentration of biliary cholesterol are sufficient to initiate gallstone formation.

If the gallstone is small, it may pass through the bile duct into the intestine with no complications. A larger stone may become lodged in the gallbladder at its connection with the common bile duct, causing contractile spasms of the smooth muscle, and pain. A more serious complication arises when a gallstone lodges in the common bile duct, thereby preventing bile from entering the intestine. The absence of bile in the intestine markedly decreases the rate of fat digestion and absorption so that about 50 percent of ingested fat passes on to the large intestine and eventually appears in the feces. Furthermore, bacteria in the large intestine convert some of this fat into fatty acid derivatives which act upon the wall of the large intestine, altering salt and water movements and causing a net flow of fluid into the large intestine; the result is diarrhea and increased fluid loss from the body.

The buildup of pressure in a blocked bile duct inhibits further secretion of bile. As a result, bilirubin, the bile pigment formed from the degradation of the heme portion of hemoglobin, accumulates in the blood and tissues, producing the yellowish coloration in the skin known as *jaundice*. Jaundice can also occur, in the absence of a blocked bile duct, if the liver cells are diseased and therefore fail to secrete bilirubin into the bile.

Since in most individuals, the duct from the pancreas joins the common bile duct just before it enters the duodenum, a gallstone that becomes lodged at this point prevents both bile and the pancreatic secretions from entering the intestine, resulting in failure both to neutralize acid from the stomach and to digest adequately most organic nutrients, not just fat.

Why some individuals develop gallstones and others do not is still unclear. Women, for example, have about twice the incidence of gallstone formation as do men, and American Indians have a very high incidence compared to other ethnic groups in America. Increased cholesterol secretion by the liver may be one of the contributing factors in these susceptible individuals, but the factors responsible for the increased secretion are unknown. Recently, attempts to lower blood cholesterol (in the hopes of decreasing the incidence of coronary artery disease) by drugs which increase the liver's secretion of cholesterol into the bile, have resulted in an increased incidence of gallstone formation.

Lactose deficiency (milk intolerance)

Lactose is the major carbohydrate of milk. It cannot be absorbed until digested into its components—glucose and galactose—which are readily absorbed by carrier-mediated active-transport processes. This digestion is performed by the enzyme, lactase, embedded in the plasma membrane of the intestinal epithelium. Lactase is present at birth (except in those rare cases where individuals have a genetic defect and are unable to form the active enzyme) and then its concentration declines in many individuals between 18 and 36 months of age. (This does not occur in approximately 75 percent of white Americans and in most northern Europeans.) In the absence of lactase, lactose remains in the lumen of the small intestine, where it partially prevents water absorption (since the absorption of water requires prior absorption of solute to provide an osmotic gradient for osmosis). This unabsorbed lactose-containing fluid is passed on to the large intestine where bacteria, which do have the enzymes capable of metabolizing lactose, produce from it large quantities of gas (which distends the colon, producing pain) and organic products which inhibit active-transport processes and increase osmolarity. The result is the accumulation of fluid in the lumen of the large intestine. Thus, the diarrhea associated with lactose deficiency is a consequence both of diminished fluid absorption in the small and large intestine and of fluid secretion into the colon. Fluid loss in infants with congenital lack of lactose may be so large as to be fatal if they are not transferred to a diet lacking lactose. The response of adults whose lactase levels have diminished during development, to milk ingestion, varies from mild discomfort to severely dehydrating diarrhea, according to the volume of milk ingested and the amount of lactase present in the intestine.

Constipation and diarrhea

Many people have a mistaken belief that unless they have bowel movements every day, the absorption of toxic substances from fecal material in the large intestine will somehow poison them. Attempts to identify such toxic agents in the blood following prolonged periods of fecal retention have been unsuccessful. In unusual cases where defecation has been prevented for a year or more by blockage of the rectum, no serious ill effects were noted except for the discomfort of carrying around the extra weight of 50 to 100 lb of feces. There appears to be no physiological necessity for having bowel movements regulated by a clock; whatever maintains a person in a comfortable state is physiologically adequate, whether this means a bowel movement after every meal, once a day, or only once a week.

There are, on the other hand, symptoms—headache, loss of appetite, nausea, and abdominal distension—which may arise when defecation has not occurred for long periods of time. The symptoms of constipation appear to be caused, not by toxins, but by the distension of the rectum and large intestine, since inflating a balloon in the rectum of a normal individual produces similar sensations. In addition, the longer fecal material remains in the large intestine, the more water is absorbed, and the harder and drier the feces become, making defecation more difficult and sometimes painful.

Decreased motility of the large intestine is the primary factor causing constipation. This often occurs in old age or may result from damage to the nerve plexuses of the colon which coordinate motility. Emotional stress can also effect the motility of the large intestine. One of the factors increasing the motility of the colon, and thus opposing the development of constipation, is distension. Certain materials in the diet which are not absorbed and thus remain in the lumen perform this function; these include cellulose and fiber (certain undigested substances which bind water).

Although laxatives are used to promote defecation, in some cases the excessive use of laxatives in attempting to maintain a preconceived notion of regularity leads to a decreased responsiveness of the colon to the normal defecation-promoting signals initiated by distension. In such cases, a long period without defecation may occur following cessation of laxative intake, appearing to confirm the subject's belief in the necessity of taking laxatives to promote regularity.

Laxatives act through a variety of mechanisms. Some, such as mineral oil, simply lubricate the feces, making defecation easier and less painful. Others, containing magnesium and aluminum salts, are substances that are absorbed slowly and thus lead to the retention of water in the lumen. Still others, such as castor oil, stimulate the motility of the colon as well as affecting membrane processes associated with ion movements across the wall (thus indirectly affecting water movements).

Diarrhea is characterized by an increased water content in the feces, and is usually accompanied by an increased frequency of defecation, stimulated by the distension of the colon by the increased fluid in the lumen. Diarrhea is ultimately the result of decreased fluid absorption or increased fluid secretion. It is popularly thought that the increased motility which usually accompanies diarrhea actually causes the diarrhea (by decreasing the time available for fluid absorption). However, there is no good evidence for this and it may be that the increased motility is simply the result of distension produced by increased fluid secretion.

A number of bacterial, protozoan, and viral diseases of the intestinal tract cause diarrhea. In many cases they do so by releasing toxic agents which either alter various ion transport processes in the epithelial membranes or injure the epithelium by direct penetration into the mucosa (thereby leading to increased fluid secretion). As we have described earlier, the presence of unabsorbed solutes in the lumen, as a result of decreased digestion or absorption, also results in retained fluid and diarrhea.

The major consequence of prolonged severe diarrhea is the effect of the loss of fluid upon the circulatory system. Accompanying this loss of fluid there is also a loss of various ions, especially potassium and bicarbonate, the latter ions derived from the liver and pancreatic secretions. Since a net loss of bicarbonate leaves behind an equivalent amount of acid, the extracellular fluid becomes more acid during periods of diarrhea.

15

REGULATION OF ORGANIC METABOLISM AND ENERGY BALANCE

Section A: Control and integration of carbohydrate, protein, and fat metabolism

Events of the absorptive and postabsorptive states
  Absorptive state
    Glucose • Triacylglycerols • Amino acids
  Postabsorptive state
    Sources of blood glucose • Glucose sparing (fat utilization)
Endocrine and neural control of the absorptive and postabsorptive states
  Insulin
    Effects on membrane transport • Effects on enzymes • Effects of decreased insulin • Control of insulin secretion • Diabetes mellitus
  Glucagon
  Sympathetic nervous system
  Conclusion
Fuel homeostasis in exercise
Regulation of plasma cholesterol
Control of growth
  External factors influencing growth
  Hormonal influences on growth
    Growth hormone • Thyroid hormone • Insulin •

    Androgens and estrogen
  Compensatory growth

Section B: Regulation of total body energy balance

Basic concepts of energy expenditure and caloric balance
  Determinants of metabolic rate
  Determinants of total body energy balance
Control of food intake
  Obesity
Regulation of body temperature
  Heat production
    Changes in muscle activity • Nonshivering ("chemical") thermogenesis
  Heat-loss mechanisms
    Radiation, conduction, convection • Evaporation
  Summary of effector mechanisms in temperature regulation
  Brain centers involved in temperature regulation
  Afferent input to the integrating centers
    Peripheral thermoreceptors • Central thermoreceptors
  Fever

Section A
Control and integration of carbohydrate, protein, and fat metabolism

In Chap. 5, we described the basic chemistry of living cells and their need for a continuous supply of nutrients. Although a certain fraction of these organic molecules is used in the synthesis of structural cell components, enzymes, coenzymes, hormones, antibodies, and other molecules serving specialized functions, most of the molecules in the food we eat are used by cells to provide the chemical energy required to maintain cell structure and function.

Essential for an understanding of organic metabolism is the remarkable ability of most cells, particularly those of the liver, to convert one type of molecule into another. These interconversions

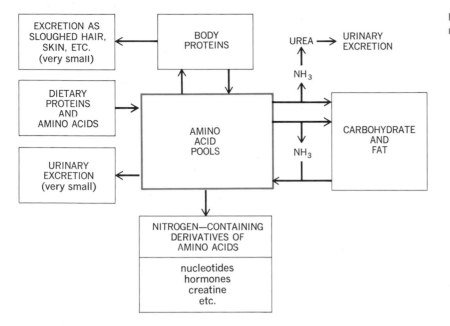

Figure 15-1. Amino acid pools and major pathways of protein metabolism.

permit the human body to utilize the wide range of molecules found in different foods, but there are limits, and certain molecules must be present in the diet in adequate amounts. Enough protein must be ingested to provide the nitrogen needed for synthesis of protein and other nitrogenous substances in the body and this protein must contain an adequate quantity of specific amino acids, called *essential* because they cannot be formed in the body by conversion from another molecule type; eight of the twenty amino acids fit this category. The other essential organic nutrients are a small group of fatty acids and the vitamins. The last group was discovered when diets adequate in both total calories and essential amino acids and fatty acids were still found to be incapable of maintaining health.

The concept of a dynamic catabolic-anabolic steady state is also a critical component of organic metabolism. With few exceptions, e.g., DNA, virtually all organic molecules are continuously broken down and rebuilt, often at a rapid rate. The turnover rate of body protein is approximately 100 g/day; i.e., this quantity is broken down into amino acids and resynthesized each day. Few of the atoms present in a person's skeletal muscle a month ago are still there today.

With these basic concepts of *molecular interconvertibility* and *dynamic steady state* as foun-

dation, we can discuss organic metabolism in terms of total body interactions. Figure 15-1 summarizes the major pathways of protein metabolism. The *amino acid pools*, which constitute the body's total free amino acids, are derived primarily from ingested protein (which is degraded to amino acids during digestion) and from the continuous breakdown of body protein. These pools are the source of amino acids for resynthesis of body protein and a host of specialized amino acid derivatives, such as nucleotides, epinephrine, etc. A very small quantity of amino acid and protein is lost from the body via the urine, skin, hair, and fingernails. The interactions between amino acids and the other nutrient types, carbohydrate and fat, are extremely important: Amino acids may be converted into carbohydrate or fat by removal of ammonia (deamination); one type of amino acid may participate in the formation of another by passing its nitrogen group to a carbohydrate. Both these processes were described in greater detail in Chap. 5 and are mentioned here to emphasize the interconvertibility of protein, carbohydrate, and fat. The ammonia, NH_3, formed during deamination is converted by the liver into urea, which is then excreted by the kidneys as the major end product of protein metabolism. Not all the events relating to amino acid metabolism occur in all cells; urea is formed in one organ (the liver)

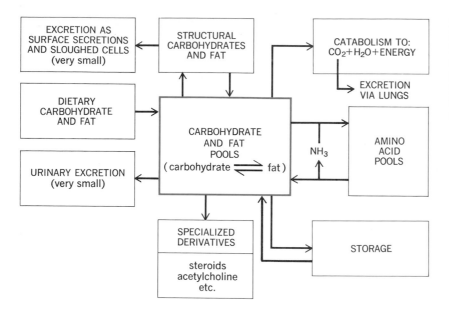

Figure 15-2. Major pathways of carbohydrate and fat metabolism. "Carbohydrate and fat pools" include the
simple unspecialized carbohydrates
and fats dissolved in the body fluids.
Note that structural carbohydrate and
fat are constantly being broken down
and resynthesized.

but excreted by another (the kidneys), but the concept of a pool is valid because all cells are interrelated by the vascular system and blood.

If any of the essential amino acids is missing from the diet, negative nitrogen balance (i.e., output greater than intake) always results. Apparently, the proteins for which that amino acid is essential cannot be synthesized and the other amino acids which would have been incorporated into the proteins are deaminated, their nitrogen being excreted as urea. It should be obvious, therefore, why a dietary requirement for protein cannot be specified without regard to the amino acid composition of that protein. Protein is graded in terms of how closely its ratio of essential amino acids approximates the ideal, which is their relative proportions in body protein. The highest-quality proteins are those found in animal products whereas the quality of most plant proteins is lower. Nevertheless, it is quite possible to obtain adequate quantities of all essential amino acids from a mixture of plant proteins alone although the total quantity of protein ingested must be larger.

Figure 15-2 summarizes the metabolic pathways for carbohydrate and fat, which are considered together because of the high rate of conversion of carbohydrate to fat (see below and Chap. 5). The similarities between Figs. 15-1 and 15-2 are obvious, but there are several critical differences: (1) The major fate of both carbohydrate and fat is

catabolism to yield energy, whereas amino acids can supply energy only after they are converted to carbohydrate or fat; (2) excess carbohydrate and fat can be stored as such, whereas excess amino acids are not stored as protein but are converted to carbohydrate and fat.

In discussing the mechanisms which regulate the magnitude and direction of these molecular interconversions, we shall see that the liver, adipose tissue (the storage tissue for fat), and muscle are the dominant effectors and that the major controlling input to them is a group of hormones and the sympathetic nerves to adipose tissue and the liver. At this point, the reader should review the biochemical pathways described in Chap. 5, particularly those dealing with glucose and the interconversions of carbohydrate, protein, and fat.

Events of the absorptive and postabsorptive states

When food is readily available human beings can get along by eating small amounts of food all day long if they wish; however, this is clearly not true for most other animals, for early man, or for most persons today, and mechanisms have evolved for survival during alternating periods of plenty and fasting. We speak of two functional states: the *absorptive state*, during which ingested nutrients are entering the blood from the gastrointestinal tract, and the *postabsorptive state*, during which the gastrointestinal tract is empty and energy must be supplied by the body's endogenous stores. Since an average meal requires approximately 4 h for complete absorption, our usual three-meal-a-day pattern places us in the postabsorptive state during the late morning and afternoon and almost the entire night. The average person can easily withstand a fast of many weeks (so long as water is provided), and extremely obese patients have been fasted for many months, being given only water and vitamins.

The absorptive state can be summarized as follows: During absorption of a normal meal, glucose provides the major energy source: only a small fraction of the absorbed amino acids and fat is utilized for energy; another fraction of amino acids and fat is used to resynthesize the continuously degraded body proteins and structural fat, respectively; most of the amino acids and fat as well as carbohydrate not oxidized for energy are transformed into adipose-tissue fat.

In the postabsorptive state, carbohydrate is synthesized in the body, but its utilization for energy is greatly reduced; the oxidation of endogenous fat provides most of the body's energy supply; fat and protein synthesis are curtailed and net breakdown occurs.

Figures 15-3 and 15-4 summarize the major pathways to be described. Although they may appear formidable at first glance, they should give little difficulty after we have described the component parts, and they should be referred to constantly during the following discussion.

Absorptive state

We shall assume an average meal to contain approximately 65 percent carbohydrate, 25 percent protein, and 10 percent fat. Recall from Chap. 14 that these nutrients enter the blood and lymph from the gastrointestinal tract primarily as monosaccharides, amino acids, and triacylglycerols,[1] respectively. The first two groups enter the blood, which leaves the gastrointestinal tract to go directly to the liver by way of the hepatic portal vein, allowing this remarkable biochemical factory

[1] Triacylglycerols were termed triglycerides until recently.

to alter the composition of the blood before it returns to the heart to be pumped to the rest of the body. In contrast, the fat droplets are absorbed into the lymph and not into the blood; the lymph drains into the systemic venous system in the neck, and, therefore, the liver does not get first crack at the absorbed fat.

Glucose. Some of the absorbed carbohydrate is galactose and fructose, but since the liver converts most of these carbohydrates immediately into glucose (and because fructose enters essentially the same metabolic pathways as does glucose), we shall simply refer to these sugars as glucose. As shown in Fig. 15-3, much of the absorbed carbohydrate enters the liver cells, but little of it is oxidized for energy, instead being built into the polysaccharide glycogen or transformed into fat. The importance of glucose as a precursor of fat cannot be overemphasized; note that glucose provides both the glycerol (as glycerophosphate) and the fatty acid components of triacylglycerols. Some of this fat synthesized in the liver may be stored there, but most is released into the blood and is then stored in adipose-tissue cells. Much of the absorbed glucose which did not enter liver cells but remained in the blood enters adipose-tissue cells, where it is transformed into fat; another fraction is stored as glycogen in skeletal muscle and certain other tissues, and a very large fraction enters the various cells of the body and is oxidized to carbon dioxide and water, thereby providing the cells' energy requirements. Glucose is usually the body's major energy source during the absorptive state.

Triacylglycerols. Almost all ingested fat is absorbed into the lymph as fat droplets (chylomicrons), containing mainly triacylglycerols. The biochemical processing of these triacylglycerols is complex; for example, the chylomicrons, themselves, cannot penetrate capillary walls and so they must be hydrolyzed into their component molecules — glycerol and free fatty acids — by enzymes located on the inner surface of the capillary endothelium. For simplicity's sake we have shown in Fig. 15-3 only the eventual overall result of this processing, namely the deposition of most of the ingested fat, once again combined into triacylglycerol, in adipose tissue. There are three major sources of adipose tissue triacylglycerol: (1) in-

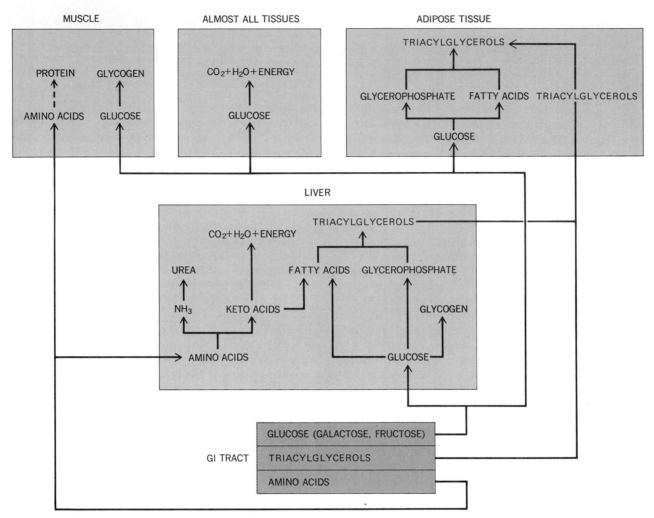

Figure 15-3. Major metabolic pathways of the absorptive phase. Glycerophosphate is the form of glycerol used to synthesize triacylglycerol.

gested fat; (2) triacylglycerol synthesized in adipose tissue; (3) triacylglycerol synthesized in the liver and transported via the blood to the adipose tissue. For simplicity, we have not shown in Fig. 15-3 that a fraction of fat is also oxidized during the absorptive state by various organs to provide energy. The actual amount utilized depends upon the content of the meal and the person's nutritional status.

Amino acids. Many of the absorbed amino acids enter liver cells and are entirely converted into carbohydrate (keto acids) by removal of the NH_3 portion of the molecule. The ammonia is converted by the liver into urea, which diffuses into the blood and is excreted by the kidneys. The keto acids formed can enter the Krebs tricarboxylic acid cycle and be oxidized to provide energy for the liver cells. Finally, the keto acids can also be converted to fatty acids, thereby participating in fat synthesis by the liver. Most of the ingested amino acids are not taken up by the liver cells and these enter other cells of the body (Fig. 15-3). Although virtually all cells require a constant supply of amino acids for protein synthesis, we have simplified the diagram by showing only muscle because it constitutes the

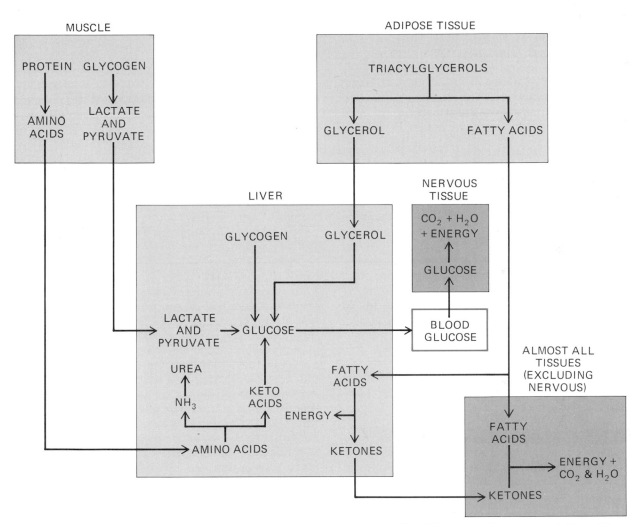

Figure 15-4. Major metabolic pathways of the postabsorptive phase or fasting. The central focus is regulation of the blood glucose concentration.

great preponderance of body mass and therefore contains the most important store, quantitatively, of body protein. Other organs of course participate, but to a lesser degree, in the amino acid exchanges occurring during the absorptive and postabsorptive states. After entering the cells, the amino acids may be synthesized into protein. This process is represented by the dotted line in Fig. 15-3 to call attention to an important fact: Excess amino acids are *not* stored as protein, in the sense that glucose and fat are stored as fat and to a lesser degree as glycogen. Eating large amounts of protein does not significantly increase body protein; the excess

amino acids are merely converted into carbohydrate or fat. On the other hand, a minimal supply of ingested amino acids is essential to maintain normal protein stores by preventing net protein breakdown in muscle and other tissues. During the usual alternating absorptive and postabsorptive states, the fluctuations in total body protein are relatively small. This discussion applies only to the adult; the growing child, of course, manifests a continuous increase of body protein.

In summary, during the absorptive period, anabolism exceeds catabolism, energy is provided primarily by glucose, body proteins are maintained, and excess calories (regardless of source) are stored mostly as fat. Glycogen constitutes a quantitatively less important storage form for carbohydrate. The use of fat to store excess calories is an excellent adaptation for mobile animals, because 1 g of triacylglycerol contains more than twice as many calories as 1 g of protein or glycogen and because there is very little water in adipose tissue.

Postabsorptive state

The essential problem during this period is that no glucose is being absorbed from the intestinal tract, yet the plasma glucose concentration must be maintained because the nervous system is an obligatory glucose utilizer; i.e., it is unable to oxidize any other nutrient for energy (there is an important exception to be described subsequently). Perhaps the most convenient way of viewing the events of the postabsorptive state is in terms of how the blood glucose concentration is maintained. These events fall into two categories: (1) sources of glucose; (2) glucose sparing (fat utilization).

Sources of blood glucose. The sources of blood glucose during the postabsorptive period (Fig. 15-4) are as follows:

1 Glycogen stores in the liver are broken down to liberate glucose but are adequate only for a short time. After the absorptive period is completed, the normal liver contains less than 100 g of glycogen; at 4 kcal/g, this provides 400 kcal, enough to fulfill the body's total caloric need for only 4 h.

2 Glycogen in muscle (and to a lesser extent other tissues) provides approximately the same amount of glucose as the liver. A complication arises because muscle lacks the necessary enzyme to form free glucose from glycogen. But glycolysis breaks the glycogen down into pyruvate and lactate, which are then liberated into the blood, circulate to the liver, and are converted into glucose. Thus, muscle glycogen contributes to the blood glucose indirectly via the liver.

3 As shown in Fig. 15-4, the catabolism of triacylglycerols yields glycerol and fatty acids. The former can be converted into glucose by the liver, but the latter cannot. Thus, a potential source of glucose is adipose-tissue triacylglycerol break-

down, in which glycerol is liberated into the blood, circulates to the liver, and is converted into glucose.

4 The major source of blood glucose during prolonged periods of fasting comes from protein. Large quantities of protein in muscle and to a lesser extent other tissues are not absolutely essential for cell function; i.e., a sizable fraction of cell protein can be catabolized, as during prolonged fasting, without serious cellular malfunction. There are, of course, limits to this process, and continued protein loss ultimately means functional disintegration, sickness, and death. Before this point is reached, protein breakdown can supply large quantities of amino acids which are converted into glucose by the liver.

To summarize, for survival of the brain, plasma glucose concentration must be maintained. Glycogen stores, particularly in the liver, form the first line of defense, are mobilized quickly, and can supply the body's needs for several hours, but they are inadequate for longer periods. Under such conditions, protein and, to a much lesser extent, fat, supply amino acids and glycerol, respectively, for production of glucose by the liver. Hepatic synthesis of glucose from pyruvate, lactate, glycerol, and amino acids is known as *gluconeogenesis*, i.e., new formation of glucose. During a 24-h fast, it amounts to approximately 180 g of glucose. The kidneys are also capable of glucose synthesis from the same sources, particularly in a prolonged fast (several weeks), at the end of which they may be contributing as much glucose as the liver.

Glucose sparing (fat utilization). A simple calculation reveals that even the 180 g of glucose per day produced by the liver during fasting cannot possibly supply all the body's energy needs: 180 g/day × 4 kcal/g = 720 kcal/day, whereas normal total energy expenditure equals 1500 to 3000 kcal/day. The following essential adjustment must therefore take place during the transition from absorptive to postabsorptive state: The nervous system continues to utilize glucose normally, but virtually all other organs and tissues markedly reduce their oxidation of glucose and depend primarily on fat as their energy source, thus sparing the glucose produced by the liver to serve the obligatory needs of the nervous system. The essential step is the catabolism of adipose-tissue triacylglycerol to liberate fatty acids into the blood. These fatty acids are picked up by virtually all tissues (excluding the nervous system), enter the Krebs cycle, and are oxidized to carbon dioxide and water, thereby providing energy. The liver, too, utilizes fatty acids for its energy source, but its handling of fatty acids during fasting is unique; it oxidizes them to acetyl CoA which is processed into a group of compounds called *ketone bodies* instead of being oxidized further via the Krebs cycle. (One of these substances is acetone, some of which is exhaled and accounts for the distinctive breath odor of persons undergoing prolonged fasting or suffering from severe untreated diabetes mellitus.) These ketone bodies are released into the blood and provide an important energy source for the many tissues capable of oxidizing them via the Krebs cycle.

The net result of fatty acid utilization during fasting is provision of energy for the body and sparing of glucose for the brain. The combined effects of gluconeogenesis and the switch-over to fat utilization are so efficient that, after several days of complete fasting, the plasma glucose concentration is reduced only by a few percent. After one month, it is decreased only 25 percent.

There also occurs an important change in brain metabolism with prolonged starvation. Apparently, after 4 to 5 days of fasting, the generalization that brain is an obligatory glucose utilizer is no longer valid, for the brain begins to utilize large quantities of ketone bodies, as well as glucose, for its energy source. The survival value of this phenomenon is very great; if the brain significantly reduces its glucose requirement (by utilizing ketones instead of glucose), much less protein need be broken down to supply the amino acids for gluconeogenesis. Accordingly, the protein stores will last longer, and the ability to withstand a long fast without serious tissue disruption is enhanced.

Thus far, our discussion has been purely descriptive; we now turn to the endocrine and neural factors which so precisely control and integrate these metabolic pathways and transformations. Without question, the most important single factor is insulin. As before, the reader should constantly refer to Figs. 15-3 and 15-4. We shall focus primarily on the following questions (Fig. 15-5) raised by the previous discussion: (1) What controls the shift from the net anabolism of protein, glycogen, and triacylglycerol to net catabolism? (2) What induces primarily glucose utilization during absorption and fat utilization during postabsorption; i.e., how do cells "know" they should start oxidizing fatty acids and ketones instead of glucose? (3) What drives net hepatic glucose synthesis (gluconeogenesis) and release during postabsorption?

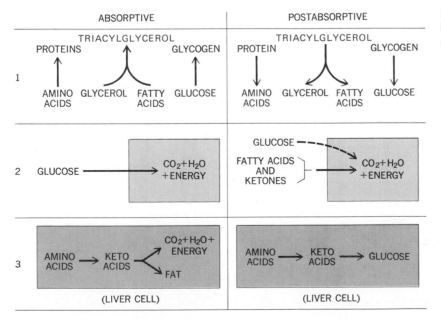

Figure 15-5. Summary of critical shifts in transition from absorptive to postabsorptive states.

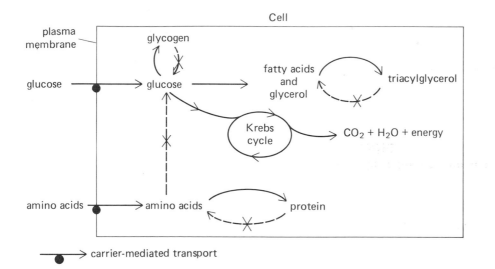

Figure 15-6. Major effects of insulin upon organic metabolism. Each solid arrow ($\rightarrow$) represents a process enhanced by insulin whereas ($-\times\rightarrow$) denotes a reaction inhibited by insulin. Conversely, when insulin concentration is low, the activity of the dashed pathways is enhanced and that of the solid pathways is decreased. The nontransport effects are mediated by insulin-dependent enzymes. All these effects are not necessarily exerted upon all cells.

Endocrine and neural control of the absorptive and postabsorptive states

Insulin

Insulin is a protein hormone secreted by the islets of Langerhans, clusters of endocrine cells in the pancreas. Appropriate histological techniques reveal three distinct types of islet cells, termed A, B, and D cells. The B cells are the source of insulin, the A cells of glucagon, and the D cells of a peptide named somatostatin (this last substance is also found in the hypothalamus and various other sites in the body, and the precise role of pancreatic somatostatin in normal physiology remains unclear).

Insulin acts directly or indirectly on most tissues of the body with the notable exception of most brain areas. The effects of insulin are so important and widespread that an injection of this hormone into a fasting person duplicates the absorptive state pattern of Fig. 15-3 (except, of course, for the absence of gastrointestinal absorption); conversely, patients suffering from insulin deficiency (diabetes mellitus) manifest the postabsorptive pattern of Fig. 15-4. From these statements alone, it might appear (correctly) that secretion of insulin is stimulated by eating and inhibited by fasting.

Insulin induces in its target cells a large number of changes (Fig. 15-6) which fall into two general categories: alteration of either membrane transport or enzyme function (Fig. 15-6). What are the biochemical mechanisms by which these changes are brought about? Being a protein, insulin combines with receptors specific for it on the outer surface of the plasma membrane, and it is this combination which triggers the sequences of events leading ultimately to the changes in membrane transport and enzyme function. Almost certainly involved in these sequences is at least one "second messenger," but its (their) identity has not been established with certainty. Insulin is clearly one hormone that does not activate adenylate cyclase; indeed it may inhibit this enzyme, thereby decreasing intracellular cyclic AMP.

Effects on membrane transport. Glucose enters most cells by the carrier-mediated mechanism which we described as facilitated diffusion in Chap. 6. Insulin stimulates this facilitated diffusion of glucose into cells, particularly muscle and adipose tissue; greater glucose entry into cells increases the availability of glucose for all the reactions in which glucose participates. (It is important to note that insulin does *not* alter glucose uptake by the brain.) Insulin also stimulates the active transport of amino acids into most cells, thereby making more amino acids available for protein synthesis.

Effects on enzymes. The enhanced membrane transport of glucose and amino acids induced by insulin favors, by mass action, glucose oxidation and net synthesis of glycogen, triacylglycerol, and protein. However, in addition, insulin alters the activities or concentrations of many of the intracellular enzymes involved in the anabolic and catabolic pathways of these substances so as to achieve the same result. Glycogen synthesis provides an excellent example: Increased glucose uptake per se stimulates glycogen synthesis but, in addition, insulin increases the activity of the rate-limiting step in glycogen synthesis and inhibits the enzyme which catalyzes glycogen catabolism; thus, insulin favors glucose transformation into glycogen by a triple-barrelled effect!

The situation for triacylglycerol storage is analogous: Increased entry of glucose into adipose tissue provides the precursors for the synthesis of fatty acid and glycerophosphate and their combination into triacylglycerol; simultaneously insulin increases the activity of certain of the enzymes which catalyze fatty acid synthesis, and most important, inhibits the enzyme which catalyzes triacylglycerol breakdown—again a triple-barrelled effect favoring, in this case, triacylglycerol storage.

And similarly for protein: Insulin stimulates uptake of amino acids while simultaneously increasing the activity of some of the ribosomal enzymes which mediate the synthesis of protein from these amino acids and inhibiting the enzymes which mediate protein catabolism.

Finally, insulin inhibits almost all the critical liver enzymes which catalyze gluconeogenesis; the net result is that insulin abolishes glucose release by the liver (and causes net uptake instead).

Effects of decreased insulin. We have thus far dealt with the positive effects of insulin. Clearly insulin deficit will have just the opposite results; when the blood concentration of insulin decreases, the metabolic pattern is shifted toward decreased glucose entry and oxidation and a net catabolism of glycogen, protein, and triacylglycerol. In other words, these metabolic pathways are in a dynamic state, capable of proceeding, in terms of net effect, in either direction. For this reason, energy metabolism can be shifted from the absorptive to the postabsorptive pattern merely by lowering the rate of insulin secretion: Glucose entry and oxidation decrease; glycogen breakdown increases; net protein catabolism liberates amino acids into the blood; net triacylglycerol catabolism liberates glycerol and fatty acids into the blood, and the resulting higher fatty acid concentration in blood facilitates

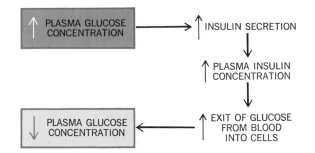

Figure 15-7. Negative-feedback nature of plasma glucose control over insulin secretion.

cellular uptake of fatty acids, which in turn stimulates fatty acid oxidation; and gluconeogenesis is stimulated not only by the increased availability of precursors (amino acids and glycerol) but by enzyme changes in the liver itself. The glucose released by the liver can be utilized by the brain since its glucose uptake is not insulin-dependent.

Control of insulin secretion. Insulin secretion is directly controlled by the glucose concentration of the blood flowing through the pancreas, a simple system requiring no participation of nerves or other hormones. An increase in blood glucose concentration stimulates insulin secretion; conversely, a reduction inhibits secretion. The feedback nature of this system is shown in Fig. 15-7. A rise in plasma glucose stimulates insulin secretion; insulin induces rapid entry of glucose into cells (as well as cessation of glucose output by the liver); this transfer of glucose out of the blood reduces the blood concentration of glucose, thereby removing the stimulus for insulin secretion, which returns to its previous level.

Figure 15-8 illustrates typical changes in plasma glucose and insulin concentrations following a normal carbohydrate-rich meal. Note the close association between the rising blood glucose level (resulting from gastrointestinal absorption) and the plasma insulin increase induced by the glucose rise. The low postabsorptive values for plasma glucose and insulin concentrations are not the lowest attainable, and prolonged fasting induces even further reductions of both variables until insulin is barely detectable in the blood.

Plasma glucose is not the sole control over insulin secretion, for insulin secretion is sensitive to numerous other types of input. One of the most

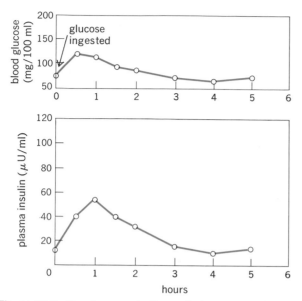

Figure 15-8. Blood concentrations of glucose and insulin following ingestion of 100 g of glucose. Study performed on normal human subjects. *(Adapted from Daughaday and Kipnis.)*

important is the plasma concentration of certain amino acids, an elevated amino acid concentration causing enhanced insulin secretion. This is easily understandable since amino acid concentrations increase after eating, particularly after a high-protein meal. The increased insulin stimulates cell uptake of these amino acids. There are also important neural and hormonal controls over insulin secretion. For example, one or more of the hormones secreted by the gastrointestinal tract (see Chap. 14) in response to the presence of food stimulates the release of insulin; this provides an "anticipatory" (or "feedforward") component to glucose regulation.

Diabetes mellitus. The name diabetes, meaning syphon or running through, was used by the Greeks over 2000 years ago to describe the striking urinary volume excreted by certain people. Mellitus, meaning sweet, distinguishes this urine from the large quantities of insipid urine produced by persons suffering from ADH deficiency (Chap. 13). This sweetness of the urine was first recorded in the seventeenth century, but in England the illness had long been called the pissing evil. Because of the marked weight loss despite huge food intake,

the body's substance was believed to be dissolving and pouring out through the urinary tract, a view not far from the truth. In 1889, experimental diabetes was produced in dogs by surgical removal of the pancreas, and 32 years later, in 1921, Banting and Best discovered insulin.

A tendency toward diabetes can be inherited. We say tendency because diabetes often is not an all-or-none disease but may develop slowly, and overt signs may all but disappear with appropriate measures, e.g., weight reduction. The cause of diabetes is relative insulin deficiency. We have described how a lowered insulin concentration induces virtually all the metabolic changes characteristic of the postabsorptive state (Fig. 15-4). The picture presented by an untreated diabetic is a gross caricature of this state (Fig. 15-9). The catabolism of triacylglycerol with resultant elevation of plasma fatty acids and ketones is an appropriate response because these substances must provide energy for the body's cells, which are prevented from taking up adequate glucose by the insulin deficiency. In contrast, glycogen and protein catabolism and the marked gluconeogenesis so important to maintain plasma glucose during fasting are completely inappropriate in the diabetic since plasma glucose is already high because it cannot enter into cells. These reactions serve only to raise the plasma glucose still higher, with disastrous consequences. Only the brain is spared glucose deprivation since its uptake of glucose is not insulin-dependent.

The elevated plasma glucose of diabetes is, itself,

Figure 15-9. Factors which elevate blood glucose concentration in insulin deficiency.

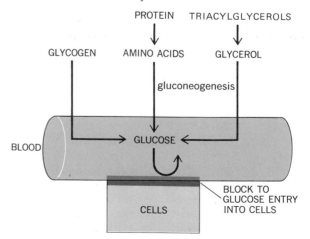

relatively innocuous,[1] but it induces changes in renal function of serious consequence. In Chap. 13, we pointed out that a normal person does not excrete glucose because all glucose filtered at the glomerulus is reabsorbed by the tubules. However, the elevated plasma glucose of diabetes may so increase the filtered load of glucose that the maximum tubular reabsorptive capacity is exceeded, and large amounts of glucose may be excreted. For the same reasons, large amounts of ketones may also appear in the urine. These urinary losses, of course, only aggravate the situation by further depleting the body of nutrients and leading to weight loss. Far worse, however, is the effect of these solutes on sodium and water excretion. In Chap. 13, we saw how tubular water reabsorption is a passive process induced by active solute reabsorption. In diabetes, the osmotic force exerted by unreabsorbed glucose and ketones holds water in the tubule, thereby preventing its reabsorption. For several reasons (the mechanisms are beyond the scope of this book) sodium reabsorption is also retarded. The net result is marked excretion of sodium and water, which leads, by the sequence of events shown in Fig. 15-10, to hypotension, brain damage, and death.

Another serious abnormality in diabetes is a markedly increased hydrogen-ion concentration, due primarily to the accumulation of ketone bodies, which as moderately strong acids generate large amounts of hydrogen ion by dissociation. The kidneys respond to this increase by excreting more hydrogen ion and are generally able to maintain balance fairly well, at least until the volume depletion described above interferes with renal function. What effect does the increased hydrogen-ion concentration have on respiration? A marked increase in ventilation occurs in response to stimulation of the medullary respiratory centers by hydrogen ion, and the resulting overexcretion of carbon dioxide further helps to keep the hydrogen-ion concentration below lethal limits.

The grim picture painted above is seen only in patients with severe uncontrolled diabetes, those with almost total inability to secrete insulin; the great majority of diabetics never develop this so-called "ketoacidosis," but they do suffer from a variety of chronic abnormalities, including athero-

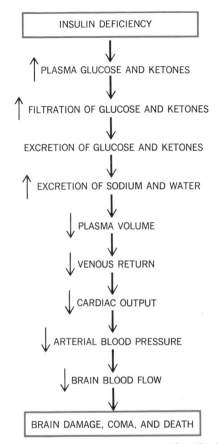

Figure 15-10. Effects of severe untreated insulin deficiency on renal function.

sclerosis, small-vessel and nerve disease, susceptibility to infection, and blindness. The mechanisms by which insulin deficiency contributes to the development of these malfunctions is the subject of much current investigation.

The treatment of diabetes is aimed at maintaining plasma glucose at a relatively normal value. The major therapy for patients with severe inability to secrete insulin is administration of insulin, which must be given by injection since as a protein it is broken down by gastrointestinal enzymes. The dose must be determined carefully since an overdose abnormally lowers the plasma glucose concentration and can cause brain malfunction. Another type of therapy used for many patients with less severe insulin deficiency is noninsulin drugs which can be taken by mouth, and act upon the islet cells to stimulate insulin secretion; thus,

---

[1] This view is presently being challenged, but no firm answer is yet available.

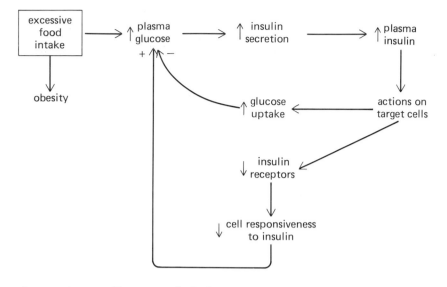

Figure 15-11. Postulated mechanism by which chronic overeating leads to chronically elevated plasma insulin and diminished cell responsiveness to insulin. Note that the increased insulin contributes to both a negative-feedback and positive-feedback effect on plasma glucose.

therapy is actually accomplished with the patient's own insulin. The effectiveness and safety of these agents remain controversial.

Finally, simple weight reduction without any other therapy is frequently sufficient to eliminate many of the manifestations of the disease in the large number of diabetics who are overweight. (Given the earlier mention of progressive weight loss as a symptom of diabetes, it may seem contradictory that many diabetics are overweight, but the paradox is resolved when one recalls that only patients with severe insulin deficiency suffer so much loss of glucose in the urine as to lose weight.) The first clue as to why diabetes is so frequently associated with obesity was the surprising discovery that many obese diabetics actually have *increased* plasma concentrations of insulin! Their elevated plasma glucose is due, therefore, not to an absolute deficiency of insulin but to a failure of insulin to influence target cells normally. This insulin insensitivity seems to have multiple causes, but one seems to be an insufficient number of insulin receptors on various target cells, the sequence of events underlying this phenomenon being as follows (Fig. 15-11): Because, as we have seen, insulin secretion is increased during ingestion and absorption of food, any person who chronically overeats secretes, on the average, increased amounts of insulin; the resulting elevation of plasma insulin induces, over time, a reduction in the number of insulin receptors (Chap. 9); thus, insulin, itself, is responsible for the decrease in target-cell re-

sponsiveness, the end result of which is a rise in plasma glucose concentration. The latter is very small in nondiabetic overeaters because the islet cells respond to it by secreting enough additional insulin to get the job done despite the reduction in available receptors; in contrast, the diabetes-prone person may also secrete additional insulin but not enough to prevent significant hyperglycemia and the appearance of diabetic symptoms.

Glucagon

Insulin unquestionably plays a central role in controlling the metabolic adjustments required for feasting or fasting, but other pathways are also involved. Glucagon, a protein hormone produced by the A cells of the pancreatic islets,[1] is probably the most important of these. The major effects of glucagon on organic metabolism (mediated by generation of cyclic AMP in its target cells) are all opposed to those of insulin (Fig. 15-12): increased glycogen breakdown in liver; increased synthesis of glucose (gluconeogenesis) by liver; increased breakdown of adipose-tissue triacylglycerol (fat mobilization). Thus, the overall results of glucagon's effects are to increase the plasma concentrations of glucose and fatty acids, all important events of the postabsorptive period.

[1] Glucagon and glucagon-like substances are also secreted by cells in the lining of the gastrointestinal tract morphologically identical to the A cells; the significance of this nonpancreatic glucagon is unclear.

From a knowledge of these effects, one would logically suppose that the secretion of glucagon should be increased during the postabsorptive period and prolonged fasting, and such is the case. The stimulus is a decreased or decreasing plasma glucose concentration. The adaptive value of such a reflex is obvious; a decreasing plasma glucose induces increased release of glucagon which, by its effects on metabolism, serves to restore normal blood glucose levels and at the same time supply fatty acids for cell utilization (Fig. 15-12). Conversely, an increased or increasing plasma glucose inhibits glucagon's secretion, thereby helping to return plasma glucose toward normal.

The pathway for this reflex is quite simple: The A cells in the pancreas respond directly to changes in the glucose concentration of the blood perfusing the pancreas.

Thus far, the story is quite uncomplicated; the B and A cells of the pancreatic islets constitute a push-pull system for regulating plasma glucose by producing two hormones (insulin and glucagon, respectively) whose actions and controlling input are just the opposite of each other. However, we must now point out a complicating feature: A second major control of glucagon secretion is the plasma amino acid concentration (acting directly on the A cells), and in this regard the effect is identical rather than opposite to that for insulin: Glucagon secretion, like that of insulin, is strongly stimulated by a rise in plasma amino acid concentration such as occurs following a protein-rich meal. Thus, during absorption of a carbohydrate-rich meal containing little protein, there occurs an increase in insulin secretion alone, caused by the rise in plasma glucose, but during absorption of a low-carbohydrate–high-protein meal, glucagon also increases, under the influence of the increased plasma amino acid concentration. The usual meal is somewhere between these extremes and is accompanied by a rise in insulin and relatively little change in glucagon, since the simultaneous increases in blood glucose and amino acids counteract each other so far as glucagon secretion is concerned. Of course, regardless of the type of meal ingested, the postabsorptive period is always accompanied by a rise in glucagon secretion (and a decrease in insulin secretion).

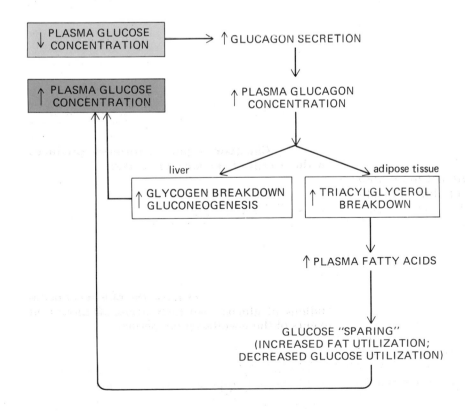

Figure 15-12. Negative-feedback nature of plasma glucose control over glucagon secretion (compare to Fig. 15-7).

What is the adaptive value of the amino acid–glucagon relationship? Imagine what might occur were glucagon not part of the response to a high-protein meal: Insulin secretion would be increased by the amino acids but, since little carbohydrate was ingested and therefore available for absorption, the increase in plasma insulin could cause a marked and sudden drop in plasma glucose. In reality, the rise in glucagon secretion caused by the amino acids permits the hyperglycemic effects of this hormone to counteract the hypoglycemic actions of insulin, and the net result is a stable plasma glucose. Thus, a high-protein meal virtually free of carbohydrate can be absorbed with little change in plasma glucose despite a marked increase in insulin secretion.

Because glucagon induces an elevation of plasma glucose, an important question is whether certain forms of diabetes might involve not only deficiency of insulin (either absolute or relative) but an excess of glucagon. One approach to this question has been to measure the concentration of glucagon in the plasma of patients with diabetes. Recall that, since glucagon secretion is inhibited by an elevated plasma glucose, one would expect to find a low plasma glucagon in diabetic persons (because they have increased plasma glucose) if glucagon had nothing to do with their disease. However, the actual finding is that most diabetic patients have plasma glucagon concentrations which are increased (either absolutely or relative to their elevated plasma glucose). The failure of glucagon to decrease seems to result from a partial loss (for unknown reasons) of glucose-sensing ability by the A cells. How much does this absolute or relative glucagon excess contribute to the carbohydrate and fat abnormalities typical of diabetes? This question is being intensively investigated using agents capable of inhibiting glucagon secretion, but no clear answer is yet available. The question is of profound clinical importance since, if glucagon were an important contributory factor, the therapy of diabetes might focus on inhibition of glucagon secretion rather than (or, much more likely, in addition to) insulin administration.

Sympathetic nervous system

Insulin and glucagon are the major hormones involved in the minute-to-minute control over the metabolic adjustments associated with feasting and fasting, but there is a third efferent pathway which may play an important role, particularly when the transition to fasting is associated with a rapidly falling plasma glucose concentration. This is the sympathetic nervous system, and the specific components of this system involved are: (1) the adrenal medulla, the endocrine gland which is an integral part of the sympathetic nervous system and which secretes mainly epinephrine into the blood; (2) the sympathetic neurons to adipose-tissue cells. The major actions of epinephrine are to stimulate glycogenolysis (by both liver and skeletal muscle) and to stimulate the breakdown of adipose-tissue triacylglycerol (fat mobilization); this latter effect is also strongly stimulated by the sympathetic neurons to adipose tissue. Thus, the overall effects on organic metabolism of enhanced activity of the sympathetic nervous system are the opposite of those of insulin and similar to those of glucagon: increased plasma concentrations of glucose, glycerol, and fatty acids (Fig. 15-13).

As might be predicted from these effects, an important stimulus leading to increased secretion of epinephrine and activity of the neurons to adipose tissue is a decreased or decreasing plasma glucose. This is the same stimulus which, as described above, leads to increased secretion of glucagon (although the receptors and pathways are totally different), and the adaptive value of the response is also the same: Blood glucose is restored toward normal and fatty acids are supplied for cell utilization. The pathway for this reflex is illustrated in Fig. 15-13. As described in Chap. 9, epinephrine release is controlled entirely by the preganglionic sympathetic fibers to the adrenal medulla; the receptors, glucose receptors in the brain (probably in the hypothalamus), initiate the reflexes which lead ultimately to increased activity in these neurons and in the sympathetic pathways to adipose tissue. This reflex is triggered when the plasma glucose concentration is rapidly decreasing, and many of the symptoms associated with such a state are accounted for by the actions of the large quantity of epinephrine, particularly those on the cardiovascular and nervous systems—palpitations, tremor, headache, nervousness, cold sweating, dizziness, etc.

It is not yet known whether the sympathetic nervous system plays any role in the metabolic adjustments to prolonged fasting (as contrasted to acute reductions in plasma glucose described

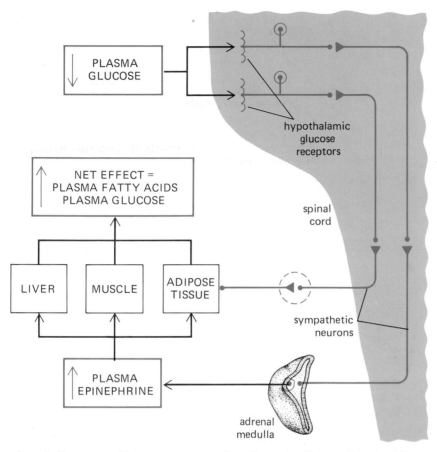

Figure 15-13. Control of epinephrine secretion and sympathetic nerves to adipose tissue by plasma glucose concentration. A decrease in plasma glucose stimulates the hypothalamic glucose receptors and, via the reflex chain shown, restores plasma glucose to normal while at the same time increasing plasma fatty acids.

above). Recent evidence suggests that the overall activity of the sympathetic nervous system is, contrary to expectations, actually decreased under such conditions; the possible adaptive significance of this change is discussed later in this chapter.

Conclusion

To a great extent insulin may be viewed as the "hormone of plenty"; it is increased during the absorptive period and decreased during postabsorption. The effects on organic metabolism of glucagon and the sympathetic nervous system are, in various ways, opposed to those of insulin and these systems are activated (the latter at least acutely) during the postabsorptive period. The influence of plasma glucose concentration is paramount in this regard, producing opposite effects on insulin secretion and on the secretion of glucagon, epinephrine, and the activity of the sympathetic nerves to adipose tissue.

It should also be noted that the secretion of glucagon and the activity of the sympathetic nervous system are stimulated not only when plasma glucose is reduced but during exercise and in a variety of nonspecific "stresses," both physical and emotional. This constitutes a mechanism for the mobilization of energy stores for coping with a fight-or-flight situation (see Chap. 17 and the section on exercise below).

Finally, in addition to the three hormones described in this section, there are others—growth hormone, cortisol, thyroxine, and the sex steroids—which have important effects (described elsewhere) on organic metabolism and the flow of nutrients. However, the secretion of these hormones is not usually keyed to the absorptive-postabsorptive phases but is controlled by other factors. Growth hormone is somewhat of an exception in that its rate of secretion does respond to prolonged fasting or rapidly occurring decreases in plasma glucose concentration; however, changes in re-

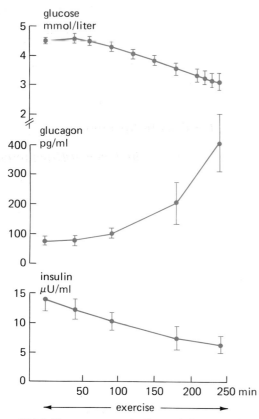

Figure 15-14. Plasma concentrations of glucose, glucagon, and insulin during prolonged exercise. *(Adapted from Felig and Wahren.)*

sponse to the usual three-meal-a-day pattern are minimal, and its short-term effects on carbohydrate and fat metabolism are relatively minor compared to its long range effects on protein metabolism and growth.

Fuel homeostasis in exercise

Exercise is another situation in which large quantities of endogenous fuels must be mobilized, in this case to provide the energy required for muscle contraction. There are at least three fuels used by exercising muscle: plasma glucose, plasma fatty acids, and the muscle's endogenous glycogen.

During the earliest phase of exercise (5 to 10 min), muscle glycogen (via glycogenolysis) is the major fuel consumed (this description applies to moderate levels of endurance-type exercise). During the middle phase (10 to 40 min) blood-borne fuels, brought to the muscle by the increased blood flow, become dominant, plasma glucose and fatty acids contributing approximately equally to the oxygen consumption of the muscle. Beyond the middle phase, fatty acids become progressively more important as glucose utilization decreases. Finally, as fatigue is approached, muscle becomes very dependent on its own glycogen, and fatigue coincides with depletion of muscle glycogen. Why glycogen depletion causes muscle to lose its ability to contract despite the availability of ample blood-borne fuels is not known. This phenomenon explains why athletes must pace themselves in a long event so as not to deplete their muscle glycogen before the event is over.

What happens to blood glucose concentration during exercise (Fig. 15-14)? It changes very little in short-term mild-to-moderate exercise (and may actually increase by 20 to 30 percent with more severe short-term activity). However, during prolonged exercise (more than 90 min) it decreases, sometimes by as much as 25 percent but rarely enough to cause symptoms of hypoglycemia. These relatively minor changes in blood glucose in the presence of marked increases in glucose utilization clearly demonstrate that glucose output by the liver must have increased approximately in proportion to the increased utilization until later stages when it begins to lag somewhat behind. The liver provides this glucose both by breakdown of its glycogen stores and by conversion of lactate, glycerol, and amino acids into glucose. The glycerol is made available by a marked increase in the catabolism of adipose-tissue triacylglycerol with resultant release of glycerol and fatty acids into the blood, the latter serving as a fuel source for the exercising muscle.

It should be evident that this combination in exercise of increased hepatic glucose production, triacylglycerol breakdown, and fatty acid utilization is similar to that seen in a fasted person, and the hormonal controls are also the same — exercise is characterized by a fall in insulin secretion, a rise in glucagon secretion, and increased activity of the sympathetic nervous system. What accounts for the changes in secretion of glucagon and insulin? One important signal during prolonged exercise is the decrease in plasma glucose, the same signal which controls these hormones in fasting. However, this cannot be the signal during less prolonged exercise since, as we have seen, plasma glucose is either unchanged or elevated. Rather, the controlling inputs appears to be enhanced activity of the sympathetic neurons supplying the cells of the pancreatic islets; this results in stimulation of glucagon release by the A cells and inhibition of insulin release by B cells (circulating epinephrine exerts similar effects on these cells). Thus, the increased sympathetic nervous system activity characteristic of exercise not only contributes directly to fuel mobilization (by enhancing glycogenolysis and fat mobilization) but indirectly by inhibiting the release of insulin and stimulating that of glucagon.

It should be noted, however, that one component of the response to exercise is quite different from fasting—in the former, glucose uptake and utilization by the muscles is enhanced, whereas in the latter, it is markedly reduced. How is it that glucose uptake can remain high in the presence of reduced plasma insulin? The most tenable hypothesis at present is that some local chemical change associated with muscle contraction occurs which stimulates the facilitated diffusion of glucose into the muscle (an interesting result of this phenomenon is that patients with diabetes require less insulin to keep their blood glucose normal when they are physically active than when they are sedentary).

Regulation of plasma cholesterol

In the previous sections, we described the flow of lipids to and from adipose tissue in the form of fatty acids and triacylglycerols. These lipids do not circulate as free molecules in the plasma, for their insolubility in water makes that impossible. Rather they are bound to particular plasma proteins, the complexes being known as lipoproteins. The various lipoproteins differ considerably in the relative amounts of fat and protein which they contain; those having little protein and much fat are categorized as *low-density lipoproteins* (since fat is less dense than protein), whereas those with the opposite distribution are *high-density lipoproteins*.

Another very important lipid, cholesterol, was not mentioned in the earlier sections because it, unlike the fatty acids and triacylglycerols, does not serve as a metabolic fuel, but rather as a precursor for bile salts, steroid hormones, plasma membranes, and other specialized molecules. It, too, circulates in combination with lipoproteins, particularly the low-density fraction. A huge number of studies have been devoted to the factors which govern its plasma concentration, because it is generally accepted that high plasma cholesterol predisposes to the development of atherosclerosis, the arterial thickening which leads to strokes, heart attacks, and other forms of damage (see Chap. 11 for a discussion of the etiology of this thickening).

Figure 15-15 summarizes the major pathways involved in the metabolism of plasma cholesterol. There are two major sources of cholesterol gain— diet and synthesis within the body; there are two major routes of loss—excretion in the feces and catabolism to bile acids. There are multiple homeostatic control points in this system. Most important is that the synthesis of cholesterol by the liver is

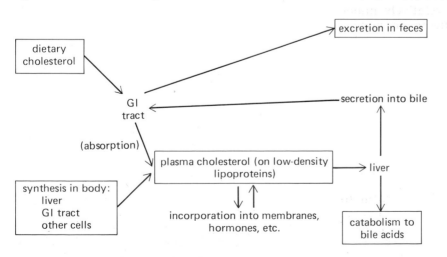

Figure 15-15. Metabolism of cholesterol.

controlled in a negative-feedback manner by dietary cholesterol, i.e., ingestion of cholesterol inhibits hepatic synthesis. Conversely, when dietary cholesterol is reduced, hepatic synthesis is increased. For this reason, reduction of dietary cholesterol alone produces only very small reductions in plasma cholesterol. Thus, the explanation for the elevated plasma cholesterol exhibited by persons ingesting large amounts of animal fat is not simply an increase in cholesterol ingestion.

However, another component of such a diet does seem to be of great importance, namely the amount of *saturated fatty acids* ingested. These substances, found mainly in animal sources, somehow stimulate the synthesis of cholesterol while at the same time reducing its excretion. Accordingly, reducing the saturated fat content of the diet may lower plasma cholesterol 15 to 20 percent. In contrast, *unsaturated fatty acids* (found mainly in vegetable oils and other plant products) enhance cholesterol excretion and catabolism so that an increase in their dietary consumption helps to reduce plasma cholesterol. Just how the various fatty acids exert these effects on cholesterol metabolism is not completely known.

Factors other than diet also influence cholesterol metabolism. For example, stress and cigarette smoke have been implicated as factors which elevate plasma cholesterol, but the quantitative contributions of such factors have not been fully documented.

Control of growth

A simple gain in body weight does not necessarily mean true growth since it may represent retention of either excess water or fat. In contrast, true growth usually involves lengthening of the long bones and increased cell division and enlargement, but the real criterion is increased accumulation of protein. People manifest two periods of rapid growth (Fig. 15-16), one during the first 2 years of life, which is actually a continuation of rapid fetal growth, and the second during adolescence. Note that total body growth may be a poor indicator of the rate of growth of specific organs (Fig. 15-16). An important implication of differential growth rates is that the so-called critical periods of development vary from organ to organ. Thus, a period of severe malnutrition during infancy when the brain

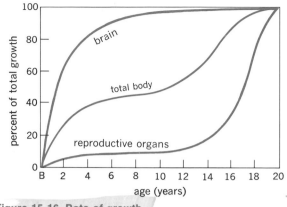

Figure 15-16. Rate of growth.

is growing extremely rapidly may produce stunting of brain development which is irreversible, whereas reproductive organs would be little affected.

External factors influencing growth

An individual's growth capacity is genetically determined, but there is no guarantee that the maximum capacity will be attained. Adequacy of food supply and freedom from disease are the primary external factors influencing growth. Lack of sufficient amounts of any of the essential amino acids, essential fatty acids, vitamins, or minerals interferes with growth, and total protein and total calories must be adequate. No matter how much protein is ingested, growth cannot be normal if caloric intake is too low, since the protein is simply oxidized for energy.

Many studies have demonstrated that the growth-inhibiting effects of malnutrition are most profound when they occur very early in life. Indeed, malnutrition during infancy may cause irreversible stunting of total body growth (and brain development, as mentioned above). The individual seems "locked in" to a younger developmental age. A related problem is that maternal malnutrition may cause retardation of growth in the fetus. One reason for the great importance of this phenomenon is that low birth weight is strongly associated with increased infant mortality; accordingly, prenatal malnutrition causes increased numbers of prenatal and early postnatal deaths. Moreover, irreversible stunting of brain development may be caused by prenatal malnutrition.

On the other hand, it is important to realize that one cannot stimulate growth beyond the genetically determined maximum by eating more than adequate vitamins, protein, or total calories; this produces obesity, not growth.

Sickness can stunt growth, but if the illness is temporary, upon recovery the child manifests a remarkable growth spurt which rapidly brings him up to his normal growth curve. The mechanisms which control this important phenomenon are unknown, but the process illustrates the strength and precision of a genetically determined sequence.

Hormonal influences on growth

The hormones most important to human growth are growth hormone, thyroxine, insulin, androgens, and estrogen, which exert widespread effects. In addition, ACTH, TSH, prolactin, and FSH and LH selectively influence the growth and development of their target organs, the adrenal cortex, thyroid gland, breasts, and gonads, respectively.

Growth hormone. Removal of the pituitary in young animals arrests growth. Conversely, administration of large quantities of growth hormone to young animals causes excessive growth. When excess growth hormone is given to adult animals after the actively growing cartilaginous areas of the long bones have disappeared, it cannot lengthen the bones further, but it does produce the disfiguring bone thickening and overgrowth of other organs known as *acromegaly*. These experiments have spontaneously occurring human counterparts, as shown in Fig. 15-17. Thus growth hormone is essential for normal growth and in abnormally large amounts can cause excessive growth.

Growth hormone has a variety of effects on organic metabolism, but its growth-promoting effects are due primarily to its ability to stimulate protein synthesis in most tissues and organs. This it does by increasing membrane transport of amino acids into cells, and by increasing the activity of ribosomes, events essential for protein synthesis. Growth hormone also causes large increases of mitotic activity and cell division, the other major components of growth.

The effects on cartilage and bone are dramatic. Bone is a living tissue consisting of a protein matrix, upon which calcium salts (particularly phosphates) are deposited. The cells responsible for laying down this matrix are *osteoblasts.* Growth of a long bone depends upon actively proliferating layers of cartilage (the *epiphyseal plates*) near the ends of the bone. The osteoblasts at the edge of the epiphyseal plates replace the cartilaginous tissue with bone while new cartilage is simultaneously formed in the plates (Fig. 15-18). Growth hormone promotes this lengthening by stimulating protein synthesis in both the cartilaginous center and bony edge of the epiphyseal plates as well as by increasing the rate of osteoblast mitosis. It is presently uncertain as to whether growth hormone, itself, exerts these growth-promoting effects on cartilage, bone, and other tissues, or whether growth hormone causes the release from the liver of so-called *somatomedins* which then are carried by the blood to the target tissues and induce the effects.

Growth hormone offers an excellent example of a hormone whose secretion occurs in episodic bursts and manifests a striking diurnal rhythm (Chap. 9). During most of the day, there is essentially no growth hormone secreted although a small burst may occur occasionally (and larger bursts may be elicited by certain stimuli such as stress or hypoglycemia). In contrast, shortly after a person falls asleep, one or more large prolonged bursts of secretion may occur. Apparently neuronal output from "sleep centers" in the brain synaptically influence the hypothalamic neurons which secrete growth-hormone releasing hormones (see Chap. 9) into the portal capillaries connecting the hypothalamus and anterior pituitary; changes in the concentrations of these releasing hormones then stimulates the secretion of growth hormone by the anterior pituitary cells.[1]

At what stages of growth is growth hormone important? It is generally accepted that growth hormone has little or no influence on the growth of the fetus or very young infant, but exerts an important influence thereafter. The total 24-h secretion rate of growth hormone (almost all during sleep) is highest during adolescence (the period of most rapid growth), next highest in children, and lowest in adults.

---

[1] It is not certain which of the two hypothalamic releasing hormones for growth hormone described in Chap. 9 is most important in this response.

Figure 15-17. Progression of acromegaly. *(Top left)* **Normal, age nine years;** *(top right)* **age sixteen years, with possible early coarsening of features;** *(bottom left)* **age thirty-three years, well-established acromegaly;** *(bottom right)* **age fifty-two years, end stage, acromegaly with gross disfigurement.** [*From William H. Daughaday, in Robert H. Williams (ed.), "Textbook of Endocrinology," 4th ed., p. 74, fig. 2-28, W. B. Saunders Company, Philadelphia, 1968.*]

Thyroid hormone. Infants and children with deficient thyroid function manifest retarded growth, which can be restored to normal by administration of physiological quantities of thyroid hormone (TH) (this term refers collectively to two hormones—thyroxine and triiodothyronine—secreted by the thyroid gland which have very similar effects). Administration of excess TH, however, does not cause excessive growth (as was true of growth hormone) but marked catabolism of protein and other nutrients, as will be explained in the section on energy balance. The essential point is that normal amounts of TH are necessary for normal growth, the most likely explanation apparently being that TH promotes the effects of growth hormone on protein synthesis; certainly the absence of TH significantly reduces the ability of growth hormone to stimulate amino acid uptake, ribosomal activation, and RNA synthesis. TH also plays a crucial role in the closely related area of organ development, particularly that of the central nervous system. Hypothyroid infants *(cretins)* are mentally retarded, a defect that can be completely repaired by adequate treatment with TH although if the infant is untreated for long, the developmental failure is largely irreversible. The defect is probably due to failure of nerve myelination which occurs as a result of thyroid deficiency.

Insulin. It should not be surprising that adequate amounts of insulin are necessary for normal growth since insulin is, in all respects, an anabolic hormone. Its stimulatory effects on amino acid uptake and protein synthesis are particularly important in favoring growth.

Androgens and estrogen. In Chap. 16 we shall describe in detail the various functions of the sex hormones in directing the growth and development of the sexual organs and the obvious physical characteristics which distinguish male from female. Here we are concerned only with the effects of these hormones on general body growth.

Sex hormone secretion begins in earnest at about the age of eight to ten and progressively increases

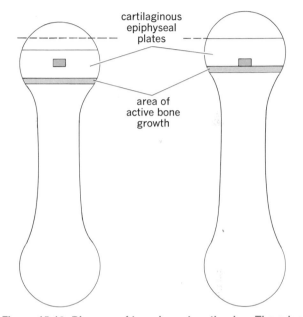

□ – "MARKER" PIECE OF CARTILAGE

cartilaginous
epiphyseal
plates

area of
active bone
growth

Figure 15-18. Diagram of long-bone lengthening. The edge of the cartilaginous plate is converted into bone, but the cells of the plate keep producing more cartilage to take its place. Thus the marker, once in the center of the plate, is now at the edge and about to be converted into bone; yet the marker itself has not moved.

to reach a plateau within 5 to 10 years. The testicular hormone, testosterone, is the major male sex hormone, but other androgens similar to it are also secreted in significant amounts by the adrenal cortex of both sexes. Females manifest a sizable increase of adrenal androgen secretion during adolescence. However, the adrenal androgens are generally not nearly so potent as testosterone. During adolescence the large increases in secretion of estrogen, the dominant female sex hormone, are virtually limited to the female. Thus the relative quantities of androgen and estrogen are very different between the sexes.

Androgens strongly stimulate protein synthesis in many organs of the body, not just the reproductive organs, and the adolescent growth spurt in both sexes is due, at least in part, to these anabolic effects. Similarly, the increased muscle mass of men compared with women may reflect their greater amount of more potent androgen. Androgens stimulate bone growth but also ultimately stop bone growth by inducing complete conversion of the epiphyseal plates to bone. This accounts for the pattern seen in adolescence, i.e., rapid lengthening of the bones culminating in complete cessation of growth for life, and explains several clinical situations: (1) Unusually small children treated before puberty with large amounts of androgens may grow several inches very rapidly but then stop completely; (2) eunuchs may be very tall because bone growth, although slower, continues much longer due to persistence of the epiphyseal plates. Estrogen profoundly stimulates growth of the female sexual organs and sexual characteristics during adolescence (Chap. 16), and is also required for normal growth of bone and tissues.

Compensatory growth

We have dealt thus far only with growth during childhood. During adult life, maintenance of the status quo is achieved by the mechanisms described earlier in this chapter. In addition, a specific type of organ growth, known as *compensatory growth,* can occur in many human organs and is actually a type of regeneration. For example, within 24 h of the surgical removal of one kidney, the cells of the other begin to manifest increased mitotic activity, ultimately growing until the total mass approaches the initial mass of the two kidneys combined. What causes this compensatory growth? It certainly does not depend upon the

nerves to the organ since it still occurs after their removal or destruction. Several types of experiments indicate that unidentified blood-borne agents are responsible.

Section B
Regulation of total body energy balance

Basic concepts of energy expenditure and caloric balance

The breakdown of organic molecules liberates the energy locked in their intramolecular bonds (Chap. 5). This is the source of energy utilized by cells in their performance of the various forms of biological work (muscle contraction, active transport, synthesis of molecules, etc.). As described in Chap. 5, the first law of thermodynamics states that energy can be neither created nor destroyed but can be converted from one form to another. Thus, internal energy liberated (ΔE) during breakdown of an organic molecule can either appear as heat (H) or be used for performing work (W).

$$\Delta E = H + W$$

In all animal cells, most of the energy appears immediately as heat, and only a small fraction is used for work. (As described in Chap. 5, the energy used for work must first be incorporated into molecules of ATP, the subsequent breakdown of which serves as immediate energy source for the work.) It is essential to realize that the body is not a heat engine since it is totally incapable of converting heat into work. The heat is, of course, valuable for maintaining body temperature.

It is customary to divide biological work into two general categories: (1) *external work,* i.e., movement of external objects by contracting skeletal muscles; (2) *internal work,* which comprises all other forms of biological work, including skeletal muscle activity not moving external objects. As we have seen, most of the energy liberated from the catabolism of nutrients appears immediately as heat, only a small fraction being used for performance of external or internal work. What may not be obvious is that all internal work is ultimately transformed into heat except during periods of growth (Fig. 15-19). Several examples will illustrate this essential point:

Figure 15-19. General pattern of energy liberation in a biological system. Most of the energy released when nutrients, such as glucose, are broken down appears immediately as heat. A smaller fraction goes to form ATP, which can be subsequently broken down and the released energy coupled to biological work. Ultimately, the energy which performs this work is also completely converted into heat.

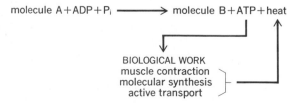

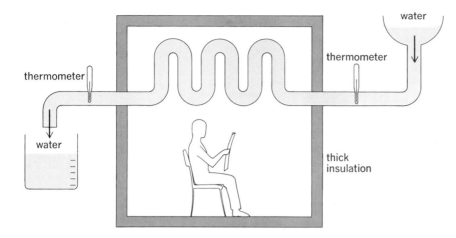

Figure 15-20. Direct method for measuring metabolic rate. The water flowing through the calorimeter carries away the heat produced by the person's body. The amount of heat is calculated from the total volume of water and the difference between inflow and outflow temperatures.

1 Internal work is performed during cardiac contraction, but this energy appears ultimately as heat generated by the resistance (friction) to flow offered by the blood vessels.
2 Internal work is performed during secretion of HCl by the stomach and $NaHCO_3$ by the pancreas, but this work appears as heat when the H^+ and HCO_3^- react in the small intestine.
3 The internal work performed during synthesis of a plasma protein is recovered as heat during the inevitable catabolism of the protein, since, with few exceptions, all bodily constituents are constantly being built up and broken down. However, during periods of net synthesis of protein, fat, etc., energy is stored in the bonds of these molecules and does not appear as heat.

Thus, the total energy liberated when organic nutrients are catabolized by cells may be transformed into body heat, appear as external work, or be stored in the body in the form of organic molecules, the latter occurring only during periods of growth (or net fat deposition in obesity). The total energy expenditure of the body is therefore given by the equation

Total energy expenditure =

(heat produced + external work + energy storage)

The units for energy are kilocalories (Chap. 5), and total energy expenditure per unit time is called the *metabolic rate*.[1]

[1] In the field of nutrition, 1 Calorie implies, by convention, 1 *large* calorie, which is actually 1 kilocalorie.

In people the metabolic rate can be measured directly or indirectly. In either case the measurement is much simpler if the person is fasting and at rest; total energy expenditure then becomes equal to heat production since energy storage and external work are eliminated. The direct method is simple to understand but difficult to perform; the subject is placed in a *calorimeter*, an instrument large enough to accommodate people, and heat production is measured by the temperature changes in the calorimeter (Fig. 15-20). This is an excellent method in that it measures heat production directly, but calorimeters are found in only a few research laboratories; accordingly, a simple, indirect method has been developed for widespread use.

Using the indirect procedure, one simply measures the subject's oxygen uptake per unit time (by measuring total ventilation and Po_2 of inspired and expired air). From this value one calculates heat production based on the fundamental principle (Fig. 15-21) that the energy liberated by the catabolism of foods in the body must be the same as when they are catabolized outside the body. We know precisely how much heat is liberated when 1 liter of oxygen is consumed in the oxidation of fat, protein, or carbohydrate outside the body; this same quantity of heat must be produced when 1 L of oxygen is consumed in the body. Fortunately, we do not need to know precisely which type of nutrient is being oxidized internally, because the quantities of heat produced per liter of oxygen consumed are reasonably similar for the oxidation of fat, carbohydrate, and protein, and average 4.8 kcal/L of oxygen. When more exact calculations

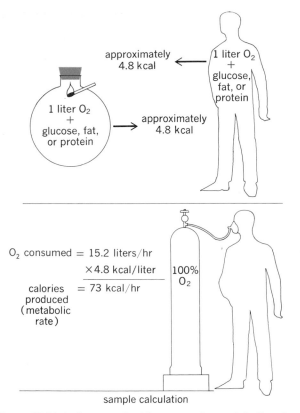

approximately
4.8 kcal

1 liter O_2
+
glucose,
fat, or
protein

1 liter O_2
+
glucose, fat,
or protein

approximately
4.8 kcal

O_2 consumed = 15.2 liters/hr

$\times$ 4.8 kcal/liter

calories
produced
(metabolic
rate)

= 73 kcal/hr

100%
O_2

sample calculation

Figure 15-21. Indirect method for measuring metabolic rate. The calculation depends upon the basic principle that when 1 L of oxygen is utilized in the oxidation of organic nutrients approximately 4.8 kcal is liberated.

are required, it is possible to estimate the relative quantity of each nutrient. Figure 15-21 presents values obtained by the indirect method for a normal fasted resting adult male. This man's rate of heat production is approximately equal to that of a single 100-W bulb.

Determinants of metabolic rate

Since many factors cause the metabolic rate to vary (Table 15-1), when one wishes to compare metabolic rates of different people, it is essential to control as many of the variables as possible. The test used clinically and experimentally to find the *basal metabolic rate* (BMR) is designed to accomplish this by standardizing conditions: The subject is at mental and physical rest, in a room at comfortable temperatures, and has not eaten for at least

Table 15-1. Factors affecting the metabolic rate

Age
Sex
Height, weight, and surface area
Growth
Pregnancy, menstruation, lactation
Infection or other disease
Body temperature
Recent ingestion of food (SDA)
Prolonged fasting
Muscular activity
Emotional state
Sleep
Environmental temperature
Circulating levels of various hormones, especially epineph-
rine and thyroid hormone

12 h. These conditions are arbitrarily designated basal, the metabolic rate during sleep being actually less than the BMR. The measured BMR is then compared with previously determined normal values for a person of the same weight, height, age, and sex.

BMR is often appropriately termed the metabolic *cost of living.* Under these conditions, most of the energy is expended, as might be imagined, by the heart, liver, kidneys, and brain. Its magnitude is related not only to physical size but to age and sex as well. The growing child's resting metabolic rate, relative to size, is considerably higher than the adult's because the child expends a great deal of energy in net synthesis of new tissue. On the other end of the age scale, the metabolic cost of living gradually decreases with advancing age, for unknown reasons. The female's resting metabolic rate is generally less than that of the male (even taking into account size differences) but increases markedly, for obvious reasons, during pregnancy and lactation. The greater demands upon the body by infection or other disease generally increase total energy expenditure; moreover the presence of fever increases metabolic rate.

The ingestion of food also increases the metabolic rate, as shown by measuring the oxygen consumption or heat production of a resting man before and after eating; the metabolic rate is 10 to 20 percent higher after eating. This effect of food on metabolic rate is known as the *specific dynamic action* (SDA). Protein gives the greatest effect, carbohydrate and fat less. The cause of SDA is not what one might expect, namely, the energy expended in the digestion and absorption of ingested food. These processes account for only a small fraction of the increased metabolic rate (intravenous administration of amino acids produces almost the same SDA effect as oral ingestion of the same material). Most of the increased heat production appears to be secondary to the processing of the exogenous nutrients by the liver, since it does not occur in an animal whose liver has been removed. In contrast to eating, prolonged fasting causes a decrease in metabolic rate. This is due, in part, simply to reduction of body mass, but even when expressed on a per weight basis, metabolic rate is reduced. The mechanism is unclear, but the adaptive value of this change is considerable since it decreases the amount of nutrient stores which must be catabolized each day. One possible mechanism may be a reduction in sympathetic nervous system activity.

All these influences on metabolic rate are small compared with the effects of *muscular activity* (Table 15-2). Even minimal increases in muscle tone significantly increase metabolic rate, and severe exercise may raise heat production more than fifteenfold. Changes in muscle activity also explain part of the effects on metabolic rate of sleep (decreased muscle tone), reduced environmental temperature (increased muscle tone and shivering), and emotional state (unconscious changes in muscle tone).

Metabolic rate is strongly influenced by the hormones epinephrine and thyroid hormone. The intravenous injection of epinephrine may promptly increase heat production by more than 30 percent. As we have seen, epinephrine has powerful effects on organic metabolism, and its calorigenic, i.e., heat-producing, effect may be related to its stimulation of glycogen and triglyceride catabolism since ATP splitting and energy liberation occur in both the breakdown and the subsequent resynthesis of these molecules. Regardless of the mechanism, whenever epinephrine secretion is stimulated, the metabolic rate rises. This probably accounts for part of the greater heat production associated with emotional stress, although increased muscle tone is also contributory.

Thyroid hormone (TH) also increases the oxygen consumption and heat production of most body tissues, a notable exception being the brain. The mechanism of this calorigenic effect remains controversial. Long-term excessive TH, as in patients

Table 15-2. Energy expenditure during different types of activity for a 70-kg man

Form of activity	kcal/h
Awake, lying still	77
Sitting at rest	100
Typewriting rapidly	140
Dressing or undressing	150
Walking level at 2.6 mi/h	200
Sexual intercourse	280
Bicycling on level, 5.5 mi/h	304
Walking 3 percent grade at 2.6 mi/h	357
Sawing wood or shoveling snow	480
Jogging (5.3 mi/h)	570
Rowing 20 strokes/min	828
Maximal activity (untrained)	1440

with hyperthyroidism, induces a host of effects secondary to the hypermetabolism which well illustrate the interdependence of bodily functions. The increased metabolic demands markedly increase hunger and food intake; the greater intake frequently remains inadequate to meet the metabolic needs, and net catabolism of endogenous protein and fat stores leads to loss of body weight; excessive loss of skeletal muscle protein results in muscle weakness; catabolism of bone protein weakens the bones and liberates large quantities of calcium into the extracellular fluid, resulting in increased plasma and urinary calcium; the hypermetabolism increases the requirement for vitamins, and vitamin deficiency diseases may occur; respiration is increased to supply the required additional oxygen; cardiac output is also increased and, if prolonged, the enhanced cardiac demands may cause heart failure; the greater heat production activates heat-dissipating mechanisms, and the patient suffers from marked intolerance to warm environments. These are only a few of the many results induced by the calorigenic effect of TH. The important effects of TH relating to growth and development, described earlier, appear to be quite distinct from the calorigenic effect.

Determinants of total body energy balance

Using the basic concepts of energy expenditure and metabolic rate as a foundation, we can consider total body energy balance in much the same way as any other balance, i.e., in terms of input and output. The laws of thermodynamics dictate that, in the steady state, the total energy expenditure of the body equals total body energy intake. We have already identified the ultimate forms of energy expenditure: internal heat production, external work, and net molecular synthesis (energy storage). The source of input, of course, is the energy contained in ingested food. Therefore the energy-balance equation is

$$\begin{matrix} \text{Energy} \\ \text{(from food intake)} \end{matrix} = \begin{matrix} \text{internal} \\ \text{heat} \\ \text{produced} \end{matrix} + \begin{matrix} \text{external} \\ \text{work} \end{matrix} + \begin{matrix} \text{energy} \\ \text{storage} \end{matrix}$$

Our equation includes no term for loss of fuel from the body via urinary excretion of nutrients. In a normal person, almost all the carbohydrate and amino acids filtered at the glomerulus are reabsorbed by the tubules, so that the kidneys play no significant role in the regulation of energy balance. In certain diseases, however, the most important being diabetes, urinary losses of organic molecules may be quite large and would have to be included in the equation. In all normal persons very small losses occur via the urine, feces, and as sloughed hair and skin, but we can ignore them as being negligible.

As predicted by this energy-balance equation, three states are possible:

Energy intake = internal heat production
+ external work
(body weight constant)
Energy intake > internal heat production
+ external work
(body weight increases)
Energy intake < internal heat production
+ external work
(body weight decreases, i.e., negative energy storage)

In most adults, body weight remains remarkably constant over long periods of time, implying that precise physiological regulatory mechanisms operate to control (1) food intake or (2) internal heat production plus external work or (3) both. Actually, all these variables are subject to control in humans, but the amount of food intake is the dominant factor. Control mechanisms for heat production are aimed primarily at regulating body temperature, rather than total energy balance. For example, when someone is cold, her body produces additional heat by shivering even if she is starving; conversely, a fat person is not automatically impelled by his

hypothalamus to run around the block—quite the reverse in most cases. It is essential to understand that as shown by the energy-balance equation, an individual's degree of activity, i.e., heat production plus external work, *is* one of the essential determinants of total body energy balance, but its automatic physiological control is not aimed at achieving such a balance. Moreover, a man's total activity generally reflects the kind of work he does, his inclination toward sports, etc. The important generalization is that caloric intake is the major factor being automatically controlled so as to maintain energy balance and constant body weight. To alter the two examples cited above: When exposure to cold or running around the block causes increased energy expenditure, the individual automatically increases his food intake by an amount sufficient to match the additional energy expended (this example, however, will be qualified later in the section on obesity).

Despite the fact that the control of food intake is the major mechanism by which energy balance is *normally* regulated, it should be emphasized that balance can also be achieved in the presence of a markedly deficient intake by means of several physiological adaptations. These responses to partial chronic malnutrition offer an excellent summary of many of the factors which determine metabolic rate. Perhaps the most striking experiment designed to study this problem was the reduction of caloric intake in normal volunteers from 3492 kcal/day to 1570 kcal/day for 24 weeks (these men were soldiers performing considerable physical activity). Initially, weight loss was very rapid and ultimately averaged 24 percent of the men's original body weights, but the important fact is that the weights did stabilize, i.e., energy balance was reestablished despite the continued caloric intake of 1570 kcal/day. Clearly, metabolic rate must have decreased an identical amount. This was due to a reduction in both BMR and physical activity as the men became apathetic and reluctant to engage in any activity. The decrease in BMR was due both to a decreased total body mass and to a decrease in metabolism of the liver and other organs out of proportion to their changes in weight (the mechanism of the latter change is unknown). Thus, people can adjust to marked caloric restriction but only at a price. Moreover, in children some of the changes caused by malnutrition are irreversible.

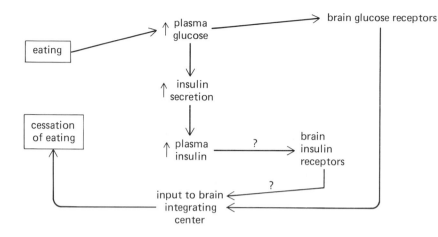

Figure 15-22. Hypothesized negative-feedback inhibition of eating by glucose and insulin.

Control of food intake

The control of food intake can be analyzed in the same way as any other biological control system; i.e., in terms of its various components (regulated variable, receptor sensitive to this variable, afferent pathway, etc.). As our previous description emphasized, the variable which is being maintained relatively constant in this system is total body energy content. Here we have a problem similar to that described in Chap. 13 for total body sodium balance: What kind of receptors could possibly detect the *total body content* of a particular variable, in this case calories? It is very unlikely that any such receptors exist and so, instead, the system must depend on signals which are intimately related to total energy storage — the plasma concentrations of glucose, fatty acids, glycerol, and amino acids.

These signals provide both short-term and long-term information concerning the body's energy stores. For example (Fig. 15-22), plasma glucose concentration (and the rate of cellular glucose utilization) rises during eating; detection of this increase by glucose receptors (in the brain) constitutes a signal to those portions of the brain which controls eating behavior and leads to cessation of eating. Conversely, fasting would decrease glucose utilization, remove this signal, and promote eating. Plasma concentrations of fatty acids or glycerol could serve as a long-term indicator of the body's total adipose-tissue mass (itself an excellent indicator of total body energy stores) if their basal rates of release from adipose tissue

were in direct proportion to the size of the adipose tissue cells; this seems to be the case for glycerol.

In addition to the plasma concentrations of these metabolic fuels, the concentrations of the hormones which regulate them may be important negative-feedback signals to the brain's integrating centers. For example, insulin (which increases during food absorption) has been shown to suppress hunger (Fig. 15-22), whereas glucagon (which increases during fasting) is known to stimulate hunger. Epinephrine and the gastrointestinal hormones released during meal ingestion may also contribute important signals.

Finally, a nonchemical correlate of food ingestion — body temperature — may be yet another signal. The specific dynamic action of food, i.e., the increase in metabolic rate induced by eating, tends to raise body temperature and this constitutes a signal inhibitory to eating. Such a mechanism would also be consistent with the fact that people eat more in colder climates than in warm ones.

It would certainly be simpler if we could quantitate the precise contributions of each of these signals to the control of food intake and describe the location and nature of the receptors stimulated by each, but this information is simply not available at present. Moreover, the confusion does not lessen when we consider the brain areas which integrate all these afferent inputs and cause the individual either to seek out and ingest food or to desist from doing so. For many years, the evidence favoring two specific interacting hypothalamic nuclei as the integrators of feeding behavior seemed quite convincing, but a large number of recent experiments have destroyed this neat picture; areas of the hypothalamus may well be involved but there are other brain areas which play equally important roles, and the ways in which these various neuronal collections interact remains to be determined.

Thus far we have described the control of food intake in a manner identical to that for any other homeostatic control system, such as those regulating sodium or potassium. Thus, the automatic "involuntary" nature of the system was emphasized. However, although total energy balance unquestionably reflects, in large part, the reflex input from some combination of glucose receptors, glycerol receptors, thermostats, etc., it also is strongly influenced by the reinforcement (both positive and negative) of such things as smell, taste, texture, psychological associations, etc. Thus, the behavioral concepts of reinforcement, drive, and motivation to be described in a later chapter must be incorporated into any comprehensive theory of food-intake control. Obviously, these psychological factors having little to do with energy balance are of very great importance in obese persons. It should be emphasized, however, that most people whose obesity is ascribed to psychological factors do not continuously gain weight; their automatic homeostatic control mechanisms are operative but maintain total body energy content at supranormal levels.

Obesity

Obesity has been called the most common disease in America. Despite adverse effects on health, the social stigma of being obese, and a bewildering array of new diets and treatments, the prevalence of obesity in the United States is actually increasing. The term disease is perfectly justified since obesity is correlated with illness and premature death from a multitude of causes. The seriousness of being overweight is underlined by statistics which show a mortality rate more than 50 percent greater than normal in overweight persons in the same age groups.

Ultimately, all obesity represents failure of normal food-intake control mechanisms, although what causes this failure (and how best to treat it) is unknown for the great majority of obese persons. It is widely assumed that psychological factors, habit, and social custom are the predominant causes of obesity, but many investigators believe

that this view is not really well supported by experimental data.

The relationship between food intake and physical activity has also received recent attention. For example, a study of obese high school girls revealed that they ate, on the average, *less* than a control group of normal weight girls. The obese girls had much less physical activity than the control group and were eating "too much" only in relationship to their physical activity. This example stresses the fact, so easily forgotten, that there are two sides to the energy-balance equation. Moreover, low levels of physical activity may cause *increased* eating. The caloric intakes and body weights of large numbers of workers in the same factory in India were studied after grouping the men according to the physical exertion required by their jobs. Levels of activity below a certain arbitrary minimum were classified as sedentary. As shown in Fig. 15-23, men performing work loads above the sedentary range displayed the expected pattern; caloric intake was directly proportional to work level, and body weights for all groups of men were similar. The unexpected finding was that for men performing small work loads in the sedentary range, caloric intake varied inversely with work load, i.e., the less physical activity the men performed, the more they ate. Accordingly, these men were considerably fatter, on the average, than the other men. Since this kind of study is difficult to interpret in terms of cause and effect, a series of experiments was performed on rats forced to remain sedentary or to exercise. Figure 15-24 shows data similar to those for the factory workers. It is therefore apparent that very low levels of activity do not induce similar reductions of food intake but actually stimulate eating. The implications of these findings for energy balance in a society where so many of us fall into the sedentary category are obvious, and the elucidation of the factors responsible will be of considerable importance.

Another important question concerns the possibility that influences early in life may induce physiological changes which predispose to adult obesity. Adipose-tissue growth is due to both an increased size and number of cells. The number of cells stabilizes sometime during early adulthood and does not change thereafter. Present evidence suggests that overfeeding early in life, particularly during infancy, causes the generation of an ab-

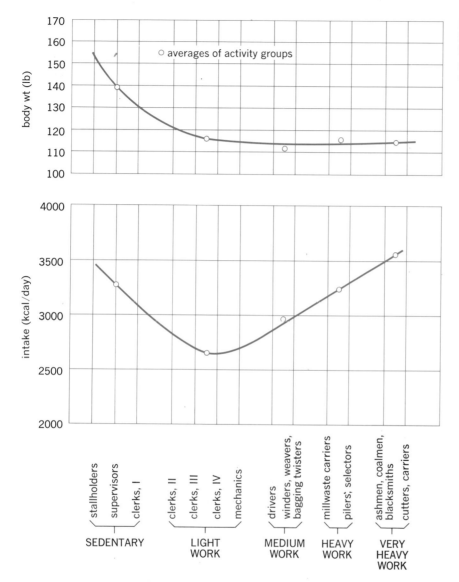

Figure 15-23. Body weight and caloric intake as functions of physical activity in workers at an Indian factory. *(Adapted from Mayer et al.,* Am. J. Clin. Nutrition.*)*

normally large number of adipose-tissue cells which persists throughout later life; the presence of these extra cells may well act as a stimulus for increased food intake (possibly through the glycerol signal described earlier). In any case, the view that "a fat baby is a healthy baby" requires reevaluation in light of the potential effects of early obesity on later obesity with its attendant consequences.

As an exercise in energy balance, let us calculate how rapidly a person can expect to lose weight on a reducing diet. Suppose a woman whose metabolic rate per 24 h is 2000 kcal goes on a 1000 kcal/day diet. How much of her own body fat will be required to supply this additional 1000 kcal/day? Almost all of the organic nutrients lost are fat, and fat contains 9 kcal/g; therefore,

1000 kcal/day $\div$ 9 kcal/g = 111 g/day or 777 g/week

Approximately another 77 g of water is lost from the adipose tissue along with this fat (adipose tissue is 10 percent water) so that the grand total for one week's loss equals 854 g or 1.8 lb. Thus, during her diet she can reasonably expect to lose approxi-

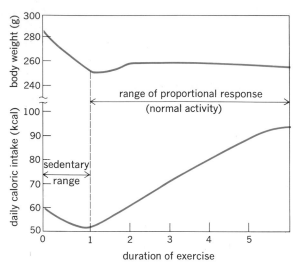

Figure 15-24. **Voluntary food intake and body weight as functions of duration of controlled exercise (hours per day) in normal adult rats.** *(Adapted from Mayer et al.,* Am. J. Physiol.*)*

mately this amount of weight each week. Actually, the amount of weight lost during the first week might be considerably greater since, for reasons poorly understood, a large amount of water may be lost early in the diet, particularly when the diet contains little carbohydrate. This early loss, which is really of no value so far as elimination of excess fat is concerned, often underlies the wild claims made for fad diets (indeed, to enhance this effect, drugs which cause the kidneys to excrete even more water are sometimes included in the diet). Clearly, weight loss is a slow process requiring patience and a meaningful reshaping of eating patterns.

Regulation of body temperature

Animals capable of maintaining their body temperatures within very narrow limits are termed *homeothermic.* The adaptive significance of this ability stems primarily from the marked effects of temperature upon the rate of chemical reactions in general and enzyme activity in particular. Homeothermic animals are spared the slowdown of all bodily functions which occurs when the body temperature falls. However, the advantages obtained by a relatively high body temperature

impose a great need for precise regulatory mechanisms since even moderate elevations of temperatures begin to cause nerve malfunction, protein denaturation, and death. Most people suffer convulsions at a body temperature of 41°C, and 43°C is the absolute limit for life. In contrast, most body tissues can withstand marked cooling (to less than 7°C), which has found an important place in surgery when the heart must be stopped, since the dormant cold tissues require little nourishment.

Figure 15-25 illustrates several important generalizations about normal human body temperature: (1) Oral temperature averages about 0.5°C less than rectal; thus, all parts of the body do not have the same temperature. (2) Internal temperature is not absolutely constant but varies several degrees in perfectly normal persons in response to activity pattern and external temperature, and in addition, there is a characteristic diurnal fluctuation, so that temperature is lowest during sleep and slightly higher during the awake state even if the person remains relaxed in bed. An interesting variation in women is a higher temperature during the last half of the menstrual cycle (Chap. 16).

If temperature is viewed as a measure of heat "concentration," temperature regulation can be studied by our usual balance methods. In this case, the total heat content of the body is determined by net difference between heat produced and heat lost from the body. Maintaining a constant body temperature implies that, overall, heat production must equal heat loss. Both these variables are subject to precise physiological control.

Temperature regulation offers a classic example of a biological control system; its generalized components are shown in Fig. 15-26. The balance between heat production and heat loss is continuously being disturbed, either by changes in metabolic rate (exercise being the most powerful influence) or by changes in the external environment which alter heat loss. The resulting small changes in body temperature reflexly alter the output of the effector organs, which drive heat production or heat loss and restore normal body temperature.

Heat production

The basic concepts of heat production have already been described. Recall that heat is produced by virtually all chemical reactions occurring in the body and that the cost-of-living metabolism by all

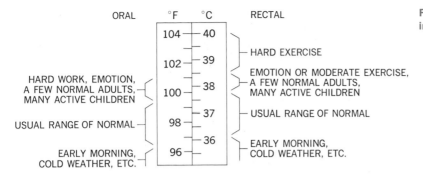

organs sets the basal level of heat production, which can be increased as a result of skeletal muscular contraction or the action of several hormones.

Changes in muscle activity. The first muscle changes in response to cold are a gradual and general increase in skeletal *muscle tone.* This soon leads to *shivering,* the characteristic muscle response to cold, which consists of oscillating rhythmic muscle tremors occurring at the rate of about 10 to 20 per second. So effective are these contractions that body heat production may be increased severalfold within seconds to minutes. Because no external work is performed, all the energy liberated by the metabolic machinery appears as internal heat. As always, the contractions are directly controlled by the efferent motor neurons to the muscles. During shivering these nerves are controlled by descending pathways under the primary control of the hypothalamus. It is important to note that this "shivering pathway" can be suppressed, at least in part, by input from the cerebral cortex since a cold man ceases to shiver when he starts to perform voluntary activity. Besides increased muscle tone and shivering, which are completely reflex in nature, people also use voluntary heat-production mechanisms such as foot stamping, hand clapping, etc.

Thus far, our discussion has focused primarily on the muscular response to cold; the opposite reactions occur in response to heat. Muscle tone is reflexly decreased and voluntary movement is also

Figure 15-26. Summary of temperature regulation. Heat loss from the body depends directly upon the external environment and upon changes controlled by temperature-regulating reflexes. In certain environments, heat gain rather than heat loss may actually occur.

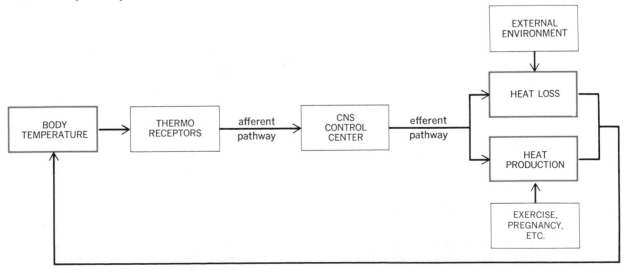

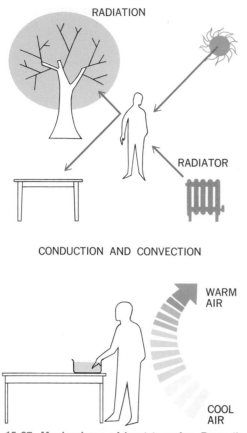

Figure 15-27. Mechanisms of heat transfer. By radiation, heat is transferred by electromagnetic waves (solid arrows); in conduction, heat moves by direct transfer of thermal energy from molecule to molecule (dashed arrow).

diminished ("It's too hot to move"). However, these attempts to reduce heat production are relatively limited in capacity both because muscle tone is already quite low normally and because of the *direct* effect of a body temperature increase on metabolic rate.

Nonshivering ("chemical") thermogenesis. In most experimental animals chronic cold exposure induces an increase in metabolic rate, which is not due to increased muscle activity. Indeed, as this so-called nonshivering thermogenesis increases over time, it is associated with a decrease in the degree of shivering. The cause of nonshivering thermogenesis has been the subject of considerable controversy; present evidence suggests that it

is due mainly to an increased secretion of epinephrine. (Thyroxine may also be involved but this is much less clear.) Equally controversial is the question whether nonshivering thermogenesis is a significant human phenomenon. Regardless of the outcomes of these debates, it seems clear that human hormonal changes and nonshivering thermogenesis are of secondary importance and that changes in muscle activity constitute the major control of heat production for temperature regulation, at least in the early response to temperature changes.

Heat-loss mechanisms

The surface of the body exchanges heat with the external environment by radiation, conduction and convection (Fig. 15-27), and water evaporation.

Radiation, conduction, convection. The surface of the body constantly emits heat, in the form of electromagnetic waves. Simultaneously, all other dense objects are radiating heat. The rate of emissions is determined by the temperature of the radiating surface. Thus, if the surface of the body is warmer than the *average* of the various surfaces in the environment, net heat is lost, the rate being directly dependent upon the temperature difference. The sun, of course, is a powerful radiator, and direct exposure to it greatly decreases heat loss by radiation or may reverse it.

Conduction is the exchange of heat not by radiant energy but simply by transfer of thermal energy from atom to atom or molecule to molecule. Heat, like any other quantity, moves down a concentration gradient, and thus the body surface loses or gains heat by conduction only through direct contact with cooler or warmer substances, including, of course, the air.

Convection is the process whereby air (or water) next to the body is heated, moves away, and is replaced by cool air (or water), which in turn follows the same pattern. It is always occurring because warm air is less dense and therefore rises, but it can be greatly facilitated by external forces such as wind or fans. Thus, convection aids conductive heat exchange by continuously maintaining a supply of cool air. In the absence of convection, negligible heat would be lost to the air, and conduction would be important only in such unusual circumstances as immersion in cold water.

(Because of the great importance of air movement in aiding heat loss, attempts have been made to quantitate the cooling effect of combinations of air speed and temperature; the most useful tool has been the wind-chill index.) Henceforth we shall also imply convection when we use the term conduction.

It should now be clear that heat loss by radiation and conduction is largely determined by the temperature difference between the body surface and the external environment. It is convenient to view the body as a central core surrounded by a shell consisting of skin and subcutaneous tissue (for convenience, we shall refer to the complex shell of tissues simply as skin) whose insulating capacity can be varied. It is the temperature of the central core which is being regulated at approximately 37°C; in contrast, as we shall see, the temperature of the outer surface of the skin changes markedly. If the skin were a perfect insulator, no heat would ever be lost from the core; the outer surface of the skin would equal the environmental temperature (except during direct exposure to the sun), and net conduction or radiation would be zero. The skin, of course, is not a perfect insulator, so that the temperature of its outer surface generally lies somewhere between that of the external environment and the core. Of great importance for temperature regulation of the core is that the skin's effectiveness as an insulator is subject to physiological control by changing the blood flow to the skin. The more blood reaching the skin from the core, the more closely the skin's temperature approaches that of the core. In effect, the blood vessels diminish the insulating capacity of the skin by carrying heat to the surface (Fig. 15-28). These vessels are controlled primarily by vasoconstrictor sympathetic nerves. Vasoconstriction may be so powerful

that the skin of the finger, for example, may undergo a 99 percent reduction in blood flow during exposure to cold.

Exposure to cold increases the difference between core and environment; in response, skin vasoconstriction increases skin insulation, reduces skin temperature, and lowers heat loss. Exposure to heat decreases (or may even reverse) the difference between core and environment; in order to permit the required heat loss, skin vasodilation occurs, the difference between skin and environment increases, and heat loss increases. Although we have spoken of skin temperature as if it were uniform throughout the body, certain areas participate much more than others in the vasomotor responses; accordingly, skin temperatures vary with location.

What are the limits of this type of process? The lower limit is obviously the point at which maximal skin vasoconstriction has occurred; any further drop in environmental temperature increases the environment-skin temperature difference and causes excessive heat loss. At this point, the body must increase its heat production to maintain temperature. The upper limit is set by the point of maximal vasodilation, the environmental temperature, and the core temperature itself. At high environmental temperatures, even maximal vasodilation cannot establish a core-environment temperature difference large enough to eliminate heat as fast as it is produced. Another heat-loss mechanism, therefore, is brought strongly into play, sweating. Thus, the skin vasomotor contribution to temperature is highly effective in the midrange of environmental temperature (20–28°C), but the major burden is borne by increased heat production at lower temperatures and by increased heat loss via sweating at higher temperatures.

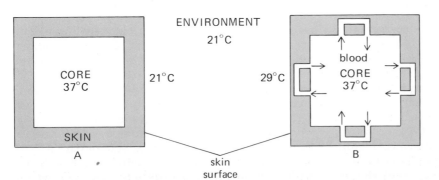

Figure 15-28. Relationship of skin's insulating capacity to its blood flow. (A) Skin as a perfect insulator, i.e., with zero blood flow, the temperature of the skin outer surface equaling that of the external environment. When the skin blood vessels dilate (B) the increased flow carries heat to the body surface, i.e., reduces the insulating capacity of the skin, and the surface temperature becomes intermediate between that of the core and the external environment.

Two other important mechanisms for altering heat loss by radiation and conduction remain: changes in surface area and clothing. Curling up into a ball, hunching the shoulders, and similar maneuvers in response to cold reduce the surface area exposed to the environment, thereby decreasing radiation and conduction. In human beings, clothing is also an important component of temperature regulation, substituting for the insulating effects of feathers in birds and fur in other mammals. The principle is similar in that the outer surface of the clothes now forms the true "exterior" of the body surface. The skin loses heat directly to the air space trapped by the clothes; the clothes in turn pick up heat from the inner air layer and transfer it to the external environment. The insulating ability of clothing is determined primarily by its type as well as by the thickness of the trapped air layer. We have spoken thus far only of the ability of clothing to reduce heat loss; the converse is also desirable when the environmental temperature is greater than body temperature, since radiation and conduction then produce heat gain. People therefore insulate themselves against temperatures which are greater than body temperature by wearing clothes. The clothing, however, must be loose so as to allow adequate movement of air to permit evaporation, the only source of heat loss under such conditions. White clothing is cooler since it reflects more radiant energy, which dark colors absorb. Contrary to popular belief, loose-fitting light-colored clothes are far more cooling than going nude during direct exposure to the sun.

Evaporation. Evaporation of water from the skin and lining membranes of the respiratory tract is the second major process for loss of body heat. Thermal energy is required to transform water from the liquid to the gaseous state. Thus, whenever water vaporizes from the body's surface, the heat required to drive the process is conducted from the surface, thereby cooling it. Even in the absence of sweating, there is still loss of water by diffusion through the skin, which is not completely waterproof. A like amount is lost during expiration from the respiratory lining. This insensible water loss amounts to approximately 600 mL/day in human beings and accounts for a significant fraction of total heat loss. In contrast to this passive water diffusion, *sweating* requires the active secretion of fluid by sweat glands and its extrusion into ducts, which carry it to the skin surface. The sweat is pumped to the surface by periodic contraction of cells resembling smooth muscle in the ducts. Production and delivery of sweat to the surface are stimulated by the sympathetic nerves. Sweat is a dilute solution containing primarily sodium chloride. The loss of this salt and water during severe sweating can cause diminution of plasma volume adequate to provoke hypotension, weakness, and fainting. It has been estimated that there are over 2.5 million sweat glands spread over the adult human body, and production rates of over 4 liters/h have been reported. This is 9 lb of water, the evaporation of which would eliminate almost 2400 kcal from the body! It is essential to recognize that sweat must evaporate in order to exert its cooling effect. The most important factor determining evaporation is the water vapor concentration of the air, i.e., the *humidity*. The discomfort suffered on humid days is due to the failure of evaporation; the sweat glands continue to secrete, but the sweat simply remains on the skin or drips off. Most other mammals differ from human beings in lacking sweat glands. They increase their evaporative losses primarily by panting, thereby increasing pulmonary air flow and increasing water losses from the lining of the respiratory tract, and they deposit water for evaporation on their fur or skin by licking.

Heat loss by evaporation of sweat gradually dominates as environmental temperature rises since radiation and conduction decrease as the body-environment temperature gradient diminishes. At environmental temperatures above that of the body, heat is actually gained by radiation and conduction, and evaporation is the sole mechanism for heat loss. A person's ability to survive such temperatures is determined by the humidity and by the maximal sweating rate. For example, when the air is completely dry, a person can survive a temperature of 130°C (266°F) for 20 min or longer, whereas very moist air at 46°C (115°F) is not bearable for even a few minutes.

Changes in sweating determine people's chronic adaptation to high temperatures. A person newly arrived in a hot environment has poor ability to do work initially, body temperature rises, and severe weakness and illness may occur. After several days, there is great improvement in work tolerance with little increase in body temperature, and the

Table 15-3. Summary of effector mechanisms in temperature regulation

Stimulated by cold	
Decrease heat loss	Vasoconstriction of skin vessels; reduction of surface area (curling up, etc.); behavioral (put on warmer clothes, raise thermostat setting, etc.)
Increase heat production	Shivering and increased voluntary activity; (?) increased secretion of thyroxine and epinephrine; increased appetite

Stimulated by heat	
Increase heat loss	Vasodilation of skin vessels; sweating; behavioral (put on cooler clothes, turn on fan, etc.)
Decrease heat production	Decreased muscle tone and voluntary activity; (?) decreased secretion of thyroxine and epinephrine; decreased appetite

person is said to have *acclimatized* to the heat. Body temperature is kept low because there is an earlier onset of sweating and because of increased rates. The sodium content of the sweat is reduced, thereby minimizing the loss of sodium from the body. The mechanisms of these changes remain unknown although heightened aldosterone secretion plays an important role in reducing the sodium.

Summary of effector mechanisms in temperature regulation

Table 15-3 summarizes the mechanisms regulating temperature, none of which is an all-or-none response but calls for a graded progressive increase or decrease in activity. As we have seen, heat production via skeletal muscle activity becomes extremely important at the cold end of the spectrum, whereas increased heat loss via sweating is critical at the hot end.

Brain centers involved in temperature regulation

Neurons in the hypothalamus and other brain areas, via descending pathways, control the output of somatic motor nerves to skeletal muscle (muscle tone and shivering) and of sympathetic nerves to skin arterioles (vasoconstriction and dilation), sweat glands, and the adrenal medulla.

In animals in which thyroxine is an important component of the response to cold, these centers also control the output of hypothalamic TSH-releasing hormone (TRH).

Afferent input to the integrating centers

The final component of temperature-regulating systems we must describe is really the first component, i.e., the afferent input. Obviously, these temperature-regulating reflexes require receptors capable of detecting changes in the body temperature. There are two groups of receptors, one in the skin (*peripheral thermoreceptors*) and the other in deeper body structures (*central thermoreceptors*).

Peripheral thermoreceptors. In the skin (and certain mucous membranes) are nerve endings usually categorized as *cold* and *warm receptors*. In one sense, these are misleading terms since cold is not a separate entity but a lesser degree of warmth. Really there are two populations of temperature-sensitive skin receptors, one stimulated by a lower and the other by a higher range of temperatures (Fig. 15-29). Information from these receptors is transmitted via the afferent nerves and ascending pathways to the hypothalamus and other integrating areas, which respond with appropriate efferent output; in this manner, the firing of cold receptors stimulates heat-producing and heat-conserving mechanisms.

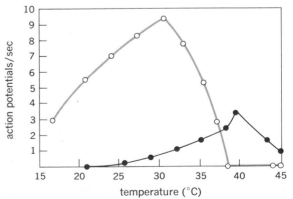

Figure 15-29. Discharge rates of a typical skin cold receptor (open circles) and warm receptor (closed circles) in response to changes in temperature. *(Adapted from Dodt and Zotterman.)*

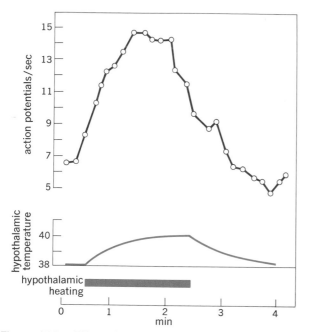

Figure 15-30. Effect of local heating of a discrete area of hypothalamus on the discharge rate of a single thermosensitive hypothalamic neuron. *(Adapted from Nakayama et al.)*

Central thermoreceptors. It should be clear that the skin thermoreceptors alone would be highly inefficient regulators of body temperature for the simple reason that it is the core temperature, not the skin temperature, which is actually being regulated. On theoretical grounds alone it was apparent that core, i.e., central, receptors had to exist somewhere in the body, and numerous experiments have localized them to the hypothalamus, spinal cord, abdominal organs, and other internal locations. In unanesthetized dogs, local warming of hypothalamic neurons (through previously implanted thermodes) causes them to fire rapidly (Fig. 15-30) and reproduces the entire picture of the dog's usual response to a warm environment: He becomes sleepy, stretches out to increase his surface area, pants heavily, salivates, and licks his fur. Conversely, local cooling induces vasoconstriction, intensive shivering, fluffing out of the fur, and curling up. These hypothalamic thermoreceptors have synaptic connections with hypothalamic and other integrating centers, which also receive input from other central thermoreceptors as well as from the skin thermoreceptors. The precise rela-

tive contributions of the various thermoreceptors remain the subject of considerable debate. Temperature-regulating reflexes are summarized in Fig. 15-31.

Fever

The elevation of body temperature so commonly induced by infection is due not to a breakdown of temperature-regulating mechanisms but to a "resetting of the thermostat" in the hypothalamus or other brain area. Thus, a person with a fever regulates her temperature in response to heat or cold but always at a higher set point. The onset of fever is frequently gradual but it is most striking when it occurs rapidly in the form of a chill. It is as though the thermostat were suddenly raised; the person suddenly feels cold, and marked vasoconstriction and shivering occur; the person also curls up and puts on more blankets because she feels cold. This association of heat conservation and increased heat production serves to drive body temperature up rapidly. Later, the fever breaks as the thermostat is reset to normal; the person feels hot, throws off the covers, and manifests profound vasodilation and sweating.

What is the basis for the resetting? As is described in the chapter on resistance to infection, certain endogenously produced chemicals known as *pyrogens* are released from white blood cells in the presence of infection or inflammation. These pyrogens act directly upon the thermoreceptors in the hypothalamus (and perhaps other brain areas), altering their rate of firing and their input to the integrating centers. This effect of pyrogens may be mediated via local release of prostaglandins which then directly alter thermoreceptor function. Consistent with this hypothesis is the fact that aspirin, which reduces fever by restoring thermoreceptor activity toward normal, inhibits the synthesis of prostaglandins, but this theory still remains unproven.

In addition to infection and inflammation, in which fever is induced by pyrogens, as described above, there are other situations in which hyperthermia is produced by quite different mechanisms. Excessive blood levels of epinephrine or thyroxine resulting from diseases of the adrenal medulla or thyroid gland elevate the body temperature by direct actions on heat-producing metabolic reactions rather than by altering the hypothalamic

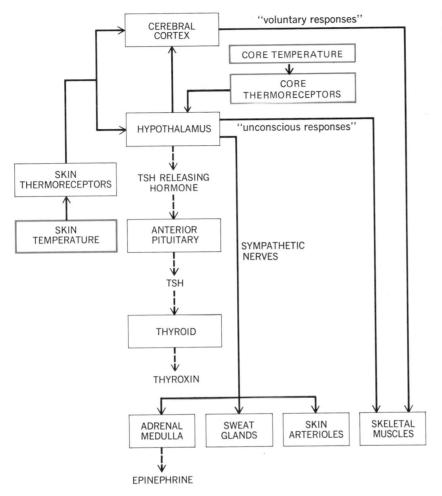

Figure 15-31. Summary of temperature-regulating mechanisms. The dashed lines are hormonal pathways, which are probably of minor importance in human beings. Not shown are other nonhypothalamic integrating areas.

thermoreceptor setting. Certain lesions of the brain do not reset the hypothalamus but rather completely destroy its normal regulatory capacity; under such conditions lethal hyperthermia may occur very rapidly. *Heat stroke* is also characterized by a similar breakdown in function of the regulatory centers. It is frequently a positive-feedback state in which, because of inadequate balancing of heat loss and production, body temperature becomes so high that the brain regulatory centers are put out of commission and body temperature therefore rises even higher. Thus a patient suffering from heat stroke manifests a dry skin (absence of sweating) despite a markedly elevated body temperature.

In this regard, it is instructive to calculate how rapidly body temperature can rise when heat-loss mechanisms are completely shut down. The amount of heat required to raise the body temperature 1°C is 0.83 kcal/kg. Therefore, in a 70-kg person, the body temperature is increased 1°C by the retention of 0.83 kcal × 70 kg = 58 kcal. Thus, a lethal 6°C rise in body temperature occurs when 6°C × 58 kcal/1°C rise = 348 kcal has been retained. Normal metabolism produces this much heat in approximately 5 h (actually less since metabolic rate is greatly increased as body temperature rises).

Heat stroke, the attainment of a body temperature at which vital bodily functions are endangered, should be distinguished from *heat exhaustion*. The former is due to a breakdown in heat-regulating mechanisms, whereas the latter is not the result of failure of heat regulation but rather is the inability to meet the price of heat regulation. Heat exhaustion is a state of collapse due to hypotension brought on by depletion of plasma volume (secondary to sweating) and by extreme dilation of skin blood vessels, i.e., by decreases in both cardiac output and peripheral resistance. Thus, heat exhaustion occurs as a direct consequence of the activity of heat-loss mechanisms; because these mechanisms have been so active, the body temperature is only modestly elevated. In a sense, heat exhaustion is a safety valve which, by forcing cessation of work when heat-loss mechanisms are overtaxed, prevents the larger rise in body temperature which would precipitate the far more serious condition of heat stroke.

16

REPRODUCTION

Section A: Male reproductive physiology

Spermatogenesis
  Delivery of sperm
  Erection
  Ejaculation
Hormonal control of male reproductive functions
  Effects of testosterone
    Spermatogenesis ● *Accessory reproductive organs* ● *Behavior* ● *Secondary sex characteristics* ● *Mechanism of action of testosterone*
  Anterior pituitary and hypothalamic control of testicular function

Section B: Female reproductive physiology

Ovarian function
  Ovum and follicle growth
  Ovum division
  Formation of corpus luteum
  Ovarian hormones
  Cyclical nature of ovarian function
Control of ovarian function
  Control of follicle and ovum development
  Control of ovulation
  Control of corpus luteum
  Summary
Uterine changes in the menstrual cycle
Nonuterine effects of estrogen and progesterone
Female sexual response
Pregnancy
  Ovum transport
  Sperm transport
  Entry of the sperm into the ovum
  Early development, implantation, and placentation
  Hormonal changes during pregnancy
  Parturition (delivery of the infant)
  Lactation
  Fertility control

Section C: The chronology of sex development

Sex determination
Sex differentiation
  Differentiation of the gonads
  Differentiation of internal and external genitalia
  Sexual differentiation of the central nervous system
  Puberty
  Menopause

Before beginning detailed descriptions of male and female reproductive systems, it is worthwhile to summarize some of the important terminology. The primary reproductive organs are known as the *gonads*, the *testes* in the male and the *ovaries* in the female. In both sexes, the gonads serve dual functions: (1) production of the reproductive cells, *sperm* or *ova*; (2) secretion of the so-called *sex hormones*. The systems of ducts through which the sperm or ova are transported and the glands lining or emptying into the ducts are termed the *accessory reproductive organs* (in the female the breasts are also usually included in this category).

Finally, the *secondary sexual characteristics* comprise the many external differences (hair, body contours, etc.) between male and female which are not directly involved in reproduction. The gonads and accessory reproductive organs are present at birth but remain relatively small and nonfunctional until the onset of *puberty*, at about 10 to 14 years of age. The secondary sexual characteristics are virtually absent until puberty. The term puberty signifies the attainment of sexual maturity in the sense that conception becomes possible; as commonly used, it refers to the 3 to 5 years of sexual development culminating in the

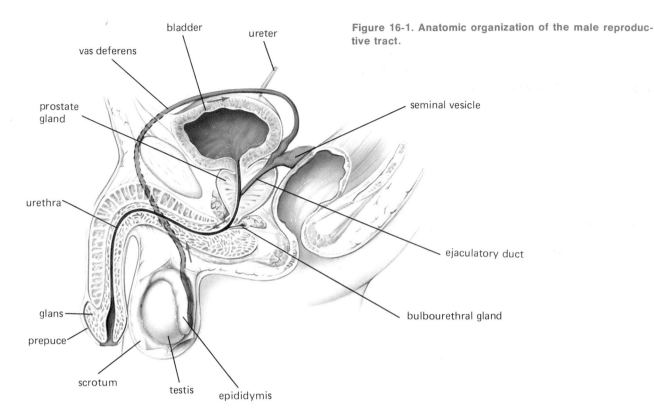

bladder

ureter

vas deferens

Figure 16-1. Anatomic organization of the male reproductive tract.

prostate
gland

seminal vesicle

urethra

ejaculatory duct

glans

prepuce

bulbourethral gland

scrotum   testis   epididymis

attainment of sexual maturity. The term *adolescence* has a much broader meaning and includes the total period of transition from childhood to adulthood in all respects, not just sexual.

Section A:
Male reproductive physiology

The essential male reproductive functions are the manufacture of sperm (*spermatogenesis*) and the deposition of the sperm in the female. The organs which carry out these functions are the testes, epididymides (singular, epididymis), vas deferens, ejaculatory ducts, and penis together with the accessory glands: the seminal vesicles, prostate, and bulbourethral glands (Fig. 16-1). The *testes* are surrounded by a connective-tissue capsule and suspended outside the body in a sac, the *scrotum*. During embryonic development, the testes lie in the rear wall of the abdomen, and during the seventh month of intrauterine development, they descend into the scrotum, carrying with them the arteries, veins, and nerves which supply

them, and the ducts (vas deferens) connecting them with the urethra. This descent is essential for normal sperm production during adulthood since sperm formation requires a temperature lower than normal internal body-temperature.

Each testis contains many tiny convoluted *seminiferous tubules*, the combined lengths of which are 250 m; they are lined with the sperm-producing *spermatogenic cells*. The testes also serve an endocrine function, the manufacture of the primary male sex hormone *testosterone*. The endocrine-secreting *interstitial cells* (Leydig cells) lie in the small connective tissue spaces surrounding the seminiferous tubules (Fig. 16-2). Thus, the sperm-producing and testosterone-producing functions of the testes are carried out by different cells.

Spermatogenesis

In Fig. 16-2, a microscopic section of an adult human testis, it can be seen that the seminiferous tubules contain many cells, the vast majority of which are in various stages of division; these are the spermatogenic cells. Each seminiferous tubule is surrounded by a basement membrane, and only the outermost layer of spermatogenic cells is in contact with this membrane; these are undifferentiated germ cells termed *spermatogonia,* which, by dividing mitotically, provide a continuous source of new cells. Some spermatogonia move away from the basement membrane and increase markedly in size. Each of these large cells, now termed a *primary spermatocyte,* divides to form two *secondary spermatocytes,* each of which in turn divides into two *spermatids,* the latter ultimately being transformed into mature *spermatozoa (sperm)* (Fig. 16-3).

The division of the primary and secondary spermatocytes by *meiosis* differs from the ordinary

mitotic division. During mitosis (Chap. 4) each daughter cell receives the full number of chromosomes; in meiosis, which is really two divisions in succession, each final cell receives only half of the chromosomes present in the original cell. The primary spermatocyte contains 46 chromosomes, 23 from each parent. Prior to its division, each of these 46 chromosomes becomes duplicated and during the condensation into chromosomes, the duplicated maternal and paternal chromatin threads pair with each other, forming a four-stranded chromosome (Fig. 16-4). Thus, there are 23 four-stranded chromosomes instead of 46 two-stranded chromosomes. When the primary spermatocyte divides, the two maternal strands pass into one of the secondary spermatocytes and the pa-

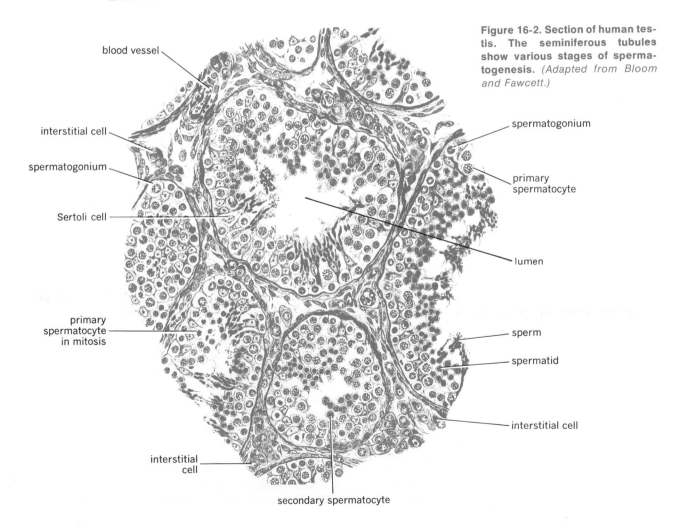

Figure 16-2. Section of human testis. The seminiferous tubules show various stages of spermatogenesis. *(Adapted from Bloom and Fawcett.)*

blood vessel

interstitial cell

spermatogonium

Sertoli cell

primary spermatocyte in mitosis

interstitial cell

secondary spermatocyte

spermatogonium

primary spermatocyte

lumen

sperm

spermatid

interstitial cell

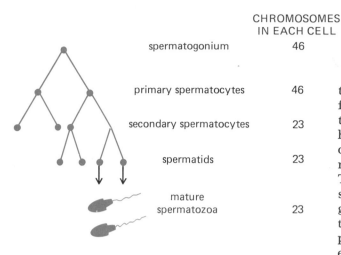

	CHROMOSOMES IN EACH CELL
spermatogonium	46
primary spermatocytes	46
secondary spermatocytes	23
spermatids	23
mature spermatozoa	23

Figure 16-3. Summary of spermatogenesis. Each spermatogonium yields eight mature sperm, each containing 23 chromosomes.

Figure 16-4. Separation of chromosomes during cell divisions of meiosis.

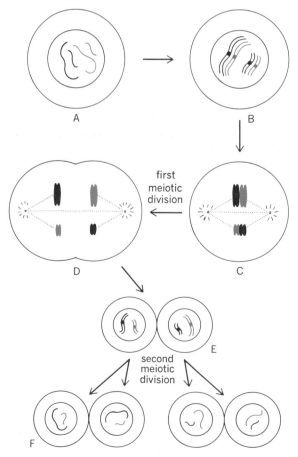

ternal strands into the other. Since each of the 23 four-stranded chromosomes becomes attached to the spindle fibers at random, one chromosome may have its maternal strands oriented toward one pole of the cell and the next chromosome have its maternal strands oriented toward the opposite pole. Thus, when the maternal and paternal strands separate, the two secondary spermatocytes, in general, receive a mixture of maternal and paternal chromosomes. It would be extremely improbable that all 23 maternal chromosomes would end up in one cell and all 23 paternal chromosomes in the other.

The random distribution of maternal and paternal chromosomes during meiosis provides the basis for much of the variability in the genetic constitution of the offspring from a single set of parents. Over 8 million (2^{23}) different combinations of maternal and paternal chromosomes can result from the distribution of the 23 chromosomes during meiosis. This is only the minimum number of possible combinations of genetic material since segments of the maternal and paternal chromosomes may exchange with each other in a process known as *crossing over*, which occurs during chromosomal pairing before the first meiotic division. This forms chromosomes containing both maternal and paternal genes in the same chromosome. Thus all the genes initially present in a paternal or maternal chromosome do not necessarily pass into the same cell during meiosis.

When the secondary spermatocytes divide (the second division of meiosis), the chromosomes separate from their duplicates so that the resulting cells, the spermatids, have 23 nonduplicated chromosomes. Since development of the female germ cell follows a similar pattern, the eventual combination of sperm and ovum results in the reestablishment of the normal complement of 46 chromosomes.

The final phase of spermatogenesis, the transformation of the spermatids into mature spermatozoa, involves no further cell divisions. The head of a mature spermatozoon (Fig. 16-5) consists almost entirely of the nucleus, a dense mass of DNA bearing the sperm's genetic information. The

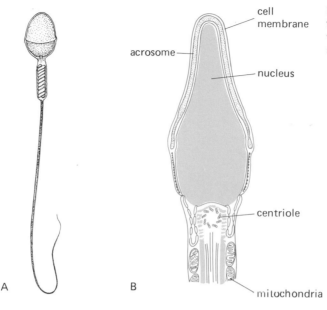

cell membrane

acrosome

nucleus

centriole

mitochondria

Figure 16-5. (A) Human mature sperm seen in frontal view. (B) The close-up is a side view.

A   B

tip of the nucleus is covered by the *acrosome*, a protein-filled vesicle containing several lytic enzymes which enable the sperm to enter the ovum. The tail comprises a group of contractile filaments which produces a whiplike movement capable of propelling the sperm at a velocity of 1 to 4 mm/s. The mitochondria of the spermatid form the midpiece of the tail and probably provide the energy for its movement.

Throughout the entire process of maturation, the developing germ cells remain intimately associated with another type of cell, the *Sertoli cell* (Fig. 16-6). Each Sertoli cell extends from the basement membrane of the seminiferous tubule all the way to the fluid-filled lumen, and is joined to adjacent Sertoli cells by means of tight junctions. Thus the Sertoli cells form an unbroken ring around the outer circumference of the seminiferous tubule and divide the tubule into two compartments—a basal compartment (between the basement membrane and the tight junction), and a central (adluminal) compartment including the lumen. This arrangement has several very important results: First, the ring of interconnected Sertoli cells forms a "barrier" which limits the movement of chemicals from blood into the lumen of the seminiferous tubule (the membranes surrounding the entire tubule form a second component of the so-called blood-testis barrier); second, mitosis of the spermatogonia takes place entirely

in the basal compartment, and the primary spermatocytes must move through the tight junctions of the Sertoli cells (which open to make way for them) to gain entry into the central compartment. In this compartment, the spermatids are contained in recesses formed by invaginations of the Sertoli-cell luminal membranes until mature enough for release. Apparently, the Sertoli cells serve as the route by which nutrients and chemical signals reach the developing germ cells. As we shall see, they may also produce hormones and other chemicals important for the control of spermatogenesis.

The mechanisms which guide the remarkable cellular transformation from spermatid to mature sperm remain uncertain. In any small segment of seminiferous tubules, the entire process of spermatogenesis proceeds in a regular sequence. For example, at any given time, virtually all the primary spermatocytes in one portion of the tubule are undergoing division, whereas in an adjacent segment, the secondary spermatocytes may be dividing. The entire process in a single area takes approximately 72 days. In mammals which breed seasonally, spermatogenesis is periodic, activity being followed by degeneration of the spermatogonia and shrinking of the seminiferous tubules. In contrast, nonseasonal breeders, such as human beings, manifest continuous activity, the cycles following each other without a break. Perhaps the most amazing characteristic of spermatogenesis

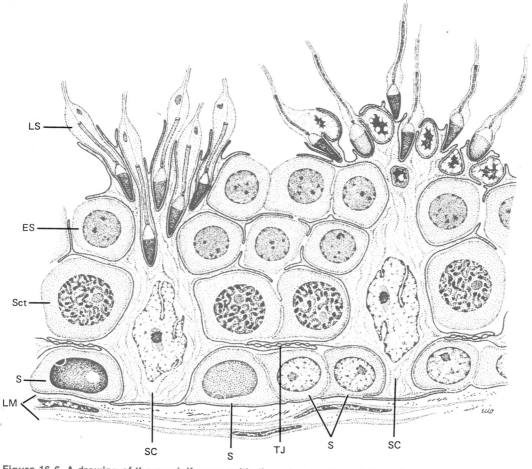

LS

ES

Sct

S

LM

SC   S   TJ   S   SC

Figure 16-6. A drawing of the seminiferous epithelium showing the relation of the Sertoli cells and germ cells. Spermatocytes (Sct); early spermatids (ES); late spermatids (LS); spermatogonia (S); tight junction between two Sertoli cells (TJ); lining membrane (LM); Sertoli cell (SC). *(From R. O. Greep and L. Weiss, "Histology," 3d ed., McGraw-Hill Book Company, 1973.) (Courtesy of Y. Clermont.)*

is its sheer magnitude: the normal human male may manufacture several hundred million sperm per day.

Delivery of sperm

From the seminiferous tubules, the sperm pass through a network of interconnected highly coiled ducts which join to form a single duct within the *epididymis* (Fig. 16-7), which in turn leads to the large thick-walled *vas deferens.* Movement through these ducts is accomplished by two means

(the sperm themselves are nonmotile at this time): the pressure created by the continuous formation of sperm and fluid; a peristaltic-like action exerted by the smooth muscle cells in the duct walls.

Besides serving as a route for sperm exit, this system performs several other important functions: (1) The epididymis and first portion of the vas deferens store sperm prior to ejaculation; (2) during passage through the epididymis or storage there (approximately 9 to 14 days are required for a sperm to traverse the epididymis), a maturation process occurs, without which the sperm would be nonmotile and infertile when they enter the female tract; (3) during ejaculation, the sperm are expelled from the epididymis and vas deferens by strong contractions of the smooth muscle lining the duct walls.

A pair of large glands, the *seminal vesicles,* drain into the two vas deferens which now become the ejaculatory ducts which pass into the *prostate gland* and join the urethra (Fig. 16-1). The prostate and seminal vesicles, as well as the bulbourethral glands just below the prostate, secrete the bulk of the fluid to be ejaculated, the *semen,* which contains a large number of different chemical substances, the functions of which are presently being worked out. For example, the seminal vesicles secrete large quantities of the carbohydrate fructose, utilized by the sperm contractile apparatus for energy. We shall describe later the possible contributions of the prostaglandins, present in very large concentrations in seminal fluid. Another function of the seminal fluid is that of sheer dilution of the sperm (in human beings, sperm constitutes only a few percent of the total ejaculated semen); without such dilution, motility is impaired.

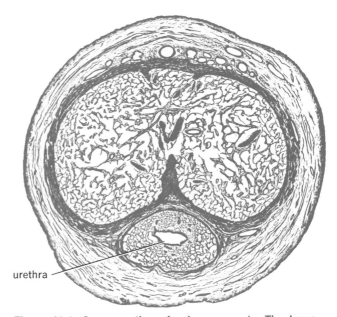

urethra

Figure 16-8. Cross section of a human penis. The large central area and that surrounding the urethra are vascular spaces which become filled with blood to cause erection. *(Adapted from Bloom and Fawcett.)*

Figure 16-7. Diagrammatic cross section of a human testis. The highly coiled seminiferous tubules are the sites of sperm production.

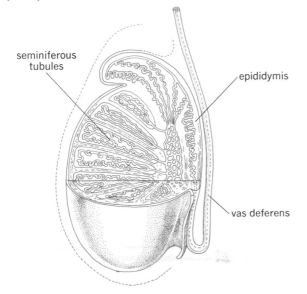

seminiferous tubules

epididymis

vas deferens

Erection

The primary components of the male sexual act are *erection* of the penis, which permits entry into the female vagina, and *ejaculation* of the sperm-containing semen into the vagina. Erection is a vascular phenomenon which can be understood from the structure of the penis (Fig. 16-8). This organ consists almost entirely of three cylindrical cords of *erectile tissue,* which are actually vascular spaces. Normally the arterioles supplying these chambers are constricted so that they contain little blood and the penis is flaccid; during sexual excitation, the arterioles dilate, the chambers become engorged with blood, and the penis becomes rigid. Moreover, as the erectile tissues expand, the veins emptying them are passively compressed, thus minimizing outflow and contributing to the engorgement. This entire process occurs rapidly, complete erection sometimes taking only 5 to 10 s. The vascular dilation is accomplished by stimulation of the parasympathetic nerves and inhibition of the sympathetic nerves to the arterioles of the penis (Fig. 16-9). This ap-

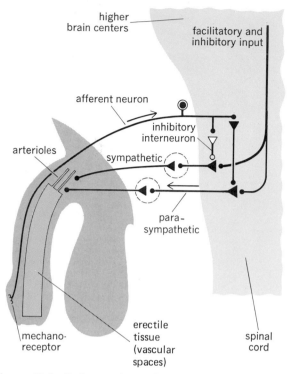

Figure 16-9. Reflex pathway for erection. The reflex is initiated by mechanoreceptors in the penis. Input from higher centers can facilitate or inhibit this reflex.

pears to be one of the few cases of direct parasympathetic control over high-resistance blood vessels. In addition to these vascular effects, the parasympathetic nerves stimulate urethral glands to secrete a mucuslike material which aids in lubrication of the head of the penis. What receptors and afferent pathway initiate these reflexes? The primary input comes from highly sensitive mechanoreceptors located in the head of the penis. The afferent fibers carrying the impulses synapse in the lower spinal cord and trigger the efferent outflow. It must be stressed, however, that higher brain centers, via descending pathways, may exert profound facilitative or inhibitory effects upon the efferent neurons. Thus, thoughts or emotions can induce erection in the complete absence of mechanical stimulation of the penis; conversely, failure of erection (impotence) may frequently be due to psychological factors. The ability of alcohol to inhibit erection is probably due to its effects on higher brain centers.

Ejaculation

This process is basically a spinal reflex, the afferent pathways apparently being identical to those described for erection. When the level of stimulation reaches a critical level, a patterned automatic sequence of efferent discharge is elicited to both the smooth muscle of the genital ducts and the skeletal muscle at the base of the penis. The precise contribution of various pathways is complex, but the overall response can be divided into two phases: (1) The genital ducts and glands contract, as a result of sympathetic stimulation to them, emptying their contents into the urethra (emission); (2) the semen is then expelled from the penis by a series of rapid muscle contractions. During ejaculation the sphincter at the base of the bladder is closed so that sperm cannot enter the bladder nor can urine be expelled. The rhythmical contractions of the muscles which occur during ejaculation are associated with intense pleasure and many systemic physiological changes, the entire event being termed the *orgasm.* A marked skeletal muscle contraction throughout the body is followed by the rapid onset of muscular and psychological relaxation, and there is also a marked increase in heart rate and blood pressure. Once ejaculation has occurred, there is a so-called latent period during which a second erection is not possible; this period is quite variable but may be hours in perfectly normal men.

The average volume of fluid ejaculated is 3 mL, containing approximately 300 million sperm. However, the range of normal values is extremely large, and older ideas of the minimal concentration of sperm required for fertility are now being reevaluated. Although quantity is important, it is obvious that the quality of the sperm is another critical determinant of fertility.

Hormonal control of male reproductive functions

Virtually all aspects of male reproductive function are either directly controlled or indirectly influenced by testosterone or the anterior pituitary gonadotropins, *follicle-stimulating hormone* (FSH) and *luteinizing hormone* (LH). These pituitary hormones were named for their effects in the female, but their molecular structures are precisely the same in both sexes. (LH, in the male, is frequently called interstitial-cell-stimulating hormone, ICSH.) FSH and LH exert their effects only upon the testes, whereas testosterone manifests a broad spectrum of actions not only on the testes but on the accessory reproductive organs, the secondary sexual characteristics, sexual behavior, and organic metabolism in general.

Effects of testosterone

Testosterone, the hormone produced by the interstitial cells of the testes, is the major male sex hormone. As pointed out in Chap. 9, other steroids with actions similar to those of testosterone are produced by the adrenal cortex and are, with testosterone, collectively known as *androgens*. Although the adrenally produced androgens constitute a large fraction of total blood androgens in men, they are much less potent than is testosterone and are unable to maintain testosterone-dependent functions should testosterone secretion be decreased or eliminated by disease or castration. Accordingly, we shall not further discuss adrenal androgens in our description of male reproductive function.

Spermatogenesis. Adequate amounts of testosterone are essential for spermatogenesis, and sterility is an invariable result of testosterone deficiency. The cells which secrete testosterone are the *interstitial cells;* as shown in Fig. 16-2, they lie scattered between the seminiferous tubules. It seems likely that the stimulatory effects of testosterone on spermatogenesis are exerted locally by the hormone diffusing from the interstitial cells into the seminiferous tubules. Testosterone is not the only hormone required for spermatogenesis; the pituitary gonadotropins are also required, and the relationship with testosterone will be described subsequently.

It must be emphasized that although testosterone is required for the process of spermatogenesis, testosterone production does *not* depend upon spermatogenesis. In other words, testosterone deficiency produces sterility by interrupting spermatogenesis, but interference with the function of the seminiferous tubules does not alter normal testosterone production by the interstitial cells. The great importance of this relationship is that a simple, effective method of sterilizing the male is

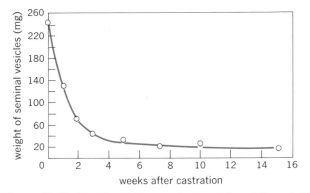

Figure 16-10. Atrophy of the seminal vesicles of the adult mouse after castration. *(Adapted from Deanesley and Parkes.)*

vasectomy, surgical ligation and removal of a segment of the vas deferens, which carry sperm from the testes. This procedure prevents the delivery of sperm but does not appear to alter secretion of testosterone.

Accessory reproductive organs. The morphology and function of the entire male duct system, glands, and penis all depend upon testosterone. Following removal of the testes *(castration)* in the adult, all the accessory reproductive organs decrease in size (Fig. 16-10), the glands markedly reduce their rates of secretion, and the smooth muscle activity of the ducts is inhibited. Erection and ejaculation may be deficient. These defects disappear upon the administration of testosterone.

Behavior. Most of our information comes from experiments on animals other than human beings, but even from our fragmentary information about man, there is little doubt that the development and maintenance of normal sexual drive and behavior in men are testosterone-dependent and may be seriously impaired by castration. However, it is a mistake to assume that deviant male sexual behavior must therefore be due to testosterone deficiency or excess. For example, most (but not all) male homosexuals have normal rates of testosterone secretion; although administration of exogenous testosterone may sometimes increase sexual activity in these men, they remain homosexual. To date, no clear-cut correlation has been established between homosexuality or hypersexuality and hormonal status in either men or women.

A question which has recently become the subject of enormous controversy is whether testosterone influences other human behavior in addition to sex, i.e., are there any inherent male-female differences or are the observed differences in behavior all socially conditioned. There is little doubt that behavioral differences based on sex do exist in other mammals; for example in mice aggression is clearly greater in males and is testosterone-dependent. Obviously, it will be difficult to answer such questions with respect to human beings but attempts are now being made to study them in a controlled scientific manner.

Secondary sex characteristics. In the animal world secondary sex characteristics range from the exotic courtship dances of salamanders to the mane of the lion to the stag's antlers. In many species they are important for normal sexual functions; antlers used in fighting for the female and the deep attracting voice of the tree toad are two examples from literally thousands.

In man, virtually all the obvious masculine secondary characteristics are testosterone-dependent. For example, a male castrated before puberty does not develop a beard or axillary or pubic hair. A strange and unexplained finding is that baldness, although genetically determined in part, does not occur in castrated men (castration will not, of course, reverse baldness). Other testosterone-dependent secondary sexual characteristics are the deepening of the voice (resulting from growth of the larynx), skin texture, thick secretion of the skin oil glands (predisposing to acne), and the masculine pattern of muscle and fat distribution. This leads us into an area of testosterone effects usually described as *general metabolic effects* but very difficult to separate from the secondary sex characteristics. It is obvious that the bodies of men and women (even excepting the breasts and external genitals) have very different appearances; a woman's curves are due in large part to the feminine distribution of fat, particularly in the region of the hips and lower abdomen but in the limbs as well. A castrated male gradually develops this pattern; conversely, a woman treated with testosterone loses it. A second very obvious difference is that of skeletal muscle mass; testosterone

exerts a profound effect on skeletal muscle to increase it size. The overall relationship of testosterone to general body growth was described in Chap. 15.

Mechanism of action of testosterone. Testosterone, like other steroid hormones, crosses cell membranes readily and, within target cells, combines with specific cytoplasmic receptors. The hormone-receptor complex then moves into the nucleus where the testosterone-receptor complex influences the transcription of certain genes into messenger RNA. The result is a change in the rate at which the target cell synthesizes the proteins coded for by these genes, and it is the changes in the concentrations of these proteins which underlie the cell's overall response to the hormone. For example, testosterone induces increased synthesis of enzymes in the prostate gland which then catalyze the formation of the gland's secretions.

In Chap. 9 we mentioned that hormones sometimes must undergo transformation in their target cells in order to be effective, and this is true of testosterone in certain of its target cells. In some, after its entry to the cytoplasm, testosterone undergoes an enzyme-mediated transformation into another steroid—*dihydrotestosterone*—and it is this molecule which then combines with a specific receptor and moves into the nucleus to exert its effect (Fig. 16-11). There are men whose cells lack the enzyme required for formation of dihydrotestosterone, and these men display normal masculinization of those cells which respond directly to testosterone (skeletal muscles, for example) but failure of normal growth of those in which dihydrotestosterone is the actual messenger (the prostate, for example).

Quite startling has been the recent discovery that, in still other target cells, notably neurons in certain areas of the brain, testosterone is transformed not to dihydrotestosterone but to estrogen, which then combines with cytoplasmic receptor, moves into the nucleus . . . etc. (Fig. 16-11). Thus, a "male" sex hormone must first be transformed into a "female" sex hormone to be able to influence those neurons, which, in turn, mediate "male" behavior!

Testosterone is not a uniquely male hormone, but is found in very low concentration in the blood of normal women (as a result of production by the

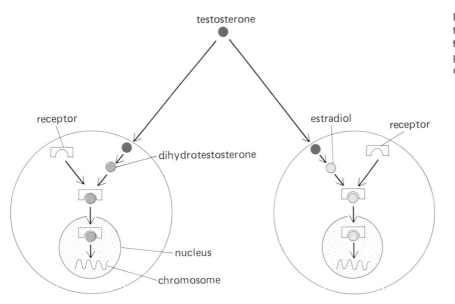

testosterone

receptor

estradiol

receptor

dihydrotestosterone

nucleus

chromosome

Figure 16-11. Transformation of testosterone either to dihydrotestosterone or to estradiol is accomplished by enzymes in many target cells. *(Adapted from McEwen.)*

ovaries and adrenal glands). Of greater importance is the fact that androgens other than testosterone are found in quite significant concentrations in the blood of women. The sites of production are mainly the adrenal glands, which contribute these same nontestosterone androgens in the male; in contrast to their lack of significance in the male, adrenal androgens do play several important roles in the female, specifically stimulation of general body growth (Chap. 15) and maintenance of sexual drive. In several disease states, the female adrenals may secrete abnormally large quantities of androgen, which produce a remarkable virilism; the female fat distribution disappears, a beard appears along with the male body-hair distribution, the voice lowers in pitch, the skeletal muscle mass enlarges, the clitoris (homologue of the male penis) enlarges, and the breasts diminish in size. These changes illustrate the sex-hormone dependency of secondary characteristics.

Anterior pituitary and hypothalamic control of testicular function

FSH and LH are essential for normal spermatogenesis and testosterone secretion (following removal of the anterior pituitary, the testes decrease greatly in weight, and spermatogenesis and testosterone secretion almost cease). FSH stimulates spermatogenesis by an action on the seminiferous tubules. However, this effect is not exerted directly on the spermatogenic cells but rather via FSH stimulation of Sertoli cells, which as we have seen, are in intimate contact with the spermatogenic cells at all stages of development. LH stimulates testosterone secretion by a direct action on the interstitial cells, and because testosterone is required for spermatogenesis, LH is indirectly involved in this latter process as well (Fig. 16-12).

No discussion of the anterior pituitary is complete without inclusion of the hypothalamus. As described in Chap. 9, all anterior pituitary hormones are controlled by releasing hormones secreted by discrete areas of the hypothalamus and reaching the pituitary via the hypothalamo-pituitary portal blood vessels. This input is essential for sexual function since destruction of the relevant hypothalamic areas stops spermatogenesis and markedly reduces testosterone secretion. Figure 16-12 summarizes the hypothalamic–anterior pituitary–testicular relationships; note that the hypothalamus exerts no direct effects on the testes but rather influences them indirectly via its control of the anterior-pituitary gonadotropins.

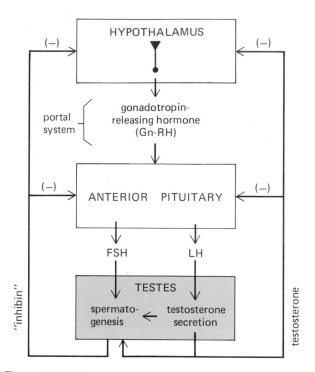

Figure 16-12. Summary of hormonal control of testicular function. The negative signs indicate that testosterone inhibits LH secretion via both the hypothalamus and the anterior pituitary. Testosterone reaches the seminiferous tubules to stimulate spermatogenesis both by local diffusion and by release into the blood and recirculation to the testes. FSH influences spermatogenesis via an action on the Sertoli cells and is, in turn, inhibited by a substance, inhibin, released by the Sertoli cells.

Several of the paths shown in Fig. 16-12 remain controversial. A major unsettled question is whether there exist two distinct hypothalamic hormones, one for each gonadotropin. Following the purification of LH-releasing hormone it was found that this peptide also possessed the capacity to stimulate FSH release, and although the evidence is not yet conclusive it favors the hypothesis that there is only a single releaser of both LH and FSH, and so LH-releasing hormone was appropriately rechristened *gonadotropin-releasing hormone* (Gn-RH).

Given that Gn-RH controls the secretion of both FSH and LH, we must now ask what controls Gn-RH. Recall from Chap. 9 that the hypothalamic cells which produce releasing hormones

are neurons which secrete the releasing hormones into the hypothalamo-pituitary portal vascular system in response to action potentials generated in them. This electrical activity in the neurons which secrete Gn-RH may be in part spontaneous, but it also reflects both synaptic input from other brain areas (these neural inputs are not indicated in Fig. 16-12) and inhibitory influences of testicular hormones reaching the hypothalamus via the blood.

The precise nature of this latter negative-feedback inhibition also remains unsettled. The observed fact is that testosterone, at physiological levels, inhibits the secretion of LH but has little effect on FSH secretion. This is not what would be expected were testosterone to exert its major inhibitory effect on the hypothalamic secretion of Gn-RH since this latter hormone influences the secretion of both FSH and LH. Therefore, it is likely that testosterone's major negative-feedback effect is directly on the anterior pituitary rather than on the hypothalamus, i.e. testosterone directly inhibits the pituitary's LH-releasing mechanism but has little or no effect on FSH.

How then do the testes inhibit FSH secretion, if not via testosterone? It is most likely that the inhibitory chemical signal to the hypothalamus and anterior-pituitary is a protein hormone termed *inhibin,* which is released into the blood from the seminiferous tubules themselves, probably from the Sertoli cells. That the Sertoli cells are the source of the negative-feedback inhibition of FSH secretion makes sense since, as described above, the stimulatory effect of FSH on spermatogenesis is exerted via the Sertoli cells; thus, the Sertoli cells are, in all ways, the link between FSH and spermatogenesis. The solution to this problem of how the seminiferous tubules exert an inhibitory influence on FSH secretion is not merely of academic interest since an agent which inhibits FSH only and not LH would constitute an ideal male contraceptive; spermatogenesis, but not testosterone secretion, would be eliminated.

Despite all this complexity, one should not lose sight of the fact that the secretion of LH and FSH in the male normally proceeds at a rather fixed, continuous rate during adult life; accordingly, spermatogenesis and testosterone secretion also occur at relatively unchanging rates. This is completely different from the large cyclical swings of activity so characteristic of the female reproductive hormones. A word of caution may be in order here, however, for recent work in nonprimate mammals has indicated that testosterone levels can be made to vary in response to various sexual and social stimuli; the relevance of these studies for men is unknown.

Section B:
Female reproductive physiology

The ovaries and female reproductive duct system—uterine tubes, uterus, and vagina—are located in the pelvic region and constitute the internal genitalia (Fig. 16-13A and B). In the female, the urinary and reproductive duct systems are entirely separate.

The uterine tubes (also known as *oviducts* or *Fallopian tubes*) are not directly connected to the ovaries but open into the abdominal cavity close to them. This opening of each uterine tube is a trumpet-shaped expansion surrounded by long fingerlike projections (the *fimbriae*) which are lined with ciliated epithelium. The other ends of the uterine tubes empty directly into the cavity of the *uterus* (womb), which is a hollow thick-walled muscular organ lying between the bladder and rectum. The upper portion of the uterus is the *corpus*, or *body*, and the lower portion is the *cervix*. A small opening in the cervix leads from the uterus to the vagina.

The external genitalia (Fig. 16-13C) include the mons pubis, labia majora and minora, the clitoris, the vestibule of the vagina, and the vestibular glands. The term *vulva* includes all these parts. The *labia majora*, the female analogue of the scrotum, are two prominent skin folds. The *labia minora* are small skin folds lying between the labia majora; they surround the urethral and vaginal openings, and the area thus enclosed is the *vestibule*. The vaginal opening lies behind that of the urethra. Partially overlying the vaginal opening is a thin fold of mucous membrane, the *hymen*. The *clitoris* is an erectile structure homologous to the penis.

Unlike the continuous sperm production of the male, the maturation and release of the female germ cell, the *ovum*, is cyclic and intermittent. This pattern is true not only for ovum development but for the function and structure of virtually the entire female reproductive system. In

human beings and other primates these cycles are called *menstrual cycles*.

Ovarian function

The ovary, like the testis, serves a dual purpose: (1) production of ova; (2) secretion of the female sex hormones, *estrogen* and *progesterone*.

Ovum and follicle growth

In Fig. 16-14, a cross section of an ovary, note the discrete cell clusters known as *primary follicles*. Each is composed of one ovum surrounded by a single layer of follicular *(granulosa)* cells (Fig. 16-14A). (As will be described in a subsequent section, the ovum is referred to by different names at each stage of development, but for clarity, we shall simply use the term ovum.) At birth, normal human ovaries contain an estimated one million such follicles, and no new ones appear after birth. Thus, in marked contrast to the male, the newborn female already has all the germ cells she will ever have. Only a few, perhaps 400, are destined to reach full maturity during her active reproductive life. All the others degenerate, so that few, if any, remain by the time she reaches menopause at approximately 50 years of age. One result of this is that the ova which are released (ovulated) near menopause are 30 to 35 years older than those ovulated just after puberty; it has been suggested that certain congenital defects much commoner among children of older women are the result of aging changes in the ovum.

The development of the follicle (Fig. 16-15) is characterized by an increase in size of the ovum and a proliferation of the surrounding granulosa cells (we do not know what mechanisms stimulate development of certain primary follicles, leaving most unstimulated to degenerate without ever showing a growth phase). The ovum becomes separated from the granulosa cells by a thick membrane, the *zona pellucida*; however, despite the presence of the zona pellucida, the granulosa cells remain intimately associated with the ovum by means of cytoplasmic processes which traverse the zona pellucida and form gap junctions with the ovum. Whether estrogen or other chemical messengers cross these gap junctions (which become uncoupled at the time of ovulation) is uncertain, but

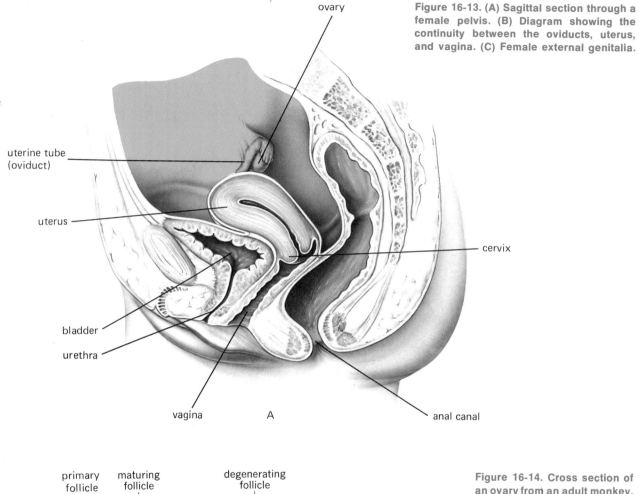

ovary

Figure 16-13. (A) Sagittal section through a female pelvis. (B) Diagram showing the continuity between the oviducts, uterus, and vagina. (C) Female external genitalia.

uterine tube (oviduct)

uterus

cervix

bladder

urethra

vagina

A

anal canal

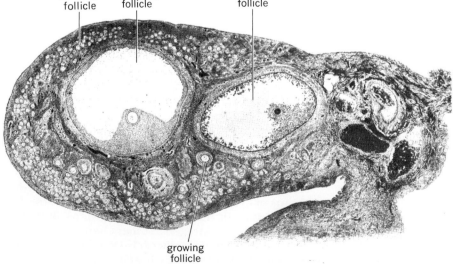

primary follicle

maturing follicle

degenerating follicle

Figure 16-14. Cross section of an ovary from an adult monkey. *(Adapted from Bloom and Fawcett.)*

growing follicle

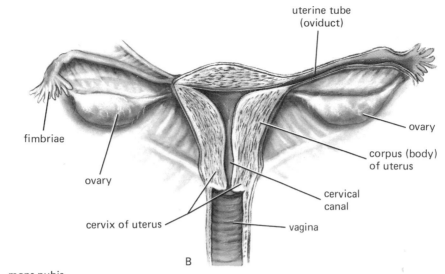

uterine tube
(oviduct)

fimbriae

ovary

ovary

cervix of uterus

cervical
canal

corpus (body)
of uterus

vagina

B

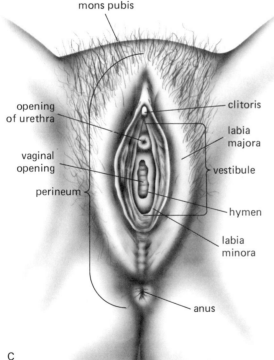

mons pubis

opening
of urethra

vaginal
opening

perineum

clitoris

labia
majora

vestibule

hymen

labia
minora

anus

C

it may be that the granulosa cells play a role in mediating the actions of gonadotropic hormones on the ovum analogous to the role of the Sertoli cells in the male.

As the follicle grows, new cell layers are formed, not only from mitosis of the original follicle cells but from the growth of specialized ovarian connec-

tive-tissue cells. Thus, the follicle, originally composed of the ovum and its surrounding layers of granulosa cells, becomes invested with additional outer layers of cells known as the *theca*. When the follicle reaches a certain diameter, a fluid-filled space, the *antrum*, begins to form in the midst of the granulosa cells as a result of fluid they secrete.

By the time the antrum begins to form, the ovum has reached full size. From this point on, the follicle grows in part because of continued follicular cell proliferation but largely because of the expanding antrum. Ultimately, the ovum, surrounded by the zona pellucida and several layers of granulosa cells (the cumulus), occupies a ridge projecting into the antrum. The antrum becomes so large (about 1.5 cm) that the completely mature follicle actually balloons out on the surface of the ovary. *Ovulation* occurs when the wall at the site of ballooning ruptures and the ovum, surrounded by its tightly adhering zona pellucida and cumulus, is carried out of the ovary by the antral fluid (Fig. 16-16). Many women experience varying degrees of abdominal pain at approximately the midpoint of their menstrual cycles, which has generally been presumed to represent abdominal irritation induced by the entry of follicular contents at ovulation. However, precise timing of ovulation has indicated that this time-honored concept may be wrong, and the cause of discomfort remains unclear.

In the human adult ovary, there are always several antrum-containing follicles of varying sizes,

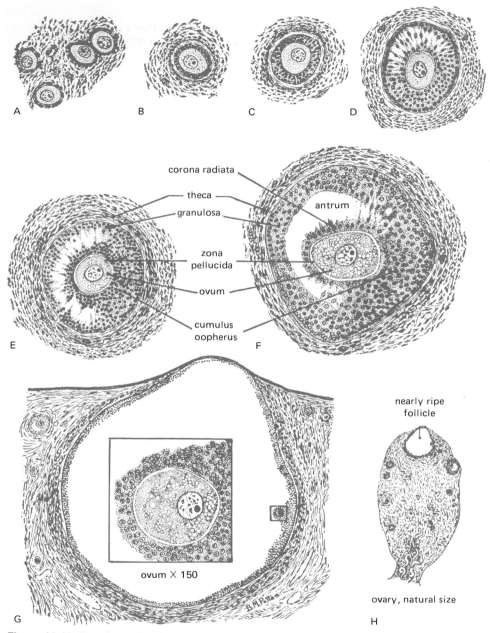

Figure 16-15. Drawings showing a series of stages in development of the human ovum and ovarian follicle. (Projection drawings in A–F, ×150; in G the follicle is ×15 but the inset detail of the ovum is ×150, like the other drawings in the series.) *(From B. M. Patten and B. M. Carlson, "Human Embryology," 3d ed., McGraw-Hill Book Company, New York, 1968.)*

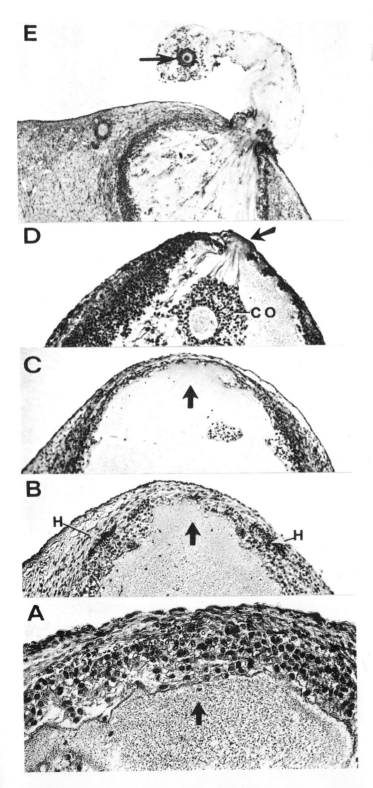

but during each cycle, normally only one follicle reaches the complete maturity just described, the process requiring approximately 2 weeks. All the other partially matured antral follicles undergo degeneration at some stage in their growth, the mechanism being unknown. On occasion (1 to 2 percent of all cycles), two or more follicles reach maturity, and more than one ovum may be ovulated. This is the commonest cause of multiple births; in such cases the siblings are fraternal, not identical.

Ovum division

The ova present at birth are the result of numerous mitotic divisions of the primitive ova, the *oogonia* (a term analogous to spermatogonia in the male), which occurred during early fetal development (Fig. 16-17). At some point the oogonia cease dividing in the fetus and then develop into primary oocytes (analogous to primary spermatocyte). These primary oocytes all begin the first division of meiosis but do not complete it; accordingly, all the ova present at birth are primary oocytes (thus, the name primary follicle) containing 46 replicated DNA strands and in a state of meiotic arrest. This division is completed only just before an ovum is about to be ovulated and is analogous to the division of the primary spermatocyte because each daughter cell receives 23 replicated chromosomes. However, in this division one of the two resulting cells, the *secondary oocyte*, retains virtually all the cytoplasm, the other (the "polar body") being very small. In this manner, the already full-size ovum loses half of its chromosomes but almost none of its nutrient-rich cytoplasm. The second cell division of meiosis occurs in the oviduct after ovulation (indeed, after penetration by a sperm), and the

Figure 16-16. Stages in ovulation. (A) Several hours before ovulation the granulosa is still quite thick, as is the theca. (B) One-half hour before ovulation decreased blood flow appears in the region of the bulge. There is a significant thinning out of the follicular cells and connective tissue. (C) A few minutes before rupture the follicular cells have almost disappeared, as has the connective tissue in the region of the bulge. (D) The ovarian wall ruptures, and the free cumulus oophorus (CO) streams toward the opening. (E) Ovulation is completed but the viscous antral fluid still adheres to the site of rupture. *(From R. O. Greep and L. Weiss, "Histology," 3d ed., McGraw-Hill Book Company, 1973.)*

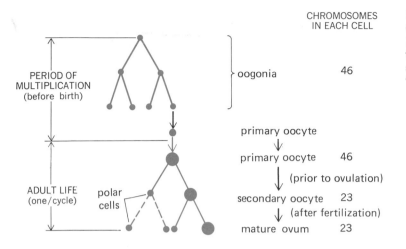

CHROMOSOMES
IN EACH CELL

oogonia 46

primary oocyte

primary oocyte 46

(prior to ovulation)

secondary oocyte 23

(after fertilization)

mature ovum 23

PERIOD OF
MULTIPLICATION
(before birth)

ADULT LIFE
(one/cycle)

polar
cells

Figure 16-17. Summary of ovum development. Compare with the male pattern of Fig. 16-3. Each primitive ovum produces only one mature ovum containing 23 chromosomes. It is a semantic oddity that the ovum is not termed "mature" until after fertilization occurs.

daughter cells each retain 23 chromosomes. Once again, one daughter cell, now the mature ovum, retains nearly all the cytoplasm. The net result is that each primary oocyte is capable of producing only one mature fertilizable ovum (Fig. 16-17); in contrast, each primary spermatocyte produces four viable spermatozoa.

Formation of corpus luteum

After rupture of the follicle and discharge of the antral fluid and the ovum, a transformation occurs within the follicle, which collapses, and the antrum fills with partially clotted fluid. The follicular cells enlarge greatly, and the entire glandlike structure is known as the *corpus luteum*. If the discharged ovum is not fertilized, i.e., if pregnancy does not occur, the corpus luteum reaches its maximum development within approximately 10 days and then rapidly degenerates. If pregnancy does occur, the corpus luteum grows and persists until near the end of pregnancy.

Ovarian hormones

Just as "androgen" refers to a group of hormones with similar actions, the term "estrogen" denotes not a single specific hormone but rather a group of steroid hormones which have similar effects on the female reproductive tract. These include estradiol, estrone, and estriol, of which the first is the major estrogen secreted by the ovaries. It is common to refer to any of them simply as estrogen and we shall follow this practice.

Estrogen is secreted to some extent by various ovarian cell types but primarily by the follicle cells (*not* the ovum) and by the corpus luteum. Progesterone, the second ovarian hormone, may be secreted in very minute amounts by the follicle cells, but its major source is the corpus luteum. The detailed physiology of these hormones will be described subsequently.

Cyclical nature of ovarian function

The length of a menstrual cycle varies considerably from woman to woman, averaging about 28 days. Day 1 is the first day of menstrual bleeding, and in a typical 28-day cycle, ovulation occurs around day 14. In terms of ovarian function, therefore, the menstrual cycle may be divided into two approximately equal phases: (1) the *follicular phase,* during which a single follicle and ovum develop to full maturity and (2) the *luteal phase,* during which the corpus luteum is the active ovarian structure. It must be stressed that the day of ovulation varies from woman to woman and frequently in the same woman from month to month.

Control of ovarian function

The basic factors controlling ovum development, ovulation, and formation of the corpus luteum is analogous to the controls described for testicular function in that the anterior pituitary gonado-

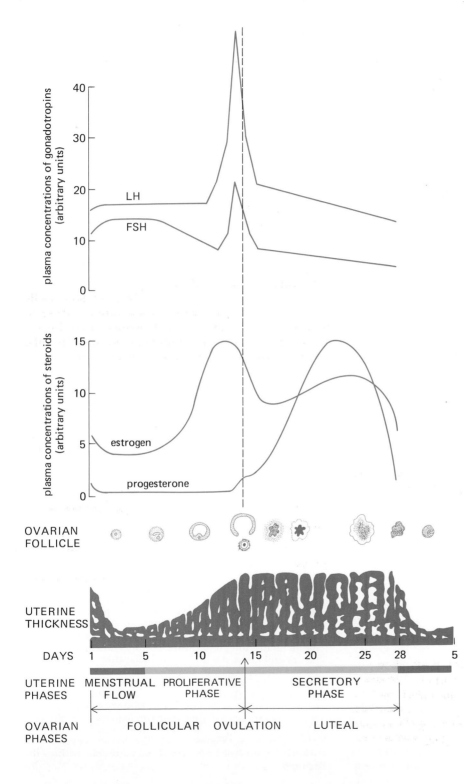

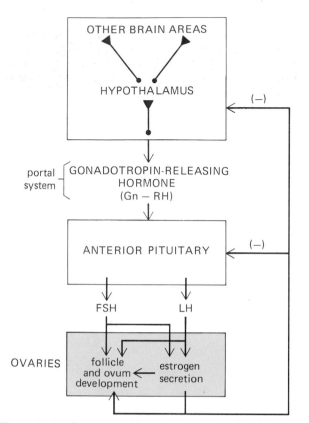

Figure 16-19. Summary of hormonal control of follicle and ovum development and estrogen secretion during the follicular phase of the menstrual cycle. Compare with the analogous pattern for the male (Fig. 16-12). The negative signs indicate that estrogen inhibits both the hypothalamus and the anterior pituitary. Estrogen reaches the developing ovum and follicle both by local diffusion and by release into the blood and recirculation to the ovaries.

tropins, FSH and LH, and the gonadal sex hormone, estrogen, play primary roles. However, the overall schema is more complex in the female since it includes a second important gonadal hormone (progesterone) and a hormonal cycling, quite different from the more stable continuous rates of male hormone secretion.

For purposes of orientation, let us look first at the changes in the blood concentrations of all four participating hormones during a normal menstrual cycle (Fig. 16-18). Note that FSH is slightly elevated in the early part of the follicular phase of the menstrual cycle and then steadily decreases throughout the remainder of the period except for a small transient midcycle peak. LH is quite con-

stant during most of the follicular phase but then shows a very large midcycle surge (peaking approximately 24 h before ovulation) followed by a progressive slow decline during the luteal phase. The estrogen pattern is more complex. After remaining fairly low and stable for the first week (as the follicle develops), it rises to reach a peak just before LH starts off on its surge. This peak is followed by a dip, a second rise (due to secretion by the corpus luteum), and, finally, a rapid decline during the last days of the cycle. The progesterone pattern is simplest of all; virtually no progesterone is secreted by the ovaries during the follicular phase, but very soon after ovulation, the developing corpus luteum begins to secrete progesterone, and from this point the progesterone pattern is similar to that for estrogen. It is hoped that, after the following discussion, the reader will understand how these changes are all interrelated to yield a self-cycling pattern.

**Control of follicle
and ovum development**

Growth of the follicle and ovum maturation depend directly upon the pituitary gonadotropins—follicle-stimulating hormone (FSH) and luteinizing hormone (LH)—and on estrogen, which may act, in large part, locally within the ovary. Because estrogen is secreted mainly by the follicle cells, its secretion rate progressively increases as the follicle enlarges. This secretion requires stimulation by FSH and LH.[1] Thus, both FSH and LH are directly required for normal follicle and ovum development as well as for estrogen secretion (Fig. 16-19), unlike the separation of function of the gonadotropins in the male.

---

[1] The interaction of FSH and LH in promoting estrogen secretion by the cells of the follicle offers an excellent illustration of endocrine complexity: LH acts upon the thecal cells of the follicle to stimulate the synthesis of androgens; these androgens diffuse across the basement membrane separating the thecal and granulosa layers and enter the granulosa cells; there they are converted into estrogen by enzymes whose activity is under the control of FSH. Thus, the overall process of estrogen production by the follicle requires the interplay of both types of follicle cells and both pituitary gonadotropins. It also points up the fact that the ovaries actually produce large quantities of "male" sex hormone; however, little of this androgen enters the blood because it is almost completely converted to estrogen within the follicle.

Table 16-1. Summary of important feedback effects of estrogen and progesterone

1 Estrogen, in *low* plasma concentrations, inhibits the hypothalamic neurons which secrete Gn-RH and the anterior pituitary cells which secrete FSH.
  Result: Negative-feedback inhibition of FSH secretion.
2 Estrogen, in *high* plasma concentrations, stimulates the hypothalamic neurons which secrete Gn-RH and the anterior pituitary cells which secrete LH.
  Result: Positive-feedback stimulation of LH secretion.
3 Estrogen and progesterone *together* inhibit the hypothalamic neurons which secrete Gn-RH and possibly the anterior pituitary as well.
  Result: Negative-feedback inhibition of FSH and LH secretion.

As in the male, the secretion of these pituitary hormones, in turn, requires stimulation by hypothalamic Gn-RH. The rate at which this releasing hormone is secreted by specific hypothalamic neurons during this first part of the cycle is probably determined by a combination of spontaneous neuronal automaticity, synaptic input from other brain areas, and negative-feedback inhibition by estrogen reaching the brain via the blood (Table 16-1). This latter input is responsible for the decline in FSH secretion seen during the second week of the cycle (Fig. 16-18). Estrogen exerts a negative-feedback inhibition of FSH secretion not only by an action on the hypothalamus but directly on the anterior pituitary cells, and this accounts for the fact (Fig. 16-18) that during these first two weeks, FSH and LH secretion are not completely parallel, as would otherwise have been expected, given the existence of only a single hypothalamic releasing hormone for the two pituitary hormones.

Control of ovulation

If one administers small quantities of FSH and LH each day to a woman whose pituitary has been removed (because of disease), she manifests normal follicle development, ovum maturation, and estrogen secretion, but she does not ovulate. If, on the other hand, after approximately 14 days of this therapy, she is given one or two larger injections of LH, ovulation occurs. This is precisely what happens in the normal woman; the follicle and ovum mature for 2 weeks under the influence of FSH, LH, and estrogen, and ovulation is triggered

by a rapid brief outpouring from the pituitary of larger quantities of LH (Fig. 16-18). This LH then causes ovulation by inducing increased synthesis of enzymes which catalyze the dissolution of the thin ovarian membrane at the bulge of the mature follicle. Thus, the midcycle surge of LH emerges as perhaps the single most decisive event of the entire menstrual cycle; indeed, it is the presence of this surge which most distinguishes the female pattern of pituitary secretion from the relatively unchanging pattern of the male.

What causes the LH surge? In the previous section we described how estrogen exerts a negative-feedback inhibition on the hypothalamus and pituitary; the fact is that this inhibitory effect occurs only when blood estrogen concentration is relatively low, as during the first part of the follicular phase. In contrast, high blood concentrations of estrogen, as exist during the estrogen peak of the late follicular phase, *stimulate* the hypothalamus to secrete Gn-RH and also act directly

Figure 16-20. Ovulation and corpus luteum formation are induced by the markedly increased LH, itself induced by the stimulatory effects of high levels of estrogen.

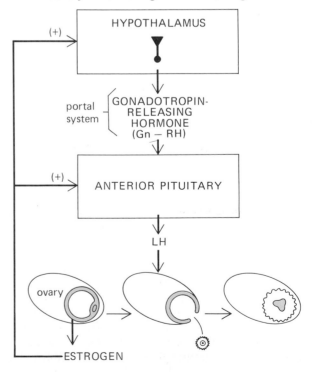

upon the pituitary to *enhance* the sensitivity of LH-releasing mechanisms to this Gn-RH (Table 16-1); these stimulatory effects are often called "positive feedback" effects. The net result is that, as estrogen secretion rises rapidly during the last half of the follicular phase, its blood concentration eventually becomes high enough to cause an outpouring of LH, which, in turn, induces ovulation.

Why is there only one surge per cycle, i.e., why does not the elevated estrogen existing throughout most of the luteal phase keep inducing LH surges? The answer is that, as a result of ovulation and succeeding corpus luteum formation, both induced by the LH surge, progesterone secretion begins and the progesterone-estrogen combination (unlike estrogen alone) exerts only inhibitory effects on the hypothalamic-pituitary mechanisms for LH release (Table 16-1).

Control of the corpus luteum

Once formed, maintenance of the corpus luteum requires some stimulatory support from LH, but the amount of LH needed is quite small. What causes the corpus luteum to degenerate if no pregnancy results? The blood concentration of LH shows no sudden decrease during the late luteal phase so that it is difficult to blame regression on any sudden withdrawal of LH. In some mammalian species, it seems that regression is actively induced by some substance (perhaps prostaglandin) produced by the nonpregnant uterus, but such seems not to be the case in women. There are other hypotheses, including the idea that the corpus luteum has a "built-in" life span of approximately 10 to 14 days and that it "self-destructs" unless prevented from doing so by the onset of pregnancy. How pregnancy does this will be described subsequently.

During its short life, the corpus luteum secretes large quantities of estrogen and progesterone. As mentioned above, these hormones exert a powerful negative-feedback inhibition (via the hypothalamus and anterior pituitary) of FSH and LH secretion. Accordingly, during the luteal phase of the cycle, pituitary gonadotropin secretion is reduced, which explains the diminished rate of follicular maturation during this second half of the cycle. With degeneration of the corpus luteum, blood estrogen and progesterone concentrations decrease, FSH and LH increase, and a new follicle is stimulated to mature.

Summary

The events described thus far in this section are summarized in Fig. 16-18, showing the ovarian and hormonal changes during a normal nonpregnant menstrual cycle:

1 Under the influence of FSH, LH, and estrogen, a single follicle and ovum reach maturity at about 2 weeks.
2 During the second of these two weeks, under the influence of LH and FSH, estrogen secretion by the follicle cells progressively increases.
3 For several days near midperiod, production of LH (and FSH, to a lesser extent) increases sharply as a result of the stimulatory (positive-feedback) effects of high levels of estrogen on the brain and the pituitary.
4 The high concentration of LH induces rupture of the ovarian membranes, and ovulation occurs (it is not known what role if any the increased FSH plays).
5 The reason for the midcycle decrease in estrogen secretion is not certain but may be due in part to disruption of the follicle.
6 The ruptured follicle is rapidly transformed into the corpus luteum, which secretes large quantities of both estrogen and progesterone.
7 The high blood concentrations of these hormones inhibit the release of LH and FSH, thereby lowering their blood concentrations and preventing the development of a new follicle or ovum during the last two weeks of the period; in addition, they prevent any additional LH surge.
8 Failure of ovum fertilization is associated with the degeneration of the corpus luteum during the last days of the cycle.
9 The disintegrating corpus luteum is unable to maintain its secretion of estrogen and progesterone, and their blood concentrations drop rapidly.
10 The marked decrease of estrogen and progesterone removes the inhibition of FSH (and LH) secretion.
11 The blood concentration of FSH (and LH) begins to rise, follicle and ovum development are stimulated, and the cycle begins anew.

Uterine changes in the menstrual cycle

Profound changes in uterine morphology occur during the menstrual cycle and are completely attributable to the effects of estrogen and progesterone (Table 16-2). Estrogen stimulates growth of the uterine smooth muscle (*myometrium*) and the glandular epithelium (*endometrium*) lining its cavity, and in addition, induces the synthesis of receptors for progesterone. Progesterone acts upon this estrogen-primed endometrium to convert it to an actively secreting tissue: The tubular glands become coiled and filled with secreted glycogen; the blood vessels become spiral and more numerous; various enzymes accumulate in the glands and connective tissue of the lining. All these changes are ideally suited to provide a hospitable environment for implantation of a fertilized ovum.

Estrogen and progesterone also have important effects on the mucus secreted by the *cervix*. Under the influence of estrogen alone, this mucus is abundant, clear, and nonviscous; all these characteristics are most pronounced at the time of ovulation and facilitate penetration by sperm. In contrast, progesterone causes the cervical mucus to become thick and sticky, in essence a "plug" which may constitute an important blockade against the entry of bacteria from the vagina—a further protection for the fetus should conception occur.

The endometrial changes throughout the normal menstrual cycle should now be readily understandable (Fig. 16-18).

1 The fall in blood progesterone and estrogen, which results from regression of the corpus luteum, deprives the highly developed endometrial lining of its hormonal support; the immediate result is profound constriction of the uterine blood vessels, which leads to diminished supply of oxygen and nutrients. Disintegration starts, the entire lining begins to slough, and the *menstrual flow* begins, marking the first day of the cycle.
2 After the initial period of vascular constriction, the endometrial arterioles dilate, resulting in hemorrhage through the weakened capillary walls; the menstrual flow consists of this blood mixed with endometrial debris. (Average blood loss per period equals 50 to 150 mL of blood.)

Table 16-2. Effects of female sex steroids

Effects of estrogens

1 Growth of ovaries and follicles
2 Growth and maintenance of the smooth muscle and epithelial linings of the entire reproductive tract
Also: *a* Oviducts: increased motility and ciliary activity
b Uterus: increased motility
secretion of abundant, clear cervical mucus
c Vagina: increased "cornification" (layering of epithelial cells)
3 Growth of external genitalia
4 Growth of breasts (particularly ducts)
5 Development of female body configuration: narrow shoulders, broad hips, converging thighs, diverging arms
6 Stimulation of fluid sebaceous gland secretions ("anti-acne")
7 Pattern of pubic hair (actual growth of pubic and axillary hair is androgen-stimulated)
8 Stimulation of protein anabolism and closure of the epiphyses (? due to stimulation of adrenal androgens)
9 (?) Sex drive and behavior
10 Reduction of blood cholesterol
11 Vascular effects (deficiency → "hot flashes")
12 Feedback effects on hypothalamus and anterior pituitary
13 Fluid retention

Effects of progesterone

1 Stimulation of secretion by endometrium; also induces thick, sticky cervical secretions
2 Stimulation of growth of myometrium (in pregnancy)
3 Decrease in motility of oviducts and uterus
4 Decrease in vaginal "cornification"
5 Stimulation of breast growth (particularly glandular tissue)
6 Inhibition of effects of prolactin on the breasts
7 Feedback effects on hypothalamus and anterior pituitary

3 The *menstrual phase* continues for 3 to 5 days, during which time blood estrogen levels are low.
4 The menstrual flow ceases as the endometrium repairs itself and then grows under the influence of the rising blood estrogen concentration; this period, the *proliferative phase*, lasts for the 10 days between cessation of menstruation and ovulation.
5 Following ovulation and formation of the corpus luteum, progesterone, acting in concert with estrogen, induces the secretory type of endometrium described above.
6 This period, the *secretory phase*, is terminated by disintegration of the corpus luteum, and the cycle is completed.

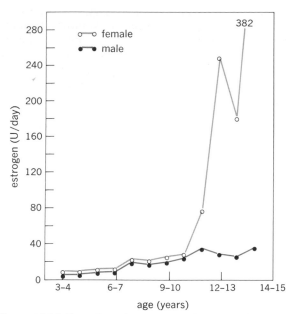

Figure 16-21. Excretion of estrogen in the urine of children, an indicator of the blood concentration of estrogen. *(Adapted from Nathanson et al.)*

It is evident that the phases of the menstrual cycle can be named either in terms of the ovarian or uterine events. Thus, the ovarian follicular phase includes the uterine menstrual and proliferative phases; the ovarian luteal phase is the same as the uterine secretory phase. The essential point is that the uterine changes simply reflect the effects of varying blood concentrations of estrogen and progesterone throughout the cycle. In turn, the secretory pattern of these hormones reflects the complex hypothalamic–anterior pituitary–ovarian interactions described previously.

Nonuterine effects of estrogen and progesterone

The uterine effects of the sex hormones described above represent only one set of a wide variety exerted by estrogen and progesterone (Table 16-2); all these effects are discussed in this chapter and are listed here for reference. The effects of estrogen in the female are analogous to those of testosterone in the male in that estrogen exerts dominant control over all the accessory sex organs and secondary sex characteristics. Estrogenic stimulation maintains the entire female genital tract—uterus, oviducts, vagina—the glands lining the tract, the external genitalia, and the breasts. It is responsible for the female body-hair distribution and the general female body configuration: narrow shoulders, broad hips, and the characteristic female "curves," the result of fat deposition in the hips, abdomen, and other places. Estrogen has much less general anabolic effect on nonreproductive tissues than testosterone but probably contributes to the general body growth spurt at puberty. Finally, as described above, estrogen is required for follicle and ovum maturation; its increased secretion at puberty, in concert with that of the pituitary gonadotropins, permits ovulation and the onset of menstrual cycles. For the rest of the woman's reproductive life, estrogen continues to support the ovaries, accessory organs, and secondary characteristics. Because its blood concentration varies so markedly throughout the cycle, associated changes in all these dependent functions occur, the uterine manifestations being the most striking.

As was true for testosterone, estrogen acts on the cell nucleus, and its biochemical mechanism of action appears to be at the level of the genes themselves. It should be reemphasized that estrogen is not a uniquely female hormone. Small and usually insignificant quantities of estrogen are secreted by the male adrenal and the testicular interstitial cells (Fig. 16-21), the latter probably being responsible for the breast enlargement so commonly observed in pubescent boys; apparently, the rapidly developing interstitial cells release significant quantities of estrogen along with the much larger amounts of testosterone.

Progesterone is present in significant amounts only during the luteal phase of the menstrual cycle, and its effects are less widespread than those of estrogen, the endometrial changes being the most prominent. Progesterone also exerts important effects on the breasts, the oviducts, and the uterine smooth muscle, the significance of which will be described later. Progesterone also causes a transformation of the cells lining the vagina (decreased cornification), and the microscopic examination of some of these cells provides an indicator that ovulation has or has not occurred. Note that in this regard, and others in Table 16-2, progesterone exerts an "anti-estrogen effect" (it probably does so by inhibiting the synthesis of estrogen receptors).

Another indicator that ovulation has occurred is a small rise (approximately 0.5°C) in body temperature that usually occurs at this time and persists throughout the luteal phase. This change was previously ascribed to an action of progesterone on temperature regulatory centers in the brain, but this is probably incorrect and the actual cause is not known.

Female sexual response

The female response to sexual intercourse is very similar to that of the male in that it is characterized by marked vasocongestion and muscular contraction in many areas of the body. For example, mounting sexual excitement is associated with engorgement of the breasts and erection of the nipples, resulting from contraction of muscle filaments in them. The *clitoris* (Fig. 16-13), which is a homologue of the penis and is composed primarily of erectile tissue and endowed with a rich supply of sensory nerve endings, also becomes erect. During intercourse, the vaginal epithelium becomes highly congested and secretes a mucuslike lubricant. The final stage of female sexual excitement may be the process of orgasm, as in the male; if no orgasm occurs, there is a slow resolution of the physical changes and sexual excitement. The female has no counterpart to male ejaculation, but with this exception, the physical correlates of orgasm are very similar in the sexes: there is a sudden increase in skeletal muscle activity involving almost all parts of the body; the heart rate and blood pressure increase; the female counterpart of male genital contraction is transient rhythmical contraction of the vagina and uterus. Psychologically, the woman feels a sudden instant of release, followed immediately by an intense awareness of clitoral and pelvic sensation radiating upward, often reported as a feeling of opening up. Finally, a feeling of warmth and relaxation ensues. Orgasm seems to play no essential role in assuring fertilization.

A final question related to the female sexual response is sex drive. Incongruous as it may seem, sexual desire in adult women is more dependent upon androgens than estrogen. Thus libido is usually not altered by removal of the ovaries (or the physiological analog, menopause). In contrast, sexual desire is greatly reduced by adrenalectomy, since these glands are the major source of androgens in women. Finally, women receiving large doses of testosterone (for the treatment of breast cancer) generally report a large increase in sexual desire.

This completes our survey of normal reproductive physiology in the nonpregnant female. In weaving one's way through this maze, it is all too easy to forget the prime function subserved by this entire system, namely, reproduction. Accordingly, we must now return to the mature ovum we left free in the abdominal cavity, find it a mate, and carry it through pregnancy, delivery, and breast feeding.

Pregnancy

Following ejaculation into the vagina, the sperm live approximately 48 h; after ovulation, the ovum remains fertile for 10 to 15 h. The net result is that for pregnancy to occur, sexual intercourse must be performed no more than 48 h before or 15 h after ovulation (these are only average figures, and there is probably considerable variation in the survival time of both sperm and ovum). However, even these short time limits are probably too generous since, although fertile, the older ova manifest a variety of malfunctions after fertilization, frequently resulting in their death.

Ovum transport

At ovulation, the ovum is extruded from the ovary, and its first mission is to gain entry into the oviduct. The end of the oviduct has long, fingerlike projections lined with ciliated epithelium (Fig. 16-13). At ovulation, the smooth muscle of these projections causes them to pass over the ovary while the cilia beat in waves toward the interior of the duct; these motions sweep in the ovum as it emerges from the ovary and start it on its trip toward the uterus.

Once in the oviduct, the ovum moves rapidly for several minutes, propelled by cilia and by contractions of the duct's smooth muscle coating; the contractions soon diminish, and ovum movement becomes so slow that it takes several days to reach the uterus. Thus, fertilization must occur in the oviduct because of the short life span of the unfertilized ovum.

Sperm transport

Transport of the sperm to the site of fertilization within the oviduct is so rapid that the first sperm arrive within 30 min of ejaculation. This is far too rapid to be accounted for by the sperm's own motility; indeed, the movement produced by the sperm's tail is probably essential only for the final stages of approach and penetration of the ovum. The act of intercourse itself provides some impetus for transport out of the vagina into the uterus because of the fluid pressure of the ejaculate and the pumping action of the penis during orgasm. After intercourse, the primary transport mechanism may be the contractions of the uterine and oviduct musculature. The factors controlling these muscular contractions remain obscure but may involve prostaglandins in the semen. The mortality rate of sperm during the trip is huge; of the several hundred million deposited in the vagina, only a few thousand reach the oviduct. This is one of the major reasons that there must be so many sperm in the ejaculate to permit pregnancy.

In addition to aiding transport of sperm, the female reproductive tract exerts a second critical effect on them, namely, the conferring upon them of the capacity for fertilizing the egg. Although, as we have mentioned, sperm gain some degree of maturity during their stay in the epididymis, they are still not able to penetrate the zona pellucida surrounding the ovum until they have resided in the female tract for some period of time. The mechanism by which this process, known as *capacitation*, occurs is still very poorly understood, but the result is a change in the sperm membrane which permits release of the acrosomal enzymes upon contact.

Entry of the sperm into the ovum

The sperm makes initial contact with the cells surrounding the ovum presumably by random motion, there being no good evidence for the existence of "attracting" chemicals. Having made contact, the sperm rapidly moves between these adhering cells of the cumulus and through the zona pellucida by releasing from its acrosomal cap enzymes which break down cell connections and intermolecular bonds.

Once through the zona pellucida, the sperm makes contact with the ovum plasma membrane, fuses with it, and slowly passes through into the cytoplasm (frequently losing its tail in the process). Upon penetration by the sperm, the ovum completes its last division, and the one daughter cell with practically no cytoplasm—the second polar body—is extruded. The nuclei of the sperm and ovum then unite; the cell now contains 46 chromosomes, and *fertilization* is complete. However, viability depends upon stopping the entry of additional sperm. The mechanism of this "block to polyspermy" is as follows: The fusion of the sperm and ovum membrane causes secretory vesicles located around the ovum's periphery to fuse with the ovum's plasma membrane and release their contents into the space between the plasma membrane and zona pellucida; some of these molecules are enzymes which break down binding sites for sperm in the zona pellucida, thereby preventing additional sperm from moving through the zona; others may cause the zona or plasma membrane of the ovum, itself, to become impenetrable. The immediate trigger for the release of these vesicles is a transient increase in cytoplasmic calcium resulting from membrane changes induced by fusion of sperm and ovum; i.e., calcium is acting as a "second messenger" in this system (the sperm being the first). This calcium very likely also triggers activation of ovum enzymes required for the ensuing cell divisions and embryogenesis.

The fertilized egg is now ready to begin its development as it continues its passage down the oviduct to the uterus. If fertilization had not occurred, the ovum would slowly disintegrate and usually be phagocytized by the lining of the uterus. Rarely a fertilized ovum remains in the oviduct, where implantation may take place; such tubal pregnancies cannot succeed because of lack of space for the fetus to grow, and surgery may be necessary. Even more rarely a fertilized ovum may move in the wrong direction and be expelled into the abdominal cavity where implantation may take place and (rarely) proceed to term, the infant being delivered surgically.

Early development, implantation, and placentation

During the leisurely 3- to 4-day passage through the oviduct, the fertilized ovum undergoes a number of cell divisions (identical twins result when a single fertilized ovum at a very early stage of de-

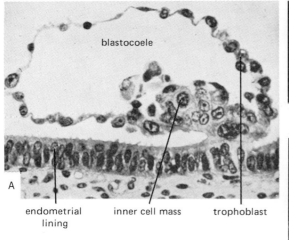

blastocoele

A

endometrial   inner cell mass   trophoblast
lining

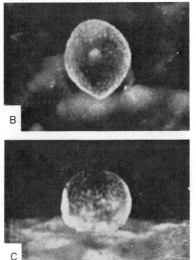

B

C

Figure 16-22. Photomicrographs showing initial implantation of 9-day monkey blastocysts. (A) Cross section of the embryo and uterine lining. (B) An entire blastocyst attached to the uterus, viewed from above. (C) The same embryo viewed from the side. [*From C. H. Heuser and G. L. Streeter,* Carnegie Contrib. Embryol., **29**:15 *(1941)*.]

velopment becomes completely divided into two independently growing cell masses) and after reaching the uterus, it floats free in the intra-uterine fluid (from which it receives nutrients) for several more days, all the while undergoing cell division. This entire time span corresponds to days 14 to 21 of the typical menstrual cycle; thus while the ovum is undergoing fertilization and early development, the uterine lining is simultaneously being prepared by estrogen and progesterone to receive it. On approximately the twenty-first day of the cycle, i.e., 7 days after ovulation, *implantation* occurs. The fertilized ovum has by now developed into a ball of cells surrounding a recently formed central fluid-filled cavity and is known as a *blastocyst.*

A section of the blastocyst is shown in Fig. 16-22; note the disappearance of the zona pellucida, an event necessary for implantation. The inner cell mass is destined to develop into the fetus itself, whereas the outer lining of cells is already differentiating into specialized cells, *trophoblasts,* which will form the nutrient membranes for the fetus, as described below. Once the zona pellucida has disintegrated, the trophoblastic layer rapidly enlarges and makes contact with the uterine wall. The trophoblast cells of blastocysts recovered from the uterus have been found to be quite sticky, particularly in the region overlying the inner cell mass; it is this portion which adheres to the endometrium upon contact and initiates implantation.

This initial contact somehow induces rapid development of the trophoblasts, tongues of which penetrate between endometrial cells. By this means, the *embryo* is soon completely embedded within the endometrium (Fig. 16-23), the nutrient-rich cells of which provide the metabolic fuel and raw materials for the developing embryo. This system, however, is adequate to provide for the embryo only during the first few weeks when it is very small. The system taking over after this is the fetal circulation and *placenta* (after the end of the second month, the embryo is known as a *fetus*).

The placenta is a combination of interlocking fetal and maternal tissues which serves as the organ of exchange between mother and fetus. The expanding trophoblastic layer breaks down endometrial capillaries, allowing maternal blood to ooze into the spaces surrounding it; clotting is prevented by the presence of some anticoagulant substance produced by the trophoblasts. Soon the trophoblastic layer completely surrounds and projects into these oozing areas (Fig. 16-23). By this time, the developing embryo has begun to send out into these trophoblastic projections blood vessels which are all branches of larger vessels, the *umbilical arteries and veins,* communicating with the main intraembryonic arteries and veins via the *umbilical cord* (Fig. 16-24). Five weeks after implantation, this system has become well established, the fetal heart has begun to pump blood, and the entire mechanism for nutrition of the fetus

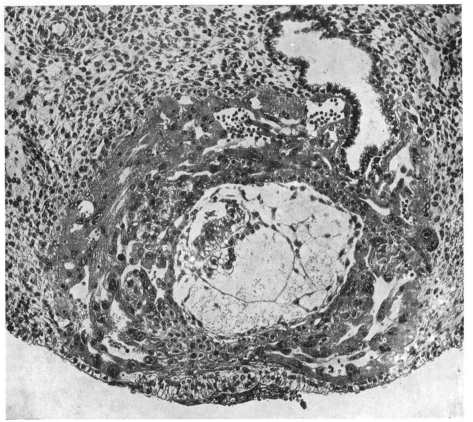

Figure 16-23. Eleven-day human embryo, completely embedded in the uterine lining. [*From A. T. Hertig and J. Rock,* Carnegie Contrib. Embryol., **29:** *127 (1941).*]

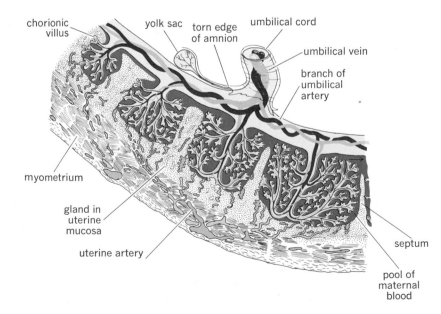

chorionic villus

yolk sac

torn edge of amnion

umbilical cord

umbilical vein

branch of umbilical artery

myometrium

gland in uterine mucosa

uterine artery

septum

pool of maternal blood

Figure 16-24. Schematic diagram: interrelations of fetal and maternal tissues in formation of placenta. The placenta becomes progressively more developed from left to right. *(From B. M. Patten and B. M. Carlson "Human Embryology," 3d ed.,* © *1968 by McGraw-Hill, Inc., New York.)*

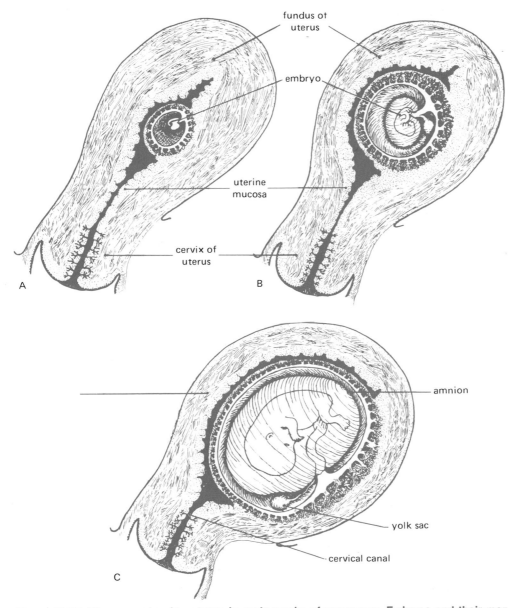

Figure 16-25. Diagrams showing uterus in early weeks of pregnancy. Embryos and their membranes are drawn to actual size. Uterus is within actual size range. (A) At fertilization age of 3 weeks; (B) 5 weeks; (C) 8 weeks. *(From B. M. Patten and B. M. Carlson, "Human Embryology," 3d ed., McGraw-Hill Book Company, New York, 1968.)*

is in operation. Waste products move from the fetal blood across the placental membranes into the maternal blood; nutrients move in the opposite direction. Many substances, such as oxygen and carbon dioxide, move by simple passive diffusion, whereas other substances are carried by active-transport mechanisms in the placental membranes.

At first, as described above, the trophoblastic projections simply lie in endometrial spaces filled with blood, lymph, and some tissue debris. This basic pattern is retained throughout pregnancy, but many structural alterations have the net effect of making the system more efficient; e.g., the trophoblastic layer thins, the distance between maternal and fetal blood thereby being reduced. It must be emphasized that there is exchange of materials between the two blood streams but no actual mingling of the fetal and maternal blood; the maternal blood enters the placenta via the uterine artery, percolates through the spongelike endometrium, and then exits via the uterine veins; similarly, the fetal blood never leaves the fetal vessels. In several ways, the system is analogous to the artificial kidney described in Chap. 13 with the endometrial vascular spaces serving as the bath through which the fetal vessels course like the dialysis tubing. A major difference is that active transport is an important component of placental function. Moreover, the placenta must serve not only as the embryo's kidney but as its gastrointestinal tract and lungs as well. Finally, as will be discussed below, the placenta (probably the trophoblasts) secretes several hormones of crucial importance for maintenance of pregnancy.

The fetus, floating in its completely fluid-filled cavity and attached by the umbilical cord to the placenta (Fig. 16-25), develops into a viable infant during the next 9 months (babies born prematurely, often as early as 7 months, frequently survive). (Description of intrauterine development is beyond the scope of this book but can be found in any standard textbook of human embryology.)

A point of great importance is that the developing embryo and fetus is subject to considerable influence by a host of factors (noise, chemicals, viruses, etc.) affecting the mother. Via the placenta, drugs taken by the mother can reach the fetus and influence its growth and development. The thalidomide disaster was our major reminder of this fact in recent years, as is the growing number of babies who suffer heroin withdrawal symptoms after birth as a result of their mothers' drug use during pregnancy. Two other examples: Lead and DDT cross the placenta very easily; we do not know the potential effects, if any, on the fetus of many agents present in the environment.

Hormonal changes during pregnancy

Throughout pregnancy, the specialized uterine structures and functions depend upon high concentrations of circulating estrogen and progesterone (Fig. 16-26). During approximately the first 2 months of pregnancy, almost all these steroid hormones are supplied by the extremely active corpus luteum formed after ovulation. Recall that if pregnancy had not occurred, this glandlike structure would have degenerated within 2 weeks after ovulation; in contrast, continued corpus luteum growth and steroid secretion occur when pregnancy occurs. Persistence of the corpus luteum is essential since continued secretion of estrogen and progesterone is required to sustain the uterine lining and prevent menstruation (which does not occur during pregnancy). We have mentioned our lack of knowledge of the factors causing corpus luteum degeneration during a nonpregnant cycle; its persistence during pregnancy is due, at least in part, to a hormone from the blastocyst and placenta called *chorionic gonadotropin* (CG). Almost immediately after beginning their endometrial invasion, the trophoblastic cells start to secrete

Figure 16-26. Urinary excretion of estrogen, progesterone, and chorionic gonadotropin during pregnancy. Urinary excretion rates are an indication of blood concentrations of these hormones.

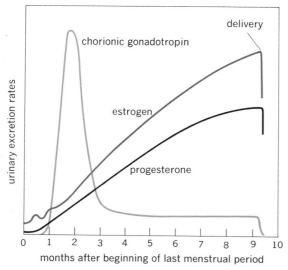

CG into the maternal blood. This protein hormone has properties very similar to those of LH (although it is chemically different) and it strongly stimulates steroid secretion by the corpus luteum. Recall that secretion of both LH and FSH is powerfully inhibited by estrogen and progesterone; therefore, the blood concentrations of these pituitary gonadotropins remain extremely low throughout pregnancy; by this means, further follicle development, ovulation, and menstrual cycles are eliminated for the duration of the pregnancy.

The detection of CG in urine or plasma is the basis of most pregnancy tests. The secretion of CG increases rapidly during early pregnancy, reaching a peak at 60 to 80 days after the end of the last menstrual period; it then falls just as rapidly, so that by the end of the third month, it has reached a low but definitely detectable level which remains relatively constant for the duration of the pregnancy. Associated with this falloff of CG secretion, the placenta itself begins to secrete large quantities of estrogen and progesterone. The very marked increases in blood steroids during the last 6 months of pregnancy are due almost entirely to the placental secretion. The corpus luteum remains, but its contribution is dwarfed by that of the placenta; indeed, removal of the ovaries during the last 7 months has no effect at all upon the pregnancy, whereas removal during the first 2 months causes immediate loss of the fetus (abortion).

An important and clinically useful aspect of placental steroid secretion is that the placenta does not have the enzymes required for the complete synthesis of the major estrogen of pregnancy, estriol. However, the enzymes it lacks are present in the adrenal cortex of the fetus; therefore, the placenta and fetal adrenals, working together, with intermediates transported between them via the fetal circulation, produce the estriol required to maintain the pregnancy. Since the fetus is necessary for estriol production, measurement of this hormone in maternal blood provides a means for monitoring the well-being of the fetus.

Finally, the placenta not only produces steroids and CG but several other hormones as well. The best-documented one at present is a hormone which has effects very similar to those of growth hormone and prolactin. This hormone (chorionic somatomammotropin) may play an important role in the mother—maintaining a positive protein balance, mobilizing fats for energy, stabilizing plasma glucose at relatively high levels to meet the needs of the fetus, and facilitating development of the breasts.

Of the numerous other physiological changes, hormonal and nonhormonal, in the mother during pregnancy, many, such as increased metabolic rate and appetite, are obvious results of the metabolic demands placed upon her by the growing fetus. Of great importance is salt and water metabolism. During a normal pregnancy, body sodium and water increase considerably, the extracellular volume alone rising by approximately 1 L. At present, it appears that important factors causing fluid retention are renin, aldosterone, ADH, and estrogen, all acting upon the kidneys. Some women retain abnormally great amounts of fluid and manifest protein in the urine and hypertension, which if severe enough may cause convulsions. These are the symptoms of the disease known as *toxemia of pregnancy*. Since it can usually be well controlled by salt restriction, in the United States it is seen primarily in the lower socioeconomic segments of the population which do not obtain adequate medical care during pregnancy. Despite the obvious association with salt retention, all attempts to determine the factors responsible for the disease have failed.

Parturition (delivery of the infant)

A normal human pregnancy lasts approximately 40 weeks, although many babies are born 1 to 2 weeks earlier or later. Delivery of the infant, followed by the placenta, is produced by strong rhythmical contractions of the uterus. Actually, beginning at approximately 30 weeks, weak and infrequent uterine contractions occur, gradually increasing in strength and frequency. During the last month, the entire uterine contents shift downward so that the baby is brought into contact with the outlet of the uterus, the cervix. In over 90 percent of births, the baby's head is downward and acts as the wedge to dilate the cervical canal. By the onset of labor, the uterine contractions have become coordinated and quite strong (although usually painless at first) and occur at approximately 10- to 15-min intervals. Usually during this period or before, the membrane surrounding the fetus ruptures, and the intrauterine fluid escapes

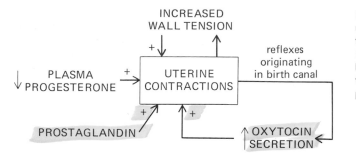

Figure 16-27. Factors stimulating uterine contractions during parturition. Note the positive-feedback nature of several of the inputs. What initiates parturition is not known.

out the vagina. As the contractions, which begin in the upper portion and sweep down the uterus, increase in intensity and frequency, the cervical canal is forced open to a maximum diameter of approximately 10 cm. Until this point, the contractions have not moved the fetus out of the uterus but have served only to dilate the cervix. Now the contractions move the fetus through the cervix and vagina. At this time the mother, by bearing down to increase abdominal pressure, can help the uterine contractions to deliver the baby. The umbilical vessels and placenta are still functioning, so that the baby is not yet on its own, but within minutes of delivery both the infant's and mother's placental vessels completely contract, the entire placenta becomes separated from the underlying uterine wall, and a wave of uterine contractions delivers the placenta (the afterbirth). Ordinarily, the entire process from beginning to end proceeds automatically and requires no real medical intervention, but in a small percentage of cases, the position of the baby or some maternal defect can interfere with normal delivery. The position is important for several reasons: (1) If the baby is not oriented head first, another portion of his body is in contact with the cervix and is generally a far less effective wedge; (2) because of its large diameter compared with the rest of the body, if the head went through the canal last, it might be obstructed by the cervical canal, leading to obvious problems when the baby attempts to breathe; (3) if the umbilical cord becomes caught between the birth canal and the baby, mechanical compression of the umbilical vessels can result. Despite these potential difficulties, however, most babies who are not oriented head first are born normally and with little difficulty.

What mechanisms control the events of parturition? Let us consider a set of fairly well-established facts:

1 The uterus is composed of smooth muscle capable of autonomous contractions and having inherent rhythmicity, both of which are facilitated by stretching the muscle.
2 The efferent neurons to the uterus are of little importance in parturition since anesthetizing them in no way interferes with delivery.
3 Progesterone exerts a powerful inhibitory effect upon uterine contractility (apparently by decreasing its sensitivity to estrogen and oxytocin). Shortly before delivery, the secretion of progesterone sometimes drops, perhaps due to "aging" changes in the placenta.
4 *Oxytocin,* one of the hormones released from the posterior pituitary, is an extremely potent uterine-muscle stimulant. Oxytocin is reflexly released as a result of afferent input into the hypothalamus from receptors in the uterus, particularly the cervix.
5 The pregnant uterus contains several prostaglandins, at least one of which is a profound stimulator of uterine smooth muscle; an increase in the release of this substance has been demonstrated during labor.

These facts can now be put together in a unified pattern, as shown in Fig. 16-27. The precise contributions of each of these factors are unclear; moreover, we cannot answer the crucial question: Which factor (if any) actually *initiates* the process? Once started, the uterine contractions exert a positive-feedback effect upon themselves via reflex stimulation of oxytocin and local facilitation of their inherent contractility; but what *starts* the contractions? The decrease in progesterone cannot be essential since it simply does not occur in most women. Nor is uterine distension or the presence of a fetus a requirement, as attested by the remarkable fact that typical "labor" begins at the expected time in some animals from which the fetus has been removed weeks previously. A primary role for prostaglandin has been touted but the evidence is far from conclusive. Regardless of their relative contributions to normal parturition, both prostaglandins and oxytocin are useful clinically in artificially inducing labor.

Another candidate for initiator of parturition is the hormone, cortisol, secreted by the adrenal cortex. There is a marked rise in cortisol secretion during the last days of pregnancy, and cortisol is known to enhance the action on the myometrium of several of the factors shown in Fig. 16-27.

Lactation

Perhaps no other process so clearly demonstrates the intricate interplay of various hormonal control mechanisms as milk production. The endocrine control has been established by numerous investigations and observations, none more striking than that in 1910 of Siamese twins: when one twin became pregnant, both women lactated after delivery.

The *breasts* are formed of epithelium-lined ducts which converge at the *nipples*. These ducts branch all through the breast tissue and terminate in saclike glands, called *alveoli*. The alveoli, which secrete the milk, look like bunches of grapes with stems terminating in the ducts. The alveoli and the ducts immediately adjacent to them are surrounded by specialized contractile cells called *myoepithelial cells*. Before puberty, the breasts are small with little internal glandular structure. With the onset of puberty, the increased estrogen causes a marked enhancement of duct growth and branching, but little development of the alveoli, and much of the breast enlargement at this time is due to fat deposition. Progesterone secretion also commences at puberty (during the luteal phase of each cycle) and this hormone also contributes to breast growth.

During each menstrual cycle, the breasts undergo fluctuations in association with the changing blood concentrations of estrogen and progesterone, but these changes are small compared with the marked breast enlargement which occurs during pregnancy as a result of the stimulatory effects of high plasma concentrations of estrogen, progesterone, prolactin, and chorionic somatomammotropin. This last hormone, as described earlier, is secreted by the placenta whereas pro-

lactin is secreted by the anterior pituitary. Under the influence of all these hormones, the alveolar structure becomes fully developed.

Prolactin is unique among the anterior pituitary hormones in that its secretion is controlled mainly by a hypothalamic inhibiting hormone (prolactin inhibiting hormone, PIH)[1]; thus an increase in prolactin secretion is achieved by inhibiting the hypothalamic neurons which release PIH into the portal vessels. Prolactin secretion is low prior to puberty (presumably because of a high rate of PIH release) but increases considerably at puberty in girls (but not in males—the function of prolactin in men is unknown). This increase is due to the stimulatory effects of estrogen (it is not known whether the estrogen acts on the hypothalamus to inhibit PIH secretion or on the anterior pituitary to reduce the sensitivity of the prolactin-secreting cells to PIH). During pregnancy, there is a marked increase in prolactin secretion (due to the elevation in plasma estrogen) beginning at about eight weeks, and rising throughout the remainder of the pregnancy.

Prolactin is the single most important hormone promoting milk production. Yet, despite the fact that prolactin is elevated and the breasts markedly enlarged and fully developed as pregnancy progresses, there is no secretion of milk. This is because estrogen and progesterone, in large concentration, prevent milk production by inhibiting the action of prolactin on the breasts (Table 16-2). Thus, although estrogen causes an increase in the secretion of prolactin and acts with it in promoting breast growth and differentiation, it (along with progesterone) is antagonistic to prolactin's ability to induce milk secretion. Delivery removes the sources of the large amounts of sex steroids and, thereby, the inhibition of milk production.

The major factor maintaining the secretion of prolactin during lactation is reflex input to the hypothalamus from receptors in the nipples which are stimulated by suckling (Fig. 16-28). After several months, however, plasma prolactin concentrations return to prepregnancy levels and lacta-

[1] As mentioned in Chap. 9, prolactin secretion is controlled not only by PIH but by one or perhaps even two stimulatory hypothalamic releasing hormones. Present evidence indicates, however, that PIH is normally the single most important input and we have, for simplicity, ignored any possible role of the stimulatory hormones in the control of prolactin during lactation.

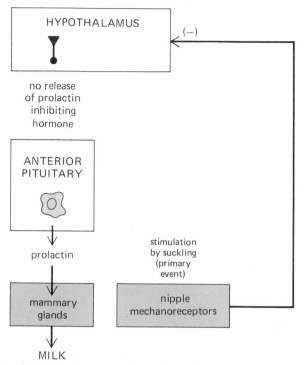

Figure 16-28. Nipple suckling reflex. The neural pathway is schematic; actually multiple interneurons are involved. Start reading this figure with the primary event—stimulation of the nipple mechanoreceptors.

tion is no longer dependent upon this hormone. Milk production ceases soon after the mother stops nursing her infant but continues uninterrupted for years if nursing is continued.

One final reflex process is essential for nursing. Milk is secreted into the lumen of the alveoli, but because of their structure the infant cannot suck the milk out. It must first be moved into the ducts, from which it can be sucked. This process is called *milk let-down* and is accomplished by contraction of the myoepithelial cells surrounding the alveoli; the contraction is directly under the control of oxytocin, which is reflexly released very rapidly (within one minute) by suckling (Fig. 16-29), just like prolactin. Higher brain centers also exert important influence over oxytocin release: A nursing mother may actually leak milk when hearing her baby cry. In view of the central role of the nervous system in lactation reflexes, it is no wonder that psychological factors can interfere with a woman's ability to nurse.

The end result of all these processes, the milk, contains four major constituents: water, protein, fat, and the carbohydrate lactose. The mammary alveolar cells must be capable of extracting the raw materials—amino acids, fatty acids, glycerol, glucose, etc.—from the blood and build them into the higher-molecular-weight substances. These synthetic processes require the participation of prolactin, insulin, growth hormone, cortisol, and probably other hormones—an amazing coordination.

Another important neuroendocrine reflex triggered by suckling (and mediated, in part, by prolactin) is inhibition of the hypothalamic-pituitary-ovarian chain at a variety of steps, with resultant block of ovulation. This inhibition apparently is relatively short-lived in many women, and approximately 50 percent begin to ovulate despite continued nursing. Pregnancy is common in women lulled into false security by the mistaken belief that failure to ovulate is always associated with nursing.

Fertility control

Table 16-3 summarizes possible means of preventing fertility. We present it because it should

Figure 16-29. Suckling-reflex control of oxytocin secretion and milk let-down. As with Fig. 16-28, begin following the figure with the nipple mechanoreceptors.

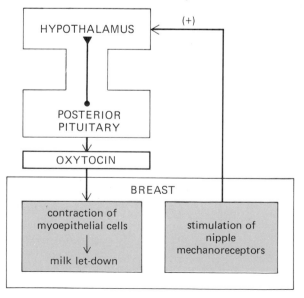

Table 16-3.

Possible means of preventing fertility in men*

I Interference with sperm survival
   a Prevention of maturation process in epididymis
   b Prevention of function of accessory glands
      1 Prevention of androgen action on accessory glands
      2 Prevention of formation of accessory-gland secretion
      3 Prevention of accessory-gland secretion from entering uretha
   c Creation of hostile environment to sperm in vas deferens ing urethra

II Interference with testicular function
   a Prevention of androgen action on seminiferous tubules
   b Prevention of action of FSH on seminiferous tubules
   c Prevention of sperm division

III Interference with hypothalamic or pituitary function
   a Prevention of FSH secretion
   b Prevention of action of Gn-releasing hormone on pituitary
   c Prevention of secretion of Gn-releasing hormone
   d Use of inhibin to inhibit FSH secretion

Possible means of preventing fertility in women†

I Interference with ovarian function
   a Prevention of initiation of follicle growth (no response to early FSH rise)
   b Prevention of response to ovulatory LH surge
   c Prevention of maturation of follicle or maturation of ova (e.g., inhibition of meiotic division)
   d Prevention of corpus luteum formation
   e Prevention of estrogen secretion
   f Prevention of progesterone secretion
   g Prevention of the maintenance of the corpus luteum during early pregnancy

II Interference with hypothalamic or pituitary function
   a Prevention of FSH or LH secretion, or both
   b Prevention of action of Gn-releasing hormone on pituitary
   c Prevention of secretion of Gn-releasing hormone
   d Alteration of estrogen or progesterone action on hypothalamus or pituitary (too much or too little)
   e Abolishment of extrahypothalamic central nervous factors required for normal reproductive hypothalamic function

III Interference with sperm action
   a Prevention of sperm entrance to vagina
   b Prevention of sperm entrance to uterus
   c Prevention of sperm entrance to oviducts
   d Creation of hostile environment to sperm in vagina, uterus, or oviducts
   e Prevention of sperm capacitation
   f Prevention of sperm penetration of ovum by action on sperm

IV Interference with ovum action
   a Prevention of ova release from ovary
   b Prevention of ova entrance to oviducts
   c Creation of hostile environment for ova in oviducts
   d Prevention of sperm penetration of ovum by action on ova

V Prevention of survival of fertilized ova
   a Prevention of fertilized ova from undergoing division
   b Alteration of migration along oviduct (too fast or too slow)
   c Prevention of implantation of blastocyst in uterine wall
   d Destruction or expulsion of embryo after implantation in uterine wall
   e Prevention of CG secretion by blastocyst and placenta
   f Prevention of sex steroid secretion by placenta

* Not listed are surgical interventions removing the source of sperm, such as castration, or preventing egress of sperm, such as vasectomy.

† Not listed is surgical removal of any of the component organs (e.g., uterus, oviducts, ovaries).

serve to summarize a majority of the information presented in this chapter (there are no new facts in the table). For each possibility listed there are many possible preventive aspects; for example, implantation is extremely complex and requires a long sequence of events. Thus, for each possibility, scientific investigation revolves around the question: What are the normal physiological events that make the process possible and how can we intervene to prevent them?

Until recently, techniques of birth control (*contraception*) were primarily those which prevent sperm from reaching the ovum; vaginal diaphragms, sperm-killing jellies, and male condoms. Each of these methods has drawbacks, and, more

Table 16-4. Mean effectiveness of contraceptive methods

Method	Pregnancies per 100 women per year
None	115
Douche	31
Rhythm	24
Jelly alone	20
Withdrawal	18
Condom	14
Diaphragm	12
Intrauterine device	5
Oral contraceptive correctly used	1
	0

important, they are sometimes ineffective (Table 16-4). Another widely used method is the so-called rhythm method, in which couples merely abstain from sexual intercourse near the time of ovulation. Unfortunately, it is difficult to time ovulation precisely even with laboratory techniques; e.g., the small rise in body temperature or change in cervical mucus and vaginal epithelium, all of which are indicators of ovulation, occur only *after* ovulation. This problem, combined with the marked variability of the time of ovulation in many women, explains why this technique is only partially effective.

Since 1950, an intensive search has been made for a simple, effective contraceptive method, the first fruit of these studies being "the pill," the *oral contraceptive.* Its development was based on the knowledge that combinations of estrogen and progesterone inhibit pituitary gonadotropin release, thereby preventing ovulation. The most commonly used agents, at least at first, were combinations of an estrogen- and a progesterone-like substance (progestogens). Each month, the pill is taken for 20 days, then discontinued for 5 days; this steroid withdrawal produces menstruation, and the net result is a menstrual cycle without ovulation. The monthly withdrawal is required to avoid "breakthrough" bleeding which would occur if the steroids were administered continuously. Another type of regimen is the so-called sequential method in which estrogen is administered alone for 15 days followed by estrogen plus progestogen for 5 days, followed by withdrawal of both steroids. As with the combination pills, this regimen interferes with the orderly secretion of gonadotropins and prevents ovulation; no LH surge occurs, at least in part because of the constant estrogen levels, i.e., the absence of a rising estrogen level capable of exerting positive feedback.

It has now become clear that the oral contraceptives do not always prevent ovulation, yet still are effective because they have multiple antifertility effects. In other words, the hormonal milieu required for normal pregnancy is such that these exogenous steroids interfere with many of the steps between intercourse and implantation of the blastocyst. Taken correctly, they are almost 100 percent effective. Serious side effects, such as intravascular clotting, have been reported but only in a relatively small number of women; nonetheless, only time can show how many undesirable

effects will ultimately appear as a result of chronic alteration of normal hormonal balance.

Another type of contraceptive which is highly effective (although not 100 percent) and which illustrates the dependency of pregnancy upon the right conditions is the *intrauterine device* (Table 16-4). Placing one of these small objects in the uterus prevents pregnancy, perhaps by somehow interfering with the endometrial preparation for acceptance of the blastocyst.

The search goes on for an effective method which will reduce even further the possibility of unwanted side effects and will still be easy to use. Almost every possible process shown in Table 16-3 is worth following up. Two recent developments are post-intercourse medications: prostaglandins and the estrogenlike diethylstilbestrol (DES). Both cause increased contractions of the female genital tract and may, therefore, cause expulsion of the fertilized ovum or failure to implant. Enthusiasm for DES has been tempered by the possibility that it may have cancer-producing properties.

Fertility prevention is of great importance, but the other side of the coin is the problem of unwanted infertility. Approximately 10 percent of the married couples in the United States are infertile. There are many reasons—some known, some unknown—for infertility; indeed, Table 16-3 also serves as a list of possible causes of infertility. Careful investigation of infertile couples frequently permits diagnosis and therapy of the basic problem.

Section C
The chronology of sex development

Sex determination

Sex is determined by the genetic inheritance of the individual, specifically by two chromosomes called the *sex chromosomes.* With appropriate tissue-culture techniques all the chromosomes in human cells can be made visible; such studies have demonstrated the presence of 46 chromosomes, 22 pairs of somatic (nonsex) chromosomes and 1 pair of sex chromosomes. The larger of the sex chromosomes is called the X chromosome and the smaller the Y chromosome. Genetic males possess one X and one Y, whereas females have two X chromosomes. Thus the genetic difference between male and female is simply the difference in one chromosome. The reason for the approximately equal sex distribution of the population should be readily apparent (Fig. 16-30): the female can contribute only an X chromosome, whereas the male, during meiosis, produces sperm, half of which are X and half of which are Y. When the sperm and ovum join, 50 percent should have XX and 50 percent XY.

Interestingly, however, sex ratios at birth are not 1:1. Rather, there tends to be a slight preponderance of male births (in England, the ratio of male to female births is 1.06). Even more surprising, the ratio at the time of conception seems to be much higher. From various types of evidence, it has been estimated that there may be 30 percent more male conceptions than female. There are several implications of these facts. First, there must be a considerably larger *in utero* death rate for males. Second, the "male," i.e., XY, sperm must have some advantage over the "female" sperm in reaching and fertilizing the egg. It has been suggested, for example, that since the Y chromosome is lighter than the X, the "male" sperm might be able to travel more rapidly. There are numerous other theories, but we are far from an answer. Moreover, it should be pointed out that conception and birth ratios show considerable variation in different parts of the world and, indeed, in rural and urban areas of the same country.

As easy method for distinguishing between the sex chromosomes was found quite by accident; the cells of female tissue (scrapings from the cheek mucosa are convenient) contain a readily detected nuclear mass which is a condensed X chromosome (in the normal female only one X chromosome functions in each cell). This has been called the *sex chromatin* and is not usually found in male cells. The method has proved valuable when genetic sex was in doubt. Its use and that of the more exacting tissue-culture visualization have revealed a group of genetic sex abnormalities characterized by such bizarre chromosomal combinations as XXX, XXY, X, and many others. Just how these combinations arise remains obscure, but the end result is usually tragic, the failure of normal anatomical and functional sexual development. For example, patients with only one sex chromosome, an X, show no gonadal development.

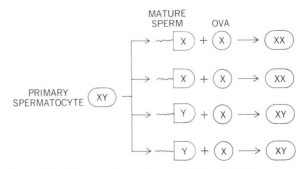

Figure 16-30. Basis of genetic sex determination.

Sex differentiation (Fig. 16-31)

It is not surprising that persons with abnormal genetic endowment manifest abnormal sexual development, but careful study has also revealed patients with normal chromosomal combinations but abnormal sexual appearance and function. For example, a genetic male (XY) may have testes and female internal genitalia (vagina and uterus); such people are termed male pseudohermaphrodites. This kind of puzzle leads us into the realm of sex differentiation, i.e., the process by which the fetus develops the male or female characteristics directed by its genetic makeup. The genes *directly* determine only whether the individual will have testes or ovaries; virtually all the rest of sexual differentiation depends upon this genetically determined gonad.

Differentiation of the gonads

The male and female gonads derive embryologically from the same site in the body. Until the sixth week of life, there is no differentiation of this site. During the seventh week, in the genetic male, the testes begin to develop; in the genetic female, several weeks later, ovaries begin to develop instead. The embryonic gonad, testis or ovary, then regulates the remainder of the individual's sexual development.

Differentiation of internal and external genitalia

The very early fetus is sexually bipotential so far as its internal duct system and external genitalia are concerned. During differentiation, either the

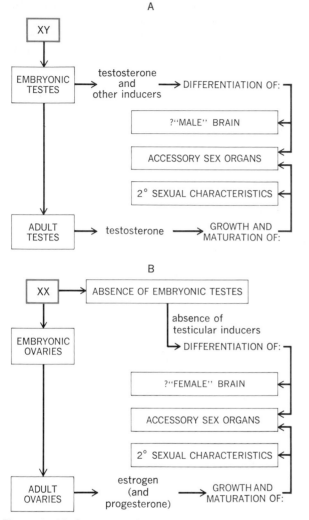

Figure 16-31. Summary of sex-organ differentiation in (A) male and (B) female.

ducts. If the gonad differentiates into a testis, the Wolffian duct remains and develops into the male duct system (epididymis, vas deferens, seminal vesicles). In contrast, if the gonad differentiates into an ovary, the Müllerian system develops into the female duct system (oviducts and uterus). In each case, most of the other set of ducts degenerates. In many species it is possible to remove or damage the area from which the gonads derive without interrupting the pregnancy. When this is done, regardless of its genetic sex, the gonadless embryo develops the female duct system; i.e., the Müllerian ducts persist and differentiate into oviducts and uterus, whereas the Wolffian ducts degenerate. The conclusion is that normally the presence of functioning testes represses the development of the female duct system and induces development of the male organs; in contrast, a female gonad need not be present for the female organs to develop. The primitive testes exert these effects by secreting two substances, both of which act locally: Müllerian regression factor is a protein made by the Sertoli cells, and as its name suggests, it causes regression of the Müllerian duct; testosterone is secreted by the Leydig cells and induces differentiation of the Wolffian duct into the epididymis, vas deferens, and seminal vesicles.

A similar analysis holds for the later development of the external genitals and vagina or prostate (these all derive embryologically not from the Müllerian or Wolffian ducts but from other structures). The absence of functioning testes results in female development, whereas the presence of testes, through the inducing effects of testosterone, results in male differentiation of these structures. Once again, female development seems to be the "natural" course unless testosterone is present to prevent it and cause male development instead. To return to our previously described XY patient with the female duct system, it seems reasonable to suspect a failure of his gonadal hormone secretion during the period of duct differentiation. Note that, depending upon the timing of gonadal failure, one could develop external and internal genitalia of opposite sexes.

Sexual differentiation of the central nervous system

As we have seen, the male and female hypothalamus differ in that there is cyclical secretion of

male or female duct system will fully develop, the other becoming vestigial. Similarly, externally, either a penis or clitoris will develop, and the tissue near it will either fuse, in the male, to become a scrotum or remain separate, in the female, as the labia. Let us look at these events in more detail.

Shortly after the gonadal development, the genital ducts begin to differentiate. Until the sixth or seventh week of embryonic life, the primitive undifferentiated gonad makes connections to two distinct sets of ducts, the *Wolffian ducts* (a part of the developing urinary system) and the *Müllerian*

Gn-releasing hormone in the female but rather fixed continuous release in the male. In primates it appears that this difference does not reflect any qualitative difference in hypothalamic development between male and female; rather it is due simply to the fact, described previously, that large amounts of the dominant female sex hormone, estrogen, can stimulate LH release whereas testosterone can only inhibit it. Thus, it has been demonstrated that the administration of estrogen to male castrated monkeys elicited LH surges indistinguishable from those shown by females.

The situation may be quite different for sexual behavior in that qualitative differences may be formed during development; genetic female monkeys given testosterone during late *in utero* life manifest not only masculinized external genitalia but pronounced masculine sex behavior as adults. Related to this is the question of whether exposure to androgens during *in utero* existence is necessary for development of other behavior patterns in addition to sex. Again, for other primates, the answer seems a clear-cut *yes*; for example, virilized female monkey offspring manifest a high degree of male-type play behavior during growth.

The evidence for human beings is very scanty; perhaps the best-studied group has been 10 young women who were accidentally virilized during late *in utero* development because their mothers were given synthetic hormones (to prevent abortion) not known to be androgenic at the time. At birth they had male external genitalia (enlarged clitoris and fused empty scrotum) but normal internal genitalia; this was corrected surgically very early in life, and their behavior for the next 5 to 15 years was studied carefully and compared with a control group matched to them in every possible way. Certain of the behaviors of these young girls differed from those of the girls in the control group; for example, they took part in more rough-and-tumble outdoor activities. Nonetheless, on the basis of this study and several others like it, the present tentative conclusions are that the most powerful factors determining gender identity and behavior are experiential and social; these influences play upon a range of potentials which is, in some way, influenced by prenatal exposure to (in the male) or lack of exposure to (in the normal female) testosterone.

In summary (Fig. 16-31), it is evident that the male gonadal secretions play a crucial role in determining normal sexual differentiation *in utero* or just after birth. Shortly after delivery, testosterone secretion becomes very low until puberty, when it once again increases to stimulate the organs it had previously helped to differentiate. In contrast, estrogen probably takes little active part in *in utero* development, female differentiation requiring only the absence of testicular secretion. Estrogen is, of course, the stimulating agent for the female sex organs during puberty and adult life. Finally, the X and Y chromosomes appear to dictate directly only whether testes or ovaries develop.

Puberty

Puberty is the period, usually occurring sometime between the ages of ten and fourteen, during which the reproductive organs mature and reproduction becomes possible. In the male, the seminiferous tubules begin to produce sperm; the genital ducts, glands, and penis enlarge and become functional; the secondary sex characteristics develop; and sexual drive is initiated. All these phenomena are effects of testosterone, and puberty in the male is the direct result of the onset of increased testosterone secretion by the testes (Fig. 16-32). Although significant testosterone secretion occurs during late fetal life, within the first days of birth testosterone secretion becomes very low and remains so until puberty.

The critical question is: What stimulates testosterone secretion at puberty? Or conversely: What inhibits it before puberty? Experiments with testis or pituitary transplantation in other mammals and studies of hormone injections in human beings have suggested the hypothalamus as the critical site of control. Before puberty, the hypothalamus fails to secrete significant quantities of Gn-RH to stimulate secretion of FSH and LH (Fig. 16-33); deprived of these latter gonadotropic hormones, the testes fail to produce sperm or large amounts of testosterone. Puberty is initiated by an unknown alteration of brain function which permits increased secretion of the hypothalamic Gn-RH; the mechanism of this change remains unknown. One likely hypothesis is that the hypothalamus prior to puberty is so sensitive to the negative feedback effects of testosterone that even the extremely low blood concentration of this hormone in prepubescent

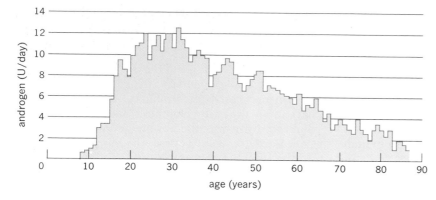

Figure 16-32. Excretion of androgen in the urine of normal boys and men, an indicator of the blood concentration of androgen. *(Adapted from Pedersen-Bjergaard and Tonnesen,* Acta Med. Scand.*)*

boys is adequate to block secretion of the releasing hormone; of course, this simply raises the question of what brain change occurs at puberty to lower responsiveness to the negative feedback. In any case, the process is not abrupt, but develops over several years as evidenced by slowly rising plasma concentrations of the gonadotropins and testosterone. It is interesting that children with brain tumors or other lesions of the hypothalamus or pineal may undergo precocious puberty, i.e., sexual maturation at an unusually early age, sometimes within the first 5 years of life, but the interpretation of these "experiments of nature" remains controversial. In particular, the possible role of that mysterious gland, the pineal, in control of reproduction is presently the subject of much investigation.

The picture for the female is analogous to that for the male. Throughout childhood, estrogen is secreted at very low levels (Fig. 16-21). Accordingly, the female accessory sex organs remain small and nonfunctional; there are minimal secondary sex characteristics, and ovum maturation does not occur. As for the male, prepuberal dormancy is probably due mainly to deficient secretion of hypothalamic Gn-RH, and the onset of puberty is occasioned by an alteration in brain function which raises secretion of this releasing hormone which, in turn, stimulates secretion of pituitary gonadotropins. The increased estrogen stimulated by the gonadotropins then induces the striking changes associated with puberty and, through its stimulatory effect on the brain and pituitary, permits menstrual cycling to begin. "Maturation" of the positive-feedback mechanism also occurs at puberty, for it is not possible to elicit an LH surge

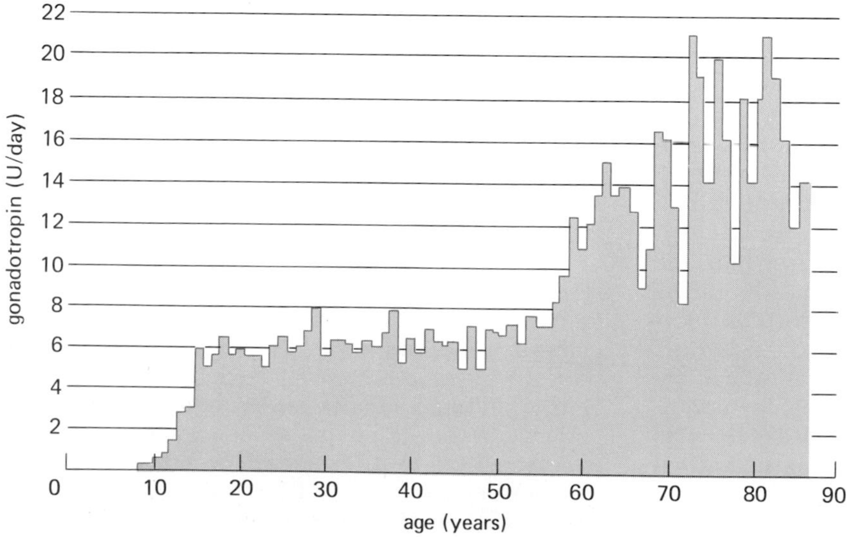

Figure 16-33. Excretion of anterior pituitary gonadotropins (FSH and LH) in the urine of normal boys and men, an indicator of the blood concentration of gonadotropins. *(Adapted from Pedersen Bjergaard and Tonnesen,* Acta Med. Scand.*)*

in prepubescent girls by giving estrogen. Precocious puberty also occurs in females as well as males; the youngest mother on record gave birth to a full-term healthy infant by cesarean section at 5 years, 8 months.

It should be recognized that the maturational events of puberty usually proceed in an orderly sequence but that the ages at which they occur may vary among individuals. In boys, the first sign of puberty is acceleration of growth of testes and scrotum; pubic hair appears a trifle later, and axillary and facial hair still later. Acceleration of penis growth begins on the average at 13 (range 11 to 14.5 years) and is complete by 15 (13.5 to 17). But note that, because of the overlap in ranges, some boys may be completely mature whereas others at the same age (say 13.5) may be completely prepubescent. Obviously, this can lead to profound social and psychological problems. In girls, appearance of "breast buds" is usually the first event (average age = 11) although pubic hair may, on occasion, appear first. *Menarche*, the first menstrual period, is a later event (average = 13) and occurs almost invariably after the peak of the total body growth spurt has passed. The early menstrual cycles are usually not accompanied by ovulation so that conception is generally not possible for 12 to 18 months after menarche. One of the most striking facts concerning menarche is the remarkable decrease over the past 150 years in the age at which it occurs in all industrialized countries. For example, the age of menarche in

Norway has decreased from near 17.5 in 1830 to 13 at present. Improved nutrition has probably played an important causal role in this phenomenon but other factors, as yet unknown, may also contribute. Again, one can imagine the social and psychological impact of such a change on young people.

Menopause

Ovarian function declines gradually from a peak usually reached before the age of thirty, but significant problems, if they occur at all, do not usually arise until the forties. Figure 16-34 demonstrates that the cause of the decline is decreasing ability of the aging ovaries to respond to pituitary gonadotropins, in large part because of the diminished number of follicles. Estrogen secretion drops despite the fact that the gonadotropins, partially released from the negative-feedback inhibition by estrogen, are secreted in greater amounts. Ovulation and the menstrual periods become irregular and ultimately cease completely. Some ovarian secretion of estrogen generally continues beyond these events but gradually diminishes until it is inadequate to maintain the estrogen-dependent tissues:[1] the breasts and genital organs gradually atrophy; the decrease in protein

[1] Even after cessation of ovarian secretion of estrogen, some estrogen is found in plasma due to conversion of adrenally secreted androgens into estrogen by nonovarian tissues.

Figure 16-34. Excretion of estrogen in the urine of women from puberty to senescence, an indicator of blood concentration of estrogen: (1) before menopause and (2) menopause. *(Adapted from Pedersen-Bjergaard and Tonnesen.)*

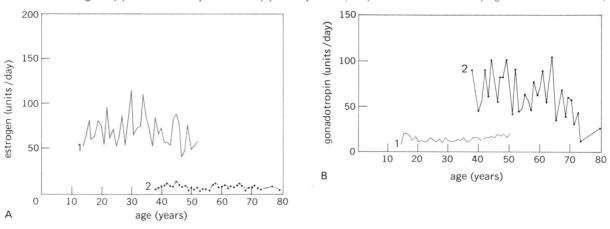

anabolism causes thinning of the skin and bones; however, sexual drive is frequently not diminished and may even be increased. Severe emotional disturbances are not uncommon during menopause and are generally ascribed not to a direct effect of estrogen deficiency but to the disturbing nature of the entire period. The hot flashes, so typical of menopause, result from dilation of the skin arterioles, causing a feeling of warmth and marked sweating; why estrogen deficiency causes this is unknown. Many of these symptoms of the menopause can be reduced by the administration of estrogen, but the safety of such administration is being seriously questioned because of the possibility that estrogen may facilitate development of breast or uterine cancers.

Another important aspect of menopause is the relationship between estrogen and plasma cholesterol. Estrogen significantly lowers the plasma cholesterol, and this or some related effect on lipid metabolism may explain why women have much less atherosclerosis than men until after the menopause, when the incidence becomes similar in both sexes.

Male changes with aging are much less drastic. Once testosterone and pituitary gonadotropin secretions are initiated at puberty, they continue throughout adult life (Figs. 16-32 and 16-33). A steady decrease in testosterone secretion in later decades apparently reflects slow deterioration of testicular function. The mirror-image rise in gonadotropin secretion is due to diminishing negative-feedback inhibition from the decreasing plasma testosterone concentration. Despite the significant decrease, testosterone secretion remains high enough in most men to maintain sexual vigor throughout life, and fertility has been documented in men in their eighties. Thus, there is usually no complete cessation of reproductive function analogous to menopause.

17
DEFENSE MECHANISMS OF THE BODY: IMMUNOLOGY, FOREIGN CHEMICALS, STRESS, AND AGING

Section A: Immunology: The body's defenses against foreign matter

Effector cells of the immune system
Nonspecific immune responses
  External anatomic and chemical "barriers"
  Inflammatory response
    Vasodilation and increased permeability to protein • Chemotaxis • Phagocytosis • Tissue repair
  Chemical mediators of the nonspecific inflammatory response
    Histamine • The kinin system • The complement system
  Systemic manifestations of inflammation
  Interferon
Specific immune responses
  Humoral immunity
    Antibodies • Functions of antibodies • Antibody production • Active and passive humoral immunity • Summary
  Cell-mediated immunity
    Immune surveillance: Defense against cancer • Rejection of tissue transplants

Transfusion reactions and blood types
Factors which alter the body's resistance to foreign cells
Hypersensitivity and tissue damage
  Atopic allergy
  Autoimmune disease

Section B: Metabolism of foreign chemicals

Absorption
Storage sites
Excretion
Biotransformation
  Alcohol: An example
Inorganic elements

Section C: Resistance to stress

Functions of cortisol in stress
  Cortisol's pharmacological effects and disease
Functions of the sympathetic nervous system in stress
Other hormones released during stress

Section D: Aging

Section A:
Immunology: The body's defenses against foreign matter

Immunity constitutes all the physiological mechanisms which allow the body to recognize materials as foreign or abnormal and to neutralize or eliminate them; in essence, these mechanisms maintain uniqueness of "self." Classically, immunity referred to the resistance of the body to microbes: viruses, bacteria, and other unicellular and multicellular organisms. It is now recognized, however, that the immune system has more diverse functions than this. It is involved both in the elimination of "worn-out" or damaged body cells (such as old erythrocytes) and in the destruction of abnormal or mutant cell types which arise within the body. This last function, known as *immune surveillance*, constitutes one of the body's major defenses against cancer.

It has also become evident that immune responses are not always beneficial but may result in serious damage to the body. In addition, the immune system seems to be involved in the process of aging. Finally, it constitutes the major obstacle to successful transplantation of organs. Because of these broad relationships, few other research areas of biology have grown so rapidly or have produced such a wealth of new and exciting, albeit sometimes bewildering, information.

Immune responses may be classified into two categories: *specific* and *nonspecific*. Specific immune responses, the function of lymphocytes (and cells derived from lymphocytes), depend upon prior exposure to a specific foreign material, recognition of it upon subsequent exposure, and reaction to it. In contrast, the nonspecific immune responses do not require previous exposure to the particular foreign material; rather they nonselectively protect against foreign materials without having to recognize their specific identities. They are particularly important during the initial exposure to a foreign organism before the specific immune responses have been activated.

Immune responses can be viewed in the same way as other homeostatic processes in the body, i.e., as stimulus-response sequences of events. In such an analysis, the groups of cells which mediate the final responses are effector cells. Before we begin our detailed description of immunology by introducing the various effector cells involved, let us first at least mention some of their major opponents—bacteria and viruses.

Bacteria are unicellular organisms which have not only a cell membrane but also an outer coating, the cell wall. Most bacteria are self-contained complete cells in that they have all the machinery required to sustain life and to reproduce themselves. In contrast, the *viruses* are essentially nucleic acid cores surrounded by a protein coat. They lack both the enzyme machinery for energy production and the ribosomes essential for protein synthesis. Thus, they cannot survive by themselves but must "live" inside other cells whose biochemical apparatus they make use of; the viral nucleic acid

directs the synthesis by the host cell of the proteins required for viral replication, nucleotides and energy sources also being supplied by the host cell. Other types of microorganisms and multicellular parasites are potentially harmful to human beings, but we shall devote most of our attention to the body's defense mechanisms against the bacteria and viruses.

How do microorganisms cause damage and endanger health? There are many answers to this question, depending upon the specific bacterium or virus involved. Some bacteria themselves cause cellular destruction by locally releasing enzymes which break down cell membranes and organelles. Others give off toxins which disrupt the functions of organs and tissues. The effect of viral habitation and replication within a cell depends upon the type of virus. Some viral particles, entering a cell, multiply very rapidly and kill the cell by depleting it of essential components or by directing it to produce toxic substances; with death of the host cell, the viral particles leave and move on to another cell. In contrast, other viral particles replicate very slowly, and the viral nucleic acid may even become associated with the cell's own DNA molecules, replicating along with them and being passed on to the daughter cells during cell division; such a virus may remain in the cell or its offspring for many years. As we shall see, cells infected with these so-called "slow" viruses may be damaged by the body's own defense mechanisms turned against the cells because they are no longer "recognized" as "self." Finally, it is likely that certain viruses cause transformation of their host cells into cancer cells.

Effector cells of the immune system

The major effector-cell types of the immune system (Table 17-1) are the white blood cells (leukocytes), plasma cells, and macrophages. The white blood cells were described in Chap. 11 and should be reviewed at this time. Plasma cells are derived from lymphocytes (a white blood cell type) and are the cells which secrete antibodies. Macrophages are derived from monocytes (another white blood cell type), have as their major function phagocytosis, and are found scattered throughout the tissues of the body; the structure of these cells varies from place to place, but their common distinctive fea-

Table 17-1. Major effector-cell types of the immune system

	White blood cells (leukocytes)						
	Polymorphonuclear granulocytes						
	Neutrophils	Eosinophils	Basophils	Lymphocytes	Monocytes	Plasma cells	Macrophages
Percent of total leukocytes	50–70%	1–4%	0.1%	20–40%	2–8%		
Primary site of production	Bone marrow	Bone marrow	Bone marrow	Bone marrow, thymus, and other lymphoid tissues	Bone marrow	Derived from B lymphocytes in lymphoid tissue	Most are formed from monocytes
Primary known function	Phagocytosis of bacteria, cell debris, and antibody-primed foreign matter; release of chemicals (endogenous pyrogen, chemotaxic factors, etc.)	? ? ?	Release of histamine and other chemicals (similar to tissue mast cells)	B cells: production of antibodies (after transformation into plasma cells); T cells: responsible for cell-mediated immunity; influence B cells	Transformed into tissue macrophages with high phagocytic activity; release of granulopoietin	Production of antibodies	Phagocytosis of bacteria, cell debris, and antibody-primed foreign matter; assist in antibody formation and T-cell sensitization; release of granulopoietin

tures are numerous cytoplasmic granules and the ability to ingest almost any kind of foreign particle. Some macrophages are capable of mitotic activity, but the major means of repletion of tissue macrophages is by influx of blood monocytes and their differentiation locally.

The effector cells of the immune system are distributed throughout the organs and tissues of the body, most being housed in the so-called lymphoid tissues (Fig. 17-1 and Table 17-2): lymph nodes, spleen, thymus, tonsils, and aggregates of lymphoid follicles such as those in the lining of the gastrointestinal tract.

Lymph nodes function as filters along the course of the lymph vessels, lymph flowing through them before being returned to the general circulation. Lymph enters the node via afferent lymphatic vessels, trickles through the lymphatic sinuses of the node, and leaves via efferent lymphatic vessels

Table 17-2. Some functions of lymphoid tissues

Lymph Nodes (including tonsils and lymphoid follicles in gastrointestinal tract)	1 Remove and store lymphocytes 2 Form and add new lymphocytes to lymph flow through nodes 3 Macrophages phagocytize particulate matter
Spleen	1 Production of red blood cells in fetus but not adult 2 Production of new lymphocytes 3 Macrophages phagocytize products of red cell degradation and other foreign matter 4 Store red blood cells, which can be added to circulation by contraction of the spleen
Thymus	1 Production of T lymphocytes 2 Secretion of hormones (thymosin)

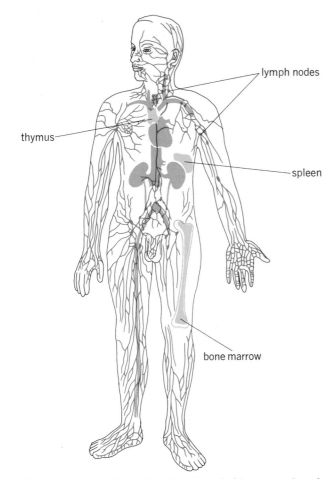

Figure 17-1. Location of various lymphoid organs: lymph nodes, thymus, and spleen. Not shown are the lymphoid patches in the various epithelial linings of the body. The bone marrow is not a lymphoid organ but is the site of production of many of the cells which come to reside in the lymphoid organs.

on the other side. The lymphatic sinuses are relatively open channels and as the lymph flows through them, some lymphocytes are removed to be stored temporarily in the node and others are added to the lymph. Some of the lymphocytes released into the lymph are those previously stored, but others have been newly formed in the node. The lymphatic sinuses are lined with macrophages, which phagocytize particulate matter, such as dust (inhaled into the lungs), cellular debris, bacteria, and other microorganisms.

The *spleen* is the largest of the lymphoid organs and lies in the left part of the abdominal cavity between the stomach and diaphragm. The interior of the spleen is filled with a reticular meshwork, the *red pulp* and the *white pulp*. Blood, rather than lymph, percolates through the red pulp (thus its name), and erythrocytes as well as lymphocytes and macrophages are collected in the spaces in the meshwork. In the human fetus, the spleen is an important red-cell-forming organ, but in the adult only lymphocytes are formed there. They are produced in the *white pulp* of the spleen, areas of lymphoid tissue which surround the arterioles. The macrophages of the spleen phagocytize many of the products of red cell degradation as well as various kinds of foreign matter.

The *thymus* lies in the upper part of the chest, and its size varies with age; it is relatively large at birth and grows until puberty when it gradually atrophies and is replaced by fatty tissue. The thymus is an important lymphocyte-producing organ; it also secretes a group of hormones presently known collectively as *thymosin*. The role of the thymus in immunity will be described below.

Nonspecific immune responses

External anatomic and chemical "barriers"

The body's first lines of defense against infection are the barriers offered by surfaces exposed to the external environment. Very few microorganisms can penetrate the intact skin, and the sweat, sebaceous, and lacrymal glands secrete chemical substances which are highly toxic to certain forms of bacteria. The mucous membranes also contain antimicrobial chemicals, but more important, mucus is sticky. When particles adhere to it, they can be swept away by ciliary action, as in the upper respiratory tract, or engulfed by phagocytic cells. Other specialized surface "barriers" are the hairs at the entrance to the nose, the cough reflexes, and the acid secretion of the stomach. Finally, a major "barrier" to infection is the normal microbial flora of the skin and other linings exposed to the external environment; these microbes suppress the growth of other potentially more virulent microorganisms.

Inflammatory response

Despite the effectiveness of the external barriers, small numbers of microorganisms penetrate them every day. Think of all the small breaks produced in the skin or mucous membranes by tooth brushing, shaving, tiny scratches, etc. In addition, we now recognize that many viruses are able to penetrate seemingly intact healthy skin or mucous membranes.

Once the invader has gained entry, it triggers off *inflammation*, the response to injury. The local manifestations of the inflammatory response are a complex sequence of highly interrelated events, the overall functions of which are to bring phagocytes into the damaged area so that they can destroy (or inactivate) the foreign invaders and set the stage for tissue repair. The sequence of events which constitute the inflammatory response varies, depending upon the injurious agent (bacteria, cold, heat, trauma, etc.), the site of injury, and the state of the body, but the similarities are in many respects more striking than the differences. It should be emphasized that, in this section, we describe inflammation in its most basic form, i.e., the nonspecific innate response to foreign material. As we shall see, inflammation remains the basic scenario for the acting out of specific immune responses as well, the difference being that the entire process is amplified and made more efficient by the participation of antibodies and sensitized lymphocytes, i.e., agents of specific immune responses.

The sequence of events in an inflammatory response is briefly as follows, using bacterial infection as our example:

1 Initial entry of bacteria
2 Vasodilation of the vessels of the microcirculation leading to increased blood flow
3 A marked increase in vascular permeability to protein
4 Filtration of fluid into the tissue with resultant swelling
5 Exit of neutrophils (and later, monocytes) from the vessels into the tissues
6 Phagocytosis and destruction of the bacteria
7 Tissue repair

All these events are mediated by chemical substances released or generated locally and these will be described in the last part of this section. The familiar gross manifestations of this process are redness, swelling, heat, and pain, the latter being the result both of distension and of the effect of local substances on afferent nerve endings.

Vasodilation and increased permeability to protein. Immediately upon microbial entry, chemical mediators dilate most of the vessels of the microcirculation in the area and somehow alter the material between the endothelial cells so as to make it quite leaky to large molecules. Tissue swelling is directly related to these changes: The arteriolar dilation increases capillary hydrostatic pressure, thereby favoring filtration of fluid out of the capillaries; in addition, the protein which leaks out of the vessels as a result of increased permeability builds up locally in the interstitium, thereby diminishing the difference in protein concentration between plasma and interstitium (recall from Chap. 11 that this difference is mainly responsible for fluid movement from interstitium into capillaries).

The adaptive value of all these vascular changes is twofold: (1) The increased blood flow to the inflamed area increases the delivery of phagocytic leukocytes and plasma proteins crucial for immune responses (to be described below), (2) the increased capillary permeability to protein ensures that the relevant plasma proteins—all normally restrained by the capillary membranes—can gain entry to the inflamed area.

Chemotaxis. Within 30 to 60 min after the onset of inflammation, a remarkable interaction occurs between the vascular endothelium and circulating neutrophils. First, the blood-borne neutrophils begin to stick to the inner surface of the endothelium. The process is quite specific since erythrocytes show no tendency to stick and other leukocytes do so only later, if at all. How the endothelium is made sticky by injury remains unknown.

Following their surface attachment, the neutrophils begin to manifest considerable amoebalike activity. Soon a narrow amoeboid projection is inserted into the space between two endothelial cells (Fig. 17-2), and the entire neutrophil then squeezes into the interstitium. The alterations of vessel structure described above may facilitate this process by loosening the intercellular connections, or the neutrophil may simply pry the connection apart by the force of its amoeboid movement. By this process (known as neutrophil exudation), huge numbers of neutrophils migrate into the inflamed areas of tissue and move toward the microbes. This entire response of the neutrophils is known as *chemotaxis* and is induced by chemi-

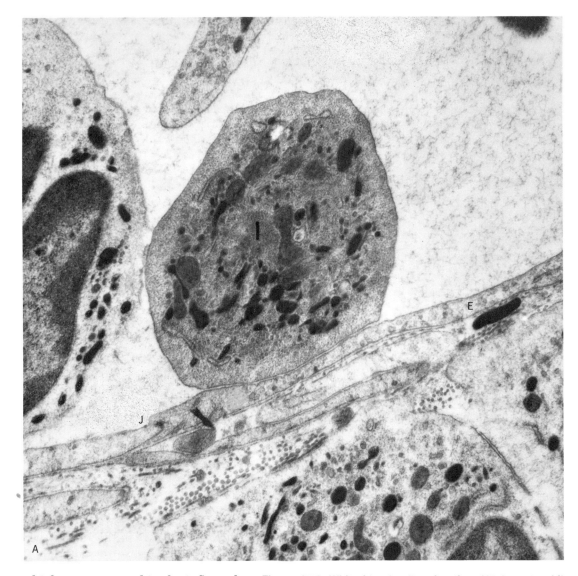

cal mediators which are generated in the inflamed area and move out by diffusion; therefore, there exists a concentration gradient for them, the area immediately around the microbes containing the greatest concentration. The chemotactic chemicals polarize the motion of the neutrophils along this gradient, but precisely how they do so is not known.

Movement of leukocytes into the tissue is usually not limited only to neutrophils. Monocytes follow, but much later, and once in the tissue are transformed into macrophages; meanwhile some of the macrophages normally present in the tissue

Figure 17-2. White blood cell emigration. (A) A neutrophil (at center) in the lumen of a capillary has just adhered to the endothelium (E); note the intact intercellular junction (J). (B) A neutrophil is protruding a pseudopod through the intercellular junction and out into the interstitial space. The entire cell will move through the wall in this manner. [*From V. T. Marchesi, Q. J. Exp. Physiol., 46:115 (1961).*]

have begun to multiply by mitosis and to become motile. Thus, all the phagocytic cell types are present in the inflamed area. Usually the neutrophils predominate early in the infection but tend to die off more rapidly than the others, thereby yielding a predominantly macrocytic picture later.

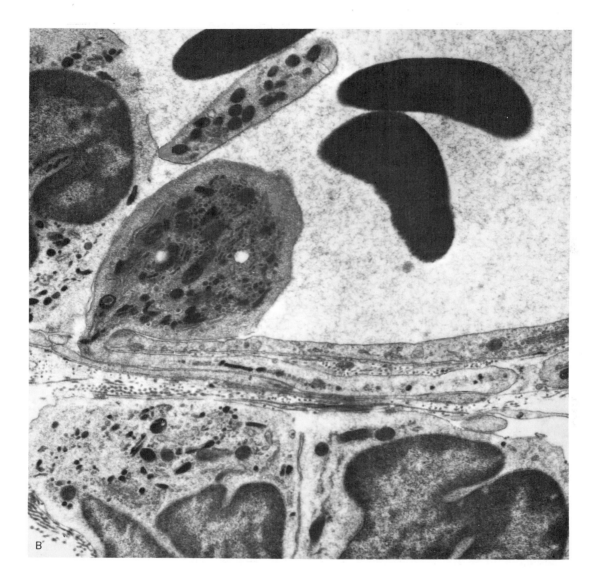

B

In contrast, in certain types of allergies and inflammatory responses to parasites the eosinophils are in striking preponderance. Thus, one of the tests for allergically induced "runny nose" is to study a small amount of nasal discharge for the presence of eosinophils.

Phagocytosis. Phagocytosis is the primary function of the inflammatory response, and the increased blood flow, vascular permeability, and leukocyte exudation serve only to ensure the presence of adequate numbers of phagocytes and to provide the milieu required for the performance of their function. Having arrived at the site, phagocytes must "pick out" what to attack by discerning certain characteristics on the surfaces of the foreign cells (or damaged native cells); in the case of nonspecific inflammation, what these characteristics are and how they trigger phagocytosis is not clear.

The process of ingestion is illustrated in Fig. 17-3. The phagocyte engulfs the organism by endocytosis, i.e., membrane invagination and pouch (*phagosome*) formation. Once inside, the microbe remains in the phagosome, a layer of phagocyte plasma membrane separating it from the phagocyte

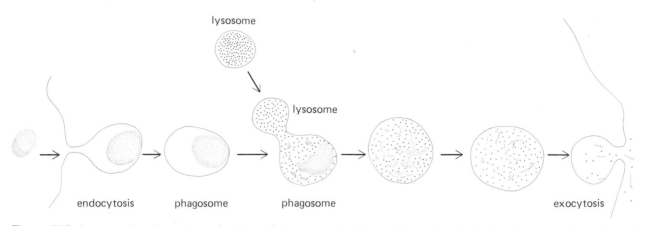

lysosome

lysosome

endocytosis   phagosome   phagosome   exocytosis

Figure 17-3. Large molecules and particulate substances enter the cell by endocytosis, forming a membrane-bound phagosome (at left in diagram). A phagosome may then merge with a lysosome, which brings together the digestive enzymes of the lysosome and the contents of the phagosome. After digestion has taken place, the contents may be released to the outside of the cell by exocytosis.

cytoplasm. The next step is known as *degranulation* (Fig. 17-3): The membrane surrounding the phagosome makes contact with one of the phagocyte's lysosomes (these appear microscopically as "granules") which are filled with a variety of digestive enzymes. The likelihood of contact is facilitated by the fact that the lysosomes are stimulated to undergo increased movements in the vicinity of the phagosome. These movements, like those responsible for the original engulfment, are due to contractions of actin and myosin filaments within the phagocytes. Once contact between the phagosome and lysosome is made, the membranes fuse (the disappearance of the granules during this fusion leads to the term "degranulation"), and the microbe is exposed to the lysosomal enzymes capable of killing it by breaking down its macromolecules.

The lysosomal enzymes are not the only mechanism within the phagosome for killing bacteria. The phagosome produces a high concentration of hydrogen peroxide which is extremely destructive to macromolecules. This production is mediated by the activity of enzymes in the phagosomal membrane itself, in response to the presence of the foreign particle.

The lysosomal enzymes and hydrogen peroxide not only kill the microorganism but they catabolize it into low-molecular-weight products which can then be safely released from the phagocyte or actually utilized by the cell in its own metabolic processes. This entire process need not kill the phagocyte, which may repeat its function over and over before dying. Nondegradable foreign particles (such as wood, tattoo dyes, or metal) and certain species of microorganisms may be retained indefinitely within macrophages.

The neutrophils may also release lysosomal granules into the extracellular fluid; the enzymes released from these granules attack extracellular debris at the injury site, making it easier for the macrophages to phagocytize it at the battle's end, thus paving the way for repair of the damaged area.

Thus far we have presented the role of the phagocytes, but what are the microorganisms doing all this time to protect themselves? Most of them do very little. Certain kinds, however, release substances which diffuse into the phagocyte cytoplasm and disrupt the membranes of the lysosomes, thereby allowing these potent chemicals to destroy the phagocyte itself.

Finally, it should be noted that, by mechanisms described below, some microbes are killed during inflammation without having been previously phagocytized.

Tissue repair. The final stage of the inflammatory process is tissue repair. Depending upon the tissue involved, regeneration of organ-specific cells may or may not occur (for example, regeneration occurs in skin and liver but not in muscle or the central nervous system; the latter two tissues are incapable of cell division in adults). In addition, fibroblasts in the area divide rapidly and begin to secrete large quantities of collagen.

The end result may be complete repair (with or without a scar), *abscess* formation, or *granuloma* formation. An abscess is basically a bag of pus (microbes, leukocytes, and liquefied debris) walled off by fibroblasts and collagen. This occurs when tissue breakdown is very severe and when the microbes cannot be eliminated but only contained. When this stage has been reached, the abscess must be drained for it will not resolve spontaneously.

A granuloma occurs when the inflammation has been caused by certain microbes (such as the bacteria causing tuberculosis) which are engulfed by phagocytes but survive within them. It also occurs when the inflammatory agent is a nonmicrobial substance which cannot be digested by the phagocytes. The granuloma consists of numerous layers of phagocytic-type cells, the central ones of which contain the offending material. The whole thing is, itself, usually surrounded by a fibrous capsule. Thus, a person may harbor live tuberculosis-producing bacteria within the body for many years and show no ill effects as long as the microbes are contained within the granuloma and not allowed to escape.

Chemical mediators of the nonspecific inflammatory response

The events described in the previous section are elicited by chemical substances of varying origins: Some are released from tissue cells present in the area prior to the invasion; others are brought into the area by circulating leukocytes from which they are then released; still others are newly generated in the interstitial fluid of the area as a result of enzyme-mediated reactions. There are a large number of these chemical mediators and no quantitative assessment of their roles is presently available. We describe here three of the mediators which have been most studied to date: histamine, the kinin system, and the complement system. (See Table 17-3.)

Histamine. Histamine is present in many tissues of the body but is particularly concentrated in mast cells, circulating basophils, and platelets. Release

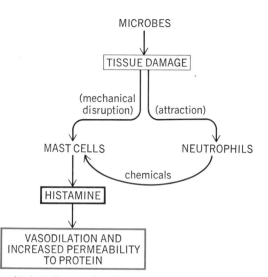

Figure 17-4. Pathways for histamine release in nonspecific inflammatory reactions.

of histamine is induced by a variety of factors. In an inflammatory response in which specific immune responses are absent, two of the major factors are simple mechanical disruption of histamine-containing cells (due to injury) and chemicals secreted by neutrophils attracted to the site (Fig. 17-4). The released histamine acts directly on vascular smooth muscle and endothelial cells to cause vasodilation and increased permeability, respectively. In addition, histamine has profound effects on nonvascular smooth muscle, perhaps the most significant being its constriction of the respiratory airways.

The kinin system. The kinins are small polypeptides whose vascular effects are similar to those of histamine and which, in addition, are powerful

Table 17-3. Summary of major mediators in nonspecific inflammation

Effect	Mediators
Vasodilation and increased permeability	Histamine Kinins Complement
Chemotaxis	Kinins Complement
Enhancement of phagocytosis	Complement
Direct microbial killing	Complement

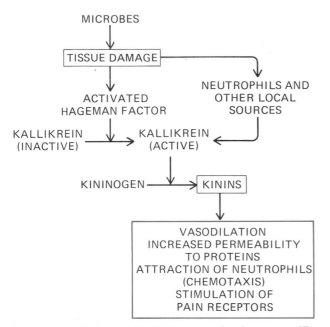

Figure 17-5. Pathways for kinin generation in nonspecific inflammatory reactions.

chemotaxic agents. The kinins are generated in plasma in the following ways (Fig. 17-5). There is present in plasma an enzyme, known as *kallikrein,* which exists normally in an inactive form but which, when activated, catalyzes the splitting off of the kinins from another normally occurring plasma protein known appropriately as *kininogen.* What is it that activates this enzyme? Again we find that multiple factors are capable of doing so but that, in the absence of specific immune responses, the most important is a chemical substance we already met in the section of Chap. 11 on hemostasis: activated Hageman factor. Here is one of many connections between the hemostatic and immune systems. Recall that Hageman factor is, itself, inactive until activated locally by contact with altered vascular surfaces; once activated, it catalyzes the first steps in the cascade sequences leading to both clotting and plasmin formation. Now we see that, in addition, it leads to the activation of the key enzyme of the kinin-generating system, thereby contributing to the first phases of the inflammatory process.

In addition to being generated from its inactive form in blood, active kallikrein is also found in many different tissues as well as in neutrophils. Once inflammation has begun, the kallikrein present in these sources contributes to the generation of kinin. In essence, this is a positive feedback since plasma kininogen is usually protected from the action of tissue kallikrein because the former is too large a molecule to pass through capillaries; with enhanced vascular permeability, it penetrates into the tissues, thereby permitting the generation of more kinin and more inflammation. It is a positive feedback in a second respect, also, since the kinins attract more neutrophils to the area, which in turn facilitate the generation of more kinin. It should also be noted that, by their effects on afferent neuron terminals, the kinins may account for much of the pain associated with inflammation.

The complement system. The *complement* system is yet another example (the clotting, anticlotting, and kinin systems are others) of a "system" consisting of a group of plasma proteins which normally circulate in the blood in an inactive state; upon activation of the first protein of the group, there occurs a sequential cascade in which active molecules are generated from inactive precursors. In each of the other cascade systems we have described, only the final activated molecule (fibrin in clotting, plasmin in anticlotting, kinins in the kinin system) is the direct mediator of the biological response of that system; in contrast, in the complement system, certain of the different proteins activated along the cascade may function as mediators of a particular response (vasodilation, chemotaxis, etc.) as well as activators of the next step. Moreover, several of the protein fragments split off in the process of activation are also biologically active (Fig. 17-6). Since this system consists of eleven distinct proteins (as well as the several active protein fragments split during their enzymatic activation), the overall system is enormously complex and we will make no attempt to identify the roles of individual complement proteins except for purposes of illustration or clarity.

As summarized in Fig. 17-7, one or more of the activated complement molecules is capable of mediating virtually every step of the inflammatory response. Certain of them enhance vasodilation and increased permeability by direct effects on the vasculature as well as by stimulating release of histamine from mast cells (and platelets) and activating plasma kallikrein. They also facilitate

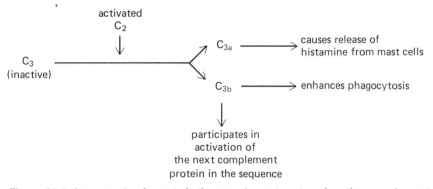

Figure 17-6. An example of events in the complement system. Inactive complement protein C_3 is split by complement protein C_2 (itself having been activated by the preceding step in the cascade) into two proteins, C_{3a} and C_{3b}. C_{3a} functions as one of the mediators of the inflammatory response, in large part by stimulating mast cells to release histamine. C_{3b} attaches to the microbe where it not only facilitates phagocytosis but participates in the activation of the next complement protein in the sequence.

neutrophil exudation by acting as powerful chemotaxic agents. Another complement component enhances phagocytosis by coating the microbe and permitting the phagocyte to ingest it more efficiently, most likely by acting as receptors to which the phagocyte can bind. This is a particularly important function because many virulent bacteria have a thick polysaccharide capsule which strongly resists engulfment by phagocytes. Finally, yet another complement protein mediates direct killing of microbes without prior phagocytosis. The complement becomes embedded in the microbial surface and somehow forms a tunnel making the microbe leaky and killing it.

In describing the actions of the complement system we have emphasized the sequential or cascade nature of the reactions required for the generation of active mediators, but we have so far ignored the

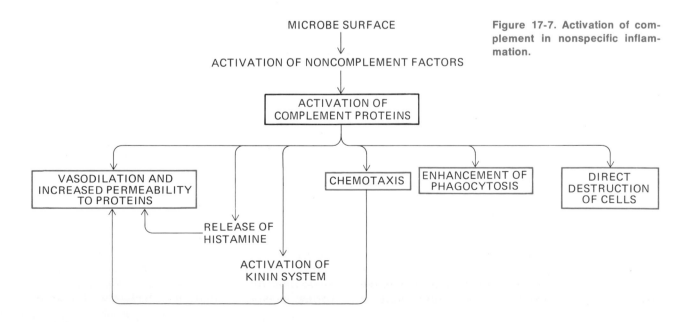

Figure 17-7. Activation of complement in nonspecific inflammation.

problem of just how the sequence is initiated. As we shall see in a later section, antibody is required to activate the very first molecules in the complete complement sequence, but antibody is not present in nonspecific inflammatory responses. Fortunately, the first (antibody-dependent) steps can be bypassed; i.e., alternative mechanisms exist for initiating the sequence at a point beyond these first steps. Little biological activity is lost in bypassing the first steps because no inflammatory mediators are generated by them.

What are the initiators in this alternate antibody-independent complement pathway? The major one is a normally circulating plasma protein which is not itself a member of the complement family and is capable of binding to certain types of carbohydrate chains commonly present on bacterial or viral surfaces (but not present on the body's own cells). This binding changes the conformation so as to confer upon it the capacity to initiate the enzymatic splitting of complement C_3.

Systemic manifestations of inflammation

We have thus far described the *local* inflammatory response. What are the systemic, overall responses? Probably the single most common and striking systemic sign of injury is fever. The substance primarily responsible for the resetting of the hypothalamic thermostat, as described in Chap. 15, is a protein (endogenous pyrogen) released by the neutrophils (and perhaps other cells) participating in the inflammatory response.

One would expect fever, being such a consistent concomitant of infection, to play some important protective role, and recent evidence, at least for nonmammalian species, strongly suggests that such is the case (this is a very important question in light of the widespread use of aspirin to suppress fever). In contrast to the possible benefits of fever, there is no question that an extremely high fever may be quite harmful, particularly its effects on the functioning of the central nervous system, and convulsions are not infrequent in highly febrile young children.

Endogenous pyrogen has effects in addition to eliciting fever; in particular, it causes movement of iron (and other trace metals) out of the plasma and into the liver. The net result is to reduce the extracellular concentration of iron, a phenomenon which is beneficial since bacteria require the presence of high concentrations of iron to multiply. Thus, endogenous pyrogen may exert a spectrum of effects having the common denominator of enhancing resistance to infection.

Another systemic manifestation of many bacterial diseases is a marked increase in the synthesis and release of neutrophils by the bone marrow (in contrast, viral infections frequently are associated with decreased numbers of circulating neutrophils). This stimulation of neutrophil ("granulocyte") production is mediated by a circulating chemical named *granulopoietin* (just as the hormone which stimulates erythrocyte production is called erythropoietin). Granulopoietin seems to be produced and released mainly by monocytes and macrophages, and its release is induced by the presence of bacterial products. Thus, the mononuclear phagocytic cells not only participate in the destruction of bacteria but, in the process, release chemicals which elicit increased production of the other major class of phagocytes, the neutrophils.

Interferon

Interferon is a nonspecific defense mechanism against viral infection (Fig. 17-8). It is a protein which inhibits viral multiplication and is produced by several different cell types in response to a viral infection. Its production (or lack of production) can be understood from basic principles of protein synthesis (Chap. 4). When there is no virus in the host cell, the potential for interferon synthesis exists but the actual synthesis is repressed. Entry of a virus into the cell induces interferon synthesis by the usual DNA-RNA-ribosomal mechanisms; the inducer seems to be viral nucleic acid. Once synthesized, interferon may leave the cell, enter the circulation, and bind to the plasma membrane of other cells. Thus, interferon produced by one cell can protect other cells from viral infection. The interferon molecule does not, itself, have antiviral activity; rather its binding to the plasma membrane triggers the synthesis of "antiviral" proteins by the cell to which it is bound. These antiviral proteins act by blocking synthesis within the cell of macromolecules the virus requires for its multiplication. The fact that interferon itself is not antiviral but rather stimulates the cells to make their own antiviral proteins is another example of amplification —a single molecule of interferon can trigger the synthesis of a lot of antiviral protein so that only a

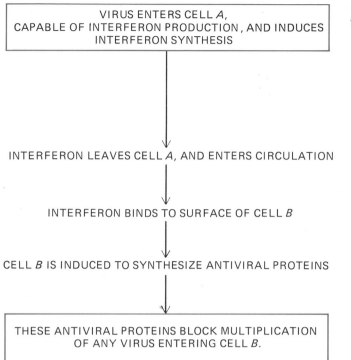

Figure 17-8. Role of interferon.

few interferon molecules seem to be required to protect a cell.

To reiterate, interferon is *not* specific; all viruses induce the same kind of interferon synthesis, and interferon in turn can inhibit the multiplication of many different viruses. For many years the hope has been to administer interferon to patients suffering from viral infections in order to enhance their resistance. The few studies now completed with exogenous interferon have been successful in conferring increased resistance against several different viruses and, in at least one case, against a form of cancer.

Specific immune responses

Traditionally, specific immune responses are placed in one of two categories, according to the nature of the effector mechanisms employed in the response. One category has been termed *humoral,* or *antibody-mediated immunity,* in recognition of the central role of circulating antibodies in the destructive process; the antibodies are secreted by the so-called B *lymphocyte* (to be more precise, by

the *plasma cells* into which B lymphocytes differentiate upon being stimulated). The second category of specific immune defenses is *cell-mediated immunity,* and it is mediated not by antibodies but by intact cells, in this case a second population of lymphocytes, the *T lymphocytes,* which are distinct from the B lymphocytes.

In the adult, new lymphocytes arise by mitosis mainly in lymph nodes, thymus, and spleen. However, the ancestors of these lymphocytes were cells from either the bone marrow or thymus; after leaving these sites, they took up residence in the spleen, lymph nodes, and other lymphoid aggregations, but their bone-marrow or thymus residence is what conferred upon them B cell or T cell characteristics. At some point in their development, cells destined to be T cells enter (or arise within) the thymus (thus the name T cell) which in some manner confers upon them the ability to differentiate and mature into cells competent to act as effectors in cell-mediated immunity. The T cells then leave the thymus and take up residence in the other lymphoid tissues, but the thymus continues to stimulate them and their offspring by

Table 17-4. Sequence of events in humoral immunity against bacteria

1 Bacterial antigen is carried to spleen or lymph nodes, where it binds to surface of specific B lymphocytes. (Macrophages "help" this process; T cells may "help" or "suppress" it.)
2 B lymphocyte differentiates into plasma cell and secretes antibody specific for that antigen into the blood.
3 Antibody circulates to site of infection and combines with antigen on surface of bacteria.
4 Presence of antibody bound to antigen facilitates phagocytosis and activates complement system which further enhances phagocytosis and also directly kills bacteria.
5 Certain of the B lymphocytes differentiate into memory cells capable of responding very rapidly should the bacteria be encountered again.

means of hormones (*thymosin*) which its epithelial cells secrete. In contrast, the B cells are derived from the bone marrow prior to their taking up residence in lymph nodes and other nonthymic lymphoid tissues.

Unlike the nonspecific defense mechanisms, B-cell and T-cell responses depend upon the cells' "recognizing" the specific foreign matter to be attacked or neutralized. This recognition is made possible by the fact that molecular components of foreign cells (and other foreign matter) combine specifically with receptor sites on the surface of B cells and/or T cells, triggering off the attack. These foreign molecular components which stimulate a specific immune response are known as *antigens*. The B-cell system is characterized by an ability to recognize an enormous variety of different specific antigens, and its major function is to confer, via its secreted antibodies, resistance against most bacteria (and their toxins) and some viruses. By contrast, the T-cell system appears to recognize a more limited number of antigens, but these antigens, as we shall see, are the crucial cell-surface markers which label cells "self" or "non-self"; the T cells confer major resistance against fungi, viruses (once they have become intracellular), parasites, the few bacteria which must live inside cells to survive, cancer cells, and solid-tissue transplants.

B cells and T cells influence each other in a variety of ways. On the one hand, antibodies (a B-cell product) may either facilitate or decrease the ability of an attacking T cell to destroy a foreign cell. On the other hand, T cells may either enhance

or suppress the secretion of antibody by B cells. This last phenomenon led to the discovery that T cells do not constitute a homogenous population but are of three kinds: (1) *cytolytic T cells*, which upon activation perform the role classically ascribed to T cells—the killing of those foreign cells listed in the previous paragraph; (2) *"helper" T cells*, which enhance antibody production; (3) *"suppressor" T cells*, which reduce antibody production. Thus, so interrelated are T cells and B cells that two of the three classes of T cells have as their function the regulation of B cells. (It should be emphasized that T cells, themselves, never secrete antibodies, they alter the B cells' capacity to do so.)

Humoral immunity

Table 17-4 summarizes the sequence of events which result in antibody-directed destruction of bacteria. We withhold discussion of step 1, induction of antibody synthesis, until we have described the nature of antibodies and their mechanisms of action.

Antibodies. An *antibody* is a specialized protein capable of combining chemically with the specific antigen which stimulated its production. The word specific is essential in the definition since an antigenic substance reacts only with the type of antibodies elicited by its own kind or an extremely closely related kind of molecule. Specificity is thus related to the chemical structure of the antigen and its antibody.

Antibodies are all proteins, each composed of four interlinked polypeptide chains (Fig. 17-9). The two long chains are called heavy chains, and the two short ones are light chains (these chains are all joined by disulfide bridges). The amino acid sequences along the chains are constant from one antibody to the next except for relatively short sequences in portions of opposing light and heavy chains (Fig. 17-9). These sequences are unique for each of the extremely large number of antibodies that an individual can produce, and constitute the binding sites (two per antibody) for the antigen specific for that antibody. The constant sequences in the "stem" of the heavy chains also contain nonspecific binding sites for molecules and cells which function as the effectors for antibody action (see below). The high degree of specificity of an antibody binding site for an antigen is the result of two factors: First, its shape is complementary to that of the antigen; second, the amino acid side chains within the site are precisely positioned to maximize electrostatic forces and hydrogen bonding between the antibody and antigen. [It should be evident that in this binding, the antigen is a ligand and the antibody a binding site (Chap. 4).] Thus, only the antibody containing a unique amino acid sequence can "recognize" i.e., combine with, a given antigen.

Antibodies belong to the family of proteins known as immunoglobulins, which may be subdivided into five classes according to the types of light and heavy chains they contain. It is very easy to misunderstand this concept, which states that there are five *classes* of immunoglobulins (antibodies), not five antibodies; each class contains many thousands of unique antibodies. These classes are designated by the letters G, A, M, D, and E after the symbol Ig (for immunoglobulin). IgG (the G stands for gamma; thus "gamma globulin" is a synonym for immunoglobulin G, the most abundant plasma antibodies) and IgM antibodies provide the bulk of specific immunity against bacteria and viruses. The other class we shall be concerned with is IgE, for these antibodies mediate certain allergic responses. IgA antibodies are produced by lymphoid tissue lining the gastrointestinal, respiratory, and genitourinary tracts and exert their major activities in the secretions of these tracts. The function of the IgD class is presently uncertain.

Many small chemicals injected into the body do not induce antibody formation because an essential determinant of a molecule's capacity to serve as an antigen is size (most antigens have molecular weights greater than 10,000). However, many smaller molecules can act as antigens after first attaching themselves to one of the host's proteins, thus forming a complex large enough to induce antibody formation. Still other low-molecular-weight substances incapable of inducing antibody synthesis because of their small size can combine with antibodies induced by another antigen; in such cases the structural unit of the large true antigen which was critical in the induction process must have been similar or identical to that of the small molecule. These last two phenomena explain why many small molecules can cause allergic attacks.

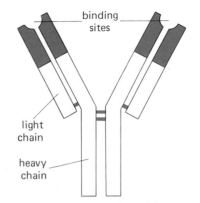

Figure 17-9. Antibody structure. The links represent disulfide bonds.

The antigens we shall be most concerned with are large protein or carbohydrate molecules which are in the outer surface of microbes or are microbial products, such as bacterial toxins. However, components of almost any foreign cell or molecule not normally present in the body may act as an antigen (moreover, even normal body components can induce antibody formation under unusual circumstances). Once the antibodies have been formed (the mechanism will be discussed in a subsequent section), they are released into the blood, reach the site where the antigen is located, and combine with it.

Functions of antibodies. In a previous section we described how a local inflammatory response is induced nonspecifically by any tissue damage. Now we shall see that the presence of antigen-antibody complexes triggers off events which profoundly *amplify* the inflammatory response. In other words, the major function of humoral immune mechanisms is to enhance and make more efficient the inflammatory response and cell elimination already initiated in a nonspecific way by the invaders. Thus, vasodilation, increased vascular permeability to protein, neutrophil exudation, phagocytosis, and killing of microbes without prior phagocytosis are all markedly enhanced.

Activation of complement system. There are several mechanisms by which the presence of antigen-antibody complexes enchances inflammation, but the single most important involves the complement system. As described earlier, this system, which provides mediators for virtually every process in inflammation, is involved to some extent in the nonspecific inflammatory response, but the presence of antibody attached to antigen profoundly increases its participation because the antibody-antigen complex is a powerful activator of the first step in the complement sequence. Precisely how this activation is brought about is not completely clear, but it is certain that the critical event is combination of the first complement molecule in the sequence to the antibody of the antibody-antigen complex.

It is important to note that the complement molecule binds not to the unique antigen-specific binding sites in the antibody's "prongs," but rather to binding sites in the antibody's stem. Since the latter are the same in virtually all antibodies of the IgG and IgM classes, the complement molecule will bind to *any* antibodies belonging to these classes. In other words, there is only one set of complement molecules, and, once activated, they do essentially the same thing, regardless of the specific identity of the invader. In contrast, the formation of antibodies to antigens on the invader and their subsequent combination are highly specific.

To reiterate, the function of the antibodies is to "identify" the invading cells as foreign by combining with antibody-specific antigens on the cell's surface; the complement system is subsequently activated when the first complement molecules in the sequence combine with this antigen-bound antibody, and the cascade of activated complement molecules then mediate the actual attack. Once activated, the complement components which facilitate phagocytosis or kill the microbes outright are so nonspecific in their ability to bind to cells that they are potentially capable of attacking the body's own cells as well; this does not normally occur because their period of activation is extremely short-lived and they do not have time to get to an "innocent" host cell from the microbe on whose surface the original activation had occurred (not only the first antibody-induced step but several others in the complement sequence occur on the surface of the microbe).

Direct enhancement of phagocytosis. Activation of complement is not the only mechanism by which antibodies enhance phagocytosis. Merely the presence of antibody attached to antigen on the microbe's surface has some enhancing effect. This

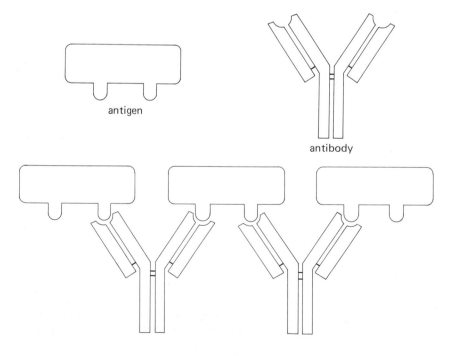

Figure 17-10. Interlocking complex of antigens and antibodies.

antigen

antibody

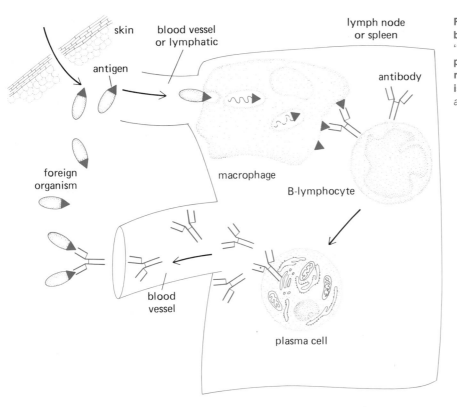

Figure 17-11. Induction of antibody synthesis by a microbe. The "processing" of antigen by macrophage is still not clear. The possible role of helper or suppressor T cells is not shown. *(Adapted from Singer and Hilgard.)*

combination of antibody with microbial antigen increases the activity of phagocytes bound to the same microbe by somehow enhancing release of membrane-bound calcium into the phagocyte's cytosol, thereby triggering the interactions between actin and myosin leading to membrane invagination.

Direct neutralization of bacterial toxins and viruses. Bacterial toxins and certain viral components act as antigens to induce antibody production. The antibodies then combine chemically with the toxins and viruses to "neutralize" them. Neutralization in reference to a virus means that the combined antibody prevents attachment of the virus to host cell membranes, thereby preventing virus entry into the cell. Similarly, antibodies neutralize bacterial toxins by combining chemically with them, thus preventing the interaction of the toxin with susceptible cell-membrane sites. In both cases, since each antibody has two binding sites for combination with antigen, chains of antibody-antigen complexes are formed (Fig. 17-10) which are then phagocytized.

Antibody production. When a foreign antigen reaches lymphoid tissues it triggers off antibody synthesis (Table 17-4 and Fig. 17-11). The antigen may reach the spleen via the blood, but much more commonly it is carried from its site of entry via the lymphatics to lymph nodes. There it stimulates a tiny fraction of the B lymphocytes to enlarge and undergo rapid cell division, most of the progeny of which then differentiate into plasma cells, which are the active antibody producers. The most striking aspect of this transformation is a marked expansion of the cytoplasm, which consists almost entirely of the granular type of endoplasmic reticulum (Fig. 17-12) found in other cells which manufacture protein for export; after synthesis, the antibodies are released into the blood or lymph. Some of the B-cell progeny do not fully differentiate into plasma cells but rather into "memory" cells, ready to respond rapidly should the antigen ever reappear at a future time; thus, the presence of the "memory" cells avoids much of the delay which occurred during the initial infection.

Note that we stated that only a tiny fraction of the total B cells respond to any given antigen. It

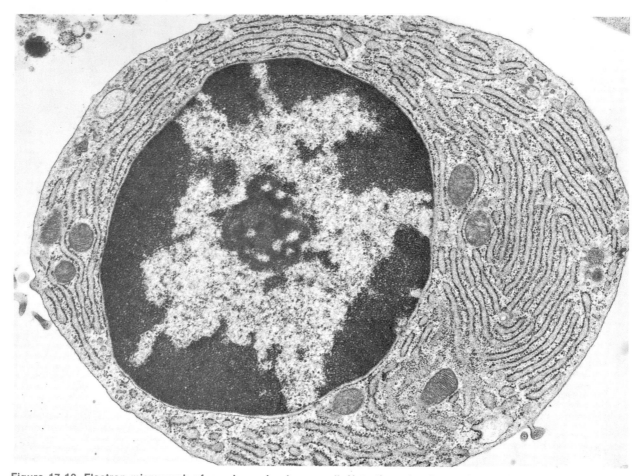

Figure 17-12. Electron micrograph of a guinea pig plasma cell. Note the extensive endoplasmic reticulum. *(From W. Bloom and D. W. Fawcett, "A Textbook of Histology," 9th ed., W. B. Saunders Company, Philadelphia, 1968.)*

can be demonstrated that different antigens stimulate entirely different populations *(clones)* of B cells. This is because the cells of any one lymphocyte clone (and the plasma cells it gives rise to) are capable of secreting only one kind of antibody. This limited synthetic capacity was probably determined by random mutations of the genes coding for the variable amino acids in the specific portions of the antibody chains. (Antibody-synthesizing cells replicate and die so often that randomly occurring mutations could account for the observed degree of diversity.) Thus, according to this clonal theory, different antigens do not direct a single cell to produce different antibodies; rather each specific antigen triggers activity in the clone of cells already predetermined to secrete only anti-

body specific to that antigen. The antigen selects this particular clone and no other because the B cell displays on its surface immunoglobulin binding sites identical to those of the antibodies which it is capable of producing (after differentiation into a plasma cell). These surface immunoglobulins act as receptor sites with which the antigen can combine, thereby triggering off the entire process of division, differentiation, and antibody secretion just described. The staggering but statistically possible implication is that there must exist millions of different clones, one for each of the possible antigens an individual *might* encounter during life.

In concluding this section on antibody synthesis, it should be pointed out that interaction between antigen and B-cell receptor site is considerably more complex than that just described above, for it seems clear that macrophages also play a crucial role. During antigen activation of the B cells, there occurs a clustering of macrophages around the relevant B-cell clone. It is likely that the macrophages "process" and "present" the antigen in some way so as to allow it to activate the B-cell receptor sites, but the actual events remain unknown. As mentioned earlier, T cells also play a role, for they may facilitate or inhibit B-cell antibody formation in response to an antigenic stimulus.

Active and passive humoral immunity. We have been discussing antibody formation without regard to the course of events in time. The response of the antibody-producing machinery to invasion by a foreign antigen varies enormously, depending upon whether it has previously been exposed to that antigen. Antibody response to the first contact with a microbial antigen occurs slowly over several days, with some circulating antibody remaining for long periods of time, but a subsequent infection elicits an immediate and marked outpouring of additional antibody (Fig. 17-13). It is evident that this type of "memory" confers a greatly enhanced resistance toward subsequent infection with a particular microorganism. This resistance, built up as a result of actual contact with microorganisms and their toxins or other antigenic components, is known as *active immunity.* Until modern times, the only way to develop active immunity was actually to suffer an infection, but now a variety of other medical technics are used, i.e., the injection of vaccines or microbial derivatives. The actual material injected may be small quantities of living or weakened microbes, e.g., polio vaccine, small quantities of toxins, or harmless antigenic materials derived from the microorganism or its toxin. The general principle is always the same: Exposure of the body to the agent results in the induction of the antibody-synthesizing machinery required for rapid, effective response to possible future infection by that particular organism. However, for many microorganisms the memory component of the antibody response does not occur, and antibody formation follows the same time course regardless of how often the body has been infected with the particular microorganism.

A second kind of immunity, known as *passive*

immunity, is simply the direct transfer of actively formed antibodies from one person (or animal) to another, the recipient thereby receiving preformed antibodies. This exchange normally occurs between fetus and mother across the placenta and is an important source of protection for the infant during the first months of life when the antibody-synthesizing capacity is relatively poor. The same principle is used clinically when specific antibodies or pooled gamma globulin are given a person exposed to or actually suffering from certain infections, such as measles, hepatitis, or tetanus. The protection afforded by this transfer of antibodies is relatively short-lived, usually lasting only a few weeks. The procedure is not without danger since the injected antibodies (often of nonhuman origin) may themselves serve as antigens, eliciting antibody production by the recipient and possibly severe allergic responses.

Summary. We may now summarize the interplay between nonspecific and specific immune mechanisms in resisting a bacterial infection. When a bacterium is encountered for the first time, nonspecific defense mechanisms resist its entry and, if entry is gained, attempt to eliminate it by phagocytosis (and, to some extent, by nonphagocytic killing). Simultaneously, the bacterial antigens induce the differentiation of specific B-cell clones into plasma cells capable of antibody production. If the nonspecific defenses are rapidly successful, these specific immune responses may never play an important role. If only partly successful, the infection may persist long enough for significant amounts of antibody to reach the scene; antibody activates its chemical amplification system—complement—which both enhances phagocytosis and directly destroys the foreign cells. In either case all subsequent encounters with that type of bacteria will be associated with the same sequence of events, with the crucial difference that the specific immune responses are brought into play much sooner and with greater force; i.e., the person would enjoy active immunity against that type of bacteria.

Cell-mediated immunity

The T lymphocytes are responsible for cell-mediated immunity. T cells, like B cells, are clonal in that each clone of T cells bears on its cell surface

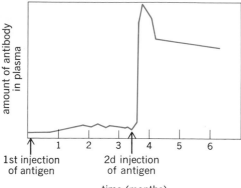

Figure 17-13. Rate of antibody production following the initial contact with an antigen and a subsequent contact with the same antigen.

genetically determined receptor sites for a specific foreign antigen. Upon exposure to and combination with the appropriate antigen (in the presence of macrophages), the T cell becomes "sensitized," i.e., undergoes enlargement and mitosis, with differentiation of the daughter cells. So far, the story sounds similar to that described earlier for B cells, but the crucial difference has to do with the activities of these differentiated daughter cells: Instead of secreting antibodies (as do the plasma-cell progeny of B cells), these cells actually combine with the antigen on the surface of the foreign cell (Fig. 17-13), and this triggers the release locally of a powerful battery of chemicals (known collectively as *lymphokines*) (Fig. 17-13). We must emphasize the important geographic difference between plasma cell and sensitized T-cell function: Antibodies are secreted by plasma cells located in lymph nodes and other lymphoid organs often far removed from the invasion site and reach the site via the blood; in contrast, a sensitized T cell travels to the invasion site where, upon combination with antigen bound to the foreign cell, it releases its chemicals (Fig. 17-14).

As is true for the B system, some of the sensitized T cells do not actually participate in the immune response but serve as a "memory bank" which greatly speeds up and enhances the immune

Figure 17-14. Mechanism of rejection of renal transplant by sensitized T cells. *(Adapted from Hume.)*

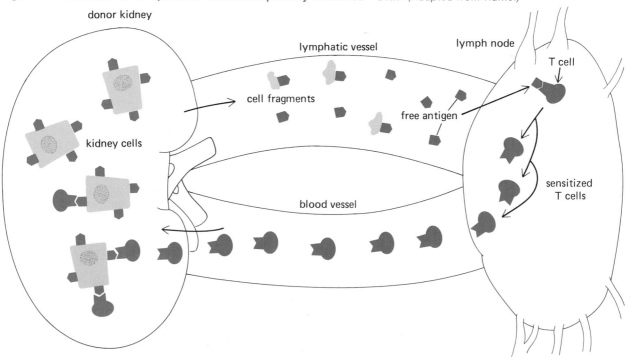

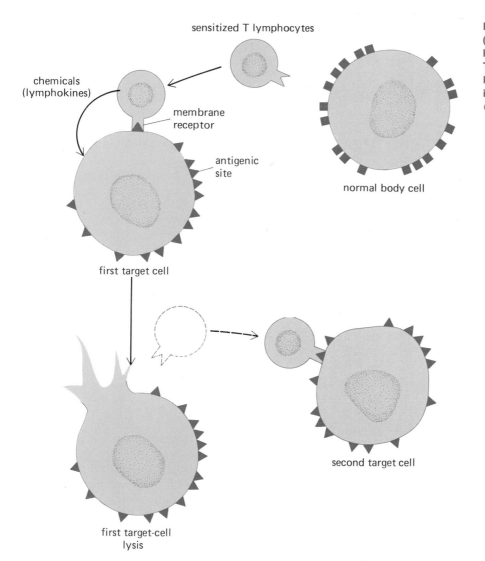

Figure 17-15. Killing of foreign (target) cells by sensitized T lymphocytes. The nature of the T-lymphocyte binding site is not known for certain, but may be an immunoglobulin. *(Adapted from Cerottini.)*

response if the person is ever exposed to the specific antigen again. Thus, active immunity exists for cell-mediated immune responses just as for antibody responses. Passive immunity also exists in this system and can be conferred by administering sensitized lymphocytes taken from a previously infected person (or animal).

The chemicals released when sensitized lymphocytes combine with specific antigen kill cells directly (Fig. 17-15) and, in addition, act as an amplification system for the facilitation of the inflammatory response and phagocytosis. Thus, cell-mediated immunity is analogous to humoral im-

munity in that it serves, in large part, to enhance and make more efficient the nonspecific defense mechanisms already elicited by the foreign material. The major difference is that humoral immunity utilizes a circulating group of plasma proteins (the complement system) as its major amplification system whereas the T cells literally produce and secrete their own chemical amplification system.

As might be predicted, some of these chemicals are chemotaxic factors. These serve to attract some neutrophils but many more monocytes to the area. The monocytes are converted to macrophages

Table 17-5. Summary of host responses to viruses

	Main cells involved	Comment on action
Nonspecific responses		
Anatomic barriers	Surface linings of body	Simple barrier; antiviral chemicals
Inflammation	Tissue macrophages	Phagocytosis of extracellular virus
Interferon	Multiple cell types	Prevention of viral replication inside host cells
Specific responses		
Humoral	Plasma cells derived from B lymphocytes secrete antibodies which:	Neutralize virus, preventing entry to cell. Activate complement leading to both enhanced phagocytosis and direct destruction of extracellular virus.
Cell-mediated	Sensitized T lymphocytes secrete chemicals which:	Destroy host cell, inducing release of virus so that it can be phagocytized. Prevent viral replication (this secreted chemical is interferon).

and begin their job of phagocytosis. Not only do the T cells secrete chemotaxic factors to attract the macrophages-to-be; they secrete another substance which keeps the macrophages in the area and stimulates them to greater phagocytic activity (indeed, such revved-up macrophages are known as "angry" macrophages). In addition to facilitating the killing of target cells by phagocytosis, the lymphocytes secrete so-called cytotoxins which are able to kill target cells directly, i.e., without phagocytosis. Here is another analogy to the complement system, with its ability to destroy cells directly as well as to facilitate phagocytosis. Interferon is also released in large quantity from T cells.

The cytolytic T cells we have been describing in this section belong to only one of three distinct T-cell populations mentioned earlier. Cells of the other two classes play totally different roles in specific immunity, namely the facilitation or suppression of B-cell production of antibody. At present we know little about how these "helper" and "suppressor" cells perform their functions but it is likely that they combine with antigenic sites on the microbe, and release chemicals which act upon the B cells to influence their differentiation into plasma cells. Their possible roles in both the excessive and deficient antibody production seen in many human diseases are presently being intensively studied.

One aspect of T-lymphocyte function against viruses requires emphasis (Table 17-5). As stated several times, viruses take up residence within the body's cells; accordingly, in order to attack a virus once it has gained cellular entry, the lymphocyte must destroy the cell, itself. Such destruction occurs because the viral nucleic acid codes for a protein foreign to the body, and once this protein has been synthesized by the host cell, it becomes located on the cell's surface where it can act as an antigen to elicit a T-cell attack. The result is destruction of the host cell with release of the viruses which can then be directly attacked. Generally, only a few host cells must be sacrificed in this way, but once viruses have had a chance to replicate and spread from cell to cell, so many host cells may be attacked by the body's own defenses that serious malfunction may result.

Immune surveillance: Defense against cancer. A major function of cell-mediated immunity is to recognize and destroy cancer cells. This is made possible by the fact that virtually all cancer cells have some surface antigens different from those of other body cells and can, therefore, be recognized as "foreign." It is likely that cancer arises as a result of genetic alteration (by viruses, chemicals, radiation, etc.) in previously normal body cells. One manifestation of the genetic change is the appearance of the new surface antigens. Circulating T cells encounter and become sensitized to these foreign cells, combine with the antigens on their surface, and release the effector chemicals which destroy the cells by the mechanisms described above. It is presently believed that such transformations occur very frequently, i.e., that we may "get cancer once a day" (one expert's estimate), but that the cells are destroyed as fast as they arise. According to this view, only when the cell-mediated system is ineffective in either recognizing or destroying the cells do they multiply and produce clinical cancer.

In this last regard, there may occur an important interaction between the humoral and cell-mediated systems which actually protects the cancer cell. We have just pointed out that the tumor-specific antigens stimulate the development of sensitized lymphocytes against themselves. Simultaneously, they may also stimulate the production by B cells of circulating antibodies (of the IgG class). These antibodies combine with the cell-surface antigen sites, and the next events may be quite variable; in some cases, the antibody may facilitate the destruction of the tumor cell whereas in most cases it may actually protect the cell by preventing T cells from combining with the antigen. In the latter cases, the antibodies are called "blocking" antibodies and the process is called *immune enhancement*. Whether facilitation or blocking occurs seems to depend upon precisely how the antibody interacts with the antigen, but many unanswered questions remain (for example, why do the blocking antibodies not activate complement which would destroy the tumor cells?).

Clearly, the relative amounts of blocking antibody and sensitized T cells elicited by the antigenic stimulus are a major determinant of whether the emergent cancer cell is destroyed or not. Of great potential importance is the finding that dietary protein deficiency markedly impairs the production of IgG antibodies and, therefore, may favor destruction of tumor cells. Does our large protein intake favor tumor growth by enhancing IgG synthesis? Can growth of certain tumors be medically controlled by restriction of certain amino acids?

Such questions have obvious practical significance, and the hope is that an understanding of the immune responses to cancer may make possible the development of more effective anticancer weapons. Some avenues presently being studied are the possibilities of vaccination with tumor antigens (active immunity) and transfer of immunity with sensitized T lymphocytes (passive immunity). These approaches attempt to enhance defenses against specific types of tumors, but another quite different approach is to induce nonspecific stimulation of the T-cell system. It has been found that injections of certain substances (for example, an attenuated strain of a certain bacterium known as BCG) somehow "rev" up the entire T-cell system so that defenses against any cancer cell might be enhanced in the process.

Rejection of tissue transplants. The cell-mediated immune system is also mainly responsible for the recognition and destruction, i.e., *rejection*, of tissue transplants (Fig. 17-14). On the surfaces of all nucleated cells of an individual's body are antigenic protein molecules known as *histocompatibility antigens*. The genes which code for these proteins are, of course, inherited from one's parents, so that the offspring's group of antigens are, in part, similar to the parents' but not identical. Clearly, the more closely related two people are the more similar these antigens will be, but no two people (other than identical twins) have identical groups.

When tissue is transplanted from one individual to another, those surface antigens which differ from the recipient's are recognized as foreign and are destroyed by sensitized circulating T cells. As was true for the response to cancer cells, the foreign cells may also stimulate the secretion of circulating "blocking" antibodies; if this occurs, the chances for graft survival are enhanced. Note that, in immunology, whether a phenomenon is desirable or undesirable depends on the point of view; tumor enhancement is undesirable whereas graft enhancement is desirable.

Some of the most valuable tools aimed at reducing graft rejection are radiation and drugs which kill actively dividing lymphocytes and, thereby, decrease the T-cell population. Unfortunately, this also results in depletion of B cells as well so that antibody production is diminished and the patient becomes highly susceptible to infection. A more discriminating method presently being tried is to prepare and inject into the recipient antibodies against the T cells; by this means, the T cells would be destroyed but not the B cells. Many other tools are presently being developed, all of which depend upon an understanding of the basic physiology of specific immune responses. As might be predicted from the previous section, graft recipients receiving medication to reduce their T-cell activity manifest an increased tendency to develop certain types of cancer.

Related to the general problem of graft rejection is one of the major unsolved questions of immunology: How does the body avoid producing antibodies or sensitized lymphocytes to its own cells; i.e., how does it distinguish self-antigens from nonself-antigens? In general, it appears that any antigens present during embryonic and very early neonatal life are recognized as self and no antibodies or sensitized lymphocytes are formed against them later in life, following maturation of specific immune mechanisms. This can be shown by fooling the embryo in the following manner: Foreign mouse cells are injected into an embryo mouse during intrauterine life; months later, when the mature mouse is given a graft from the same foreign species, the graft is not rejected.

Of course, the generalization stated above is

empirical fact, not explanation. Present evidence warrants the generalization that the thymus, as might be predicted, is very much involved in this "imprinting" of self-recognition but just how is simply not known. The question is clearly of more than academic interest since, if we understood the mechanism by which tolerance for one's own tissues is established, then we might be able to confer tolerance for transplants.

Transfusion reactions and blood types

Transfusion reactions are a special example of tissue rejection, which illustrates the fact that antibodies rather than sensitized T cells can sometimes be the major factor in leading to the destruction of nonmicrobial cells. Among the large numbers of erythrocyte membrane antigens, we still recognize those designated A, B, and O as most important. These antigens are inherited, A and B being dominant. Thus, an individual with the genes for either A and O or B and O will develop only the A or B antigen. Accordingly, the possible blood types are A, B, O, and AB. If the typical pattern of antibody induction were followed, one would expect that a type A person would develop antibodies against type B cells only if the B cells were introduced into the body. However, what is atypical of this system is that even without initial exposure the type A person always has a high plasma concentration of anti-B antibody. The sequence of events during early life which lead to the presence of the so-called natural antibodies in all type A persons is unknown. Similarly, type B persons have high levels of anti-A antibodies; type AB persons obviously have neither anti-A nor anti-B antibody; type O persons have both; anti-O antibodies are usually not present in anyone.

With this information as background, what will happen if a type A person is given type B blood? There are two incompatibilities: (1) The recipient's anti-B antibody causes the transfused cells to be attacked; (2) the anti-A antibody in the transfused plasma causes the recipient's cells to be attacked. The latter is generally of little consequence, however, because the transfused antibodies become so diluted in the recipient's plasma that they are ineffective. It is the destruction of the transfused cells which produces the problems. The range of possibilities is shown in Table 17-6. It

Table 17-6. Summary of ABO blood-type interactions

Recipient	Donor	Compatible ?
A	O, A	yes
B	O, B	yes
AB	O, A, B, AB	yes
O	O	yes
A	AB, B	no
B	AB, A	no
AB	– – –	– – –
O	AB, A, B	no

should be evident why type O people are frequently called universal donors whereas type AB people are universal recipients. These terms, however, are misleading and dangerous since there are a host of other incompatible erythrocyte antigens and plasma antibodies besides those of the ABO type. Therefore, except in dire emergency, the blood of donor and recipient must be carefully matched.

Another antigen of medical importance is the so-called *Rh factor* (because it was first studied in rhesus monkeys) now known to be a group of erythrocyte membrane antigens. The Rh system follows the classic immunity pattern in that no one develops anti-Rh antibodies unless exposed to Rh-type cells (usually termed Rh-positive cells) from another person. Although this can be a problem in an Rh-negative person, i.e., one whose cells have no Rh antigen, subjected to multiple transfusions with Rh-positive blood, its major importance is in the mother-fetus relationship. When an Rh-negative mother carries an Rh-positive fetus, some of the fetal erythrocytes may cross the placental barriers into the maternal circulation, inducing her to synthesize anti-Rh antibodies. Because the movement of fetal erythrocytes into the maternal circulation occurs mainly during separation of the placenta at delivery, a first Rh-positive pregnancy rarely offers any danger to the fetus, since delivery occurs before the antibodies can be made. In future pregnancies, however, these anti-Rh antibodies will already be present in the mother and can cross the placenta to attack the erythrocytes of an Rh-positive fetus. The risk increases with each Rh-positive pregnancy as the mother becomes more and more sensitized. Fortunately, Rh disease can be prevented by giving any Rh-negative mother gamma globulin against Rh erythrocytes within 72 h after she has delivered an Rh-positive infant. These exogenous antibodies bind to the antigenic sites on any Rh erythrocytes entering the mother's blood during delivery and prevent them from inducing antibody synthesis by the mother (both the cells and exogenous antibody are soon destroyed).

Factors which alter the body's resistance to foreign cells

Let us examine two seemingly opposed statements: (1) Tuberculosis is *caused* by the tubercle bacillus; (2) tuberculosis is caused by malnutrition. The first statement seems the more accurate in the sense that the disease, tuberculosis, will not occur in the absence of infection by tubercle bacilli. Yet we also know that many people harbor these bacteria, but never develop tuberculosis. There must exist, therefore, other factors which, by upsetting the balance between host and microbe, permit invasion, multiplication, and production of symptoms. In a sense, then, malnutrition "causes" tuberculosis by doing just this. In fact, there need be no quibbling over semantics once one realizes that the presence of the microbe is the necessary but frequently not sufficient cause of the disease. Therefore, it becomes very important to define those influences which determine the body's capacity to resist infection (as well as cancer cells and transplants). We offer here only a few examples.

A person's general nutritional status is extremely important, but there is no indication that *excess* vitamins, etc., can confer increased resistance. A preexisting disease (infectious or noninfectious) can also predispose the body to infection. Diabetics, for example, suffer from a propensity to numerous infections, at least partially explainable on the basis of defective leukocyte function. Moreover, any injury to a tissue lowers its resistance, perhaps by altering the chemical environment or interfering with blood supply.

In numerous examples one of the basic resistance mechanisms itself is deficient. A striking case is that of congenital deficiency of plasma gamma globulin, i.e., failure to synthesize antibodies; these patients cannot survive without frequent intravenous injections of gamma globulin. Similarly, a lethal complement deficiency may exist. A decrease in the production of leukocytes is also an important cause of lowered resistance, as, for example, in patients given drugs specifically to inhibit rejection of tissue or organ transplants. The total quantity of leukocytes circulating is not necessarily critical; patients with leukemia, for example, may have tremendous numbers of blood neutrophils or monocytes, but these cells are almost all immature or otherwise incapable of normal function; such patients are extremely prone to infection. Reduced functional activity of lymphocytes is also seen in the elderly and may account for their decreased resistance.

Finally, we must mention the most important of the external agents we employ in altering resistance to infection, the *antibiotics*, such as penicillin. Use of these antibacterial agents is made possible because they are harmful to microbes but relatively innocuous to the body's cells. This characteristic distinguishes them from the *disinfectants*, which are highly effective antibacterial agents but equally toxic to the body. The "relatively" is quite important since all antibiotics are toxic to a lesser or greater degree and must not be used indiscriminately. A second reason for judicious use is the problem of drug resistance. Most large bacterial populations contain mutants which are not sensitive to the drug and which are thus selected out by the drug. These few are capable of multiplying into large populations resistant to the effects of that particular antibiotic. Perhaps even more important, resistance can be transferred from one microbe directly to another previously nonresistant microbe by means of chemical agents ("resistance factors") passed between them. A third reason for the judicious use of antibiotics is that these agents may actually contribute to a new infection by altering the normal flora so that overgrowth of an antibiotic-resistant species occurs. Antibiotics exert a wide variety of effects, but the common denominator is interference with the synthesis of one or more of the bacteria's essential macromolecules.

Hypersensitivity and tissue damage

Immune responses obviously evolved to protect the body against invasion by foreign matter. Unfortunately, they frequently cause malfunction or damage to the body itself. The term *hypersensitivity* refers to an acquired reactivity to an antigen which can result in bodily damage upon subsequent exposure to that particular antigen. Hypersensitivity responses may be due to activation of either the humoral or cell-mediated system. There are a variety of types of hypersensitivity responses, and we shall describe only two categories: atopic allergy and autoimmune disease.

Atopic allergy

A certain portion of the population is susceptible to sensitization by environmental antigens such as pollen, dusts, foods, etc. Initial exposure to the antigen leads to some antibody synthesis but, more important, to the memory storage which characterizes active immunity. Upon reexposure, the antigen elicits a more powerful antibody response. So far, none of this is unusual. What is it then that leads to body damage? The fact is that these particular antigens stimulate the production of the IgE class of antibodies which, upon their release from plasma cells, circulate to various parts of the body and attach themselves to mast cells (and basophils). When the antigen then combines with the IgE attached to the mast cell, the complex triggers entry of calcium into the cell and a resulting release of the mast cell's histamine and other vasoactive chemicals (complement is not involved in IgE effects). These chemicals then initiate a local inflammatory response. Thus, the symptoms of atopic allergy are due to the various effects of these chemicals and the body site in which the antigen–IgE–mast cell combination occurs. For example, when a previously sensitized person inhales ragweed pollen, the antigen combines with IgE–mast cells in the respiratory passages. The chemicals released cause increased mucus secretion, increased blood flow, leakage of protein, and contraction of the smooth muscle lining the airways. Thus, there follow the symptoms of congestion, running nose, sneezing, and difficulty in breathing which characterize hayfever.

In this manner, the symptoms of atopic allergy may be localized to the site of entry of the antigen. However, sometimes systemic symptoms may result if very large amounts of the vasoactive chemicals released enter the circulation and cause severe hypotension and bronchiolar constriction. This sequence of events (anaphylactic shock) can actually cause death and can be elicited in some sensitized people by the antigen in a single bee sting.

A major puzzle to biologists is the inappropriate nature of most atopic allergic responses, which are usually far more damaging to the body than the antigen triggering them. In other words, we clearly see the maladaptive nature of antigen-IgE reactions, but we do not know why such a system should have evolved, i.e., what normal physiological function is subsumed by IgE antibodies.

As may well be imagined, antihistamines offer some relief in atopic allergies; it is usually incomplete, however, since other vasoactive agents are also released from the mast cells. In severe cases, the anti-inflammatory powers of large doses of cortisol are employed. The therapy known as *desensitization* is also of great interest because it offers another example of so-called blocking antibodies. After the specific antigen to which a person is sensitive has been identified, it is injected frequently in minute but increasing quantities. This procedure induces the synthesis of IgG antibodies against the antigen, so that when the antigen is subsequently encountered in the normal way, it will be complexed by these IgG antibodies, thereby preventing it from combining with the IgE. Because the IgG antibodies are not fixed to mast cells, no allergic response results.

Autoimmune disease

We must qualify a generalization made previously by pointing out that the body does, all too often, produce antibodies or sensitized T cells against its own tissues, the result being cell damage or alteration of function. A growing number of human diseases are being recognized as *autoimmune* in origin.

There are multiple causes for the body's failure to recognize its own cells: (1) Normal antigens may be altered by combination with drugs or environmental chemicals; (2) the cell may be infected by a virus whose nucleic acid codes for a new protein (antigen); (3) genetic mutations may yield new antigens; (4) the body may encounter microbes whose antigens are so close in structure to certain self-antigens that the antibodies or sensitized lymphocytes produced against these antigens cross-react with the self-antigens; (5) components of certain tissues might never be exposed during embryonic life to whatever organs (? the thymus) must recognize and memorize the self-antigens; if they appear in the blood later in life, as the abnormal result of tissue disruption following injury or infection, they are treated as foreign; (6) many investigators are coming to the conclusion that suppressor T cells normally prevent the production of autoantibodies and that a deficiency of these cells contributes to the development of autoimmunity in many cases. This list of possibilities is by no means complete, but whatever the cause, a breakdown in self-recognition results in turning the body's immune mechanisms against its own tissues.

The above description centers on the production of antibodies or sensitized lymphocytes against the body's own cells. However, autoimmune damage may also be brought about in several other quite different ways. An overzealous response (too much generation of complement or release of chemicals from platelets, neutrophils, or sensitized lymphocytes) may cause damage not only to invading foreign cells but to neighboring normal cells or membranes as well. For example, were a circulating antigen-antibody complex to be trapped within capillary membranes, the generation of complement or release of chemicals into the area might cause damage to the adjacent membranes. As might be predicted, the kidney glomeruli, with their large filtering surface, are prime targets for such autoimmune destruction.

Another important question in the field of autoimmunity concerns the relationship between mother and fetus: Why does not the mother's immune system reject the fetus (half of the fetal antigens are, of course, paternal and therefore foreign to the mother)? No generally accepted explanation is available at present.

Section B
Metabolism of foreign chemicals

The body is exposed to a huge number of environmental chemicals, including inorganic nonnutrient elements, naturally occurring fungal and plant toxins, and synthetic chemicals. This last category is by far the largest, since there are now more than 10,000 foreign chemicals being commercially synthesized (over 1 million have been synthesized at one time or another); these are "foreign" in the sense that they are not normally found in nature. These foreign chemicals inevitably find their way into the body, either because they are purposely administered, as drugs (medical or "recreational"), or simply because they are in the air, water, and food we use.

As described in section A of this chapter, foreign materials can induce inflammation and specific immune responses. However, these defenses are directed mainly against foreign cells and although noncellular foreign chemicals can also elicit certain of them (as in atopic allergy, for example), such immune responses do not constitute the major defense mechanisms against most foreign chemicals. Rather, molecular alteration (biotrans-

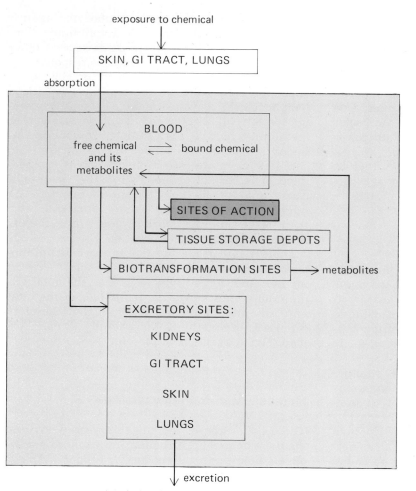

exposure to chemical

SKIN, GI TRACT, LUNGS

absorption

BLOOD

free chemical and its metabolites ⇌ bound chemical

SITES OF ACTION

TISSUE STORAGE DEPOTS

BIOTRANSFORMATION SITES → metabolites

EXCRETORY SITES:

KIDNEYS

GI TRACT

SKIN

LUNGS

excretion

chemical and its metabolites

Figure 17-16. Metabolic pathways for foreign chemicals.

formation) and excretion constitute the primary mechanisms.

A central focus for the body's handling of any foreign chemical is those factors which determine the effective concentration of the chemical at its sites of action (Fig. 17-16). First, the chemical must gain entry to the body through the gastrointestinal tract, lungs, or skin (or placenta in the case of a fetus). Accordingly, its ability to move across these barriers will have an important influence on its blood concentration. But, as Fig. 17-16 illustrates, the rate of entry into the body is only one of many factors determining the concentration of the chemical at its site of action. Once in the blood, the chemical may become bound reversibly to plasma proteins or to erythrocytes; this

lowers its free concentration and, thereby, its ability to alter cell function. It may be transported into storage depots (for example, DDT into fat tissue) or it may undergo biotransformation. The metabolites resulting from these latter enzyme-mediated reactions themselves enter the blood and are subject to the same fates as the parent molecules. Finally, the foreign chemical and/or its metabolites may be eliminated from the body in the urine, expired air, skin secretions, or feces (having been deposited in the latter by biliary secretion). All these processes serve to determine the free blood concentration of chemical, which in turn is one of the major determinants of the chemical's ability to reach its sites of action, and we now discuss each of them.

Absorption

In practice, most organic molecules move through the lining of some portion of the gastrointestinal tract fairly readily, either by simple diffusion or by carrier-mediated transport. This should not be surprising, since the gastrointestinal tract evolved to favor absorption of the wide variety of nutrient molecules in the environment; the nonnutrient synthetic organic chemicals are the beneficiaries of these relatively nondiscriminating transport mechanisms.

The lung alveoli are also highly permeable to most organic chemicals and therefore offer an easy entrance route for airborne chemicals. They are also an important entry site for airborne metals, which generally penetrate the gastrointestinal tract very poorly. One important aspect of absorption through the lungs is that the liver does not get first crack at the chemical (as it does when entry is via the gastrointestinal tract); by the same token, any chemical which is toxic to the liver is not as dangerous when it enters via the lungs.

Lipid solubility is all-important for entry through the skin, so that this route is of little importance for charged molecules but can be used by oils, steroids, and other lipids.

The penetration of the placental membranes by foreign chemicals is one of the most important fields in toxicology, since the effects of environmental agents on the fetus during critical periods of development may be quite marked and, in many cases, irreversible (thalidomide offers a tragic example). We are still relatively ignorant of placental transport mechanisms but it is clear that diffusion is an important mechanism for lipid-soluble substances and that carrier-mediated systems (which evolved for the carriage of endogenous nutrients) may be usurped by foreign chemicals to gain entry into the fetus.

Storage sites

The major storage sites for foreign chemicals are cell proteins, bone, and fat. The chemical bound in these sites is in equilibrium with the free chemical in the blood so that an increase in blood concentration causes more movement into storage; conversely, as the chemical is eliminated from the body and its blood concentration falls, movement occurs out of storage sites.

These storage sites obviously are a source of protection but they also permit the multiplier effect in food chains. Moreover, it sometimes happens that the storage sites accumulate so much chemical that they become damaged, themselves. Finally, there is the possibility of rapid release with potentially toxic effects.

Excretion

To appear in the urine, a chemical must either be filtered through the glomerulus or secreted across the tubular epithelium (Chap. 13). Glomerular filtration is, as emphasized in Chap. 13, a bulk flow process so that all low-molecular-weight substances in plasma undergo filtration; accordingly, there is considerable filtration of most environmental chemicals except for those which are mainly bound to plasma proteins (note that protein binding is, therefore, a mixed blessing in that it reduces toxicity but impedes excretion). In contrast, tubular secretion is by discrete transport processes, and many environmental chemicals (penicillin is a good example) utilize the mediated transport systems available for naturally occurring substances.

Once in the tubular lumen, either via filtration or tubular secretion, the foreign chemical may still not be excreted for it may be reabsorbed back across the tubular epithelium into the blood. This is a major problem since so many foreign chemicals are highly lipid-soluble; as the filtered fluid moves along the renal tubules, these molecules passively diffuse along with reabsorbed water through the tubular epithelium and back into the blood. The net result is that little is excreted in the urine, and the chemical is retained in the body. If these chemicals could be transformed into more polar (and, therefore, less lipid-soluble) molecules, their passive reabsorption from the tubule would be retarded, and they would be excreted more readily. This type of transformation is precisely what occurs in the liver, as described in the next section.

An analogous problem exists for those foreign molecules (and trace metals) excreted into the bile. Many of these substances having reached the lumen of the small intestine are absorbed back into the blood, thereby escaping excretion in the feces. This cyclic enterohepatic circulation was described in Chap. 14.

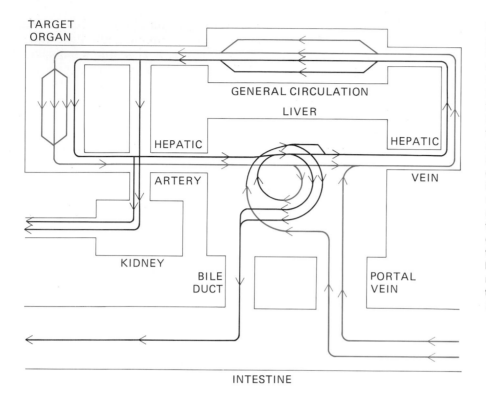

Figure 17-17. Pathway of an ingested drug that is transformed in the liver is shown schematically. After its absorption, the drug *(colored arrows)* enters the liver through the portal vein, may then pass into the general circulation, exert its effect on the target organ, and eventually return to the liver. On each passage through the liver a fraction of the drug is converted, usually into inactive less lipid-soluble metabolites *(black arrows)*. The metabolites may be carried by the bile into the intestines for excretion or may pass through the circulation to the kidneys, to be excreted into the urine; they may also exert actions on the target cells. *(Adapted from Kappas and Alvares.)*

Biotransformation

The metabolic alteration of foreign molecules occurs mainly in the liver (but to some extent also in kidney, skin, placenta, and other organs). A large number of distinct enzymes and pathways are involved, but the common denominator of most of them is that they transform chemicals into more polar, less lipid-soluble substances. One consequence of this transformation is that the chemical may be rendered less toxic but this is not always so. The second, more important, consequence is that its tubular reabsorption is diminished and urinary excretion facilitated (Fig. 17-17). Similarly, for substances handled by biliary excretion, gut absorption of the metabolite is less likely so that fecal excretion is also enhanced (Fig. 17-17).

The hepatic enzymes which perform these transformations were first discovered in the context of drug metabolism and were called "drug-metabolizing enzymes," but their spectrum of action is much wider than this. The preferred name is *microsomal enzyme system* (MES) to denote the

fact that they are found mainly in the smooth endoplasmic reticulum. One of the most important facts about this enzyme system is that it is easily inducible, i.e., the activity of the enzymes can be greatly increased by exposure to a chemical which acts as a substrate for the system.

However, all is not really so rosy, for the hepatic biotransformation mechanisms vividly demonstrate how an adaptive response may, under some circumstances, turn out to be maladaptive. These enzymes all too frequently "toxify" rather than "detoxify" a drug or pollutant; in fact many foreign chemicals are quite nontoxic until the liver enzymes biotransform them. Of particular importance is the likelihood that many, if not all, chemicals which cause cancer do so only after biotransformation. These enzymes can also cause problems in another way, because they evolved primarily not to defend against foreign chemicals (which were much less prevalent during our evolution) but rather to metabolize endogenous substrates, particularly steroids and other fat-soluble molecules. Therefore, their induction by a drug or pollutant increases metabolism not only of that drug or pollutant, but of the endogenous substrates as well. The result is a decreased concentration in the body of that normal substrate.

Another fact of great importance concerning the microsomal enzyme system is that, just as certain chemicals induce it, others inhibit it. The presence of such chemicals in the environment could have deleterious effects on the system's capacity to protect against those chemicals it transforms. (Just to illustrate how complex this picture can be, note that any chemical which inhibits the microsomal enzyme system may actually confer protection against those other chemicals which must undergo transformation in order to become toxic.)

Alcohol: An example

Alcohol (specifically one type—ethanol) offers an excellent example of the role biotransformation plays in determining a substance's toxicity and its influence on the body's responses to other chemicals. The overuse of alcohol is associated with liver damage, and for many years it was thought that this damage was due to the malnutrition so frequently accompanying alcoholism. It is now clear that, although severe malnutrition may play some role, the toxic damage to the liver is caused mainly by the metabolites of alcohol, produced by the liver cells, themselves.

Alcohol is initially broken down by liver cells to hydrogen and acetaldehyde, and these seem to be the major culprits. Acetaldehyde is a very toxic chemical which damages mitochondria. The excess hydrogen exerts its damaging effect more subtly by causing increased accumulation of fat (in the form of triacylglycerol) in the liver cells by two pathways: (1) By mass action it drives the synthesis of fatty acids and glycerol (see Chap. 5), the building blocks of triacylglycerol; (2) the hydrogen is "burned" in the mitochondria (see oxidative phosphorylation, Chap. 5), to supply the liver's energy requirement, and this allows fat, the usual source of hydrogen for the liver's mitochondria, to accumulate. The accumulation of fat leads to enlargement of liver cells and damage to them. (We have restricted our discussion to the toxic effects of hydrogen and acetaldehyde on liver cells; other organs and tissues are also damaged by them in a variety of ways.)

The metabolism of alcohol not only damages the liver cells but leads to marked changes in the metabolism of other drugs and foreign chemicals, all explainable by the fact that these other agents share certain of alcohol's metabolic pathways. First, when alcohol is taken at the same time as another drug such as a barbiturate, enhanced effects of the two drugs are observed. This is because the alcohol and barbiturate complete for the same hepatic microsomal enzyme system, resulting in a decreased rate of catabolism of both (and, therefore, increased blood concentrations).

The situation just described was for simultaneous administration of the two drugs and is independent of whether the individual is a chronic overuser of alcohol or not. Let us look now at the long-term effects of chronic overuse. Alcohol is a powerful inducer of the microsomal enzyme system so that its chronic presence causes an increase in the system's activity. The result is an increase in the rate of catabolism of other chemicals, such as barbiturates, which share the enzymes. Thus, the chronic use of alcohol decreases the potency of any given dose of barbiturate by causing its blood concentration to fall more rapidly.

Putting these last two paragraphs together, one can see that the effect of alcohol on the blood concentrations of other chemicals depends both on whether the two drugs are taken simultaneously and whether the person is a chronic overuser of alcohol. But the situation is even more complex, for if the person has suffered serious liver damage due to chronic alcoholism, then the microsomal enzyme system, instead of being induced, may be inadequate simply because the cells have been damaged. At this stage, then, the metabolism of the other agents will be diminished and their effects will be exaggerated.

Finally, we have presented the chronic effects of alcohol only in terms of the effects on the metabolism of other chemicals. It should be clear, however, that similar effects are exerted on the metabolism of alcohol itself. Thus, chronic overuse, but before significant damage has occurred, results in increased catabolism (because of the induction of the microsomal enzymes), and this accounts for much of the "tolerance" to alcohol, i.e., the fact that increasing doses must be taken to achieve a given magnitude of effect. Once severe damage has occurred, the decline in rate of catabolism may lead to increased sensitivity to a given dose, just the opposite of tolerance.

Inorganic elements

Thus far our discussion of environmental chemicals has dealt mainly with organic molecules. Many potentially dangerous substances are not organic chemicals but are normally occurring inorganic elements present in excess because of human activity. As is true for organic drugs and pollutants, the concentrations of inorganic elements at their sites of action depend on rates of entry, excretion, and storage. The gastrointestinal tract offers an important first line of defense since absorption of them is usually quite limited. In contrast, airborne inorganic elements gain entry to the blood more readily through the lungs. Little is known of the mechanisms by which the kidneys and liver handle potentially harmful trace elements; specifically, it is not known whether adaptive increases in excretion are induced by exposure to the element.

Section C
Resistance to stress

Much of this book has been concerned with the body's response to stress in its broadest meaning of an environmental change which must be adapted to if health and life are to be maintained. Thus, any change in external temperature, water intake, etc., sets into motion mechanisms designed to prevent a significant change in some physiological variable. In this section, however, we de-

scribe the basic stereotyped response to stress in the more limited sense of noxious or potentially noxious stimuli. These comprise an immense number of situations, including physical trauma, prolonged heavy exercise, infection, shock, decreased oxygen supply, prolonged exposure to cold, pain, fright, and other emotional stresses. It is obvious that the overall response to cold exposure is very different from that to infection, but in one respect the response to all these situations is the same: Invariably secretion of cortisol is increased (Fig. 17-18); indeed, the term stress has come to mean to physiologists any event which elicits increased cortisol secretion. Also, sympathetic nervous activity is usually increased.

Historically, activation of the sympathetic nervous system was the first overall response to stress to be recognized and was labeled the fight-or-flight response. Only later did further work clearly establish the contribution of the adrenal cortical response. The increased cortisol secretion is mediated entirely by the hypothalamus-anterior pituitary system (Fig. 17-19) and does not occur in animals lacking a pituitary. Thus, afferent input to the hypothalamus induces secretion of ACTH-releasing hormone, which is carried by the hypothalamopituitary portal vessels to the anterior pituitary and stimulates ACTH release. The ACTH, in turn, circulates to the adrenal and stimulates cortisol release. As described in Chap. 9, the hypothalamus receives input from virtually all areas of the brain and receptors of the body, and the pathway involved in any given situation depends upon the nature of the stress; e.g., ascending pathways from the arterial baroreceptors carry the input during hypotension, whereas pathways from other brain centers mediate the response to emotional stress. The destination is always the same, namely, synaptic connection with the hypothalamic neurons which secrete ACTH-releasing hormone. These same pathways also converge on the hypothalamic areas which control sympathetic nervous activity (including release of epinephrine from the adrenal medulla).

Functions of cortisol in stress

Many of cortisol's most important effects are on organic metabolism. Cortisol (1) stimulates protein catabolism, (2) stimulates liver uptake of amino acids and their conversion to glucose (gluconeo-

Figure 17-18. Effects of trauma on plasma cortisol concentration.

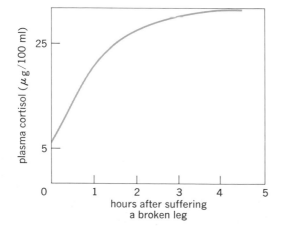

genesis), (3) is permissive for stimulation of gluconeogenesis by other hormones (glucagon, growth hormone, etc.), and (4) inhibits glucose uptake and oxidation by many body cells ("insulin antagonism") but not by the brain. Indeed, so striking are these effects that cortisol is often called a glucocorticoid to distinguish it from the other major adrenal steroid, aldosterone, called a mineralcorticoid because its major effects are on sodium and potassium metabolism.

These effects are ideally suited to meet a stressful situation. First, an animal faced with a potential threat is usually forced to forgo eating, and these metabolic changes are essential for survival during fasting—indeed, an adrenalectomized animal rapidly dies of hypoglycemia and brain dysfunction during fasting. Second, the amino acids liberated by catabolism of body protein stores provide not only energy, via gluconeogenesis, but also constitute a potential source of amino acids for tissue repair should injury occur.

A few of the many medically important implications of these cortisol-induced metabolic effects associated with stress are as follows: (1) Any patient ill or subjected to surgery catabolizes considerable quantities of body protein; (2) a diabetic who suffers an infection requires much more insulin than usual; (3) a child subjected to severe stress of any kind manifests retarded growth. The explanations for these phenomena should be evident.

Cortisol has important effects other than those on organic metabolism. One of the most important is that of enhancing vascular reactivity. A patient lacking cortisol faced with even a moderate stress may develop hypotension and die if untreated. This is due primarily to a marked decrease in total peripheral resistance. For unknown reasons, stress induces widespread arteriolar dilation, despite massive sympathetic nervous system discharge, unless large amounts of cortisol are present. A large part of its counteracting effect is ascribable to the fact that moderate amounts of cortisol permit norepinephrine to induce vasoconstriction, but this can be only part of the story since considerably larger amounts of cortisol are required to prevent stress-induced hypotension completely. In other words, the normal cardiovascular response to stress requires *increased* cortisol secretion, not just permissive quantities.

Thus far we have presented the adaptive value of

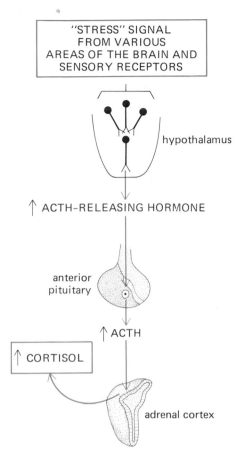

Figure 17-19. Pathway by which stressful stimuli elicit increased cortisol secretion.

the stress-induced cortisol increase mainly in terms of its role in preparing the body physically for fight or flight, and there is no doubt that cortisol does function importantly in this way. However, in recent years, it has become apparent that cortisol may have other important functions. Table 17-7 is a partial list of the large variety of psychosocial situations demonstrated to be associated with increased cortisol secretion. Common denominators of many of them are novelty and challenge. Of great interest, therefore, are recent experiments which suggest that cortisol affects memory in experimental animals, most likely through direct actions on the brain. Even more striking, ACTH (independent of its stimulation of cortisol secretion) facilitates learning and memory. (ACTH seems to be only one of many peptides which exert such actions on the brain [see Chap. 8].) Thus, it

Table 17-7. Psychosocial situations shown to be associated with increased plasma concentration or urinary excretion of adrenal cortical steroids

Experimental animals

1 Any "first experience" characterized by novelty, uncertainty, or unpredictability.
2 Conditioned emotional responses; anticipation of something previously experienced as unpleasant.
3 Involvement in situations in which the animal must master a difficult task in order to avoid or forestall aversive stimuli. The animal must really be "trying."
4 Situations in which long-standing rules are suddenly changed so that previous behavior is no longer effective in achieving a goal.
5 Socially subordinate animals. (Dominant animals have decreased cortisol.)
6 Crowding (increased social interactions).
7 Fighting or merely observing other animals fighting.

Human beings

I Normal persons
  a Acute situations
    1 Aircraft flight
    2 Awaiting surgical operation
    3 Final exams (college students)
    4 Novel situations
    5 Competitive athletics
    6 Anticipation of exposure to cold
    7 Workdays, compared to weekends
    8 Many job experiences
  b Chronic life situations
    1 Predictable personality-behavior profile: aggressive, ambitious, time-urgency
    2 Discrepancy between levels of aspiration and achievement
  c Experimental techniques
    1 "Stress" or "shame" interview
    2 Many motion pictures
II Psychiatric patients
  a Acute anxiety
  b Depression, but only when patient is aware of and involved in a struggle with it

may well be that the rise in ACTH secretion induced by psychosocial stress helps one to cope with the stress by facilitating the learning of appropriate responses.

Cortisol's pharmacological effects and disease

There are several situations in which adrenal corticosteroid levels in human beings become ab-

normally elevated. Patients with excessively hyperactive adrenals (there are several causes of this disease) represent one such situation, but the common occurrence is that of steroid administration for medical purposes. When cortisol is present in very high concentration, the previously described effects on organic metabolism are all magnified, but in addition there may appear one or more new effects, collectively known as the *pharmacological effects* of cortisol. The most obvious is a profound reduction in the inflammatory response to injury or infection (indeed, reducing the inflammatory response to allergy, arthritis, or other diseases is the major reason for administering the cortisol to patients). Large amounts of cortisol inhibit almost every step of inflammation (vasodilation, increased vascular permeability, phagocytosis) and may decrease antibody production as well. As might be expected, this decreases the ability of the person to resist infections. In addition, large amounts of cortisol may accelerate development of hypertension, atherosclerosis, and gastric ulcers, and may interfere with normal menstrual cycles.

As emphasized above, these pharmacological effects are known to be elicited when cortisol levels are extremely elevated. Yet an unsettled question of great importance is whether long-standing lesser elevations of cortisol may do the same thing, albeit more slowly and less perceptibly. Put in a different way, do the psychosocial stresses, noise, etc., of everyday life contribute to disease production via increased cortisol?

Functions of the sympathetic nervous system in stress

A list of the major effects of increased general sympathetic activity almost constitutes a guide on how to meet emergencies. Since all these actions have been discussed in other sections of the book, they are listed here with little or no comment:

1 Increased hepatic and muscle glycogenolysis (provides a quick source of glucose)
2 Increased breakdown of adipose tissue triacylglycerol (provides a supply of glycerol for gluconeogenesis and of fatty acids for oxidation)

3 Increased central nervous system arousal and alertness
4 Increased skeletal muscle contractility and decreased fatigue
5 Increased cardiac output secondary to increased cardiac contractility and heart rate.
6 Shunting of blood from viscera to skeletal muscles by means of vasoconstriction in the former beds and vasodilation in the latter
7 Increased ventilation
8 Increased coagulability of blood

The adaptive value of these responses in a fight-or-flight situation is obvious. But what purpose do they serve in the psychosocial stresses so common to modern life when neither fight nor flight is appropriate? As for cortisol, a question yet to be answered is whether certain of these effects, if prolonged, might not enhance the development of certain diseases, particularly atherosclerosis and hypertension. For example, one can easily imagine the increased blood fat concentration and cardiac work contributing to the former disease. Considerable work remains to be done to evaluate such possibilities.

Other hormones released during stress

Other hormones which are usually released during many kinds of stress are aldosterone, antidiuretic hormone, and growth hormone. The increases in ADH and aldosterone ensure the retention of sodium and water within the body, an important adaptation in the face of potential losses by hemorrhage or sweating (ADH may also influence learning by an action on the brain). Growth hormone reinforces the insulin antagonism effects of cortisol and the fat-mobilizing effects of epinephrine. Moreover, it probably stimulates the uptake of amino acids by an injured tissue and thereby facilitates tissue repair if needed; but since it cannot counteract the generalized protein catabolic effects of the increased cortisol, gluconeogenesis is not hampered.

Finally, recent evidence suggests that this list of hormones whose secretion rates are altered by stress is by no means complete. It is likely that the secretion of almost every known hormone may be influenced by stress. For example, prolactin, thy-roxine, and glucagon are often increased whereas the pituitary gonadotropins (LH and FSH), insulin, and the sex steroids (testosterone or estrogen) are decreased. The adaptive significance of many of these changes is unclear but their possible contribution to stress-induced disease processes may be very important.

Section D
Aging

Old age is not an illness; it is a continuation of life with decreasing capacities for adaptation. This view of aging in terms of a progressive failure of the body's various homeostatic adaptive responses has gained wide acceptance only recently, for there had been a strong tendency to confuse what we now recognize as a distinct aging process with those diseases frequently associated with aging. However, these diseases—notably atherosclerosis and cancer—are not *necessary* accompaniments of old age, even though their incidence does increase with age; rather, they seem to interact with the aging process in a positive-feedback cycle, each accelerating the other. For example, one theory of aging focuses on the fact that the collagen molecules of connective tissue develop increased cross-linkages with age, and that the resulting rigidity decreases tissue functioning. It is not difficult to imagine how this aging change in collagen, were it to occur in the large arteries, would also enhance the rate of development of atherosclerosis, which, in turn, might enhance the rate of cross-linkage formation by altering the chemical environment of the collagen.

What is the nature of the aging process itself (in contrast to the "diseases of aging")? The physiological manifestations of aging are a gradual deterioration in function and in the capacity to respond to environmental stresses. Thus, such quantitative parameters of function as glomerular filtration rate and basal metabolic rate decrease (Fig. 17-20), as does the ability to maintain the internal environment constant in the face of changes in temperature, diet, oxygen supply, etc. These manifestations of aging are related both to a decrease in the actual number of cells in the body (for example, we lose an estimated 100,000 brain cells each day) and to the disordered functioning of many of the cells which remain.

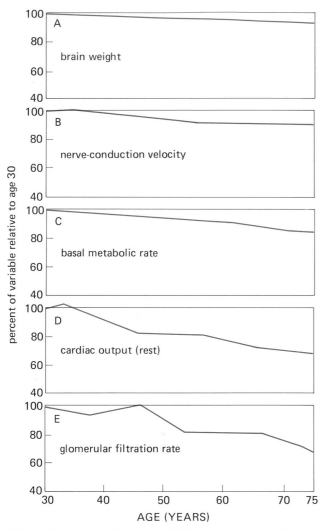

percent of variable relative to age 30

A

brain weight

B

nerve-conduction velocity

C

basal metabolic rate

D

cardiac output (rest)

E

glomerular filtration rate

AGE (YEARS)

Figure 17-20. Loss of function with increasing age. *(Adapted from Leaf.)*

In thinking of how the total number of cells in the body diminishes, it is crucial to recognize that this number reflects the balance between new cell generation (by cell division) and the death of old cells. It had been known for many years that, in the adult human, certain specialized cells—notably nerve and muscle—lose their ability to divide, but it was not until recently that a limitation on cell division of other cell types was firmly established. In the crucial experiment, cells, when grown outside of the body, divided only a certain number of

times and then stopped; moreover, the number of divisions correlated with the age of the donor. The fact that the number of divisions also correlated with the normal life span of the different species from which the cells were obtained is strong evidence for the idea that cessation of mitosis is a normal, genetically programmed event. However, other experiments have demonstrated that manipulation of the chemical environment of some cells (in this case, by the addition of large quantities of vitamin E) can result in the cells dividing 120 times (rather than the 50 usually observed) before mitosis ceases. Thus, as in any gene-environment interaction, it is likely that environmental factors determine just how many divisions actually occur, within the limits set by the genetic program.

In addition to inherent degeneration of macromolecules and external-environmental factors ("wear-and-tear," radiation, chemicals, etc.), it is very likely that cell death may be the result of the influence of one tissue or organ upon another. One of the most interesting possibilities in this last regard concerns the role of the immune system in aging. The concept of autoimmunity was discussed in an earlier section of this chapter; as emphasized there, the immune system all too frequently attacks the body's own cells and it seems likely that, over time, forbidden clones might arise, or the macromolecules of the body's cells might undergo subtle changes such that they would no longer be recognized as "self," with the failure of recognition leading to their destruction by the immune system. A decline in suppressor T cells is also a likely possibility.

The ultimate failure of mitosis may be due to an accumulation of copying errors in a cell's DNA molecules. This emphasis on the importance of DNA in aging applies not only to cell division but to all aspects of cell function. As mentioned above, aging is expressed not only by a decrease in total number of cells but also by the deterioration of the functional capacity of those cells which remain. There is fairly general agreement that the immediate cause of this deterioration is an interference in the function of the cells' macromolecules—not just DNA, but RNA, cell proteins, and the flow of information between these macromolecules as well. There are probably many factors responsible for these macromolecular disturbances; as one recent reviewer put it, this is a field in which there are as many theories as there are investigators.

18
PROCESSING SENSORY INFORMATION

Basic characteristics of sensory coding
  Stimulus modality
  Stimulus intensity
  Stimulus localization
  Control of afferent information
The sensory systems
  Somatic sensation
    Touch-pressure • *Joint position* • *Temperature* •
    Pain
  Vision

    Light • *The optics of vision* • *Receptor cells and
the retina* • *Visual system coding* • *Color vision
• Eye-movement control*
Hearing
  Sound localization
Vestibular system
Chemical senses
  Taste • *Smell*
Further perceptual processing

A human being's awareness of the world is determined by the physiological mechanisms involved in the processing of afferent information, the initial steps of which are the conversion of stimulus energy into action potentials in nerve fibers. This code represents information from the external world even though, as is frequently the case with symbols, it differs vastly from what it represents. Afferent information may or may not have a conscious correlate; i.e., it may or may not give rise to a conscious awareness of the physical world. Afferent information which does have a conscious correlate is called *sensory information,* and the conscious experience of objects and events of the external world which we acquire from the neural processing of sensory information is called *perception.*

Intuitively, it might seem that sensory systems operate like familiar electrical communications equipment, but this is true only up to a point. As an example, let us compare telephone transmission with our auditory sensory system. The telephone changes sound waves into electric impulses, which are then transmitted along wires to the receiver; thus far the analogy holds. (Of course, the mechanisms by which electric currents and action potentials are transmitted are quite different, but this does not affect our argument.) The telephone then changes the coded electric impulses *back into sound waves.* Here is the crucial difference, for our brain does not physically translate the code into sound; rather the coded information itself or some correlate of it is what we perceive as sound. At present there is absolutely no understanding of how coded action potentials or composites of them are perceived as conscious sensations.

Basic characteristics of sensory coding

It is worthwhile restating the fact that all the information transmitted by the nervous system over distances greater than a few millimeters is signaled in the form of action potentials traveling

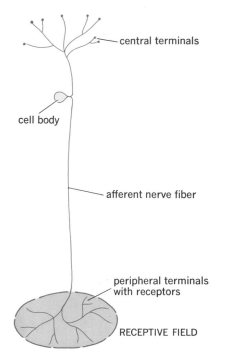

Figure 18-1. Sensory unit and receptive field.

over discrete neural pathways. Several different kinds of information must be relayed by this code: stimulus modality, intensity, and localization.

Stimulus modality

As described in Chap. 8, different receptors have different sensitivities; i.e., each receptor type responds more readily to one form of energy than to others. These different forms of stimulus energy are called *modalities*. Therefore, the type of receptor activated by a stimulus constitutes the first step in the coding of different types (modalities) of stimuli. The afferent nerve fibers from these receptors and at least some of the ascending spinal cord and brain pathways activated by them retain the same specificity in that they carry information that pertains to only one sensory modality. As expected, therefore, there are *specific pathways* (labeled "lines," as it were) for the different modalities. In tracing these pathways we shall begin at the receptor.

A single afferent neuron plus all the receptors it innervates make up a *sensory unit*. In a few cases

the afferent neuron innervates a single receptor, but generally the peripheral end of an afferent neuron divides into many fine branches, each terminating at a receptor (Fig. 18-1); all the receptors of a sensory unit are preferentially sensitive to the same stimulus modality. The *receptive field* of a neuron is that area which, if stimulated, leads to activity in the neuron (Fig. 18-1).

The central processes of the afferent neurons terminate in the central nervous system, *diverging* to terminate on several (or many) interneurons (Fig. 18-2A) and *converging* so that the processes of many afferent neurons terminate upon a single interneuron (Fig. 18-2B). Many of the parallel chains of interneurons are grouped together to form the specific ascending pathways of the central nervous system. Each chain consists of three to five synaptically connected neuronal links. Several sensory units may converge upon a given chain of neurons, but all these sensory units respond to the same stimulus modality so that specificity is maintained. For example, certain brainstem interneurons which form links in the specific pathways fire action potentials only when hair-follicle receptors are stimulated, others only when vibratory stimuli are applied to pacinian corpuscles, and still others only when a specific joint is moved.

The specific pathways (except for the olfactory pathways) pass to the brainstem and thalamus, the final neurons in the pathways going to different areas of the cerebral cortex (Fig. 18-3). The fibers transmitting the *somatic sensory modalities* (touch, temperature, etc.) synapse at cortical levels in *somatosensory cortex*, a strip of cortex which lies in the parietal lobe just behind the junction between the parietal and frontal lobes. (The word "somatic" in this context refers to the framework or outer walls of the body, including skin, skeletal muscle, tendons, and joints, as opposed to the viscera, i.e., the organs in the thoracic and abdominal cavities.) The specific pathways which originate in receptors of the taste buds, after synapsing in brainstem and thalamus, probably pass to cortical areas adjacent to the face region of the somatosensory strip. The specific pathways from the ears, eyes, and nose do not pass to the somatosensory cortex but go instead to other primary cortical receiving areas (Fig. 18-3). The pathways subserving olfaction are different from all the others in that they do not pass through thalamus and have no representation in cerebral

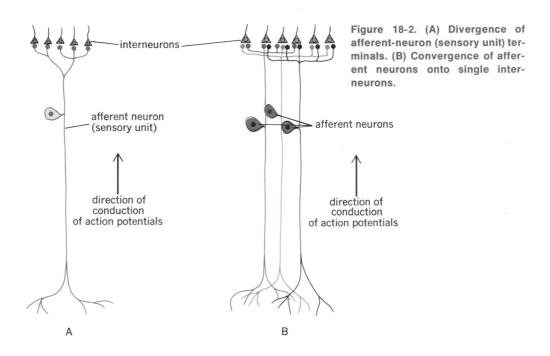

Figure 18-2. (A) Divergence of afferent-neuron (sensory unit) terminals. (B) Convergence of afferent neurons onto single interneurons.

interneurons

afferent neuron (sensory unit)

direction of conduction of action potentials

afferent neurons

direction of conduction of action potentials

A   B

cortex; they pass instead into parts of the limbic system (Fig. 8-47).

To reiterate, stimulus modality is indicated by the specific sensitivity of individual receptors and the pathways conveying the information to the primary sensory areas of the brain. However, these specific pathways just described are not the only ascending pathways. In contrast to the specific pathways, chains of neurons in other *nonspecific pathways* are activated by sensory units of several different modalities and therefore signal only general information about the level of excitability; i.e., they indicate that *something* is happening, usually without specifying just what (or where). A given cell in the nonspecific pathways may respond, for example, to maintained skin pressure, heating, cooling, skin stretch, and other stimuli applied to afferent nerves from deep tissues. Such cells, which respond to different kinds of stimuli, are called *polymodal.* The nonspecific pathways feed into the brainstem reticular formation and regions of the thalamus and cerebral cortex which are not highly discriminative.

Stimulus intensity

The second kind of information contained in the action-potential code indicates the intensity of the

stimulus. As described in Chap. 8, one important mechanism for signaling intensity is the number of sensory units activated; generally the greater the intensity of the stimulus, the greater is the number

Figure 18-3. Primary sensory areas of the cerebral cortex.

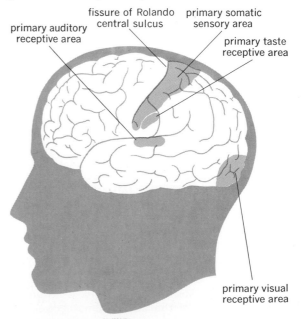

fissure of Rolando central sulcus

primary auditory receptive area

primary somatic sensory area

primary taste receptive area

primary visual receptive area

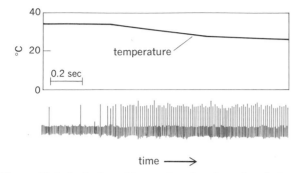

Figure 18-4. A single cold receptor signals a drop in temperature from 34 to 26°C with an increase in firing rate of action potentials in its afferent-nerve fiber. *(Adapted from Hensel and Bowman.)*

of sensory units activated. A second mechanism is the frequency at which any given sensory unit fires; an example is illustrated in Fig. 18-4, which shows the impulses in a single afferent fiber as a temperature receptor in the skin is gradually cooled from 34 to 26°C (normal body temperature is close to 37°C). Action-potential frequency correlates with stimulus intensity for afferent pathways leading to the sensory experiences not only of temperature, but of touch, limb position (joint extension or flexion), taste, sound (loudness), and light (brightness), but the relationship between stimulus intensity and action-potential frequency is generally more complicated than the one seen in temperature-sensitive systems.

Stimulus localization

A third type of information to be signaled is the location of the stimulus. Since only sensory units from a restricted area converge upon any one interneuron of the specific pathways, the pathway which begins with that particular interneuron transmits information about that restricted area. Thus, the specific pathway indicates stimulus location as well as stimulus modality.

Despite the fact that the sizes of the receptive fields of individual sensory units may vary considerably, e.g., from 2 to 200 mm² in skin, and that receptive fields of adjacent neurons overlap considerably so that stimulation of only a single receptor almost never occurs, it is nevertheless possible for a person to pinpoint the location of a stimulus because of the interactions between the

activated sensory units. Let us examine more closely how such interaction occurs. The density of the receptors of a single sensory unit (i.e., the number of receptors per unit area) varies within its receptive field, usually being greatest at the geometric center. Thus, a stimulus of a given intensity activates more receptors and generates more action potentials if it occurs at the center of the receptive field (point *A*, Fig. 18-5) than at the periphery (point *B*). Since the peripheral terminations of afferent neurons overlap to a great extent (Fig. 18-6), the placement of a stimulus determines not only the rate at which one nerve fiber fires but also the activity in sensory units with overlapping receptive fields. In the example in Fig. 18-6, neurons *A* and *C*, stimulated near the edge of their receptive fields where the receptor density is lower, fire at a lower frequency than neuron *B*, stimulated at the center of its receptive field. Because of this gradient of receptor density across the receptive field, the information content of the pattern of activity in a population of afferent neurons is great.

As we have seen, stimulus strength is related to the firing frequency of the afferent neuron, but a

Figure 18-5. Two stimulus points, *A* and *B*, in the receptive field of a single afferent neuron.

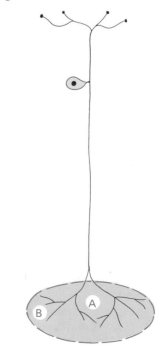

high frequency of impulses in the single afferent fiber of Fig. 18-5 could mean either that a stimulus of moderate intensity was applied at the center of the receptive field (point A) or that a strong stimulus was applied at the periphery (point B). Neither the intensity nor the localization of the stimulus can be detected precisely. But in a group of sensory units (Fig. 18-6), a high frequency of action potentials in neuron B arriving simultaneously with a lower frequency of action potentials in neurons A and C permits more accurate localization of the stimulus near the center of neuron B's receptive field. Once the location of the stimulus within the receptive field of neuron B is known, the firing frequency of neuron B can be used as a meaningful measure of stimulus intensity.

Thus, the precision with which a stimulus can be localized and differentiated from an adjacent stimulus depends both on the size of the receptive field covered by a single afferent neuron and on the amount of overlap of nearby receptive fields. For example, the ability to discriminate between two adjacent mechanical stimuli to the skin is greatest on the thumb, fingers, lips, nose, and cheeks, where the sensory units are small and overlap considerably. The localization of visceral sensations is less precise than that of somatic stimuli because there are fewer afferent fibers and each has a larger receptive field.

Control of afferent information

All incoming afferent information is subject to extensive control before it reaches higher levels of the central nervous system. Much of it is reduced or even abolished by inhibition from other neurons. Some of the neural elements mediating these controls are collaterals from afferent neurons, interneurons in the local vicinity, and descending pathways from higher regions, particularly the reticular formation and cerebral cortex. They do so by synapsing mainly upon two sites: (1) axon terminals of the afferent neurons; (2) the second-order neurons, i.e., the interneurons directly activated by the afferent neurons. The afferent terminals are influenced by presynaptic inhibition (Chap. 8), whereas the second-order cells receive both pre- and postsynaptic activity.

In some cases afferent input is continually, i.e., tonically, inhibited to some degree. This provides the flexibility of either removing the inhibition

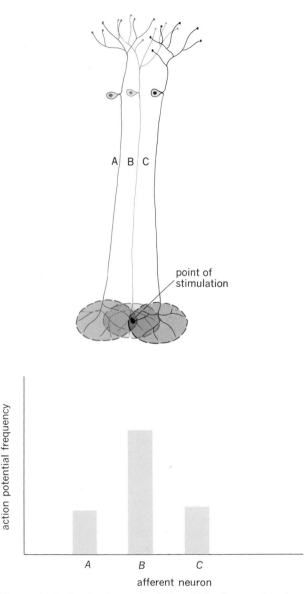

point of stimulation

action potential frequency

A   B   C

afferent neuron

Figure 18-6. A stimulus point falls within the overlapping receptive fields of three afferent neurons.

(disinhibition) so as to allow a greater degree of signal transmission or of increasing the inhibition so as to block the signal more completely. The thin myelinated and unmyelinated fibers that play an important role in pain seem to be under such tonic influence.

Afferent information can also be modified by central facilitation as well as by inhibition. Both methods increase the contrast between "wanted"

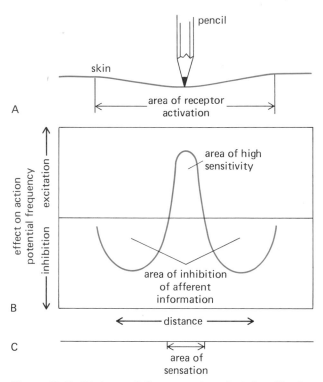

Figure 18-7. (A) A pencil tip pressed against the skin depresses surrounding tissues. Receptors are activated under the pencil tip and in the adjacent tissue. (B) Because of lateral inhibition, the central area of excitation is surrounded by an area where the afferent information is inhibited. (C) The sensation is localized to a more restricted region than that in which mechanoreceptors were actually activated.

and "unwanted" information, thereby increasing the effectiveness of selected pathways and focusing sensory-processing mechanisms on "important" messages.

In some afferent systems, the control is organized in such a way that the stronger inputs are enhanced and the weaker inputs of adjacent sensory units simultaneously inhibited. Such *lateral inhibition* can be demonstrated in the following way. While pressing the tip of a pencil against the finger with one's eyes closed, one can localize the pencil point quite precisely, even though the region around the pencil tip is also indented and mechanoreceptors within this entire area are activated (Fig. 18-7); this is because the information from the peripheral region is removed by mechanisms of lateral inhibition. Lateral inhibition occurs in the pathways of most sensory modalities but is utilized

to the greatest degree in the pathways providing the most accurate localization. For example, a stimulus can be localized very precisely in the pathways relaying information about hair deflection, whereas stimuli activating temperature receptors, whose pathways lack lateral inhibition, are localized only poorly.

Lateral inhibition also occurs between pathways relaying information about different sensory modalities. One example may explain why we rub an area to stop a pain there: Afferent fibers whose receptors are activated by mechanical stimuli can to a certain extent inhibit the output of afferent fibers whose receptors are stimulated by irritating or painful stimuli; therefore, a mechanical stimulus (rubbing) can inhibit an irritating or painful stimulus.

With these general principles as background, we now turn to the specific sensory systems, their receptor mechanisms, and the way the resulting neural signals are processed by the central nervous system.

The sensory systems

Somatic sensation

Somatic receptors respond to mechanical stimulation of the skin or hairs and underlying tissues, rotation or bending of joints, temperature changes, and possibly some chemical changes. Their activation gives rise to the sensations of touch, pressure, heat, cold, the awareness of the position and movement of the parts of the body, and pain. Recalling that by *receptor* we mean the specialized peripheral ending of an afferent nerve fiber or the specialized receptor cell associated with it, we can say that each of these sensations is probably associated with a specific type of receptor; i.e., there are distinct receptors for heat, cold, touch, pressure, joint position, and pain.

After entering the central nervous system, the afferent fibers from the somatic receptors synapse onto interneurons which form the specific pathways which go to somatosensory cortex via the brainstem and thalamus. The specific pathways cross from the side of afferent-neuron entry to the opposite side of the central nervous system in the spinal cord or brainstem; thus the sensory pathways from receptors on the left side of the body go

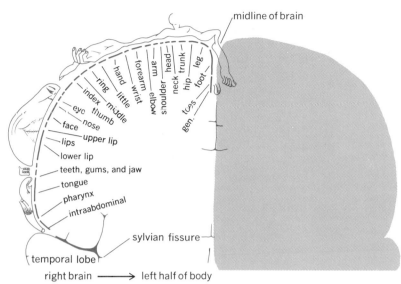

midline of brain

ring middle
index thumb
eye nose
face
lips upper lip
lower lip
teeth, gums, and jaw
tongue
pharynx
intraabdominal
hand little
forearm
wrist
arm elbow
shoulder
neck trunk
head
hip leg
foot
toes
gen.

sylvian fissure

temporal lobe

right brain ⟶ left half of body

Figure 18-8. Location of pathway termina- tions for different parts of the body in the somatosensory cortex. This pattern is dupli- cated on the opposite cerebral hemisphere. The left half of the body is represented on the right hemisphere of the brain, and the right half of the body is represented on the left cerebral hemisphere.

to the somatosensory cortex of the right cerebral hemisphere and vice versa. In somatosensory cor- tex, the terminations of the individual fibers of the specific somatic pathways are grouped according to the location of the receptors. The pathways which originate in the foot end nearest the longi- tudinal dividing line between the two cerebral hemispheres. Passing laterally over the surface of the brain, one finds the terminations of the path- ways from leg, trunk, arm, hand, face, tongue, throat, and viscera (Fig. 18-8). The parts with the greatest sensitivity (fingers, thumb, and lips) are represented by the largest areas of somatosensory cortex.

Touch-pressure. One of the best examples of how receptor specificity can be determined by the char- acteristics of the surrounding tissue is the pacinian corpuscle, discussed in Chap. 8. The nerve terminal of the pacinian corpuscle is surrounded by alter- nating layers of cells and extracellular fluid in such a way that a mechanical stimulus reaches the central nerve terminal only after displacing the layers of the surrounding capsule (Fig. 18-9). Rapidly applied pressures are transmitted to the nerve terminal without delay and give rise to a re- ceptor potential by the mechanisms described in Chap. 8. The energy of slowly administered or sus- tained forces is, in part, absorbed by the elastic tissue components of the capsule and is therefore partially dissipated before it reaches the nerve

terminal at the core. Thus, the receptor fires only at the fast onset—and perhaps again at the release —of the mechanical stimulus but not under sus- tained pressure. What the pacinian corpuscle sig- nals is not pressure, but *changes* in pressure with time. It can effectively discriminate stimuli vi- brating up to frequencies of 300 Hz (cycles per second). Other mechanoreceptors which adapt more slowly than the pacinian corpuscles provide information about both the *rate* of pressure ap- plication and the stimulus *intensity.*

Joint position. The activation of mechanoreceptors in the joints, associated ligaments, and tendons and their pathways through the nervous system give rise to the conscious awareness of the position and movement of the joints and the stresses acting upon them. Input from these receptors is integrated

Figure 18-9. Pacinian corpuscle.

area of pressure stimulation

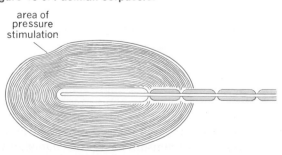

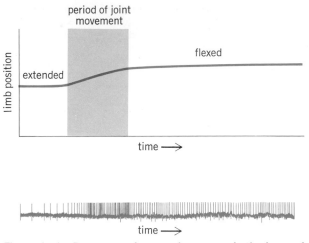

Figure 18-10. Response of a stretch receptor in the knee of a cat as the limb is flexed. *(Redrawn from Boyd and Roberts.)*

with visual and other information to provide awareness of the position of the body in space. The different receptors in the joints and ligaments are activated by mechanical stimuli such as stretching, twisting, or compressing or by painful stimuli caused by damage. Their combined sensitivities signal movement and final position of the joint. Some receptors fire rapidly at the initiation of the movement, the action-potential frequency indicating the speed of the movement, and then slow their rate of firing down to a frequency dependent upon the final joint position (Fig. 18-10). Figure 18-11 shows the response of a single joint receptor during flexion and extension of the joint. Other receptors respond oppositely, firing faster during extension and slower during flexion.

Temperature. The thermoreceptors in the skin are classified, according to their responses, as cold and warm receptors even though distinguishing anatomical structures have not been identified for either type. Warm receptors increase their discharge upon warming and show a temporary inhibition upon cooling, whereas cold receptors respond in the opposite way. Also, the temperature at which the sensory unit fires maximally is much higher for warm than for cold receptors. Several mechanisms have been proposed for thermoreceptor responsiveness, and such receptors might work in the following way: Changes in temperature and in the associated thermal agitation of molecular

bonds cause configurational changes in protein molecules in the nerve ending; these alter the membrane permeability, causing a receptor potential, which leads to the generation of action potentials.

Thermoreceptors are sensitive not only to temperature but to some chemicals as well; for example, menthol stimulates cold receptors with the result that the skin feels cooler than it actually is, and some substances in spices stimulate warm receptors resulting in sensations of heat and even burning pain.

Pain. A stimulus which causes or is on the verge of causing tissue damage often elicits a sensation of pain and a reflex escape or withdrawal response as well as the gamut of physiological changes mediated by the sympathetic nervous system similar to those elicited during fear, rage, and aggression.

Figure 18-11. Two different responses of a single joint stretch receptor to movements in opposite directions, the upper curve during flexion and the lower curve during extension. *(Redrawn from Boyd and Roberts.)*

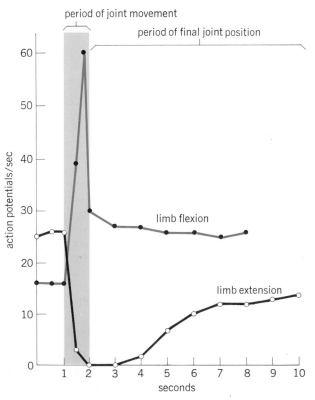

These changes usually include faster heart rate, higher blood pressure, greater secretion of epinephrine into the bloodstream, increased blood sugar, less gastric secretion and motility, decreased blood flow to the viscera and skin, dilated pupils, and sweating. In addition, the experience of pain includes an emotional component of fear, anxiety, and sense of unpleasantness as well as information about the stimulus's location, intensity, and duration. And probably more than any other type of sensation, the experience of pain can be altered by past experiences, suggestion, emotions (particularly anxiety), and the simultaneous activation of other sensory modalities.

To reiterate, the stimuli which give rise to pain result in a sensory experience *plus* a reaction to it, the reaction including the emotional response (anxiety, fear) and behaviorial response (withdrawal or other defensive behavior). Both the sensation and the reaction to the sensation must be present for tissue-damaging stimuli to cause suffering. The sensation of pain can be dissociated from the emotional and behavioral reactive component by drugs, e.g., morphine, or by selective brain operations which interrupt pathways connecting the frontal lobe of the cerebrum with other parts of the brain. When the reactive component is no longer associated with the sensation, pain is felt, but it is not necessarily disagreeable; the patient does not mind as much. Thus, pain relief can be obtained even though the perception of painful stimuli is not reduced.

The receptors whose stimulation gives rise to pain are at the ends of certain small unmyelinated or lightly myelinated afferent neurons. These receptors, known as *nociceptors*, all appear the same when viewed with an electron microscope, yet different endings respond preferentially to harmful mechanical, chemical, or thermal stimuli. It is hypothesized that a chemical is released from the damaged tissue. Upon release, the chemical depolarizes the nerve ending, initiating action potentials in the afferent nerve fiber. The chemical has not been identified, although there are many candidates.

Different nociceptors have different thresholds; some respond only to severe procedures such as cutting the skin, pricking it with needles, or heating it to 50°C, whereas receptors having lower thresholds respond to firm pressure or a temperature as low as 42°C, neither of which damages the skin. These latter receptors signal impending harm, and the sensation they invoke is described as threatening or warning rather than actual pain. Both types of receptors fire more vigorously as stimulus intensity increases.

The primary afferents coming from nociceptors synapse onto two different types of interneurons after entering the central nervous system (one transmitter thought to be released at this synapse is substance P) such that information about pain is transmitted to higher centers via the specific ascending pathways for pain and via nonspecific pathways. It is hypothesized that the specific pathways, which go to the thalamus and cerebral cortex, convey information about where, when, and how strongly the stimulus was applied and are thought to convey information about the sharp, localized aspect of pain. The nonspecific pathways, which go to the brainstem reticular formation and a part of the thalamus different from that supplied by the specific pathways, are believed to convey information about the aspect of pain which is duller, longer lasting, and less well localized. Neurons of the reticular formation and thalamus which are activated by the nonspecific pathway are interconnected with the hypothalamus and other areas of the brain which play major roles in integrating autonomic and endocrine stress responses and in generating the behavioral patterns of aggression and defense.

Descending pathways capable of altering the transmission of information in the afferent neurons, spinal pathways, or brain centers, are particularly important in pain. One powerful inhibitory pathway that controls transmission in pain fibers descends from a part of the brainstem reticular formation; the descending axons of these neurons end on the second-order neurons in the pain pathways and on the terminals of the afferent fibers themselves. Electrical stimulation of this area of brainstem reticular formation produces a profound reduction of pain (analgesia), a phenomenon called stimulus-produced analgesia (SPA). SPA can also be produced by stimulating other areas of the nervous system and is used in people in an attempt to control severe, long-lasting pain. It has been hypothesized that acupuncture analgesia may result from stimulation of certain nerve pathways which feed into these brainstem regions and activate the powerful inhibitory control mechanisms.

At some of the synaptic links in the pain-suppres-

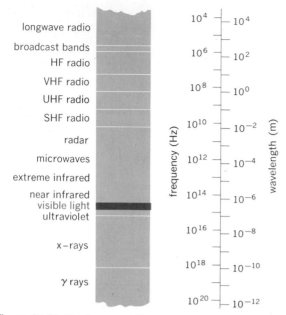

Figure 18-12. Electromagnetic spectrum.

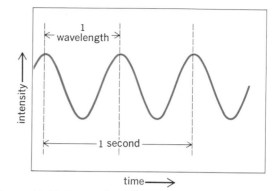

Figure 18-13. Properties of a wave. The frequency of this wave is 2 Hz.

The optics of vision. Light can be represented most simply by a ray or line drawn in the direction in which the wave is traveling. Light waves are propagated in all directions from every point of a visible object. These divergent light waves must

Figure 18-14. Human eye. The blood vessels depicted run along the back of the eye between the retina and the vitreous humor.

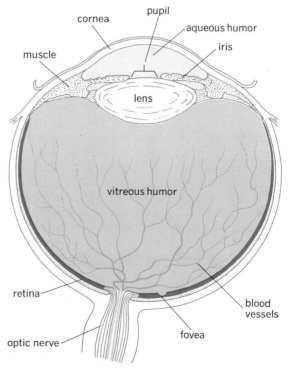

sion pathways the neurotransmitters, enkephalin and endorphin are thought to be involved, and morphine, which activates the receptor sites for these transmitters, is thought to relieve pain by its ability to activate this inhibitory system. Since the dull, long-term, poorly localized type of pain responds more to morphine than does the sharp, well-localized type, this inhibitory pathway presumably affects mainly the nonspecific pathways.

Vision

Light. The receptors of the eye are sensitive to only that tiny portion of the vast spectrum of electromagnetic radiation which we call light (Fig. 18-12). Radiant energy is described in terms of wavelengths and frequencies. The *wavelength* is the distance between two successive wave peaks of the electromagnetic radiation (Fig. 18-13) and varies from several kilometers at the top of the spectrum to minute fractions of a millimeter at the bottom end. Those wavelengths capable of stimulating the receptors of the eye (the visible spectrum) are between 400 and 700 nm (nanometers), and light of different wavelengths in this band is associated with different color sensations.

pass through an optical system which focuses them back into a point before an accurate image of the object is achieved. In the eye itself, the image of the object being viewed must be focused upon the *retina*, a thin layer of neural tissue lining the back of the eyeball (Fig. 18-14), where the light-sensitive receptor cells of the eye are located. The *lens* and *cornea* of the eye (Fig. 18-14) are the optical systems which focus the image of the object upon the retina. At a boundary between two substances, such as the cornea of the eye and the air outside it, the rays are bent so that they travel in a new direction. The degree of bending depends in part upon the angle at which the light enters the second medium. The cornea plays the larger role in focusing light rays because the rays are bent more in passing from air into the cornea than in passing into and out of the lens.

The surface of the cornea is curved so that light rays coming from a single point source hit the cornea at different angles and are bent different amounts, but all in such a way that they are directed to a point after emerging from the lens (Fig. 18-15A). Notice what happens to the image when the object being viewed has more than one dimension (Fig. 18-15B); the image on the retina is upside down relative to the original light source. It is also reversed left to right.

The shape of the cornea and lens and the length of the eyeball determine the point where light rays reconverge. Moreover, light rays from objects close to the eye strike the cornea at greater angles (are more divergent) and have to be bent more in order to reconverge on the retina. Although the cornea performs the greater part quantitatively of focusing the visual image on the retina, all adjustments for distance are made by changing the shape of the lens. Such changes are called *accommodation*. The shape of the lens is controlled by a muscle which flattens the lens when distant objects are to be focused upon the retina and allows it to assume a more spherical shape to provide ad-

Figure 18-15. Refraction (bending) of light by the lens system of the eye. *A* illustrates the focusing of light rays from **a single** point, and *B* the focusing of light rays from more than one point to form the image of an object on the retina.

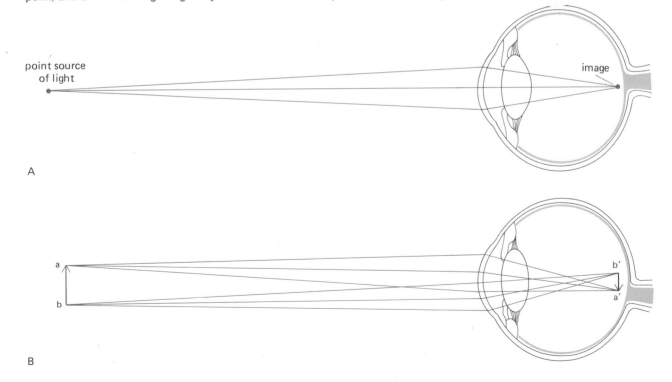

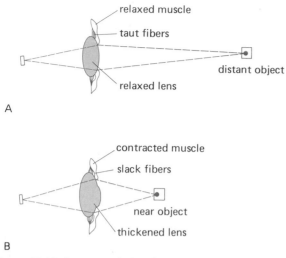

relaxed muscle

taut fibers

distant object

relaxed lens

A

contracted muscle

slack fibers

near object

thickened lens

B

Figure 18-16. Accommodation for distant and near vision by the pliable lens. (A) The lens is stretched for distant vision so that it adds the minimum amount of focusing power. (B) The lens thickens for near vision to provide greater focusing power.

ditional bending of the light rays when near objects are viewed (Fig. 18-16). These muscles are controlled by parasympathetic nerve fibers.

Cells are added to the lens throughout life but only to the outer surface. This means that cells at the center of the lens are both the oldest and the farthest away from the nutrient fluid which bathes the outside of the lens (if capillaries ran through the lens, they would interfere with its transparency). These central cells age and die first, and with death they become stiff, so that accommodation of the lens for near and far vision becomes more difficult; this impairment is known as *presbyopia*. This is one reason why many people who never needed glasses before start wearing them in middle age.

Cells of the lens can also become opaque so that detailed vision is impaired; this is known as *cataract*. The defective lens can usually be removed surgically from persons suffering from cataract, and with the addition of compensating eyeglasses, or even an artificial lens, effective vision can be restored although the ability to accommodate will be lost.

Defects in vision occur if the eyeball is too long in relation to the lens size, for then the images of near objects fall on the retina but the images

of far objects are focused in front of the retina. This is a *nearsighted*, or *myopic*, eye, which is unable to see distant objects clearly. If the eye is too short for the lens, distant objects are focused on the retina while near objects are focused behind it (Fig. 18-17); this eye is *farsighted*, or *hyperopic*, and near vision is poor. Defects in vision also occur where the lens or cornea does not have a smoothly spherical surface. The improperly shaped eyeball or irregularities in the cornea (*astigmatism*) or lens can usually be compensated for by eyeglasses.

The amount of light entering the eye is controlled by a ringlike pigmented muscle known as the *iris*, the color being of no importance as long as the tissue is sufficiently opaque to prevent the passage of light. The hole in the center of the iris through which light enters the eye is the *pupil*. The iris muscle reflexly contracts in bright light, decreasing the diameter of the pupil; this not only reduces the amount of light entering the eye but also directs the light to the central and most optically accurate part of the lens. Conversely, the iris relaxes in dim light, when maximal sensitivity is needed.

Receptor cells and the retina. The receptor cells in the retina (the photoreceptors) are called *rods* and *cones* because of their microscopic appearance (Fig. 18-18). Photoreceptors contain light-sensitive molecules called *photopigments* which absorb light. There are four different photopigments in the retina, each of which is made up of a protein (*opsin*) bound to a *chromophore* molecule. The chromophore is always *retinal* (a slight variant of vitamin A) but the opsin differs in each of the four types. This difference causes each of the four photopigments to absorb light most effectively at a different part of the visual spectrum. For example, one photopigment absorbs wavelengths in the range of red light better than those in blue or green ranges whereas another absorbs green light better than red or blue.

The events leading to receptor activation have been most studied for the rods. The rod-shaped ending of these receptor cells contains many stacked, flattened membrane discs parallel to the light-receiving surface of the retina. The disc membranes contain a high concentration of the photopigment *rhodopsin*. Its opsin is covalently bonded to the chromophore, 11-*cis*-retinal, which

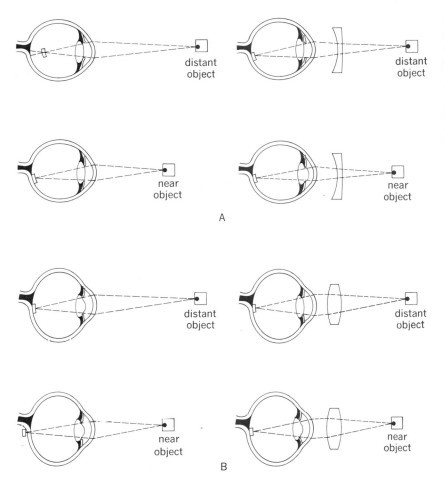

Figure 18-17. (A) In the nearsighted eye, light rays from a distant source are focused in front of the retina. A concave lens placed before the eye bends the light rays out sufficiently to move the focused image back onto the retina. When near objects are viewed through concave lenses, the eye accommodates to focus the image on the retina. (B) The farsighted eye must accommodate to focus the image of distant objects upon the retina. (The normal eye views distant objects with a flat, stretched lens.) The accommodating power of the lens of the eye is sufficient for distant objects, and these objects are seen clearly. The lens cannot accommodate enough to keep images of near objects focused on the retina, and they are blurred. A convex lens converges light rays before they enter the eye and allows the eye's lens to work in a normal manner.

upon exposure to light, changes its configuration into the all-*trans* form of the molecule (enzymes are not required for this reaction, for it is catalyzed by light). The initial change in molecular shape, caused by the absorption of light, leads to an unstable molecule which spontaneously undergoes further changes resulting in the dissociation of the chromophore from the opsin. A substance (possibly calcium) is released and causes a decrease in the sodium permeability of the cell membrane, producing a receptor potential. Since the change is a *decrease* in sodium permeability, the receptor potential in this case is a hyperpolarization. Thus the receptors are relatively depolarized in the dark and hyperpolarized in the light, and transmitter release, which occurs with depolarization, increases in the dark and is reduced by illumination. Needless to say, this is still an effective signaling

mechanism. After dissociation of the photopigment, the all-*trans*-retinal is then changed back to the 11-*cis* form, which combines with opsin to regenerate rhodopsin. These latter reactions do not depend on light, but are enzyme mediated.

The receptor cells (whether rods or cones) synapse upon *bipolar cells* in the retina (Fig. 18-19), which in turn synapse upon *ganglion cells.* Generally, cone receptor cells have relatively direct lines to the brain; i.e., there is little convergence along the neural pathway. This relative lack of convergence provides precise information about the area of the retina that was stimulated, and cone visual acuity is very high. Because cones are concentrated in the center of the retina, it is that part that we use for finely detailed vision. However, the lack of convergence in the cone visual pathways offers little opportunity for the summa-

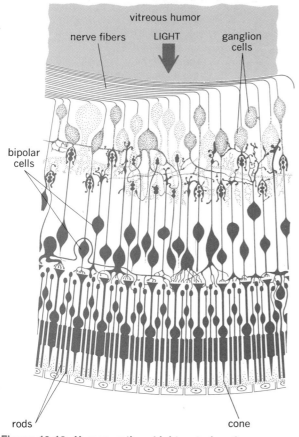

Figure 18-18. Human retina. Light entering the eye must pass through the fibers and cells of the retina before reaching the sensitive tips of the rods and cones. *(Adapted from Gregory.)*

mation are good. Therefore, a relatively low-intensity light stimulus that would cause only a subthreshold response in a cone ganglion cell can cause an action potential in a rod ganglion cell. Thus, the difference in acuity and light sensitivity between rod and cone vision is due, at least in part, to the anatomical wiring patterns of the retina.

These differences explain why objects in a darkened theater are indistinct; with such low illumination the cones do not generate effective signals, so that all vision is supplied through the more sensitive but less accurate rod vision. Moreover, rods contain only one type of photopigment (rhodopsin) and, as we shall see later, cannot give rise to color vision; to perceive color at least two types of receptors with different photopigments must be activated. Thus, objects in a darkened theater appear in shades of gray.

The sensitivity of the eye is decreased under conditions of bright illumination *(light adaptation)* and improves after being in the dark for some time *(dark adaptation)*. The state of adaptation is partly related to the amount of intact rhodopsin (the 11-*cis* form); thus visual sensitivity improves as more 11-*cis* rhodopsin is regenerated. Adaptation is mainly determined, however, by a neural mechanism which alters the "gain" of the system so that in the dark-adapted condition activation of a rhodopsin molecule produces a larger electrical signal than in the light-adapted state.

Visual system coding. In most of the experiments dealing with coding in the visual system, simple visual shapes such as white bars against a black background were projected onto a screen in front of the anesthetized animal while the activity of single cells in the visual system was recorded. (This discussion is based on research done chiefly on frogs, cats, and monkeys, but it almost certainly applies to people as well.) Different parts of the retina could be stimulated by varying the position

tion of subthreshold events to fire action potentials in the ganglion cells, and high levels of illumination are needed to activate these pathways. Thus, the cone pathways are useful only in "daytime" vision. In contrast, there is much convergence in the rod visual pathway, and, although acuity is poor, opportunities for spatial and temporal sum-

Figure 18-19. Diagrammatic representation of the cells in the visual pathway.

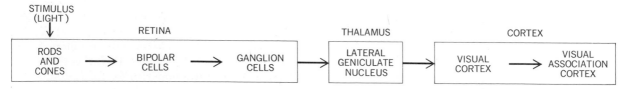

of the bar on the screen. We shall constantly refer to receptive fields of neurons within the visual pathway and to the responses of these neurons to light. It is essential to recognize that only the rods and cones respond directly to light; all other components of the pathway are influenced only by the synaptic input to them. Thus when we speak of the receptive field of a neuron in the visual pathway, we really mean that area of the retina which when stimulated can influence the activity of that neuron. Similarly the neuron's "response to light" is really its response to neural activity within the visual pathway initiated by light falling upon the rods and cones. We shall follow the information processing elucidated by these experiments through the stages of the visual pathway, starting at the level of the ganglion cells (Fig. 18-19).

In the retina an amazing amount of data processing has occurred. The retinal ganglion cells discharge spontaneously; i.e., they fire in the absence of any light stimulus. This spontaneous activity gives the cell an important second signal with which to work; it can either increase or decrease its rate of firing. Each converging receptor-bipolar-ganglion-cell chain is synaptically connected to other similar chains by cells which conduct laterally through the retina. These interconnections occur at both bipolar- and ganglion-cell levels.

The receptive fields of the ganglion cells are circular; i.e., any light falling within a specific circular area of the retina influences the activity of a given ganglion cell (by way of the receptor-bipolar-cell chain). The response of the ganglion cell varies markedly, depending on the region of its receptive field stimulated. Some ganglion cells speed up their rate of firing when a spot of light is directed at the center of their receptive field and slow down their firing when the periphery is stimulated. Such a cell is said to have an *on* center. The activity of a ganglion cell of this type is shown in Fig. 18-20A. Other ganglion cells, such as cell 2 (Fig. 18-20), have just the opposite response, decreasing activity when the center of its receptive field is stimulated (Fig. 18-20B) and increasing activity when the light stimulates the periphery (Fig. 18-20C). Notice the spontaneous activity of the *off*-center cell and the abrupt inhibition of its activity when the light is turned on.

Moreover, the basic pattern of ganglion-cell activity can be greatly modified. An *on* response increases, i.e., the frequency of firing action poten-

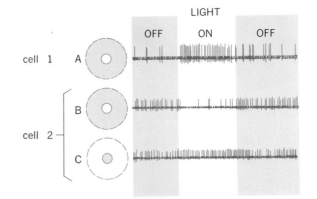

Figure 18-20. Recordings of the activity of a single ganglion cell. (A) Response of an *on*-center ganglion cell (cell 1) which increased its activity when light stimulated the center of its receptive field. (B) Activity of an *off*-center cell (cell 2) is suppressed when the center of its receptive field is stimulated. (C) The same *off*-center cell increases its activity when the light is restricted to the periphery. *(Adapted from Hubel and Wiesel,* J. Physiol., **154.***)*

tials increases, if the intensity of the light spot is greater or if the diameter of the spot is larger but it decreases if the diameter of the spot becomes so much larger that it encroaches upon adjacent *off* regions or if a second spot is shown simultaneously on a nearby *off* region.

The axons of the ganglion cells form the optic nerve, which passes to the brain. The optic nerves from the two eyes meet near the center of the head where some of the fibers cross over to the opposite side of the brain. This partial crossover provides both cerebral hemispheres with input from both eyes.

After entering the brain, the visual pathways pass to the *lateral geniculate nucleus* in the thalamus. The receptive field of a single cell there, i.e., that area of the retina which when stimulated can influence the activity of the lateral geniculate neuron, resembles that of a retinal ganglion cell in being concentric with either an *on*-center–*off*-periphery pattern or vice versa.

The partially processed visual information is transmitted along the axons of the lateral geniculate neurons to *primary visual cortex*, where the processing continues. Although the receptive fields of retinal ganglion cells and lateral geniculate neurons are usually concentric, with *on* or *off*

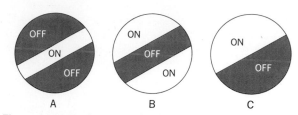

Figure 18-21. Retinal receptive fields of simple cortical cells are no longer arranged concentrically but are organized to provide information about lines and borders. *(Adapted from Hubel and Wiesel,* J. Physiol., **160.***)*

centers, the receptive fields of cells in the visual cortex vary widely in organization. The cortical cells are classified as concentric, simple, complex, or hypercomplex according to the stimuli to which they respond. The *concentric* cells have a receptive-field organization similar to that of the lateral geniculate cells. The *simple* cells have receptive fields which are divided into *off* and *on* regions having a side-by-side arrangement of excitatory and inhibitory areas with straight boundaries rather than circular ones (Fig. 18-21). Diffuse light over the entire receptive field generally gives little or no response because the effects of the simultaneously stimulated *on* and *off* areas cancel out. The most effective stimulus is one which covers the *on* area but does not encroach upon the *off* area, e.g., long, narrow slits of light; dark, rectangular bars against a light background (lines); or straight line borders between areas of different brightness. The orientation of the optimum stimulus varies from cell to cell.

Some of the *complex* cells respond optimally to a line in a particular orientation across the receptive field but, unlike simple cells, the complex cells can respond when lines of that orientation are presented at different places in the receptive field (Fig. 18-22). These cells have no separation of their receptive fields into excitatory and inhibitory parts. Most cells at cortical levels can be influenced from either eye with the most effective stimulus form, orientation, and rate of movement similar for both eyes. The cortical response increases when the two eyes are stimulated simultaneously, although some respond best to input from one of the eyes. The complex cells, in turn, activate others which are hypercomplex.

The multiple interconnections in the visual path-

ways are there to provide for active data processing rather than the simple transmission of action potentials. By means of these intricate cellular hookups, cells of the visual pathways respond only to selected features of the visual world. They are organized to handle information about line, contrast, movement, and color, but they are not very good intensity detectors. Note, also, that the visual system does not form a picture in the brain but through the simultaneous activation of many neurons forms a specifically coded electrical statement.

Color vision. Light is the source of all colors. Pigments, such as those mixed by a painter, serve only to reflect, absorb, or transmit different wavelengths of light, yet the nature of the pigments determines how light of different wavelengths will react. For example, an object appears red because

Figure 18-22. Complex cortical cells may increase their rate of firing when a properly oriented bar of light appears at any position in the visual field. (The white bars represent the location of the bar of light and the dark bars represent different positions in the visual field.) (A) The cell responds regardless of the bar's position as long as it is horizontal. (B) A vertical bar causes no response. Complex cells such as this help indicate movement of a visual stimulus.

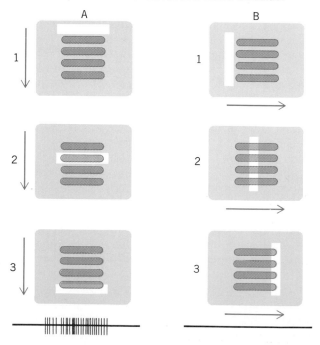

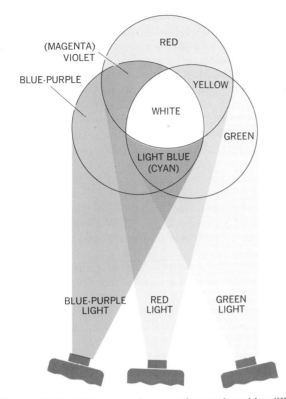

Figure 18-23. All known colors can be produced by different combinations of the three primary wavelengths of light, those giving rise to the color sensations of red, blue, and green. *(Adapted from R. L. Gregory, "Eye and Brain: The Psychology of Seeing," McGraw-Hill Book Company, New York, 1966.)*

the shorter wavelengths are absorbed by the material whereas the long wavelengths are reflected to excite the photopigment of the retina most sensitive to them. Light perceived as white is a mixture of all wavelengths, and black is the absence of all light. Sensation of any color can be obtained by the appropriate mixture of three lights, red, blue, and green (Fig. 18-23). Light and pigments are properties of the physical world, but color exists only as a sensation in the mind of the beholder. The problem for the scientist is to discover how the perception of a brilliantly colored world results from packets of photic energy of varying wavelengths.

Color vision begins with the activation of the photopigments in the cone receptor cells. Human retinas have three kinds of cones, which contain either yellow-, green-, or blue-sensitive photopigments, responding optimally to light of 570-, 535-,

and 445-nm wavelengths, respectively. (The 570 pigment sensitivity extends far enough to sense the long, red wavelengths and is sometimes called the red photopigment.) Although each type of cone is excited most effectively by light of one particular wavelength, it responds to other wavelengths as well; thus, for any given wavelength, the three cone types are excited to different degrees. For example (Fig. 18-24), in response to a light of 535-nm wavelength, the green cones respond maximally, the red cones less, and the blue cones hardly at all. Our sensation of color depends upon the ratios of these three cone outputs and their comparison by higher-order cells in the visual system.

The fact that there are three different kinds of cone cells explains the various types of color blindness. Most people (over 90 percent of the male population and over 99 percent of the female population) have normal color vision; i.e., their color vision is determined by the differential activity of the three types of cones. Most color-blind (or better, color-defective) people appear to lack one of the three photopigments, and their color vision is therefore formed by the differential activity of the remaining two types of cones. For example, people with green-defective vision see as if they have only red- and blue-sensitive cones.

The pathways for color vision follow those described earlier for the line-contrast processors (Fig. 18-19); i.e., the cones synapse upon bipolar cells, and the bipolar cells synapse upon ganglion

Figure 18-24. Sensitivities of the three photopigments found in human cones. One pigment senses blue (445 nm), another green (535 nm), and the third yellow (570 nm). The yellow pigment sensitivity extends far enough to sense red (about 650 nm) and is, in fact, called the red photopigment. *(Redrawn from Michael.)*

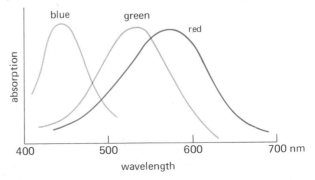

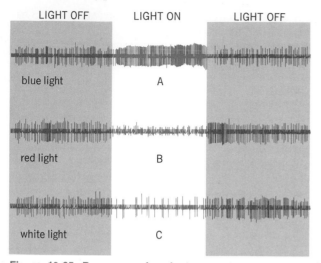

Figure 18-25. Response of a single ganglion cell to blue light, red light, and white light. *(Adapted from Hubel and Wiesel, J. Physiol.,* **154.***)*

cells, etc. Single ganglion cells receive simultaneously input (via bipolar cells) from two types of cones. For example, one group of ganglion cells might receive excitatory input from red and green cones. The inputs are additive so that such a cell will respond more briskly when it receives input from red and green cones together than it does to input from red or green cones alone. These cells code brightness rather than specific colors. A second type of ganglion cell codes specific colors and is called the *opponent color cell;* these cells receive excitatory input from one of the three cone types and inhibitory input from another. For example, the cell in Fig. 18-25 increased its rate of firing when stimulated by a blue light but decreased it when a red light replaced the blue. The cell gave a weak response when stimulated with a white light because the light contained both blue and red wavelengths.

This opponent aspect of color vision means that perception of one member of the pair is associated with decreased sensitivity to the other. For example, the neuronal activity which results in the perception of green inhibits the pathways to the perception of red. As mentioned earlier, the relatively common red color-blindness in people has been attributed to a photopigment that is abnormally formed or lacking in one of the cone cells, but it may be due to a paucity of red-green opponent cells.

Eye-movement control. The cones are most concentrated in a specialized area of the retina known as the *fovea,* and images focused there are seen with the greatest acuity. In order to get the visual image focused on the fovea and keep it there, the six muscles attached to the outside of each eyeball perform two basic movements, fast and slow.

The fast movements include *saccades,* small jerking movements which rapidly bring the eye from one fixation point to another to allow search of the visual field. In addition, saccades move the visual image over the foveal receptors, thereby preventing adaptation. In fact, if saccades are prevented, all color and most detail fade away in a matter of seconds. These movements also occur during certain periods of sleep when the eyes are closed, and perhaps they are associated with "watching" the visual imagery of dreams. Saccades are among the fastest movements in the body.

Slow eye movements are involved in tracking visual objects as they move throughout the visual field. In tracking an object as it moves up and down or side to side, both eyes make similar movements, e.g., turning upward or to the right together. In tracking a visual object in depth, however, the eyes turn inward (convergence) as the object comes nearer and outward (divergence) as it moves farther away. Slow movements are also used during compensation for movements of the head; if a stationary visual object is focused on the fovea and the head is moved to the left, the eyes must be moved an equal distance to the right if the object's image is to remain focused on the fovea; if the head moves up, the eyes must move down. These compensating movements obtain their information about the movement of the head from the semicircular canals of the vestibular system, which will be described shortly. Control systems for the other slow movements require the continual feedback of visual information about the moving object.

Hearing

Sound energy is transmitted through air by a movement of air molecules. When there are no air molecules, as in a vacuum, there can be no sound. The disturbance of air molecules that makes up a sound wave consists of regions of compression, in which the air molecules are close together and

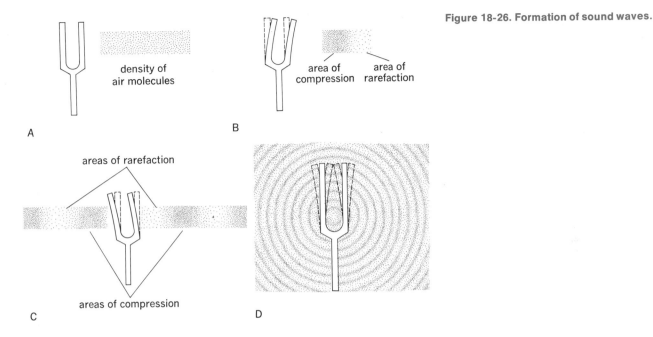

Figure 18-26. Formation of sound waves.

A

density of
air molecules

B

area of   area of
compression   rarefaction

C

areas of rarefaction

areas of compression

D

the pressure is high, alternating with areas of rarefaction, where the molecules are farther apart and the pressure is lower. Anything capable of creating such disturbances can serve as a sound source. A tuning fork at rest emits no sound (Fig. 18-26), but if it is struck sharply, it vibrates at a single fixed frequency and gives rise to a pure tone. As the arms of the tuning fork move, they push air molecules ahead of them, creating a zone of compression, and leave behind a zone of rarefaction (Fig. 18-26B). As they move in the opposite direction, they again create pressure waves of compression and rarefaction (Fig. 18-26C).

The molecules in an area of compression, pushed together by the vibrating prong of the tuning fork, bump into the molecules ahead of them, push them together, and create a new region of compression. Individual molecules travel only short distances, but the disturbance passed from one molecule to another can travel many miles; and it is by these disturbances (sound waves) that sound energy is transmitted. The sound dies out only when so much of the original sound energy has been dissipated that one sound wave can no longer disturb the air molecules around it. The tone emitted by the tuning fork is said to be *pure* because the waves of rarefaction and compression are regularly spaced. The waves of speech and many other common

sounds are not regularly spaced but are complex waves made up of many frequencies of vibration.

The sounds heard most keenly by human ears are those from sources vibrating at frequencies between 1000 and 4000 Hz (Fig. 18-27), but the entire range of frequencies audible to human beings extends from 20 to 20,000 Hz. The *frequency* of vibration

Figure 18-27. Human audibility curve. The threshold of hearing varies with sound frequency. The conversational voice of an average male is about 120 Hz, that of an average female about 250 Hz.

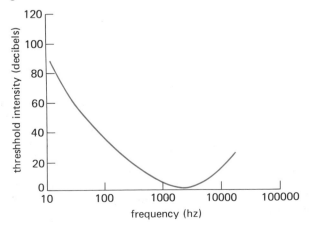

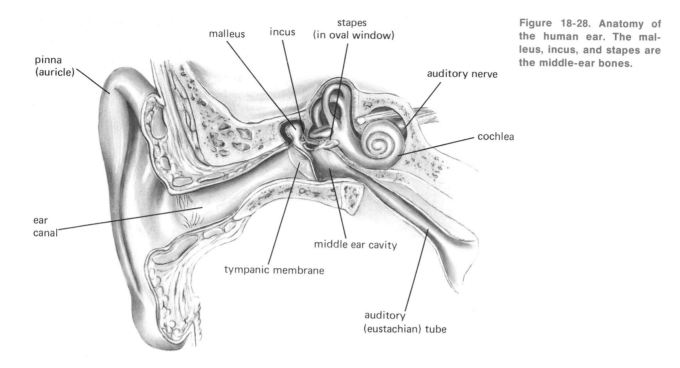

malleus   incus   stapes
(in oval window)

pinna
(auricle)

auditory nerve

cochlea

ear
canal

middle ear cavity

tympanic membrane

auditory
(eustachian) tube

Figure 18-28. Anatomy of the human ear. The malleus, incus, and stapes are the middle-ear bones.

of the sound source is related to the pitch we hear; the faster the vibration, the higher the pitch. We can also detect loudness and tonal quality, or timbre, of a sound. The difference between the packing (or pressure) of air molecules in a zone of compression and a zone of rarefaction, i.e., the *amplitude* of the sound wave, is related to the loudness of the sound that we hear. The number of sound frequencies in addition to the fundamental tone, i.e., the lack of *purity* of the sound wave, is related to the timbre of the sound. We can distinguish some 400,000 different sounds; we can distinguish the note A played on a piano from the same note played on a violin, and we can identify voices heard over the telephone. We can also selectively *not* hear sounds, tuning out the babel of a party to concentrate on a single voice.

The first step in hearing is usually the entrance of pressure waves into the *ear canal* (Fig. 18-28). The waves reverberate from the side and end of the ear canal, filling it with the continuous vibrations of pressure waves. The *tympanic membrane (eardrum)* is stretched across the end of the ear canal. The air molecules, under higher pressure during a wave of compression, push against the membrane, causing it to bow inward. The distance

the membrane moves, although always very small, is a function of the force and velocity with which the air molecules hit it and is therefore related to the loudness of the sound. During the following wave of rarefaction, the membrane returns to its original position. The exquisitely sensitive tympanic membrane responds to all the varying pressures of the sound waves, vibrating slowly in response to low-frequency sounds and rapidly in response to high tones. It is sensitive to pressures to which the most delicate touch receptors of the skin are totally insensitive.

The tympanic membrane separates the ear canal from the *middle-ear cavity* (Fig. 18-28). The pressures in these two air-filled chambers are normally equal. The ear canal is at atmospheric pressure, and the middle ear is exposed to atmospheric pressure only through the *auditory (eustachian) tube*, which connects the middle ear to the pharynx. The slitlike ending of this tube in the pharynx is normally closed, but during yawning, swallowing, or sneezing, when muscle movements of the pharynx open the entire passage, the pressure in the middle ear equilibrates with atmospheric pressure. A difference in pressure can be produced with sudden changes in altitude, as in an elevator or airplane, when the pressure outside the ear changes while the pressure within the middle ear remains constant because of the closed eustachian tube. This difference distorts the tympanic membrane and causes pain.

The second step in hearing is the transmission of sound energy from the tympanic membrane, through the cavity of the middle ear, and then to the fluid-filled chambers of the *inner ear*. Because the liquid in the inner ear is more difficult to move than air, the pressure transmitted to the inner ear must be amplified. This is achieved by a movable chain of three small middle-ear bones which couple the tympanic membrane to the membrane-covered opening (the oval window) separating the middle and inner ear. The *total* force of the tym-

panic membrane is transferred to the oval window, but because the oval window is so much smaller, the *force per unit area* (i.e., pressure) is increased 15 to 20 times. Additional advantage is gained through the lever action of the three middle-ear bones. The amount of energy transmitted to the inner ear can be modified by the contraction of two small muscles in the middle ear which alter the tension of the tympanic membrane and the position of the third middle-ear bone (*stapes*) in the oval window. These muscles protect the delicate receptor apparatus from intense sound stimuli and possibly aid intent listening over certain frequency ranges.

Thus far, the entire system has been concerned with the transmission of the sound energy into the

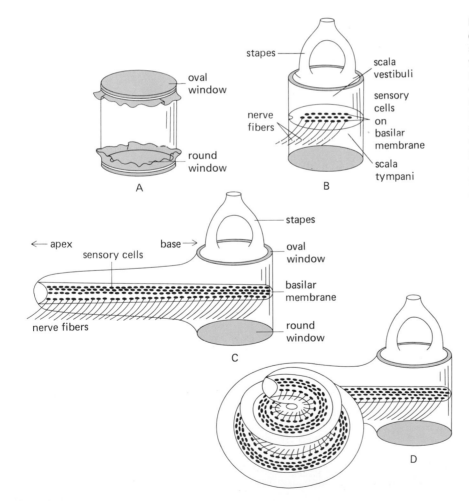

Figure 18-29. Basic plan of the cochlea. (A) The cochlea is represented as a fluid-filled container with an elastic membrane (the oval and round window membranes) covering each end. The rigid wall of the container corresponds to the bony walls of the cochlea. (B) A driving piston (the stapes) and an elastic partition (the basilar membrane) are added. Sensory hair cells are placed on the basilar membrane. The nerve fibers from these cells will form the auditory nerve. A hole is placed in one end of the basilar membrane. (C) The wall of the cochlea near the hole is extended and the basilar membrane is elongated. (D) The elongated cochlea is coiled along its length. The base of the cochlear duct, i.e., the basilar membrane, is shown, but the rest of the duct is not. (*Redrawn from Kiang.*)

inner ear, where the receptors are located. The inner ear, or *cochlea*, is a coiled passage in the temporal bone. It is almost completely divided lengthwise by the *cochlear duct*, the base of which is the *basilar membrane* (Figs. 18-29 and 18-30). As the pressure wave pushes in on the tympanic membrane, the chain of bones rocks the footplate of the stapes against the membrane covering the oval window, causing it to bow into the scala vestibuli compartment of the cochlea (Fig. 18-29) and create there a wave of pressure. The wall of the scala vestibuli is largely bone, but there are two paths by which the pressure waves can be dissipated. One path is to the end of the scala vestibuli, where the waves pass around the end of the

cochlear duct into a second compartment, the *scala tympani*, and back to a second membrane-covered window, which they bow out into the middle-ear cavity. However, most of the pressure waves do not follow this route but are transmitted to the cochlear duct and thereby to the basilar membrane, which is deflected into the scala tympani.

The pattern by which the basilar membrane is deflected is important because this membrane contains the spiral organ of Corti with its sensitive receptor cells which transform sound energy, i.e., the pressure wave, into action potentials. At the end of the cochlea closest to the middle-ear cavity, the basilar membrane is narrow and relatively

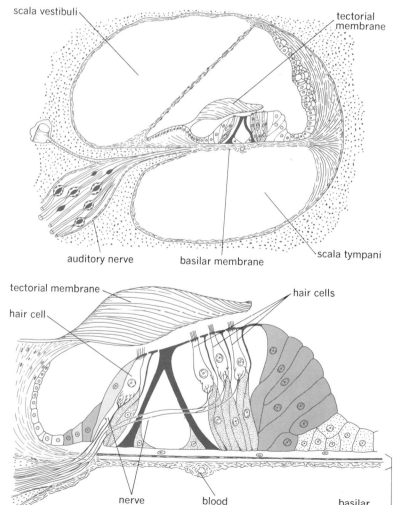

Figure 18-30. Cross section of the membranes and compartments of the inner ear with detailed view of the hair cells and other structures upon the basilar membrane. *(Adapted from Rasmussen.)*

Figure 18-31. Wave motion of the basilar membrane in response to pressure changes in the inner ear. *(From M. Alpern, M. Lawrence, and D. Wolsk, "Sensory Processes," Brooks/Cole Publishing Company, Belmont, Calif., 1968.)*

is displaced by pressure waves, the hair cells move in relation to the tectorial membrane and the fluid surrounding the hairs and, consequently, the hairs are displaced. The resulting mechanical deformation of the hair cells generates a receptor potential, possibly by opening channels for ion flow across the hair-cell membrane; the receptor potential in

Figure 18-32. Point along the basilar membrane where the traveling wave peaks is different with different sound frequencies. The region of maximal displacement of the basilar membrane occurs near the end of the membrane for low-pitched (low-frequency) tones and near the oval window and middle ear for high-pitched tones. *(Adapted from Kim and Molnar.)*

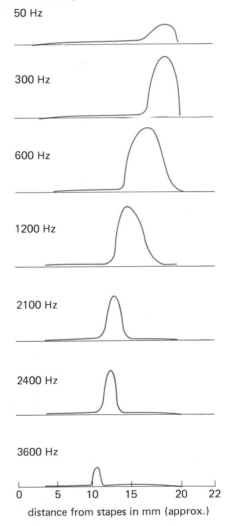

stiff, but it becomes wider and more elastic as it extends throughout the length of the cochlear spiral. The stiff end nearest the middle-ear cavity vibrates immediately in response to the pressure changes transmitted to the scala vestibuli, but the responses of the more distant parts are slower. Thus, with each change in pressure in the inner ear, a wave of vibrations is made to travel down the basilar membrane (Fig. 18-31).

The region of maximal displacement of the basilar membrane varies with the frequency of vibration of the sound source. The properties of the membrane nearest the oval window and middle ear are such that this region resonates best with high-frequency tones and undergoes the greatest amplitude of vibration when high-pitched tones are heard. The vibration of the basilar membrane in response to high-frequency sound waves soon dies out once it is past this region. Lower tones also cause the basilar membrane to vibrate near the middle-ear cavity, but the vibration wave travels out along the membrane for greater distances. The more distant regions of the basilar membrane vibrate maximally in response to low tones. Thus the frequencies of the incoming sound waves are in effect sorted out along the length of the basilar membrane (Fig. 18-32).

Vibration of the basilar membrane serves to stimulate the receptor cells *(hair cells)* of the organ of Corti (Fig. 18-30), which ride upon the membrane (the greatest stimulation occurs wherever displacement of the basilar membrane is greatest). Some of the fine hairs on the top of the receptor cells are in contact with the overhanging *tectorial membrane*, which projects inward from the side of the cochlea. As the basilar membrane

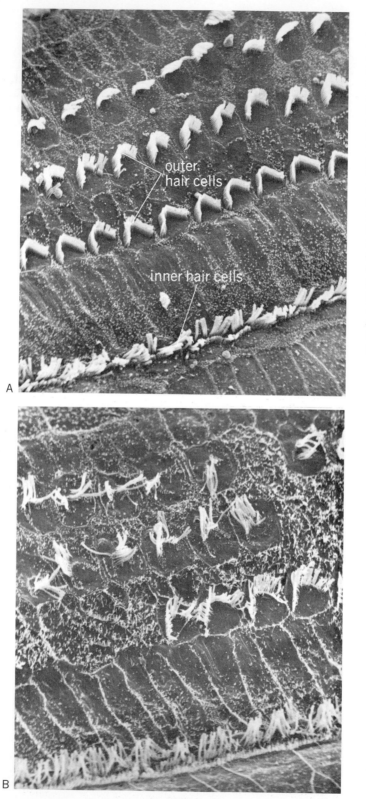

Figure 18-33. Injury to the inner ear by intense noise. (A) Normal organ of Corti (guinea pig) showing the three rows of outer hair cells and single row of inner hair cells. (B) Injured organ of Corti after 24-h exposure to noise levels typical of very loud rock music (2000-Hz-octave band at 120 dB). Several outer hair cells are missing, and the cilia of others no longer form the orderly W pattern of the normal ear. Note also the increased number and size of small villi on the cell surfaces. *(Scanning electron micrograph by Robert E. Preston. Courtesy Joseph E. Hawkins, Kresge Hearing Research Institute.)*

turn causes the release of a chemical transmitter from the hair cell. The transmitter diffuses across an extracellular gap and activates receptor sites on the afferent neuron which underlie the hair cell, and the neuron is depolarized.

The hair cells are easily damaged by exposure to high-intensity noises such as the typical live amplified rock music concerts, engines of jet planes, and revved-up motorcycles. The damaged sensory hairs form giant, abnormal hair structures or are lost altogether, and in cases of long exposure to loud sounds, hair cells and their supporting cells completely degenerate (Fig. 18-33). Much lesser noise levels also cause damage if exposure is chronic.

In normal hearing, the depolarizations formed in the endings of the afferent neurons trigger bursts of action potentials which are transmitted into the central nervous system. The greater the energy of the sound wave (loudness), the greater is the movement of the basilar membrane, the greater the depolarization of the afferent neuron, and the greater the frequency of action potentials in it. Nerve pathways from different parts of the basilar membrane are connected to specific sites along the strip of auditory cortex in much the same way that signals from different regions of the body are represented at different sites in somatosensory cortex; it is as though the basilar membrane were unrolled and spread along a strip of cortex. Each neuron along the auditory pathways responds at threshold levels of sound intensity (loudness) to a very small range of sound frequencies, its so-called *best frequency*. Different neurons have different best frequencies, depending on the location of their receptive fields in the basilar membrane.

It is tempting to explain completely our ability to distinguish sounds of different pitches (frequencies) by the specialization of the auditory cells; however, the fact is that we can differentiate various sound frequencies much more precisely than can single neurons stimulated by moderate sound intensities. This is because, as sound intensities increase to those typical of our everyday surroundings, the frequency band to which the cells respond widens markedly. The precision of our auditory perceptions must, therefore, be explained in part by other factors, a major one being inhibitory mechanisms, many of which are under the control of descending pathways. One such inhibitory mechanism works as follows: Neurons in the audi-

tory pathways firing in response to sounds at their best frequency fire faster than do other neurons having a slightly different best frequency. In the nuclei along the auditory pathways, neurons are interconnected by inhibitory processes so that the fastest-firing cells, i.e., those stimulated at their best frequencies, exert the greatest inhibition on their neighbors; therefore, signals from only those cells firing at their best frequency are retained for perceptual processing.

Sound localization. In hearing (as in vision and smell) there is the stimulus-localization problem of projecting the stimulus to an external source. Although the receptors stimulated by sound are located in the cochlea, the source of the sound is perceived to be external, for example, the ringing telephone across the room. Localization of high-frequency (high pitch) tones depends mainly on the difference in sound intensity at the two ears, the intensity being greater in the ear nearer the sound source; the sound is thus localized on the side where it is louder. A simple experiment can demonstrate this effect. One localizes a quietly hummed tone at the middle of the head. If, while you continue to hum, you lightly plug one ear, the sound intensity in that ear increases because of the greater sound reverberations in the blocked ear canal, and the sound becomes localized in the closed ear. Thus, even a slight imbalance in intensity is sufficient to cause large changes in localization of the stimulus. The first step in this comparison occurs in the brainstem before there has been much chance for synaptic alteration and delay to modify the information. Certain neurons in the brainstem are excited by sound stimuli applied to one ear but inhibited when the opposite ear is stimulated; if the intensity is the same in both ears, the effects cancel out and these neurons do not change their level of activity, whereas if the sound intensities are different the activity is altered.

In contrast to high-frequency tones, localization of low-frequency tones depends mainly on comparison of the times of onset of the tone at the two ears rather than on differences in intensity. A sound originating on the left stimulates the left ear slightly before it stimulates the right (Fig. 18-34), periods of condensation and rarefaction of each sound wave occurring slightly earlier on the left. Certain cells in the brainstem auditory path-

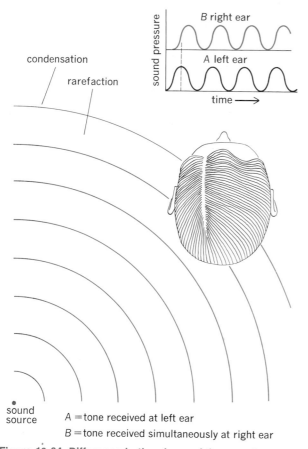

Figure 18-34. Difference in the phase of the sound wave at the two ears helps to localize the sound source.

way fire most rapidly when the sounds at the two ears are "phase-locked"; i.e., when they occur at the same time, and their firing rate progressively slows as the times of onset become increasingly out of phase. Turning the head, which is frequently done to help locate a sound source, alters both intensity and time-of-onset differences between the two ears.

Vestibular system

Changes in both the motion and position of the head are detected by mechanoreceptors which are part of the *vestibular apparatus*, membranous sacs filled with fluid (*endolymph*) and lying in tunnels in the temporal bone of the skull, one on each side of the head. The vestibular apparatus consists of three *semicircular canals* and a slight bulge for the *utricle* and *saccule* (Fig. 18-35A).

The three semicircular canals are arranged at right angles to each other (Fig. 18-35B). The mechanoreceptors of the semicircular canals are hair cells, the hairs of which are closely ensheathed by a gelatinous mass which extends into the channel of the semicircular canal at one point. Whenever the head is moved, the bony tunnel wall of the semicircular canal, its enclosed membranous sac, and the attached bodies of the hair cells all, of course, turn with it. The endolymph fluid filling the membranous sac, however, is not attached to the skull; therefore, because of inertia, the fluid tends to retain its original position, i.e., to be "left behind" and pushes against the gelatinous mass, bending the hairs within it and thereby stimulating the hair cells (Fig. 18-36). The speed and magnitude of the movement of the head determine the way in which the hairs are bent and the hair cells stimulated. As the inertia is overcome and the endolymph fluid begins to move at the same rate as the rest of the head, the hairs slowly return to their resting position. For this reason, the hair cells are stimulated only during *changes* in the rate of motion, i.e., during acceleration of the head. In contrast, during motion at a constant speed, stimulation of the hair cells ceases.

The hair cells are functionally connected to the afferent nerve fibers underlying them by chemically mediated synapses. Some transmitter is released from the hair cells in the absence of stimulation so that even when the head is motionless, the afferent neurons fire at a relatively low rate. Thus, the vestibular receptors can signal information by either increasing or decreasing the frequency of action potentials in the afferent-nerve fiber. The frequency of action potentials in the afferent nerve is related to the shearing force bending the hairs on the receptor cells; when the hairs are bent one way, the rate of firing speeds up, and when the hairs are bent in the opposite direction, the firing frequency slows down (Fig. 18-36B and C).

Whereas the semicircular canals signal the rate of change of motion of the head, the utricle and saccule contain receptors which provide information about the position of the head relative to the direction of the forces of gravity and about any linear acceleration of the head. The receptor cells here, too, are mechanoreceptors sensitive to the dis-

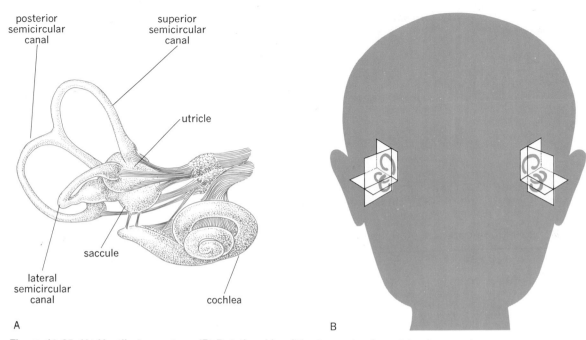

posterior
semicircular
canal

superior
semicircular
canal

utricle

saccule

lateral
semicircular
canal

cochlea

A

B

Figure 18-35. (A) Vestibular system. (B) Relationship of the two sets of semicircular canals.

placement of projecting hairs. The hair cells of the utricle and saccule are collected into groups from which the hairs protrude into a gelatinous substance. Tiny calcium carbonate stones, or *otoliths*, are embedded in the gelatinous covering of the hair cells, making it heavier than the surrounding endolymph, or attached to the ends of some of the hairs. When the head is tipped, the gelatinous-otolith material changes its position, pulled by gravitational forces to the lowest point in the utricle or saccule. The shearing forces of the gelatinous-otolith substance against the hair cells bend the hairs and stimulate the receptor cells.

The information from the vestibular apparatus is used for two purposes. One is to control the muscles which move the eyes so that, in spite of changes in the position of the head, the eyes remain fixed on the same point. As the head is turned to the left, for example, impulses from the vestibular nuclei activate the ocular muscles which turn the eyes to the right and inhibit their antagonists; the eyes therefore turn toward the right and remain fixed on the point of interest.

The second use of vestibular information is in reflex mechanisms for maintaining upright posture. In monkeys, cats, and dogs, for example, the

vestibular apparatus plays a definite role in the postural fixation of the head, orientation of the animal in space, and reflexes accompanying locomotion. However, in man very few postural re-

Figure 18-36. Relation between position of hairs and activity in afferent neurons. Movement of the hairs in one direction (B) increases the action potential frequency in the afferent nerve fiber activated by the hair cell, whereas movement in the opposite direction (C) decreases the rate relative to the resting state (A). *(Adapted from Wersall, Gleisner, and Lundquist.)*

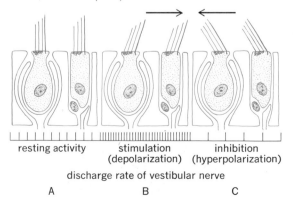

resting activity   stimulation   inhibition
(depolarization) (hyperpolarization)

discharge rate of vestibular nerve

A   B   C

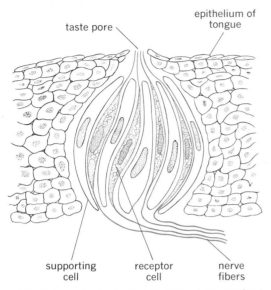

Figure 18-37. Structure and innervation of a taste bud.

flexes are known to depend primarily on vestibular input despite the fact that the vestibular organs are sometimes called the sense organs of balance.

Chemical senses

Receptors sensitive to certain chemicals in the environment are *chemoreceptors.* Some of these respond to chemical changes in the internal environment (e.g., the oxygen and hydrogen receptors in certain of the large blood vessels); others respond to external chemical changes, and in this category are the receptors for taste and smell. Although taste and smell affect a person's appetite, flow of saliva and gastric secretions, and the avoidance of harmful substances, they do not exert strong or essential influence in human beings, and are less important than the receptors for vision or hearing.

Taste. The specialized receptor organs for the sense of taste are the 10,000 or so *taste buds* which are located on the tongue, roof of the mouth, pharynx, and larynx. In the taste buds the receptor cells are arranged like segments of an orange so that the multifolded upper surfaces of the receptor cells extend into a small pore at the surface of the

taste bud, where they are bathed by the fluids of the mouth (Fig. 18-37).

Taste sensations are traditionally divided into four basic groups: sweet, sour, salt, and bitter, but different types of taste buds or receptor cells which would support this specificity have not been identified. In fact, a single receptor cell can respond in varying degrees to many different chemical substances falling into more than one of the basic categories.

The mechanisms by which the taste receptors are stimulated are not known. It has been suggested that the first step is a loose binding of the individual molecules of the chemical substance with specific sites on the receptor cell membrane. The fact that a single receptor can respond to more than one basic taste quality could be explained if a single receptor cell had several different sites, each capable of binding with a different type of molecule. It is known that the receptors depolarize when stimulated chemically.

Afferent nerve fibers enter the buds to end on the receptor cells, separated from them by a cleft across which transmitters released from the receptors diffuse. One nerve fiber may innervate several receptor cells, and one receptor may be innervated by several different neurons. There is clearly no one-to-one relationship by which each receptor cell has a direct line into the central nervous system. The frequency of action potentials in single nerve fibers increases in response to increasing concentrations of the chemical stimulant; therefore, frequency signals the quantity, but what signals the quality? How can we distinguish so many different taste sensations when the receptor cells lack specificity both in terms of the kind of chemical to which they respond and the way in which they are connected to the brain? The afferent fibers involved in taste show different firing patterns in response to different substances; e.g., one fiber may fire very rapidly when the stimulatory substance is salt but only sporadically when it is sugar, and another fiber may have just the opposite reaction. Therefore, awareness of the specific taste of a substance probably depends upon the pattern of firing within a group of neurons rather than that in a specific neuron. Identification of the substance is also aided by information about its temperature and texture which is transmitted to the central nervous system from other receptors on the tongue and surface of the oral cavity. The odor of the substance clearly helps, too, as is attested by the common experience that food lacks taste when one has a stuffy head cold.

Smell. The olfactory receptors which give rise to the sense of smell lie in a small patch of mucus-secreting membrane (the *olfactory mucosa*) in the upper part of the nasal cavity (Fig. 18-38A). The olfactory receptors are not separate cells but are specialized areas of the afferent neurons. The cell body of these neurons bears a knob from which several long ciliary processes extend out to the surface of the olfactory mucosa (Fig. 18-38B). The knob and cilia of the afferent neurons contain the receptor sites, and the axons project to the brain as the olfactory nerve.

Before an odorous substance can be detected, it must release molecules which diffuse into the air and pass into the nose to the region of the olfactory mucosa, dissolve in the layer of mucus covering the receptors, establish some sort of relation with the receptors, and depolarize the membrane enough to initiate action potentials in the afferent-nerve fiber. Upon stimulation of the olfactory mucosa with odorous substances, receptor potentials can be recorded which change with stimulus quality and intensity.

The physiological basis for discrimination between the tens of thousands of different odor qualities is speculative. There are no apparent differences in receptor cells, at least on a microscopic level, but meaningful differences between receptors are thought to exist at molecular levels. One theory is that odor discrimination depends upon a large number of different types of interactions between the odor-substance molecules and receptor sites. This theory is based on the supposition that the receptor cells probably have 20 or 30 different types of receptor sites, the receptor-site populations varying from one cell to another. For example, one receptor cell may have mainly receptor sites of types A, B, and C, whereas another cell has types B, D, and E.

The molecules of the odor substance combine with most types of receptor sites, but the "fit" varies. A good contact occurs when the odor-producing molecule and receptor sites' size, shape, and polarity match; in such a case the resulting depolarization of the receptor cell is large. In cases of poorer fit, the receptor cell still depolarizes but to a smaller degree. All depolarizations occurring

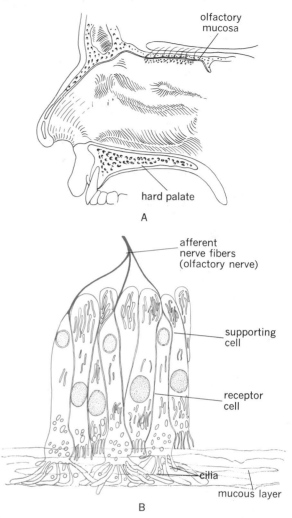

Figure 18-38. Location and structure of the olfactory receptors.

together in one cell are summed, forming a receptor potential, which determines the firing rate in the afferent-nerve fiber. According to this theory, receptor cells with different receptor-site populations respond to the same odor substance with different firing rates. Thus, it is the simultaneous yet differential stimulation of many receptor cells that provides the basis for odor discrimination.

It is known that olfactory discrimination depends only partially upon the action-potential pattern generated in the different afferent neurons; it also varies with attentiveness, state of the olfactory mucosa (acuity decreases when the mucosa is congested, as in a head cold), hunger (sensitivity is greater in hungry subjects), sex (women in general have keener olfactory sensitivities than men), and smoking (decreased sensitivity has been repeatedly associated with smoking). And just as the awareness of the taste of an object is aided by other senses, so the knowledge of the odor of a substance is aided by the stimulation of other receptors. This is responsible for the description of odors as pungent, acrid, cool, or irritating.

Further perceptual processing

Our actual perception of the events around us often involves areas of brain other than primary sensory cortex. Information from the primary sensory areas achieves further elaboration through the neural activity in *cortical association areas.* These brain areas lie outside the classic primary cortical sensory or motor areas but are connected to them by association fibers (Fig. 18-39). Although it has not often been possible to elucidate the specific roles performed by the association areas, they are acknowledged to be of the greatest importance in

Figure 18-39. Areas of association cortex.

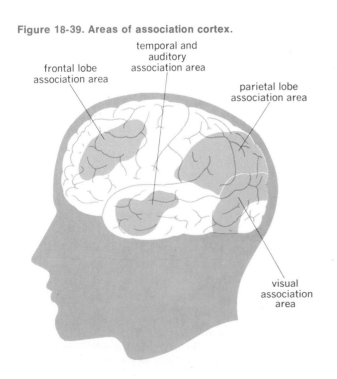

frontal lobe
association area

temporal and
auditory
association area

parietal lobe
association area

visual
association
area

the maintenance of higher mental activities in human beings. For example, if areas of primary visual cortex are stimulated (when the brain surface is exposed under local anesthesia during neurosurgical procedures and the patient is awake), the patient "sees" a flash of light. Upon stimulation of the association areas surrounding visual cortex, the patient reports seeing more elaborate visual sensations such as "brilliant colored balloons floating around in an infinite sky." Upon stimulation of association areas still farther from primary visual cortex, the patient might report visual memories which seem to be reenacted before his eyes. Another example of the embellishment of the sensory experience provided by association areas can be found in persons who have undergone removal of parts of cortex because of tumors or accidents. A person who has no primary visual cortex is blind; if a chair is placed in his path, he walks into it because he does not see it. In contrast, a person who has functional primary visual cortex but has no visual association areas sees the chair and therefore does not walk into it, but he is not able to say that the object in his path is a chair or to explain its function. Similarly, a patient with damaged auditory association areas may hear a spoken word but not comprehend its meaning. A patient with damaged parietal association areas can feel a small, cool object in his hand but know neither that it is called a key nor that it is used to open locks.

Further perceptual processing involves arousal, attention, learning, memory, language, and emotions, as well as comparing the information presented via one sensory modality with that of another. For example, we may hear a growling dog, but our perception of the actual event taking place varies markedly, depending upon whether our visual system detects that the sound source is an angry animal or a loudspeaker.

Sometimes the objects we view are ambiguous and have more than one logical interpretation. Such a figure is the drawing entitled "My wife and my mother-in-law," illustrated in Fig. 18-40. When first trying to identify such a figure, we pick out one detail, say the curved line at the midportion of the left side of the drawing. If it reminds us of a bent nose, we immediately seek to verify our impression by looking for the expected eyes, mouth, hair, etc. An old woman's face is perceived, but as we look at it, the image suddenly shifts. The line that was a nose is now a chin; the image of a stylish young woman appears. Since both images are equally plausible, our interpretation of the image shifts back and forth between them; it is impossible to see them as both plausible at the same time.

We put great trust in our sensory-perceptual processes despite the inevitable modifications we know to exist. Some factors known to distort our perceptions of the real world are as follows:

1 Afferent information is distorted by receptor mechanisms and by its processing along afferent pathways, e.g., by adaptation.
2 Such factors as emotions, personality, and social background can influence perceptions so that two people can witness the same events and yet perceive them differently.
3 Not all information entering the central nervous system gives rise to conscious sensations. Actually, this is a very good thing because many unwanted signals, generated by the extreme sensitivity of our receptors, are canceled out. The afferent systems are very sensitive. Under ideal conditions the rods of the eye can detect the flame of a candle 17 mi away. The hair cells of the ear can detect vibrations of an amplitude much lower than that caused by the flow of blood through the vascular system and can even detect molecules in random motion bumping against the tympanic membrane. Olfactory receptors respond to the presence of only four to eight odorous molecules. It is possible to detect one action potential generated by a pacinian corpuscle. If no mechanisms existed to select, restrain, and organize the barrage of impulses from the periphery, life would be unbearable. Information in some receptors' afferent pathways is not canceled out; it simply does not give rise to a conscious sensation. For example, stretch receptors in the muscles detect changes in the length of the muscles, but activation of these receptors does not lead to a conscious sense of anything. Similarly, stretch receptors in the walls of the carotid sinus effectively monitor both absolute blood pressure and its rate of change, but people have no conscious awareness of their blood pressure.
4 We lack suitable receptors for many energy forms. For example, we can have no direct information about radiation and radio or television

Figure 18-40. Ambiguous figure, "My wife and my mother-in-law," in which the young girl's chin is the woman's nose, created by cartoonist W. E. Hill in 1915.

waves until they are converted to an energy form to which we are sensitive. Many regions of the body are insensitive to touch, pressure, and pain because they lack the appropriate receptors. The brain itself has no pain or pressure receptors, and brain operations can be performed painlessly on patients who are still awake, provided that the cut edges of the sensitive brain coverings are infused with local anesthetic.

However, the most dramatic examples of a clear difference between the real world and our perceptual world can be found in illusions and drug- and disease-induced hallucinations when whole worlds can be created.

Any sense organ can give false information; e.g., pressure on the closed eye is perceived as light in darkness, and electrical stimulation of the afferent nerve fibers from any sense organ produces the sensory experience normally arising from the activation of that receptor. Why do such illusions appear? The two bits of false information just men-

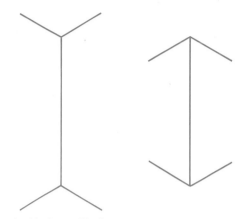

Figure 18-41. Arrow illusion.

tioned arise because most afferent fibers transmit information about one modality, and action potentials in a given afferent pathway going to a certain area of the brain signal the kind of information normally carried in that pathway. The explanations in most other illusions are hypothetical. In the arrow illusion (Fig. 18-41) the shaft of the arrow with the outgoing fins looks longer than that of the other arrow even though the two shafts are exactly the same length. Like many other illusions this one probably occurs because our perceptual systems try to group information into meaningful patterns. This particular illusion appears to those of us brought up in a rectangular world of lines and corners, the two-dimensional arrows possibly suggesting the three-dimensional arrangement of a cube in very familiar situations such as the corner of a room or corner of a building. The arrow with outward-going fins suggests the distant corner of a room whereas the arrow with the inward-going fins suggests the nearby corner of a building. We realize that distant objects appear smaller than they really are and mentally adjust for the apparent difference in distance between the two corners. It is suggested that because of this adjustment we perceive the arrow with the outward-going fins to be longer than it really is. People brought up in curvilinear societies (round buildings, etc.) do not have this illusion and report that the shafts of both arrows appear to be the same length.

In conclusion, the two processes of transmitting data through the nervous system and interpreting it cannot be separated. Information is processed at each synaptic level of the afferent pathways. There is no one point along the afferent pathways or one particular level of the central nervous system below which activity cannot be a conscious sensation and above which it is a recognizable, definable sensory experience. Perception has many levels, and it seems that the many separate stages are arranged in a hierarchy, with the more complex stages receiving input only after it is processed by the more elementary systems. Every synapse along the afferent pathways adds an element of organization and contributes to the sensory experience.

19
CONTROL OF BODY MOVEMENT

Local control of motor neurons
   Length-monitoring systems and the stretch reflex
     Alpha-gamma coactivation
   Tension-monitoring systems
Descending pathways and the brain centers which control
   them
   Corticospinal pathway

Multineuronal pathways
Cerebral cortex
Subcortical nuclei (including the basal ganglia) and
   brainstem nuclei
Cerebellum
Maintenance of upright posture and balance
Walking

The execution of a coordinated movement is a complicated process. Consider reaching to pick up an object. The fingers are extended (straightened) and then flexed (bent), the degree of extension depending upon the size of the object to be grasped, and the force of flexion depending upon the weight and consistency of the object. Simultaneously, the wrist, elbow, and shoulder are extended, and the trunk is inclined forward, the exact movements depending upon the distance of the object and the direction in which it lies. The shoulder must be stabilized to support the weight first of the arm and then of the object. Upright posture must be maintained in spite of the body's continually shifting center of gravity.

The building blocks for this simple action—as for all movements—are active motor units, each comprising one motor neuron together with all the skeletal muscle cells it innervates (Chap. 10). Thus, anything that affects the movement of skeletal muscle does so by means of synaptic input to the motor neurons. Neural inputs from many sources converge upon the motor neurons to control their activity, and the precision of coordinated muscle movement depends upon the *balance* of their influence. If for example, an inhibitory sys-

tem is damaged and its input to the motor neurons is lessened, the still-normal excitatory input will be unopposed; the motor neurons will fire excessively and the muscle will be hyperactive. In cases of abnormal synaptic input, the motor units do not operate with the same degree of coordination or under the wide range of conditions typical of a normal motor system. No one source of input to the motor neuron is essential for movement, but to provide the precision and speed of normally coordinated movements, the balanced input from all sources is necessary.

Each of the myriad coordinated body movements is characterized by a set of motor-unit activities occurring over space and time, and the interrelating systems which converge upon the motor neurons to control their activity are the subject of this chapter. We present here a summary of a model of motor-system functioning (Fig. 19-1) and then describe each component in detail.

First, a general "command" such as "pick up sweater" or "write signature" or "answer telephone" is generated, but it is not known where such a command is generated or how the neurons involved in this function are activated. Regions of association cortex and cerebellum show the earli-

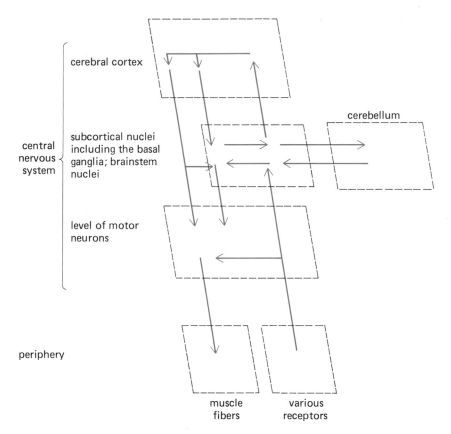

Figure 19-1. Diagram of motor system function. All motor system control of the muscles is exerted via the alpha motor neurons; thus, they are called the *final common pathway*.

est signs of coordinated electrical activity preceding and related to motor movements, and if these areas do not themselves house the "command neurons," they are certainly important early coordinating centers. Certain neurons in these brain areas, as well as in the basal ganglia, continually receive information from receptors in the muscles and joints, skin, vestibular apparatus, and eyes. These brain areas, therefore, have information about the starting position of the body part that is to be moved. Integrating this information with that from the "command neurons," they produce an initial program of motor unit activation needed to perform the desired movement. This overall program is, however, then influenced by program subsets contributed by other brain regions; such subset programs exist because of inherent genetically determined connections between neurons and the connections formed later as a result of learning.

The complete program arrived at is relayed via nuclei in the thalamus to the motor cortex where

it results in changes in the firing patterns of the motor-cortex cells. Action potentials in the axons of these cells then descend to alter the activity of the appropriate motor units; via axon collaterals, they also inform the basal ganglia and cerebellum of the signals that are being sent. This starts the process over again; the basal ganglia and cerebellum compare the new directives with ongoing information about the limb position, and compute adjustments, which are then sent back to the motor cortex. Thus, the basal ganglia and cerebellum receive a constant stream of information from the cortex about what actions are supposed to be taking place, while they simultaneously receive reports from the periphery about the actions that actually are taking place; any discrepancies between the intended and actual movements are detected, and program corrections are sent to the motor cortex. This loop from the motor cortex to the cerebellum and basal ganglia and back again continues throughout the course of the movement.

However, although activation of this loop takes only 1/50th s, very rapid movements do not provide enough time for continual correction, and the entire course of the action in such movements is completely preprogrammed. The cerebellum deals more with these movements, whereas the basal ganglia are more involved with slow, continuous movements.

We have acknowledged that we know very little concerning the location or activation of the command neurons. Note that our system as described also does not take into account the mechanisms by which the "decision" to make a particular movement is reached. What neural events actually occur in the brain to cause one to "decide" to pick up an object in the first place? Presently we have no insight on this question.

Given such a model, it is difficult to use the word *voluntary* with any real precision, but we shall use it to refer to those actions which are characterized as follows:

1 These are actions we think about. The movement is accompanied by a conscious awareness of what we are doing and why we are doing it rather than the feeling that it "just happened," a feeling that often accompanies reflex responses.
2 Our attention is directed toward the action or its purpose.
3 The actions are the result of learning. Actions known to have disagreeable consequences are less likely to be performed voluntarily.

In the previous example of reaching to pick up an object, the activation of some of the motor units, such as those actually involved in grasping the object, can be classified clearly as voluntary; but most of the muscle activity associated with the act is initiated without any conscious, deliberate effort. In fact, almost all motor behavior involves both conscious and unconscious components, and the distinction between the two cannot be made easily.

Even a highly conscious act such as threading a needle involves the unconscious postural support of the hand and arm and inhibition of antagonistic muscles (those muscles whose activity would oppose the intended action, in this case the finger extensor muscles which straighten the fingers). Moreover, unconscious basic reflexes such as dropping a hot object can be influenced by conscious effort. If the hot object is something that took a great deal of time and effort to prepare, one probably would not drop it but would try to inhibit the reflex, holding on to the object until it could be put down safely. Most motor behavior is neither purely voluntary nor purely involuntary but falls at some point on a spectrum between these two extremes. But even this statement is of little help because patterned muscle movements shift along the spectrum according to the frequency with which they are performed.

For example, when a person first learns to drive a car with standard transmission, stopping is a fairly complicated process involving the accelerator, clutch, and brake. The sequence and force of the various operations depend upon the speed of the car, and their correct implementation requires a great deal of conscious attention. With practice the same actions become automatic. If a child darts in front of the car of an experienced driver, he does not have to think about the situation and decide to remove his foot from the accelerator and depress the brake and clutch. Upon seeing the child, he immediately and automatically stops the car. A complicated pattern of muscle movements is shifted from the highly conscious end of the spectrum over toward the involuntary end by the process of learning. Such a shift presumably occurs because relevant programs have been established, presumably by altering the synaptic connections between neurons (see Chap. 20). During early stages of learning, movements are slow and clumsy and are characterized by hesitant exploration; such movements depend greatly upon sensory feedback for guidance. With repetition of the movement, programming occurs (i.e., learning takes place). Thus, with later attempts at the movement, the altered synaptic connections are utilized, the dependence on sensory feedback is decreased, and the movement is carried out with greater precision and efficiency.

Whether activated voluntarily or involuntarily, given motor units are frequently called upon to serve many different functions. For example, one demand upon the muscles of the limbs, trunk, and neck is made by postural mechanisms; these muscles must support the weight of the body against gravity, control the position of the head and different parts of the body relative to each other to maintain equilibrium, and regain stable, upright posture after accidental or intentional shifts in position. Superimposed upon these basic postural

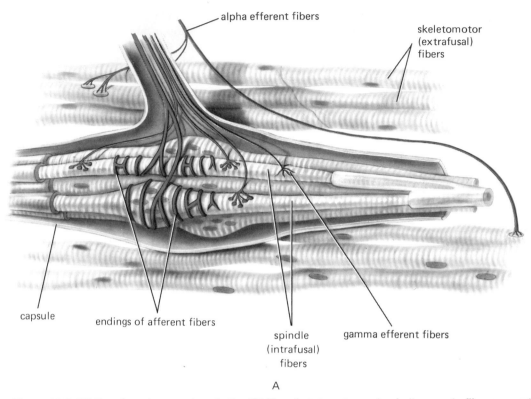

alpha efferent fibers

skeletomotor
(extrafusal)
fibers

capsule

endings of afferent fibers

spindle
(intrafusal)
fibers

gamma efferent fibers

A

Figure 19-2. (A) Drawing of a muscle spindle. (B) The skeletomotor and spindle muscle fibers are relaxed and the spindle receptors are under slight tension. (C) The skeletomotor fibers are contracting, the spindle fibers are relaxed, and the tension has been removed from the spindle receptors. (D) The skeletomotor and spindle muscle fibers are contracting and tension has been restored to the spindle stretch receptors. *(Part A adapted from Merton.)*

requirements are the muscle movements associated with locomotion. For these purposes, the muscles must be capable of transporting the body from one place to another under the coordinated commands of neural mechanisms for alternate stepping movements and shifting the center of gravity. And added to the requirements of posture and locomotion can be the highly skilled movements of a ballerina or hockey player. The motor units are activated and the sometimes conflicting demands are settled, usually without any conscious, deliberate effort.

We now turn to an analysis of the individual components of the motor-control-system model, beginning with local control mechanisms because their activity serves as a base upon which the descending pathways frequently exert their influence.

Local control of motor neurons

Much of the synaptic input to the motor neurons arises from neurons at the same level of the central nervous system as the motor neurons. Indeed, some of these neurons are activated by receptors in the very muscles controlled by the motor neurons, in other nearby muscles, and in tendons associated with the muscles. These receptors monitor muscle length and tension and pass this information via afferent nerve fibers into the central nervous system. This input forms the afferent component of purely local reflexes which provide negative-feedback control over muscle length and tension. In addition, it is transmitted to cerebral cortex, cerebellum, and the basal ganglia where it can be integrated with input from other types of receptors.

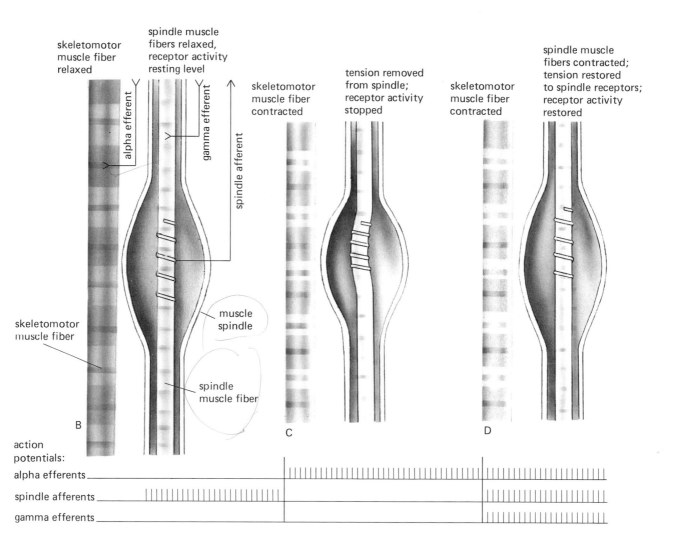

skeletomotor
muscle fiber
relaxed

spindle muscle
fibers relaxed,
receptor activity
resting level

alpha efferent

gamma efferent

spindle afferent

skeletomotor
muscle fiber
contracted

tension removed
from spindle;
receptor activity
stopped

skeletomotor
muscle fiber
contracted

spindle muscle
fibers contracted;
tension restored
to spindle receptors;
receptor activity
restored

skeletomotor
muscle fiber

muscle
spindle

spindle
muscle fiber

B

C

D

action
potentials:

alpha efferents

spindle afferents

gamma efferents

Length-monitoring systems and the stretch reflex

Embedded within skeletal muscle are stretch receptors which are made up of afferent-nerve endings wrapped around modified muscle cells, both partially enclosed in a fibrous capsule. The entire structure is called a *muscle spindle*. The modified muscle fibers within the spindle are known as *spindle fibers;* the typical skeletal muscle cells outside the spindle are the *skeletomotor* (or *extrafusal*) *fibers* (Fig. 19-2).

The muscle spindles are parallel to the muscle such that passive stretch of the muscle pulls on the spindle fibers, stretching them and activating their receptors; the greater the stretch the faster the rate of firing. In contrast, contraction of the skeletomotor fibers and the resultant shortening of the muscle releases tension on the muscle spindle and slows down the rate of firing of the stretch receptor. There are different kinds of spindle receptors, one responding to the magnitude of the stretch, another to both the absolute magnitude of the stretch and the speed with which it occurs. The importance of the kind of information relayed by the first receptor is apparent: It tells the central nervous system about the length of the muscle. However, by indicating the rate of change of the muscle length, the second type of

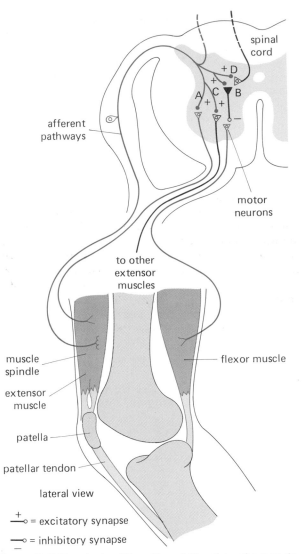

Figure 19-3. Terminals of the afferent fiber from the muscle spindle involved in the knee jerk.

When the afferent neurons from the muscle spindle enter the central nervous system, they divide into branches which can take several different paths. One group of terminals (*A* in Fig. 19-3) directly forms excitatory synapses upon the motor neurons going back to the muscle that was stretched, thereby completing a reflex arc known as the *stretch reflex*. This reflex is probably most familiar in the form of the knee jerk, which is tested as part of routine medical examinations. The physician taps on the patellar tendon, which stretches over the knee and connects muscles in the thigh to a bone in the foreleg. As the tendon is depressed, the muscles to which it is attached are stretched, and the receptors within the muscle spindles are activated. Information about the change in length of the muscles is fed back to the motor neurons controlling the same muscles. The motor units are excited, the thigh muscles shorten, and the patient's foreleg is raised to give the familiar knee jerk. The proper performance of the knee jerk tells the physician that the afferent limb of the reflex, the balance of synaptic input to the motor neuron, the motor neuron itself, the neuromuscular junction, and the muscle are functioning normally. In usual circumstances, of course, the stretch receptors are activated neither simultaneously nor so strongly, and the response is not a jerk. Rather, the stretch reflex maintains resting tension and smooths motions.

Because the group of afferent terminals mediating the stretch reflex synapses directly with the motor neurons without the interposition of any interneurons, the stretch reflex is called *monosynaptic*. Stretch reflexes are the only known monosynaptic motor reflex arcs in people; all other reflex arcs are polysynaptic, having at least one interneuron (and usually many) between the afferent and efferent pathways.

A second group of afferent terminals (*B* in Fig. 19-3) ends on interneurons which, when excited, inhibit the motor neurons controlling antagonistic muscles whose contraction would interfere with the reflex response. For example, the normal response to the knee-jerk reflex is straightening of the knee to extend the foreleg. The antagonists to these extensor muscles are a group of flexor muscles which, when activated, draw the foreleg back and up against the thigh. If both opposing groups of muscles are activated simultaneously, the knee

receptor allows the central nervous system to anticipate the magnitude of the stretch. If the rate of stretch is increasing very rapidly, the stretch itself cannot stop immediately, and an additional change in length can be predicted. Although the two kinds of stretch receptors are separate entities, they will be referred to collectively as the *muscle-spindle stretch receptors*.

joint is immobilized and the leg becomes a stiff pillar. This is certainly what is required in some situations, but if the foreleg is to be extended from a flexed position, the motor neurons which activate the flexor muscles must be inhibited as the motor neurons controlling the extensor muscles are activated. The excitation of one muscle and the simultaneous inhibition of its antagonistic muscle is called *reciprocal innervation.*

A third group of terminals (*C* in Fig. 19-3) ends on interneurons which, when excited, activate *synergistic muscles,* i.e., muscles whose contraction assists the intended motion. For example, in the knee jerk, interneurons facilitate motor neurons which control other leg extensor muscles.

A fourth group of afferent terminals (*D* in Fig. 19-3) synapses with interneurons which convey information about the muscle length to areas of the brain (particularly cerebellum and basal ganglia) dealing with coordination of muscle movement. Although the muscle stretch receptors initiate activity in pathways eventually reaching cerebral cortex, the information relayed by these action potentials does not have a strong conscious correlate; rather the conscious awareness of the position of a limb or joint comes more from the joint, ligament, and skin receptors.

Alpha-gamma coactivation. Because the muscle spindles are parallel to the large skeletomotor muscle fibers, stretch on them is removed when the skeletomotor fibers contract. If the spindle stretch receptors were permitted to go slack at this time they would stop firing action potentials and this important afferent information would be lost. To prevent this, the spindle muscle fibers themselves are frequently made to contract during the shortening of the skeletomotor fibers, thus maintaining tension in the spindle and firing in the receptors. The spindle fibers are not large and strong enough to shorten whole muscle and move joints; their sole job is to produce tension on the spindle stretch receptors. The muscle fibers in the spindles shorten in response to motor neuron activity (Fig. 19-4), but these motor neurons which activate the spindle fibers are usually not the same motor neurons which activate the skeletomotor muscle fibers. The motor neurons controlling the skeletomotor muscle fibers are larger and are classified as *alpha motor neurons;* the smaller neurons whose axons innervate the spindle fibers are known as the *gamma motor neurons.* The latter neurons are activated primarily by synaptic input from descending pathways. The overall route—descending pathway, gamma motor neu-

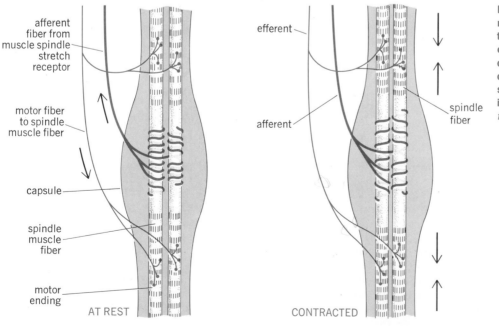

afferent fiber from muscle spindle stretch receptor

motor fiber to spindle muscle fiber

capsule

spindle muscle fiber

motor ending

AT REST

efferent

afferent

spindle fiber

CONTRACTED

Figure 19-4. (A) Diagram of muscle-spindle innervation. (B) As the two striated ends of a spindle fiber contact, they pull on the center of the fiber and stretch the receptor, which is located there. *(Adapted from Merton.)*

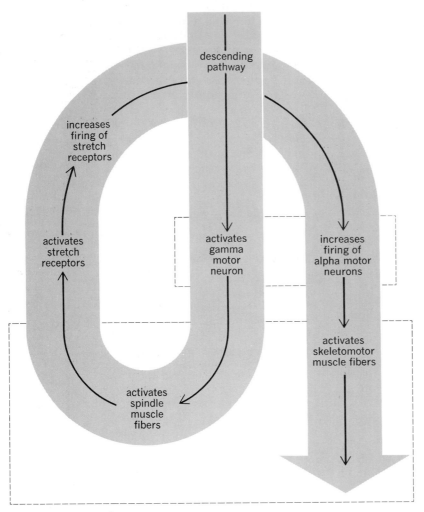

Figure 19-5. Gamma loop. The small rectangle at the center indicates those events occurring in the brainstem or spinal cord; the large rectangle, the events in the muscle.

ron, spindle muscle fiber, stretch receptor and afferent neuron, alpha motor neuron—is known as the *gamma loop* (Fig. 19-5).

In many voluntary and involuntary movements alpha and gamma motor neurons are *coactivated,* i.e., fired at almost the same time. To understand the usefulness of alpha-gamma coactivation, consider picking up a book whose weight is unknown. The position of the involved joints is determined by the length of the muscles. Suppose that the initially programmed strength of alpha motor-unit firing is not sufficient to lift the book. The skeletomotor muscle fibers will be unable to shorten but the spindle fibers, activated simultaneously by the gamma motor neurons, will shorten and the spindle receptors will be stretched. By way of the stretch

reflex, the excitatory synaptic input to the alpha motor units will increase, causing summation of contraction, recruitment of additional motor units, and greater muscle tension. Thus coactivation and the gamma loop provide a mechanism by which motor commands and muscle performance can be compared at the local level and compensation brought about on the spot. Moreover, information that the spindles are longer than expected (i.e., that the skeletomotor fibers have not shortened enough) will be transmitted to those higher brain centers involved in programming and controlling motor behavior so that they can alter their output as well.

Alpha-gamma coactivation works the other way too. If the initial program caused too intense alpha motor-unit activity, the book would be lifted too rapidly. The faster-than-expected shortening of the spindle fibers would remove tension from the spindles, stopping the receptors' firing. This would reflexly remove a component of excitatory input from the alpha motor neurons, automatically slowing the muscle movement to a more desirable rate. Thus, coactivation of alpha and gamma motor neurons can lead to fine degrees of regulation of muscle activity.

Tension-monitoring systems

A second component of the local motor-control apparatus monitors tension rather than length. The receptors employed in this system are the *Golgi tendon organs*, which are located in the tendon near its junction with the muscle. Endings of afferent-nerve fibers are wrapped around collagen bundles of the tendon, which are slightly bowed in the resting state. When the skeletomotor fibers of the attached muscle contract, they pull on the tendon, straightening the collagen bundles, and distorting the receptor endings of the afferent nerves. The receptors fire in proportion to the increasing force or tension generated by the contracting muscle. The afferent neuron's activity causes inhibitory postsynaptic potentials in the motor neurons of the contracting muscle (Fig. 19-6). Some of the Golgi tendon organs have high thresholds and respond only when the tension is very great. These high-threshold receptors may function as safety valves, inhibiting the muscle when the force it generates is great enough to damage the tendon or bone. The remainder of the Golgi tendon organs have lower thresholds, comparable to the receptors of the muscle spindle, and they supply the motor control systems with continuous information about the tension generated. This information is necessary for effective movement because a given input to a group of motor neurons does not always provide the same amount of tension. The tension developed by a contracting muscle depends on the muscle length and the degree of muscle fatigue, as well as the number of activated motor neurons and the rate at which they are firing (Chap. 10). Because one set of inputs to the motor neurons can lead to a large number of different tensions, feedback of information is necessary to inform the motor control systems of the tension actually achieved.

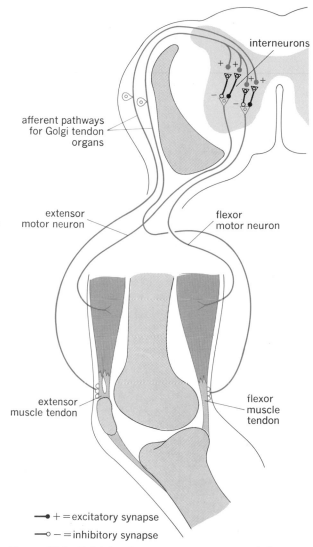

Figure 19-6. Golgi tendon organ component of the local control system.

Descending pathways and the brain centers which control them

The cerebral cortex, basal ganglia and other subcortical nuclei, brainstem nuclei, and cerebellum influence the motor neurons by descending pathways. There are three mechanisms by which these pathways alter the balance of synaptic input converging upon the alpha motor neurons:

1 By synapsing directly upon the alpha motor

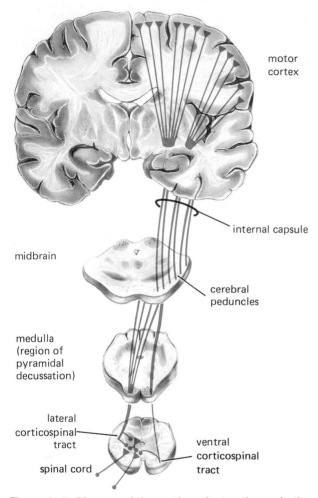

motor
cortex

internal capsule

midbrain

cerebral
peduncles

medulla
(region of
pyramidal
decussation)

lateral
corticospinal
tract

ventral
corticospinal
tract

spinal cord

Figure 19-7. Diagram of the corticospinal pathway. In the medulla most of its fibers cross to descend in the opposite side of the spinal cord. *(Adapted from Eccles.)*

motor neurons, it has the advantage of the co-ordination built into the interneuron network as described earlier (e.g., recruitment of synergistic muscles, reciprocal innervation).

The degree to which each of these three mechanisms is employed varies, depending upon the nature of the descending pathway, of which there are two major categories: the *corticospinal pathway* and the *multineuronal pathways.*

Corticospinal pathway

The fibers of the corticospinal pathway, as the name implies, have their cell bodies in the cerebral cortex. The axons of these cortical neurons pass without any additional synapsing to end in the immediate vicinity of the motor neurons (Fig. 19-7). The group of fibers innervating muscles of the eye, face, tongue, and throat branch away from these descending pathways in the brainstem to contact motor neurons whose axons travel out with the cranial nerves; the rest descend to their various terminations in the spinal cord, traveling down the spinal cord to innervate the motor neurons controlling the muscles of the extremities which are associated with fine-skilled movements. Near the junction of the spinal cord and brainstem, most of these fibers cross the spinal cord to descend on the opposite side. Thus, the skeletal muscles on the left side of the body are controlled largely by neurons in the right half of the brain. The corticospinal pathway is also called the *pyramidal tract* or *pyramidal system,* perhaps because of its shape in some parts of the brain or because it was formerly thought to arise solely from the giant pyramidal neurons of the cortex.

The corticospinal pathway is the major mediator of fine, intricate movements. However, it is not the sole mediator of these movements, since surgical section of this pathway in people does not completely eliminate such movements, although it does make them weaker, slower, and less well coordinated. Clearly, the multineuronal pathways (to be described next) must also contribute to the performance of delicate movements.

The fibers of the corticospinal pathway end in all three of the ways described, i.e., on alpha motor neurons, gamma motor neurons, and interneurons. In addition they end presynaptically on the terminals of afferent neurons and, through collateral

neurons themselves. This has the advantage of speed and specificity.

2 By synapsing on the gamma motor neurons, which, via the gamma loop, influence the alpha motor neurons. This pathway and number 1 above usually operate together (alpha-gamma coactivation). As described earlier, this has the advantage of maintaining output from the stretch receptors and providing a means for local on-the-spot compensation.

3 By synapsing on interneurons, often the same ones subserving the local reflexes. Although this route is not as fast as directly influencing the

branches, on neurons of the ascending afferent pathways. The overall effect of their input to afferent systems is to limit the area of skin, muscle, or joints allowed to influence the cortical neurons, thereby sharpening the focus of the afferent signal and improving the contrast between important and unimportant information. The collaterals also convey the information that a certain motor command is being delivered and possibly give rise to the sense of effort. Because of this descending (motor) control over ascending (sensory) information, there is clearly no real functional separation of these two systems.

Multineuronal pathways

Some of the neurons of cerebral cortex do not go directly to the region of the motor neurons; rather, they form the first link in neuronal chains which pass through the basal ganglia, several other subcortical nuclei, and the brainstem, forming multiple synapses along their course (Fig. 19-8). At each successive neuron, the information carried by the pathways is altered according to the balance of excitatory and inhibitory input to them. Some of these chains descend to the level of the motor neurons, ending there either on interneurons or gamma motor neurons; thus, the motor neurons (or the interneurons which directly influence them) receive input from both the corticospinal and multineuronal pathways. Other neuronal chains of

the multineuronal pathways do not descend to the level of the motor neurons, but rather, at some point, loop back to earlier way stations, including cerebral cortex. These loops constitute the paths by which the programs and program corrections described in our original model modify the activity of motor cortex, including the neurons of the corticospinal pathway. These multineuronal pathways and the structures which they connect are often called the *extrapyramidal system* to distinguish them from the corticospinal *(pyramidal)* pathways.

Whereas the corticospinal neurons have greater influence over the motor neurons which control muscles in the more distal ends of the limbs, e.g., the fingers and hands, the multineuronal pathways have more effect on the muscles of the trunk and more proximal parts of the limbs. However, it is wrong to imagine a complete separation of function between these two pathways, for the distinctions between them are not at all clear cut; moreover, some areas of motor cortex give rise to neurons of both pathways. All movements, whether automatic or voluntary, require the continual coordinated interaction of both pathways.

Cerebral cortex

Many areas of the cortex give rise to the two types of descending pathways described above, but a large number of the fibers come from the posterior

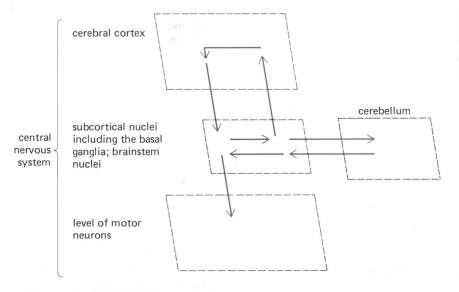

Figure 19-8. Diagram of the multineuronal pathways. Note that this is basically Fig. 19-1 without the corticospinal pathway, muscles, and receptors.

cerebral cortex

central nervous system

subcortical nuclei including the basal ganglia; brainstem nuclei

cerebellum

level of motor neurons

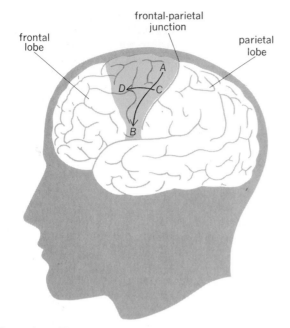

Figure 19-9. The motor cortex (shaded).

part of the frontal lobe, which is therefore called the *motor cortex* (Fig. 19-9). The function of the neurons in the motor cortex varies with position in the cortex. As one starts at the top of the brain and moves down along the side (A to B in Fig. 19-9), the cortical neurons affecting movements of the toes and feet are at the top of the brain, followed (as one moves laterally along the surface of the brain) by neurons controlling leg, trunk, arm, hand, fingers, neck, and face. The size of each of the individual body parts in Fig. 19-10 is proportional to the amount of cortex devoted to its control; clearly, the cortical areas representing hand and face are the largest. The great number of cortical neurons for innervation of the hand and face is one of the factors responsible for the fine degree of motor control that can be exerted over those parts.

The neurons also contribute to different pathways as one explores from the back of the motor cortex forward (C to D in Fig. 19-9). Those in the back portion, i.e., closest to the junction of the frontal and parietal lobes, mainly contribute to the corticospinal pathway. Moving anterior (forward) in the motor cortex, this zone of neurons gradually blends into the group which forms the first link in the multineuronal pathways. However,

none of these cortical neurons functions as an isolated unit. Rather, they are interconnected so that those cortical neurons controlling motor units having related functions fire together. The more delicate the function of any skeletal muscle, the more nearly one-to-one is the relationship between cortical neurons and the muscle's motor units.

Colonies of cells having similar functions are arranged in columns perpendicular to the surface of motor cortex, each column containing hundreds of neurons. The neurons in the columns function together, and this has several advantages: (1) Because as many as 100 motor neurons must sometimes fire in order to evoke a movement, activity in isolated neurons being ineffective, having the cells which are activated together near to each other minimizes relay time. (2) The cells in one column inhibit activity in adjacent columns so that the relative effectiveness of strongly active columns is enhanced; such heightened contrast clarifies and focuses the movement. (3) Because the columns receive afferent input from the skin and tissue areas overlying the muscle fibers activated by the column, one can visualize the cortical cells as integrators of long reflex arcs initiated by the stimulation of receptors, and, indeed, some movement is explainable in these terms. For example, skin receptors on the palm of the hand feed into columns controlling the flexor muscles of the area;

Figure 19-10. Arrangement of the motor cortex.

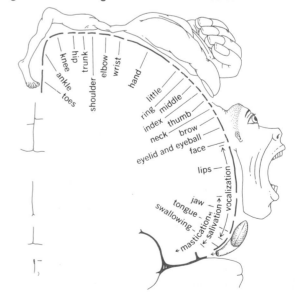

thus, when the palm is stimulated, the fingers and thumb flex to approach the stimulus. This mechanism most likely underlies the *grasp reflex* seen, for example, in an infant holding tightly to an object placed in its hand.

Throughout this description we have presented the cerebral cortex as the origin of the descending pathways to motor neurons, but the neurons of cerebral cortex manifest no spontaneous generation of action potentials. Rather, their firing pattern, like that of almost all neurons, is determined solely by the balance of synaptic activity impinging upon them. In other words, as we emphasized in the earlier description of the overall model of motor system control, the motor cortex is not the prime initiator of movement; it is simply a tremendously important relay station. What then are the inputs which drive the neurons of motor cortex? Again as emphasized in our model, where the "command" neurons are located is not known, but regions of association cortex and cerebellum have been implicated. For example, electrodes placed on the skull to record the electrical activity of the brain during the "decision making period" pick up over wide areas of association cortex distinctive activity patterns (the so-called *readiness potential*) about 800 ms (0.8 s) before the movement begins. Only afterwards, 50 to 60 ms prior to the movement, does sharper electrical activity (the *motor potential*) appear over the motor-cortex region of the hemisphere controlling the involved muscles.

Another source of important input to the motor cortex is the pathways from the receptors listed earlier in the chapter. Yet such input does not generally account for the actual initiation of movement. For example, if the neurons of the motor cortex are stimulated in conscious patients during surgical exposure of the cortex, the patient moves but not in a purposefully organized way. The movement made depends upon the part of the motor cortex stimulated. The patient is aware of the movement but says, "You made me do that," recognizing that it is not a voluntary movement. Stimulation of parts of cortex in the parietal lobe sometimes causes patients to say that they want to make a certain movement, but the movement does not occur.

Finally, the input to the cortex neurons not only must be directed toward some purpose but must select from a variety of ways of achieving that purpose since learned motor programs can be executed in many ways with many sets of muscles. For a simple example, a rat trained to press a lever whenever a signal light flashes will do so quite consistently but, depending upon its original position in the cage, its movements can be quite varied. If the rat is to the left of the lever, it moves to the right; if it is to the right, it moves to the left. If its paw is on the floor, it raises the paw; if its paw is above the lever, it lowers the paw. The only consistent act is pressing the lever. The movements performed to achieve the purpose are variable and seem almost inconsequential. Yet, although it is the end result that matters, only the intervening acts or movements can be programmed by the nervous system. How a given program is selected is not known.

Subcortical nuclei (including the basal ganglia) and brainstem nuclei

As mentioned above, the multineuronal pathways descending from cerebral cortex synapse in the basal ganglia, other subcortical nuclei, and brainstem nuclei; collateral branches from axons in the corticospinal pathway also end in these regions. These nuclei and the pathways that interconnect them serve to correlate fine, detailed voluntary movements with the appropriate postural mechanisms upon which these movements are superimposed. In addition, the basal ganglia play a special role in the control of slow, smooth, voluntary movements.

Perhaps the roles these regions play in the control of muscle activity can be better understood by learning what happens when they are damaged by accident or disease. Such conditions usually involve one or more of the following: (1) There can be an increase in muscle tone, i.e., a resistance to stretch, so that movements become stiffer. (2) Unwanted, purposeless, uncontrolled movements are often present. The latter can take the form of rapid contractions of opposing muscle groups (*tremor*), writhing, continuous swings between two positions (*athetosis*, e.g., hyperextension of the fingers with the palm facing upward alternating with clenching the fingers with the palm facing downward), or sudden, random, coordinated movements of the distal part of the arms or legs that are relatively mild (*chorea*) or more violent (*ballis-*

mus). (3) There is usually a decrease or loss of associated movements, such as arm-swings during walking or the facial movements which normally give vitality and emotional expression to one's appearance.

One particular subcortical nucleus is the *substantia nigra* (black substance), which gets its name from the deep pigment in its cells. It is this group of neurons, whose fibers synapse in the basal ganglia, that is implicated in the disease *parkinsonism,* characterized by tremor, rigidity, and a delay in the initiation of movement. The neurons of the substantia nigra liberate the transmitter dopamine at their axon terminals in the basal ganglia. When they degenerate, as they do in parkinsonism, the amount of dopamine delivered to the postsynaptic cells is reduced. Normal functioning of the basal ganglia neurons seems to depend upon the balance of input from such dopamine-releasing neurons, which are inhibitory, and excitatory acetylcholine-releasing neurons. Decrease of the inhibitory transmitter allows excessive excitation which, in turn, leads to oscillating bursts of activity in basal-ganglia neurons feeding back to motor cortex. This activates neurons there whose axons descend to the level of the motor neurons and drives them in the abnormal patterns of rhythmic activity.

The dopamine receptors on the basal-ganglia neurons are normal in parkinsonism; it is the quantity of transmitter that is deficient. Therefore, L-dopa, which is a precursor of dopamine (Chap. 8), is given to these patients, enters their blood stream, crosses the blood-brain barrier, and enters the neurons where it is converted to dopamine (dopamine itself isn't used because it cannot cross the blood-brain barrier). The newly formed dopamine activates the receptors in the basal ganglia and relieves the symptoms of the disease. On the other hand, drugs which block dopamine receptors (and are sometimes used in the treatment of schizophrenia) have as one of their unwanted side effects symptoms characteristic of parkinsonism.

Cerebellum

The cerebellum sits on top of the brainstem, as can be seen in Fig. 8-45. It does not initiate movement but acts by influencing other regions of the brain responsible for motor activity. Destruction

of the cerebellum does not cause the loss of any specific movement; instead it is associated with a general inadequacy of that movement. The main problems of people with cerebellar damage are as follows:

1 They cannot perform movements smoothly. If they try to grasp an object, the movement is jerky and is accompanied by oscillating, to-and-fro tremors which become more marked as the hand approaches the object. In contrast, when their limbs are at rest, they are steady and motionless, but if they try to move them for any reason, even to help maintain balance, they go into oscillations which are sometimes quite wild. This oscillating tremor is known as *intention tremor* and is classically associated with cerebellar damage. Contrast this to the resting tremor that occurs with basal ganglion disease.
2 They walk awkwardly with the feet well apart. They have such difficulty maintaining their balance that their gait appears drunken.
3 They cannot start or stop movements quickly or easily and, if asked to rotate the wrist back and forth as rapidly as possible, their motions are slow and irregular.
4 They may not be able to combine the movements of several joints into a single, smooth, coordinated motion. To move the arm, they might first move the shoulder, then the elbow, and finally the wrist.

In cases of severe cerebellar damage, the combined difficulties of poor balance and unsteady movements may become so great that the person is incapable of walking or even standing alone. Since speech depends on the intricately timed coordination of many muscle movements, cerebellar damage is accompanied by speech disturbances. The person with cerebellar damage shows no evidence whatsoever of sensory or intellectual deficits.

From this discussion, the reader will deduce that the cerebellum is involved in the control of muscles utilized both in maintaining steady posture and in effecting coordinated, detailed movements. As stated earlier, it receives input both from cortex and subcortical centers with information about what the muscles *should* be doing and from many afferent systems with information about what the muscles *are* doing. If there is a discrepancy between the two, an error signal is sent from cerebellum to the cortex and subcortical centers where new commands are initiated to decrease the discrepancy and smooth the motion.

The cerebellum plays a special role in the control of rapid movements (and thus serves as the counterpart to the basal ganglia, which influence slow movements). Rapid movements are largely "preprogrammed" in their entirety, rather than being modified during their course as slower movements are. Such preprogramming requires calculation of the time needed for the movement (taking into account the particular amount of muscle force necessary in any special situation) and integration of these data with information about the moved structure's resting position and final location. The cerebellum performs this function and, like the basal ganglia, projects the information to the motor cortex.

The afferent inputs to the cerebellum come from the vestibular system, eyes, ears, skin, muscles, joints, and tendons, i.e., from the major receptors affected by movement. Inputs from receptors in a single small area of the body end in the same region of the cerebellum as do the inputs from the higher brain centers controlling the motor units in that same area. Thus information from the muscles, tendons, and skin of the arm arrive at the same area of cerebellar cortex as the motor commands for "arm" from the cerebral cortex. This permits the cerebellum to compare motor commands with muscle performance.

This completes our analysis of the components of motor control systems. We now analyze the interactions of these components in two situations: maintenance of upright posture and walking.

Maintenance of upright posture and balance

The skeleton supporting the body is a system of long bones and a many-jointed spine which cannot stand alone against the forces of gravity. Even when held together with ligaments and covered with flesh, it cannot stand erect unless there is coordinated muscular activity. This applies not only to the support of the body as a whole but also to the fixation of segments of the body on adjoining segments, e.g., the support of the head, which is held erect without any conscious effort and without fatigue in a normal awake person. A decrease in postural fixation of the head is seen when

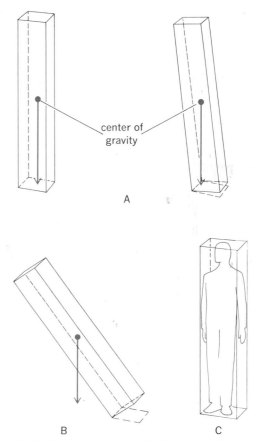

Figure 19-11. The center of gravity is the point in a system at which, if a string were attached and pulled upward, all the downward forces due to gravity would be exactly balanced. The center of gravity must remain within the vertical projections of the base of the system if stability is to be maintained. (A) Stable conditions. (B) Unstable conditions. (C) The vertical projections of the base, in which the center of gravity must remain for stable posture in human beings.

someone falls asleep sitting up; the head nods until the chin is on or near the chest.

Added to the problem of supporting his own weight against gravity is that of maintaining equilibrium. A person is a very tall structure balanced on a relatively small base, and the center of gravity is quite high, being situated just above the pelvis. For stability, the center of gravity must be kept within the small area determined by the vertical projection of the base of the body (feet) in contact with the surface upon which a person stands (Fig. 19-11). Yet human beings are almost always in motion, swaying back and forth and side to side even when standing still. Clearly, they often operate under conditions of unstable equilibrium and would be toppled easily by physical forces in the environment if their equilibrium were not protected by reflex postural mechanisms.

The maintenance of posture and balance is accomplished by means of complex counteracting reflexes, all the components of which we have met previously. The efferent arc of the reflexes is, of course, the alpha motor neurons to the skeletal muscles. The major coordinating centers are, as usual, the basal ganglia, brainstem nuclei, and reticular formation, all of which influence the motor neurons mainly via the descending multineuronal pathways. What is the source of afferent input? One might predict the existence of "center-of-gravity" receptors but, in fact, no such receptors exist. Rather, information about the location of the center of gravity is given by the integration of all the afferent signals from muscles, joints, skin, vestibular system, and eyes. This integration provides the coordinating centers with a "map" of the position of the whole body in space.

Figure 19-12. Interaction of the vestibular, skin, joint, and tendon receptors.

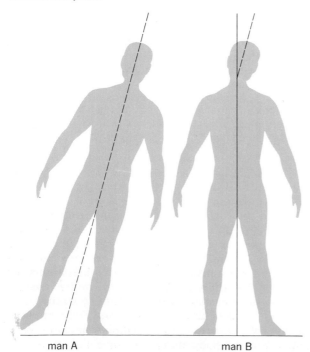

man A   man B

In such a system it is extremely difficult to assign a certain percentage of importance to any one afferent system, even when the exact conditions are specified. However, it does seem that in people, under conditions which permit vision, visual information is probably most important. Yet, so influential are the other inputs and so adaptable is the overall system that a blind person maintains balance quite well, with only slight loss of precision. Moreover, with changing circumstances, the dependence on different afferent inputs may change considerably.

Let us take the vestibular input as an example. Despite the fact that the vestibular organs are called the sense organs of balance, persons whose vestibular mechanisms have been completely destroyed may have very little disability in everyday life (one such man was even able to ride a motorcycle). Such persons are not seriously handicapped as long as their visual system, joint position receptors, and cutaneous receptors are functioning. However, they do have difficulty walking in darkness over uneven ground or walking down stairs, where they cannot see a point immediately in front of their feet for visual reference. Thus, in a normal person vestibular input must increase in importance under such circumstances. Finally, vestibular information provides the only clue to orientation with respect to gravity when one is swimming under water, where visual, skin, and joint input is inappropriate.

Skin receptors sensing contact of the body with other surfaces also play a role in regulation of body posture as can be shown by the following test. Walking in the dark, one's gait is halting and uncertain; stability and confidence are greatly improved by the simple act of running a fingertip along a wall. The fingertip certainly provides no physical support, but it adds significant afferent information to the coordinating centers in the subcortical centers and cerebellum.

The joint-position receptors add information too. The two figures in Fig. 19-12 illustrate this point. Man A is clearly in danger of falling and is in need of muscle movements to keep himself upright. Consider the synaptic input to the central nervous system: The vestibular inputs from the two sets of semicircular canals are different, the inputs from the joints and ligaments of the limbs on the two sides of his body are different, and the joint receptors from the section of vertebral column that runs through his neck indicate that his neck is not bent. The final integration of all these synaptic inputs leads to the initiation of postural reflex responses. Man B is not in danger of falling even though the input from his vestibular systems is exactly the same as that of man A, because the total synaptic input to the central nervous system is quite different. There is no imbalance of input from his limb joint receptors, but there is imbalance from his neck joint receptors. The information in man A indicates instability; that in man B does not. The reactions which would be initiated by vestibular stimulation do not occur unless the input from joint receptors indicates that the body is unstable.

Finally, we must point out the special contribution of the muscle-spindle (stretch) receptors to the maintenance of upright posture against the force of gravity. These length-monitoring receptors provide information via afferent and ascending pathways to the higher brain centers but, in addition, they initiate the locally occurring stretch reflex. Imagine a person standing upright; gravity causes his knees to begin to buckle. As this occurs the patellar tendon and extensor muscles are stretched and the stretch reflex is elicited, resulting in increased contraction of the extensor muscles to straighten the knee and prevent him from falling.

From the above description, it might seem (incorrectly) that the stretch reflexes, by themselves, could maintain upright position against the forces of gravity. In fact, their input is effective in enhancing motor activity only when there is a simultaneously occurring facilitation of the alpha motor neurons by descending pathways via alpha-gamma coactivation. Indeed, the providing of such facilitation is a major function of the multineuronal pathways. As might be predicted, these pathways generally facilitate preferentially the motor neurons to the so-called antigravity muscles, for example, the extensor muscles of the legs.

Finally, what happens if all the postural reflexes, once initiated, are unsuccessful and the person loses his balance? If a person is tilted, the arm on the lower side is pulled toward his body, but if the tilting goes so far that he is in danger of losing his balance, the arm on the lower side is quickly extended to break the fall. The first reflex pattern in response to tilt breaks up and a different set of reflexes takes over as soon as there is no hope of restoring balance. Indeed, it is just such a transformation that accounts for stepping in walking.

Walking

Postural fixation of the body is intimately related to the problems of locomotion and maintaining equilibrium in the face of movement. In fact, the mechanisms for posture and movement are the same, and a disturbance of one is almost always associated with a disturbance of the other. In considering locomotion, the need for some structure or mechanism capable of carrying the body along is added to the basic requirement of antigravity support of the body. In walking, the human body is balanced on the very small base provided by one foot. The weight of the body is supported on each leg alternately, and to accomplish this the body moves from side to side in such a way that the center of gravity is alternately poised over the right and then the left leg. Only when the center of gravity is shifted over the right leg can the left foot be raised from the ground and advanced. As the left foot is lifted, the trunk of the body sways to the right to counterbalance the weight of the left leg. It must be apparent that strict and delicate control of the center of gravity is essential to permit these movements without loss of equilibrium.

Effective stepping can be evoked in newborn infants if the child is held with his feet set on a surface and his legs supporting the weight of his body. The stepping movements can be initiated by simply tilting the child's body forward and rocking him slightly from side to side. Thus, small shifts in the center of gravity start the coordinated, alternate flexion and extension of the leg involved in purposeful stepping.

The stimulus necessary to trigger the stepping is a slight forward tilt of the body. When a child or adult takes a step, the weight of the body is shifted to one foot, and the opposite foot is lifted from the ground. The body is allowed to fall forward and loses its equilibrium until it is caught on the leg which has swung forward. During this process, the center of gravity has moved both sideways and forward from its original position over one leg to a similar position over the other leg. This action is repeated rhythmically, and its continuation depends on both components of the shift in the center of gravity, the forward shift, which causes the fall forward, and the sideways shift, which allows one foot to be lifted and advanced. Thus, there are four necessary components for locomotion: antigravity support of the body, stepping, control of the center of gravity to provide equilibrium, and a means of acquiring forward motion. All four of these components must be present simultaneously and continuously. It would obviously be futile to apply forward motion without an adequate stepping mechanism. An interesting point is that the forward fall serves to both provide the advance in position and evoke stepping. In fact, if a person leans forward beyond a certain point, he must either take a step or fall, and in this case the stepping is part of a protective reflex.

In persons with diseases of the basal ganglia, disturbances in locomotion often occur, and occasionally the disorders may become so great that the patients become immobilized. Such patients can stand and can make rhythmical, alternate stepping movements, but they cannot walk. One such patient with diseased basal ganglia, who walks only poorly, is able to get along very well if she carries a 14-lb chair in front of her. Her disorder is in the basal ganglion mechanisms which control the forward shift in the center of gravity, and she cannot lean forward to acquire the forward progression necessary for locomotion. The weight of the chair has the effect of bringing the center of gravity forward. Her disorder is only in the ability to tilt her body forward; she does not lack the ability to rock her body from side to side. Other patients have good front-to-back control but lack adequate lateral control.

20
CONSCIOUSNESS AND BEHAVIOR

States of consciousness
  Electroencephalogram
  The waking state
  Sleep
  Neural substrates of states of consciousness
    Pathways for the waking state, arousal, and attention • *The sleep centers*
Conscious experience
Motivation and emotion
  Motivation
  Emotion
  Chemical mediators for emotion-motivation

Altered states of consciousness
  Schizophrenia
  The affective disorders: Depressions and manias
  Psychoactive drugs, tolerance, and addiction
    Alcohol: An example
  Coma and brain death
Neural development and learning
  Evolution of the nervous system
  Development of the individual nervous system
  Learning and memory
    Short-term memory • *Memory consolidation* • *Long-term memory* • *Forgetting*
Cerebral dominance and language
Conclusion

The term *consciousness* includes two distinct concepts, *states of consciousness* and *conscious experience.* The second concept refers to those things of which a person is aware—thoughts, feelings, perceptions, ideas, dreams, reasoning—during any of the states of consciousness. In contrast, a person's state of consciousness, i.e., whether awake, asleep, drowsy, etc., is defined both by behavior, covering the spectrum from coma to maximum attentiveness, and by the pattern of brain activity that can be recorded electrically, usually as the electric-potential difference between two points on the scalp. This record is the *electroencephalogram* (EEG).

States of consciousness

Electroencephalogram

The EEG is such an important tool in identifying the different states of consciousness that we begin with it. The wavelike pattern of the EEG is about 100 times smaller than the amplitude of an action potential, and the frequency of the waves may vary from 3 to 22 Hz, the patterns being distinguished from one another by both their frequencies and amplitudes. Correlated with changes in EEG pattern are changes in behavior spanning the entire normal range from attentive alertness to sleep.

Each EEG pattern is the result of varying degrees of intermittent synchronization of membrane potential changes in groups of neurons located in the cerebral cortex. These potential changes are thought to be synaptic potentials (and other graded potential changes) rather than action potentials. The involved synapses manifest more or less synchronous electrical changes because of intimately coordinated activities of the thalamus and cortex. Small areas of thalamus act as pacemakers, or rhythm generators, each of which projects to and influences the activity of a specific

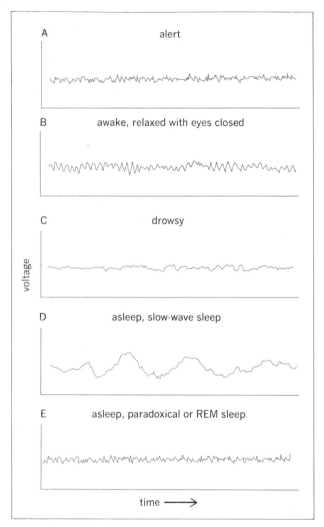

A   alert

B   awake, relaxed with eyes closed

C   drowsy

D   asleep, slow-wave sleep

E   asleep, paradoxical or REM sleep

voltage

time ──────▶

Figure 20-1. EEG patterns corresponding to various states of consciousness.

small region of cortex.[1] Because the cells within a given small thalamic unit are interconnected, they tend to fire together; moreover, these individual units are affected by both excitatory and inhibitory feedback loops with other thalamic units as well as by loops with the cells in the cortex to which the thalamic units project. These loops lead to the synchronization of large areas of cortex, and it is these bursts of cortical activity that are recorded

[1] Although other regions of the brain can affect cortical synchrony, these thalamocortical interactions form the basis of the EEG-generating mechanism.

as the EEG. Cortical neurons other than those synchronized at any given moment also have fluctuating membrane potentials due to synaptic input, but their activity is not synchronized and, therefore, the potentials tend to cancel each other out. Periodically, new groups of neurons are synchronized.

We have been speaking thus far mainly of the highly synchronized EEG waves characteristic of slow-wave sleep (Fig. 20-1D). The EEG pattern becomes less synchronized during arousal, and it becomes completely desynchronized with full alertness. This breakdown in synchrony occurs as regions of the midline thalamus are activated. Their input to cortex causes, in turn, greater activation of the excitatory feedback loops that pass from cortex back to thalamus, thereby counteracting the IPSPs that are largely responsible for strong synchronization of the pacemakers.

The EEG is a useful clinical tool because the normal patterns are altered over brain areas that are diseased or damaged. It is also useful in defining states of consciousness, but it is not known what function, if any, this electric activity serves in the brain's task of information processing. We do not know whether these electric waves actually influence brain activity or whether they are merely epiphenomena. (An epiphenomenon is a phenomenon which occurs with an event but is not causally related to it, for example, the sound of a baseball bat striking a ball – the sound results from the impact but does not influence how far the ball will travel.)

The waking state

Behaviorally, the waking state is far from homogeneous, comprising the infinite variety of things one can be doing. The prominent EEG wave pattern of an awake relaxed adult whose eyes are closed is a slow oscillation of 8 to 12 Hz, known as the *alpha rhythm* (Fig. 20-1B) (each region of the brain has a characteristic alpha rhythm). The alpha rhythm is associated with decreased levels of attention, and when alpha rhythms are being generated, subjects commonly report that they feel relaxed and happy. A high degree of alpha rhythm is also associated with meditational states. However, people who normally experience high numbers of alpha episodes have not been shown to be psychologically different from others with lower levels, and the relation between brain-wave activity and subjective mood is obscure. People have been trained to increase the amount of alpha brain rhythms by providing a feedback signal such as a tone whenever alpha rhythm appears in their EEG.

When people are attentive to an external stimulus (or are thinking hard about something), the alpha rhythm is replaced by lower, faster oscillations (Fig. 20-1A). This transformation is known as *EEG arousal* and is associated with the act of attending to stimuli rather than with the perception itself; for example, if people open their eyes in a completely dark room and try to see, EEG arousal occurs. With decreasing attention to repeated stimuli, the EEG pattern reverts to the alpha rhythm.

Sleep

Although adults spend about one-third of their time sleeping, we know little of the functions served by it. We do know that sleep is an active process and not a mere absence of wakefulness. Moreover, it is not a single simple phenomenon; there are distinct states of sleep characterized by different EEG and behavior patterns.

The EEG pattern changes profoundly in sleep. As a person becomes drowsy, the alpha rhythm is gradually replaced (Fig. 20-1C), and as sleep deepens, the EEG waves become slower and larger (Fig. 20-1D). This *slow-wave sleep* is periodically interrupted by episodes of *paradoxical sleep*, during which the subject still seems asleep but has an EEG pattern similar to that of EEG arousal, i.e., an awake alert person (Fig. 20-1E).

It is difficult to tell precisely when a person passes from drowsiness into slow-wave sleep, for in slow-wave sleep there is considerable tonus in postural muscles and only a small change in cardiovascular or respiratory activity. The sleeper can be awakened fairly easily and if awakened, rarely reports dreaming. Slow-wave sleep has a characteristic kind of mentation described by subjects as "thoughts" rather than "dreams." The thoughts are more plausible and conceptual; they are more concerned with recent events of everyday life and more like waking-state thoughts than are true dreams.

At the onset of paradoxical sleep, there is an abrupt and complete inhibition of tone in the postural muscles, although periodic episodes of twitching of the facial muscles and limbs and rapid eye movements behind the closed lids occur. (Paradoxical sleep is therefore also called *rapid-eye-movement* or *REM sleep*.) Respiration and heart rate are irregular, and blood pressure may go up or down. When awakened during paradoxical sleep, 80 to 90 percent of the time subjects report that they have been dreaming.

Continuous recordings show that the two states of sleep follow a regular 30- to 90-min cycle, each episode of paradoxical sleep lasting 10 to 15 min. Thus, slow-wave sleep constitutes about 80 percent of the total sleeping time in adults, and paradoxical sleep about 20 percent. The time spent in paradoxical sleep increases toward the end of an undisturbed night. Normally it is not possible to pass directly from the waking state to an episode of paradoxical sleep; it is entered only after at least 30 min of slow-wave sleep. Thus, subjects wakened at the beginning of every period of paradoxical sleep can be prevented from spending much time in that state although their total sleeping time remains approximately normal. After being deprived of paradoxical sleep for several nights, all subjects spend a greater than usual proportion of time in paradoxical sleep the next time they sleep.

What is the functional significance of sleep, i.e., what happens to the brain during sleep? The brain, as a whole, does not rest during sleep and there is no generalized inhibition of activity of cerebral neurons; however, there is a reorganization of neuronal activity, some individual neurons being less active during sleep than during waking and others showing just the opposite pattern. The total blood flow and oxygen consumption of the brain, signs of its metabolic activity, do not decrease in sleep.

Although sleep is not a period of generalized rest for the whole brain, it may represent a period of rest for certain specific elements, during which they can replenish substrates necessary for their generation of action potentials. Yet, when isolated neural tissue is exposed to extreme rates of stimulation far exceeding those occurring under physiological circumstances, neurons recover within a period of minutes. Alternatively, it has been suggested that the functional significance of sleep lies not in short-term recovery but in the relatively long-term chemical and structural changes that the brain must undergo to make learning and memory possible.

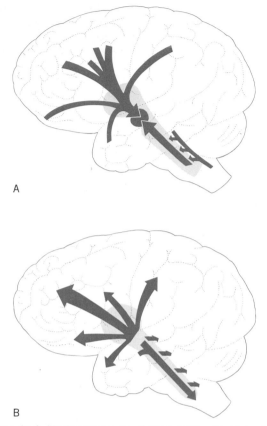

A

B

Figure 20-2. (A) Convergence of descending, local, and ascending influences upon reticular formation (shaded area). (B) Projections from reticular formation to spinal cord, brainstem and cerebellum, and cerebrum. *(Adapted from Livingston.)*

Neural substrates of states of consciousness

The prevailing state of consciousness is mainly the resultant of the interplay between three neuronal systems, one causing arousal and the other two sleep. All three are parts of the *reticular formation.* The reticular formation proper lies in the central core of the brainstem in the midst of the neural pathways ascending and descending between the brain and spinal cord, but neurons in the midline thalamus also function as an extension of this brainstem system. Neurons of the reticular formation are influenced by and influence virtually all areas of the central nervous system (Fig. 20-2).

The reticular formation is not homogeneous, and discrete areas frequently have specific functions, several of which we have already discussed: It helps to coordinate skeletal muscle activity (Chap. 19); it contains the primary cardiovascular and respiratory control centers (Chaps. 12 and 13); it monitors the huge number of messages ascending and descending through the central nervous system (Chaps. 8 and 18). In this chapter we are concerned with its role in determining states of consciousness.

Pathways for the waking state, arousal, and attention. In 1934, it was discovered that after the cerebrum is surgically isolated from the spinal cord and lower three-fourths of the brainstem, the EEG loses the wave patterns typical of an awake animal, indicating that some neural structures within the separated brainstem or spinal cord are essential for the maintenance of a waking EEG. The relevant neural structures actually lie within the reticular formation and form part of the *reticular activating system* (RAS), destruction of which produces coma and the EEG characteristic of the sleeping state.

Single neurons in the reticular formation may be activated by any afferent modality—a flash of light, a ringing bell, a touch on the skin. As they pass from the brainstem into the central core of the cerebrum, fibers of the RAS activate the *diffuse thalamic projection system.* Fibers of these thalamic neurons synapse in the cortex, but unlike the specific thalamic projections described in Chap. 18, they are not involved in the transmission of information about specific sensory modalities; rather, they maintain the behavioral characteristics of the awake state.

The RAS is crucial not only for maintenance of a waking state but for arousal and attention. The important generalization to keep in mind during this analysis is that human beings are conscious of a stimulus only when the nervous system is oriented and appropriately receptive toward it, and it is the neurons of the reticular activating system which arouse the brain and facilitate information reception by the appropriate neural structures. However, the sensitivity of this system is selective. A mother may awaken instantly at her baby's faintest whimper whereas she can sleep peacefully through the roar of a jet plane passing overhead.

Arousal can be studied in experimental animals by the presentation of a novel stimulus, which causes a change in the EEG pattern from the quiet resting alpha rhythm to that characteristic of arousal. This response is called the *orienting response* because the animal stops whatever it is doing and looks around or listens intently, orienting itself toward the stimulus source. What pathway mediates this response? Information about the stimulus is relayed to the cortex via both specific and nonspecific pathways, the nonspecific pathways traveling via the reticular formation and nonspecific areas of thalamus. The distribution of synaptic endings of these two pathways on given cortical neurons is different; those of the specific pathways are clustered together close to the site at which action potentials are initiated in the cortical neuron whereas those of the nonspecific pathways are scattered around other portions of the neuron. Thus, one can say that the specific pathways "control" the cortical neurons, whereas the nonspecific pathways exert a more general "bias" (i.e., various degrees of facilitation or inhibition). It is hypothesized that the interplay between these two inputs to cortical neurons forms the attention-focusing mechanism in which a biasing of certain of the cortical neurons by the nonspecific system affects their response to input over the specific pathways.

But since attention is directed only to meaningful stimuli, it seems that at some point in the nervous system there must be a comparison of present stimuli with those that have gone before to answer questions such as: "Is this new information?" "Is it relevant in terms of past events?" The neural outcome of this questioning biases (or fails to bias) the relevant cortical neurons so that attention is paid to the stimulus or it is ignored. In other words, a "yes" answer to either of the above questions results in the orienting response, and a "no" answer leads to a progressive decrease in response *(habituation)*. For example, when a loud bell is sounded for the first time, it may evoke an orienting response (or startle response) in the animal; but after several ringings, the animal makes progressively less response and eventually may ignore the bell altogether. An extraneous stimulus of another modality or the same stimulus at a different intensity restores the original response *(dishabituation)*. Habituation is not due to receptor fatigue or adaptation.

Not unexpectedly, the neuronal mechanisms that "select" the response (the orientating response or habituation) are thought to be located in the thalamus and reticular formation, a hypothesis supported in part by the finding there of "novelty detectors," which will be described shortly.

The attention-focusing mechanism just described is a type of reflex action, and it is suggested that in humans (and probably other higher animals) there exists a second, "voluntary," attention-directing mechanism involving temporal and frontal cortex. This mechanism would permit one to "choose" to pay attention to selected stimuli, but it still requires much of the organizational function of the reticular formation.

It should be evident from this description that both the cortex and reticular formation are required for sustained wakefulness and normal arousal and attention-focusing. In addition, certain areas of the limbic system (see Chap. 8) are implicated in the EEG and behavioral aspects of the waking state. For example, during the orienting response a portion of the limbic system known as the *hippocampus* manifests low-frequency, high-amplitude waves, the so-called *theta rhythm* (or hippocampal "arousal"); note that the theta rhythm is essentially opposite to the EEG pattern of the cerebral cortex during arousal. Injury to the hippocampus results in faulty orienting responses as well as to the faulty learning described in a subsequent section.

The sleep centers. The control of sleep is exerted by two neuronal systems which oppose the tonic activity of the RAS. The two neuronal clusters, one in the central core of the brainstem, the other in the pons, are also part of the reticular formation (Fig. 20-3). The sleep-wake cycle is thought to occur because the brainstem-core neurons tonically release the transmitter serotonin; when serotonin levels become high enough, the neurons of RAS are inhibited. This results in the loss of awake conscious behavior and its EEG manifestations and the replacement of these by the behavior and EEG characteristics of slow-wave sleep.

These brainstem-core neurons also facilitate the sleep center of the pons, whose activity induces paradoxical sleep. Pathways ascending from this paradoxical-sleep center of the pons establish the low-voltage, fast EEG pattern and activate the muscles of the eyes; descending pathways inhibit

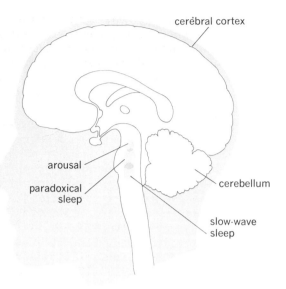

Figure 20-3. Brainstem structures involved in arousal, paradoxical sleep, and slow-wave sleep. *(Adapted from Jouvet.)*

motor-neuronal activity, which results in loss of muscle tone.

Of great importance is the fact that, while instituting the EEG and behavioral manifestations of paradoxical sleep, the neurons of the sleep center in the pons also influence the central core of the brainstem in a true feedback fashion. We have said that the release of serotonin by these central core neurons causes slow-wave sleep and facilitates the paradoxical-sleep center. It is thought that the paradoxical-sleep-center neurons feed back and stimulate the *uptake* of serotonin by the endings of the brainstem-core neurons. This decrease in free (i.e., extracellular) serotonin concentration lessens the inhibition on the RAS and permits a return of the awake state. Waking continues until sleep is again triggered by the release of sufficient serotonin to inhibit the RAS. Thus, the cycling of the sleeping and waking states of consciousness is probably due, at least in part, to the slow accumulation and dissipation of chemical transmitters.

As presented above, sleep is basically the result of the cyclical inhibition of the RAS by brainstem-core neurons. However, other brain regions also affect sleep. Many of them are in the limbic system, brain regions associated with emotions. This provides a neural basis for the common experience that thoughts, worries, fears, etc., can interfere with sleep.

In summary, two opposing systems exist in the brain, one a sleep-producing system and the other an arousal system; the two are connected in such a way that activity in one suppresses activity in the other, and vice versa. It is the interaction of these two systems which produces the sleep-waking cycles. However, the periodic inhibition of the arousal systems (which would normally cause sleep) can be overridden by input from afferent pathways or other brain centers so that RAS activity is maintained sufficiently high to keep one awake or interrupt sleep. In fact, the waking mechanisms seem to be more easily activated than those causing sleep. An example familiar to all parents is that it is much easier to arouse a sleeping child than to get an alert, attentive child to sleep.

Conscious experience

All subjective experiences are popularly attributed to the workings of the mind. This word conjures up the image of a nonneural "me," a phantom interposed between afferent and efferent impulses, with the implication that mind is something more than neuronal activity. The truth of the matter is that physiologists and psychologists have absolutely no idea of the mechanisms which give rise to conscious experience. Nor are there even any scientifically meaningful hypotheses concerning the problem.

Conscious experiences are difficult to investigate because they can be known only by verbal report. Such studies lack scientific objectivity and must be limited to people. In an attempt to bypass these difficulties scientists have studied the behavioral correlates of mental phenomena in other animals. For example, a rat deprived of water performs certain actions to obtain it. These actions are the behavioral correlates of thirst. But it must be emphasized that we do not know whether the rat consciously experiences thirst; this can only be inferred from the fact that human beings are conscious of thirst under the same conditions.

However, one crucial question which cannot be investigated in experimental animals is whether conscious experiences actually influence behavior. Although, intuitively, it might seem absurd to question this, the fact is that the answer is crucial for the development of one's concept of humanity. It is possible that conscious experience is an epiphenomenon.

Consider the following sequence of events: The ringing of a telephone reminds a student that he had promised to call his mother; he finishes the page he has been reading and makes the call. What causes him to do so? The epiphenomenon view holds that the conscious awareness accompanies but does not influence the passage of information from afferent to motor pathways. Thus, in this view, behavior occurs automatically in response to a stimulus, memory stores supplying the direct link between afferent and efferent activity. In contrast, the processing of the afferent information (the sound of the bell), acting through memory stores, could result in the conscious awareness of his promise, which, in turn, leads to the relevant activity in motor pathways descending from the cortex. There is no way of choosing between these two views at present.

In contrast to this presently unapproachable question, some aspects of conscious experience have yielded, at least in part, to experimentation in human beings. In general, these can be described by two questions: What is the relationship between the conscious experience and information arriving over the principal afferent pathways? Is there an anatomically distinct area of the central nervous system involved in conscious experience as opposed to those areas of the brain engaged in the unconscious processing of information or the execution of automatic movements?

The first question has been approached by studies performed in conscious human subjects whose brains are exposed for neurosurgery. Tiny electrodes lowered into different areas of the thalamus record the activity of single cells. In certain thalamic cells a given somatic stimulus, such as movement of a joint, regularly produces a given response which contains in coded form the precise location and intensity of the stimulus. The electric response recorded is unchanged regardless of whether the patients are keenly aware of each stimulus or whether their attention is diverted so that they are unaware that they have been stimulated; i.e., the electric activity is the same whether or not information about the stimulus is incorporated into conscious experience. The thalamic cells which respond in this manner are in nuclei known to form part of the specific afferent pathway. Thus, information relayed in the specific ascending pathways does not necessarily become part of conscious experience.

There are other neurons in nonsensory parts of the thalamus which have quite different properties and patterns of activity, which are more closely related to conscious experience. Each of these cells, called *novelty detectors*, responds to input from many parts of the body, but it rapidly *ceases* responding to repeated stimuli of the same kind as the person's attention to them wanes. The firing pattern is more closely related to the degree to which the person is aware of a given stimulus than to precise information about a certain sensory modality.

Other evidence also suggests that conscious experience is determined by central structures as well as by peripheral stimuli. Clinical examples of conscious sensory experiences existing in the absence of neural input are not unusual. For example, after a limb has been amputated, the patient sometimes feels as though it were still present. The nonexistent limb, called a phantom limb, can be the "site" of severe pain.

In this context, it is interesting to study the conscious experiences of persons undergoing periods of sensory deprivation. Student volunteers lived 24 h a day in as complete isolation as possible—even to the extent that their movements were greatly restricted. External stimuli were almost completely absent, and stimulation of the body surface was relatively constant. At first the students slept excessively, but soon they began to be disturbed by vivid hallucinations which sometimes became so distorted and intense that the students refused to continue the experiment. The neural bases of these hallucinations are poorly understood, but it has been suggested that central structures may generate patterns of activity corresponding to those normally elicited by peripheral stimuli when varied sensory input is absent and that conscious experience is not solely dependent upon the senses.

There seems to be an optimal amount of afferent stimulation necessary for the maintenance of the normal, awake consciousness. Levels of stimulation greater or less than this optimal amount can lead to trances, hypnotic states, hallucinations, "highs," or other altered states of consciousness. In fact, alteration of sensory input is commonly used to induce such experiences intentionally.

The answer to the second question: Is there a specific brain area in which conscious experience resides? can be gleaned from conscious persons undergoing neurosurgical procedures and from persons who have accidental or disease-inflicted damage to parts of the brain; but the answer is still far from clear. We have mentioned that it is possible that conscious experience may be just another aspect of the neural activity in those brain centers which receive and process afferent information. Or, on the other hand, conscious experience may depend upon the transmission of the processed afferent information to special parts of the brain whose function is to "release" the contents of conscious experience. The only available clues as to how or where this might be done have been gained from inference. For example, with evolution comes (we assume) greater complexity of conscious experience. Since the human brain is distinguished anatomically from that of other mammals by a greatly increased volume of cerebral cortex, this is a logical place to look for the seat of conscious experience. The cortex has been stimulated when patients on the operating table are fully alert and the brain is exposed. If a certain area of association cortex in temporal lobe is stimulated, the subject may report one of two types of changes in his conscious experience. Either he is aware of a sudden change in his interpretation of the present situation, i.e., what he is seeing or hearing suddenly becomes familiar or strange or frightening or coming closer or going away, or he has a sudden flashback or awareness of an earlier experience. Although he is still aware of where he is, an earlier experience comes to him and repeats itself in the same order and detail as the original experience. It may have been a particular occasion when he was listening to music. If asked to do so, he can hum an accompaniment to the music. If in the past he thought the music beautiful, he thinks so again. During such electric stimulation, visual or auditory experiences are recalled only if the patient was attentive to them when they originally occurred. Other experiences which are also part of conscious experience have never been produced by such stimulation. No one has ever reported periods

when he was trying to make a decision or solve a problem or add up a row of figures. It is also interesting that no brain area other than temporal association cortex has been found from which complete memories have been activated. However, one cannot assume that association cortex is the site of the stream of consciousness, since stimulation of these cortical neurons causes the propagation of action potentials to many other parts of the brain.

It is best to take a different view of the matter; for, in the words of one famous neurosurgeon, "Consciousness is not something to be localized in space." It is a function of the integrated action of the brain. Sensations and perceptions form part of conscious experience, and yet there is no one point along the ascending pathways or one particular level of the central nervous system below which activity cannot be a conscious sensation and above which it is a recognizable, definable sensory experience. Every synapse along the ascending pathways adds an element of meaning and contributes to the sensory experience.

On the other hand, consciousness and, we presume, the accompanying conscious experiences, are inevitably lost when the function of regions deeper within the cerebrum or the nerve fibers passing to association cortex from the reticular formation are interrupted by injury. Although the matter is far from settled, evidence suggests that neuronal systems in the reticular formation of the brainstem and regions deep within the cerebrum are involved in brain mechanisms necessary for perceptual awareness. It has been suggested that this system of fibers has widespread interactions with various areas of the cortex and somehow determines which of these functional areas is to gain temporary dominance in the on-going stream of the conscious experience.

The concept we want to leave as an answer to the question of where the conscious experience resides is perhaps best presented in the following analogy. In an attempt to say which part of a car is responsible for its controlled movement down a highway, one cannot specify the wheels or axle or engine or gasoline. The final performance of an automobile is achieved only through the coordinated interaction of many components. In a similar way, the conscious experience is the result of the coordinated interaction of *many* areas of the nervous system. One neuronal system would be incapable of creating a conscious experience without the effective interaction of many others.

Motivation and emotion

Motivation

Motivation is presently undefinable in neurophysiological terms, but it can be defined in behavioral terms as the processes responsible for the goal-directed quality of behavior. Much of this behavior is clearly related to homeostasis, i.e., the maintenance of a stable internal environment, an example being putting on a sweater when one is cold. In such homeostatic goal-directed behavior specific bodily needs are being satisfied, the word "needs" having a physicochemical correlate. Thus, in our example the correlate of need is a drop in body temperature, and the correlate of need satisfaction is return of the body temperature to normal. The neurophysiologic integration of much homeostatic goal-directed behavior has been discussed earlier (thirst and drinking, Chap. 13; food intake and temperature regulation, Chap. 15; reproduction, Chap. 16).

However, many kinds of motivated behavior, e.g., the selection of a particular sweater on the basis of style, have little if any apparent relation to homeostasis. Clearly, much of human behavior fits this latter category. Nonetheless, the generalization that motivated behavior is induced by needs and is sustained until the needs are satisfied is a useful one despite the inability to understand most needs in physicochemical terms.

A concept inseparable from motivation is that of *reward* and *punishment,* rewards being things that organisms work for or things which strengthen behavior leading to them, and punishments being the opposite. They are related to motivation in that rewards may be said to satisfy needs. Many psychologists believe that rewards and punishments constitute the incentives for learning. Because virtually all behavior is shaped by learning, reward and punishment become crucial factors in directing behavior. Although some rewards and punishments have conscious correlates, many do not. Accordingly, much of human behavior is influenced by factors (rewards and punishments) of which the individual is unaware.

We have thus far described motivation without regard to its neural correlates. Nothing is known of the mechanisms which underlie the subjective components of this phenomenon, nor is it known how rewards and punishments influence learning and behavior. Present knowledge is limited to recognition of some of the brain areas (and their interconnecting pathways) which are important in motivated behavior.

It should not be surprising that the brain area most important for the integration of motivated behavior related to homeostasis is the hypothalamus, since it contains the integrating centers for thirst, food intake, temperature regulation, and many others. Much information concerning the reinforcing effects of rewards and punishments on hypothalamic function has been obtained through *self-stimulation* experiments, in which an unanesthetized experimental animal regulates the rate at which electric stimuli are delivered through electrodes previously implanted in discrete brain areas. The animal is placed in a box containing a lever it can press (Fig. 20-4). If no stimulus is delivered to the animal's brain when the bar is pressed, he usually presses it occasionally at

Figure 20-4. Apparatus for self-stimulation experiments. *(Adapted from Olds.)*

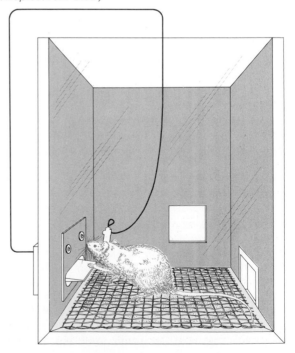

random. However, if a stimulus is delivered to the brain as a result of the bar press, a different behavior can result, depending upon the location of the electrodes. If the animal increases his bar-pressing rate above control, the electric stimulus is, by definition, rewarding; if he decreases it, the stimulus is punishing. Thus, the rate of bar pressing is taken to be a measure of the effectiveness of the reward (or punishment). Bar pressing that results in self-stimulation of the sensory and motor systems produces response rates not significantly different from the control rate. Brain stimulation through electrodes implanted in certain areas of hypothalamus serves as a positive reward. Animals with electrodes in these areas bar-press to stimulate their brains from 500 to 5,000 times per hour. In fact, electric stimulation of some areas of hypothalamus is more rewarding than external rewards; e.g., hungry rats often ignore available food for the sake of electrically stimulating their brains.

This rewarding effect of self-stimulation is not found in all areas of the hypothalamus but is most closely associated with those areas which normally mediate highly motivated behavior, e.g., feeding, drinking, and sexual behavior. Consistent with this is the fact that the animal's rate of self-stimulation in some areas increases when he is deprived of food; in other areas, it is decreased by castration and restored by administration of sex hormones. Thus, it appears that neurons controlling homeostatic goal-directed behavior are themselves intimately involved in the reinforcing effects of reward and punishment.

Although it has been generally assumed that such feeding and drinking behaviors are associated with the underlying feelings of hunger and thirst, it is puzzling as to why an animal would self-stimulate in order to experience "thirst" or "hunger."

Emotion

Related to motivation are the complex phenomena of *emotion*. Scientists are presently trying to understand the operation of the chain of events leading from the perception of an emotionally toned stimulus, i.e., the subjective feeling of fear, love, anger, joy, anxiety, hope, etc., to the complex display of emotional behavior, and they are beginning to find some answers. Most experiments point to the involvement of the limbic system, which has been studied in experimental animals, using electric stimulation of specific areas within it. The physiological results of these procedures vary markedly but justify the concept that three distinct neural systems mediate the various emotional behaviors. Of course, in these experiments there was no way to assess the subjective emotional feelings of the animals; instead they were observed for behaviors which usually are associated with emotions in human beings. As different areas of the limbic system were electrically stimulated in awake animals three types of behavior resulted.

During stimulation of one particular area the animal actively approaches a situation as though expecting a reward. Stimulation of a second area causes the animal to stop the behavior he is performing, as though he knew it would lead to punishment. Stimulation of a third area of the limbic system causes the animal to arch its back, puff out its tail, hiss, snarl, bare its claws and teeth, flatten its ears, and strike. Simultaneously, its heart rate, blood pressure, respiration, salivation, and concentrations of plasma epinephrine and fatty acids all increase. Clearly, this behavior typifies that of an enraged or threatened animal. In fact, an animal's behavior can be changed from quiet to savage or from savage to docile simply by electrically stimulating different areas of the limbic system.

Limbic areas have also been stimulated in awake human beings undergoing neurosurgery. These patients, relaxed and comfortable in the experimental situation, report vague feelings of fear or anxiety during periods of stimulation to certain areas even though they are not told when the current is on. Stimulation of other areas induces pleasurable sensations which the subjects find difficult to define precisely.

Surgical damage to parts of the limbic system in experimental animals is another commonly used tool; it leads to a great variety of changes in behavior, particularly that associated with emotion. Destruction of a nucleus in the tip of the temporal lobe produces docility in an otherwise savage animal, whereas surgical damage to an area deep within the brain produces vicious rage in a tame animal; and the rage caused by this lesion can be counteracted by a lesion in the tip of the temporal lobe. A rage response can also be caused by destruction of part of the hypothalamus. Lesioned animals sometimes manifest bizarre sexual behavior in which they attempt to mate with animals of other species; females frequently assume male positions and attempt to mount other animals.

Self-stimulation experiments have shown the presence of reward and punishment responses in various parts of the limbic system. When the electrodes are in certain midline areas, the animal presses the bar once and never goes back, indicating that stimulation of these brain areas has a punishing effect. These are the same brain areas which, when stimulated, give rise to behavioral activity signifying avoidance, rage, or escape. In contrast, self-stimulation of other limbic areas has a strong rewarding effect.

Stimulation of certain hypothalamic areas (like the stimulation of other limbic structures described above) elicits behavior that *seems* to have a strong subjective emotional component; yet if the hypothalamus is isolated from the other portions of the limbic system, the emotional component of the behavior is lacking. For example, stimulation of a cat's hypothalamus under such conditions can cause enraged, aggressive behavior complete with attack directed at any available object, but as soon as the stimulation ends, the animal immediately reverts to its usual friendly behavior. It seems as though the actions lacked emotionality and purpose, representing only the motor component of the behavior. We use the word "purpose" but could have said that, except for the experimentally induced stimulus, the behavior lacked "motivation."

What is the relationship between the hypothalamus and the rest of the limbic system? The three components of the limbic system described above converge in the medial hypothalamus, which acts as an integrating center. For example, in cats the medial hypothalamus exerts a tonic inhibition on the neural pathways which lead to fight-or-flight behavior. However, upon receipt of appropriate environmental stimuli, the nuclei of the temporal lobe inhibit the medial hypothalamus, thus decreasing its inhibitory influence over the fight-or-flight system and allowing activity in that system to increase. The resulting emotional behavior, then, seems to result from the balance of input to the medial hypothalamic integrating centers. Notice that, although the structures involved in the control of emotional behavior are predominantly located in the limbic system, and (as described in previous chapters) the main controlling centers for consummatory behavior related to

homeostasis are located in the hypothalamus, the two meet and interact at the level of the medial hypothalamus.

Finally the subjective aspects, or feelings, that make up part of an emotional experience possibly also involve the cortex, particularly cortex of the frontal lobes, which is implicated because changes in emotional states frequently occur following damage there. These alterations in mood and character are described as fear, aggressiveness, depression, rage, euphoria, irritability, or apathy. There are indications that frontal regions may exert inhibitory influences upon the hypothalamus and other areas of the limbic system. There may be facilitatory frontal regions as well. Anatomical connections between frontal cortex and hypothalamus exist to support the suggested interrelationship of these two areas in motivated and emotional behaviors. Excitatory and inhibitory influences from the limbic system and possibly from nonlimbic areas of cortex are of great importance in determining and patterning the level of excitability of hypothalamic and brainstem neurons. How activity in the limbic system is initiated and influences other brain areas is still poorly understood.

Chemical mediators for emotion-motivation

The pathways which underlie the brain reward systems and motivation (Fig. 20-5) are, of course, made up of chains of synaptically interconnected neurons, and the neurotransmitters and the many drugs that affect the synapses in these pathways are fairly well-defined. A role for norepinephrine and dopamine (both of which belong, along with epinephrine, to the family of catecholamines) is suggested by several kinds of evidence. First, the anatomical sites that give high rates of self-stimulation contain catecholamine-mediated pathways. Second, rates of self-stimulation by experimental animals are altered by drugs that affect transmission at catecholamine-mediated synapses; for example, drugs (such as amphetamine) which increase synaptic activity in the catecholamine pathways, increase self-stimulation rates. As mentioned below, these drugs also elevate mood in humans. Conversely, drugs (such as chlorpromazine, an antipsychotic agent) that lower activity in the catecholamine pathways decrease self-stimulation and depress mood. (Note that the fact that

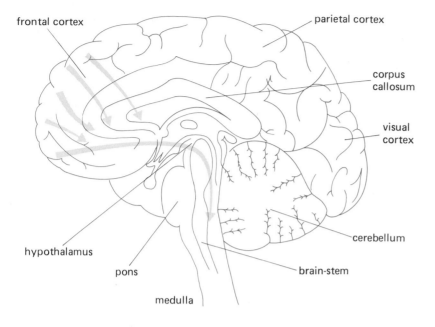

frontal cortex

parietal cortex

corpus
callosum

visual
cortex

cerebellum

hypothalamus

pons

brain-stem

medulla

catecholamines correlate both with the brain reward-system and moods is further evidence for a neural relation between motivation and emotion.)

The catecholamines are also implicated in the pathways subserving learning. This association is not unexpected since we have just stated that they are involved in the neural systems underlying reward and punishment, and many psychologists believe that rewards and punishments constitute the incentives for learning.

Altered states of consciousness

Schizophrenia

Schizophrenia means simply a fragmentation of mental functioning. It is uncertain whether it is a single disease or a family of disorders, but the general symptoms are clear and include altered motor behavior, perceptual distortions, disturbed thinking, altered mood, and abnormal interpersonal behavior. The motor behavior can range from total immobilization (catatonia) to wild purposeless activity. Perceptual distortions can include hallucinations, particularly auditory ones, such as hearing voices or hearing one's own thoughts out loud. The disturbed thinking can include delusions (illogical thought and beliefs that

are false or improbable but cannot be changed by contrary evidence or argument); e.g., the belief that one has been chosen for a special mission or is persecuted by others.

Schizophrenia is a fairly common disease; the chance of becoming schizophrenic is about one percent, and persons suffering from this disease occupy more than one-fourth of all hospital beds in the United States, even though less than 20 percent of the diagnosed schizophrenics are hospitalized.

The cause of schizophrenia is unknown and both genetic and environmental factors have been suspected. The suspicion of genetic factors immediately suggests that there is something biochemically different about people with schizophrenia since genes operate only through biochemical mechanisms. Consequently, an amazing variety of tissues have been analyzed for changes, but few differences have been found that relate to schizophrenia in an understandable way. One of the most promising explanations for the disease, the *dopamine hypothesis*, suggests that there is excessive activity of dopamine-mediated synapses, especially in structures of the limbic system, but even this is not supported by tissue analyses; rather, it is suggested by the fact that virtually all drugs that are effective in the therapy of schizophrenia interfere with transmission at dopamine-mediated synapses (for example, one blocks post-synaptic dopamine receptors).

Abnormalities in other neurotransmitters and neuromodulators, including serotonin, histamine, acetylcholine, the endorphins, and norepinephrine have been implicated as well. Norepinephrine deficiency is particularly suspect, and could be associated with excess dopamine (suggested by the dopamine hypothesis) if there were a deficiency in the hydroxylase enzyme which normally converts dopamine to norepinephrine (Chap. 8); this would lead to the release of dopamine, rather than norepinephrine, at the synaptic endings.

An alternative hypothesis concerning the cause of schizophrenia is that potent hallucinogens may be formed from neurotransmitters which are normally in the brain. For example, the hallucinogenic drugs mescaline and psilocin are methylated derivatives of dopamine and serotonin, respectively, and enzymes exist in the body (in the lung and brain) for this methylation step.

Despite the presence of these interesting hypotheses and the availability of antipsychotic drugs which relieve some of the symptoms of the disease, the fact remains: The exact nature of the cause of schizophrenia and a means of producing a lasting cure in all schizophrenics are unknown.

The affective disorders: Depressions and manias

The affective disorders are characterized mainly by serious, prolonged disturbances of mood and generally include the *depressions,* the *manias,* or swings between these two states, so-called *bipolar affective disorder.* In the depressive disorders, the prominent feature is a pervasive sadness or loss of interest or pleasure, whereas the essential feature of the manias is an elated mood, sometimes with euphoria (i.e., an exaggerated sense of well being), overconfidence, and irritability. The affective disorders are also accompanied by changes in thought patterns and behaviors. Along with schizophrenia, they form the major psychiatric illnesses today.

Two of the monoamine neurotransmitters, norepinephrine and serotonin, are implicated in the affective disorders. One current hypothesis states that the depressions are due to deficiencies or decreased effectiveness of norepinephrine and serotonin in certain pathways, whereas the manias are caused by functional hyperactivity of these two neurotransmitters. Variations in the degree of involvement of the transmitters would then account for differences in the subgroups which can be distinguished clinically and biologically within the two major classifications of the affective disorders.

Much of the evidence supporting the norepinephrine-serotonin hypothesis is circumstantial and controversial; it comes largely from studies of the actions of those drugs used in the affective disorders and the excretion rates of the various neurotransmitters and their metabolites. First, the drugs used to combat depressions generally potentiate the effects of the monoamines. Some (the monoamine oxidase inhibitors) inhibit one of the enzymes responsible for monoamine breakdown, thereby increasing the concentrations of these transmitters in the central nervous system; others (the tricyclic antidepressants) interfere with the reuptake of norepinephrine and serotonin by presynaptic endings. Thus, both drug types result in an increased concentration of transmitter available to activate receptors on the postsynaptic membrane.

Second, in persons suffering from depression (or going through the depressive phase of a manic-depressive bipolar disorder) there is a decrease in the urinary excretion of the monoamines and their metabolic biproducts, suggesting a decreased synaptic release. In contrast, the excretion of these compounds increases during manias. However, here one encounters the kind of difficulty met when trying to analyze the cause of a mental disorder: Do the changes reflect disturbances which are the cause of the disease or do they reflect the altered behaviors and mental states resulting from the disease?

Other hypotheses concerning depression suggest that the problem lies in the balance between norepinephrine and serotonin rather than in the absolute quantities of either one, or that acetylcholine is involved or even that the problem lies with the receptors or neuromodulators rather than the neurotransmitters.

Lithium carbonate is the most effective drug currently used in the treatment of the manias; it is highly specific for the manias, normalizing mood and slowing down thinking and motor behavior without causing sedation. In addition, it decreases the severity of the swings between mania and depression that occur in the bipolar disorder and, in some cases, is effective in depression not associated with manias. Lithium decreases the release of norepinephrine from the presynaptic terminals and enhances its reuptake, actions opposite those of the tricyclic antidepressants. This explanation fits nicely with the hypothesis that norepinephrine is overactive in mania, but it does not explain why lithium is occasionally effective in treatment of the depressions.

Psychoactive drugs, tolerance, and addiction

In the previous sections we mentioned several drugs that are used in an attempt to combat altered states of consciousness. These as well as other psychoactive drugs are also used in a deliberate attempt to elevate mood (euphorigens) and produce unusual states of consciousness ranging from meditative states to hallucinations. All these drugs seem to act on neuronal membranes or synaptic mechanisms in one way or another (neurotransmitter release or reuptake, membrane permeability, etc.).

Actually, as we have seen, psychoactive drugs are often chemical forms related to neurotransmitters such as serotonin, dopamine (Fig. 20-6) and norepinephrine; moreover, it is possible that the endorphins and enkephalins, in conjunction with their effects in the pain-inhibiting pathways (Chap.

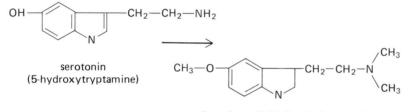

Figure 20-6. Molecular similarities between neurotransmitters and some euphorigens. At high doses these euphorigens can cause hallucinations.

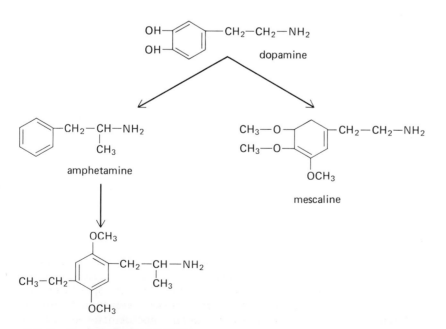

18), serve as natural euphorigens. As we have seen, several of the naturally occurring euphorigens have been implicated in diseases such as schizophrenia.

In general, there seems to be a tendency for the neurotransmitter-receptor interaction to stay at some baseline state and to regain that state if moved from it. Perhaps this balance is upset by some malfunction in disease states such as schizophrenia, the depressions, and the manias. In any case, it may be this tendency to return to a baseline that underlies the dual phenomena of tolerance and dependence (addiction) to administered psychoactive drugs.

Tolerance to a drug occurs when increasing doses of that drug are required to achieve effects that initially occurred in response to a smaller dose, i.e., it takes more drug to do the same job. One cause of tolerance is that the presence of the drug may stimulate the manufacture (especially, in the liver) of those enzymes that degrade it; as drug concentrations increase so do the concentrations of the degratory enzymes so that more drug must be administered to produce the initial effect.

However, another possible cause for tolerance has nothing to do with drug degradation but results from the drug's actions as a neurotransmitter. The drug's effects add to those of the normally-occurring neurotransmitter, thereby producing an increased effect which, by feedback mechanisms, decreases the release of the naturally-occurring neurotransmitter. For example, tolerance to opiate drugs such as morphine, which react with some types of the receptors normally activated by enkephalin, may occur in the following way (Fig. 20-7). Enkephalin is normally continued released, at least to some degree, by those neurons which use it as their transmitter so that the enkephalin receptors are always exposed to a certain amount of stimulation. When morphine is present, it binds to any receptors not already occupied by enkephalin and thereby increases the analgesic-euphorigenic effects of the enkephalin pathways. As the enkephalin receptors are stimulated more and more strongly (by the morphine-enkephalin combination), a feedback system (still hypothetical) decreases the firing rate of the enkephalin-releasing neurons, thereby decreasing the synthesis and release of the transmitter and the number of receptors activated by it. The receptors, no longer receiving their usual amount of enkephalin, can

therefore accept more morphine. In other words, to regain the previous analgesic-euphorigenic effects of full receptor activation, increased amounts of the drug are required.

This type of reasoning can also explain the physical symptoms associated with cessation of drug use, i.e., *withdrawal.* When the drug is stopped, the receptors receive for some time neither the drug nor the normal neurotransmitter (because the latter's synthesis and release had been inhibited by the previous prolonged drug use).

Alcohol: An example. Ethyl alcohol is an example of a depressant drug even though lay people often think of it as a stimulant. Medically, the central nervous system depressants are used to produce sedation, sleep, and anesthesia; nonmedicinally they are used for relaxation and relief of anxiety, to produce a mild euphoria, to promote sleep, and for suicidal purposes. With repeated usage they produce tolerance and both the physical and psychological forms of drug addiction.

Excess ingestion of alcohol causes a variety of central nervous system symptoms, for example, poor coordination, sluggish reflexes, emotional lability, and out-of-character behavior (such as belligerent, aggressive behavior in a person typically shy, retiring, and mild mannered). Sometimes, amnesia for events that happen during the intoxicated period (blackouts) occurs. The degree of intoxication relates to body size, type of beverage consumed, rapidity of drinking, tolerance, the presence or absence of food in the stomach, and many other factors. Individual response to alcohol consumption varies greatly, but signs of intoxication such as those just described are almost always present at blood levels of 2000 mg/L, unconsciousness usually occurs when blood levels are between 4000–5000 mg/L, and death between 6000–8000 mg/L. Unconsciousness usually occurs before the person can drink enough to die, but when death does occur, it is usually because of respiratory depression or vomiting followed by aspiration of the vomitus into the respiratory tract.

Alcohol acts on the lipid matrix of cell membranes rather than on specific receptors. Many attempts have been made to find a metabolic basis for the effects on the central nervous system but there has been little success (see Chap. 17 for effects on the liver). Alcohol does block the conduction of action potentials in peripheral nerves by

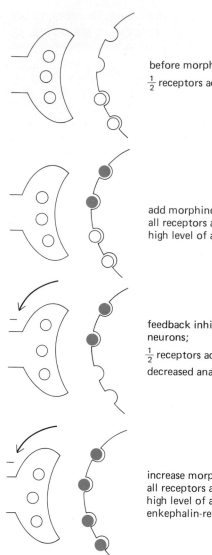

before morphine;
$\frac{1}{2}$ receptors activated

add morphine;
all receptors activated;
high level of analgesic-euphorigenic effect

feedback inhibition of enkephalin-releasing neurons;
$\frac{1}{2}$ receptors activated;
decreased analgesic-euphorigenic effect

increase morphine;
all receptors activated;
high level of analgesic-euphorigenic effect;
enkephalin-releasing neurons inhibited

○ enkephalin

● morphine

) enkephalin receptor

altering both sodium and potassium permeabilities and depolarizing the nerve membrane, but these effects are seen only at alcohol concentrations much higher than those needed to affect the central nervous system.

Recent research on the basis of alcohol addiction suggests that, in the presence of acetaldehyde (a product of alcohol oxidation), abnormal products are formed during the metabolism of some of the neurotransmitters. These presumably act as "false" neurotransmitters and interfere with the function or metabolism of the normal neurotransmitters in neural pathways that mediate the craving for alcohol.

The length of the period of intoxication ranges from several hours to 8 to 12 h after alcohol consumption stops. More than 95 percent of the alcohol consumed is metabolized in the liver by oxidation to acetaldehyde and acetyl coenzyme A (Chap. 17), and this occurs at a relatively slow rate; for example, it takes 5 to 6 h for an average-sized person to metabolize the alcohol in 120 mL (4 oz) of whisky or 1.2 1 (1.25 qt) of beer.

Withdrawal symptoms following cessation of prolonged, heavy drinking are tremors (the "shakes"), nausea and vomiting, a dry mouth (even though the person is not dehydrated), headache and muscular weakness, nightmares, and feelings of anxiety, guilt, and irritability. More severe symptoms such as hallucinations and seizures can also occur.

Coma and brain death

There have been great advances in the past 100 years in the development of techniques of resuscitation; for example, now it is fairly common for artificial respiration to supply gases to the lungs and cardiac massage to restore a nonfunctioning circulation. However, the brain is particularly susceptible to anoxia and it is likely to suffer damage even though resuscitative techniques have restored the functioning of the heart and lungs. Techniques are available to maintain persons in such brain-damaged, comatose states (regardless of their cause), resulting occasionally in a body with adequate cardiovascular and respiratory function but an inactive nervous system—in other words, a living body which contains a dead brain.

The question "When is a person actually dead?" often has urgent medical, legal, and social consequences. For example, with the advent of organ transplantation and the need for viable tissues, it became imperative to know "Is a person in a coma dead or alive?" *Brain death* is widely accepted by doctors, lawyers, and the general public as the criterion for death, the term being taken to mean "a brain that no longer functions and has no possibility of functioning again." The problem now becomes practical: "How does one know when brain death has occurred? What criteria are valid determinants of brain death?" There is certainly no agreement about these criteria, but the list below gives one set of conservative standards frequently accepted. These criteria are tested only after appropriate therapeutic procedures have been performed, and they must be present for at least 30 min, beginning 6 h after the onset of suspected brain damage.

1 The person is unconscious and unresponsive, obeying no commands and making no spontaneous sounds or movements.
2 The person is apneic (not breathing). Apnea is considered to have occurred if there is no sign of a spontaneous respiration for a period of 15 min. (Once a patient is on a respirator, it is of course unadvisable to remove him for the 15-min test period because of the danger of further anoxic damage occurring. Therefore, apnea is diagnosed if the patient's reflexes do not drive respiration at a depth or rhythm different from those of the respirator and if there is no spontaneous attempts to fight the respirator.)
3 There is electrocerebral silence, i.e., a flat EEG with wave amplitudes less than two microvolts. The EEG is a most significant test since its outcome correlates highly with survival or death; the greater the activity on the EEG recording, the greater the chances of survival. However, this correlation does not hold in those cases when electrocerebral silence follows drug intoxication or periods of severely lowered body temperature.
4 Cortical and brainstem reflexes, such as the pupillary, corneal, cough, swallowing, and gag reflexes are absent; in some cases reflexes mediated by the spinal cord, such as the tendon and stretch reflexes, may be present.
5 Absence of cerebral blood flow for a period of at least 30 min.

Neural development and learning

Evolution of the nervous system

During an early stage of evolution, the nervous system was probably a simple system with a limited number of interneurons interposed between the afferent and efferent nerve cells. The interneuronal component expanded rapidly until it came to be by far the largest part. The interneurons formed networks of increasing complexity, at first involved mainly with the stability of the internal environment and position of the body in space. Those cells with increasing specialization of function came to be localized at one end of the primitive nervous system, and the brain began to evolve.

The brainstem, which is the oldest part of the brain in this evolutionary sense, retains today many of the anatomical and functional charac-

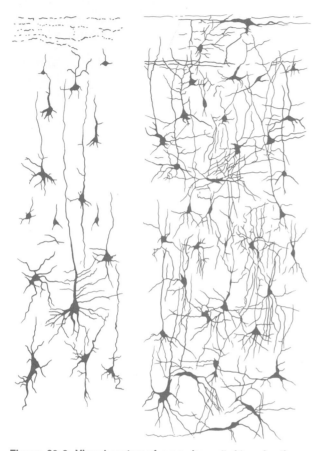

Figure 20-8. Visual cortex of a newborn *(left)* and a three-month-old child *(right)*. *(Redrawn from J. L. Conel, "The Postnatal Development of the Human Cerebral Cortex," vol. 1, The Cortex of the Newborn, Harvard University Press, Cambridge, Mass., 1939.)*

teristics typical of those most primitive brains. With continued evolution, newer, increasingly complex structures were added on top of (or in front of) the older ones. They developed as paired symmetrical tissues, the cerebral hemispheres, and reached their highest degree of sophistication with the formation of the cerebal cortex.

The newer structures served in part to elaborate, refine, modify, and control already existing functions. For example, it is possible to perceive the somatic stimuli of pain, touch, and pressure with only a brainstem, but the stimulus cannot be localized without a functioning cerebral cortex. Perhaps even more important is that these newer, more anterior parts of the brain came to be in-

volved in the perception of goals, the ordering of goal priorities, and the patterning of behaviors to serve in pursuit of these goals.

Development of the individual nervous system

All neurons are present at birth, but during postnatal development there occur many important changes in these neurons: The cells enlarge, their numbers of dendrites increase, axons elongate, myelinization increases, the number of synaptic contacts increases, neurons migrate to new locations, etc. Notice the changes in visual cortex cells in just the first three months of life (Fig. 20-8) and the gradual change in electric activity (Fig. 20-9). These developmental patterns are reflected in behavioral changes. It is believed that the cortex is relatively nonfunctional at birth, a notion supported by the fact that infants born without a cortex have almost the same behavior and reflexes as shown by a normal infant. As cortical function develops, it seems gradually to exert an inhibitory control over the lower (and phylogenetically older) structures.

With cortical development, some of the reflexes whose integrating centers are in subcortical structures come under at least a degree of cortical control. Examples are the rooting and sucking reflexes present in infancy. (The stimulus is a touch on the infant's cheek; in response, the head turns toward the stimulus, the mouth opens, the stimulating object is taken into the mouth, and sucking begins.) The reflex is essential for the survival of the young, but it is superseded by other eating behaviors. As the cerebral hemispheres develop and autonomy increases, allowing the elaboration of hand feeding, the primitive rooting and sucking reflexes are inhibited. The basic reflex arc does not disappear; it is simply inhibited. For example, while examining a patient with damaged frontal lobes of the brain, a physician standing behind the patient and out of his view quietly reached over and touched the patient's cheek. In response, the patient nuzzled the physician's finger until it was in his mouth and even began sucking on it. Upon realizing what he had unconsciously and automatically done, he was quite embarrassed; but his behavior, released from its normal inhibition because of the brain damage, serves as a perfect illustration of the above point.

During development of the brain and spinal cord,

nerve growth and synaptic contact occur with remarkable precision and selectivity. How do the right nerve connections get established in the first place? The best answer presently available is the *chemoaffinity theory,* in which the complicated neural circuits are said to grow and organize themselves according to selective attractions between neurons, which are determined by chemical codes under genetic control. Thus, during early stages of their development, neurons acquire individual chemical "identifications" by which they can recognize and distinguish each other. The chemical specificity is precise enough to determine not only the particular postsynaptic cell to which the axon tip will grow but even the precise area on the postsynaptic cell that will be contacted. (Recall that synapses closer to the postsynaptic cell's initial segment have greater influence over its activity.)

Several other factors also influence the final neural organization: timing of the development of the individual cells (for example the axons that reach a postsynaptic neuron earliest have the greatest chance of forming large numbers of synapses with it); the presence of neighboring neurons and glia (which form structural constraints and guides for the developing nerve processes); chemical events involving neurotransmitters and enzyme systems in the tip of the developing fiber and adjacent cells.

Of course, adequate nutritive environments are required for normal brain development. The developing brain is extremely sensitive to the effects of malnutrition (i.e., a deficiency of calories, protein, trace metals, or vitamins), and some of the damage produced may be permanent. Malnutrition in the prenatal period, during which neurons are dividing, causes a decrease in the rate of cell division, and the final number of cells is less than normal. This change is permanent and cannot be reversed by improving nutrition at a later time after the period for cell division has passed. Postnatally, malnutrition prevents the normal enlargement of cells, but this growth is restored with the resumption of an adequate diet.

The level of hormones is also important; for example, decreased thyroid hormone concentrations during development cause a type of mental retardation known as *cretinism.* Although it was formerly believed that the developing fetus "took what it needed" nutritionally from the mother and that, in cases of malnutrition after birth, the brain

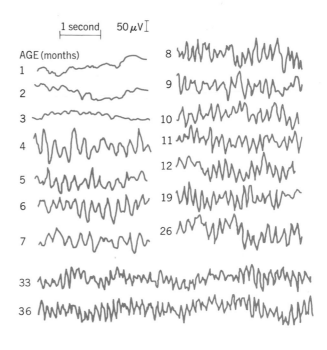

Figure 20-9. EEGs from the same individual, showing the onset of the alpha rhythm at four months, the attainment of adult wave frequencies at ten years of age, and little change thereafter. *(Redrawn from D. B. Lindsley, Attention, Consciousness, Sleep, and Wakefulness, in "Handbook of Physiology; Neurophysiology," vol. 3, pp. 1553-1593, American Physiological Society, Washington, D.C., 1960.)*

was "spared" at the expense of the rest of the body, there is increasing evidence that nutritional deprivation both before and after birth has serious and irreversible effects on the chemical and structural maturation of the brain. Intermittent deprivation affects those cells that are still developing at that time more than cells already mature.

Despite the evidence indicating the selective growth and interconnectivity of nerve cells, the nature-nurture controversy, i.e., how much of development is due to genetically determined patterns and how much to experience and subsequent learning, must also be considered, for, in fact, the course of neural development can be altered. However, the period during which this alteration can occur is genetically built into the neuron's developmental timetable. The modifiability of some neurons is limited to so-called *critical periods* early in development but in others it persists longer. Thus, some neuronal organizational patterns are highly specified early in development and are unmodifiable thereafter. These periods of maximal structural and functional growth depend upon the availability of proper internal and external environmental conditions for their full development. As an example of such a critical period it was found that the visual systems of sheep dogs deprived of sight in the first 5 weeks of life are anatomically, biochemically, and electrophysiologically retarded; this does not occur if the deprivation occurs from 5 to 10 weeks of age or during a 5-week period in adult dogs.

Other neurons remain uncommitted and are modifiable by function and experience until late in development. But in the case of either limited or lengthy periods of modifiability, it can occur only within the constraints imposed by the neurons' genetic code and its experiential history. The genes specify the capacity whereas the environmental stimuli determine its specific expression and content.

Learning and memory

Learning is the increase in the likelihood of a particular response to a stimulus as the consequence of experience. There are many kinds of learning: habituation (defined classically as a reversible decrease in the strength of a response upon repeated stimulation), conditioning, and other forms typical only of higher animals in which learning may occur following a single stimulus and in the absence of direct reinforcement. Generally, rewards or punishments, as mentioned earlier, are crucial ingredients of learning, as is contact with and manipulation of the environment. *Memory* is the relatively permanent form of this change in responsiveness, and the postulated neural correlate of memory is called the *memory trace.*

Memory occurs in two forms, *short-term* and *long-term*, and the process of transferring memories from the short- to the long-term form is called *memory consolidation.* The existence of two forms is borne out by the behavior of patients with bilateral lesions in the hippocampal region of the limbic system. Such patients can recall information immediately after it has been presented, but they cannot retain the material as well as normal people do, particularly if it involves verbal clues. It is as though they were unable to transfer the information from short- to long-term memory storage. Although short-term memory involves cerebral cortex, and long-term memory involves the limbic system, there is no exclusive site for memory storage; removal of various parts of the brain does not remove specific memories, and practically every part of the nervous system is actually capable of undergoing some form of learning.

Short-term memory. Short-term memory is a limited-capacity storage process lasting as little as a few seconds and serving as the initial depository of information. After the registration of events in short-term memory, they may either pass away or be retained in long-term memory stores depending upon factors such as attention and motivation. This distribution is not all-or-none, since some fragments of an event may be retained and others lost.

One theory suggests that the memory trace during the early phases of learning is a reverberating neural circuit in which electric activity passes around and around in closed neural loops until the mechanisms by which it is consolidated into long-term stores are activated. There is certainly evidence for the existence of such neuronal pathways in the brain, and activity once started in such loops, could be maintained to keep the "memory" of the input for some time. That such reverberating circuits could be responsible for the temporary storage of acquired information in short-term memory is supported by evidence that conditions which interfere with the electric activity of the brain (for example, coma, deep anesthesia, electroconvulsive shock, and insufficent blood supply to the brain), also interfere with the retention of recently acquired information. These same states do not usually interfere with previously laid-down long-term memory. For example, when a man becomes unconscious from a blow on the head, he often cannot remember anything that happened for about 30 min before he was hit, so-called *retrograde amnesia.* As recovery from retrograde amnesia occurs, memories of the most recent events often return last and least completely (although islands of memory can return in a discontinuous fashion). The loss of consciousness in no way interferes with memories of experiences that were learned before the period of amnesia. There are no known chemicals that selectively block the formation of short-term memories.

Memory consolidation. Memory consolidation is a labile process during which the fixation of an experience is susceptible to external interference, the lability lasting for periods of a few seconds to days or even weeks. The amount of time required to consolidate an experience is fairly constant in a given species, although it may be markedly increased or decreased by various drugs which presumably alter the chemical reactions involved in transferring an experience to a more stable form of memory. Agents that enhance learning also generally increase attentiveness, activity, and sensory responsiveness, any of which would increase learning, but these substances are effective even when given after the learning experience and are therefore thought to act directly on consolidation. According to the "reverberating circuit" theory, these drugs facilitate consolidation because they accelerate or prolong the reverberations of activity through the neural circuit which in turn facilitates the permanent changes required for long-term memory of the event.

All experiences do not register in long-term memory. What determines whether consolidation will occur? It is as though, for memory to become fixed, a "fix signal" must follow the event that is to be remembered. Such a signal is probably not specific for a particular event but rather may indicate "whatever just happened, remember it". Thus, the signal for memory consolidation may be likened to the fixation step in photographic developing; the latent image on the film will fade rapidly unless a chemical fixative is applied. The same fixative is used for photographs of any subject and itself contains no specific information. We have no real idea as to the nature of the physiological "fix" signal.

Long-term memory. Behavioral investigations and common experience indicate that memories of past events and well-learned behavior patterns normally can have very long life-spans. They may be changed or suppressed by other experiences, but, contrary to popular opinion, memories do not usually fade away or decay with time. This stability and durability, combined with the fact that removing parts of the brain does not remove specific memories, suggests that memory is stored in widespread chemical form or that the memory trace is an alteration in structure of some elements of the brain.

Any physicochemical explanation of learning must be able to account for the following phenomena:

1 Learning can occur very rapidly. In fact, under some situations, learning can occur in one trial.
2 Learning must be translated from an initial form involving action potentials to some permanent form of storage that can survive deep anesthesia, trauma, or electroconvulsive shock, which disrupt the normal patterns of neural conduction in the brain.
3 Information can be retained over long periods of time during which most components of the body have been renewed many times. Learning and memory must therefore reside either in systems that do not turn over rapidly or in systems which are self-perpetuating.
4 Information can be retrieved from memory stores after long periods of disuse. The common notion that memory, like muscle, always atrophies with lack of use is largely wrong.
5 Learning opposing responses interferes with the memory of initial responses to the same stimulus.
6 When learning a specific task, after an initial period of rapid learning, the process seems to slow down and proceed at a rate that offers diminishing returns.
7 Many memories become more vivid and fixed with time.

It is safe to say that no single theory can presently explain all these phenomena. However, it is generally assumed that the neuronal changes underlying learning and memory are due to changes in synaptic transmission in the absence of major structural changes in nervous tissue, i.e., without the establishment of new connections by the sprouting and growth of neurons such as occurs during development and repair of injuries. This ability of neural tissue to change its responsiveness to stimulation because of its past history of activation is known as *plasticity*.

The synaptic theory of learning and memory is based on two generally accepted views: (1) The overall plan of neuronal connections is determined by heredity alone, but in order to function optimally, the connections must be brought out by (or, once brought out, maintained by) interactions with the environment; (2) the environmental interactions that result in learning modify communication between neurons at synapses that already exist due to heredity and maturation, i.e., preexisting synapses are thought to become more efficient. The critical modifications of synaptic relationships might be a change in the area of synaptic contact, an increase or decrease in the concentration of synaptic vesicles, changes at the presynaptic or postsynaptic membranes, etc. In any case, present evidence indicates that whatever the changes are, they are ultimately ascribable to alterations in the neuron's macromolecules, i.e., that large, stable molecules within neurons are changed during learning and that information is stored in the specific configuration of these molecules.

We are already familar with information-storage mechanisms of this kind, e.g., the coding of genetic information by the nucleotide sequences in chromosomal DNA and the transfer of this information to RNA and proteins (Chap. 4). Experiments have shown that drugs which block protein synthesis interfere with the laying down or retention of learned behavior, strongly supporting the hypothesis that RNA or protein synthesis is involved. For example, goldfish were placed in a tank partially divided by a barrier and were given learning trials of 20 s of light followed by 20 s of light paired with shock. The fish learned to swim over the hurdle from the light to the dark end of the tank to avoid the shock and, when tested 4 days after the learning situation, remembered the experience. If a drug which blocks protein synthesis was injected into the brains of the fish immediately after the learning trials, the fish had no memory of the experience 4 days later (they could perform immediately after learning the task). The blocking agent did not interfere with long-term memory per se; it interfered with the laying-down of long-term memory, i.e., with consolidation. This is evidenced by the fact that, if injected only a few hours after the learning situation, the fish did remember days later, for consolidation had occurred during that time. Moreover, after fish have been injected with the protein-synthesis blocker, they still remember what they had learned many days before.

One of the most interesting new developments concerning learning and memory is the possible role of certain hormones or other neuromodulators in these processes. Several peptide hormones (ACTH and melanocyte stimulating hormone) and B-lipotropin all contain similar segments seven amino acids in length. These substances (or simply their shared segment) enhance learning of several different kinds of tasks and in some cases protect against amnesia. It has been suggested that these substances may be released in increased amounts during learning situations and affect learning and memory by temporarily increasing the motivational value of certain cues in the environment, perhaps via enhanced arousal of portions of the limbic system. The effects of these substances on learning and memory last 2–4 h and are clearly distinct from their other known effects.

The posterior-pituitary peptide hormones, ADH (vasopressin) and oxytocin, also affect learning. Their effects last longer than those of the ACTH-related peptides and are thought to somehow enhance memory consolidation.

Forgetting. Forgetting denotes an inability to retrieve certain memory traces, but it does not necessarily mean that the traces no longer exist in the brain, for memory traces may be inaccessible to certain stimuli whereas they can be reached by others. For example, one may recognize the voice of a telephone caller but not be able to recall his name, which is immediately remembered upon seeing him. Our understanding of the mechanism of forgetting, like that of learning, is still mainly at the level of theories. One, the *interference theory of forgetting*, states that forgetting results from competition between responses at the time of recall. Competition comes from conflicting information stored both before and after the storage of the particular item that is being recalled. Thus we carry with us (in the form of prior learning) the source of much of our forgetting.

Cerebral dominance and language

To the casual observer, the two cerebral hemispheres appear to be symmetrical, but they both have functional specializations, the most marked of which is language. All facets of language—the conceptualization of what one wants to say or write, the neuronal control of the act of speaking or writing, and the comprehension of what has been spoken or written—are mediated by structures of the left cerebral hemisphere in 97 percent of the population. This is also the dominant hemisphere for fine motor control in the large majority of people, and so most people are right-handed (recall that the left hemisphere controls the movements of the right side of the body). Control of both-handedness and language in the left hemisphere suggests that this hemisphere is specialized for controlling successive changes in the precise positioning of muscle groups. However, the correlation between handedness and language function is not perfect, and for many left-handed people, language functions are controlled by the left hemisphere (in the remaining left-handed people, language is represented bilaterally or in the right hemisphere).

Just as the left hemisphere is dominant for language in most people, the right hemisphere appears to be dominant for many nonlanguage processes, particularly those requiring spatial ability. The right hemisphere also predominately processes sounds such as musical patterns. Memories are asymmetrically stored, too, verbal memories being greater in the left hemisphere and nonverbal memories, e.g., visual patterns, being greater in the right. It may be that even the emotional responses of the two hemispheres are different, some authors claiming that when electroconvulsive shock is administered in the treatment of depression, better effects are obtained when both electrodes are placed over the right hemisphere.

Much of the evidence about the specific functions of the two cerebral hemispheres has been obtained from studies on patients whose main commissures (nerve fiber bundles) joining the hemispheres have been cut to relieve uncontrollable epilepsy (neuronal activity, which starts at a cluster of abnormal cortical neurons, spreads throughout adjacent cortex, and gives rise to seizures and convulsions). In essence, this operation leaves two separate cerebral hemispheres with a single brainstem—two separate mental domains within one head. Events experienced, learned, and remembered by one hemisphere remain unknown to the other because the memory processing of one hemisphere is inaccessible to the other. Because control of language production resides in the left hemisphere, only that hemi-

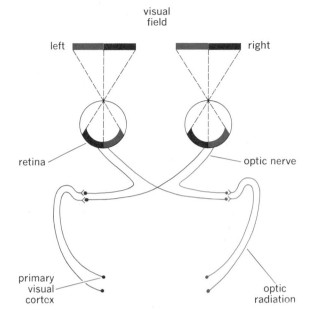

Figure 20-10. Visual pathways. Information about objects in the left half of the visual field is projected to the right cerebral hemisphere and vice versa.

sphere can communicate orally or in writing about its conscious experience. When vision is experimentally limited so that only that part of the retina whose fibers pass to the right hemisphere is excited (Fig. 20-10), the left hemisphere, which controls speech, is unaware of the visual experience and the patient is unable to describe it either orally or in writing. In contrast, when the portion of retina projecting to the left hemisphere is activated alone, the patient can describe the experience without difficulty. However, the right silent hemisphere does understand some language even though it is itself incapable of producing speech, for if the word for an object is flashed only to that hemisphere and the patient is told to point to the object or to demonstrate the use of the object, the patient complies (responding with the left hand since the right hemisphere controls the muscles of the opposite side of the body).

The left hemispheric specialization for processing speech is clearly in evidence in five-year-olds and, in fact, seems to be present and effectively operating (for speech sounds at least) in the first few months of life. This indicates that the left hemisphere is slated from birth to be the speech hemisphere, a hypothesis supported by the fact that the anatomical and physiological asymmetries in the two hemispheres are present at birth. This fundamental asymmetry is more flexible in early years than in later life. For example, accidental damage to the left hemisphere of children under the age of two causes no impediment of future language development, and language develops in the intact right hemisphere. Even if the left hemisphere is traumatized after the onset of language, language is reestablished in the right hemisphere after transient periods of loss. The prognosis becomes rapidly worse as the age at which damage occurs increases, so that after the early teens, language is interfered with permanently. The dramatic change in the possibility of establishing language (or a second language) in the teens is possibly related to the fact that the brain attains its final structural, biochemical, and functional maturity at that time. Apparently, with maturation of the brain, language functions are irrevocably assigned and the utilization of language propensities of the right hemisphere is no longer possible. Interestingly, if the left-hemisphere damage occurs early enough in the child's development so that language functions can be transferred to the right hemisphere, defects occur in those functions in which the right hemisphere normally plays a role, i.e., language develops at the expense of other right-hemisphere functions.

Different areas of left cerebral cortex are related to specific aspects of language. Areas in the frontal lobe near motor cortex are involved in the articulation of speech, whereas areas in the parietal and temporal lobe are involved in sensory functions and language interpretations. Such cortical specialization is demonstrated by the *aphasias*, specific language deficits not due to mental retardation or paralysis of the muscles involved in speech. Damage to the parietal and temporal lobe areas result in aphasias related to conceptualization—the patients cannot understand spoken or written language even though their hearing and vision are unimpaired. In contrast, damage to the areas near motor cortex result in expressive aphasias—the patients are unable to carry out the coordinated respiratory and oral

movements necessary for language even though they can move their lips and tongue; they understand spoken language and know exactly what they want to say but are unable to speak. Expressive aphasias are often associated with an inability to write.

Conclusion

Until recently it was commonly thought that control of most complex behavior such as thinking, remembering, learning, etc., was handled almost exclusively by the cerebral cortex. Actually, damage of cortical areas outside of motor and sensory areas produces behavioral results that are subtle rather than obvious (the one exception, language, is highly sensitive to cortical damage), and stimulation of the cortex causes little change in the orientation or level of excitement of the animal.

In general, it is best to consider that particular behavioral functions are not controlled exclusively by any one area of the nervous system but that the control is shared or influenced by structures in other areas. Cortex and the subcortical regions — particularly limbic and reticular systems — form a highly interconnected system in which many parts contribute to the final expression of a particular behavioral performance. Moreover, the nervous system is so abundantly interconnected that it is difficult to know where any particular subsystem begins or ends.

In the early seventeenth century Descartes taught that all things in nature, including human beings, are machines. The brain's mode of operation was compared to that of a clock. When computers became widely used, the brain was compared to a computer. The most recent analogy compares the brain to a hologram, a photographic process which records specially processed lightwaves themselves, rather than the image of an object. These widely divergent analogies only emphasize how little we know of how the brain really functions.

APPENDIX
ENGLISH AND METRIC UNITS

	English	Metric
Length	1 foot = 0.305 meter 1 inch = 2.54 centimeters	1 meter = 39.37 inches 1 centimeter (cm) = 1/100 meter 1 millimeter (mm) = 1/1000 meter 1 micrometer (μm) = 1/1000 millimeter 1 nanometer (nm) = 1/1000 micrometer * (1 angstrom (Å) = 1/10 nanometer)
Mass	** 1 pound = 433.59 grams 1 ounce = 27.1 grams	1 kilogram (kg) = 1,000 grams = 2.2 pounds 1 gram (g) = 0.037 ounce 1 milligram (mg) = 1/1000 gram 1 microgram (μg) = 1/1000 milligram 1 nanogram (ng) — 1/1000 microgram 1 picogram (pg) = 1/1000 nanogram
Volume	1 gallon = 3.785 liters 1 quart = 0.946 liter	1 liter = 1,000 cubic centimeters = 0.264 gallon 1 liter = 1.057 quarts 1 milliliter (ml) = 1/1000 liter 1 microliter (μl) = 1/1000 milliliter

* The angstrom unit of length is not a true metric unit, but has been included because of its frequent use, until recently, in the measurement of molecular dimensions.

** A pound is a unit of force not mass. When we write 1 kg = 2.2 pounds, this means that one kilogram of *mass* will have a *weight* under standard conditions of gravity at the Earth's surface of 2.2 pounds *force*.

GLOSSARY

A band: one of the transverse bands making up the repeating striated pattern of cardiac and skeletal muscle; located in the middle of a sarcomere; corresponds to the length of the thick filaments

abortion: removal or expulsion of a fetus (or embryo) from the uterus before it is viable

abscess: microbes, leukocytes, and liquified tissue debris walled off by fibroblasts and collagen

absolute refractory period: the time during which an excitable membrane is unable to generate an action potential in response to any stimulus

absorption: movement of materials across a layer of epithelial cells from a body cavity or compartment toward the blood; e.g., movement from interstitial space into a capillary, from the lumen of the digestive tract into the interstitial space, or from the lumen of a nephron into the interstitial space

absorptive state: the period during which nutrients are entering the bloodstream from the gastrointestinal tract

acceptor: a protein associated with DNA; binds specifically to a receptor-hormone complex

accessory reproductive organs: the system of ducts through which the sperm or ova are transported and the glands emptying into these ducts (in the female the breasts are usually also included in this category)

acclimatization: an adaptive change induced during an individual's lifetime by an environmental stress with no alteration in the person's genetic endowment

accommodation: adjustment of the eye for viewing various distances by changing the shape of the lens

acetone: a ketone body produced from acetyl CoA during prolonged fasting or untreated severe diabetes mellitus

acetyl coenzyme A: a metabolic intermediate that transfers acetyl groups to the Krebs cycle and to various synthetic pathways such as those for fatty acid synthesis

acetyl group: $CH_3\overset{\overset{O}{\|}}{C}-$; transferred from one reaction to another by way of coenzyme A

acetylcholine: a chemical transmitter released from many peripheral nerve endings (e.g., from postganglionic parasympathetic fibers and at neuromuscular junctions) and from some neurons in the central nervous system

acetylcholinesterase: an enzyme that breaks down acetylcholine into acetic acid and choline; located near the acetylcholine receptors on the postsynaptic membrane

ACh: *see* acetylcholine

acid: any compound capable of releasing hydrogen ions to a solution; *see also* strong acid, weak acid

acidity: a description of the hydrogen-ion concentration of a solution; the greater the acidity, the higher the hydrogen-ion concentration

acidosis: any situation in which the hydrogen-ion concentration of arterial blood is elevated

acromegaly: a disease characterized by disfiguring bone thickening and overgrowth of other organs; occurs in adults having excess growth hormone

acrosome: a vesicle containing digestive enzymes located at the head of a sperm

ACTH: *see* adrenocorticotropic hormone

actin: the globular contractile protein to which the myosin cross bridge binds; located in the thin filaments of muscle

action potential: the electric signal propagated over long distances by excitable cells, e.g., nerve and muscle; it is characterized by an all-or-none reversal of the membrane potential in which the inside of the cell temporarily becomes positive relative to the outside; has a threshold and is conducted without decrement

activation energy: the energy necessary to disrupt existing chemical bonds holding atoms or molecules together so that new chemical bonds can be formed

active hyperemia: the increased blood flow through a tissue resulting from an increase in its metabolic activity

active immunity: a resistance to reinfection acquired as the result of contact with microorganisms, their toxins, or other antigenic components

active site: the region of an enzyme to which the ligand (substrate) binds; a term equivalent to binding site

active state: the tension at any instant generated by the muscle sliding filaments as a result of cross-bridge activity; this tension may differ from the external tension exerted by a muscle, depending on the extent to which the series elastic component has been stretched

active transport: a carrier-mediated transport system in which molecules can be transported across a membrane from a region of low to a region of high concentration at the expense of energy provided by metabolism

acuity: acuteness or clarity of perception

adaptation: (neural) a decrease in the frequency of action potentials fired by a neuron despite a stimulus of constant magnitude

adenine: one of the purine bases found in the nucleotides which constitute the subunits of DNA and RNA

adenosine diphosphate (ADP): the two-phosphate product of ATP breakdown; it is formed when ATP releases its stored energy, ATP $\rightarrow$ ADP + P_i + Energy

adenosine monophosphate (AMP): the monophosphate derivative of ATP; one of the nucleotides in DNA and RNA

adenosine triphosphate (ATP): the major energy carrier that transfers chemical potential energy from one molecule to another; energy is transferred to ATP during its formation from ADP and phosphate as a result of the catabolism of carbohydrates, fat, and protein, and it is released to perform work by the cell during the subsequent breakdown of ATP into ADP and phosphate

adenylate cyclase: the enzyme which catalyzes the transformation of ATP to cyclic AMP

ADH: *see* antidiuretic hormone

adipocytes: cells specialized for the synthesis and storage of triacylglycerol (neutral fat) during periods of food uptake; fat cells

adipose: fatty; e.g., adipose tissue is tissue composed largely of fat-storing cells

adolescence: the period of transition from childhood to adulthood in all respects, not just sexual ones

ADP: *see* adenosine diphosphate

adrenal cortex: the endocrine gland forming the outer shell of each adrenal gland; secretes mainly cortisol, aldosterone, and androgens

adrenal glands: a pair of endocrine glands located just above each kidney; each gland consists of two regions: an outer layer, the adrenal cortex, and an inner core, the adrenal medulla

adrenal medulla: the endocrine gland forming the inner core of each adrenal gland; secretes mainly epinephrine

adrenaline: the British name for epinephrine

adrenergic: pertaining to those nerve fibers that release norepinephrine from their synaptic terminals; a compound that acts like norepinephrine; neuroendocrine pathways in which either epinephrine or norepinephrine is the transmitter

adrenocorticotropic hormone (ACTH): a polypeptide hormone secreted by the anterior pituitary; stimulates some of the cells of the adrenal cortex; also called corticotropin

affective disorders: the manias and depressions; diseases characterized by serious, prolonged disturbances of mood as well as changes in thought patterns and behaviors

afferent arteriole: vessels in the kidney which convey blood from arteries to glomeruli

afferent neuron: a neuron whose cell body lies outside the brain or spinal cord; carries information to the central nervous system from the receptors at its peripheral endings; sometimes called primary afferent or first-order neuron

afferent pathway: the component of a reflex arc that transmits information from a receptor to an integrating center; any pathway conveying information toward the central nervous system (or toward the brain)

affinity: the degree of attraction between a ligand and its binding site; the strength with which a ligand binds to its binding site; at high-affinity sites the binding is tight, whereas at low-affinity sites the binding is weak

after-hyperpolarization: a temporary increase in the transmembrane potential following an action potential, so that the inside of the cell is more negative than its resting state

afterbirth: the placenta and associated membranes expelled from the uterus after delivery of an infant

aggregation factor: a protein which contains several binding sites capable of interacting with specific binding sites on plasma membranes; aggregation factors form cross links between specific types of cells, holding them together to form tissues

albumin: one of the major proteins found in blood plasma; more abundant than the globulins and usually of smaller molecular weight

aldosterone: a mineralocorticoid steroid hormone secreted by the adrenal cortex; regulates electrolyte balance

alkalosis: any situation in which the hydrogen-ion concentration of arterial blood is reduced

all or none: an event that either occurs maximally or it does not occur at all, e.g., an action potential

all-*trans*-retinal: the form of retinal, in the receptor cells of the eye, that results from the action of light upon all-*cis*-retinal; dissociates from opsin

allosteric enzyme: an enzyme whose activity can be altered by modulator molecules acting at a site other than the substrate binding site

allosteric regulation: control of the properties of a protein binding site by modulator molecules that act at locations other than the binding site

alpha-adrenergic receptor: a type of receptor for norepinephrine and epinephrine that responds to certain drugs and can be distinguished from a beta-adrenergic receptor

alpha helix: a coiled conformation of a polypeptide chain found in many proteins, resulting from the electrical attraction between repeating peptide bonds along the chain

alpha motor neuron: a large motor neuron which innervates skeletomotor muscle fibers

alpha rhythm: the prominent 8- to 13-Hz oscillation in the EEG of awake, relaxed adults with their eyes closed; associated with decreased levels of attention

alveolar dead space: the volume of inspired air that reaches the alveoli but cannot undergo gas exchange with the blood

alveolar duct: a thin-walled airway branching from the end of the terminal bronchiole; alveoli extend from the end and sides of this duct

alveolar pressure: the air pressure within the pulmonary alveoli

alveolar ventilation: the volume of fresh atmospheric air entering the alveoli each minute (respiratory rate times the tidal volume minus the anatomic dead space); also called alveolar ventilatory volume

alveolus: a small cavity; the thin-walled, air-filled outpocketing from the alveolar duct in the lungs; the cluster of cells at the end of a duct in secretory glands

amino acid: a molecule containing an amino group, a carboxyl group, and a side chain all attached to a single carbon atom; 20 different amino acids (distinguished by their side chains) form the structural subunits of proteins

amino acyl synthetase: one of 20 different enzymes, each of which catalyzes the linkage of one type of amino acid to its particular type of tRNA during protein synthesis

aminopeptidases: enzymes located in the intestinal epithelial membrane; they break the peptide bond of the amino acid located at the end of a polypeptide that has a free amino group

ammonia: NH_3; substance produced during the breakdown of amino acids; converted in the liver to urea

amnesia: loss of memory; *see also* retrograde amnesia

AMP: *see* adenosine monophosphate

amphetamine: a drug that increases transmission at catecholamine-mediated synapses in the brain

amphipathic molecule: a molecule containing polar or ionized groups in one region and nonpolar groups in another

amylase: an enzyme that breaks down starch, producing disaccharides and small polysaccharides; secreted by the salivary glands and exocrine portion of the pancreas; also called alpha amylase or, if secreted by the salivary glands, ptyalin

anabolic reactions: synthetic chemical reactions; reactions which put small molecular fragments together to form larger molecules

anaerobic: without oxygen

analgesia: removal of pain

anatomic dead space: the space within the airways of the respiratory tract whose walls do not permit gas exchange

androgen: any chemical with actions similar to those of testosterone

anemia: a decreased concentration of hemoglobin in the blood

anemic hypoxia: hypoxia in which the arterial Po_2 is normal, but the total oxygen content of the blood is reduced because of inadequate numbers of erythrocytes, deficient or abnormal hemoglobin, or competition for the hemoglobin molecule by carbon monoxide

angina pectoris: chest pain associated with inadequate blood flow to the heart muscle

angiotensin I: a small polypeptide (10 amino acid residues) hormone generated in the blood by the action of renin on a plasma protein, angiotensinogen

angiotensin II: an octapeptide formed by the enzymatic removal of two amino acids from angiotensin I; stimulates the secretion of aldosterone from the adrenal cortex and the contraction of vascular smooth muscle

angiotensin III: a heptapeptide formed by the enzymatic removal of an amino acid from angiotensin II

angiotensinogen: a plasma protein synthesized in the liver; the precursor of the hormone angiotensin

angiotensins: angiotensin I, II, and III collectively

"angry" macrophages: macrophages whose activity has been facilitated by chemicals released from T cells

anion: a negatively charged ion, e.g., Cl^-, SO_4^{2-}

anoxia: the absence of oxygen

antagonist: a muscle whose action opposes the intended movement; e.g., the finger flexors are antagonists to the finger extensors

anterior pituitary: the anterior portion of the pituitary gland which synthesizes, stores, and releases ACTH, GH, TSH, prolactin, FSH, and LH

antibody: a specialized protein secreted by plasma cells and capable of combining with the specific antigen that stimulated its production

antibody-mediated immunity: *see* humoral immunity

anticodon: the three-nucleotide sequence in tRNA that is exposed and able to base-pair with a complementary codon in mRNA during protein synthesis

antidiuretic hormone (ADH): a peptide hormone synthesized in the hypothalamus and released from the posterior pituitary; also called vasopressin; it increases the permeability of the kidney collecting ducts to water

antigen: a foreign protein or protein-polysaccharide complex that stimulates a specific immune response against itself when introduced into the body

antihistamine: a chemical that blocks the action of histamine; e.g., it relieves the airway constriction and increased mucus secretion caused by histamine release during allergic attacks

antrum: (gastric) the lower portion of the stomach, i.e., the region closest to the pyloric sphincter; (ovarian) the fluid-filled cavity in the maturing ovarian follicles

aorta: the largest artery in the body; carries blood from the left ventricle to the upper portion of the thoracic cage where it turns and descends to the lower portion of the abdomen

aortic body: *see* aortic chemoreceptor

aortic chemoreceptor: chemoreceptor located near the arch of the aorta; sensitive to oxygen, carbon dioxide, and hydrogen-ion concentrations

aortic valve: the valve between the left ventricle of the heart and the aorta

aphasia: a specific language deficit not due to mental retardation; deficiency can be in understanding or formulating oral or written language

apnea: cessation of respiration

apneic: without breathing

appendix: appendix vermiformis; a small, finger-like projection from the cecum in the large intestine; has no known digestive function in human beings

arrhythmia: any variation from the normal heartbeat rhythm

arterial baroreceptors: nerve endings sensitive to stretch or distortion of the blood vessels (carotid sinus and arch of the aorta) in whose walls they lie

arteriolar resistance: the resistance offered by the arterioles to blood flow from arteries to capillaries

arteriole: a narrow vessel, surrounded by several layers of smooth muscle, that regulates the flow of blood from an artery into a capillary network; the primary site at which vascular resistance is regulated

artery: a thick-walled, elastic vessel that carries blood away from the heart

artificial kidney: an apparatus that eliminates by

dialysis the excess ions and wastes which accumulate in the blood when the kidneys fail

ascending limb: (of Henle's loop) the portion of the renal nephron leading to the distal convoluted tubule

ascorbic acid: vitamin C

aspartic acid: an amino acid; a possible neurotransmitter at some excitatory synapses

asthma: a disease characterized by severe airway smooth muscle constriction and by plugging of the airways with mucus

astigmatism: a defect in vision that occurs because of irregularities in the shape of the cornea

atherosclerosis: a disease characterized by a thickening of the arterial wall with abnormal smooth muscle cells, deposits of cholesterol, and connective tissue resulting in narrowing of the vessel lumen

athetosis: unwanted, purposeless, uncontrolled movements taking the form of writhing, continuous swings between two positions, e.g., hyperextension of the fingers with the palm facing upward alternating with clenching the fingers with the palm downward

atmospheric pressure: the air pressure of the external environment (760 mmHg at sea level); barometric pressure

atom: the smallest unit of matter that has chemical characteristics; consists of a complex arrangement of electrons moving around a positively charged nucleus, but the numbers of positive and negative charges are equal so an atom has no net charge; atoms combine to form molecules

atomic nucleus: the dense region at the center of an atom consisting of positively charged protons and uncharged neutrons; contains virtually all the mass of the atom

atomic number: the distinguishing characteristic of specific atoms; equal to the number of protons in the nucleus of the atom; e.g., calcium atoms, which have 20 protons, have the atomic number 20

atomic weight: a relative value that indicates how much heavier a given atom is than a hydrogen atom; approximately equal to the number of protons and neutrons, each of which is assigned the weight "1"

atopic allergy: the responses to IgE antibodies (and the histamine and other vasoactive chemicals released by them) induced by an antigen

ATP: *see* adenosine triphosphate

atrioventricular (AV) node: a region, at the base of the right atrium near the interventricular septum, containing specialized cardiac muscle cells through which electrical activity must pass to reach the ventricles from the atria

atrioventricular (AV) valve: a valve between an atrium and a ventricle of the heart; the AV valve on the right side of the heart is the tricuspid valve, and that on the left is the mitral valve

atrium: a chamber of the heart that receives blood from the veins and passes it on to the ventricle on the same side of the heart

auditory: pertaining to the sense of hearing

auditory cortex: the region of cerebral cortex which receives the nerve fibers from the auditory pathways

auditory tube: the tube that connects the middle-ear cavity to the pharynx; eustachian tube

autoimmune disease: a disease produced by antibodies or T cells sensitized against the body's own cells, which results in damage or alteration of cell function

automaticity: capable of self-excitation

autonomic nervous system: a component of the efferent division of the peripheral nervous system; innervates cardiac muscle, smooth muscle, and glands; consists of sympathetic and parasympathetic subdivisions

autoregulation: (of blood flow) the ability of individual organs to alter their vascular resistance and, thereby, to control (self-regulate) their blood flows independent of external neural and hormonal influence

AV node: *see* atrioventricular node

AV valve: *see* atrioventricular valve

axon: (also nerve fiber) a single process extending from the cell body of a neuron; propogates action potentials away from cell body

axon collateral: a branch of an axon leading to a set of axon terminals that differ in location from the main axon terminals

axon terminals: the network of fine branches at the end of the axon, each branch ending at a synaptic or neuroeffector junction

B cells: lymphocytes capable of becoming antibody-secreting plasma cells

B lymphocytes: *see* B cells

bacteria: unicellular organisms which have an outer coating, the cell wall, in addition to a cell membrane; are self-contained complete cells in that they have all the machinery required to sustain life and to reproduce themselves

balance concept: the statement that if the quantity of any substance in the body is to be maintained at a constant level over a period of time, the total amounts ingested and produced must equal the total amounts excreted and consumed

ballismus: sudden, violent, purposeless movements of the distal part of the arms or legs

barometric pressure: *see* atmospheric pressure

baroreceptor: a receptor sensitive to pressure and rate of change in pressure; e.g., the arterial baroreceptors respond to mean arterial pressure and pulse pressure

basal: resting level

basal ganglia: several nuclei in the cerebral hemispheres which code and relay information associated with the control of muscle movements; specifically, the caudate nucleus, globus pallidus, and putamen

basal metabolic rate (BMR): a measurement by a clinical test in which the metabolic rate is determined under standardized conditions, i.e., the subject is at mental and physical rest (but not sleeping), at comfortable temperature, and fasting for at least 12 hours, and compared with values for a normal person of the same weight, height, age, and sex; the metabolic cost of living

base: any molecule that can combine with a hydrogen ion; *see also* purine bases, pyrimidine bases

basement membrane: a thin proteinaceous layer of extracellular material associated with the plasma membranes of many cells, especially epithelial cells where it is located on the basal side of the cell

basic electrical rhythm: the spontaneous depolarization-repolarization cycles of pacemaker cells, in the longitudinal smooth muscle layer of the stomach and intestines, which coordinate the muscular activity of the GI tract

basilar membrane: the membrane in the inner ear separating the cochlear duct and scala tympani; supports the organ of Corti

basophil: a polymorphonuclear granulocytic leukocyte whose granules stain with basic dyes

BCG: an attenuated strain of a certain bacterium, the injection of which activates the entire T-cell system so that defenses against any cancer cell might be enhanced

best frequency: the sound frequency that causes the greatest response in neurons in the auditory pathway, different neurons having different best frequencies

beta-adrenergic receptor: a type of receptor for norepinephrine and epinephrine that responds to certain drugs and can be distinguished from an alpha-adrenergic receptor

biceps: muscle that causes flexion of the elbow

bile: a yellow-green fluid containing bile salts, cholesterol, lecithin (a phospholipid), bile pigments and other end products of organic metabolism, and certain trace metals; secreted by the liver

bile canaliculi: small ducts adjacent to the liver cells; receive bile secreted by these cells

bile pigment: a pigmented substance, derived from the breakdown of hemoglobin, secreted in bile; gives bile its color

bile salts: a group of steroid molecules secreted by the liver and passed via bile ducts into the intestinal lumen where they promote the solubilization and digestion of fat

bilirubin: a yellow substance resulting from the breakdown of heme; excreted in bile as a bile pigment

binding site: a region on a protein that has chemical groups which are able to interact with a specific ligand, binding it to the surface of the protein

"biological clock": cells which function on a rhythmic basis in the absence of apparent external stimulation

biotin: one of the B-complex vitamins

biotransformation: alteration of foreign molecules by an organism's metabolic pathways

bipolar affective disorder: an affective disorder characterized by swings in mood between mania and depression; formerly called manic depression

bipolar cells: the neurons in the retina that are postsynaptic to the rods and cones

blackout: a period of temporary amnesia

blastocyst: a ball of developing cells surrounding a central cavity; stage of embryonic development

blood-brain barrier: a complex group of anatomical barriers and physiological transport systems which closely controls the kinds of substances that enter the extracellular space of the brain from the blood and the rate at which they enter

blood-testis barrier: a barrier, formed by Sertoli cells and the tight junctions between them, which limits the movements of chemicals between the blood and the lumen of the seminiferous tubules

blood type: classification of blood determined by the presence of antigens of either A, B, O, or AB types on the plasma membranes of erythrocytes and by the presence in the plasma of the anti-A or anti-B antibodies (or both or neither); mixing bloods of different types may result in the agglutination of the erythrocytes

blue-sensitive cones: cones containing photopigment most sensitive to light of 445-nm wavelengths

BMR: *see* basal metabolic rate

bolus: a lump of food that has been chewed and mixed with saliva

bone marrow: a highly vascular, cellular substance in the central cavity of some bones; site of the synthesis of erythrocytes, some types of leukocytes, and platelets

botulinus toxin: a chemical which blocks the release of acetylcholine from motor neuron terminals, thereby preventing the excitation of the muscle membrane; produced by the bacterium *Clostridium botulinum*

Bowman's capsule: the blind sac at the beginning of the tubular component of a kidney nephron; closely associated with the glomerulus of the nephron

brain death: the state, determined by established criteria, in which the brain no longer functions and has no possibility of functioning again; widely accepted as the criterion for death

brainstem: a subdivision of the brain consisting of medulla, pons, and midbrain and located between the spinal cord and the cerebrum

bronchiole: a small division of a bronchus (lung airway)

bronchus: the branch of the air passage that enters the substance of the lung (the main bronchus); a second- or third-order branch of the main bronchus (secondary or tertiary bronchi)

buffering: the reversible binding of hydrogen ions by various compounds, e.g., bicarbonate; excess hydrogen ions are bound to the buffer compounds thereby decreasing the hydrogen-ion concentration, whereas a low hydrogen-ion concentration results in a release of hydrogen ions from the buffer compounds; these reversible reactions tend to maintain the acidity of a buffered solution nearly constant

bulk flow: movement of fluids or gasses down a pressure gradient

Ca^{2+}: calcium ion

calcitonin: a polypeptide hormone secreted by nonfollicular cells of the thyroid; involved in calcium regulation

calorie (cal): the unit in which heat energy is measured; the amount of heat needed to raise the temperature of 1 g of water 1°C; one large calorie, used to represent the caloric equivalent of various nutrients, equals 1000 calories

calorimeter: an instrument which measures heat changes

cAMP: *see* cyclic AMP

cancer: an uncontrolled growth of cells

capacitation: final maturation of the sperm

capillary: smallest of the blood vessels; water, nutrients, dissolved gases, and other molecules are able to diffuse across capillary walls; composed of a single layer of endothelial cells

carbamino compound: (e.g., $HbCO_2$) a compound resulting from the combination of carbon dioxide with protein amino groups, particularly hemoglobin

carbohydrate: a substance composed of atoms of carbon, hydrogen, and oxygen according to the general formula $C_n(H_2O)_n$, where n is any whole number, e.g., glucose, $C_6H_{12}O_6$

carbon monoxide: CO; a gas which reacts with the same iron binding sites on hemoglobin as does oxygen but with a much greater affinity than oxygen

carbonic acid: H_2CO_3; an acid formed from H_2O and CO_2

carbonic anhydrase: the enzyme that catalyzes the reversible reaction by which CO_2 and H_2O combine to form carbonic acid (H_2CO_3)

carboxyl group: the -COOH grouping in a molecule

carboxypeptidases: two enzymes secreted by the exocrine pancreas as precursors, the procarboxypeptidases; break the peptide bond of the amino acid located at the end of a protein having a free carboxyl group

cardiac: pertaining to the heart

cardiac cycle: one contraction-relaxation episode of the heart

cardiac output: the volume of blood pumped by each ventricle per minute (not the total output pumped by both ventricles)

cardiovascular center: a group of neurons in the brainstem medulla which serves as a major integrating center for reflexes affecting heart rate, cardiac contractility, vascular resistance, and vascular capacity in the control of blood pressure

carrier: a protein binding site in a membrane capable of combining with a transported molecule and enabling it to pass through the membrane

carotid: pertaining to the two major arteries (the carotid arteries) in the neck which convey blood to the head

carotid bifurcation: the point at which the common carotid artery divides to form the internal and external carotid arteries

carotid body: *see* carotid chemoreceptors

carotid chemoreceptors: chemoreceptors located near the carotid bifurcation; sensitive to oxygen, carbon dioxide, and hydrogen-ion concentrations

carotid sinus: a dilatation of the internal carotid artery just above the carotid bifurcation; location of the carotid baroreceptors

castration: removal of the gonads, i.e., testes or ovaries

catabolic reaction: a degradative chemical reaction which results in the fragmentation of a molecule into smaller and smaller parts

catalyst: a substance that accelerates chemical reactions but does not itself undergo any net chemical change during the reaction

cataract: an opaque area of the lens of the eye

catatonia: abnormal behavior characterized by alternate periods of stupor, impulsive or stereotyped activity, and muscular rigidity

catechol-O-methyl transferase: an enzyme that degrades the catecholamine neurotransmitters

catecholamine: a neurotransmitter, either dopamine, epinephrine, or norepinephrine, containing a catechol ring (a six-sided carbon ring with two adjacent hydroxyl groups) and an amine (NH_2) group

CCK: *see* cholecystokinin

cecum: the dilated pouch at the beginning of the large intestine into which open the ileum, colon, and appendix

cell: the basic structural and functional unit of living things, i.e., the smallest structural unit into which an organism can be divided and still retain the characteristics we associate with life

cell division: the process by which a single cell forms two daughter cells; consists of nuclear division with replication of chromosomes (mitosis) followed by division of the entire cell

cell-mediated immunity: a type of specific immune response mediated by T lymphocytes

cell membrane: *see* membrane

cellulose: a straight-chain polysaccharide composed of thousands of glucose subunits; found in plant cells

center of gravity: the point in a body at which the mass of the body is in perfect balance; if the body were suspended from a string attached to this point, there would be no movement

central chemoreceptors: receptors in the brainstem medulla which respond to changes in the H^+ and CO_2 concentrations of the cerebrospinal and brain extracellular fluid

central dogma: (of molecular biology) the doctrine that in all organisms genetic information flows from DNA to RNA and then to proteins

central nervous system (CNS): brain plus spinal cord

central thermoreceptors: temperature receptors in the hypothalamus, spinal cord, abdominal organs, and other internal locations

centrioles: two small bodies composed of nine fused sets of microtubules located in the cell cytoplasm; participate in nuclear and cell division

centromere: the point in a chromosome at which the two chromatin threads are attached to each other

cephalic phase: (of gastrointestinal control) initiation of the neural and hormonal reflexes by stimulation of receptors in the head (cephalic)—sight, smell, taste, and chewing

cerebellum: a subdivision of the brain that lies behind the forebrain and above the brainstem and deals with control of muscle movements

cerebral: pertaining to the brain

cerebral cortex: the 3-mm-thick, cellular layer covering the cerebrum

cerebrospinal fluid (CSF): the clear fluid that fills the cerebral ventricles and the subarachnoid space surrounding the brain and spinal cord

cervix: the lower portion of the uterus; connects the lumen of the uterus with that of the vagina

CG: *see* chorionic gonadotropin

cGMP: *see* cyclic GMP

channel: (membrane) a small passage through a membrane through which small-diameter molecules and ions can diffuse; a membrane pore

chemical bonds: the interactions between the electrons of adjacent atoms which hold the atoms together in a molecule

chemical element: a given type of atom; e.g., carbon atoms form one type of chemical element, oxygen atoms another; there are over 100 different chemical elements, each characterized by a distinct number of subatomic particles

chemical equilibrium: a situation in which the rates of the forward and reverse reactions of a chemical reaction are equal, and thus no net change in the concentrations of the reactants or products occurs

chemical formula: a term listing the numbers and kinds of atoms in a given molecule; e.g., H_2O is the chemical formula for water, the molecules of which contain two hydrogen atoms and one oxygen atom

chemical specificity: only certain ligands (molecules, ions) can bind with a given type of binding site; the fewer the kinds of ligands capable of binding to that site, the greater its specificity; determined by the shape of the ligand and binding site

chemical synapse: a synapse at which chemical mediators (neurotransmitters) released in response to action potentials in one neuron diffuse across an extracellular gap to influence the activity of the second neuron

chemoaffinity theory: the suggestion that developing neural circuits form the proper connections because of selective attractions, determined by individual chemical codes, between the neurons

chemoreceptors: afferent nerve endings or cells associated with them that are sensitive to concentrations of certain chemicals

chemotaxis: the orientation and movement of cells in a specific direction in response to a chemical stimulus

chief cells: gastric gland cells which secrete pepsinogen, precursor of the enzyme pepsin

cholecystokinin (CCK): a polypeptide hormone secreted by cells in the upper small intestine; regulates several gastrointestinal activities including motility and secretion of the stomach, gall bladder contraction, and enzyme secretion of the pancreas; formerly called pancreozymin

cholesterol: a steroid molecule; precursor of the steroid hormones and bile acids

cholinergic: pertaining to a nerve fiber that releases acetylcholine; a compound that acts like acetylcholine

cholinesterase: *see* acetylcholinesterase

chorea: relatively mild, sudden, purposeless motions of the distal part of the arms or legs

chorionic gonadotropin (CG): a protein hormone secreted by the trophoblastic cells of the blastocyst and placenta; maintains the secretory activity of the corpus luteum during the first three months of pregnancy

chorionic somatomammotropin: a hormone secreted by the placenta with effects (in the mother) similar to those of growth hormone and prolactin

choroid plexus: a highly vascular, epithelial structure lining portions of the cerebral ventricles; responsible for the formation of much of the cerebrospinal fluid

chromatin: a substance in the nucleus consisting of fibrous threads of DNA, nuclear proteins, and small amounts of RNA

chromatin protein: *see* nuclear protein

chromophore: the retinal component of a photopigment molecule

chromosome: a highly coiled, condensed form of chromatin that is formed in the cell nucleus during the process of mitosis and meiosis

chyle: the fatty material taken up by the central lacteals in the villi of the duodenum after a fatty meal

chylomicron: a small (0.1- to 3.5-μm diameter) lipid droplet, consisting of triacylglycerol, phospholipid, cholesterol, free fatty acids, and protein, that is released from the intestinal epithelial cells and enters the lacteals during fat absorption

chyme: the solution of partially digested food in the lumen of the stomach and intestines

chymotrypsins: a family of enzymes secreted by the exocrine pancreas as precursors, the chymotrypsinogens; break certain peptide bonds in proteins and polypeptides

cilia: hairlike projections from the surface of specialized epithelial cells; sweep back and forth in a synchronized way to propel material along the cell surface

circadian: daily; approximately once every 24 hours

circular muscle layer: the layer of smooth muscle surrounding the submucosal layer in the stomach and intestinal walls, and whose muscle fibers are oriented circumferentially around these organs

circulatory shock: a condition resulting from a decrease in cardiac output severe enough to impair seriously the blood supply to various organs and tissues

"circus" movement: the abnormal traveling of waves of cardiac excitation via circular routes through cardiac muscle resulting in fibrillation

11-*cis*-retinal: the form of retinal in the receptor cells of the eye that is bound to opsin and responds to light

citric acid: a six-carbon molecule which is an intermediate in the Krebs cycle

citric acid cycle: *see* Krebs cycle

clearance: the volume of plasma from which a particular substance has been removed

clitoris: a small body of erectile tissue in the female external genitalia; homologous to the penis

clone: a group of genetically identical objects

clot retraction: the tightening up or contraction of a fibrin meshwork which squeezes serum out of a blood clot and strengthens it

CNS: *see* central nervous system

CO$_2$: carbon dioxide

coactivation: (of alpha and gamma motor neurons) almost simultaneous stimulation of alpha and gamma motor neurons which maintains tension on the spindle stretch receptors during skeletal muscle fiber shortening

cobalamine: vitamine B$_{12}$; cyanocobalamine

cochlea: the inner ear; the fluid-filled, spiral-shaped compartment that contains the organ of Corti

cochlear duct: a fluid-filled, membranous bag that extends almost the entire length of the inner ear, dividing it into two compartments, the scala vestibuli and the scala tympani; contains the organ of Corti

code word: the three-nucleotide sequence in DNA which signifies a given amino acid

codon: the sequence of three nucleotide bases in mRNA that corresponds to a given code word in DNA

coenzyme: nonprotein organic molecule which functions as a cofactor; many coenzymes are derived from vitamins; generally serves as a carrier molecule which transfers atoms (such as hydrogen) or small molecular fragments from one reaction to another; is not consumed in the reaction in which it participates and can be reused until it is degraded

coenzyme A: a derivative of the B-vitamin pantothenic acid; the coenzyme that transfers acetyl groups from one reaction to another

cofactor: nonprotein substance (e.g., metal ions or coenzymes) which binds to specific regions of an enzyme and thereby either maintains the shape of the active site or participates directly in the binding of substrate to the active site; is not consumed in the reaction in which it participates and can be reused until it is degraded

collagen: a fibrous protein which has great strength but no active contractile properties; functions as a structural element in the interstitium of various types of connective tissue, e.g., tendons and ligaments

collateral ventilation: movement of air between alveoli through pores in the wall separating the alveoli

collecting duct: a portion of the tubular component of a kidney nephron; ducts from several nephrons combine situated between the distal convoluted tubule and the renal pelvis

colloid: a large molecule to which the capillaries are relatively impermeable, e.g., a plasma protein

colon: the large intestine, specifically that part extending from the cecum to the rectum

colony: (of neurons) a group of cerebral-cortex neurons with related function lying in a column perpendicular to the cortical surface

color blindness: a defect in color vision due to the absence of one or even two of the three cone photopigments or to a defect in color-processing neurons in the visual pathway

colostrum: the thin, milky fluid secreted by the mammary glands for a few days after birth

coma: a deep, prolonged unconsciousness not fulfilling all the criteria for brain death, e.g., there may be spontaneous movement and some intact reflexes

"command" neurons: neurons or groups of neurons whose activity initiates the series of neural events resulting in a voluntary action; the identity of these neurons is unknown

commissure: a large bundle of myelinated nerve

fibers that connects the two cerebral hemispheres

compensatory growth: a type of regeneration present in many organs after tissue damage

complement system: a group of plasma proteins that normally circulate in the blood in an inactive state but, upon activation of the first protein, take part in a "cascade" of reactions, the active proteins generated during the cascade mediating certain events of the immune response

complex cell: a visual cortex neuron that responds to lines at particular orientations

compliance: *see* lung compliance

concentration: the amount of material present in a unit volume; expressed as mass/volume, e.g., g/L or mmols/L

concentration gradient: a graded concentration between two regions, e.g., across a cell membrane or epithelial layer of cells

conducting portion: (of the respiratory airways) those air passageways having walls too thick to allow the exchange of gas between blood and air

conducting system: (of the heart) a network of cardiac muscle fibers specialized to conduct electrical activity to different areas of the heart

conductor: material having a low resistance to current flow

cone: one of the two receptor types for photic energy in the retina

congestive heart failure: a set of signs and symptoms associated with a decreased contractility of the heart and engorgement of the heart, veins, and capillaries with blood

conscious experience: the things of which a person is aware; the thoughts, feelings, perceptions, ideas, and reasoning during any of the states of consciousness

conscious movements: *see* voluntary actions

consciousness: *see* conscious experience, state of consciousness

constipation: infrequent or difficult defecation

contractility: the component of contractile force that depends directly on cross-bridge activity rather than on the extent of filament overlap (fiber length)

contraction: the active process of generating force in a muscle

contraction time: the time interval between the stimulus and the development of peak tension by a muscle

control system: a collection of interconnected components which functions to keep a physical or chemical parameter within a predetermined range of values

convection: the process by which air (or water) next to a warm body is heated, moves away, and is replaced by colder air, which, in turn, follows the same cycle

convergence: synapses from many presynaptic neurons acting upon a single postsynaptic neuron; the turning of the eyes inward, i.e., toward the nose, to view near objects

core temperature: the temperature of the inner organs as opposed to that of the skin

cornea: the transparent structure covering of the front of the eye; forms part of the optical system of the eye and helps focus an image of the object on the retina

corneal reflex: closure of the eyelid in response to irritation of the cornea

cornification: layering of flat (squamous) epithelial cells

coronary artery: an artery branching from the aorta to carry blood to the heart muscle

corpus callosum: a large, wide band of myelinated nerve fibers that connects the two cerebral hemispheres; one of the brain commissures

corpus luteum: an ovarian structure formed from the follicle after ovulation; secretes estrogen and progesterone

cortex: *see* adrenal cortex, cerebral cortex

cortical association areas: regions of the parietal and frontal lobes of cerebral cortex which receive input from the different sensory modalities, memory stores, etc., and perform further perceptual processing

corticospinal pathway: a descending motor pathway having its nerve cell bodies of origin in the cerebral cortex; the axons pass without synapsing to the region of the motor neurons; also called the pyramidal tract

corticotropin: *see* adrenocorticotropic hormone

corticotropin releasing hormone (CRH): the hypothalamic hormone that stimulates ACTH (corticotropin) secretion by the anterior pituitary

cortisol: a glucocorticoid steroid hormone secreted by the adrenal cortex; regulates various aspects of metabolism; has many other actions

countercurrent: flow in opposing directions through the two limbs of a hairpin loop

countercurrent multiplier system: the mechanism

associated with the loops of Henle which creates in the medulla of the kidney a region having a high interstitial-fluid osmolarity

covalent bond: a chemical bond between two atoms in a molecule in which each atom shares one of its electrons with the other; e.g., 0:0 indicates that the two oxygen atoms of an oxygen molecule interact via a covalent bond

CP: *see* creatine phosphate

cranial nerve: one of the 24 (12 pairs) peripheral nerves that joins the brainstem or forebrain

craniosacral division: (of the autonomic nervous system) *see* parasympathetic nervous system

creatine phosphate: a molecule that transfers phosphate and energy to ADP to generate ATP

creatinine: a waste product derived from muscle creatine

cretin: a mentally retarded person whose defect is due to insufficient thyroid hormone during the critical periods of brain development

CRH: *see* corticotropin releasing hormone

critical period: a time during the developmental history when a system is most readily influenced by both favorable and adverse environmental factors

cross bridge: a projection extending from a thick filament in muscle; a portion of the myosin molecule capable of exerting force on the thin filament and causing it to slide past the thick one

crossed-extensor reflex: increased activation of the extensor muscles on the side opposite the limb flexion

crossing over: a process in which segments of maternal and paternal chromosomes exchange with each other during chromosomal pairing before the first meiotic division

crystalloid: a low-molecular-weight solute, e.g., Na^+, glucose, or urea

CSF: *see* cerebrospinal fluid

cumulus: the granulosa cells surrounding the ovum as it projects into the antrum of the ovarian follicle

curare: a drug which binds strongly to acetylcholine receptors at neuromuscular junctions preventing their activation by acetylcholine and thereby causing paralysis

current: the movement of electric charge; in biological systems, this is achieved by the movement of ions

cutaneous: pertaining to the skin

cyanide: a poison that works by reacting with the final cytochrome in the cytochrome system associated with oxidative phosphorylation, thereby preventing electron transfer to oxygen and the production of ATP

cyclic AMP (cAMP): cyclic 3′,5′-adenosine monophosphate; a cyclic nucleotide which serves as second messenger for many nonsteroid hormones, neurotransmitters, and paracrine agents

cyclic GMP (cGMP): cyclic 3′,5′-guanosine monophosphate; a cyclic nucleotide which acts as second messenger in some cells, possibly in opposition to cyclic AMP

cytochrome: one of a group of iron-containing proteins which transfers electrons from one molecule to another during oxidative phosphorylation

cytochrome system: a chain of enzymes which transfers hydrogen from coenzyme hydrogen carriers, such as NADH, to oxygen, forming water; the enzymes are located in the inner membranes of mitochondria and are associated with oxidative phosphorylation

cytolytic T cells: a type of T cell which upon activation kills fungi, intracellular viruses, parasites, the few bacteria that can survive only inside cells, cancer cells, and solid-tissue transplants

cytoplasm: the region of a cell located outside of the nucleus

cytosine: one of the pyrimidine bases present in DNA and RNA

cytosol: the watery medium inside cells but located outside of cell organelles

dark adaptation: the improvement in the sensitivity of vision after one is in the dark for some time

daytime vision: the vision possible only under conditions of high illumination; color vision

deamination: the removal of an amino ($—NH_2$) group from a molecule, such as an amino acid

decremental: decreasing in amplitude

defecation: expulsion of feces from the rectum

degradative reaction: *see* catabolic reaction

degranulation: the fusion of the phagosome and lysosome membranes; the lysosomes, which appear as microscopic "granules," disappear

delusion: illogical thought that is false or improbable but cannot be changed by argument or evidence to the contrary

dendrite: a highly branched process of a neuron

which receives synaptic inputs from other neurons

denervation atrophy: a decrease in the size of muscle fibers whose nerve supply is destroyed

dense body: a cytoplasmic structure to which actin filaments are anchored in smooth muscle cells

deoxyhemoglobin: *see* reduced hemoglobin

deoxyribonucleic acid (DNA): the nucleic acid that stores and transmits genetic information; consists of a double strand (double helix) of nucleotide subunits; the sequence of nucleotides in DNA (three nucleotides specifying one amino acid) determines the sequence of amino acids in the proteins synthesized by a cell

depolarize: to change the value of the membrane potential toward zero so the inside of the cell becomes less negative

depression: (as an affective disorder) a serious, prolonged disturbance in mood characterized by a pervasive sadness or loss of interest or pleasure

DES: *see* diethylstilbestrol

descending limb: (of Henle's loop) the segment of a renal nephron into which the proximal tubule drains

desensitization: the frequent injection of an antigen to which a person is hypersensitive into that person in order to induce a state of reduced responsiveness to that antigen

desmosome: a type of cell junction whose function is to hold cells together; consists of two opposed plasma membranes that remain separated by a 20-nm extracellular space; fibers extend into the cytoplasm from the inner surface of the desmosome and appear to be linked to other desmosomes of the cell

diabetes insipidus: a disease resulting from defective ADH control of urine concentration; marked by great thirst and the excretion of a large volume of urine

diabetes mellitus: a disease in which the hormonal control of plasma glucose is defective because of an absolute or relative deficiency of insulin

diaphragm: the dome-shaped sheet of skeletal muscle that separates the thoracic and abdominal cavities; the principle muscle of respiration

diarrhea: liquid feces

diastole: the period of the cardiac cycle when the ventricles are not contracting

diastolic pressure: the minimum blood pressure during the cardiac cycle, i.e., the pressure just prior to ventricular ejection

diethylstilbestrol (DES): an estrogenlike compound which causes increased contractions of the female genital tract and may, therefore, cause expulsion of a fertilized ovum

differentiation: the process by which cells undergo alteration and acquire specialized structural and functional properties during growth and development

diffuse thalamic projection system: fibers of thalamic neurons which synapse in the cortex and are concerned with maintaining the behavioral characteristics of the awake state and with direction of attention to selected events

diffusion: the movement of molecules from one location to another because of random thermal molecular motion; net diffusion always occurs from a region of high concentration to a region of lower concentration

diffusion coefficient (k_D): a proportionality constant determining the rate at which a substance diffuses through a solution; its magnitude depends on molecular weight, temperature, and the medium through which the molecule is diffusing

diffusion equilibrium: the state during which the diffusion fluxes in opposite directions are equal (i.e., the net flux = 0 and there is no concentration gradient)

diffusion potential: a voltage created by the separation of charge that results from the diffusion of charged particles in a solution

digestion: the process of breaking down large particles and high-molecular-weight substances into small molecules

dihydrotestosterone: a steroid formed by the enzyme-mediated alteration of testosterone; the active form of testosterone in certain of its target cells

2,3-diphosphoglycerate (DPG): a substance produced by erythrocytes during glycolysis; binds reversibly with hemoglobin, causing it to change conformation and release oxygen

disaccharidase: one of a family of enzymes located in the plasma membrane of the intestinal epithelium and capable of splitting disaccharides into monosaccharides

disaccharide: a carbohydrate molecule composed of two monosaccharides; e.g., the disaccharide sucrose (table sugar) is composed of glucose and fructose

dishabituation: restoration of an orienting response when a novel stimulus (or the same stimulus at a different intensity) replaces a stimulus to which an animal has become habituated

disinhibition: the removal of inhibition from a neuron, thereby allowing its activity to increase

distal convoluted tubule: a portion of the tubular component of a kidney nephron; situated between Henle's loop and the collecting duct

disulfide bond: R-S-S-R, where R is the remaining portion of the molecule

disuse atrophy: a decrease in the size of muscle fibers which are not used for a long period of time

diuretic: any substance that causes an increase in volume of urine excreted

diurnal: daily; occurring in a 24-hour cycle

divergence: (neuronal) the process in which a single presynaptic neuron synapses upon and thereby influences the activity of many postsynaptic neurons; (eyes) the movement in which the eyes are turned outward to view distant objects

DNA: *see* deoxyribonucleic acid

DNA-dependent-RNA-polymerase: *see* transcriptase

DNA polymerase: the enzyme which during replication joins together nucleotides base paired with the nucleotides of one strand of DNA to form a new double-stranded molecule of DNA

dominant hemisphere: the cerebral hemisphere that controls the hand used most frequently for intricate tasks, e.g., the left hemisphere in a right-handed person

dopa: dihydroxyphenylalanine; an intermediary in the catecholamine synthetic pathway

dopamine: a catecholamine neurotransmitter; a precursor of epinephrine and norepinephrine

dopamine hypothesis: the suggestion that schizophrenia is due to excessive activity of dopamine-mediated synapses, especially in the limbic system

dorsal root: a group of afferent nerve fibers that enters the left and right sides of the region of the spinal cord facing toward the back of the body

double bond: two chemical bonds formed between the same two atoms; symbolized by $=$, e.g., the carbon dioxide molecule (CO_2) has two double bonds, $O=C=O$

DPG: *see* 2,3-diphosphoglycerate

dual innervation: innervation of an organ or gland by both sympathetic and parasympathetic nerve fibers

duodenum: first portion of the small intestine; that portion of the intestine between the stomach and jejunum

ear canal: the passageway that leads from the outside of the skull inward to end at the tympanic membrane

eardrum: *see* tympanic membrane

ECG: *see* electrocardiogram

ectopic focus: a region of the heart (other than the SA node) that assumes the role of cardiac pacemaker

edema: accumulation of excess fluid in the interstitial space

EDV: *see* end-diastolic volume

EEG: *see* electroencephalogram

EEG arousal: the transformation of the EEG pattern from alpha rhythm to a lower, faster oscillation; associated with increased levels of attention

effector: a cell or cell collection which changes its activity in response to a neural or hormonal signal; specifically, muscle and gland cells, but technically all cells, are effectors

efferent arteriole: a vessel which conveys blood from a renal glomerulus to a set of peritubular capillaries

efferent neuron: a neuron which carries information away from the central nervous system to muscle cells, gland cells, or postganglionic neurons

efferent pathway: that component of a reflex arc that transmits information from the integrating center to the effector; any pathway conveying information out of the central nervous system (or away from the brain within the central nervous system)

ejaculation: a reflex consisting of contraction of the genital ducts and skeletal muscle at the base of the penis with resulting expulsion of semen from the penis

ejaculatory duct: the continuation of the vas deferens after it receives the duct from the seminal vesicle; joins the urethra within the prostate gland

EKG: *see* electrocardiogram

electric charge: one of the fundamental units of

measurement; labeled positive or negative depending upon its interaction with other, known electric charges

electric force: that force which causes the movement of charged particles toward regions having an opposite charge and away from regions having a similar charge

electric potential: (or electric potential difference) *see* potential

electrical synapse: a synapse at which local currents resulting from action potentials in one neuron flow through gap junctions, joining two neurons, to influence the activity of the second neuron

electrocardiogram (ECG, EKG): a recording of the electrical currents generated by action potentials of cardiac muscle cells

electrocerebral silence: a flat EEG, i.e., without electrical indication of brain activity

electrochemical gradient: the force determining the magnitude of the net diffusion of charged particles; a combination of the electrical gradient (as determined by the voltage difference between two points) and the chemical gradient (as determined by the concentration difference between the same two points)

electroconvulsive shock: an electrically or chemically induced shock to the brain resulting in a seizure

electrode: a probe to which electric charges can be added (or from which they can be removed) to cause changes in the electric current flowing to a recorder which usually transforms this input into terms of voltage

electroencephalogram (EEG): a recording of the electric potential differences between two points on the scalp (or between one point on the scalp and a distant electrode)

electrogenic pump: an active transport system which directly separates electric charge thereby producing a potential difference

electrolyte: mineral ions, e.g., Na^+ and K^+

electromagnetic radiation: radiation comprised of waves with electric and magnetic components that travel through matter and vacuum; generally caused by the motion of charged particles and includes gamma rays, x-rays, ultraviolet, visible, and infrared light, and television and radio waves

electron: a subatomic particle that carries one negative charge and revolves in an orbit around the nucleus of the atom; may be shared between two atoms in the formation of a chemical bond or captured by another atom in the formation of ions

embryo: a developing organism during the early stages of development; in human beings, the first two months of intrauterine life

emission: movement of the contents of the male genital ducts into the urethra

emotion: subjective feelings such as fear, joy, anxiety, amusement, etc.

emphysema: a disease in which the internal structures of the lung become more stretchable, resulting in a markedly increased resting-lung size but a decreased surface area for gas exchange with the blood

emulsification: a fat-solubilizing process in which large lipid droplets are broken into smaller droplets

emulsion: a suspension of small (1-μm diameter) lipid droplets in an aqueous medium

end-diastolic volume (EDV): the amount of blood in the ventricle of the heart just prior to systole

end-plate potential (EPP): a depolarization of the motor end plate of skeletal muscle fibers in response to acetylcholine which is released from the overlying axon terminal in response to an action potential in the motor neuron; initiates an action potential in skeletal muscle plasma membrane

end-product inhibition: inhibition of a metabolic pathway by the action of the final product upon an allosteric site of some enzyme (usually the rate-limiting enzyme) in the pathway; in this case, the end product is the modulator molecule

end-systolic volume (ESV): the amount of blood remaining in the ventricle after ejection

endocrine: pertaining to a ductless (i.e., an endocrine) gland

endocrine gland: a group of epithelial cells specialized for secretion but having no ducts, so they secrete their products into the extracellular space around the gland cells from which the products diffuse into the bloodstream; a ductless gland whose secretory products are hormones

endocrine system: all the hormone-secreting glands of the body

endocytosis: the process in which the plasma membrane of a cell invaginates, and the invaginations become pinched off forming small, intracellular, membrane-bound vesicles which enclose a volume of material; when the extracellular substance is fluid, the process is *fluid endocytosis* (pinocytosis); when specific molecules bind to sites on the plasma membrane and are carried into the cell as the membrane invaginates, the process is *adsorptive endocytosis* (phagocytosis)

endolymph: the fluid that fills the membranous sacs of the vestibular apparatus and the cochlear duct of the inner ear

endometrium: glandular epithelium lining the uterine cavity

endoplasmic reticulum: a cell organelle consisting of an interconnected network of membrane-bound tubules, vesicles, and flattened sacs located in the cytoplasm; two types are distinguished: granular, which has ribosomes embedded in its membranes, and agranular, which is smooth surfaced

endorphins: a group of peptides; some are probably neurotransmitters at the synapses activated by opiate drugs; "endogenous morphines"

endothelium: the thin layer of cells that lines the cavities of the heart and the blood vessels

energy: the ability to perform work; energy is redistributed in all physical and chemical change, and its presence is revealed only when change is occurring; the total energy content of any physical object consists of two components: potential energy and kinetic energy

enkephalin: a peptide which possibly functions as a neurotransmitter at the synapses activated by opiate drugs

enterohepatic circulation: a "recycling" pathway for bile salts by reabsorption from the intestines, passage to the liver (via the hepatic portal vein), and back to the intestines (via the bile duct)

enterokinase: an enzyme embedded in the luminal plasma membrane of the intestinal epithelial cells; activates pancreatic trypsinogen

enzymatic activity: the rate at which an enzyme converts substrate to product

enzyme: a protein that accelerates specific chemical reactions but does not itself undergo any net change during the reaction; a biochemical catalyst

eosinophil: polymorphonuclear granulocytic leukocyte whose granules take up the red dye eosin

epididymis: a portion of the duct system of the male reproductive system located between the seminiferous tubules and the vas deferens

epiglottis: a thin, triangular flap of cartilage that folds down and back covering the trachea; deflects food away from the respiratory passages during swallowing

epilepsy: neuronal activity that starts at a cluster of abnormal cortical neurons and spreads to adjacent neural tissue, giving rise to convulsions

epinephrine: a hormone secreted by the adrenal medulla and involved in the regulation of organic metabolism; a catecholamine neurotransmitter; also called adrenaline

epiphenomenon: an event that accompanies some action but is not causally related to it

epiphyseal plate: an actively proliferating zone of cartilage near the ends of bones; the region of bone growth.

epithelial transport: *see* transcellular transport

epithelium: tissue that covers all body surfaces and lines all body cavities; most glands are formed from epithelial tissue

EPP: *see* end-plate potential

EPSP: *see* excitatory postsynaptic potential

equilibrium: a situation in which no net change occurs within a system with time; no energy input is required to maintain the state of equilibrium

equilibrium potential: the voltage difference that causes a movement of a particular ion species equal in magnitude but opposite in direction to the concentration gradient causing net diffusion of that same ion species; at the equilibrium potential there is no net movement of the ion species

erectile tissue: spongy vascular tissues in the penis and clitoris which become engorged with blood and cause the tissue to become erect

erection: a vascular phenomenon in which the penis or clitoris becomes stiff and erect

erythrocyte: a red blood cell

erythropoiesis: formation of erythrocytes

erythropoietin: a glycoprotein hormone secreted mainly by cells in the kidney; stimulates red-blood-cell production

esophagus: the portion of the digestive tract that connects the mouth (pharynx) and stomach

essential amino acids: the amino acids that cannot be formed by the body at a rate adequate to meet metabolic requirements and, therefore, must be obtained from the diet

essential hypertension: hypertension of unknown cause

essential nutrients: substances which are required for normal or optimal body function but are synthesized by the body either not at all or in amounts inadequate to achieve balance

estradiol: a steroid hormone of the estrogen family; the major estrogen secreted by the ovaries

estriol: a steroid hormone of the estrogen family; the major estrogen of pregnancy

estrogen: a group of steroid hormones which have similar effects on the female reproductive tract

estrone: a steroid hormone of the estrogen family

ESV: *see* end-systolic volume

euchromatin: uncoiled region of chromatin; site of most gene transcription

euphoria: an exaggerated sense of well being

euphorigen: a drug that elevates mood

eustachian tube: *see* auditory tube

excitability: the ability to produce action potentials

excitation-contraction coupling: the mechanisms in muscle fibers linking depolarization of the plasma membrane with generation of force by the cross bridges

excitatory postsynaptic potential (EPSP): a depolarizing graded potential which arises in the postsynaptic neuron in response to the activation of excitatory synaptic endings upon it

excitatory synapse: a synapse which, when activated, either increases the likelihood that the membrane potential of the postsynaptic neuron will reach threshold and undergo action potentials, or increases the firing frequency of existing action potentials

excretion: the appearance of a substance in the urine or feces

exocrine gland: a group of epithelial cells specialized for secretion and having ducts, which lead to a specific compartment or surface, into which the products are secreted

exocytosis: the process in which the membrane of an intracellular vesicle fuses with the plasma membrane, the vesicle opens, and the vesicle contents are liberated into the extracellular fluid

expiration: the bulk flow of air out of the lungs to the atmosphere

expiratory reserve volume: the volume of air that can be exhaled by maximal active contraction of the expiratory muscles at the end of a normal expiration

expressive aphasia: a language deficit, not due to mental retardation, in which persons cannot perform the coordinated movements necessary for speech even though they are not paralyzed, understand what is spoken to them, and know what they want to say; often involves writing as well

extension: the straightening of a joint

extensor muscle: a muscle whose activity causes straightening of a joint

external anal sphincter: a band of skeletal muscle around the lower end of the rectum; generally under voluntary control

external environment: the environment that immediately surrounds the external surfaces of an organism

external genitalia: (female) the mons pubis, labia majora and minora, clitoris, vestibule of the vagina, and vestibular glands; (male) the penis and scrotum

external urethral sphincter: a ring of skeletal muscle that surrounds the lower end of the urethra, preventing micturition when contracted

external work: movement of external objects by the contraction of skeletal muscles

extracellular fluid: the watery medium that surrounds each of the body's cells; the environment in which the cells live; comprises the interstitial fluid and plasma; accounts for about one-third of total body water

extrafusal fibers: *see* skeletomotor fibers

extrapyramidal system: *see* multineuronal pathways

extrasystole: a heartbeat that occurs before the normal time in the cardiac cycle

extrinsic clotting pathway: the formation of fibrin clots by a pathway using tissue thromboplastin

facilitated diffusion: a carrier-mediated transport system in which binding sites on carrier proteins are accessible to extra- and intracellular fluids, permitting the transported molecules to be moved in either direction across the membrane; net movement always occurs from high to low concentration, and at equilibrium the concentrations on the two sides of the membrane become equal

facilitation: (neuronal) the general depolarization of a nerve cell membrane potential when excitatory synaptic input to the cell exceeds inhibitory input

factor XII: *see* Hageman factor

fallopian tube: *see* uterine tubes

farsighted: pertaining to the defect in vision that occurs because the eyeball is too short for the lens, so that near objects are focused behind the retina

fat mobilization: the increased breakdown of triacylglycerols and release of glycerol and fatty acids into the blood

fat-soluble vitamins: vitamins A, D, E, and K that are soluble in nonpolar solvents and insoluble in water

fatty acid: a chain of carbon atoms 14 to 22 carbons long (16 and 18 are most common) with a carboxyl group at one end through which it can be linked to glycerol in the formation of triacylglycerol (neutral fat); *see also* polyunsaturated fatty acid, saturated fatty acid, unsaturated fatty acid

Fe^{2+}: ferrous ion

Fe^{3+}: ferric ion

feces: waste material consisting primarily of water, bacteria, and ingested matter that failed to be digested and absorbed in the intestines; the material expelled from the large intestine during defecation

feedback: a situation in which the ultimate effects of a system themselves influence the system; *see also* negative feedback, positive feedback

feedforward: an aspect of some control systems which allows the system to anticipate changes of stimuli to its sensor

ferritin: an iron-binding protein which is the storage form for iron

fertilization: union of a sperm with an ovum

fetus: an organism during the later stages of development prior to birth, in human beings the period from the second month of gestation until birth

fever: an increased body temperature due to the setting of the "thermostat" of the temperature-regulating mechanisms at a higher-than-normal level

fibrillation: extremely rapid contractions of cardiac muscle in an unsynchronized, repetitive way that prevents effective pumping of blood

fibrin: a protein fragment resulting from the enzymatic cleavage of fibrinogen and having the ability to bind to other similar fragments, i.e., to polymerize, and to turn blood into a solid gel (clot)

fibrinogen: the plasma protein precursor of fibrin

filtration: movement of essentially protein-free plasma across the walls of a capillary out of its lumen as a result of a pressure gradient across the capillary wall

fimbria: fingerlike projections surrounding the trumpet-shaped expansion at the ovarian end of each uterine tube

first law of thermodynamics: the maxim that energy is neither created nor destroyed during any chemical or physical process; law of the conservation of energy

"first" messenger: the extracellular hormone, paracrine agent, or neurotransmitter that combines with a plasma-membrane or cytosol receptor

first-order neuron: the afferent neuron

flatus: intestinal gas

flexion: the bending of a joint

flexion reflex: flexion of those joints that withdraw an injured part away from a painful stimulus

flexor muscle: a muscle whose activity causes bending of a joint

fluid mosaic model: (membrane structure) molecular structure of cell membranes consisting of proteins embedded in a bimolecular layer of phospholipids; the phospholipid layer has the physical properties of a fluid, allowing the membrane proteins to move laterally within the lipid layer

flux: the amount of material crossing a surface in a unit of time; *see also* net flux, unidirectional flux

folic acid: pteroylglutamic acid; a water-soluble vitamin of the B-complex group essential for the formation of DNA; also called folacin

follicle: *see* ovarian follicle

follicle cell: *see* granulosa cells

follicle-stimulating hormone (FSH): a polypeptide hormone secreted by the anterior pituitary in both males and females; one of the gonadotropins whose target organs are the gonads

follicular phase: that portion of the menstrual cycle during which a single follicle and ovum develop to full maturity prior to ovulation

forebrain: a subdivision of the brain consisting of the right and left cerebral hemispheres (the cerebrum) and the diencephalon

fovea: the area near the center of the retina in which cones are most concentrated and which, therefore, gives rise to the most acute vision

frontal lobe: a region of the anterior cerebral cortex, where motor cortex, speech centers, and some association cortices are located

fructose: a five-carbon sugar found in the dissaccharide sucrose (table sugar)

FSH: *see* follicle-stimulating hormone

GABA: *see* gamma-aminobutyric acid

gag reflex: contractions of the constrictor muscle of the pharynx, and gagging in response to touching the back of the pharynx

galactose: a six-carbon monosaccharide found in the dissaccharide lactose (milk sugar)

gall bladder: a small sac underneath the liver; concentrates bile and stores it between meals; contraction of the gall bladder ejects bile into the bile duct and small intestine

gallstone: a precipitate of cholesterol (and occasionally other substances as well) in the gall bladder or bile duct

gametes: the germ cells or reproductive cells; the sperm cells in a male and the ova in a female

gamma-aminobutyric acid (GABA): a possible neurotransmitter at some inhibitory synapses

gamma globulin: immunoglobulin G (IgG), the most abundant plasma antibody; a plasma protein

gamma loop: an indirect pathway leading to the facilitation of alpha motor neurons; the route from descending motor pathway to gamma motor neuron to spindle muscle fiber to stretch receptor and afferent neuron to alpha motor neuron

gamma motor neuron: a small motor neuron which controls spindle (intrafusal) muscle fibers

ganglion: (pl. ganglia) a cluster of neuronal cell bodies located outside the central nervous system

ganglion cells: the neurons in the retina that are postsynaptic to the bipolar cells; their axons form the optic nerve

gap junction: a type of cell junction allowing direct electric current flow between adjacent cells; the two opposing plasma membranes come within 2 to 4 nm of each other and are joined by small channels or tubes through which ions and small molecules can flow between the cytoplasms of the two cells

gas equilibration: the condition in which the partial pressure of a gas in the liquid phase and the partial pressure of that gas in the gaseous phase are equal

gastric: pertaining to the stomach

gastric-inhibitory peptide (GIP): a polypeptide hormone secreted by cells of the upper small intestine; one action is to inhibit gastric motility

gastric phase: (of gastrointestinal control) the initiation of neural and hormonal reflexes by stimulation of the wall of the stomach by distension, acidity, and peptides

gastrin: a polypeptide hormone secreted by the antral region of the stomach; stimulates gastric acid secretion

gastrocnemius: muscle in the calf of the leg whose contraction can cause both extension of the ankle and flexion of the leg

gastroilial reflex: a reflex increase in contractile activity of the ileum during periods of gastric emptying

gastrointestinal system: gastrointestinal tract plus the salivary glands and portions of the liver and pancreas

gastrointestinal tract: mouth, esophagus, stomach, small and large intestines

gene: a unit of hereditary information; the portion of a DNA molecule which contains the information, coded in its nucleotide sequence, required to determine the amino acid sequence of a single polypeptide chain; a single DNA molecule contains many genes

generator potential: *see* receptor potential

germ cells: cells which give rise to the male and female gametes, the sperm and ova

GFR: *see* glomerular filtration rate

GH: *see* growth hormone

GIP: *see* gastric-inhibitory peptide

gland: a group of epithelial cells specialized for the function of secretion; *see also* endocrine gland, exocrine gland

glial cell: a nonneuronal cell type in the brain and spinal cord; helps regulate the extracellular environment of the central nervous system; also neuroglial cell

globin: the polypeptide chains of a hemoglobin molecule

globulin: one of the types of protein found in blood plasma, less abundant that albumin and usually of larger molecular weight

glomerular filtration: the movement of an essentially protein-free plasma through the capillaries of the renal glomerulus into Bowman's capsule

glomerular filtration rate (GFR): the volume of fluid filtered through the renal glomerular capillaries into the renal tubules per unit time

glomerulus: a capillary tuft which forms the vascular component at the beginning of a kidney nephron; intimately associated with Bowman's capsule

glottis: the opening between the vocal cords

glucagon: a polypeptide hormone secreted by the A (or alpha) cells of the islets of Langerhans of the pancreas; its action on several target cells leads to a rise in plasma glucose

glucocorticoid: one of several hormones produced by the adrenal cortex having major effects on glucose metabolism, e.g., cortisol

gluconeogenesis: synthesis by the liver of glucose from pyruvate, lactate, glycerol, or amino acids; new formation of glucose

glucose: the major monosaccharide (carbohydrate) in the body; a six-carbon sugar, $C_6H_{12}O_6$; the catabolism of glucose during the absorptive state provides most of the energy for the cells of the body

glucose-6-phosphate: a carbohydrate formed by transferring a phosphate from ATP to glucose; the first step in the pathways of glycolysis and glycogen synthesis

glucose sparing: the switch from glucose to fat utilization by most cells during the postabsorptive state

glutamic acid: an amino acid; a possible neurotransmitter at some excitatory synapses

glycerol: a three-carbon carbohydrate molecule; forms the backbone of triacylglycerol (neutral fat)

glycerol phosphate: an important intermediate in glycolysis and triacylglycerol synthesis

glycine: an amino acid; possible neurotransmitter at some inhibitory synapses

glycogen: a highly branched polysaccharide composed of thousands of glucose subunits; the major form of carbohydrate storage in the body; also called animal starch

glycogenolysis: the breakdown of glycogen

glycolipid: lipid with a covalently bound carbohydrate group; present in small amounts in the plasma membrane oriented with the carbohydrate at the extracellular surface

glycolysis: the breakdown of glucose to form two molecules of lactic acid; occurs in the absence of oxygen and gives rise to the net synthesis of two molecules of ATP; enzymes located in the cytoplasm

glycoprotein: protein with a covalently linked carbohydrate group; present in plasma membranes with the carbohydrate facing the extracellular surface

GnRH: *see* gonadotropin releasing hormone

goiter: an enlarged thyroid

Golgi apparatus: a cellular organelle consisting of membranes and vesicles, usually located near the nucleus; processes newly synthesized proteins for secretion; contains the enzymes for adding carbohydrates to proteins to form glycoproteins

Golgi tendon organs: mechanoreceptor endings of afferent nerve fibers wrapped around collagen bundles in tendons; activated by distortion when tension is applied to the tendon

gonad: gamete-producing reproductive organ, i.e., the testes in the male and the ovaries in the female

gonadotropin releasing hormone (GnRH): the hypothalamic hormone that controls LH and FSH secretion by the anterior pituitary

gonadotropins: hormones secreted by the anterior pituitary which control the function of the gonads; FSH and LH

granuloma: a mass consisting of numerous layers of phagocytic-type cells (the central one of which contains a microbe or nonmicrobial substance) enclosed in a fibrous capsule, which isolates the potentially harmful substance from healthy tissues

granulopoietin: the hormone, produced and released mainly by macrophages, that stimulates neutrophil ("granulocyte") production

granulosa cells: cells which surround the ovum in the ovarian follicle

grasp reflex: flexion of the fingers and thumb in response to stimulation of the palm; present in infants

gray matter: that portion of the brain and spinal cord which appears gray in unstained specimens and consists mainly of cell bodies and the unmyelinated portions of nerve fibers

green-sensitive cones: cones containing photopigment most sensitive to light of 535-nm wavelengths

GRH: *see* growth-hormone releasing hormone

growth: an increase in body size, i.e., lengthening of the long bones, increased cell division and cell volume, and accumulation of protein; *see also* compensatory growth

growth hormone (GH): a polypeptide hormone secreted by the anterior pituitary; also called somatotropin or somatotropic hormone (STH); stimulates body growth by means of its actions on carbohydrate and protein metabolism

growth-hormone-release inhibiting hormone: *see* somatostatin

growth-hormone releasing hormone (GRH): the hypothalamic hormone that stimulates the secretion of growth hormone by the anterior pituitary

GTP: *see* guanosine triphosphate

guanine: one of the purine bases in DNA and RNA

guanosine triphosphate (GTP): a high-energy molecule similar to ATP except that it contains the base guanine rather than adenine

H^+: hydrogen ion; the concentration of hydrogen ions in a solution determines the acidity of a solution

H zone: one of the transverse bands making up the repeating striated pattern of cardiac and skeletal muscle; a light region that bisects the A band of striated muscle; corresponds to the space between the ends of the thin filaments; thus, it consists only of thick filaments

habituation: a reversible decrease in the strength of a response upon repeatedly administered stimulation

Hageman factor: factor XII; the initial factor in the sequence of reactions that results in fibrin formation

hair cells: mechanoreceptors in the organ of Corti and vestibular apparatus

hallucination: an altered state of consciousness in which whole worlds are created in the mind and accepted as real

haploid: having a chromosome number equal to that of the gametes, i.e., half the diploid number found in most somatic cells

Hb: *see* reduced hemoglobin

$HbCO_2$: *see* carbamino compound

HbO_2: *see* oxyhemoglobin

H_2CO_3: carbonic acid

head: (of the penis) the relatively firm, smooth region at the tip of the penis

heart attack: damage to or death of cardiac muscle, i.e., a myocardial infarct

heart-lung preparation: an experimental setup in which the performance of a heart lacking any hormonal or neural input can be determined in response to varying venous pressures

heart murmur: the abnormal heart sounds caused by the turbulent flow of blood through narrowed or leaky valves or a hole in the interventricular or interatrial septum

heart rate: the number of contractions of the heart per minute

heart sounds: the noises which result from vibrations caused by closure of the atrioventricular valves (first heart sound) and pulmonary and aortic valves (second heart sound)

heartburn: pain that seems to occur in the region of the heart but is due either to spasms of the esophageal muscle following stimulation by acid stomach contents refluxed into the esophagus or to the direct stimulation of pain receptors in the esophageal wall by the acid

heat exhaustion: a state of collapse due to hypotension brought about both by depletion of plasma volume secondary to sweating and by extreme dilation of skin blood vessels; the thermoregulatory centers are still functioning

heat stroke: a positive-feedback situation in which a heat gain greater than heat loss causes body temperature to become so high that the brain thermoregulatory centers do not function, allowing temperature to rise even closer to the point at which vital bodily functions are endangered

"helper" T cells: a type of T cell that regulates B cells in a way that leads to enhanced antibody production

hematocrit: the percentage of the total blood volume occupied by blood cells

hematoma: a local, extravascular accumulation of blood

heme: the iron-containing organic molecule bound to each of the four polypeptide chains of a hemoglobin molecule; the iron-containing portion of a cytochrome

hemoglobin: a red-colored protein, located in erythrocytes, which transports most of the oxygen in the blood; an oligomeric protein composed of four polypeptide chains, each of which contains a heme group having a single atom of iron, with which oxygen reversibly combines

hemorrhage: bleeding

hemostasis: prevention of the loss of blood; stopping the flow of blood through a particular vessel

Henle's loop: the hairpinlike segment of the tubular component of a kidney nephron; situated between the proximal and distal tubules

heparin: an anticlotting agent found in various cells and tissues; an anticoagulant drug

hepatic: pertaining to the liver

hepatic portal vein: the vein that conveys blood from the capillary beds in the intestines (as well as portions of the stomach and pancreas) to the capillary beds in the liver

hertz (Hz): the measure used for wave frequencies; cycles per second

heterochromatin: region of chromatin that remains tightly coiled; stains more darkly than the uncoiled regions; genes located in these regions are inactive

hippocampal "arousal": *see* theta rhythm

hippocampus: a portion of the limbic system in the brain

histamine: a paracrine agent found in many tissues of the body but particularly in mast cells, circulating basophils, and platelets; a possible neurotransmitter

histocompatibility antigens: antigenic protein molecules on the surface of all nucleated cells; differ from person to person

histone: one of a group of positively charged proteins associated with DNA to form chromatin; interacts with the phosphate groups of the nucleotides to cause supercoiling of the chromatin threads

histotoxic hypoxia: hypoxia in which the quantity of oxygen reaching the tissues is normal, but the cell is unable to utilize the oxygen because a toxic agent has interfered with its metabolic machinery

homeostasis: the relatively stable physical and chemical composition of the internal environment of the body which results from the actions of compensating regulatory systems

homeostatic system: a control system consisting of a collection of interconnected components which functions to keep a physical or chemical parameter of the internal environment relatively constant

homeotherm: an animal capable of maintaining its body temperature within very narrow limits

hormone: a chemical substance synthesized by an endocrine gland and secreted into the blood which carries it to other cells in the body where its actions are exerted

HR: *see* heart rate

5-HT: *see* serotonin

humoral immunity: a type of specific immune response in which circulating antibodies play a central role in the destructive process

hydrocephalus: a disease in which blockage of the outflow paths for cerebrospinal fluid cause the fluid to build up, resulting in considerable damage to the brain tissue

hydrocholric acid: HCl; a strong acid secreted into the stomach lumen by gland cells lining the stomach

hydrogen bond: a weak bond between two molecules in which a negative region of one polarized substance is electrostatically attracted to a positively polarized hydrogen atom in the other

hydrogen peroxide: H_2O_2; a chemical produced by the phagosomes and highly destructive to the macromolecules within it

hydrolysis: the breakdown of a chemical bond with the addition of the elements of water (—H and —OH) to the products formed

hydrostatic pressure: the pressure exerted by a fluid

hydroxyl group: R—OH, a polar chemical group where R is the remaining portion of the molecule

5-hydroxytryptamine: *see* serotonin

hymen: a thin fold of mucous membrane partially overlying the vaginal opening

hyperemia: increased blood flow; *see also* active hyperemia

hyperglycemia: increased plasma glucose concentration above normal levels

hyperopia: *see* farsighted

hyperosmotic: having an osmolarity greater than normal plasma

hyperpolarize: to change the transmembrane potential so the inside of the cell becomes more negative than its resting state

hypersensitivity: (to an antigen) an acquired reactivity to an antigen which can result in bodily damage upon subsequent exposure to that particular antigen, e.g., allergies

hypertension: chronically increased arterial blood pressure

hyperthyroidism: the secretion of excess thyroid hormone

hypertonic solution: a solution containing a higher concentration of membrane-impermeable solute particles than cells contain; not synonymous with hyperosmotic; cells placed in a hypertonic solution will shrink due to the diffusion of water (osmosis) out of the cell

hypertrophy: an increase in size; enlargement of

a tissue or organ that results from an increase in cell size rather than in cell number

hyperventilation: ventilation greater than that needed to maintain plasma P_{CO_2} normal

hypocalcemic tetany: a disease characterized by skeletal muscle spasms due to a low calcium concentration

hypoosmotic: having an osmolarity less than normal plasma

hypotension: decreased blood pressure

hypothalamic releasing factors: *see* hypothalamic releasing hormones

hypothalamic releasing hormones: hormones released from hypothalamic neurons into the hypothalamo-pituitary portal vessels to control the release of the anterior-pituitary hormones; formerly called hypothalamic releasing factors

hypothalamo-pituitary portal vessels: vessels which transport blood from a capillary network in the hypothalamus to one in the anterior pituitary

hypothalamus: a brain region below the thalamus and responsible for the integration of many basic behavioral patterns which involve correlation of neural and endocrine function, especially those concerned with regulation of the internal environment

hypothyroid: insufficient thyroid hormone

hypotonic solution: a solution containing a lower concentration of membrane-impermeable solute particles that cells contain; not synonymous with hypoosmotic; cells placed in a hypotonic solution will swell due to the diffusion of water (osmosis) into the cell

hypoventilation: ventilation insufficient to maintain plasma P_{CO_2} normal

hypoxia: a deficiency of oxygen at the tissue level

hypoxic hypoxia: hypoxia due to decreased arterial P_{O_2}

Hz: *see* hertz

I band: one of the transverse bands making up the repeating striated pattern of cardiac and skeletal muscle; located between the A bands of adjacent sarcomeres and bisected by the Z line; extends the length of the thin filaments up to but not including regions of thick and thin filament overlap

ICSH: *see* luteinizing hormone

IgA: the class of immunoglobulin antibodies produced by lymphoid tissue lining the gastrointestinal, respiratory, and genitourinary tracts; exert their major effects on the materials encountered by the secretions of these tracts

IgD: a class of immunoglobulin antibodies whose function is unknown

IgE: the class of immunoglobulin antibodies that mediates certain allergic responses

IgG: gamma globulin; the most abundant family of plasma antibodies

IgM: a class of immunoglobulin antibodies that provides specific immunity against bacteria and viruses

ileocecal sphincter: a ring of smooth muscle separating the ileum from the cecum, i.e., the small and large intestines

ileum: the final, longest segment of the small intestine

iliogastric reflex: a reflex decrease in gastric motility in response to distension of the ileum

immune enhancement: the protection of a foreign cell from destruction by the combination of antibodies with the foreign antigens, which prevents T cells from combining with the foreign antigens

immune responses: *see* nonspecific immune responses, specific immune responses

immune surveillance: the recognition and destruction of abnormal or mutant cell types which arise within the body

immunity: the physiological mechanisms which allow the body to recognize materials as foreign or abnormal and to neutralize or eliminate them; the mechanisms which maintain the uniqueness of "self"

immunoglobulins: a family of plasma proteins to which the antibodies belong; subdivided into five classes: IgG, IgA, IgD, IgM, and IgE (Ig stands for immunoglobulin)

implantation: the event during which the fertilized ovum adheres to the uterine wall and becomes embedded in it

impotence: failure of erection of the penis, thus, inability for the male to perform sexual intercourse

inborn errors of metabolism: inherited diseases, often due to a single gene, in which there is failure to produce an active enzyme or else an abnormally active one is produced

inducer: the molecule which, when combined with the repressor protein, inactivates the repressor thereby freeing the operator gene from inhibition and initiating synthesis of the proteins coded by the operon

induction: (enzyme) stimulation of enzyme synthesis by action of the enzyme substrate on the genetic mechanisms controlling enzyme synthesis

infarct: an area of dead tissue resulting from local ischemia due to diminished circulation supplying the area

inferior vena cava: the large vein that carries blood from the lower half of the body to the right atrium of the heart

inflammation: the local response to injury characterized by swelling, pain, and increased temperature and redness in the region of injury due to increased local blood flow

inflammatory response: *see* inflammation

inhibin: a protein hormone secreted from seminiferous tubules (probably by the Sertoli cells)

inhibitory postsynaptic potential (IPSP): a hyperpolarizing graded potential which arises in the postsynaptic neuron in response to the activation of inhibitory synaptic endings upon it

inhibitory synapse: a synapse which, when activated, decreases the likelihood that the membrane potential of the postsynaptic neuron will reach threshold and fire an action potential or decreases the firing frequency of existing action potentials

initial segment: the first portion of the axon of a neuron plus the part of the cell body where the axon is joined; the region where an action potential is initiated in some postsynaptic neurons

inner ear: the cochlea; contains the organ of Corti

inorganic ion: any ion that does not contain carbon, e.g., Na^+, Cl^-, and K^+

inosine: a nucleotide base present in tRNA; base-pairs with cytosine

insensible loss: the water loss of which a person is unaware, i.e., the loss by evaporation from the cells of the skin and the lining of the respiratory passages

inspiration: the bulk flow of air from the atmosphere into the lungs

inspiratory muscles: those muscles whose contraction contributes to inspiration; the diaphragm is the major inspiratory muscle

inspiratory neuron: a neuron whose cell body is in the brainstem medulla and which fires in synchrony with inspiration and ceases firing during expiration

inspiratory reserve volume: the volume of air that can be inspired over and above the resting tidal volume

insulin: a polypeptide hormone secreted by the B (or beta) cells of the islets of Langerhans of the pancreas; stimulates the carrier-mediated movement of glucose into most cells and the enzymes that transform glucose into fat and glycogen

integral proteins: membrane proteins which are embedded in the phospholipid bilayer of the membrane and cannot be removed without disrupting the lipid components of the membrane; may span the entire membrane or be located at only one side

integrator: a cell whose output is determined by many often-conflicting bits of incoming information

integument: the skin

intention tremor: a tremor that occurs only when the person is trying to perform a movement, not when the person is at rest; associated with cerebellar disease

intercostal muscles: skeletal muscles lying between the ribs; contraction results in elevation or compression of the rib cage during inspiration or expiration

interference theory: the theory that things are forgotten because of competition with conflicting information stored before or after the storage of the particular item being recalled

interferon: a protein that stimulates cells to synthesize "antiviral" proteins, which in turn interfere with the multiplication of viruses in those cells

internal anal sphincter: a ring of smooth muscle around the lower end of the rectum

internal environment: the extracellular fluid immediately surrounding cells in the body; thus, the environment in which these cells live; includes the interstitial fluid and blood plasma

internal urethral sphincter: part of the smooth muscle wall of the urinary bladder which, by its contraction and relaxation, opens and closes the outlet from the bladder to the urethra

internal work: energy requiring activities (work) in the body, e.g., cardiac contraction, internal secretions, membrane active transport, synthetic chemical reactions, smooth muscle contractions, and the muscular movement associated with respiration

interneuron: a neuron whose cell body and axon lie entirely within the central nervous system (also called internuncial neuron)

interphase: the period of the cell-division cycle between the end of one division and the beginning of the next; DNA replication occurs in this phase

interstitial-cell stimulating hormone (ICSH): *see* luteinizing hormone

interstitial cells: (of the testes) testosterone-secreting cells which lie between the seminiferous tubules in the testes

interstitial fluid: the extracellular fluid surrounding the cells of a tissue; plasma, which surrounds blood cells, is not considered interstitial fluid

interstitium: interstitial space; the space between tissue cells

interventricular septum: the partition in the heart separating the cavities of the right and left ventricles

intestinal phase: (of gastrointestinal control) the initiation of neural and hormonal reflexes by stimulation of the walls of the intestinal tract by distension, acidity, osmolarity, and the various products of carbohydrate, fat, and protein digestion

intestino-intestinal reflex: a reflex cessation of contractile activity of the intestines in response to large distensions of the intestine, injury to the intestinal wall, or various bacterial infections in the intestine

intraabdominal pressure: the pressure within the abdominal cavity

intracellular fluid: the fluid in the cells; accounts for about two-thirds of total body water

intrafusal fibers: *see* spindle fibers

intrapleural fluid: the thin film of fluid in the thoracic cavity between the pleura lining the inner wall of the thoracic cage and the pleura covering the lungs

intrapleural pressure: the pressure within the pleural space generated by the tendency of the lung and chest wall to pull away from each other; also called intrathoracic pressure

intrathoracic pressure: the pressure within the intrapleural fluid of the chest cavity; intrapleural pressure

intrauterine device: a small object placed in the uterus which, by its presence, prevents pregnancy

intrinsic clotting pathway: the intravascular sequence of fibrin formation initiated by Hageman factor

intrinsic factor: a glycoprotein secreted by the epithelial lining of the stomach and necessary for absorption of vitamin B_{12}

inulin; a polysaccharide not normally present in the body, used to measure glomerular filtration rate

ion: any atom or small molecule containing an unequal number of electrons and protons and, therefore, carrying a net positive or negative electric charge, e.g., Cl^-, Na^+, and NH_4^+

ionic hypothesis: an explanation of the voltage changes during an action potential in terms of membrane permeability changes and the movements of sodium and potassium ions across the membrane

ionization: the dissociation of a substance in solution into atoms or molecules that carry a net electric charge, i.e., the formation of ions

IPSP: *see* inhibitory postsynaptic potential

iris: the ringlike structure that surrounds the pupil of the eye; consists of pigmented epithelium and muscle which by contraction adjusts the size of the pupil

irreversible reaction: a chemical reaction that proceeds predominately in only one direction because large quantities of energy are released in the reaction; may be reversed by an alternate route involving a different enzyme and an additional substrate to supply the large quantity of energy required

ischemia: a reduced blood supply to an organ or tissue

ischemic hypoxia: hypoxia in which the basic defect is too little blood flow to the tissues to meet the metabolic demands

isometric contraction: contraction of a muscle in such a way that it develops tension but is unable to shorten because the opposing load is greater than the tension developed

isotonic: *see* isotonic contraction, isotonic solution

isotonic contraction: shortening of a muscle that results in the moving of a load

isotonic solution: a solution containing the same number of nonpenetrating solute particles as cells contain; not synonymous with isoosmotic; a solution in which a cell will neither shrink nor swell

isovolumetric ventricular contraction: the early phase of systole when both the atrioventricular and aortic valves are closed

isovolumetric ventricular relaxation: the early phase of diastole when both the atrioventricular and aortic valves are closed

jaundice: a condition in which the skin is yellowish due to buildup in the blood and tissues of bilirubin as a result of its failure to be excreted by the liver into the bile

jejunum: the middle segment of the small intestine

k_D: *see* diffusion coefficient

k_p: *see* membrane permeability constant

kallikrein: the enzyme in plasma and tissues that generates kinins from their inactive precursors

kcal: kilocalorie; *see also* calorie

KCl: potassium chloride

keto acid: a molecule containing a carbonyl group

$$(-\overset{\overset{\displaystyle O}{\|}}{C}-)\text{ and a carboxyl group }(-\overset{\overset{\displaystyle O}{\|}}{C}-OH)$$

ketoacidosis: a form of metabolic acidosis due to increased concentration of ketone bodies; a symptom of severe untreated diabetes mellitus

ketone bodies: products of fatty acid oxidation, e.g., acetoacetic acid, acetone, and β-hydroxybutyric acid, that accumulate in the blood during starvation and in untreated diabetes mellitus

kilocalorie (kcal): *see* calorie

kinetic energy: that energy associated with an object because of its motion

kininogen: the protein from which the kinins are generated

kinins: small polypeptides generated in plasma from kininogen by the enzyme kallikrein

knee jerk: an example of the stretch reflex; stretch of the quadriceps femoris (thigh) muscle is elicited by depressing its tendon as the tendon passes over the knee, the reflex response (contraction of the muscle) causing extension of the knee

Krebs cycle: a metabolic pathway that utilizes fragments derived from the breakdown of carbohydrate, protein, and fat and produces carbon dioxide and hydrogen; the hydrogen, via oxidative phosphorylation, ultimately combines with oxygen to form water; also known as the citric acid cycle or the tricarboxylic acid cycle; enzymes located in the mitochondria of cells

labia majora: two prominent skin folds which are part of the female external genitalia; the female analogue of the scrotum

labia minora: two small skin folds lying between the labia majora; surround the urethral and vaginal openings

lacrimal glands: the glands which secrete tears

lactase: the enzyme that breaks lactose (milk sugar) into glucose and galactose; located in the luminal plasma membrane of the small-intestinal epithelia

lactate: the anion formed by the loss of hydrogen ion from lactic acid

lactation: the production and secretion of milk

lacteal: a blind-ended lymph vessel in the center of each intestinal villus

lactose: milk sugar; a disaccharide composed of glucose and galactose

lamina propria: a layer of connective tissue underlying the epithelial layer of the stomach and intestines

larynx: the part of the air passageway which connects the pharynx and trachea and contains the vocal cords

latent period: (muscle) the period (several ms) between initiation of an action potential in a muscle fiber and the beginning of mechanical activity; the period during which the events of excitation-contraction coupling occur

lateral geniculate nucleus: the thalamic nucleus that serves as a way station in the visual pathway

lateral inhibition: a method of refining information in neural pathways whereby the fibers inhibit each other, the most active causing the greatest inhibition of adjacent fibers

lateral sac: the enlarged region at the end of each segment of sarcoplasmic reticulum adjacent to the transverse tubule

law of conservation of energy: *see* first law of thermodynamics

law of the intestine: the fact that the short peristaltic waves (i.e., those traveling only several centimeters), occurring in response to local distension of the intestine, always proceed in the direction of the large intestine

law of mass action: the maxim according to which

an increase in the concentration of a reactant of a chemical reaction causes the reaction to proceed, forming more product

L-dopa: a precursor of dopamine; administered to patients with parkinsonism

learning: the increase in the likelihood of a particular response to a stimulus as a consequence of experience

lens: the adjustable part of the optical system of the eye which helps focus an image of the object on the retina regardless of the distance the object is from the eye

leukocyte: a white blood cell

LH: *see* luteinizing hormone

ligand: any molecule or ion which binds to the surface of any other molecule, such as protein, by noncovalent bonds

ligase: the enzyme that forms recombinant DNA by linking the fragmented ends of two DNA molecules previously split by a restriction enzyme

limbic system: an interconnected group of brain structures within the cerebrum involved with emotions and learning

lipid: any of several types of molecules composed primarily of carbon and hydrogen atoms; characterized by insolubility in water; include fatty acids, triacylglycerols, phospholipids, and steroids

lipoprotein: a complex of fat and protein, the amounts of which vary; low-density plasma lipoproteins have little protein and much fat whereas high-density plasma lipoproteins have the opposite distribution

lithium carbonate: a drug used in the treatment of the manias

load: the force exerted on a muscle by an external object

local-circuit neuron: a neuron which connects with cells in its immediate vicinity and has a short axon or, sometimes, none at all

local current flow: the movement of positive ions from a membrane region with a high positive charge through the cytoplasm or extracellular fluid toward a membrane region of more negative charge, and the simultaneous movement of negative ions in the opposite direction; leads to depolarization or hyperpolarization of adjacent segments of a membrane

local potential: a small, graded potential difference between two points; conducted decrementally,

has no threshold and no refractory period; EPSPs and IPSPs are examples of local potentials

local response: a biological response initiated by a change in the external or internal environment, i.e., a stimulus, which acts on cells in the immediate vicinity of the stimulus (without the involvement of nerves or hormones) with the net effect of counteracting the stimulus

local spinal reflex: a reflex whose afferents, efferents, and integrating center are located within only a few segments of the spinal cord

lockjaw: a disease produced by the toxin of the tetanus bacillus; spasm of the jaw muscles occures early in the disease

long reflex: a type of neural reflex in which the integrating center lies within the central nervous system

long-term memory: a memory that has a relatively long life-span

longitudinal muscle layer: the smooth muscle layer surrounding the circular muscle layer in the stomach and intestinal walls and whose fibers are oriented longitudinally along these organs

loop of Henle: *see* Henle's loop

lower esophageal sphincter: smooth muscles of the last 4 cm of the esophagus which can act as a sphincter, closing off the opening of the esophagus into the stomach

LSD: a potent hallucinogenic drug which induces a state resembling schizophrenia

lumen: the space within a hollow tube or organ

luminal: (adjective) pertaining to the lumen of an organ

lung compliance: the magnitude of the change in lung volume caused by a given change in pressure difference across the lung wall; i.e., the greater the lung compliance, the more stretchable the lung wall

luteal phase: the last half of the menstrual cycle during which the corpus luteum is the active ovarian structure

luteinizing hormone (LH): a polypeptide hormone secreted by the anterior pituitary that acts upon the gonads; one of the gonadotropins; sometimes called interstitial-cell stimulating hormone (ICSH) in the male; in the female the sudden increase in LH release during the middle of the menstrual cycle initiates ovulation

lymph nodes: small organs, containing lymphocytes, located along the course of the lymph vessels; site of lymphocyte formation and storage

lymphatic: pertaining to the system of vessels which conveys lymph from the tissues to the blood

lymphocyte: a type of leukocyte; mainly responsible for the "specific defenses" of the body against foreign invaders

lymphoid tissue: lymph nodes, spleen, thymus, tonsils, and aggregates of lymphoid follicles such as those lining the gastrointestinal tract

lymphokines: chemicals released from a T cell after combination of the T cell with its specific antigen

lysosome: spherical or oval cell organelle surrounded by a single membrane; contains digestive enzymes which break down bacteria and large molecules such as protein, polysaccharides, and nucleic acids that have entered a cell by endocytosis, and can digest damaged components of the cell itself

M line: one of the transverse bands making up the repeating striated pattern of cardiac and skeletal muscle; a thin dark band in the middle of the sarcomere; produced by linkages between the thick filaments

macrophage: a cell type which has as its major function the phagocytosis of foreign matter

malignant tumor: a mass of cancer cells which rapidly grow, divide, and invade surrounding tissues, disrupt the structure and function of organs, and eventually lead to death of the organism

mammary glands: the milk-secreting glands in the breasts

mania: a serious, prolonged disturbance of mood characterized by elation sometimes associated with euphoria, overconfidence, and irritability

mass movement: contraction of large segments of the ascending and transverse colon which propels fecal material into the rectum; occurs three to four times a day, usually after a meal

mast cell: a connective-tissue cell similar to a basophil except that it does not circulate in the blood; local injury releases histamine from mast cells

matrix: the extracellular substance surrounding connective-tissue cells; has some structural organization, such as connective-tissue fibers, as opposed to being a simple ionic solution

maximal oxygen uptake: the maximal amount of oxygen that can be utilized by the body per unit time during maximal exercise

maximal tubular capacity (T_m): the maximal rate of mediated transport of a substance across the wall of the kidney nephrons

mean arterial pressure: the mean blood pressure during the cardiac cycle; approximately equal to diastolic pressure plus one-third of the pulse pressure

mechanoreceptor: a sensory receptor that responds preferentially to mechanical stimuli such as bending, twisting, or compressing

mediated transport: movement of molecules across membranes mediated by binding to protein carrier molecules in the membrane; characterized by specificity, competition, and saturation; includes both facilitated diffusion and active transport

megakaryocyte: a large bone-marrow cell which gives rise to platelets

meiosis: a process of nuclear division leading to the formation of the gametes (sperm and ova); consists of two cell divisions in succession, the first and second meiotic divisions; the final cells receive only half of the chromosomes present in the original cell

melanocyte-stimulating hormones (MSH): two types of polypeptide hormones secreted by the intermediate lobe of the pituitary

membrane: a structural barrier composed of phospholipids and proteins associated with the cell surface and its organelles, e.g., endoplasmic reticulum, mitochondria, plasma membrane, etc., which provides a selective barrier to the movement of molecules and ions across the membrane and provides a structural framework to which enzymes, fibers, and a variety of ligands can bind

membrane permeability constant (k_p): a proportionality constant defining the rate at which a substance diffuses through a membrane given the value of the concentration gradient across the membrane; k_p is determined by the type of molecule, its molecular weight, the temperature, and the characteristics of the membrane; the larger the value of k_p, the more permeable the membrane

membrane potential: (or transmembrane potential) the voltage difference between the inside and outside of the cell

memory: the relatively permanent form of a change in responsiveness to a particular stimulus

"memory" cells: lymphocytes that were exposed to a certain antigen but did not differentiate fully into plasma cells or lymphokine-secreting T cells; rather, they serve as a rapidly responding pool of cells should that antigen be encountered again

memory consolidation: the processes by which memory is transferred from its short-term to its long-term form

memory trace: the postulated neural substrate of memory

menarche: the onset, at puberty, of menstrual cycling

menopause: the cessation of menstrual cycling, usually in the mid-forties

menstrual cycle: the cyclic rise and fall in female reproductive hormones; has a period of approximately 28 days in female human beings

menstrual fluid: blood mixed with debris from the disintegrating endometrium

menstrual phase: that stage of the menstrual cycle during which menstruation is occurring

menstruation: the flow of menstrual fluid from the uterus; also called the menstrual period

MES: *see* microsomal enzyme system

mescaline: a mind-altering drug that is a methylated derivative of dopamine

mesentery: a sheet of connective tissue that connects the outer surface of the stomach and intestines to the abdominal wall

messenger RNA (mRNA): the form of ribonucleic acid that transfers the genetic information corresponding to one gene (or, at most, a few genes) from DNA, in the nucleus, to the ribosomes, in the cytoplasm where proteins are synthesized

metabolic acidosis: an acidosis due to a metabolic production of acids other than carbon dioxide, e.g., the production of lactic acid during severe exercise or of ketones in diabetes mellitus

metabolic alkalosis: an alkalosis resulting from the removal of hydrogen ions from the body by mechanisms other than the respiratory removal of carbon dioxide, e.g., the loss of hydrogen from the stomach with persistent vomiting

metabolic cost of living: *see* basal metabolic rate

metabolic pathway: the sequence of enzyme-mediated chemical reactions leading to the formation of a particular product

metabolic rate: the total energy expenditure of the body per unit time

metabolism: all the chemical reactions that occur within a living organism

metastasis: the breaking away of cancer cells from the parent tumor and their spread by way of the circulatory or lymphatic system to other parts of the body

methyl group: $-CH_3$

Mg^{2+}: magnesium ion

micelle: a soluble cluster of amphipathic molecules (fatty acids, 2-monoglycerides, and bile salts), formed during the digestion of fat in the small intestine, in which the polar regions of the molecules line the outer surface of the micelle, and the nonpolar regions are located within the micelle

microbes: minute organisms, including bacteria, protozoans, and fungi

microcirculation: circulation in the capillaries

microliter: a unit of volume equal to 0.001 mL or 0.000001 L

micrometer (μm): 10^{-6} m; a micron; equivalent to 0.000039 in; formerly abbreviated μ

micropipette: an extremely fine, pointed tube used to sample or inject tiny volumes of fluid

micropuncture: the technique of penetrating a small region, such as the tubule of a renal nephron, with a fine pipette through which materials can be added to or removed from the region

microsomal enzyme system (MES): enzymes, found in the smooth endoplasmic reticulum of liver cells, which transform molecules into more polar, less lipid-soluble substances; formerly called "drug-metabolizing enzymes"

microtubules: tubular filaments in the cytoplasm of cells, which provide internal support for cells; may act in maintaining and changing cell shape and in producing the movements of organelles within the cell

microvilli: small fingerlike projections from the surface of epithelial cells, which greatly increase the absorptive surface area of a cell; a characteristic of the epithelium lining the small intestine and kidney nephrons

micturition: urination

middle-ear cavity: the air-filled space deep within the temporal bone that contains the three ossicles (ear bones) that conduct the sound waves from the tympanic membrane to the cochlea

milk intolerance: the inability to digest milk sugar (lactose) because of lack of the intestinal enzyme lactase; leads to the accumulation of large amounts of gas and fluid in the large intestine, which causes pain and diarrhea

milk let-down: the process by which milk is moved from the alveoli of the mammary glands into the ducts, from which it can be sucked

milliliter (mL): a unit of volume equal to 0.001 L

millimeter (mm): 10^{-3} m; equivalent to 0.03937 in

millivolt (mV): 0.001 V

mineralocorticoids: steroid hormones produced by the adrenal cortex which have their major effects on sodium and potassium balance; aldosterone is the major mineralocorticoid

mitochondria: rod-shaped or oval organelles in cell cytoplasm which produce most of the ATP used by the cells; site of Krebs cycle and oxidative-phosphorylation enzymes; site of cell production of carbon dioxide and oxygen utilization

mitosis: a process of nuclear division in which DNA is duplicated and an identical set of chromosomes is passed to each daughter cell at the time of cell division

mitral valve: the valve between the left atrium and left ventricle of the heart

mm: *see* millimeter

modality: (of a stimulus) the quality or kind of stimulus; e.g., touch and hearing are different modalities

modulator: a molecule which, by acting at a protein's regulatory site, alters the properties of other binding sites on the protein and thus regulates the functional activity of the protein

molarity: a unit of concentration; the number of moles of solute per liter of solution

mole: (also mol) a unit which indicates the number of molecules of a substance present; the number of moles = weight in grams/molecular weight; one mole of any substance contains the same number of molecules as one mole of any other substance (the number is 6×10^{23} — Avagadro's number)

molecular weight: a number equal to the sum of the atomic weights of all the atoms in a molecule; e.g., glucose $(C_6H_{12}O_6)$ has a molecular weight of 180 $[(6C \times 12) + (12H \times 1)(6O \times 16) = 180]$

molecule: a chemical structure formed by linking atoms together

monoamine oxidase: an enzyme that changes the catecholamine neurotransmitters and serotonin into inactive substances

monoamine-oxidase inhibitor: a drug that interferes with the action of monoamine oxidase thereby increasing the concentration of neurotransmitter available

monocyte: a type of leukocyte; leaves the bloodstream and is transformed into a tissue macrophage

monosaccharide: a carbohydrate consisting of a single sugar molecule contains generally five or six carbon atoms, e.g., glucose $(C_6H_{12}O_6)$

monosynaptic reflex: a reflex in which the afferent limb of the reflex arc (the afferent neuron) directly activates the efferent limb (the motor neuron); the stretch reflex is the only known monosynaptic reflex

motor: having to do with muscles and movement

motor cortex: a strip of cerebral cortex along the posterior border of the frontal lobe; gives rise to many (but certainly not all) of the axons descending in the corticospinal and multineuronal pathways to the motor neurons

motor end plate: the specialized region of a muscle-cell plasma membrane that lies directly under the axon terminal of a neuron, which contains the receptor sites for the neurotransmitter acetylcholine; analogous to the subsynaptic membrane of postsynaptic neuron

motor neuron: an efferent neuron which innervates skeletal muscle fibers

motor potential: electrical activity that can be recorded over motor cortex about 50 to 60 ms before a movement begins

motor unit: a motor neuron plus the muscle fibers it innervates

mRNA: *see* messenger RNA

MSH: *see* melanocyte-stimulating hormones

mucin: a protein which, when mixed with water, forms mucus

mucosa: (of the stomach and intestines) the three layers of the stomach and intestinal wall nearest the lumen, i.e., the epithelium, lamina propria, and muscularis mucosa

Müllerian duct: a part of the embryo which, in a female, develops into the ducts of the reproductive system, but in a male, degenerates

Müllerian regression factor: a protein secreted by the Sertoli cells of the primitive testes which causes the Müllerian ducts to regress

multineuronal pathways: the descending motor pathways that are made up of chains of neurons synapsing in the basal ganglia, several other subcortical nuclei, and the brainstem; only the final neuron of the chain reaches the region of the motor neurons; also called the multisynaptic pathways and the extrapyramidal system

multiunit smooth muscle: smooth muscle which exhibits little, if any, propagation of electrical activity from fiber to fiber and whose contractile activity is closely coupled to its neural input; *see also* single-unit smooth muscle

muscarinic receptor: an acetylcholine receptor that can be stimulated by the mushroom poison muscarine; most of these receptors are located on smooth muscle and gland cells

muscle: a number of muscle fibers bound together by connective tissue

muscle fatigue: a decrease in the mechanical response of muscle with prolonged stimulation

muscle fiber: a muscle cell

muscle spindle: a specialized receptor, located in skeletal muscle, which responds to stretch; a capsule-enclosed arrangement of afferent nerve endings wrapped around modified muscle fibers and arranged parallel to skeletal muscle fibers

muscle-spindle stretch receptors: the stretch receptors within the muscle spindles which are activated by stretch and provide information about muscle length

muscle tension: the force exerted by a contracting muscle on an object

muscle tone: the small amount of tension produced by skeletal muscles under resting conditions in the body as a result of a low-frequency discharge of the motor neurons; in smooth muscle tone may be maintained by spontaneous electrical activity within the muscle itself

muscularis mucosa: a thin layer of smooth muscle underlying the lamina propria in the stomach and intestines

mutation: any alteration in the base sequence of DNA which alters the genetic information stored in DNA

mV: *see* millivolt

myasthenia gravis: a neuromuscular disease associated with fatigue and skeletal muscle weakness; due to decreased numbers of ACh receptors at the motor end plate

myelin: insulating material covering the axons of many neurons; consists of multiple layers of myelin-forming cell plasma membranes that are wrapped concentrically around the axon; increases the velocity of action-potential propagation along the axon

myenteric plexus: a network of nerve cells lying between the circular and longitudinal muscle layers in the esophagus, stomach, and intestinal walls

myoblast: the embryological cell that gives rise to muscle fibers

myocardium: the cardiac muscle that forms the walls of the heart

myoepithelial cells: specialized contractile cells surrounding the alveoli of mammary glands

myofibrils: longitudinal bundles of thick and thin contractile filaments arranged in a repeating sarcomere pattern; located in the cytoplasm of striated muscles

myogenic: originating within a muscle

myoglobin: a protein in muscle fibers, similar to hemoglobin, that binds oxygen

myometrium: uterine smooth muscle

myopia: *see* nearsighted

myosin: the contractile protein which forms the thick filaments of muscle

myosin ATPase: an enzymatic site on the globular head of myosin which catalyzes the breakdown of ATP to ADP and inorganic phosphate with the release of chemical energy that is used to produce the force of muscle contraction

NaCl: sodium chloride

NAD: *see* nicotinamide adenine dinucleotide

Na$^+$–K$^+$ ATPase: an enzyme located in the plasma membranes of most cells; splits ATP in the presence of sodium and potassium ions and releases energy which is used to actively transport sodium out of the cell and potassium into the cell

nanometer (nm): 10^{-9} m; equivalent to 0.000000039 in; a millimicron; formerly abbreviated mμ

nearsighted: pertaining to the defect in vision that occurs because the eyeball is too long for the lens, so that images of distant objects are focused in front of the retina

negative balance: the situation in which the rates of loss of a substance from the body are greater than the rates of gain of that substance; also used for physical parameters, such as body temperature and energy

negative feedback: an aspect of control systems in which the ultimate effects of the system counteract the original stimulus

negative nitrogen balance: the state in which loss of nitrogen from the body is greater than intake of nitrogen

nephron: the basic functional unit of the kidney whose function is the formation of urine; consists of a glomerulus, Bowman's capsule, proximal tubule, Henle's loop, distal tubule, and collecting duct

nerve: a collection of nerve fibers (axons) in the peripheral nervous system, i.e., outside the brain or spinal cord

nerve fiber: *see* axon

net flux: the net amount of a substance diffusing across a surface in a unit of time; equal to the difference between two opposite unidirectional fluxes; $F = k_D(C_1 - C_2)$ where k_D is the diffusion coefficient, and C_1 and C_2 are the concentrations in compartments 1 and 2

neuroeffector junction: functional relationship between a nerve fiber and a muscle or gland cell, e.g., a neuromuscular junction

neuroendocrinology: the study of the interrelated functions of the nervous and endocrine systems

neuromodulator: a chemical which amplifies or dampens neuronal activity by altering the neurons directly or by influencing the effectiveness of a neurotransmitter; it does not act as a specific neurotransmitter

neuromuscular junction: an anatomically specialized junction between an axon terminal of a motor neuron and a skeletal muscle cell where the chemical mediator (acetylcholine), released in response to action potentials in the nerve, diffuses across an extracellular gap to initiate action potentials in the muscle membrane

neuron: a nerve cell

neurotransmitter: a chemical agent released by one neuron which acts upon a second neuron or upon a muscle or gland cell altering its electrical state or activity

neutral fat: triacylglycerol; formerly called triglyceride; consists of a molecule of glycerol to which three fatty acids are attached

neutral molecule: any molecule having no polar or ionized groups and thus having no electric charge

neutron: an uncharged subatomic particle located in the nucleus of an atom

neutrophil: a polymorphonuclear granulocytic leukocyte whose granules show preference for neither eosin nor basic dyes

neutrophil exudation: the amebalike movement of neutrophils from the lumen of capillaries out into the extracellular space of tissues

NH_3: ammonia

NH_4^+: ammonium ion

niacin: one of the B-complex vitamins; an essential ingredient of the coenzyme NAD

nicotinamide adenine dinucleotide (NAD): a coenzyme that transfers hydrogen from one reaction to another

nicotinic receptors: those acetylcholine receptors that respond to the drug nicotine; primarily, the receptors at the motor end plate and the receptors on postganglionic neurons of the autonomic nervous system

night vision: the only vision possible under conditions of low illumination, i.e., rod vision

nm: *see* nanometer

nociceptor: a sensory receptor whose stimulation causes the sensation of pain

nonpolar molecule: a molecule containing predominately neutral chemical bonds in which electrons are shared equally between atoms and thus having no polar or ionized groups

nonspecific immune responses: responses which do not require previous exposure to a particular foreign material but nonselectively protect against them without having to recognize their specific identities

nonspecific pathway: a chain of synaptically connected neurons in the brain or spinal cord, all of which are activated by sensory units of several different modalities

nonspecific thalamic nuclei: nuclei in the cerebrum whose axons project to widely dispersed areas of cerebral cortex; functionally, these nuclei are not clearly distinguishable from the specific thalamic nuclei

norepinephrine: a catecholamine neurotransmitter released at most sympathetic postganglionic endings, from the adrenal medulla, and in many regions of the central nervous system

novelty detectors: neurons in nonsensory parts of the thalamus which respond to inputs from many parts of the body but rapidly cease responding to repeated stimuli as the person's attention to them decreases

novocaine: a local anesthetic that works by preventing the increased membrane permeability to sodium ions required to produce an action potential in the axon of a neuron

noxious: poisonous; harmful to health

nuclear envelope: the double membrane surrounding the nucleus of a cell

nuclear pores: openings in the nuclear envelope through which mRNA passes from the nucleus into the cytoplasm

nuclear protein: the protein portion of chromatin; involved in the control of gene activity

nucleic acid: a straight-chain polymer of nucleotides in which the phosphate of one of the repeating nucleotide subunits is linked to the sugar of the adjacent one; functions in the storage and transmission of genetic information; includes both DNA and RNA

nucleolus: a prominent, densely staining nuclear structure consisting of granular and filamentous elements; contains the portions of DNA which code for rRNA

nucleoside: a molecule consisting of a purine or pyrimidine base and a sugar (ribose or deoxyribose)

nucleotide: the molecular subunit of nucleic acids, consisting of a purine or pyrimidine base, a sugar, and phosphoric acid

nucleus: a large spherical or oval membrane-bound cell organelle present in most cells; contains most of the cell's DNA and some of its RNA; (neural) a cluster of neuronal cell bodies within the CNS

obesity: an excessive accumulation of fat by the adipose tissues of the body

occipital cortex: the posterior region of the cerebral cortex where the primary visual cortex is located

off **response:** an increase in cell activity upon removal of the stimulus; a decrease in the background activity of a cell in response to stimulation

Ohm's law: the relationship between current (I), voltage (E), and resistance (R) such that $I = E/R$

olfactory: pertaining to the sense of smell

olfactory mucosa: the region of mucous membrane in the upper part of the nasal cavity containing the receptors for the sense of smell

olfactory nerve: cranial nerve I; consists of the axons of bipolar cells whose peripheral terminals contain the chemoreceptors for the sense of smell

oligodendroglia: a type of glial cell; responsible for the generation of myelin around nerve fibers in the central nervous system

oligomeric protein: a protein consisting of more than one polypeptide chain

on **response:** the response of a neuron or receptor to the onset of stimulation

oogonia: primitive ova which, upon mitotic division before birth, give rise to the primary oocytes

operator gene: the initial gene of an operon which controls the transcription of the genes in the operon; the site of repressor binding

operon: a collection of genes which can be induced or repressed as a unit

opponent color cell: a neuron whose activity is increased by input from cones of one color sensitivity but decreased by cones having a different sensitivity

opsin: the protein component of a photopigment molecule; the rods and three types of cones have different opsins

optic nerve: cranial nerve II; formed of the axons of the retinal ganglion cells

organ: a collection of tissues joined in a structural unit to serve a common function; e.g., the heart is an organ consisting of muscle, nerve, connective, and epithelial tissues which together serve as a pump for blood

organ of Corti: the collection of structures in the inner ear capable of transducing the energy of waves in a fluid medium into the electrochemical energy of action potentials in the auditory nerve

organ system: a collection of organs which together serve an overall function; e.g., the kidneys, ureters, urinary bladder, and urethra are the organs that comprise the urinary system; there are ten organ systems in the body

organelles: multimolecular structural components of cells which perform specialized functions; e.g., the mitochondria and endoplasmic reticulum are cell organelles

organic: pertaining to molecules containing the element carbon

organic acid: an acid molecule containing carbon atoms, e.g., carbonic and lactic acids

orienting response: an animal's behavior in response to the presentation of a novel stimulus, i.e., the animal stops what it is doing and looks around, listening intently and turning toward the stimulus source

osmolarity: the total solute concentration of a solution; a measure of water concentration in that the higher the osmolarity of a solution, the lower its water concentration

osmoreceptor: a neural receptor that responds to changes in the osmolarity of the surrounding fluid

osmosis: the net diffusion of water due to a difference in water concentration on the two sides of a membrane impermeable to solute molecules; the water always moves from the region of low solute concentration (high water concentration) to the region of high solute concentration (low water concentration)

osmotic equilibrium: a condition in which the solute (osmolar) concentrations on the two sides of a membrane are equal; thus, the water concentrations are also equal

osmotic hemolysis: the swelling and ultimate rupture of red cell membranes, which allows the hemoglobin to leak out; caused by placing the cells in a hypotonic solution

osmotic pressure: the pressure that must be applied to a solution to prevent the osmotic flow of pure water into the solution across a membrane that is impermeable to the solute molecules in the solution

osteoblast: the type of cell responsible for laying down the protein matrix of bone

otoliths: calcium-carbonate "stones" in the gelatinous mass of the utricle and saccule

oval window: the membrane-covered opening separating the inner-ear cavity from the scala vestibuli of the inner ear; receives the footplate of the stapes (one of the three middle-ear bones)

ovarian follicle: the ovum and its encasing follicular, granulosa, and theca cells at any stage of its development prior to ovulation

ovary: the gonad in the female

oviducts: see uterine tubes

ovulation: the release of an ovum, surrounded by its zona pellucida and cumulus, from the ovary

ovum: (pl. ova) the female gamete which is fertilized by a sperm cell at the time of conception

oxidative deamination: a reaction in which an amino group is removed from an amino acid as a molecule of ammonia (NH_3) and replaced by an oxygen atom to form a keto acid

oxidative phosphorylation: the process by which the energy released from the combination of hydrogen with molecular oxygen is used to form ATP from ADP and inorganic phosphate; occurs in the mitochondria through a highly specialized coupling process involving cytochromes embedded in the inner mitochondrial membrane

oxyhemoglobin: (HbO_2) hemoglobin combined with oxygen

oxyntic cells: see parietal cells

oxytocin: a peptide hormone of nine amino acids synthesized in the hypothalamus and released from the posterior pituitary; stimulates the mammary glands to release milk and the uterus to contract

P wave: the component of the electrocardiogram reflecting depolarization of the atria

pacemaker potential: the spontaneous depolarization to threshold potential of the plasma membrane of some nerve and muscle cells

pacinian corpuscle: a mechanoreceptor whose structure, which resembles onionlike layers around a central nerve fiber, is specialized to respond to vibrating stimuli

pain: a sensation resulting from tissue damage (or threatened tissue damage) and often accompanied by an emotional component such as fear, anxiety, or a sense of unpleasantness

palpitation: a rapid, often irregular beating of the heart that can be felt by the individual

pancreas: a gland in the abdomen near the stomach and connected by a duct to the small intestine; contains both endocrine gland cells, which secrete the hormones insulin and glucagon into the bloodstream, and exocrine gland cells, which secrete digestive enzymes and bicarbonate into the intestine

pancreatic duct: the large duct that carries the exocrine secretion of the pancreas into the duodenum

pancreatic lipase: an enzyme secreted by the exocrine pancreas; acts on triacylglycerols to form 2-monoglycerides and free fatty acids

pancreozymin: see cholecystokinin

pantothenic acid: one of the B-complex vitamins

paracrine agent: a chemical agent which, when released, exerts its effects on tissues located near the site of its secretion, in contrast to a hormone which travels by way of the blood to exert its effects on cells far removed from the site of its secretion; excludes neurotransmitters

paradoxical sleep: the state of sleep associated with small, rapid EEG oscillations (indistinguishable from those of EEG arousal) and with complete loss of tone in the postural muscles, irregular respiration and heart rate, and dreaming

parasympathetic nervous system: that portion of the autonomic nervous system whose preganglionic fibers leave the central nervous system at the brainstem and sacral portion of the spinal cord; most of its postganglionic fibers release the neurotransmitter acetylcholine

parasympathomimetic: a chemical that produces effects similar to those of the parasympathetic nervous system

parathormone: *see* parathyroid hormone

parathyroid glands: the four parathyroid-hormone secreting glands on the back of the thyroid gland

parathyroid hormone (PTH): a peptide hormone secreted by the parathyroid glands; also called parathormone; regulates calcium and phosphate concentrations of the internal environment

parietal cells: gastric gland cells which secrete hydrochloric acid and intrinsic factor; also called oxyntic cells

parietal lobe: the region of the cerebral cortex where sensory cortex and some association cortex are located

parkinsonism: a disease characterized by tremor, rigidity, and a delay in the initiation of movement; due to degeneration of the dopamine-liberating axons from the substantia nigra

parotid: one of the three pairs of salivary glands

partial pressure: the pressure exerted by one molecular species of a gas; e.g., P_{O_2}, the partial pressure of oxygen, is that part of the total pressure that is due to oxygen molecules; in a pure gas, the partial pressure equals the total pressure; a measure of the concentration of a gas in a mixture of gases

parturition: birth; delivery of an infant

passive immunity: a resistance to infection resulting from the direct transfer of antibodies or T cells from one person (or animal) to another

P_{CO_2}: the partial pressure of carbon dioxide

pepsin: an enzyme formed in the stomach from the precursor pepsinogen; breaks protein down to peptide fragments

pepsinogen: the inactive precursor of the enzyme pepsin; secreted by the chief cells of the gastric mucosa

peptide: a short polypeptide chain

peptide bond: the chemical bond $(-NH-\overset{\overset{O}{\|}}{C}-)$ join-

ing two amino acids, formed by a reaction between the amino group of one amino acid and the carboxyl group of a second amino acid resulting in the release of a water molecule; forms the backbone of protein molecules

perception: the conscious experience of objects and events of the external world which we acquire from the neural processing of sensory information

pericardium: the fibrous connective-tissue sac that surrounds the heart

peripheral chemoreceptors: the carotid and aortic bodies that respond to changes in blood oxygen, carbon dioxide, and hydrogen ions

peripheral nervous system: the nerve fibers extending from the brain and spinal cord

peripheral proteins: membrane proteins which are water soluble and can be removed from the membrane by solutions which alter the electric charge on the proteins; thus, they are presumably bound to the membrane surface

peripheral thermoreceptors: cold and warm receptors in the skin and certain mucous membranes

peristaltic wave: a progressive wave of muscle contraction that proceeds along the wall of a tube, compressing the lumen of the tube and causing the contents of the tube to move

peritubular capillaries: capillary network closely associated with the tubular components of the kidney nephrons

permissiveness: (of a hormone) the situation whereby small quantities of one hormone are required in order for a second hormone to exert its full effect upon a cell

pernicious anemia: a disease in which erythrocytes fail to proliferate and mature because of a deficiency of vitamin B_{12}

pH: the negative logarithm to the base 10 of the hydrogen-ion concentration; a measure of the acidity of a solution: the pH decreases as the acidity increases

phagocytosis: a form of endocytosis in which large multimolecular particles, such as bacteria, are taken into the cell by invagination of the plasma membrane

phagosome: the membrane-enclosed pouch formed when a phagocyte engulfs a microbe or other particulate matter

phantom limb: a limb that has been amputated but is perceived to be the source of sensations

pharmacological effects: (of a hormone) the effects produced by much larger quantities of a hormone in the body than are normally present

pharynx: throat; a passage common to the routes followed by food and air

phase-locked: pertaining to two oscillating events occurring at the same frequency; sounds at the two ears occurring so that the sound waves peak at exactly the same time

phasic: intermittent, as opposed to tonic

phosphodiesterase: the enzyme which catalyzes the breakdown of the second messenger cyclic AMP to AMP (adenosine monophosphate)

phospholipid: a subclass of lipid molecules with a glycerol backbone to which are attached two fatty acids and, to the third hydroxyl group of clycerol, a phosphate group ($-PO_4-$) plus a small polar or ionized nitrogen-containing molecule; a major component of cell membranes

phosphoric acid: an acid generated during the catabolism of phosphorus-containing compounds; dissociates to form phosphate ions and hydrogen ions

phosphorylation: addition of a phosphate group to an organic molecule

photopigment: a light-sensitive molecule altered by the absorption of photic energy of certain wavelengths; consists of an opsin bound to a chromophore

photosynthesis: the chemical process by which plants are able to capture the energy in sunlight, with which they form glucose and oxygen using carbon dioxide and water as substrates

phrenic nerve: a nerve that passes from the cervical region of the spinal cord to the diaphragm muscle

PIH: *see* prolactin inhibiting hormone

pinocytosis: *see* endocytosis

pitch: the degree of how low or high a sound is perceived; related to the frequency of the sound wave

pituitary: an endocrine gland which lies in a pocket of bone just below the hypothalamus; formerly called the hypophyseal gland, or hypophysis

placenta: a combination of interlocking fetal and maternal tissues in the uterus which serves as the organ of molecular exchange between maternal and fetal circulations

plasma: the fluid, noncellular portion of the blood; a component of the extracellular fluid compartment

plasma cells: cells which are derived from lymphocytes and secrete antibodies

plasma membrane: the cell membrane which forms the outer surface of the cell and separates its contents from the surrounding extracellular fluid

plasmin: a proteolytic enzyme able to decompose fibrin and, thereby, dissolve blood clots

plasminogen: the inactive precursor of the enzyme plasmin

plasticity: (neural) the ability of neural tissue to change its responsiveness to stimulation because of its past history of activation

platelet: a cell fragment that is present in the blood and plays a role in blood clotting; formed in bone marrow from megakaryocytes

pleura: a thin sheet of cells firmly attached to the entire interior of the thoracic cage and folding back upon itself to form two completely enclosed sacs within the thoracic cage; the interior of each sac is attached to the surface of the lung

pneumothorax: air in the intrapleural space

P_{O_2}: the partial pressure of oxygen

polar body: the small cell resulting from the unequal distribution of cytoplasm during the divisions of the primary and secondary oocytes; once formed, the polar body has no further biological function and soon dies

polar molecule: substance containing a number of electrically polarized chemical bonds in which electrons are shared unequally between atoms, the region of the molecule to which the electrons are drawn carrying a negative charge and the region from which they are drawn carrying a positive charge; polar molecules are electrically attracted to polar water molecules and are thus soluble in water

polarized: having two electric poles, one negative and one positive

polarized covalent bond: a chemical bond in which two electrons are shared unequally between two atoms, the atom to which the electrons are drawn carrying a negative charge and the other atom carrying a positive charge

polymer: a very large molecule formed by linking together many smaller similar subunits

polymodal: a neuron that responds to stimuli of more than one modality

polymorphonuclear granulocyte: a subclass of leukocytes; there are three types: neurophils, eosinophils, and basophils

polypeptide: polymer consisting of amino acid subunits joined in sequence by peptide bonds between the amino group of one amino acid and the carboxyl group of the adjacent one

polyribosome: one strand of mRNA to which are attached a number of ribosomes; each ribosome translates the coded message in mRNA into a polypeptide chain as it moves along the strand of mRNA

polysaccharide: a large carbohydrate molecule formed by linking many monosaccharide subunits together; for example, glycogen and starch

polysynaptic reflex: a reflex employing one or more interneurons in its reflex arc

polyunsaturated fatty acid: a fatty acid that contains more than one double bond

pool: the body's readily available quantity of a particular substance; frequently identical to the quantity present in the extracellular fluid

portal vessels: blood vessels which link two capillary networks

positive balance: occurs when the rates of gain of a substance by the body are greater than the rates of loss of that substance; also used for physical parameters, such as temperature, and energy

positive feedback: an aspect of control systems in which the ultimate effects of the system influence it in such a way that these effects are increased, i.e., that the original stimulus is strengthened; an unstable situation because the output of the system will rapidly and progressively increase in magnitude

postabsorptive state: the period during which nutrients are not being absorbed by the gastrointestinal tract and energy must be supplied by the body's endogenous stores

posterior pituitary: that portion of the pituitary from which oxytocin and antidiuretic hormone are released

postganglionic neuron: a neuron of the autonomic nervous system whose cell body lies in a ganglion and whose axon terminals form neuroeffector junctions with smooth muscle, cardiac muscle, or glands; a neuron which conducts impulses away from a ganglion toward the periphery

postsynaptic neuron: a neuron conducting information away from a synapse; a neuron acted upon by a synapse

postsynaptic potential: a local potential which arises in the postsynaptic neuron in response to the activation of synapses upon it; *see also* excitatory postsynaptic potential, inhibitory postsynaptic potential

postural reflexes: those reflexes that maintain or restore stable, upright posture

potential: (or potential difference) the voltage difference between two points

potential energy: that energy associated with an object because of its position or internal structure; chemical energy is a form of potential energy

prandial drinking: drinking associated with a meal

precapillary sphincter: a ring of smooth muscle around a "true" capillary at the point at which it exits from a thoroughfare channel or arteriole

preganglionic neuron: a neuron of the autonomic nervous system whose cell body lies in the central nervous system and whose axon terminals lie within a ganglion; a neuron which conducts impulses from the central nervous system toward a ganglion

preprogram: a sequence of neuronal activity that results in a movement independent of afferent input that may occur during the course of the movement

presbyopia: the impairment in vision resulting from stiffening of the lens, which makes accommodation difficult

pressure: the force exerted on a surface

"pressure" autoregulation: the ability of individual organs and tissues to alter their vascular resistance so as to maintain a relatively constant blood flow in the face of changing blood pressure

presynaptic inhibition: a relationship between two neurons in which the axon terminals of one neuron end on the synaptic knob of a second presynaptic neuron; action potentials in the first neuron decrease the release of neurotransmitter from the second neuron and thus decrease the magnitude of the postsynaptic potential produced in the third (postsynaptic) neuron

presynaptic neuron: a neuron conducting information toward a synapse

PRH: *see* prolactin releasing hormone

primary afferent: *see* afferent neuron

primary follicle: any cell cluster in the ovaries consisting of one ovum surrounded by a single layer of follicular (granulosa) cells

primary oocyte: germ cell of the female; undergoes the first meiotic division to form a secondary oocyte (and a polar body)

primary spermatocyte: a male germ cell derived from the spermatogonia; undergoes meiotic division to form two secondary spermatocytes

primary structure: (of a protein) the sequence of amino acids along the polypeptide chain of the molecule

primary visual cortex: the area of cerebral cortex that receives axons from lateral geniculate neurons; i.e., the first part of cortex to be activated by the visual pathways

progesterone: a steroid hormone secreted primarily by the corpus luteum and placenta; stimulates secretion by uterine glands, inhibits contraction of uterine smooth muscle, and stimulates breast growth

program: a related sequence of neural activity; *see also* subset program

projection neuron: a neuron having a long axon which connects it with distant parts of the nervous system or with muscles or glands

prolactin: a polypeptide hormone secreted by the anterior pituitary; stimulates milk secretion by the mammary glands

prolactin inhibiting hormone (PIH): the hypothalamic hormone that inhibits prolactin secretion by the anterior pituitary

prolactin releasing hormone (PRH): the hypothalamic hormone that stimulates prolactin secretion by the anterior pituitary

proliferative phase: that stage of the menstrual cycle between menstruation and ovulation during which the endometrium repairs itself and grows

prostaglandins: a group of unsaturated, modified fatty acids which function as chemical messengers (paracrine agents); synthesized in most —possibly all—cells of the body

prostate gland: a large gland encircling the urethra in the male; secretes fluid into the urethra

protease: an enzyme that breaks the peptide bonds between amino acids in polypeptide chains

protein: a large polymer consisting of one or more linear sequences of amino acid subunits joined by peptide bonds

protein binding site: *see* binding site

protein kinase: a group of enzymes, each of which catalyzes the addition of a phosphate group to a specific protein

prothrombin: the inactive precursor to the enzyme thrombin; produced by the liver and normally present in the plasma

protomer: each of the single polypeptide chains in an oligomeric protein

proton: a subatomic particle located in the nucleus of atoms and carrying one positive charge

protoplasm: an old term used to refer to the material within cells

proximal tubule: the first tubular component of a nephron after Bowman's capsule

psilocin: a mind-altering drug that is a methylated derivative of serotonin

psilocybin: a hallucinogenic agent found in some mushrooms

psychological fatigue: the mental factors that cause individuals to stop exercising even though their muscles are not depleted of ATP and are still able to contract

PTH: *see* parathyroid hormone

ptyalin: *see* amylase

puberty: the attainment of sexual maturity in the sense that conception becomes possible; as commonly used, it refers to the 3 to 5 years of sexual development culminating in the attainment of sexual maturity; onset usually occurs between the ages of 10 to 15 years

pulmonary: pertaining to the lungs

pulmonary circulation: the portion of the cardiovascular system between the pulmonary trunk artery (as it leaves the right ventricle) and the pulmonary veins (as they enter the left atrium); the circulation through the lungs

pulmonary edema: accumulation of fluid in the lung interstitium and air sacs

pulmonary stretch receptor: an afferent nerve ending (or cell closely associated with it) located in the airway smooth muscle and activated by lung inflation

pulmonary surfactant: *see* surfactant

pulmonary trunk: the large artery that carries blood from the right ventricle of the heart to the lungs

pulmonary valve: the valve between the right ventricle of the heart and the pulmonary trunk

pulmonary ventilation: *see* total pulmonary ventilation

pulse pressure: the difference between systolic and diastolic blood pressures

pupil: the black-appearing opening in the iris of the eye through which light passes to reach the retina

pupillary reflex: contraction of the pupil on exposure of the retina to light

purine bases: adenine and guanine; double-ring, nitrogen-containing subunits of nucleotides and nucleosides

pyloric sphincter: a ring of smooth muscle and connective tissue between the terminal antrum of the stomach and the first segment of the small intestine

pyramidal tract: *see* corticospinal pathway

pyridoxine: vitamin B_6

pyrimidine bases: cytosine, thymine, and uracil; single-ring, nitrogen-containing subunits of nucleotides and nucleosides

pyrogen: an endogenously produced chemical released from white blood cells in the presence of infection or inflammation; acts on thermoreceptors in the hypothalamus (and perhaps other brain areas), altering their rate of firing and their input to the thermoregulatory integrating centers

pyruvate: the anion formed when the carboxyl group of pyruvic acid dissociates, i.e., loses a hydrogen ion

pyruvic acid: the three-carbon keto-acid intermediate in the glycolytic pathway that, in the absence of oxygen, forms lactic acid or, in the presence of oxygen, forms an acetyl group (which can then combine with coenzyme A and enter the Krebs cycle or the pathways for fat synthesis)

QRS complex: the component of the electrocardiogram corresponding to the depolarization of the ventricles

radiation: the emission of heat from the surfaces of objects

rarefaction: a region in a sound wave where the density of air molecules is low

RAS: *see* reticular activating system

rate-limiting enzyme: the enzyme in a metabolic pathway most easily saturated with substrate; catalytic activity of the step involving the rate-limiting enzyme determines the flow of substrate through the entire metabolic pathway; usually the site of allosteric regulation in the pathway

reactant: a molecule which enters a chemical reaction; in enzyme-catalyzed reactions, the reactant is called the substrate molecule

readiness potential: electrical activity that can be recorded over wide areas of association cortex during the "decision-making period," about 0.8 s before a movement begins

receptive field: (of a neuron) the region which, if stimulated, causes activity in that neuron

receptive relaxation: the decrease in smooth muscle tension in the walls of hollow organs in response to distension

receptor: (in sensory systems) the specialized peripheral ending of an afferent neuron or a separate cell intimately connected to it which detects changes in some aspect of the environment; (in chemical communication) specific proteins either in the plasma membrane or cytosol of a target cell with which a chemical mediator combines to exert its effects

receptor density: the number of receptors per unit area

receptor potential: (also generator potential) a graded potential which arises in the ending of an afferent neuron or in a specialized cell intimately associated with it, in response to stimulation of the receptor

reciprocal innervation: inhibition of motor neurons activating those muscles whose contraction would oppose the intended movement; e.g., inhibition of flexor motor neurons during extensor motor neuron activation

recombinant DNA: DNA experimentally formed by joining portions of two DNA molecules previously fragmented by a restriction enzyme

recruitment: the activation of additional cells in response to a stimulus of increased magnitude

rectum: a short segment of the large intestine between the sigmoid colon and anus

reduced hemoglobin (Hb): deoxyhemoglobin; hemoglobin not combined with oxygen

reflex: the sequence of events elicited by a stimulus; a biological control system mediated by a reflex arc composed of neural, hormonal, or neural and hormonal elements

reflex arc: the components that mediate a reflex; usually, but not always, includes a receptor, afferent pathway, integrating center, efferent pathway, and effector

reflex response: that change brought about by the action of a stimulus upon a reflex arc; effector response

refractory: pertaining to the refractory period

refractory period: the time during which an excitable membrane does not respond to a stimulus whose magnitude is normally sufficient to trigger an action potential; *see also* absolute refractory period, relative refractory period

regulator site: the site on a protein at which interaction with a ligand (the modulator) allosterically alters the properties of the protein binding sites

"regulatory" appetite: the desire for a substance in proportion to the deficiency of that substance

relative refractory period: the time during which an excitable membrane will produce an action potential only in response to stimuli of greater strength than the usual threshold strength

relaxation time: the time interval from the development of peak tension by a muscle until the tension has decreased to zero

REM sleep: *see* paradoxical sleep

renal: pertaining to the kidneys

renal blood flow: the volume of blood delivered to the kidneys per unit of time

renal medulla: the inner region of the kidney

renal pelvis: a large central cavity at the base of each kidney; receives urine from the collecting ducts and empties it into the ureter

renin: an enzyme secreted by the kidney; it catalyzes the splitting off of angiotension I from angiotensionogen in the blood

replication: formation of a new molecule of DNA identical to the original one

repolarize: to restore the value of the transmembrane potential toward its resting level

repression: inhibition of enzyme synthesis in the presence of product molecules synthesized by that enzyme (or the enzyme chain of which it is a part); product molecules inhibit enzyme synthesis by effects upon the genes which code for the protein structure of the enzymes

repressor: a protein molecule which can bind to an operator gene and inhibit the transcription of the gene into mRNA

repressor gene: a portion of DNA that contains the genetic instructions for synthesis of a repressor protein

reproductive duct system: (male) seminiferous tubules, epididymis, vas deferens, ejaculatory duct, urethra; (female) uterine tubes, uterus, vagina

residual volume: the volume of air remaining in the lungs after a maximal expiration

resistance: the hinderance to movement through a particular substance or tube, e.g., the hinderance to the movement of electric charge through extracellular fluid, or the movement of blood through the blood vessels

respiration: the consumption of molecular oxygen during metabolism; the exchange of gas between the cells of an organism and the external environment

respiratory acidosis: an acidosis resulting from failure of the lungs to eliminate carbon dioxide as fast as it is produced

respiratory alkalosis: an alkalosis resulting from the elimination of carbon dioxide from the lungs faster than it is produced

respiratory distress syndrome: (of the newborn) a disease afflicting premature infants in whom the surfactant-producing cells are too immature to function adequately

respiratory "pump": the effect of the changing intrathoracic and intraabdominal pressures associated with respiration on blood flow through vessels in the thoracic and abdominal cavities

respiratory quotient (RQ): the ratio of CO_2 produced to O_2 consumed during metabolism

respiratory rate: the number of breaths per minute

respiratory system: the structures involved in the exchange of gases between the blood and the external environment, i.e., the lungs; the series of airways leading to the lungs, and the chest structures responsible for movement of air in and out of the lungs

resting membrane potential: (*also* resting potential) the voltage difference between the inside and outside of a cell in the absence of excitatory or inhibitory stimulation

resting tremor: a tremor present in persons who are completely at rest; occurs with basal ganglia disease

restriction enzyme: a bacterial enzyme that splits DNA into a number of fragments, acting at different loci in the two strands of the polymer and at points containing a particular sequence of nucleotides

retching: the events that normally accompany vomiting, i.e., deep inspiration, closure of the glottis, elevation of the soft palate, and contraction of abdominal and thoracic muscles, without the actual expulsion of the stomach's contents

reticular activating system: a collection of neurons within the brainstem reticular formation and its thalamic extension whose activity is concerned with alertness and direction of attention to selected events

reticular formation: an extensive network of finely branched neurons extending through the core of the brainstem and linked functionally to a similar group of cells in the midline of the thalamus

retina: a thin layer of neural tissue containing the receptors for vision and lining the back of the eyeball

retinal: a derivative of vitamin A; forms the chromophore component of a photopigment

retrograde amnesia: loss of memory for a time immediately preceding a memory-disturbing trauma such as a blow on the head

reversible reaction: a chemical reaction that can readily proceed in either direction because only small exchanges of energy are involved in the reaction

Rh factor: a group of erythrocyte plasma-membrane antigens which may (Rh+) or may not (Rh−) be present

rhodopsin: the photopigment in rods; a light-absorbing glycoprotein

rhythm method: a contraceptive technique in which couples refrain from sexual intercourse near the time of ovulation

riboflavin: vitamin B_2

ribonucleic acid (RNA): a nucleic acid involved in the decoding of the genetic information and its transfer to the site of protein synthesis; its nucleotides contain the sugar ribose to which is attached one of four bases (adenine, guanine, cytosine, or uracil); exists in three forms: messenger RNA, ribosomal RNA, and transfer RNA

ribosomal RNA (rRNA): a polymer of ribose-containing nucleotides which is synthsized in the nucleus but moves to the cytoplasm where it forms part of the ribosome

ribosome: a cytoplasmic particle which mediates the linking together of amino acids to form proteins; attached to the membranes of the endoplasmic reticulum or suspended in the cytoplasm as a free particle

rickets: a disease in which new bone matrix is inadequately calcified due to a deficiency of 1,25-dihydroxy vitamin D_3

rigor mortis: death rigor; a stiffness of the muscles resulting from a loss of ATP which is necessary for the dissociation of myosin cross bridges from the actin thin filaments; begins 3 to 4 hours after death, is complete after about 12 hours, and slowly disappears over the next 48 to 60 hours

RNA: *see* ribonucleic acid

rod: one of the two receptor types for photic energy; contains the photopigment rhodopsin

rooting reflex: the reactions of turning the head, opening the mouth, and suckling in response to a touch on the cheek; normally present in infants

round window: the membrane-covered opening separating the scala tympani of the inner ear from the middle-ear cavity

RQ: *see* respiratory quotient

rRNA: *see* ribosomal RNA

SA node: *see* sinoatrial node

saccade: a very fast, short, jerking movement of the eyeball

saliva: the secretory product of the salivary glands; a watery solution of various salts and proteins, including the mucins and amylase (ptyalin)

salivary gland: one of three paired glands which secretes into the mouth a solution containing mucus and the carbohydrate-digesting enzyme amylase

sarcomere: the repeating structural unit of a muscle myofibril; the region of a myofibril extending between two adjacent Z lines

sarcoplasmic reticulum: the endoplasmic reticulum in muscle fibers; forms sleevelike structures around each sarcomere in striated muscle; site of storage and release of calcium ions that bind to troponin to initiate contractile activity

saturated fatty acid: a fatty acid that does not contain any double bonds

saturation: (of binding sites) the degree to which binding sites are occupied by ligand; if all are occupied, the binding sites are fully saturated, if half are occupied, the saturation is 50 percent; determined by concentration of the free ligand in solution and the affinity of the binding site for the ligand

scala tympani: the fluid-filled inner-ear compartment that receives sound waves from the basilar membrane and transmits them to the round window

scala vestibuli: the fluid-filled inner-ear compartment that receives sound waves from the oval window and transmits them to the basilar membrane and cochlear duct

schizophrenia: a disease (or a family of diseases) characterized by altered motor behavior, distortions in perception, disturbed thinking, altered mood, and abnormal interpersonal behavior

Schwann cell: a nonneural cell which surrounds all peripheral nerve fibers; the plasma membrane of this cell forms the myelin sheath around a myelinated axon; its analogous cell in the central nervous system is the oligodendroglia

scrotum: the sac that contains the testes and epididymides

SDA: *see* specific dynamic action

sebaceous glands: glands in the skin which secrete a fatty substance

SEC: *see* series elastic component

second messenger: an intracellular substance which is formed or released as a result of combination of the original extracellular chemical messenger (i.e., the "first" messenger) with a receptor in the plasma membrane; alters some aspect of the cell's function and thereby mediates the cell's response to the first messenger

second-order neuron: the interneuron directly activated by the afferent neuron

secondary peristalsis: esophageal peristaltic waves not immediately preceded by the pharyngeal phase of a swallow

secondary sexual characteristics: the many external differences (hair, body contours, etc.) between male and female which are not directly involved in reproduction

secondary spermatocyte: a male germ cell derived from a primary spermatocyte

secondary structure: (of a protein molecule) the alpha-helical and random-coil conformations of portions of a polypeptide chain

secretin: a peptide hormone secreted by cells in the upper part of the small intestine; the first hormone discovered (in 1902); stimulates the pancreas to secrete bicarbonate into the small intestine

secretory phase: that stage of the menstrual cycle following ovulation and formation of the corpus luteum during which a secretory type of endometrium develops

segmentation: a series of stationary rhythmic contractions and relaxations of rings of intestinal smooth muscle which mixes the luminal contents of the intestine

self-stimulation: an experimental technique in which an animal can perform an action resulting in the stimulation of electrodes implanted in its brain

semen: the sperm-containing fluid of the male ejaculate

seminal vesicles: a pair of large glands which secrete fluid into the two vas deferens

seminiferous tubule: a tubule in the testis that contains the cells which differentiate into sperm cells

semipermeable membrane: a membrane permeable to some substances but not to others

sensor: first component of a control system; detects changes in some aspect of the environment; a receptor

sensory deprivation: an experimental situation affording isolation as complete as possible from sensory stimulation, even including restriction of movements

sensory information: afferent information which has a conscious correlate

sensory unit: a single afferent neuron plus all the receptors it innervates

series elastic component (SEC): the elastic structures associated with a muscle fiber through which the force generated by the cross bridges is transmitted to the load on a muscle; includes the cross bridges, thick and thin filaments, Z lines, and tendons

serosa: a connective-tissue layer surrounding the outer surface of the stomach and intestinal walls

serotonin: (*also* 5-hydroxytryptamine, 5-HT) a probable neurotransmitter; also functions as a paracrine agent in blood platelets and specialized cells of the digestive tract

Sertoli cell: a type of cell intimately associated with the developing germ cells in the seminiferous tubules; creates a "blood-testis barrier" and probably mediates hormonal effects on the tubules

serum: blood plasma from which the fibrinogen has been removed as a result of clotting

sex chromatin: a readily detected nuclear mass not usually found in male cells; consists of a condensed X chromosome

sex chromosomes: the two chromosomes, X and Y, which determine the genetic sex of an individual, genetic males having one X and one Y, and genetic females having two X chromosomes

sex determination: the genetic basis of an individual's sex, XY determining male and XX, female

sex differentiation: the development of male or female reproductive organs

sex hormones: estrogen, progesterone, and testosterone

shivering: oscillating rhythmic muscle tremors occurring at a rate of about 10 to 20 per second

shock: *see* circulatory shock

short reflex: a neural reflex in which the afferent and efferent pathways and integrating center are within the plexuses of the gastrointestinal tract

short-term memory: a limited-capacity storage process, possibly lasting only a few seconds, serving as the initial depository of information from which the memory may be transferred to long-term storage or forgotten

sickle cell anemia: a disease in which one of the 287 amino acids in the polypeptide chains of the hemoglobin molecule is abnormal and, at the low oxygen concentrations existing in many capillaries, causes the erythrocytes to assume sickle shapes or other bizarre forms that block the capillaries

sigmoid colon: the S-shaped terminal portion of the colon

simple cells: visual-cortex neurons that have receptive fields in which the "off" and "on" regions are side-by-side with straight, rather than circular, boundaries between them

single-unit smooth muscle: smooth muscle in which the fibers are joined by gap junctions allowing electrical activity to be conducted from cell to cell, i.e., the muscle responds to stimulation as a single unit; has pacemaker activity leading to the generation of action potentials in the absence of external stimulation

sinoatrial (SA) node: a region in the right atrium of the heart containing specialized cardiac muscle cells which depolarize spontaneously at a rate faster than that of other spontaneously depolarizing cells of the heart; the pacemaker of the heart

skeletal muscle "pump": the pumping effect of contracting skeletal muscles on the blood flow through vessels underlying them

skeletomotor fibers: the primary skeletal muscle fibers, as opposed to the modified fibers within the muscle spindle

sleep: *see* paradoxical sleep, slow-wave sleep

sleep centers: clusters of neurons in the brainstem whose activity periodically opposes the tonic activity of the reticular activating system and induces the cycling of slow-wave and paradoxical sleep

sliding filament model: the process of muscle contraction in which shortening occurs as a result of the sliding of thick and thin filaments past each other

slow viruses: viruses which replicate very slowly and may even become associated with the cell's own DNA and be passed along to the daughter cells upon cell division, altering the cells so they are no longer recognized as "self"

slow-wave potential: a slow oscillation of the membrane potential of some smooth muscle cells arising from spontaneous cyclical changes in the transport of ions across the membrane

slow-wave sleep: the state of sleep associated with large, slow EEG waves, considerable tonus in the postural muscles, and little change in cardiovascular and respiratory activity

sodium inactivation: the turning off of the increased sodium permeability at the peak of an action potential

soft palate: the nonbony region at the posterior part of the roof of the mouth

solubility: ability of the molecules of a substance to go into solution when placed in a given solvent

solute: molecules or ions dissolved in a liquid (solvent)

solution: a liquid (solvent) containing dissolved molecules or ions (solutes)

solvent: the liquid in which molecules or ions (solutes) are dissolved

somatic: pertaining to the body; related to the framework or outer walls of the body, including skin, skeletal muscle, tendons, and joints

somatic chromosomes: the 22 pairs (in human beings) of nonsex chromosomes

somatic nervous system: a component of the efferent division of the peripheral nervous system distinguished from the autonomic nervous system; innervates skeletal muscle

somatic receptor: a neural receptor that responds to mechanical stimulation of the skin or hairs and underlying tissues, to rotation or bending of joints, to temperature changes, or possibly to some chemical changes

somatomedins: hormones released mainly by the liver in response to growth hormone

somatosensory cortex: a strip of cerebral cortex in the parietal lobe just behind the junction between the parietal and frontal lobes in which nerve fibers transmitting somatic sensory modalities (e.g., touch, temperature, etc.) synapse

somatostatin: the hypothalamic hormone that inhibits growth hormone and TSH secretion by the anterior pituitary; a possible neurotransmitter; found also in stomach and D cells of the pancreatic islets of Langerhans

somatotropin: *see* growth hormone

sound localization: the projection of a perceived sound to a source, e.g., a ringing telephone, in the external world

sound wave: a disturbance in the air resulting from variations in densities of air molecules from regions of high density (compression) to low density (rarefaction)

SPA: *see* stimulus-produced analgesia

spatial summation: the concept that the effects of simultaneous stimuli (or synaptic inputs) to different points on a cell can be added together to produce a potential change greater than that caused by a single stimulus (or synaptic input)

specific dynamic action (SDA): the increase in metabolic rate induced by eating

specific immune responses: responses of lymphocytes and cells derived from them which depend upon prior exposure to a specific foreign material, recognition of it upon subsequent exposure, and reaction to it; *see also* cell-mediated immunity, humoral immunity

specific pathway: a chain of synaptically connected neurons in the brain or spinal cord, all of which are activated by sensory units of the same modality, e.g., a pathway conveying information about only touch or only temperature

specific thalamic nuclei: nuclei in the cerebrum whose axons project to localized, clearly defined regions of cerebral cortex, e.g., the nuclei relaying precise information about somatic stimuli

specificity: selectivity; the ability of a protein binding site to react only with one type (or a limited number of types) of molecule

sperm: (spermatozoa) reproductive cells in the male; the male gamete

spermatid: an immature sperm

spermatogenesis: the manufacture of sperm

spermatogonia: the undifferentiated germ cell that gives rise ultimately to sperm

spermatozoa: *see* sperm

sphincter of Oddi: a ring of smooth muscle surrounding the bile duct at its point of entrance into the duodenum

sphygmomanometer: a device consisting of an inflatable cuff and a pressure gauge for measuring blood pressure

spinal nerve: one of the 86 (43 pairs) of peripheral nerves that join the spinal cord

spindle apparatus: a cell structure which appears during nuclear division and consists of a number of microtubules passing from one side of the cell to the other between the centrioles or from a centriole to the centromere region of a chromosome

spindle fibers: the modified skeletal muscle fibers within the muscle spindle; also called intrafusal fibers

spleen: the largest of the lymphoid organs; located in the left part of the abdominal cavity between the stomach and diaphragm

split brain: a condition resulting from cutting of the nerve-fiber bundles connecting the two cerebral hemispheres, which prevents communication between the two halves of the cerebrum

stapes: the third of the three middle-ear bones; situated so that its footplate rests in the oval window and transmits sound waves to the scala vestibuli of the inner ear

starch: a moderately branched polysaccharide composed of thousands of glucose subunits; found in plant cells

Starling's law of the heart: the law that, within limits, an increase in the end-diastolic volume of the heart, i.e., increasing the length of the muscle fibers, increases the force of cardiac contraction during the following systole

stasis: cessation of blood flow in a vessel

state of consciousness: the degree of mental alertness; e.g., sleep and wakefulness are two states of consciousness

steady state: a situation in which no net change occurs in a system other than a continual energy input to the system that is required to prevent a change; *see also* equilibrium

stenosis: an abnormally narrowed opening

sternum: breastbone

steroid: a subclass of lipid molecules; the molecule consists of a skeleton of four interconnected carbon rings to which a few polar groups may be attached

STH: *see* growth hormone

stimulus: a change in the environment detectable by a receptor

stimulus-produced analgesia (SPA): analgesia

caused by electrical stimulation of certain areas of brainstem reticular formation or other regions of the central nervous system

stress: any environmental change that must be adapted to if health and life are to be maintained; any event that elicits increased cortisol secretion

stretch receptor: afferent nerve endings that are depolarized by stretching, e.g., the nerve endings in the muscle spindle

stretch reflex: a monosynaptic reflex in which stretch of a muscle causes contraction of that muscle; mediated by muscle spindles

striated muscle: muscle which demonstrates a characteristic transverse banding pattern when viewed with a microscope; skeletal and cardiac muscle

stroke: damage to a portion of the brain caused by stoppage of the blood supply to that region because of occlusion or rupture of a cerebral vessel

stroke volume: the volume of blood ejected by a ventricle during one beat of the heart

strong acid: an acid whose hydrogen atoms are completely dissociated to form hydrogen ions when the molecules are dissolved in water

subatmospheric pressure: a pressure less than the air pressure of the external environment, e.g., a pressure less than 760 mmHg at sea level

subatomic particles: the units that together make up an atom; e.g., electrons, protons, and neutrons are subatomic particles

subcortical nuclei: clusters of neuronal cells buried deep in the substance of the cerebrum, includes the basal ganglia

sublingual glands: one of the three pairs of salivary glands

submandibular glands: one of the three pairs of salivary glands

submucosa: a connective-tissue layer underlying the mucosa in the walls of the stomach and intestines

submucous plexus: a network of nerve cells in the submucosal layer of the esophagus, stomach, and intestinal walls

subset program: the sequence of neural activity that controls a component of a complete action

substance P: a substance thought to be the neurotransmitter at the synaptic endings of the afferent neurons in the pain pathway

substantia nigra: a subcortical nucleus whose cells are deeply pigmented

substrate: the ligand that binds to the specific binding site on an enzyme in an enzyme-catalyzed chemical reaction and undergoes a chemical reaction to form product molecules

substrate phosphorylation: one of the two major ways by which energy is transferred from carbohydrates, lipids, and proteins to other molecules (the other way is oxidative phosphorylation); occurs in the cytoplasm when a phosphate group is transferred directly from an intermediate in the glycolytic pathway to ADP to form ATP; also occurs during one of the Krebs-cycle reactions

subsynaptic membrane: that part of a postsynaptic neuron's plasma membrane which lies directly under the synaptic knob

subthreshold potentials: any depolarization less than the threshold potential

subthreshold stimulus: any agent capable of depolarizing the membrane a little but not enough to reach threshold

sucrose: a disaccharide carbohydrate molecule composed of glucose and fructose; table sugar

sulfhydryl group: R-SH, where R is the remaining portion of the molecule

sulfuric acid: an acid generated during the catabolism of sulfur-containing compounds; dissociates to form sulfate and hydrogen ions

summation: the increase in the mechanical response of a muscle to action potentials occurring in rapid succession

superior vena cava: the large vein that carries blood from the upper half of the body to the right atrium of the heart

"suppressor" T cells: a type of T cell that regulates B cells in a way that leads to reduced antibody production

suprathreshold stimulus: any agent capable of depolarizing the membrane more than (i.e., closer to zero than) its threshold

surface tension: the unequal attractive forces between water molecules at a surface which result in a net force which acts to reduce the surface area

surfactant: a phospholipoprotein complex produced by the type II cells of the pulmonary alveoli; markedly reduces the cohesive force of the water molecules lining the alveoli, i.e., reduces the surface tension

swallowing reflex: swallowing in response to stimulation of the roof of the mouth sweating: the secretion of a hypotonic salt solution onto the skin surface where it can evaporate, lowering body temperature

sympathetic nervous system: that portion of the autonomic nervous system whose preganglionic fibers leave the central nervous system at the thoracic and lumbar portions of the spinal cord; most of the postganglionic neurons of the sympathetic nervous system release the neurotransmitter norepinephrine

sympathetic trunk: one of the paired chains of interconnected sympathetic ganglia which lie on either side of the vertebral column

sympathomimetic: producing effects similar to those of the sympathetic nervous system

synapse: an anatomically specialized junction between two neurons where the activity in one neuron influences the excitability of the second; see also chemical synapse, electrical synapse, excitatory synapse, inhibitory synapse

synaptic cleft: the narrow extracellular space separating the pre- and postsynaptic neurons at a chemical synapse

synaptic delay: the less than 0.001-s period between excitation of the presynaptic axon terminal and membrane-potential changes in the postsynaptic neuron

synaptic knob: (also synaptic terminal) the slight swelling at the end of each axon terminal

synergistic muscle: a muscle whose action aids the intended motion

synthetic reactions: see anabolic reactions

systemic circulation: the circulation through all organs except the lungs; the portion of the cardiovascular system between the aorta (as it leaves the left ventricle) and the superior and inferior venae cavae (as they enter the right atrium)

systole: the period when the ventricles of the heart are contracting

systolic pressure: the maximum blood pressure reached during the cardiac cycle

T_m: see maximal tubular capacity

T_3: see triiodothyronine

T_4: see thyroxine

T cells: lymphocytes arising in spleen, lymph nodes, or other lymphoid tissue from precursors that at one time were in the thymus where in some way they acquired the ability to differentiate and mature into cells competent to act as effectors in cell-mediated immunity; see also cytolytic T cells, "helper" T cells, "suppressor" T cells

T lymphocytes: see T cells

t tubule: see transverse tubule

T wave: the component of the electrocardiogram corresponding to repolarization of the ventricles

taste buds: the sense organs for taste; contain chemoreceptors and supporting cells and are on the tongue, roof of the mouth, pharynx, and larynx

taste sensations: sensations traditionally divided into four categories: sweet, sour, salt, and bitter

tectorial membrane: a structure in the organ of Corti which is in contact with the hairs on the receptor cells

teleology: the explanation of events in terms of the ultimate purpose being served by them, e.g., the statement "urea is excreted in the urine because the body needs to get rid of such waste products"

temporal lobe: the region of cerebral cortex where primary auditory and one of the speech centers are located

temporal summation: the concept that the effects of two or more stimuli (or synaptic inputs) occurring at different times can be added together to produce a potential change greater than that caused by a single stimulus (or synaptic input)

tendon: a bundle of collagen fibers that connects muscle to bone and transmits the contractile force of the muscle to the bone

tension: see muscle tension

termination code word: one of the three nucleotide sequences in DNA which signifies the end of a gene; a "punctuation word" in the genetic code

termination codon: the sequence of three nucleotides in mRNA that corresponds to a given termination code word in DNA

tertiary structure: (of a protein) the three.dimensional shape of a polypeptide chain

testis: the gonad in the male

testosterone: a steroid hormone produced in the interstitial cells of the testes; the major male sex hormone; maintains the growth and development of the reproductive organs and is responsible for the secondary sexual characteristics of males

tetanus: the mechanical response of a muscle to high-frequency stimulation

tetanus bacillus: a bacterium (*Clostridium tetani*) that produces a toxin which blocks inhibitory synapses on motor neurons and causes the disease lockjaw

TH: *see* thyroid hormone

thalamus: a subdivision of the diencephalon that is a way station and integrating center for all sensory (except smell) input on its way to cerebral cortex; also contains motor nuclei and a central core which functions as an extension of the brainstem reticular formation

thalidomide: a sedative drug previously used particularly in pregnant women for morning sickness and subsequently found to cause severe fetal malformations

theca: a cell layer formed from follicle cells and specialized ovarian connective-tissue cells; surrounds the granulosa cells of the ovarian follicle

thermogenesis: heat generation

thermoreceptor: a sensory receptor that responds preferentially to temperature (and changes in temperature) particularly in a low (cold receptors) or high (warm receptors) range

theta rhythm: (of the hippocampus) low-frequency, high-amplitude electrical activity in recordings (hippocampal EEGs) during the orienting response; electrical manifestation of hippocampal "arousal"

thiamine: vitamin B_1

thick filament: the 12- to 18-nm filaments in muscle cells; located in the central region of a sarcomere in striated muscle fibers; consists of myosin molecules

thin filament: the 5- to 8-nm filaments in muscle cells; attached to the Z line at either end of a sarcomere in striated muscle fibers; consists of actin, troponin, and tropomyosin molecules

thoracic cage: wall of the thoracic cavity; chest wall

thoracic cavity: chest cavity

thoracolumbar division: (of the autonomic nervous system) *see* sympathetic nervous system

thoroughfare channel: (in a capillary network) a capillary which connects an arteriole and venule directly and from which branch the "true" capillaries

threshold: *see* threshold potential

threshold potential: (or threshold) the membrane potential to which an excitable membrane must be depolarized in order to initiate an action potential

threshold stimulus: any agent capable of depolarizing the membrane to threshold

thrombin: the enzyme that catalyzes the conversion of fibrinogen into fibrin

thrombosis: occlusion of a blood vessel by a clot

thromboxanes: substances closely related to the prostaglandins; synthesized from arachidonic acid and other essential fatty acids

thrombus: an intravascular clot

thymine: one of the pyrimidine bases; the one nucleotide base present in DNA but not in RNA

thymosin: a group of hormones secreted by the thymus

thymus: a lymphoid organ lying in the upper part of the chest; site of lymphocyte formation and thymosin secretion

thyroglobulin: a glycoprotein to which thyroid hormones bind; the storage form of the thyroid hormones in the thyroid

thyroid: a paired endocrine gland located in the neck

thyroid hormone: the collective term for the hormones released from the thyroid, i.e., thyroxine and triiodothyronine; increases the metabolic rate of most cells in the body

thyroid-stimulating hormone (TSH): a glycoprotein hormone secreted by the anterior pituitary; also called thyrotropin

thyrotropin: *see* thyroid-stimulating hormone

thyrotropin releasing hormone (TRH): the hypothalamic hormone that stimulates TSH (thyrotropin) secretion by the anterior pituitary

thyroxine (T_4): tetraiodothyronine; an iodine-containing amino acid hormone secreted by the thyroid

tidal volume: the volume of air entering or leaving the lungs during a single breath during any state of respiratory activity

tight junction: a type of cell junction found in epithelial tissues; extends around the circumference of the cell; consists of an actual fusing of the two adjacent plasma membranes and an obliteration of the extracellular space; prevents the diffusion of molecules across an epithelial layer by way of the extracellular space between cells

tissue: formerly defined as an aggregate of differentiated cells of a similar type; commonly used to denote the general cellular fabric of a given organ, e.g., kidney tissue, brain tissue; the four general classes of tissues are epithelial tissue, connective tissue, nerve tissue, and muscle tissue

tissue thromboplastin: an extravascular enzyme capable of initiating the formation of fibrin clots

tolerance: the state in which increasing doses of a drug are required to achieve effects that initially occurred in response to a smaller dose, i.e., when it takes more drug to do the same job

tonic: a continuous activity, as opposed to a phasic one

total energy expenditure: the energy stored plus the energy expended doing external work plus the energy represented in heat production by the body

total peripheral resistance (TPR): the total resistance of all the vessels in the systemic vascular tree to the flow of blood from the very first portion of the aorta to the very last portion of the venae cavae

total pulmonary ventilation: the volume of air that passes into or out of the respiratory system per minute; equal to the tidal volume times the number of breaths per minute

total vascular resistance: *see* total peripheral resistance

toxemia of pregnancy: a disease occurring in pregnant women and associated with retention of abnormally great amounts of fluid, the appearance of protein in the urine, hypertension, and possibly convulsions

toxicology: the study of substances that are harmful to the body

TPR: *see* total peripheral resistance

trace element: a mineral present in the body in extremely small quantities; at present 13 different essential trace elements are known which are required for normal growth and function

trachea: the single airway that connects the larynx with the two bronchi which enter the right and left lungs

tract: (neural) a large bundle of myelinated nerve fibers within the central nervous system

transamination: a reaction in which an amino group ($-NH_2$) is transferred from an amino acid to a keto acid, the keto acid thus becoming an amino acid

transcellular transport: the movement of molecules from an extracellular compartment into an epithelial cell and out the opposite side of the cell into a second extracellular compartment, i.e., in across the plasma membrane at one pole of the cell and out across the plasma membrane at its opposite pole; transepithelial transport

transcriptase: DNA-dependent-RNA-polymerase, the enzyme which joins the nucleotides of mRNA during its formation; catalyzes the splitting off of two of the three phosphate groups and the linking of the nucleotide to the adjacent one

transcription: formation of mRNA so that it contains in the linear sequence of its nucleotides the genetic information contained in the DNA gene from which the mRNA is transcribed; the first stage in the transfer of information from DNA to protein synthesis

transepithelial transport: *see* transcellular transport

transfer RNA (tRNA): a polymer of ribose-containing nucleotides; different tRNAs combine with different amino acids and contain an anticodon which binds to the codon in mRNA specific for that amino acid, thus arranging the amino acids in the appropriate sequence to form the protein specified by the codons in mRNA; smallest of the three forms of RNA

transferrin: an iron-binding protein which is the carrier molecule for iron in the plasma

translation: assembly of the proper amino acids in the correct order during the formation of a protein according to the genetic instructions carried in mRNA; occurs on the ribosomes of a cell

transmural pressure: the difference between the pressure exerted on one side of a structure and the pressure exerted on the other side

transverse tubule: (t tubule) a tubule extending from the plasma membrane of a striated muscle fiber into the fiber interior, passing between the opposed lateral sacs of adjacent sarcoplasmic-reticulum segments; conducts the muscle action potential into the muscle fiber leading to the release of calcium from the sarcoplasmic reticulum

tremor: unwanted, purposeless, uncontrolled movements taking the form of rapid contractions of opposing groups of muscles

TRH: *see* thyrotropin releasing hormone

triacylglycerol: a subclass of lipid molecules; neutral fat; composed of glycerol and three fatty acids; formerly called triglyceride

tricarboxylic acid cycle: *see* Krebs cycle

triceps: muscle that causes extension of the elbow

tricuspid valve: the valve between the right atrium and right ventricle of the heart

tricyclic antidepressants: drugs that interfere with the reuptake of norepinephrine and serotonin by presynaptic endings, thereby increasing the amount of neurotransmitter present in the synaptic cleft following release from the presynaptic endings

"trigger zone": that low-threshold region of a neuron at which action potentials are initiated; sometimes occurs at the initial segment of an axon

triglyceride: *see* triacylglycerol

triiodothyronine (T$_3$): an iodine-containing amino acid hormone secreted by the thyroid

tRNA: *see* transfer RNA

trophoblast: the outer layer of cells of the blastocyst; contributes to the placental tissues across which nutrients and wastes are exchanged between the maternal and fetal circulations

tropomyosin: a protein capable of reversibly covering the binding sites on actin, thereby preventing their combination with myosin; associated with the actin-containing thin filaments in muscle fibers

troponin: a protein bound to the actin and tropomyosin molecules of the thin filaments of striated muscle; site of calcium binding which initiates contractile activity in striated muscles

"true" capillary: a capillary across whose walls material exchange occurs, as opposed to a thoroughfare-channel type of capillary

trypsin: an enzyme secreted by the exocrine pancreas as the precursor trypsinogen; breaks certain peptide bonds in proteins and polypeptides

tryptophan: an essential amino acid

TSH: *see* thyroid-stimulating hormone

tubular epithelium: the sheet of epithelium one cell layer thick that forms the walls of the tubular component of a renal nephron

tubular reabsorption: the process by which materials are transferred from the lumen of a kidney tubule to the peritubular capillaries

tubular secretion: the process by which materials are transferred from the peritubular capillaries to the lumen of the kidney tubules

twitch: the mechanical response of a muscle to a single action potential

tympanic membrane: the membrane stretched across the end of the ear canal

type II cell: the cell type in the alveolar epithelium which produces surfactant

tyrosine: an amino acid; a precursor of the catecholamine neurotransmitters and the thyroid hormones; synthesized from the essential amino acid phenylalanine

tyrosine hydroxylase: the enzyme that catalyzes the formation of dopa from tyrosine in the catecholamine synthetic pathway; the rate-limiting enzyme in that pathway

ulcer: an erosion, or sore, as in the stomach or intestinal wall

umbilical cord: the flexible structure containing the umbilical arteries and vein, conveying them between the umbilicus of the fetus and the placenta

umbilical vessels: the arteries and veins transporting blood between the fetus and placenta

umbilicus: the depression in the middle of the abdomen marking the place where the umbilical cord was attached to the fetus; the navel or "belly button"

unidirectional flux: the amount of material crossing a surface in one direction in a unit of time; e.g., in the case of diffusion, $f_{1\text{-}2} = k_D C_1$ where $f_{1\text{-}2}$ is the flux from compartment 1 to 2, k_D is the diffusion coefficient, and C_1 is the concentration in compartment 1

universal donor: a person having type O blood which can be transfused into individuals with any other blood type

universal recipient: a person having type AB blood; such an individual can receive blood from individuals having any other blood type

unsaturated fatty acid: a fatty acid containing one or more double bonds

upper esophageal sphincter: a ring of skeletal muscle surrounding the esophagus just below the pharynx which, when contracted, closes the entrance to the esophagus

uracil: one of the pyrimidine bases; present in RNA but not DNA

urea: NH_2—$\overset{\overset{\textstyle O}{\|}}{C}$—$NH_2$: synthesized in the liver by linking two molecules of ammonia with carbon dioxide; the major nitrogenous waste product of protein breakdown

uremia: a general term for the symptoms of profound kidney malfunction

ureter: a tube, whose walls contain smooth muscle, which connects the renal pelvis to the urinary bladder

urethra: the tube which conveys urine from the urinary bladder to the outside of the body

uric acid: a waste product derived from nucleic acids

urinary bladder: a thick-walled sac which temporarily stores urine; receives urine from the ureters and empties it into the urethra

uterine tubes: paired structures which transport the ovum from the ovary to the uterus; also called oviducts, Fallopian tubes

uterus: womb; a hollow, thick-walled, muscular pear-shaped organ in the pelvic region of females; consists of the corpus (or body) above and the cervix below; the fertilized ovum implants in the uterine wall and develops within the uterus during the nine months of pregnancy

V: *see* volt

vagus nerve: cranial nerve X; consists of approximately 90 percent afferent and 10 percent parasympathetic preganglionic efferent fibers

varicose vein: an irregularly swollen superficial vein of the lower extremities

varicosity: a swollen region containing neurotransmitter-filled vesicles along the terminal portions of the axons of autonomic nerve fibers; analogous to a synaptic knob

vas deferens: the paired male reproductive ducts that connect the epididymidis to the urethra

vasectomy: the cutting and tying off of the vas deferens, which result in sterilization of the male without loss of testosterone secretion by the testes

vasoconstriction: a decrease in the diameter of blood vessels as a result of vascular smooth muscle contraction

vasodilation: an increase in the diameter of blood vessels as a result of vascular smooth muscle relaxation

vasopressin: *see* antidiuretic hormone

vein: a wide, thin-walled vessel that carries blood back to the heart from the venules

vena cava: one of two large veins that return systemic blood to the heart

venous return: the volume of blood flowing to the heart from the veins per unit of time

ventilation: the bulk-flow exchange of air between the atmosphere and the alveoli

ventral root: a group of efferent fibers that leaves the left and right side of the region of the spinal cord facing the front of the body

ventricle: a cavity, e.g., a ventricle of the heart or the brain

venule: a small, thin-walled vessel that carries blood from a capillary network to a vein

vesicle: a small, membrane-bound intracellular organelle

vestibular apparatus: membranous sacs in the bony semicircular canals and in the bony cavity adjacent to them (which contains the utricle and saccule)

vestibular nuclei: neuron clusters in the brainstem which receive primary afferent neurons from the vestibular apparatus

vestibular receptors: hair cells in the gelatinous mass in the semicircular canals, utricle, and saccule

vestibular system: a sense organ (often called the sense organ of balance) deep within the temporal bone of the skull; consists of the three semicircular canals, utricle, and saccule

vestibule: (of the female external genitalia) the area enclosed by the labia minora

villi: fingerlike projections from the highly folded surface of the small intestine; covered with a single layer of epithelial cells

virus: an "organism" that is essentially a nucleic acid core surrounded by a protein coat and lacking both the enzyme machinery for energy production and the ribosomes for protein synthesis; thus, it cannot reproduce except inside other cells whose biochemical apparatus it makes use of; *see also* slow viruses

viscera: the organs in the thoracic and abdominal cavities

viscosity: the property of a fluid caused by frictional interactions between its molecules that makes it resist flow; blood has a higher viscosity than pure water

vital capacity: the maximal amount of air that can be moved into the lungs following a maximal expiration; the sum of the inspiratory-reserve, normal tidal, and expiratory-reserve volumes

vitalist: one who believes that life processes occur only because of the presence of a "life force" rather than by physicochemical processes alone

vitamin: organic molecule which must be present in trace amounts to maintain normal health and growth but is not manufactured by the organism's metabolic pathways; over 20 vitamins are presently known; classified as water soluble (vitamins C and the B complex) and fat soluble (vitamins A, D, E, and K)

vitamin B_{12}: cyanocobalamine; a water-soluble, cobalt-containing vitamin of the B-complex group

vitamin D: a fat-soluble vitamin; a group of closely related sterols produced by the action of ultraviolet light on the skin or ingested in the diet; 1,25-dihydroxy-vitamin D_3, which is formed in the kidneys, is the active form

vocal cords: two strong bands of elastic tissue stretched across the opening of the larynx and caused to vibrate by the movement of air past them providing the tones of speech

volt (V): the unit of measurement of the electrical potential between two points

voltage: a measure of the potential of separated electric charges to do work; the amount of work done by an electric charge when moving from one point in a system to another; a measure of the electric force between two points

voluntary actions: actions we think about and are aware of; actions to which our attention is directed; actions that are the result of learning

vulva: the mons pubis, labia majora and minora, clitoris, vestibule of the vagina, and vestibular glands

wavelength: the distance between two wave peaks in an oscillating medium

weak acid: an acid whose hydrogen atoms are not completely dissociated to form hydrogen ions when the molecules of acid are dissolved in water

white matter: that portion of the brain and spinal cord which appears white in unstained specimens and consists mainly of myelinated nerve fibers

withdrawal: physical symptoms (usually unpleasant) associated with cessation of drug use

Wolffian duct: a part of the embryonic duct system which, in a male, remains and develops into the ducts of the reproductive system, but in a female, degenerates

work: a measure of the amount of energy required to produce a physical displacement of matter; formally defined as the product of force (F) times distance (X), $W = FX$; *see also* external work, internal work

xylocaine: a local anesthetic that works by preventing the increase in membrane permeability to sodium ions required for the generation of action potentials

yellow-sensitive cones: cones containing photopigment most sensitive to light of 570-nm wavelengths; often called red-sensitive cones because the range of pigment sensitivity extends closer to the red range than do the sensitivities of the two other cone photopigments

Z line: the structure running across the myofibril at each end of a striated muscle sarcomere; anchors one end of the thin filaments

zona pellucida: a thick membrane separating the ovum from the surrounding granulosa cells

zymogen granule: membrane-bound cytoplasmic vesicle containing inactive precursor enzymes prior to their secretion from a cell; the protein granules secreted by the exocrine portion of the pancreas

REFERENCES

REFERENCES FOR FIGURE ADAPTATIONS

Adolph, E. F.: "Physiology of Man in the Desert," Interscience, New York, 1947.

Anthony, C. P., and N. J. Kolthoff: "Textbook of Anatomy and Physiology," 8th ed., Mosby, St. Louis, 1971.

Bloom, W., and D. W. Fawcett: "A Textbook of Histology," 9th ed., Saunders, Philadelphia, 1968.

Boyd, I. A., and T. D. Roberts: *J. Physiol.*, **122**:38 (1953).

Brobeck, J. R., J. Tepperman, and C. N. H. Long: *Yale J. Biol. Med.*, **15**:831 (1943).

Carlson, A. J., V. Johnson, and H. M. Cavert: "The Machinery of the Body," 5th ed., University of Chicago Press, Chicago, 1961.

Chapman, C. B., and J. H. Mitchell: *Sci. Am.*, May 1965.

Comroe, J. H.: "Physiology of Respiration," Year Book, Chicago, 1965.

Daughaday, W. H., and D. M. Kipnis: *Recent Prog. Hormone Res.*, **22**:49 (1966).

Deanesley, R., and A. S. Parkes: *J. Physiol.*, **78**:442 (1933).

Dodt, E., and Y. Zotterman: *Acta Physiol. Scand.*, **26**:345 (1952).

Douglas, C. G., and J. S. Haldane: *J. Physiol.*, **38**:401 (1909).

Dubois, E. F.: "Fever and the Regulation of Body Temperature," Charles C Thomas, Springfield, Ill., 1948.

Eccles, J. C.: "The Understanding of the Brain," McGraw-Hill, New York, 1973.

Felig, P., and J. Wahren: *N. Eng. J. Med.*, **293**:1078 (1975).

Ganong, W. F.: "Review of Medical Physiology," 4th ed., Lange, Los Altos, Calif., 1969.

Goldberg, N. D.: *Hosp. Prac.*, May 1974.

Gordon, A. M., A. F. Huxley, and F. J. Julian: *J. Physiol.*, **184**:170 (1966).

Gregory, R. L.: "Eye and Brain: The Psychology of Seeing," McGraw-Hill, New York, 1966.

Guillemin, R., and R. Burgus: The Hormones of the Hypothalamus, *Sci. Am.*, October 1972.

Guyton, A. C.: "Functions of the Human Body," 3d ed., Saunders, Philadelphia, 1969.

Hensel, H., and K. K. A. Bowman: *J. Neurophysiol.*, **20**:564 (1960).

Hoffman, B. F., and P. E. Cranefield: "Electrophysiology of the Heart," McGraw-Hill, New York, 1960.

Hubel, D. H., and T. N. Wiesel: *J. Physiol.*, **154**:572 (1960); **160**:106 (1962).

Hume, D. M.: Organ Transplants and Immunity, *Hosp. Prac.*, 1968.

Ito, S.: *J. Cell Biol.*, **16**:541 (1963).

Jouvet, M.: *Sci. Am.*, February 1967.

Kappas, A., and A. P. Alvares: *Sci. Am.*, June 1975.

Kiang, N. Y. S.: in D. B. Tower (ed.), "The Nervous System," vol. 3, Raven Press, New York, 1975.

Kim, D. O., and C. E. Molnar: in D. B. Tower (ed.), "The Nervous System," vol. 3, Raven Press, New York, 1975.

Kuffler, S. W., and J. G. Nicholls: "From Neuron to Brain," Sinauer Associates, Sunderland, Mass., 1976.

Lambertsen, C. J.: in P. Bard (ed.), "Medical Physiology," 11th ed., Mosby, St. Louis, 1961.

Landauer, T. K.: "Readings in Physiological Psychology," McGraw-Hill, New York, 1967.

Leaf, A.: *Sci. Am.*, September 1973.

Livingston, R. B.: in G. C. Quarton, T. Mellinechuk, and F. O. Schmitt (eds.), "The Neurosciences: A Study Program," Rockefeller University Press, New York, 1967.

Loewenstein, W. R.: *Sci. Am.*, August 1960.

Mayer, J., N. B. Marshall, J. J. Vitale, J. H. Christensen, M. B. Mashayekhi, and F. J. Stare: *Am. J. Physiol.*, **177**:544 (1954).

Mayer, J., P. Roy, and K. P. Mitra: *Am. J. Nutr.*, **4**:169 (1956).

McEwen, B. S.: *Sci. Am.*, July 1976.

McNaught, A. B., and R. Callender: "Illustrated Physiology," Williams & Wilkins, Baltimore, 1963.

Merton, P. A.: *Sci. Am.*, May 1972.

Michael, C. R.: *N. Eng. J. Med.*, **288**:724 (1973).

Nakayama, T., H. T. Hammel, J. D. Hardy, and J. S. Eisenman: *Am. J. Physiol.*, **204**:1122 (1963).

Nathanson, I. T., L. E. Towne, and J. C. Aub: *Endocrinology*, **28**:851 (1941).

Olds, J.: *Sci. Am.*, October 1956.

Pedersen-Biergaard, K., and M. Tonnesen: *Acta Endocrinol.*, 1:38 (1948); *Acta Med. Scand.*, 131(suppl. 213):284 (1948).

Pitts, R. F.: "Physiology of the Kidney and Body Fluids," 2d ed., Year Book, Chicago, 1968.

Purdon-Martin, J.: "The Basal Ganglia and Posture," Lippincott, Philadelphia, 1967.

Rasmussen, A. T.: "Outlines of Neuroanatomy," 2d ed., Wm. C. Brown, Dubuque, Iowa, 1943.

Routtenberg, A.: *Sci Am.*, November 1978.

Rushmer, R. F.: "Cardiovascular Dynamics," 2d ed., Saunders, Philadelphia, 1961.

Singer, S., and H. R. Hilgard: "The Biology of People," Freeman, San Francisco, 1978.

Singer, S. J., and Garth L. Nicholson: *Science*, **175**:720 (1972).

Smith, H. W.: "The Kidney," Oxford University Press, New York, 1951.

Steinberger, E., and W. O. Nelson: *Endocrinology*, **56**:429 (1955).

Tepperman, J.: "Metabolic and Endocrine Physiology," 2d ed., Year Book, Chicago, 1968.

Wang, C. C., and M. I. Grossman: *Am. J. Physiol.*, **164**:527 (1951).

Wersall, J., L. Gleisner, and P. G. Lundquist: in A. V. S. de Reuck and J. Knight (eds.), "Myotatic, Kinesthetic, and Vestibular Mechanisms," Ciba Foundation Symposium, Little, Brown, Boston, 1967.

Woodburne, R. T.: "Essentials of Human Anatomy," 3d ed., Oxford University Press, New York, 1965.

Zweifach, B. W.: *Sci. Am.*, January 1959.

SUGGESTED READING

Like all scientists, physiologists report the results of their experiments in scientific journals. Approximately 1100 such journals in the life sciences publish 200,000 papers each year. Every physiologist must be familiar with the ever-increasing number of articles in his or her own specific field of research, but since these articles are highly technical for the nonexpert, with few exceptions we have not included any below for further reading.

Another type of scientific writing is the review article, a summary and synthesis of the relevant research reports on a specific subject. Such reviews, which are published in many different journals, are extremely useful and many of our suggested readings are of this type. Also, the American Physiological Society has a massive program of publishing comprehensive reviews in almost all fields of physiology. These reviews are gathered into a series of ongoing volumes known collectively as the "Handbook of Physiology." Another important series is the *Annual Review of Physiology*, which publishes yearly reviews of the most recent research articles in specific fields and provides the most rapid means of keeping abreast of developments in physiology.

Another type of review which is of great value to the nonexpert is the *Scientific American* article. Written in a manner usually completely intelligible to the lay person, these articles, in addition to reviewing a topic, often present actual experimental data so that the reader can obtain some insight into how the experiments were done and the data interpreted. Many articles published in *Scientific American* can be obtained individually as inexpensive offprints from W. H. Freeman and Company, San Francisco, Calif., 94104, which also publishes collections of *Scientific American* articles in a specific area, e.g., psychobiology. Two other magazines which frequently publish excellent reviews of physiology for the nonexpert are *New England Journal of Medicine* and *Hospital Practice*.

Another level of scientific writing is the monograph, an extended review of a subject area broader than the review articles described above. Monographs are usually in book form and generally attempt, not to cover all the relevant literature on the topic, but to analyze and synthesize the most important data and interpretations. At the last level is the textbook, which deals with a much larger field than the monograph or review article; its depth of coverage of any topic is accordingly much less complete and is determined by the audience to which it is aimed.

The books and articles listed below comprise only a tiny fraction of the readings available, but they should provide a source of additional information on the topics covered in this book. Their bibliographies will serve as a further entry into the scientific literature, particularly for original research reports.

We have suggested the relative difficulty of entries by asterisks. No asterisk indicates that, after finishing the relevant chapter in this book, the reader should be able to understand the entry with little difficulty. One asterisk indicates that the work is more difficult but should be comprehensible with some effort. Two asterisks denote a work which is highly detailed or technical and may require a strong background in mathematics, chemistry, or biology for a full understanding; this type of reading has been included for students who wish to pursue a subject in depth and are willing to expend the effort required.

We list first a group of books, mainly textbooks, which cover large areas of physiology. Since each can serve as additional reading for many chapters, they are grouped here according to field and are not mentioned again for individual chapters.

Books covering wide areas of physiology

Cell physiology

**Beck, Felix, and John B. Lloyd (eds.): "The Cell in Medical Science," 4 vols., Academic, New York, 1974–1976.

**Davson, H.: "A Textbook of General Physiology," 4th ed., vols. I and II, Little, Brown, Boston, 1970.

*DeRobertis, E. D. P., Francisco A. Saez, and E. M. F. DeRobertis: "Cell Biology," 6th ed., Saunders, Philadelphia, 1975.

*Giese, Arthur C.: "Cell Physiology," 4th ed., Saunders, Philadelphia, 1973.

McElroy, William D., and Carl P. Swanson: "Modern Cell Biology," 2d ed., Prentice-Hall, Englewood Cliffs, N.J., 1976 (paperback).

*Wolfe, Stephen L.: "Biology of the Cell," Wadsworth, Belmont, Calif., 1972.

Organ system physiology

Astrand, P., and K. Rodahl: "Textbook of Work Physiology," McGraw-Hill, New York, 1970.

*Ganong, W. F.: "Review of Medical Physiology," 9th ed., Lange, Los Altos, Calif., 1979.

*Guyton, A. C.: "Textbook of Medical Physiology," 5th ed., Saunders, Philadelphia, 1976.

**Handbook of Physiology," Williams & Wilkins, Baltimore, 1959– (continual publication of new volumes).

Comparative physiology

*Goldstein, Leon (ed.): "Introduction to Comparative Physiology," Holt, Rinehart and Winston, New York, 1977.

**Prosser, C. Ladd (ed.): "Comparative Animal Physiology," 3d ed., Saunders, Philadelphia, 1973.

Anatomy

Basmajian, J. V.: "Grant's Method of Anatomy," 9th ed., Williams & Wilkins, Baltimore, 1975.

Goss, C. M.: "Gray's Anatomy of the Human Body," 29th American ed., Lea & Febiger, Philadelphia, 1973.

Hamilton, W. J., and N. W. Mossman: "Human Embryology," 4th ed., Williams & Wilkins, Baltimore, 1972.

Hollinshead, W. H.: "Textbook of Anatomy," 3d ed., Harper & Row, New York, 1974.

Leeson, T. S., and C. R. Leeson: "Human Structure," Saunders, Philadelphia, 1972.

Luciano D. S., A. J. Vander, and J. H. Sherman: "Human Function and Structure," McGraw-Hill, New York, 1978.

*Woodburne, R. T.: "Essentials of Human Anatomy," 5th ed., Oxford University Press, New York, 1973.

Pathophysiology

**Frohlich, Edward D. (ed.): "Pathophysiology," 2d ed., Lippincott, Philadelphia, 1976.

*Roddie, Ian C., and William F. M. Wallace: "The Physiology of Disease," Year Book, Chicago, 1975.

Snively, W. D., Jr., and Donna R. Beshear: "Textbook of Pathophysiology," Lippincott, Philadelphia, 1972.

*Sodeman, William A., and William A. Sodeman, Jr.: "Pathologic Physiology: Mechanisms of Disease," 5th ed., Saunders, Philadelphia, 1974.

Suggestions for individual chapters

Chapter 1

Bernard, C.: "An Introduction to the Study of Experimental Medicine," Dover, New York, 1957 (paperback).

Brooks, C. McC., and P. F. Cranefield (eds.): "The Historical Development of Physiological Thought," Hafner, New York, 1959.

Butterfield, H.: "The Origins of Modern Science," Macmillan, New York, 1961 (paperback).

"The Excitement and Fascination of Science: A Collection of Autobiographical and Philosophical Essays," Annual Reviews, Palo Alto, Calif., 1966.

Fulton, J. F., and L. C. Wilson (eds.): "Selected Readings in the History of Physiology," 2d ed., Charles C Thomas, Springfield, Ill., 1966.

Harvey, W.: "On the Motion of the Heart and Blood in Animals," Gateway, Henry Regnery, Chicago, 1962 (paperback).

Leake, Chauncey: "Some Founders of Physiology," American Physiological Society, Washington, D. C., 1961.

Taylor, G. R.: "The Science of Life: A Pictorial History of Biology," Panther, London, 1967 (paperback).

Chapter 2

Berns, Michael W.: "Cells," Holt, Rinehart and Winston, New York, 1977 (paperback).

*Bloom, W., and D. W. Fawcett: "A Textbook of Histology," 10th ed., Saunders, Philadelphia, 1975.

Swanson, C. P., and P. L. Webster: "The Cell," 4th ed., Prentice-Hall, Englewood Cliffs, N. J., 1977 (paperback).

*Weiss, L., and Roy O. Greep.: "Histology," 4th ed., McGraw-Hill, New York, 1977.

Chapter 3

*Anfinsen, Christian B.: "The Molecular Basis of Evolution," Wiley, New York, 1963 (paperback).

Doty, P.: Proteins, Sci. Am., September 1967.

Frieden, Earl: The Chemical Elements of Life, Sci. Am., July 1972.

Kendrew, J. C.: Three-dimensional Study of a Protein, Sci. Am., December 1961.

Lambert, Joseph B.: The Shapes of Organic Molecules, Sci. Am., January 1970.

*Lehninger, A. L.: "Biochemistry: The Molecular Basis of Cell Structure and Function," 3d ed., Worth, New York, 1975.

Schmitt, F. O.: Giant Molecules in Cells and Tissues, Sci. Am., September 1957.

*Speakman, J. C.: "Molecules," McGraw-Hill, New York, 1966 (paperback).

Stein, W. H., and S. Moore: The Chemical Structure of Proteins, Sci. Am., February 1961.

Chapter 4

Baltimore, David: The Molecular Biology of Poliovirus, Sci. Am., May 1975.

Campbell, Allan M.: How Viruses Insert Their DNA into the DNA of the Host Cell, Sci. Am., December 1976.

Changeux, J.: The Control of Biochemical Reactions, Sci. Am., April 1965.

Clark, B. F. C., and K. A. Marcker: How Proteins Start, Sci. Am., January 1968.

Cohen, Stanley N.: The Manipulation of Genes, Sci. Am., July 1975.

Crick, F. H. C.: The Genetic Code, Sci. Am., October 1962; October 1966.

Darnell, James E., Warren R. Jelinek, and George R. Molloy: Biogenesis of mRNA: Genetic Regulation in Mammalian Cells, Science, **181: 1215 (1973).

German, James: Studying Human Chromosomes Today, Am. Sci., **58**:182 (1970).

Goodenough, Ursula W., and R. P. Levine: The Genetic Activity of Mitochondria and Chloroplasts, Sci. Am., November 1970.

Grobstein, Clifford: The Recombinant-DNA Debate, Sci. Am., July 1977.

*Hendler, Richard W.: "Protein Biosynthesis and Membrane Biochemistry," Wiley, New York, 1968.

Hurwitz, J., and J. J. Furth: Messenger RNA, Sci. Am., February 1962.

*Ingram, V. M.: "Biosynthesis of Macromolecules," 2d ed., Benjamin, New York, 1972 (paperback).

Kornberg, A.: The Synthesis of DNA, Sci. Am., September 1961.

Koshland, Daniel E., Jr.: Protein Shape and Biological Control, Sci. Am., October 1973.

Maniatis, T., and M. Ptashne: A DNA Operator-Repressor System, Sci. Am., January 1976.

Mazia, Daniel: The Cell Cycle, *Sci. Am.*, January 1974.

Miller, O. L., Jr.: The Visualization of Genes in Action, *Sci. Am.*, March 1973.

Mirsky, A. E.: The Discovery of DNA, *Sci Am.*, June 1968.

Monod, J.: "Chance and Necessity," Vintage Books, New York, 1971 (paperback).

Nirenberg, M. W.: The Genetic Code, II, *Sci. Am.*, March 1963.

Portugal, F. H., and J. S. Cohen: "A Century of DNA," MIT Press, Cambridge, Mass., 1978.

Rich, A.: Polyribosomes, *Sci. Am.*, December 1963.

*Watson, James D.: "Molecular Biology of the Gene," 3d ed., Benjamin, New York, 1976 (paperback).

Watson, James D.: "The Double Helix," Atheneum, New York, 1968. *Autobiographical account of Watson's role in discovering the DNA double helix.*

Yanofsky, C.: Gene Structure and Protein Structure, *Sci. Am.*, May 1967.

Chapter 5

Baker, Jeffrey J. W., and Garland E. Allen: "Matter, Energy, and Life," 3d ed., Addison-Wesley, Reading, Mass., 1974 (paperback).

*Bernhard, Sidney A.: "The Structure and Function of Enzymes," Benjamin, New York, 1968 (paperback).

Bolin, Bert: The Carbon Cycle, *Sci. Am.*, September 1970.

*Conn, E. E., and P. K. Stumpf: "Outlines of Biochemistry," 4th ed., Wiley, New York, 1976.

Frieden, Earl: The Enzyme-Substrate Complex, *Sci. Am.*, August 1969.

Green, D. E.: The Mitochondrian, *Sci. Am.*, January 1964.

Green, D. E.: The Synthesis of Fat, *Sci. Am.*, August 1960.

Hinkle, P. C., and R. E. McCarty: How Cells Make ATP, *Sci. Am.*, March 1978.

*Lehninger, Albert L.: "Biochemistry: The Molecular Basis of Cell Structure and Function," 3d ed., Worth, New York, 1975.

*Lehninger, Albert L.: "Bioenergetics: The Molecular Basis of Biological Energy Transformations," 2d ed., Benjamin, New York, 1971 (paperback).

Mott-Smith, Morton: "The Concept of Energy Simply Explained," Dover, New York, 1964 (paperback).

**Newsholme, E. A., and C. Start: "Regulation in Metabolism," Wiley, New York, 1975.

*Racker, Efraim: The Inner Mitochondrial Membrane: Basic and Applied Aspects, *Hosp. Prac.*, February 1974.

Segal, Harold L.: Enzymatic Interconversion of Active and Inactive Forms of Enzymes, *Science*, **180:25 (1973).

Chapter 6

Bretscher, Mark S.: Membrane Structure: Some General Principles, *Science*, **181:662 (1973).

Capaldi, Roderick A.: A Dynamic Model of Cell Membranes, *Sci. Am.*, March 1974.

**Dick, D. A. T.: "Cell Water," Butterworth, Washington, D. C., 1966.

Hospital Practice, **8**(1973) and **9**(1974) (One article each month on membrane structure and function).

**Kotyk, Arnost, and Karel Janacek: "Cell Membrane Transport," 2d ed., Plenum, New York, 1975.

Lodish, H. F., and J. E. Rothman: The Assembly of Cell Membranes, *Sci. Am.*, January 1979.

Neutra, M., and C. P. Leblond: The Golgi Apparatus, *Sci. Am.*, February 1969.

Rustad, R. C.: Pinocytosis, *Sci. Am.*, April 1961.

*Saier, Milton H., Jr., and Charles D. Stiles: "Molecular Dynamics in Biological Membranes," Springer-Verlag, New York, 1975 (paperback).

Satir, B.: The Final Steps in Secretion, *Sci. Am.*, October 1975.

Silverstein, S. C., Ralph M. Steinman, and Zanvil A. Cohen: Endocytosis, *Ann. Rev. Physiol.*, **46:669 (1977).

Singer, S. J., and Garth L. Nicolson: The Fluid Mosaic Model of the Structure of Cell Membranes, *Science*, **175:720 (1972).

Solomon, Arthur K.: The State of Water in Red Cells, *Sci. Am.*, February 1971.

Staehelin, L. A., and B. E. Hull: Junctions Between Living Cells, *Sci. Am.*, May 1978.

**Stein, W. D.: "The Movement of Molecules Across Cell Membranes," Academic, New York, 1967.

Weissman, Gerald, and Robert Claiborne (eds.):

"Cell Membranes: Biochemistry, Cell Biology, and Pathology," Hospital Practice Publishing, New York, 1976.

Chapter 7

Adolph, E. F.: Early Concepts of Physiological Regulations, *Physiol. Rev.*, **41:737 (1961).

Bernard, C.: "An Introduction to the Study of Experimental Medicine," Dover, New York, 1957 (paperback).

Cannon, W. B.: "The Wisdom of the Body," Norton, New York, 1939 (paperback).

Goldberg, N. D.: Cyclic Nucleotides and Cell Function, *Hosp. Prac.*, May 1974.

Jacobs, S., and P. Cuatrecasas: Cell Receptors in Disease, *N. Engl. J. Med.*, **297**:1383 (1977).

Langley, L. L. (ed.): "Homeostasis: Origin of the Concept," Dowden, Hutchinson, & Ross, Stroudsburg, Pa., 1973.

Pastan, I.: Cyclic AMP, *Sci. Am.*, August 1972.

Pike, J. E.: Prostaglandins, *Sci. Am.*, November 1971.

Rasmussen, H.: Ions as "Second Messengers," *Hosp. Prac.*, June 1974.

Rasmussen, H., and D. B. P. Goodman: Relationships Between Calcium and Cyclic Nucleotides in Cell Activation, *Physiol. Rev.*, **57:421, 1977.

Chapter 8

Baker, P. F.: The Nerve Axon, *Sci. Am.*, March 1966.

*Barchas, J. D., A. Huda, G. R. Elliott, R. B. Holman, and S. J. Watson: Behavioral Neurochemistry: Neuroregulators and Behavioral States, *Science*, **200**:964 (1978).

*Bodian, D.: The Generalized Vertebrate Neuron, *Science*, **137**:323 (1962).

*Bullock, T. H.: Neuron Doctrine and Electrophysiology, *Science*, **129**:997 (1959).

Cannon, W. B.: "Bodily Changes in Pain, Hunger, Fear, and Rage," Harper & Row, New York, 1963 (originally published 1915; paperback).

*DeRobertis, E.: Ultrastructure and Cytochemistry of the Synaptic Region, *Science*, **156**:907 (1967).

Eccles, J. C.: The Synapse, *Sci. Am.*, January 1965.

Eccles, J. C.: "The Understanding of the Brain," 3d ed., McGraw-Hill, New York, 1977 (paperback). Chapters 1–3.

Fernstron, J. D., and R. L. Wurtman: Nutrition and the Brain, *Sci. Am.*, February 1974.

*Goodman, L. S., and A. Gilman: Neurohumoral Transmission and the Autonomic Nervous System, in "The Pharmacological Basis of Therapeutics," 5th ed., pp. 402–441, Macmillan, New York, 1975.

*Guillemin, R.: Peptides in the Brain: The New Endocrinology of the Neuron, Nobel Prize Lecture, 1977, *Science*, **202**:390 (1978).

*Hodgkin, A. L.: "The Conduction of the Nervous Impulse," Charles C Thomas, Springfield, Ill., 1964.

*Katz, B.: Quantal Mechanism of Neural Transmitter Release, Nobel Prize Lecture, 1970, *Science*, **173**:123 (1971).

Katz, B.: How Cells Communicate, *Sci. Am.*, September 1961.

*Keynes, R. D.: Ion Channels in the Nerve-Cell Membrane, *Sci. Am.*, March 1979.

*Kuffler, S. W., and J. G. Nicholls: "From Neuron to Brain," Sinauer, Sunderland, Mass., 1976.

Loewenstein, W. R.: Biological Transducers, *Sci. Am.*, August 1960.

*Nathanson, J. A., and P. Greengard: "Second Messenger" in the Brain, *Sci. Am.*, August 1977.

*Patterson, P. H., D. D. Potter, and E. J. Furshpan: The Chemical Differentiation of Nerve Cells, *Sci. Am.*, July 1978.

**Ruch, T. C., H. D. Patton, W. Woodbury, and A. L. Towe: "Neurophysiology," Saunders, Philadelphia, 1965. Chapters 1 and 2.

*Schmitt, F. O., P. Dev, and B. H. Smith: Electronic Processing of Information by Brain Cells, *Science*, **193**:114 (1976).

Shepherd, G. M.: Microcircuits in the Nervous System, *Sci. Am.*, February 1978.

Tower, D. B. (ed.): "The Nervous System," vol. 1, The Basic Neurosciences, Raven Press, New York, 1975.

Chapter 9

Catt, K. J., and M. L. Dufau: Peptide Hormone Receptors, *Ann. Rev. Physiol.*, **39:529 (1977).

*Davidson, J. M., and S. Levine: Endocrine Regulation of Behavior, *Ann. Rev. Physiol.*, **35**:375 (1972).

Frohman, L. A.: Neurotransmitters as Regulators of Endocrine Function, *Hosp. Prac.*, April 1975.

Ganong, W. F., L. C. Alpert, and T. C. Lee: ACTH

and the Regulation of Adrenocortical Secretion, *N. Engl. J. Med.*, **290**:1006 (1974).

Gillie, R. B.: Endemic Goiter, *Sci. Am.*, June 1971.

Goldberg, N. D.: Cyclic Nucleotides and Cell Function, *Hosp. Prac.*, May 1974.

Guillemin, R.: Beta-lipotropin and Endorphins: Implications of Current Knowledge, *Hosp. Prac.*, November 1978.

McEwen, B. S.: Interactions between Hormones and Nerve Tissue, *Sci. Am.*, July 1976.

McEwen, B. S.: The Brain as a Target Organ of Endocrine Hormones, *Hosp. Prac.*, May 1975.

O'Malley, B. W., and W. T. Schrader: The Receptors of Steroid Hormones, *Sci. Am.*, February 1976.

Reichlin, S., et al.: Hypothalamic Hormones, *Ann. Rev. Physiol.*, **38:389 (1976).

Segal, H. L.: Enzymatic Interconversion of Active and Inactive Forms of Enzymes, *Science*, **180:25 (1973).

Stent, G. S.: Cellular Communication, *Sci. Am.*, September 1972.

*Sutherland, E. W.: Studies on the Mechanism of Hormone Action, *Science*, **177**:401 (1972).

*Tepperman, J.: "Metabolic and Endocrine Physiology," 3d ed., Year Book, Chicago, 1973.

*Turner, C. D.: "General Endocrinology," 6th ed., Saunders, Philadelphia, 1976.

Vale, W., et al.: Regulatory Peptides of the Hypothalamus, *Ann. Rev. Physiol.*, **39:473 (1977).

*Weitzman, E. D., et al.: The Relationship of Sleep and Sleep Stages to Neuroendocrine Secretion and Biological Rhythms in Man, *Rec. Prog. Horm. Res.*, **31**:399 (1975).

Chapter 10

*Bendall, J. R.: "Muscles, Molecules and Movement," American Elsevier, New York, 1969.

**Bourne, G. H. (ed.): "The Structure and Function of Muscle," 2d ed., Academic, New York, 1972 (vol. I), 1973 (vols. II and III), 1974 (vol. IV).

**Bulbring, E., and M. F. Shuba (eds.): "Physiology of Smooth Muscle," Raven Press, New York, 1976.

**Carlson, Francis D., and Douglas R. Wilkie: "Muscle Physiology," Prentice-Hall, Englewood Cliffs, N.J., 1974.

Close, R. I.: Dynamic Properties of Mammalian Skeletal Muscles, *Physiol. Rev.*, **52:129 (1972).

Cohen, Carolyn: The Protein Switch of Muscle Contraction, *Sci. Am.*, November 1975.

Ebashi, Setsuro: Excitation-Contraction Coupling, *Ann. Rev. Physiol.*, **38:293 (1976).

Endo, Makoto: Calcium Release from the Sarcoplasmic Reticulum, *Physiol. Rev.*, **57:71 (1977).

Guth, Lloyd: Tropic Influences of Nerve on Muscle, *Physiol. Rev.*, **44:645 (1968).

Holloszy, J. O., Booth, F. W.: Biochemical Adaptations to Endurance Exercise in Muscle, *Ann. Rev. Physiol.*, **38:273 (1976).

Hoyle, Graham: How is Muscle Turned On and Off?, *Sci. Am.*, April 1970.

**Huddart, Henry, and Stephen Hunt: "Visceral Muscle," Halstead, New York, 1975.

Huxley, H. E.: The Mechanism of Muscular Contraction, *Science*, **164:1356 (1969).

Huxley, H. E.: The Contraction of Muscle, *Sci. Am.*, November 1968.

Lazarides, E., and J. P. Revel: The Molecular Basis of Cell Movement, *Sci. Am.*, May 1979.

Lester, Henry A.: The Response to Acetylcholine, *Sci. Am.*, February 1977.

Margaria, R.: The Sources of Muscular Energy, *Sci. Am.*, March 1972.

Merton, P. A.: How We Control the Contraction of Our Muscles, *Sci. Am.*, May 1972.

Mommaerts, W. F. H. M.: Energetics of Muscle Contraction, *Physiol. Rev.*, **49:427 (1969).

Murphy, R. A.: Filament Organization and Contractile Function in Vertebrate Smooth Muscle, *Ann. Rev. Physiol.*, **41:737 (1979).

Murray, John M., and Annemarie Weber: The Cooperative Action of Muscle Proteins, *Sci. Am.*, February 1974.

*Needham, Dorothy M.: "Machina Carnia: The Biochemistry of Muscular Contraction in Its Historical Development," Cambridge University Press, New York, 1971.

Porter, K. R., and C. Franzini-Armstrong: The Sarcoplasmic Reticulum, *Sci. Am.*, March 1965.

Weber, Annemarie, and John M. Murray: Molecular Control Mechanisms in Muscle Contraction, *Physiol. Rev.*, **53:612 (1973).

Chapter 11

Adolph, E. F.: The Heart's Pacemaker, *Sci. Am.*, March 1967.

Baez, S.: Microcirculation, *Ann. Rev. Physiol.*, **39**:391 (1977).

Benditt, E. P.: The Origin of Atherosclerosis, *Sci. Am.*, February 1977.

*Berne, R. M., and M. N. Levy: "Cardiovascular Physiology," 3d ed., Mosby, St. Louis, 1977.

**Bevegard, B. S., and J. T. Shepherd: Regulation of the Circulation during Exercise in Man, *Physiol. Rev.*, 47:178 (1967).

Braunwald, E.: Regulation of the Circulation, *N. Engl. J. Med.*, 290:1124, 1420 (1974).

Chapman, C. B., and J. H. Mitchell: The Physiology of Exercise, *Sci. Am.*, May 1965.

**Chien, S.: Role of the Sympathetic Nervous System in Hemorrhage, *Physiol. Rev.*, 47:214 (1967).

*Cranefield, P. F.: Ventricular Fibrillation, *N. Engl. J. Med.*, 289:732 (1973).

*Folkow, B., and E. Neil: "Circulation," Oxford University Press, New York, 1971.

**Fozzard, H. A.: Heart: Excitation-Contraction Coupling, *Ann. Rev. Physiol.*, 39:201 (1977).

Gann, D. S.: Endocrine Control of Plasma Protein and Volume, *Surg. Clin. N. Am.*, 56:1135 (1976).

*Katz, A. M.: Congestive Heart Failure, *N. Engl. J. Med.*, 293:1184 (1975).

Kent, K. M., and T. Cooper: The Denervated Heart, *N. Engl. J. Med.*, 291:1017 (1974).

Mayerson, H.: The Lymphatic System, *Sci. Am.*, June 1963.

*Rushmer, R. F.: "Structure and Function of the Cardiovascular System," 2d ed., Saunders, Philadelphia, 1976.

Scher, A. M.: The Electrocardiogram, *Sci. Am.*, November 1961.

Scheuer, J., and C. M. Tipton: Cardiovascular Adaptations to Physical Training, *Ann. Rev. Physiol.*, 39:221 (1977).

Shepherd, J. T., and P. M. Vanhoutte: Role of the Venous System in Circulatory Control, *Mayo Clin. Proc.*, 53:247 (1978).

Vatner, S. F., E. Braunwald: Cardiovascular Control Mechanisms in the Conscious State, *N. Engl. J. Med.*, 293:970 (1975).

Wiggers, C. J.: The Heart, *Sci. Am.*, May 1957.

Wood, J. E.: The Venous System, *Sci. Am.*, January 1968.

Zweifach, B. J.: The Microcirculation of the Blood, *Sci. Am.*, January 1959.

Chapter 12

Avery, M. E., N. Wang, and H. W. Taeusch, Jr.: The Lung of the Newborn Infant, *Sci. Am.*, April 1973.

Baker, P. T.: Human Adaptation to High Altitude, *Science*, 163:1149 (1969).

*Berger, A. J., R. A. Mitchell, and J. W. Severinghaus, Control of Respiration, *N. Engl. J. Med.*, 297:92, 138, and 194 (1977).

Clements, J. A.: Surface Tension in the Lungs, *Sci. Am.*, December 1962.

Comroe, J. H., Jr.: The Lung, *Sci. Am.*, February 1966.

Crandall, E. D.: Pulmonary Gas Exchange, *Ann. Rev. Physiol.*, 38:69 (1976).

*Davenport, H. W.: "The ABC of Acid-Base Chemistry," 7th ed., University of Chicago Press, Chicago, 1978.

Fenn, W. O.: The Mechanism of Breathing, *Sci. Am.*, January 1960.

Finch, C. A., and C. Lenfant: Oxygen Transport in Man, *N. Engl. J. Med.*, 286:407 (1972).

Fishman, A. D., and G. P. Pietra: Handling of Bioactive Materials by the Lung, *N. Engl. J. Med.*, 291:884 (1974).

Guz, A.: Regulation of Respiration in Man, *Ann. Rev. Physiol.*, 37:303 (1975).

*Lenfant, C., and K. Sullivan: Adaptation to High Altitude, *N. Engl. J. Med.*, 284:1298 (1971).

*Mitchell, T. H., and G. Blomqvist: Maximal Oxygen Uptake, *N. Engl. J. Med.*, 284:1018 (1971).

**Morgan, T. E.: Pulmonary Surfactant, *N. Engl. J. Med.*, 284:1185 (1971).

Newhouse, M., et al.: Lung Defense Mechanisms, *N. Engl. J. Med.*, 295:990, 1045 (1976).

Perutz, M. F.: The Hemoglobin Molecule, *Sci. Am.*, November 1964.

*Wasserman, K.: Breathing during Exercise, *N. Engl. J. Med.*, 298:780 (1978).

Wyman, R. J.: Neural Generation of the Breathing Rhythm, *Ann. Rev. Physiol.*, 39:417 (1977).

Chapter 13

**Anderson, B.: Regulation of Body Fluids, *Ann. Rev. Physiol.*, 39:185 (1977).

*Davenport, H. W.: "The ABC of Acid-Base Chemistry," 7th ed., University of Chicago Press, Chicago, 1978.

*Elkinton, J. R., and T. S. Danowsky: "The Body Fluids: "Basic Physiology and Practical Therapeutics," Williams & Wilkins, Baltimore, 1955.

*Hays, R. M.: Antidiuretic Hormone, *N. Engl. J. Med.*, 295:659 (1976).

*Jamison, R. L., and R. H. Maffly: The Urinary

Concentrating Mechanism, *N. Engl. J. Med.*, **295**:1059 (1976).

Kuru, M.: Nervous Control of Micturition, *Physiol. Rev.*, **45:425 (1965).

Merrill, J. P.: The Artificial Kidney, *Sci. Am.*, July 1961.

Merrill, J. P., and C. L. Hampers: Uremia, *N. Engl. J. Med.*, **282**:953 (1970).

Peart, W. S.: Renin-Angiotensin System, *N. Engl. J. Med.*, **292:302 (1975).

*Ramsay, D. J., and W. F. Ganong: CNS Regulation of Salt and Water Intake, *Hosp. Prac.*, March 1977.

Rasmussen, H.: The Parathyroid Hormone, *Sci. Am.*, April 1961.

Rasmussen, H., and M. M. Pechet: Calcitonin, *Sci. Am.*, October 1970.

Schrier, R. W., and H. E. DeWardener: Tubular Reabsorption of Sodium Ion, *N. Engl. J. Med.*, **285:1731 (1971).

Smith, H. W.: "From Fish to Philosopher," Anchor Books, Doubleday, Garden City, N. Y., 1961 (paperback).

*Valtin, H.: "Renal Function," Little, Brown, Boston, 1973.

Vander, A. J.: "Renal Physiology," 2d ed., McGraw-Hill, New York, 1980.

Chapter 14

Bortoff, Alexander: Myogenic Control of Intestinal Motility, *Physiol. Rev.*, **56:418 (1976).

**Brooks, Frank P. (ed.): "Gastrointestinal Pathophysiology," 2d ed., Oxford University Press, New York, 1978 (paperback).

Davenport, Horace W.: Why the Stomach Does Not Digest Itself, *Sci. Am.*, January 1972.

*Davenport, Horace W.: "A Digest of Digestion," 2d ed., Year Book, Chicago, 1978.

*Davenport, Horace W.: "Physiology of the Digestive Tract," 4th ed., Year Book, Chicago, 1977.

**Grossman, M., V. Speranza, N. Basso, and E. Lezoche (eds.): "Gastrointestinal Hormones and the Pathology of the Digestive System," Plenum, New York, 1978.

Javitt, Norman B.: Hepatic Bile Formation, *N. Engl. J. Med.*, **295:1464, 1511 (1976).

*Javitt, Norman B., and Charles K. McSherry: Pathogenesis of Cholesterol Gallstones, *Hosp. Prac.*, July 1973.

**Johnson, L. R. (ed.): "Gastrointestinal Physiology," C. V. Mosby, St. Louis, 1977.

Kappas, Attallah, and Alvito P. Alvares: How the Liver Metabolizes Foreign Substances, *Sci. Am.*, June 1975.

Kretchmer, Norman: Lactose and Lactase, *Sci. Am.*, October 1972.

Neurath, H.: Protein Digesting Enzymes, *Sci. Am.*, December 1974.

*Phillips, Sidney F.: Fluid and Electrolyte Fluxes in the Gut, *Hosp. Prac.*, March 1973.

Rayford, Phillip L., Thomas A. Miller, and James C. Thompson: Secretin, Cholecystokinin and Newer Gastrointestinal Hormones, *N. Engl. J. Med.*, **294:1093, 1157 (1976).

*Stahlgren, Leroy H.: The Dumping Syndrome: A Study of Its Hemodynamics, *Hosp. Prac.*, December 1970.

**Wilson, T. H.: "Intestinal Absorption," Saunders, Philadelphia, 1962.

Wood, J. D.: Neurophysiology of Auerbach's Plexus and Control of Intestinal Motility, *Physiol. Rev.*, **55:307 (1975).

Chapter 15

*Adolph, E. F.: "Physiology of Man in the Desert," Interscience, New York, 1947.

Benzinger, T. H.: The Human Thermostat, *Sci. Am.*, January 1961.

*Bray, G. E., and L. A. Campfield: Metabolic Factors in the Control of Energy Stores, *Prog. Endo. Met.*, **24**:99 (1975).

Cabanac, M.: Temperature Regulation, *Ann. Rev. Physiol.*, **37:415 (1975).

Daughaday, W. H. et al.: The Regulation of Growth by Endocrines, *Ann. Rev. Physiol.*, **37:211 (1977).

*Dinarello, C. A., and S. M. Wolff: Pathogenesis of Fever in Man, *N. Engl. J. Med.*, **298**:607 (1978).

*Edelman, I. S.: Thyroid Thermogenesis, *N. Engl. J. Med.*, **290**:1303 (1974).

Eichenwald, H. F., and P. C. Fry: Nutrition and Learning, *Science*, **163**:644 (1969).

*Felig, P., and J. Wahren: Fuel Homeostasis in Exercise, *N. Engl. J. Med.*, **293**:1078 (1975).

Gerich, J. E., et al.: Regulation of Pancreatic Insulin and Glucagon, *Ann. Rev. Physiol.*, **38:353 (1976).

Gray, G. W.: Human Growth, *Sci. Am.*, October 1953.

Irving, L.: Adaptations to Cold, *Sci. Am.*, January 1966.

*Landsberg, L., and J. B. Young: Fasting, Feeding, and Regulation of the Sympathetic Nervous System, *N. Engl. J. Med.*, **298**:1295 (1978).

Mayer, J.: "Overweight: Causes, Cost, and Control," Prentice-Hall, Englewood Cliffs, N. J., 1968.

*Randle, P. J.: The Interrelationships of Hormones, Fatty Acid and Glucose in the Provision of Energy, *Postgrad. Med. J.*, **40**:457 (1964).

*Tepperman, J.: "Metabolic and Endocrine Physiology," 3d ed., Year Book, Chicago, 1973.

Unger, R. H., and L. Orci: Physiology and Pathophysiology of Glucagon, *Physiol. Rev.*, **56:778 (1976).

*Unger, R. H., R. E. Dobbs, and L. Orci: Insulin, Glucagon, and Somatostatin Secretion in the Regulation of Metabolism, *Ann. Rev. Physiol.*, **40**:307 (1978).

*Van Wyk, J. J., et al.: The Somatomedins, *Am. J. Dis. Child*, **126** (November 1973).

Wilkins, L.: The Thyroid Gland, *Sci. Am.*, March 1960.

Young, V. C., and N. S. Scrimshaw: The Physiology of Starvation, *Sci. Am.*, October 1971.

Chapter 16

Allen, R. D.: The Moment of Fertilization, *Sci. Am.*, July 1959.

*Bremner, W. J., and D. M. deKretser: The Prospects for New, Reversible Male Contraceptives, *N. Engl. J. Med.*, **295**:1111 (1976).

Brenner, R. M., and N. B. West: Hormonal Regulation of the Reproductive Tract in Female Mammals, *Ann. Rev. Physiol.*, **37:273 (1975).

*Chan, L., and B. W. O'Malley: Mechanism of Action of the Sex Steroid Hormones, *N. Engl. J. Med.*, **294**:1322 (1976).

Davidson, J. M.: Hormones and Sexual Behavior in the Male, *Hosp. Prac.*, September 1975.

Edwards, R. G., and R. E. Fowler: Human Eggs in the Laboratory, *Sci. Am.*, December 1970.

Epel, D.: The Program of Fertilization, *Sci. Am.*, November 1977.

Frantz, A. G.: Prolactin, *N. Engl. J. Med.*, **298**:201 (1978).

Klopfer, P. H.: Mother Love: What Turns It On?, *Am. Sci.*, **59**:404 (1971).

*Lloyd, C. W.: "Human Reproduction and Sexual Behavior," Lea & Febiger, Philadelphia, 1964.

*Macleod, J.: The Parameters of Male Fertility, *Hosp. Prac.*, December 1973.

Masters, W. H., and V. E. Johnson: "Human Sexual Response," Little, Brown, Boston, 1966.

*McCann, S. M.: Leuteinizing-hormone-releasing Hormone, *N. Engl. J. Med.*, **296**:797 (1977).

*Michael, R. P.: Hormones and Sexual Behavior in the Female, *Hosp. Prac.*, December 1975.

Money, J., and A. E. Ehrhardt: "Man and Woman, Boy and Girl. The Differentiation and Dimorphism of Gender Identity from Conception to Maturity," Johns Hopkins, Baltimore, 1973 (paperback).

*Patten, B. M., and B. M. Carlson: "Foundations of Embryology," 3d ed., McGraw-Hill, New York, 1974.

Segal, S. J.: The Physiology of Human Reproduction, *Sci. Am.*, September 1974.

*Tepperman, J.: "Metabolic and Endocrine Physiology," 3d ed., Year Book, Chicago, 1973.

*Wilson, J. D.: Sexual Differentiation, *Ann. Rev. Physiol.* **40**:279 (1978).

Chapter 17

Beer, A. E., and R. E. Billingham: The Embryo as a Transplant, *Sci. Am.*, April 1974.

Blumenthal, H. T.: Aging: Biologic or Pathologic, *Hosp. Prac.*, April 1978.

Burke, D. C.: The Status of Interferon, *Sci. Am.*, April 1977.

Capra, J. D., and A. B. Edmundson: The Antibody Combining Site, *Sci. Am.*, January 1977.

Clowes, R. C.: The Molecule of Infectious Drug Resistance, *Sci. Am.*, April 1973.

Cummingham, B. A.: The Structure and Function of Histocompatibility Antigens, *Sci. Am.*, October 1977.

*Dannenberg, A. M., Jr.: Macrophages in Inflammation and Infection, *N. Engl. J. Med.*, **293**:489 (1975).

Dubos, R.: "Man Adapting," Yale University Press, New Haven, Conn., 1965 (paperback).

Goldstein, S.: The Biology of Aging, *N. Engl. J. Med.*, **285**:20 (1971).

Holland, J. J.: Slow, Inapparent and Recurrent Viruses, *Sci. Am.*, February 1974.

Kappas, A., and A. P. Alvares: How the Liver Metabolizes Foreign Substances, *Sci. Am.*, June 1975.

Lerner, R. A., and F. J. Dixon: The Human Lymphocyte as Experimental Animal, *Sci. Am.*, June 1973.

Leung, K., and A. Munck.: Peripheral Actions of Glucocorticoids, *Ann. Rev. Physiol.*, **37:245 (1975).

Levine, S.: Stress and Behavior, *Sci. Am.*, January 1971.

Lieber, C. S.: The Metabolism of Alcohol, *Sci. Am.*, March 1976.

Mason, J. W.: Organization of Psychoendocrine Mechanisms, *Psychosom. Med.*, **30**(II) (1968).

Merigan, T. C.: Host Defenses Against Viral Disease, *N. Engl. J. Med.*, **290**:323 (1974).

*Müller-Eberhard, Hans J.: Chemistry and Function of the Complement System, *Hosp. Prac.*, August 1977.

*Notkins, A. L.: Viral Infections: Mechanisms of Immunologic Defense and Injury, *Hosp. Prac.*, September 1974.

Notkins, A. L., and H. Koprowski: How the Immune Response to a Virus Can Cause Disease, *Sci. Am.*, January 1973.

Old. L. J.: Cancer Immunology, *Sci. Am.*, May 1977.

*Parker, C. W.: Control of Lymphocyte Function, *N. Engl. J. Med.*, **295**:1180 (1976).

Raff, M. C.: Cell-surface Immunology, *Sci. Am.*, May 1976.

Ratnoff, O. D.: The Interrelationship of Clotting and Immunologic Mechanisms, *Hosp. Prac.*, April 1971.

Rensfeld, R. A., and B. D. Kahan: Markers of Biological Individuality, *Sci. Am.*, June 1972.

Ross, R.: Wound Healing, *Sci. Am.*, June 1969.

Stossel, T. P.: Phagocytosis, *N. Engl. J. Med.*, **290:717, 774, and 833 (1974).

Terne, N. K.: The Immune System, *Sci. Am.*, January 1973.

Vander, A. J. (ed.): "Human Physiology and the Environment in Health and Disease," Freeman, San Francisco, 1976.

Walker, W. A., and K. J. Isselbacher: Intestinal Antibodies, *N. Engl. J. Med.*, **297**:767 (1977).

Weiss, J. M.: Psychological Factors in Stress and Disease, *Sci. Am.*, June 1972.

Chapter 18

*Alpern, M., M. Lawrence, and D. Wolsk: "Sensory Processes," Brooks/Cole, Belmont, Calif., 1967 (paperback).

**Annual Review of Neuroscience, W. M. Cowan, Z. W. Hall, and E. R. Kandel (eds.), vol. 2, Annual Reviews, Palo Alto, Calif., 1979. *Good reviews on vision and vestibular mechanisms.*

Casey, K. L.: Pain: A Current View of Neural Mechanisms, *Am. Sci.*, **61**:194 (1973).

Daw, N. W.: Neurophysiology of Color Vision, *Physiol. Rev.*, **53:571 (1973).

*Dethier, V. G.: Other Tastes, Other Worlds, *Science*, **201**:224 (1978).

*De Valois, R. L., and G. H. Jacobs: Primate Color Vision, *Science*, **162**:533 (1968).

Fields, H. L., and A. I. Basbaum: Brainstem Control of Spinal Pain-Transmission Neurons, *Ann. Rev. Physiol.*, **40:217 (1978).

*Granit, R.: The Development of Retinal Neurophysiology, Nobel Prize Lecture, 1967, *Science*, **160**:1192 (1968).

**Granit, R.: "Receptors and Sensory Perception: A Discussion of Aims, Means, and Results of Electrophysiological Research into the Process of Perception," Yale, New Haven, Conn., 1955 (paperback).

Gregory, R. L.: "Eye and Brain: The Psychology of Seeing," 2d ed., World University Library, McGraw-Hill, New York, 1973.

*Hartline, H. K.: Visual Receptors and Retinal Interactions, Nobel Prize Lecture, 1967, *Science*, **164**:270 (1969).

Hubel, D. H.: The Visual Cortex of the Brain, *Sci. Am.*, November 1963.

*Michael, C. R.: Retinal Processing of Visual Images, *Sci. Am.*, May 1969.

Miller, W. H., F. Ratliff, and H. K. Hartline: How Cells Receive Stimuli, *Sci. Am.*, September 1961.

Moulton, D. G., and L. M. Beidler: Structure and Function in the Peripheral Olfactory System, *Physiol. Rev.*, **47:1 (1967).

Oakley, B., and R. M. Benjamin: Neural Mechanisms of Taste, *Physiol. Rev.*, **46:173 (1966).

*Pettigrew, J. D.: The Neurophysiology of Binocular Vision, *Sci. Am.*, August 1972.

Robinson, D. A.: Eye Movement Control in Primates, *Science*, **161**:1219 (1968).

Rock, I., and C. S. Harris: Vision and Touch, *Sci. Am.*, May 1967.

Rushton, W. A. H.: Visual Pigments and Color Blindness, *Sci. Am.*, March 1975.

Scientific American. Very often contains articles on perception, particularly visual perception.

Stent, G. S.: Cellular Communication, *Sci. Am.*, September 1972.

*Stevens, S. S.: Neural Events and the Psychophysical Law, *Science*, **170**:1043 (1970).

*Tower, D. B. (ed.): "The Nervous System," vol 3, Human Communication and Its Disorders, Raven Press, New York, 1975. *Many good articles on hearing.*

*Wald, G.: Molecular Basis of Visual Excitation, Nobel Prize Lecture, 1967, *Science*, **163**:230 (1968).

*Werblin, F. S.: The Control of Sensitivity in the Retina, *Sci. Am.*, September 1972.

Young, R. W.: Visual Cells, *Sci. Am.*, October 1970.

Chapter 19

Eccles, J. C.: "The Understanding of the Brain," 2d ed., McGraw-Hill, New York, 1977 (paperback). Chapter 4.

Evarts, E. V.: Brain Mechanisms in Movement, *Sci. Am.*, July 1973.

**Granit, R.: "The Basis of Motor Control," Academic, New York, 1970.

Lippold, O.: Physiological Tremor, *Sci. Am.*, May 1971.

Matthews, P. B. C.: Muscle Spindles and Their Motor Control, *Physiol. Rev.*, **44:219 (1964).

Merton, P. A.: How We Control the Contraction of Our Muscles, *Sci. Am.*, May 1972.

*Ochs, S.: "Elements of Neurophysiology," pp. 280–301, 342–363, and 493–526, Wiley, New York, 1965.

Pearson, K.: The Control of Walking, *Sci. Am.*, December 1976.

**Ruch, T. C., H. D. Patton, J. W. Woodbury, and A. L. Towe: "Neurophysiology," 2d ed., pp. 153–225, 252–300, Saunders, Philadelphia, 1965.

*Sherrington, Sir Charles: "The Integrative Action of the Nervous System," Yale University Press, New Haven, Conn., 1961 (originally published in 1906; paperback).

Chapter 20

Agranoff, B. W.: Memory and Protein Synthesis, *Sci. Am.*, June 1967.

"Altered States of Awareness," readings from *Scientific American*, Freeman, San Francisco, 1972 (paperback).

**Annual Reviews of Neuroscience, W. M. Cowan, Z. W. Hall, and E. R. Kandel (eds.), vols. 1 and 2, *Annual Reviews*, Palo Alto, Calif., 1978 and 1979.

*Baldessarini, R. J.: Schizophrenia, *N. Engl. J. Med.*, **297**:988 (1977).

Black, P. M.: Brain Death, *N. Engl. J. Med.*, **299**: 338 and 393 (1978).

*De Wied, D.: Hormonal Influences on Motivation, Learning, and Memory, *Hosp. Prac.*, January 1968.

Eccles, J. C.: "The Understanding of the Brain," 2d ed., McGraw-Hill, New York, 1977 (paperback). Chapters 5–6.

**Eccles, J. C. (ed.): "Brain and Conscious Experience," Springer-Verlag, New York, 1966.

Gazzaniga, M. S.: The Split Brain in Man, *Sci. Am.*, August 1967.

Gershon, E. S.: Genetics of the Affective Disorders, *Hosp. Prac.*, March 1979.

Geschwind, H.: Language and the Brain, *Sci. Am.*, April 1972.

*Hess, W. R.: Causality, Consciousness, and Cerebral Organization, *Science*, **158**:1279 (1967).

Horn, G., S. P. R. Rose, and P. P. G. Bateson: Experience and Plasticity in the Central Nervous System, *Science*, **181**:506 (1973).

Jacobson, M., and R. K. Hunt: The Origins of Nerve-cell Specificity, *Sci. Am.*, February 1973.

Kimura, D.: The Asymmetry of the Human Brain, *Sci. Am.*, March 1973.

Lenneberg, E. H.: On Explaining Language, *Science*, **164**:635 (1969).

Lopes da Silva, F. H., and D. E. A. T. Arnolds: Physiology of the Hippocampus and Related Structures, *Ann. Rev. Physiol.*, **40:185 (1978).

Luria, A. R.: The Functional Organization of the Brain, *Sci. Am.*, March 1970.

Mandell, A. J.: Neurobiological Barriers to Euphoria, *Am. Sci.*, **61**:565 (1973).

Meyer, D. E., and R. W. Schvaneveldt: Meaning, Memory Structure, and Mental Processes, *Science*, **192:27 (1976).

*Miller, N. E., and B. R. Dworkin: Effects of Learning on Visceral Functions — Biofeedback, *N. Engl. J. Med.*, **296**:1274 (1977).

Moskowitz, B. A.: The Acquisition of Language, *Sci. Am.*, November 1978.

Moskowitz, M. A., and R. J. Wurtman: Catecholamines and Neurological Diseases, *N. Engl. J. Med.*, **293:274 (1975).

Premack, A. J., and D. Premack: Teaching Language to an Ape, *Sci. Am.*, October 1972.

Pribram, K. H. The Neurophysiology of Remembering, *Sci. Am.*, January 1969.

"Psychobiology: The Biological Basis of Behavior," readings from *Scientific American*, Freeman, San Francisco, 1972 (paperback).

Rosenzweig, M. R., E. L. Bennett, and M. C. Diamond: Brain Changes in Response to Experience, *Sci. Am.*, February 1972.

Routtenberg, A.: The Reward System of the Brain, *Sci. Am.*, November 1978.

Scott, J. P.: Critical Periods in Behavioral Development, *Science*, **138**:949 (1962).

Siegel, R. K.: Hallucinations, *Sci. Am.*, October 1977.

*Sperry, R. W.: A Modified Concept of Consciousness, *Psychol. Rev.*, **76**:532 (1969).

Stent, G. S.: Limits to the Scientific Understanding of Man, *Science*, **187:1052 (1975).

Tart, C. T. (ed.): "Altered States of Consciousness," Anchor Books, Doubleday, Garden City, N.Y., 1972 (paperback).

*Thatcher, R. W., and E. R. John: "Foundations of Cognitive Processes," Functional Neuroscience, vol. 1, Lawrence Erlbaum, Hillsdale, N. J., 1977.

Wallace, R. K., and H. Benson: The Physiology of Meditation, *Sci. Am.*, February 1972.

Weissman, M. M.: Environmental Factors in Affective Disorders, *Hosp. Prac.*, April 1979.

INDEX

INDEX

A bands, 212, 214–215
A blood type, 544
A cells of pancreas, 448, 452
AB blood type, 544
Abcess, 529
Abdominal blood flow, 256
Abdominal muscles, 336
Abortion, 509
Absolute refractory period, 157
Absorption by small intestine, 402, 404, 407–408
Absorptive state, 442–447
Accessory reproductive organs, 479
Acclimatization, 364
Accommodation of eye, 567–569
Acetaldehyde, 551, 622–623
Acetone, 447
Acetyl coenzyme A, 82–83, 93–95, 98
Acetyl group, 82
Acetylcholine (ACh):
    autonomic nervous system transmitter, 173, 175, 186
    at neuromuscular junction, 223–225
    release from sympathetic endings, 288
Acetylcholinesterase, 175, 224
ACh (*see* Acetylcholine)
Acid, definition of, 26
Acidity, 395–396
    effect on hemoglobin-oxygen saturation, 349
    effects on protein conformation, 50
    of stomach, 423–426
    (*See also* Hydrogen ion)
Acidosis, 400
Acne, 488
Acromegaly, 459–460
Acrosome, 483, 504
ACTH (*See* Adrenocorticotropic hormone)
Actin, 214, 216–218, 528

Action potentials, 152–160
    frequency of, 164–166
    in heart, 266–270
    ionic hypothesis, 154–155
    propagation of, 158–160
    refractory periods, 156–158
    in skeletal muscle, 218, 222–225
    in smooth muscle, 244–248
    threshold, 156
Activation energy, 73, 75
Active hyperemia, 286–287
Active immunity, 539
Active site, enzyme, 75
Active state of muscle contraction, 229
Active transport, 111–113
Acupuncture, 565
Adaptation of receptor neurons, 164–165
Addiction, 621
Adenine, 39–41, 51–52
Adenohypophysis (*see* Anterior pituitary)
Adenosine, 287
Adenosine diphosphate, 80, 320–321
3′,5′-Adenosine monophosphate (*see* Cyclic AMP)
Adenosine triphosphate (ATP), 80–89, 93–94
ATP:
    role in forming cyclic AMP, 136–137
    role in muscle:
        actin and myosin interaction, 217–218
        energy for calcium pump, 220–222
        metabolism, 233–236
Adenylate cyclase, 136–138
ADH (*see* Antidiuretic hormone)
Adipocytes, 93
Adipose tissue, 93
    effect of insulin on, 449
    growth of, 468–469
    role in absorptive state, 443–444

Adipose tissue:
   role in postabsorptive state, 446–447
   sympathetic innervation, 454
Adolescence, 480
ADP (*see* Adenosine diphosphate)
Adrenal glands:
   cortex, 192–193, 207, 209, 382, 489
   medulla, 186–187, 192–193, 207
Adrenergic fibers, 186
   (*See also* Sympathetic nervous system)
Adrenergic receptors, 187
Adrenocorticotropic hormone, 192, 201–203, 205–206
   neuromodulator, 177
   role in memory, 628
   role in stress, 552–554
Adsorptive endocytosis, 119
Aerobic metabolism, 88–89
Affective disorders, 619–620
Afferent arterioles, 369
Afferent division of peripheral nervous system, 184
Afferent information, 561–562
Afferent neurons, 161–162
Afferent pathway, 130–131
Affinity, of binding sites, 45, 47
Afterbirth, 510
After-hyperpolarization, 154
Aggregation factors, 47–48
Aging, 555–556
Airways, 328
   resistance of, 336–337
Alanine, 96
Albumin, 257–258, 310–311
Alcohol, 311, 389, 425
   effects on CNS, 621–623
   effects on cough reflex, 361
   metabolism of, 551
Alcoholic fermentation, 84
Aldosterone, 192, 201, 209–210
   release during stress, 555
   role in maintaining potassium balance, 392–393
   role in maintaining sodium balance, 382–383, 390
   neuromodulator, 177
Alertness, 608, 610–612
Alkalosis, 400
All-or-none response, 156
Allergic responses:
   cardiovascular, 311, 313
   respiratory, 337–338
Allergy, 546

Allosteric regulation, 49, 78
Alpha-adrenergic receptors, 187, 287
Alpha helix, 37–38
Alpha motor neurons, 592–593, 595–596
Alpha rhythm, 608–609, 625
Alveolar dead space, 341
Alveolar ventilation, 340–341
Alveoli, 328–332
   gas pressures in, 333, 335–337, 345
   role in gas exchange, 348–349, 351–352
Amino acids:
   absorption of intestinal tract, 412
   effects on glucagon secretion, 453–454
   essential, 97, 441
   membrane transport, 112, 448, 459
   metabolism, 95–98
      during absorptive state, 444–445
      during postabsorptive state, 445–446
   pools, 441
   in protein structure, 34–39
   role in control of insulin secretion, 450
   structure of, 34–36
Amino acyl synthetase, 57
Amino group, 25–27
Aminopeptidase, 412
Ammonia, 95–97
   role in renal excretion of hydrogen ions, 399–400
Amnesia, 626–627
Amnion, 507
cAMP (*see* Cyclic AMP)
cAMP-protein-kinase, 199
Amphetamine, 617, 620
Amphipathic molecules, 28–29, 35
Amygdala, 183
Amylase, 404, 411
Anabolism, 71
Anaerobic metabolism, 84, 89
Anal canal, 492
Anal sphincters, 435
Analgesia, 565
Anaphylactic shock, 546
Anatomical dead space, 340–341
Androgens, 192, 487, 503
   effects on growth, 461–462
   in female, 498
Anemia, 262
   pernicious, 261, 414
   sickle cell, 258
Anemic hypoxia, 363

Angina pectoris, 319
Angiotensin:
    effect on aldosterone secretion, 382–383, 390
    effects on blood vessels, 288
    neuromodulator, 177
    role in hypertension, 317
Angiotensinogen, 288, 382
Antagonists, 232
Anterior pituitary (*see* Pituitary gland)
Antibiotics, 546
Antibodies, 522, 533–539
Anticoagulants, 326
Anticodons, 57–58
Antidiuretic hormone (ADH), 192, 201
    control of secretion, 384–385, 388–390
    effects on kidney, 379
    neuromodulator, 177
    release during stress, 555
    role in learning, 628
    site of production and release, 207
Antigens, 534
Antigravity muscles, 605
Antihistamines, 338, 546
Antrum, 423, 427, 493
Anus, 403–404, 493
Aorta, 254–255
Aortic arch baroreceptors, 306–308
Aortic bodies, 355–356
Aortic valve, 265, 272–274
Aphasias, 630
Apnea, 356
Appendix, 403, 434
Aqueduct of sylvius, 189
Aqueous humor, 566
Arachidonic acid, 136
Arachnoid villus, 189
Arousal, 608–612
Arterial pressure, 280, 282–284
    (*See also* Blood pressure)
Arteries, 254, 282–284
Arterioles, 255, 285
    hormonal control of, 288–289
    sympathetic control of, 287–288
Arteriosclerosis (*see* Atherosclerosis)
Artificial kidney, 401
Ascorbic acid, 76
Asparin, 476, 532
Aspartic acid, 173
Association areas of cortex, 586

Asthma, 337, 364
Astigmatism, 568
Atherosclerosis, 283, 318–319, 457, 520
Athetosis, 601
Atmospheric pressure, 333, 344–345
Atoms, 21–23
Atomic number, 22
Atomic weight, 22
Atopic allergy, 546
ATP (*see* Adenosine triphosphate)
Atria of heart, 264–265, 272
Atrioventricular node, 268–269, 274
Atrioventricular valves, 264–265, 272–274
Atrophy of skeletal muscle, 237
Attention, 608–611
Auditory cortex, 558–559, 581
Auditory nerve, 576, 578
Auditory tube, 576
Autoimmune disease, 547
Autonomic nervous system, 184–189
    (*See also* Parasympathetic nervous system,
        Sympathetic nervous system)
Autoregulation of blood flow, 285–287
AV node (*see* Atrioventricular node)
AV valves (*see* Atrioventricular valves)
Axon, 160–162
Axon terminals, 160, 162

B blood type, 544
B cells of pancreas, 448
B lymphocytes, 533–534, 537–539, 544
Bacteria, 522
    antibiotics, 546
    in GI tract, 403, 405, 435
Bacterial toxins, 537
Balance, 603–606
Balance concept, 139–143
Ballismus, 601–602
Banting, Frederick Grant, Sir, 450
Barbiturates, 551
Barometric pressure, 344
Baroreceptors, 165, 306–309, 316
Basal ganglia, 182–183, 590–591, 597, 601–603, 606
Basal metabolic rate, 464, 466
Base (*see* Nucleotides)
Base, alkali, 395
Base pairing, 51, 53–58, 63
Basic electrical rhythm, 427, 434
Basilar membrane, 577–579, 581

Basophils, 262
BCG, 543
Beard, 488
Bee sting, 546
Belching, 434
Bernard, Claude, 7
Beryllium, 346
Best, Charles Herbert, 450
Best frequency, 581
Beta-adrenergic receptors, 187, 266
Bicarbonate, 350-352
  secretion by liver, 407, 431-432
  secretion by pancreas, 407, 428-430
Biceps muscle, 232
Bile:
  composition, 429
  excretion of foreign chemicals in, 549-550
  secretion, 429-432
Bile canaliculi, 431
Bile pigments, 429
Bile salts, 407, 412-415, 429-432
Bilirubin, 259, 431, 438
Binding sites, characteristics of, 43-50
Biochemistry, 30
Biological clocks, 143
Biological rhythms, 143
Biotin, 76
Biotransformation, 548
Bipolar affective disorders, 619
Bipolar cells of retina, 569-571, 573-574
Birth (see Parturition)
Bladder (urinary), 368-369, 377-378, 492
Blastocyst, 505
Blood:
  carbon dioxide content, 343, 350-351
  coagulation, 319-326
  composition of, 256-264
  oxygen content, 343, 346
  plasma, 7-8, 256-258
  transfusion, 544-545
  types, 544, 545
  volume of, 257, 296-297, 305
Blood brain barrier, 188, 359
Blood clot, 325-326
Blood flow:
  to intestine, 410
  to organs, 256
  regulation, 285-289
  velocity, 290-291

Blood pressure:
  arterial, 282-284
  in capillaries, 294-295
  measurement of, 284
  regulation of, 300-310
  relation to total body sodium, 381
  in veins, 296-298
BMR (see Basal metabolic rate)
Body temperature, 467
  (See also Temperature)
Bonds (see Chemical bonds)
Bone, 132-133
  growth, 459-460, 462
  role in calcium homeostasis, 393-395
Bone marrow, 259-264, 523
Botulinus toxin, 225
Bowman's capsule, 368-370, 372
Brain:
  blood supply, 188-190, 256
  death, 623
  metabolism, 447
  structure of, 181-184
  (See also Brainstem; Cerebellum; Cerebral cortex;
      Hypothalamus; Limbic system; Medulla)
Brainstem, 180-182
Breasts, 501, 502, 511-512
  (See also Lactation)
Breath holding, 357
Breathing:
  neural control, 179
  neural generation of rhythmicity, 352-353
  work of, 341-342
Bronchi, 328-330
Bronchioles, 329-331
Buffers, 396-398
Bulbourethral glands, 480, 485
Bulk flow, 280-282
  of blood, 254
  of air, 333

Caffeine, 270, 425
Calcitonin, 192, 201, 395
Calcium, 22-23, 26
  absorption of gastrointestinal tract, 394-395, 416
  regulation of extracellular concentration, 393-395
  relation to bone, 132-133
  role in blood coagulation, 322-323
  role in cardiac muscle, 272, 279
  role in cell adhesion, 122

Calcium:
   role in fertilization, 504
   role in release of neurotransmitters, 167
   role as second messenger, 135–136
   role in skeletal muscle contraction, 218–222
   role in smooth muscle contraction, 242–246
   stimulus for exocytosis, 119
Caloric balance, 465–466
Calorie, 72, 463
Calorimeter, 463
Camel, 391
Cancer, 68–69, 522, 543
Cannon, W. B., 128
Capacitation, 504
Capillaries, 254–255, 289–296
   bulk flow across, 292–296
   diffusion across, 291–292, 296
   filtration, 525
   osmosis across, 293–296
Carbamino compounds, 350–351
Carbohydrate:
   body content, 93
   catabolism, 88–89
   digestion and absorption from GI tract, 411–412
   in membranes, 103
   metabolism, 442–443, 446–448
   storage, 90–91
   structure of, 29–32
   synthesis, 91–92
   (*See also* Cellulose; Disaccharide; Glucose;
      Glycogen; Polysaccharide; Starch)
Carbon, 22–24, 29–30
Carbon dioxide:
   blood content, 350–351
   in control of ventilation, 356–359
   metabolic formation of, 82–84, 98
   partial pressures in respiratory system, 345–346
   relation to hydrogen ion concentration, 397–399
   transport in blood, 343, 350–352
Carbonic acid, 350
Carbonic anhydrase, 350, 423–424
Carbon monoxide, 355–356, 363
Carboxyl group, 25–27
Carboxypeptidase, 412
Cardiac cycle, 272–275
Cardiac muscle, 211–212, 248–249, 265–266
   calcium, 393
   energetics, 280
   excitability, 266–271

Cardiac muscle:
   excitation-contraction coupling, 272
   length-tension relationship, 277–279
   refractory period, 269–271
Cardiac output, 275–280
   changes with age, 555–556
   determinants of, 280–281
   during exercise, 256, 275, 362–363
   role in control of blood pressure, 303
Cardiovascular center, 181, 305–306
Carotid artery, 355
Carotid bodies, 355–357
Carotid sinus baroreceptors, 165, 306–308
Carrier hypothesis, 110
Cartilage, 459, 461
Castor oil, 439
Castration, 487–488, 616
Catabolism, 71
Catalysts, 48, 74
Cataract, 568
Catatonia, 618
Catecholamines, 173–174, 617–618
CCK (*see* Cholecystokinin)
Cecum, 434–435
Cell-mediated immunity, 533, 539–544
Cells:
   aggregation of, 47–48
   basic units, 1–4, 7
   differentiation of, 1, 65–66
   division of, 63–65, 556
   junctions between, 122–124
   organelles of, 14–19
   structure of, 13–19
Cellulose, 31–32, 411, 439
Center of gravity, 603, 605–606
Central canal, 189
Central chemoreceptors, 358–359
Central dogma of molecular biology, 51
Central nervous system, 179–184
Central sulcus, 559
Central thermoreceptors, 476
Centrioles, 15, 19–20, 65
Centromere, 64–65
Cephalic phase, 419–420
Cerebellum, 180–182, 589, 591, 597, 599, 602–603
Cerebral cortex, 182–183, 624
Cerebral dominance, 628–630
Cerebral hemispheres, 182
Cerebrospinal fluid, 188–190, 358–359

Cerebrum, 180–183
Cervical nerves, 180
Cervix, 491–493, 501, 510
CG (*see* Chorionic gonadotropin)
Chemical bonds, 23–27
Chemical elements, 21–23
Chemical energy, 72
Chemical equilibrium, 74
Chemical messengers, 48, 135–136
Chemical reactions, 73–74, 79
Chemical specificity of binding, 43–44
Chemical synapses, 167
Chemoaffinity theory of CNS development, 625
Chemoreceptors, 584
    role in regulation of cardiovascular system, 309
    in respiratory system, 355–360
Chemotaxis, 525–527, 531, 541
Chewing, 404, 421
Chief cells, 423–426
Chloride, 22–23, 25–28, 175
    active transport of, 379
    role at inhibitory synapses, 170
Chlorpromazine, 617
Cholecystokinin, 192, 418–420, 429–430, 432
Cholesterol:
    absorption by intestines, 414
    blood levels, 501
    effects of estrogen, 520
    in plasma membrane, 100–101
    regulation of plasma levels, 457–458
    role in atherosclerosis, 318
    role in gallstone formation, 437–438
    in steroid hormone synthesis, 210
    structure of, 33
Cholinergic fibers, 186
Choking, 404
Chorea, 601
Chorionic gonadotropin, 508–509
Chorionic somatomammotropin, 509, 511
Chorionic villus, 506
Choroid plexus, 189–190
Chromatin, 17–18, 62–65
Chromophore, 568–569
Chromosomes, 64–65, 481–482
Chylomicrons, 414–415, 443
Chyme, 405
Chymotrypsin, 412
Cigarette smoke, 329, 345
Cilia, 211, 328–329, 331, 503

Cingulate gyrus, 183
Circadian rhythms, 145, 195
Cis-retinal, 568–569
Citric acid, 82–83, 98
Clearance, 373
Clitoris, 489, 491, 493, 503
Clot retraction, 324
CP (*see* Creatine phosphate)
Cobalamine, 76
Cobalt, 260
Cochlea, 576, 578–579, 583
Code words, genetic, 52, 56–58
Codeine, 175
Codons, 56–58
Coenzyme A, 82
Coenzymes, 75–77
Cofactors, 75–76
Cold receptors, 475, 564
Collagen, 320, 322, 529, 555
Collaterals of neuron, 160, 162
Collecting ducts, 369, 379, 388
Colon, 434–435
Color blindness, 573
Color vision, 572–574
Coma, 610, 623
Commissures, 182
Compensatory growth, 462
Competition for binding sites, 46
Competition for mediated transport sites, 109–110
Complement, 530–532, 535–536
Complex cortical cells, 572
Compliance of lung, 335
Concentration, 28–29
Concentration gradient, 106
Concentric cortical cells, 572
Conduction of heat, 472–474
Cones, 568–570, 573–574
Congestive heart failure, 317–318, 383
Connective tissue, 3
Consciousness, 607
    altered states of, 618–623
    states of, 607–612
        arousal, 608–612
        sleep, 609–612
        waking state, 608–609
Conscious experience, 612–618
Conscious sensation, 587
Constipation, 438–439
Contraception, 513–514

Contractile proteins (*see* Actin and Myosin)
Contractility of heart, 279
Contraction, muscle, 225–226
Control systems, 8–9, 128–130
Convection, 472–473
Convergence, 166, 168
Convulsions, 470
Cornea, 566–567
Cornification of vagina, 502
Coronary arteries, 265, 289, 438
Coronary blood flow, 280, 318
Corpus callosum, 182–183
Corpus luteum, 496
   during pregnancy, 508–509
   regulation of, 500
Cortical association areas, 586
Corticospinal motor pathway, 598–600
Corticotropin (*see* Adrenocorticotropic hormone)
Corticotropin releasing hormone, 205–206
Cortisol, 192, 201, 209–210
   effects on norepinephrine synthesis, 176
   effects on surfactant secretion, 342
   pharmacological effects, 554
   role in parturition, 511
   role in stress, 552–554
Cough, 361
Counter-current multiplier system, 385–388
Covalent bond, 23
Cranial nerves, 181, 184
Craniosacral division of the autonomic nervous
   system, 185
Creatine, 368
Creatine phosphate, 235
Cretinism, 461, 625
CRH (*see* Corticotropin releasing hormone)
Cross bridge, 215–218
Crossed-extensor reflex, 178
Crossing over, 482
Crystaloids, 294
Cumulus, 493–494
Curare, 224
Current (electric), 146, 152–153
Cyanide, 363
Cyclic AMP, 136–140, 199
Cyclic GMP, 141, 199
Cytochromes, 81, 258
Cytolytic T cells, 534
Cytoplasm, 14, 18
Cytosine, 39–41, 51–52

Cytoskeleton, 20
Cytosol, 14
Cytotoxins, 542

D cells of pancreas, 448
Dark adaptation, 570
DDT, 508
Dead space:
   anatomical, 340–341
   alveolar dead space, 341
Decremental current flow, 153
Defecation, 435–436, 438–439
Degranulation, 528
Dendrites, 160–162
Denervation atrophy, 237
Deoxyhemoglobin, 346
Deoxyribonucleic acid (DNA), 39–41
   functions of, 50–52
   genetic code, 51–53
   recombinant, 69–70
   replication of, 63–64
   role of folic acid in synthesis, 261
Deoxyribose, 39, 41
Depolarize, 151
Depression, 619–620
DES, 514
Descartes, Rene, 630
Desensitization, 547
Desmosome, 15, 123–124, 266
Diabetes insipidus, 389
Diabetes mellitus, 376, 448, 450–452, 454, 457, 545
Diaphragm, 329, 332, 335, 403, 422–423
Diarrhea, 381, 428, 435, 439
Diastole, 272–274
Diastolic blood pressure, 283
Diencephalon, 181–182
Diethylstilbestrol, 514
Diffuse thalamic projection system, 610
Diffusion, 103–108, 254, 345–346
Diffusion coefficient, 105–106
Diffusion potentials, 147–150, 152
Digestion, 402
Digitalis, 318
Dihydrotestosterone, 488–489
Diphosphoglycerate, 349–350, 364
Disaccharides, 31
Dishabituation, 611
Disinfectants, 546
Disinhibition, 561

Dissolved gas, 344
Distal convoluted tubule, 369, 379, 382
Disulfide bonds, 38–39
Disuse atrophy, 237
Diuretics, 383
Divergence, 166, 168
DNA (*see* Deoxyribonucleic acid)
DNA-dependent-RNA-polymerase, 54–55
Dopa, 173
Dopamine, 173–175, 602, 618–619, 620
Dorsal root ganglia, 180–181
Dorsal roots, 180–181
Double helix (*see* Deoxyribonucleic acid, structure
    of)
DPG (*see* Diphosphoglycerate)
Dreaming, 609
Drowning, 357
Drugs, effects on nervous system, 175–176
Drug tolerance, 621
Dual innervation, 187
Duodenum, 407–408
Dynamic equilibrium, 71

Ear canal, 576
Eardrum (*see* Tympanic membrane)
ECF (*see* Extracellular fluid)
ECG, 271–272
Ectopic focus of excitability in heart, 270
Edema, 295–296, 299, 318, 383
EEG (*see* Electroencephalogram)
Effector, 132–134
Efferent arterioles, 369–370
Efferent division of peripheral nervous system,
    184–188
Efferent neurons, 161–162
Efferent pathway, 132–134
Ejaculation, 485–487
Ejaculatory duct, 480
EKG, 271–273
Elastin fibers, 3
Electrical synapse, 167
Electric charge, 21, 145
Electric force, 145
Electrocardiogram, 271–272
Electroconvulsive shock, 627
Electrodes, 146
Electroencephalogram, 607–611, 623, 625
Electrogenic pump, 149
Electrolytes, 258

Electromagnetic spectrum, 566
Electrons, 21–22, 145
Embryo, 505–507
Emission, 486
Emotion, 184, 361, 616–618
Emphysema, 335, 345, 364
Emulsification, 412–413, 415
End diastolic volume, 274, 277–280
Endocrine glands, 134, 191–193
Endocytosis, 118–119
Endogenous pyrogen, 532
Endolymph, 582–583
Endometrium, 501
Endoplasmic reticulum, 15–18, 56
    role in protein secretion, 120–121
Endorphins, 173, 175, 177, 566, 620–621
Endothelium, 289, 320
End-plate potential, 223–224
End-product inhibition, 78–79
End-systolic volume, 274, 277
Energy, 72–73, 81–82
Energy balance, 462–478
Enkephalins, 173, 175, 177, 566, 620–622
Enterohepatic circulation, 431
Enterokinase, 429–430
Environment (*see* Internal environment, External
    environment)
Enzymes, 47, 74–80
Eosinophils, 262, 523, 527
Epidedymis, 480, 484–485
Epiglottis, 421–422
Epilepsy, 629
Epinephrine:
    from adrenal medulla, 187, 192, 201, 207
    effect on glycogen breakdown, 136–138
    effect on heart, 276–277, 279–280, 316
    effect on metabolic rate, 465
    effects on adipose tissue, 200
    effects on airway smooth muscle, 337
    effects on arterioles, 288–289
    effects on glucose and fat metabolism, 454–455
    effects on ventilation, 361
    neurotransmitter, 173–175
    receptors on heart, 199
    role in glycogen synthesis and breakdown, 198
    role in thermogenesis, 472
Epiphenomenon, 608, 613
Epiphyseal plates, 459, 461
Epiphyses, 501

Epithelial cells, 3, 408–411
Epithelial transport, 113
EPP (*see* End plate potential)
EPSP (*see* Excitatory postsynaptic potential)
Equilibrium potentials, 148–149, 152
Erectile tissue, 485
Erection, 485–487
Erythrocytes:
  characteristics of, 256–262
  production, 261–262, 524
  role in oxygen transport, 346
Erythropoiesis, 260–262
Erythropoietin, 192, 261–262, 364
Escherichia coli, 60
Esophagus, 328–329, 403, 405–406, 421–423
Essential amino acids, 98, 441–442
Essential hypertension, 317
Essential nutrients, 142–143
Estradiol, 496
Estriol, 496, 509
Estrogen:
  action on uterus, 199
  changes at menopause, 519–520
  during pregnancy, 508–510
  effects on breasts, 511
  effects on growth, 461–462
  in males, 488, 502
  neuromodulator, 177
  in oral contraceptives, 514
  secretion during menstrual cycle, 496–502
  secretion by ovaries, 192, 201–203
  secretory changes with age, 518–520
  structure of, 33
  synthesis of, 209–210
Estrone, 496
Ethanol (*see* Alcohol)
Euchromatin, 63
Eunuchs, 462
Euphoria, 619
Euphorigens, 620
Eustachian tube, 576–577
Evaporation, 474–475
Evolution, 69
Excitable membrane, 153
Excitation-contraction coupling:
  cardiac muscle, 272
  skeletal muscle, 218–225
  smooth muscle, 242–246
Excitatory postsynaptic potential, 169–171

Excitatory synapse, 168–171
Excretion, 370
Exercise:
  changes in circulation, 314–316
  effects on metabolic rate, 465
  fuel metabolism, 456–457
  maximal oxygen uptake, 361–363
  muscle metabolism, 235
  regulation of ventilation, 359–361
  role of skeletal muscle fiber types, 239–240
  role of sympathetic nervous system, 456
Exercise hyperemia, 286
Exocrine glands, 132
Exocytosis, 120–122
Expiration, 335–337
Expiratory reserve volume, 339–340
Extension, 231–232
External anal sphincter, 435–436
External environment, 4
External urethral sphincter, 377
Extracellular fluid, 6–8, 380–381, 384
Extrafusal fibers, 592–593
Extrapyramidal system, 599
Extrinsic clotting pathway, 324
Eye color, 51
Eye movement, 574

Facilitated diffusion, 110
Facilitation of postsynaptic neurons, 170
Factor XII, 321
Fainting, 310–311
Fallopian tubes (*see* Oviducts)
Farsighted vision, 568
Fast twitch-fatigable muscle fibers, 237–239
Fasting, 446–447
Fat:
  digestion and absorption by intestinal tract,
    406–407, 412–415
  distribution in body, 488, 501–502
  metabolism, 92–95, 442–447
  structure of, 32–34
  (*See also* Triacylglycerol)
Fat-soluble vitamins, 76–77
Fatigue, skeletal muscle, 228, 234, 236–239
Fatty acids:
  metabolism of, 93–95, 98
  structure of, 32–34
Feces, 402–404, 431, 436, 438
Feedback, 130

Feedforward, 130
Fermentation, 84
Ferritin, 260, 416
Fertility control, 512–514
Fertilization, 504
Fetus, 547
Fiber, 439, 464, 476–478, 532
Fibrillation of cardiac muscle, 271
Fibrin, 320–325
Fibrinogen, 257–258, 320–325
Fibroblasts, 529
"Fight or flight" response, 187
Filaments in muscle fibers, 214–215
Fimbriae, 491–493
First-order neurons, 184
Fissure of Rolando, 559
Flatus, 435
Flexion, 231–232
Flexion reflex, 177–178
Fluid compartments, 7–8
Fluid endocytosis, 118
Fluidity, of plasma membrane, 100–102
Fluid-mosaic model of membrane structure, 101–102
Flux, 104–106
Folic acid, 76, 260–262
Follicle growth, 491–495
Follicle-stimulating hormone:
  in female, 498–500
  in male, 486, 489–490
  secretion by pituitary, 192, 201–203, 205–206
  secretory changes with age, 517–519
Follicular cells, 491, 494
Follicular phase, 496–497
Food intake, control of, 467–470
Food poisoning, 225
Forebrain, 181–182
Foreign chemicals, metabolism of, 547–552
Forgetting, 628
Fornix, 183
Fourth ventricle of brain, 189
Fovea, 566, 574
Fraternal siblings, 495
Frequency of sound, 575–576
Frontal cortex, 617–618
Frontal lobe, 182, 558, 565, 630
Fructose, 30, 88, 485
FSH (see Follicle-stimulating hormone)
Functional residual capacity, 340
Fundus of stomach, 424

GABA (see Gamma-aminobutyric acid)
Galactose, 30, 88
Galactosidase, 60
Gallbladder, 403, 407, 431–432
Gallstones, 437–438
Gamma-aminobutyric acid, 173, 175
Gamma globulin, 257, 545
Gamma loop, 596
Gamma motor neurons, 592–593, 595–596, 598–599
Ganglia, 184–186
Ganglion cells of retina, 569–571, 573–574
Gap junctions:
  in cardiac muscle, 266
  between granulosa cells and ovum, 491
  in smooth muscle, 247–249
  structure of, 15, 122–124
Gas:
  properties of, 343–344
Gas transport, 342–352
Gastric inhibitory peptide, 192, 418–420, 425
Gastric phase, 419–420
Gastrin, 177, 192, 418–419, 423–425
Gastrocnemius muscle, 232–233
Gastroileal reflex, 434–435
Gastrointestinal system:
  digestion and absorption of
    carbohydrate, 411–412
    fat, 412–415
    nucleic acids, 414
    protein, 412
    vitamins, 414
    water and minerals, 414–416
  hormonal regulation, 417–419
  neural regulation, 417–418
  reflexes, 417
  structure of, 402–408
  (See also Bile; Large intestine; Liver; Pancreas;
    Saliva; Small intestine; Stomach)
Gates controlling membrane permeability to ions, 155
Generator potential, 152
Genes, 51, 56, 63, 65–68
Genetic information, 50–51
Genitalia, 515–516
GFR (see Glomerular filtration rate)
GIP (see Gastric inhibitory peptide)
Glans, 480
Glial cells, 160, 184
Globin, 257, 259
Globulins, 257–258

Glomerular capillary pressure, 372
Glomerular filtration rate, 370–374, 380–381, 555–556
Glomerulus, 368–370
Glottis, 421–422
Glucagon, 192, 201, 208, 448, 452–454, 456, 467
Gluconeogenesis, 446
  effect of insulin on, 449
  role of epinephrine, 454
Glucose:
  catabolism, 88–89, 98
  control of insulin secretion, 449–450
  metabolism: absorptive state, 443–444
    postabsorptive state, 446–447
  movement across plasma membrane, 110, 448
  reabsorption of kidney, 374–376
  receptors in hypothalamus, 454–455, 467
  role in brain metabolism, 189
  sparing, 446–447
  storage, 90–91
  structure of, 29–30
  synthesis, 91–92
Glucose-6-phosphate, 85, 90
Glutamic acid, 96, 173
Glycerol, 92–93, 98
  role in fat metabolism, 443–447
  role in glucose metabolism, 446–447
Glycine, 173, 175
Glycogen:
  breakdown, 136–138
  effect of insulin on, 449
  formation in absorptive state, 443–444
  in liver, 446
  metabolism of, 90–92, 98
  in muscle, 446
  role in muscle fatigue, 236
  role in postabsorptive state, 446–447
  secretion of uterine glands, 501
  structure of, 31–32
Glycolysis, 235, 237, 239
Glycolipids, 103
Glycoproteins, 103
cGMP (see Cyclic GMP)
Gn-RH (see Gonadotropin-releasing hormone)
Goiters, 197
Golgi apparatus, 15–18, 121
Golgi tendon organs, 597
Gonadotropic hormones, 192, 202–203, 205–206, 486

Gonadotropin-releasing hormone, 177, 205–206, 490, 489–500, 517
Gonads, 192–193, 479, 515
Graded potentials, 151–153
Graft rejection, 543–544
Granuloma, 529
Granulopoietin, 523, 532
Granulosa cells, 491, 493–494
Grasp reflex, 601
Gravity, effects on circulation, 313–314
Gray matter, 179, 181
Growth:
  effects of androgens and estrogens, 461–462
  effects of growth hormone, 459–461
  effects of insulin, 461
  effects of thyroid hormone, 461
  regulation of, 458–462
Growth hormone, 192, 201–203, 205–206, 455–456, 459–461, 555
GTP (see Guanosine triphosphate)
Guanine, 39–41, 51–52
Guanosine monophosphate, 139
Guanosine triphosphate, 84

H zone, 212, 214–215
Habituation, 611, 626
Hageman factor, 321–324, 530
Hallucinations, 613–614, 620
Handedness, 629
Hardening of the arteries, 283
Harvey, William, 254
Hay fever, 546
Hearing, 574–582
Heart, 264–280
  attacks, 318–319, 363, 435
  autonomic innervation, 266
  block, 270–272
  blood flow, 256
  conducting system, 268–269
  excitability, 266–271
  failure, 317–318
  influence of autonomic nerves, 275–280
  origin of heart beat, 267–272
  rate, 275–276, 280, 362–363
  sounds, 274–275
  (See also Cardiac muscle)
Heart-lung preparation, 276–277
Heart transplants, 316
Heartburn, 423

Heat:
  energy, 72
  exhaustion, 477–478
  loss mechanisms, 472–475
  production, 470–472
  stroke, 477
"Helper" T cells, 534
Hematocrit, 257
Hematoma, 319
Heme, 257, 259
Hemoglobin, 257, 259, 346–350
Hemolysis, 118
Hemorrhage, 297, 309–311
Hemostasis, 319–326
Henle's loop, 369
Heparin, 325–326
Hepatic portal vein, 404, 411, 443
Hereditary information, 50–51
Heroin, 508
Heterochromatin, 63
High-altitude respiratory responses, 364
High-density lipoproteins, 457
Hippocampus, 183, 611, 626
Histamine, 134, 173, 177
  in basophils, 262
  effects on airways, 337–338
  effects on gastric secretion, 425–426
  mediator of inflammatory response, 529–531
  release by lungs, 365
Histidine, 61
Histocompatibility antigens, 543
Histones, 62
Histotoxic hypoxia, 363
Hodgkin, A. L., 154
Homeostasis, 127–128
Homeothermic animals, 470
Homosexuals, 487
Hooke, Robert, 13
Hormones, 191
  activation, 195
  inactivation and excretion, 195
  mechanisms of action, 197–201
  pharmacological effects, 200–201
  receptors, 198–199
  secretion, 194–195, 201–208
  transport in blood, 195
Hot flashes, 501, 520
5-HT (see Serotonin)
Humidity, 474

Humoral immunity, 533–539
Huxley, A. F., 154
Hydrocephalus, 190
Hydrochloric acid secretion by stomach, 405, 423–426
Hydrogen, 22–23, 25, 81, 84
Hydrogen bonds, 27, 37
Hydrogen ion, 26
  chemoreceptors, 358–359
  regulation, 395–400
  transport between tissues and lungs, 351–352
Hydrogen peroxide, 528
Hydroxyl group, 26–27
5-Hydroxytryptamine, 173
Hymen, 491
Hypercalcemia, 393
Hypercomplex cortical cells, 572
Hyperemia, 286
Hyperopic vision, 568
Hyperpolarize, 151
Hypersensitivity, 546
Hypertension, 280, 316–317, 382, 391
Hyperthermia, 476–477
Hyperthyroidism, 198, 465
Hypertonic solutions, 117
Hypertrophy of skeletal muscle, 237
Hyperventilation, 356–357, 359–360
Hypocalcemic tetany, 393
Hypothalamic-releasing hormones, 204–207
Hypothalamo-pituitary portal vessels, 203–204
Hypothalamus, 182–183
  control of thirst, 389–391
  glucose receptors, 454–455
  osmoreceptors, 388–389
  regulation of shivering, 471
  role in control of circulation in exercise, 316
  role in controlling food intake, 467–468
  role in emotional behavior, 617
  role in exercise respiration, 361
  role in hormone secretion, 192–193
  role in male reproductive system, 489–490
  role in motivated behavior, 615, 618
  role in pituitary function, 204–207
  role in stress, 552–553
  role in temperature regulation, 475–477
Hypothyroid, 461
Hypotonic solutions, 118
Hypoxia, 363–364
Hypoxic hypoxia, 363–364

I band, 212, 214-215
ICSH, 487
Identical twins, 495, 504
Ig (see Immunoglobulins)
IgA, 535
IgE, 535, 546-547
IgG, 535-536, 543, 547
IgM, 535-536
Ileocecal sphincter, 435
Ileogastric reflex, 434
Ileum, 408, 414
Illusions, 587-588
Immune enhancement, 543
Immune responses:
  nonspecific, 522
  specific, 522, 533-544
Immune surveillance, 522, 543
Immune system, effector cells, 522-524
Immunity, 521-522, 539
Immunoglobulins, 535
Implantation, 505
Impotence, 486
Inactivation of sodium, 154
Inborn errors in metabolism, 68
Incus, 576
Induction, 60-63
Infarction, 318
Inferior vena cava, 255
Infertility, 514
Inflammation, 525
Inflammatory response, 524-529
Inhibin, 490
Inhibitory postsynaptic potential, 170-171
Inhibitory synapse, 169-171
Initial segment of neuron, 160, 162
Inner cell mass, 505
Inner ear, 577-578
Inosine, 57-58
Inspiration, 335-337
Inspiratory capacity, 340
Inspiratory neurons, 353
Inspiratory reserve volume, 339-340
Insulin, 448-452
  control of secretion, 449-450
  effects of decreased secretion, 449-452
  effects on growth, 461
  during exercise, 456-457
  receptors, 198
  role in glycogen synthesis and breakdown, 197-198

Insulin:
  role in hunger, 467
  secretion by pancreas, 192, 201, 208
Integral proteins, 101
Integrating center, 130-131
Intension tremor, 602
Intensity coding, 165-166
Intercostal muscles, 332, 336
Interferon, 532-533
Internal anal sphincter, 435
Internal environment, 4-8, 129, 367
Interneurons, 161-162
Interphase, 64
Interstitial cells, 480-481, 487
Interstitial-cell-stimulating hormone, 487
Interstitial fluid, 7-8
Interstitial fluid pressure, 294
Interventricular septum, 268
Intestinal epithelium, 408-411
Intestinal phase, 419-420
Intestine (see Large intestine, Small intestine)
Intestino-intestinal reflex, 434
Intraabdominal pressure, 336
Intracellular fluid, 7-8, 14
Intrafusal fibers, 592-593
Intrapleural fluid, 333-335
Intrapleural pressure, 335-337
Intrathoracic pressure, 278, 280, 335
Intrauterine device, 514
Intrinsic factor, 414, 423
Inulin, 373
Involuntary nervous system, 187
Iodine, 196-197
Ionic hypothesis of action potentials, 154-155
Ionization, 25
Ions, 25
IPSP (see Inhibitory postsynaptic potential)
Iris, 566, 568
Iron, 23
  absorption by intestinal tract, 416
  balance in body, 260
  role in bacterial growth, 532
  role in oxygen transport, 257, 259
Irreversible chemical reactions, 79
Ischemia, 287
Ischemic hypoxia, 363
Islets of Langerhans, 448
Isometric muscle contraction, 225-226
Isometric ventricular contraction, 273-274

Isotonic muscle contraction, 225–226
Isotonic solutions, 117
Isovolumetric ventricular relaxation, 273–274

Jaundice, 438
Jejunum, 408
Joint movements, effects on respiration, 361
Joint position, 563–564
Joint receptors, 604–605
Junctions between membranes, 122–124

Kallikrein, 530
Kangaroo rat, 385
Keto acids, 95–98, 444
Ketoacidosis, 451
Ketone bodies, 445, 447, 451
Kidneys:
    blood flow, 256
    disease, 400–401
    endocrine functions, 192–193
    erythropoietin, 261–262
    excretion of foreign chemicals, 549
    excretion of glucose in diabetes mellitus, 451
    growth, 461
    regulation of extracellular hydrogen ion
        concentration, 399–400
    structure, 368–370
    transplantation, 540
Kilocaloric, 72
Kinetic energy, 72
Kininogen, 530
Kinins, 529–531
Knee jerk reflex, 594–595
Krebs cycle, 82–84, 89–92, 98, 447
Krebs, Hans, 82

Labia majora, 491, 493
Labia minora, 491, 493
Labor, 510
Lacrymal glands, 524
Lactase, 438
Lactation, 511–512
Lacteal, 410, 414–415
Lactic acid, 84–87
    formation during muscle activity, 236
    role in glucose synthesis, 446
Lactose, 60, 411, 438
Lactose deficiency, 438
Lamina propria, 403, 405

Language, 628–630
Large intestine, 403–404, 407–408
    digestion, 411
    motility, 435–436
Larynx, 328–329, 488
Latent period in muscle, 219–220, 226
Lateral geniculate nucleus, 571
Lateral inhibition, 562
Lateral sacs of sarcoplasmic reticulum, 219–221
Lateral ventricle of brain, 189
Law of mass action, 74, 77
Law of the intestine, 434
Laxatives, 439
L-dopa, 602
Lead, 508
Learning, 591, 618, 626–628
Length-tension relationship:
    cardiac muscle, 277–279
    skeletal muscle, 229–230
    smooth muscle, 241–242
Lens, 566–568
Leukemia, 545
Leukocytes, 256, 262–264, 522–524
Lever action of muscles, 231–233
Leydig cells, 480
LH (see Luteinizing hormone)
Libido, 503
Ligaments, 3, 563–564
Ligand, 43
Ligase, 69–70
Light, 566
Light adaptation, 570
Limbic system, 182–183, 611–612, 616–618, 626
Lipase, 412–413
Lipids, structure of, 32–35
Lipoproteins, 457
B-Lipotropin, 628
Lithium carbonate, 619–620
Liver:
    biotransformation of foreign chemicals, 549–550
    effects of alcohol, 551
    glycogen storage, 90–91
    plasma proteins, 257
    role in amino acid metabolism, 444
    role in bile secretion, 431–432
    role in blood clotting mechanisms, 323–324
    role in carbohydrate metabolism, 443–447
    role in fat metabolism, 443–447
    role in gastrointestinal function, 403, 407–408

Liver:
    role in specific dynamic action, 465
    role in vitamin D activation, 395
    synthesis of cholesterol, 457–458
Load muscle, 225
Local anesthetics, 155–156
Local responses, 133
Lockjaw, 176
Long-term memory, 626–628
Loop of Henle, 385–388
Low-density lipoproteins, 457
Lower esophageal sphincter, 422–423
LSD, 175
Lumbar nerves, 180
Lungs:
    metabolic functions of, 365
    role in activating plasmin, 325
    structure of, 328–331
    volumes of, 339–340
Luteal phase, 496–497
Luteinizing hormone:
    in female, 498–500
    in male, 486, 489–490
    role in ovulation, 499
    secretion by pituitary, 192, 201–203, 205–206
    secretory changes with age, 517–519
Lymph, 299
Lymph flow, 299–300
Lymph nodes, 523–524
Lymphatics, 299–300
Lymphocytes, 262–264, 522–524, 533
    (*See also* B lymphocytes, T lymphocytes)
Lymphoid tissues, 523
Lymphokines, 540
Lysosomes, 15–18, 120, 528

M line, 212, 214–215
Macrophages, 259, 262–263, 330, 522–524, 526, 541–542
Magnesium, 22–23, 26
Male reproductive organs, 480
Malignant tumors, 68
Malleus, 576
Malnutrition, 458, 466, 545, 625
Maltose, 411
Manias, 619–620
Mass movement, 435
Mast cells, 262, 546
Maximal oxygen uptake, 361–363

Maximum tubular capacity, 376
Mean arterial pressure, 283, 302–303
Mechanist view of life, 9
Mechanoreceptors, 163, 563
Median eminence, 204
Mediated transport, 108–113
    (*See also* Active transport; Facilitated diffusion)
Medulla, 180–182
    cardiovascular center, 305–306
    respiratory center, 352
    swallowing center, 421
    vomiting center, 436
Megakaryocytes, 264
Meiosis, 480–482, 495–496
Melanocyte stimulating hormone, 202, 628
Melatonin, 192
Membrane:
    cell organelles, 13–19
    junctions between, 122–124
    permeability, 106–108, 155
    structure, 100–103
Membrane potentials:
    action potentials, 152–160
    diffusion potentials, 147–150
    graded potentials, 151–153
    resting potentials, 146–147, 149–150
Memory, 614
    consolidation, 626–627
    long-term, 626–628
    short-term, 626–627
Menarche, 519
Menopause, 491, 503, 519–520
Menstrual cycle, 491, 496, 501–502
Menstrual flow, 501
Mescaline, 619, 620
Mesenteries, 403
Messenger RNA, 53–60, 63
Metabolic acidosis, 400
Metabolic alkalosis, 400
Metabolic pathway, 78
    carbohydrates, 88–92
    fats, 92–95
    protein, 95–97
Metabolic rate:
    determinants of, 464–465
    measurement of, 463–464
Metabolism, 71–72
    (*See also* Carbohydrate; Fat; Protein)
Metastasis, 68

Micelles, 28–29, 412–415
Microcirculation, 290
Microbes, 524
Microorganisms, 522
Microscopes:
    electron, 13–14
    light, 13
Microsomal enzyme system, 550–551
Microtubules, 19–20, 65
Microvilli, 408–410
Micturition, 377–378
Midbrain, 181–182
Middle-ear cavity, 576
Milk:
    intolerance, 438
    let-down, 512
    production, 511–512
Millivolt, 146
Mineral elements, 22–23
Mitochondria:
    role in energy metabolism, 81–84
    role in fat metabolism, 93
    structure of, 15–19
Mitosis, 66, 459, 480
Mitral valve, 264–265, 274–275
Modality of stimulus, 558
Modulator molecules, 49
Mole, 29
Molecular weight, 29
Molecules, characteristics of, 23–27
Monoamine oxidase inhibitors, 619
Monocytes, 262, 522–523, 526, 532
Monosaccharides, 31
Monosynaptic reflex, 594
Mons pubis, 491, 493
Morphine, 175, 364, 565, 566, 621, 622
Motion sickness, 436
Motivation, 615–618
Motor control:
    descending pathways, 597–603
    local, 592–597
Motor cortex, 590, 598–601
Motor-end plate, 223–224
Motor neurons, 161, 185, 222–224, 231
    alpha, 592–593, 595–596
    central control of, 597–603
    gamma, 592–593, 595–596
    local control of, 592–598
Motor units, 223, 231, 589

Movement, control of, 589–592
MSH (see Melanocyte stimulating hormone)
Mucins, 421
Mucosa, 403, 405
Mucus secretion:
    airways, 328–329, 337
    salivary glands, 404, 421
    stomach, 423
Mullerian ducts, 516
Mullerian regression factor, 516
Multineuronal motor pathways, 599–601, 605
Multiunit smooth muscle, 245, 247–249
Murmurs, heart, 275
Muscarinic receptors, 187
Muscle: summary of characteristics, 248
    (See also Cardiac muscles; Skeletal muscle; Smooth
        muscle)
Muscle differentiation, 236–239
Muscle fatigue, 228, 234, 236–239
Muscle fibers, 212
Muscle spindle, 592–597, 605
Muscularis mucosa, 403, 405, 410
Mutation, 66–69
Myasthenia gravis, 225
Myelin, 158–160
Myelinated nerve fibers, 158–160
Myelinated receptor neurons, 164
Myoblasts, 236–237
Myocardium, 264
Myenteric plexus, 403, 405, 417
Myoepithelial cells, 511–512
Myofibrils, 212–215
Myoglobin, 38, 237, 239, 258, 364
Myometrium, 501
Myopic vision, 568
Myosin, 214–218, 528
Myosin ATPase, 217, 237–239

NAD (see Nicotinamide adenine dinucleotide)
Na-K ATPase, 112–113, 149–150
Natural selection, 69
Nearsighted vision, 568
Negative balance, 144
Negative feedback, 132–133
Negative nitrogen balance, 442
Nephron, 368–370
Nerve cells, 3
Nerve fibers, 184
Nerve gas, 224

Nerves, 184
Nervous system:
  development of, 624–626
  evolution of, 623–624
  (*See also* Autonomic nervous system, Brain,
    Central nervous system, Somatic nervous
    system, Spinal cord)
Net flux, 106–107
Neuroendocrinology, 191
Neuromodulators, 176–177
Neuromuscular junction, 221–225
Neuron structure, 160–162
Neuronal plasticity, 627
Neurotransmitters, 135, 167, 173–175
Neutral fat (*see* Triacylglycerol)
Neutrons, 21–22
Neutrophils, 262, 523, 525–526, 528, 530–532, 541
Niacin, 76
Nicotinamide adenine dinucleotide, 76, 81
Nicotinic receptors, 186
Nitrogen, 22–24
Nociceptors, 565
Nodes of Ranvier, 158–160
Nonpolar molecules, 26
Nonshivering thermogenesis, 472
Nonspecific immune responses, 522, 524–533
Nonspecific sensory pathways, 559
Norepinephrine:
  neurotransmitter, 173–175, 186–187, 192, 201
  role in affective disorders, 619
  role in schizophrenia, 619
Novelty detectors, 611, 613
Novocaine, 155
Nuclear envelope, 14–15, 18
Nuclear pores, 14–15
Nuclei in central nervous system, 179
Nucleic acids:
  digestion and absorption by small intestine, 414
  structure of, 39–41
  (*See also* Deoxyribonucleic acid; Ribonucleic acid)
Nucleolus, 14–18, 55–56
Nucleoproteins, 56, 62
Nucleotides, 39–41
Nucleus:
  atomic, 21–22
  cell, 14–15, 18, 55–56, 65

O blood type, 544
Obesity, 452, 468–470

Occipital lobe, 182
Ohm's law, 146
Olfaction, 558, 585–586
Olfactory mucosa, 585
Oligodendrocyte, 184
Oligomeric proteins, 38–39
Oogonia, 495–496
Operating point, 131
Operator gene, 60–61
Operon, 61–62
Opiate drugs, 175
Opponent color cell, 574
Opsin, 568–569
Optic nerve, 566, 571
Optics, 566–568
Oral cavity, 329
Oral contraceptives, 514
Organic molecules, 29–30, 32
Organ of Corti, 578–580
Organ systems, 3–5
Orgasm, 486, 503
Orienting response, 611
Osmolarity, 114
  regulation of, 385–389
Osmoreceptors, 388–389
Osmosis, 113–118
Osmosis across capillaries, 293–296
Osteoblasts, 459
Otoliths, 583
Oval window, 576–579
Ovaries, 192–193, 491–496
Oviducts, 491–493, 503–504
Ovulation, 493, 495, 497–498
  control of, 499–500
  following pregnancy, 512
Ovum:
  development, 491–496
  regulation of development, 498–500
  transport, 503
Oxaloacetic acid, 83, 91–92, 98
Oxidative deamination, 95–96
Oxidative phosphorylation:
  mechanism of, 81–82, 89–90
  in skeletal muscle, 235, 237, 239
Oxygen:
  binding to hemoglobin, 346–350
  caloric equivalents of, 463
  control of ventilation, 354–356
  formation of photosynthesis, 87–88

Oxygen:
    maximal uptake, 361–363
    partial pressures in respiratory system, 345–346
    properties of, 22–24
    transport in blood, 343, 346–350
    utilization by cells, 81–82, 328
Oxy-hemoglobin, 346
Oxyntic cells, 423
Oxytocin, 177, 192, 201, 207, 379, 510, 512, 628

P wave, 271–274
Pacemaker potentials, 152
    in smooth muscle, 244–248
Pacemaker, heart, 268
Pacinian corpuscle, 563
Pain, 175, 361, 562, 564–566, 587
Pancreas:
    endocrine secretions, 192–193, 209, 448, 457
    hormones effecting, 419
    intestinal secretions, 406–408, 428–430
Pantothenic acid, 76
Paracrines, 134, 177
Paradoxical sleep, 609, 611–612
Parahippocampal gyrus, 183
Parasympathetic division of the autonomic nervous
    system, 185–189
Parasympathetic nerves:
    to blood vessels of external genitals, 288
    effects on heart, 275–277
Parathormone, 132–133, 394–395
Parathyroid hormone, 192, 201, 208, 394–395
Parathyroids, 132–133, 192–193
Parietal cells, 423–426
Parietal lobe, 182, 558, 630
Parkinsonism, 602
Parotid salivary gland, 403, 421
Partial pressure, gas, 344
Parturition, 509–511
Passive immunity, 539
Penicillin, 377
Penis, 485–486
Pepsin, 406, 424–426
Pepsinogen, 423–426
Peptide bond, 35–37
Peptides, 36
Perceptual processing, 586–588
Pericardium, 264
Peripheral chemoreceptors, 357
Peripheral nervous system, 179, 184–188

Peripheral proteins, 101
Peripheral thermoreceptors, 475
Peristalsis:
    esophagus, 422
    large intestine, 435
    small intestine, 434
    stomach, 427
    ureters, 377
Peritubular capillaries, 370
Permeability of membranes, 106–108
Permissiveness of hormones, 200
Pernicious anemia, 261, 414
Pesticides, 224
Petellar tendon, 594
pH, 395
Phagocytic cells in lungs, 329–330
Phagocytosis, 119, 262, 527–528, 531
Phantom limb, 613
Pharmacological effects of hormones, 200–201
Pharynx, 328–329, 403, 405–406, 421–422
Phenylalanine, 68
Phenylketonuria, 68
Phosphate:
    in phospholipids, 33–34
    reabsorption by renal tubules, 394
    role in renal excretion of hydrogen ions, 399
Phosphodiesterase, 136
Phosphoenolpyruvic acid, 85, 91–92
Phospholipids:
    in cell membranes, 100–102
    structure of, 33–35
Phosphorus, 22–23
Photopigments, 568
Photosynthesis, 84, 87–88
Phrenic nerves, 352
PIH (*see* Prolactin-inhibitory hormone)
Pineal, 192
Pinna, 576
Pinocytosis, 118
Pitch, 576
Pituitary gland:
    anterior hormones, 192–193, 202–207
    posterior hormones, 192–193, 207
    structure of, 201–202
Placenta, 505–508, 549
Plasma, blood, 256–258
Plasma cells, 264, 522–523, 533, 547–548
Plasma membrane, 14–15, 18
    formation and removal, 120–122

Plasma membrane:
  functions of, 99–100
  permeability of, 106–108
  structure of, 100–103
Plasma proteins, 257–258
Plasmin, 324–325
Platelets, 175, 256, 264, 320–321
Pleura, 332–333
Pneumonia, 364
Pneumothorax, 334–335, 364
Polar body, 495–496
Polar molecules, 26–27
Polarized chemical bonds, 26–27
Poliomyelitis, 252, 364
Polymers, 30
Polymodal neurons, 559
Polymorphonuclear granulocytes, 262–263
Polyribosome, 59
Polysaccharides, 31
Polyspermy, 504
Polysynaptic reflexes, 178
Polyunsaturated fatty acids, 32, 34
Pons, 181–182, 611
Pores in membrane, 101, 107–108
Portal vein, 255, 431, 432
Positive balance, 144
Positive feedback, 130, 155–156
Postabsorptive state, 442, 445–447
Postganglionic fibers, 186
Postsynaptic neurons, 166
Posture, 583, 603–606
Potassium:
  absorption by intestinal tract, 416
  active transport, 111–112, 149–150
  equilibrium potential, 149
  essential element, 22–23, 26
  regulation of extracellular concentration, 391–393
  role in generation of action potential, 154–155
  secretion by kidneys, 377
Potential energy, 72
Potentials (*see* Membrane potentials, Voltage)
Prandial drinking, 391
Precapillary sphincter, 290
Preganglionic fibers, 186
Pregnancy, 503–514
  effects on metabolic rate, 464
  heartburn, 423
  hormonal changes during, 508–509
  Rh factor, 545

Prepuce, 480
Presbyopia, 568
Pressure autoregulation, 287
Pressure-flow equation, 280–282
Pressure sensation, 563
Presynaptic inhibition, 176, 561
Presynaptic neurons, 166
PRH (*see* Prolactin releasing hormone)
Primary afferent neurons, 184
Primary follicles, 491–492, 494
Primary oocytes, 495–496
Primary spermatocyte, 481–483
Primary structure of proteins, 36, 39
Primary visual cortex, 571
Products, of enzyme reactions, 47
Progesterone:
  action on uterus, 199
  during pregnancy, 508–510
  effects on breasts, 511
  in oral contraceptives, 514
  regulation of secretion, 496–502
  sites of production, 192, 201–203, 209–210
Progestogens, 514
Prolactin, 192, 201–202, 205–206, 511–512
Prolactin-inhibitory hormone, 205–206, 511–512
Prolactin-releasing hormone, 205–206
Proliferative phase of menstrual cycle, 497, 501–502
Propagation of action potentials, 158–160
Prostaglandins:
  in active hyperemia, 288–289
  chemical mediators, 136, 177
  effects on airways, 338
  removal from pulmonary circulation, 365
  role in blood hemostasis, 320
  role in parturition, 510–511
  in semen, 485, 504
Prostate gland, 480, 485
Proteases, 95
Proteins:
  addition of chemical group to amino acid side chains, 50
  binding sites, characteristics of, 43–50
  body content, 93
  conformation, 49–50
  digestion and absorption from GI tract, 411
  in membranes, 101–102
  metabolism, 95–98, 441, 444, 449
  role in gluconeogenesis, 446
  secretion, 120–121

Proteins:
    structure of, 34–38
    synthesis, 51–63
        effect of insulin on, 449
        effects of androgens on, 461–462
        effects of growth hormone on, 459, 461
        regulation of, 59–63
        role in memory, 628
Protein kinase, 136–140
Prothrombon, 321–323, 326
Protomer, 40
Protons, 20–21, 143
Protoplasm, 14
Proximal convoluted tubule, 368–369
Pseudohermaphrodites, 515
Psilocin, 619
Psilocybin, 175
Psycoactive drugs, 620–623
Puberty, 479, 517–519
Pubic hair, 501–502
Pulmonary arterial pressure, 303
Pulmonary arteries, 255
Pulmonary blood vessels, 338–339
Pulmonary circulation, 254–255, 273–274, 338–339
Pulmonary edema, 318, 339, 345–346
Pulmonary valve, 265, 272–274
Pulmonary veins, 255
Pulmonary ventilation, 339–341
Pulse pressure, 283
Punishment, 615–617
Pupil, 566, 568
Purines, 38–40
Pus, 529
Pyloric sphincter, 424, 427
Pyramidal decussation, 598
Pyramidal tract, 598
Pyridoxine, 76
Pyrimidines, 38–40
Pyrogens, 476
Pyruvic acid, 84–87, 89–92, 96, 98, 446

QRS complex in electrocardiogram, 271–274
Quadriceps femoris, 232–233
Quatenary structure of proteins, 37–38

Radiation of heat, 472–474
Rage, 616–617
Ragweed pollen, 546
Random thermal motion, 103

Rapid-eye-movement sleep, 609
RAS (*see* Reticular activating system)
Reabsorption by kidney:
    of salt, 379
    of sodium, 382–383
    of water, 379
    (*See also* Tubular reabsorption)
Receptive field, 558, 560–561
Receptive relaxation, 427
Receptor neurons, 161–166
Receptor potential, 152, 163–165
Receptors, 132–133, 136–137, 198–199
Reciprocal innervation, 178, 595
Recombinant DNA, 69–70
Recruitment of muscle fibers, 231
Recruitment of receptor neurons, 166
Rectum, 403–404, 407, 435–436, 439
Red blood cells (*see* Erythrocytes)
Red pulp, 524
Reduced hemoglobin, 346
Reflex, 130–134
Refractory periods of action potentials, 156–158
Regurgitation through heart valve, 275
Relative refractory period, 157
Relaxation time skeletal muscle, 226
REM sleep (*see* Rapid-eye-movement sleep)
Renal blood flow, 373–374
Renal hypertension, 317
Renal pelvis, 368
Renin, 192, 201, 288, 317, 382–383, 390
Repolarize, 151
Repression, 60–63
Reproduction, 479–520
Residual volume, 339–340
Resistance:
    of airways, 336–337
    to blood flow, 281–282, 285–289
    of capillaries, 290
    electrical, 146
    total peripheral, 302–303
    of veins, 298
Respiration, 328
    control of, 179, 352–364
    effects on venous return, 298
Respiratory acidosis, 400
Respiratory alkalosis, 400
Respiratory center, 181
Respiratory-distress syndrome of the newborn, 342
Respiratory quotient, 342

Respiratory system, 328
Resting membrane potential, 146–147, 149, 152
Restriction enzyme, 69
Retching, 436
Reticular activating system, 610–612, 614
Reticular formation, 181, 559, 565, 610–612
Retina, 566–571
Retinal, 76, 568–569
Retrograde amnesia, 626–627
Reversible chemical reactions, 79
Reward, 615–618
Rh factor, 545
Rhodopsin, 568–570
Riboflavin, 76
Ribonucleic acid (RNA):
    messenger RNA, 53–60, 63
    ribosomal RNA, 56
    structure of, 40
    transfer RNA, 57–58
Ribose, 40
Ribosomal RNA, 56
Ribosomes, 15, 17, 19, 56–60, 120–121
Ribs, 332
Rickets, 395
Rigor mortis, 217
RNA (see Ribonucleic acid)
    mRNA (see Messenger RNA)
    rRNA (see Ribosomal RNA)
    tRNA (see Transfer RNA)
Rods, 568–570
Round window, 577
RQ (see Respiratory quotient)

Saccades, 574
Saccule, 582–583
Sacral nerves, 180
Saliva, 421
Salivary glands, 403–406, 408
Salt appetite, 391
Salt balance, 378–391, 509
Salt, solubility of, 27–28
Saltatory conduction, 160
SA node (see Sinoatrial node)
Sarcomere, 212–215
Sarcoplasmic reticulum, 219–221, 272
Saturated fatty acid, 32, 34, 458
Saturation:
    of binding sites, 45–47
    of mediated transport flux, 108–109

Scala tympani, 577–578
Scala vestibuli, 577–579
Schizophrenia, 602, 618–619
Schwann cell, 184
Scrotum, 480
SDA (see Specific dynamic action)
Sebaceous glands, 501, 524
SEC (see Series elastic component)
Secondary oocyte, 495–496
Secondary peristalsis, 422
Secondary sex characteristics, 479, 488
Secondary spermatocyte, 481–482
Secondary structure of proteins, 39
Second messengers, 135–139
Secretin, 192, 418–420, 429–430
Secretory granules (vesicles), 17–18, 120–122
Secretory phase of menstrual cycle, 497, 501–502
Segmentation, 433–434
Semen, 485
Semicircular canals, 582–583
Seminal vesicles, 480, 485
Seminiferous tubules, 480–481, 483–484
Sensor, 129
Sensory:
    coding, 557–562
    deprivation, 613
    information, 557
Septal area, 183
Series elastic component, 229
Serosa, 403, 405
Serotonin, 173, 175, 620
    role in affective disorders, 619
    role in blood hemostasis, 320–321
    role in sleep, 611–612
Sertoli cell, 483–484, 489, 490, 516
Serum, blood, 257, 324
Sex chromosomes, 514–515
Sex chromatin, 515
Sex determination, 514–515
Sex differentiation, 515–519
Sex drive, 489, 503
Sex hormones:
    effects on CNS, 516–517
    (See also Estrogen; Progesterone; Testosterone)
Shivering, 471
Shock, 311–313
Short-term memory, 626–627
Sickle cell anemia, 258
Sigmoid colon, 435

Simple cortical cells, 572
Single-unit smooth muscle, 247–248
Sinoatrial node, 268
Skeletal muscle:
  action potentials, 218, 222–225
  blood flow, 256, 288
  differentiation and growth, 236–239
  energy metabolism, 233–236
  mechanics, 225–230
    control of whole-muscle tension, 230–233
    length-tension relationship, 229–230
    summation, 227–229
    velocity of shortening, 227, 233
  structure of, 212–215
  summary of events during contraction, 225
  testosterone effects on, 488
  types of muscle fibers, 237–239
Skeletal muscle "pump", 298
Skeletomotor fibers, 592–593
Skin, 395, 524
  blood flow, 256, 287–288
  role in temperature regulation, 473–475
Skull, 180
Sleep, 609–612
Sliding filament model of muscle contraction, 215–217, 229–230
Slow twitch-resistant to fatigue muscle fibers, 237–239
Slow viruses, 522
Slow wave potentials in smooth muscle, 244
Slow-wave sleep, 609, 612
Small intestine:
  absorption, 411–416
  blood flow, 410
  functions of, 403–404, 407
  hormones, 419
  immune system, 412
  motility, 433–434
  secretions, 433
  structure of, 408–411
Smell, 585–586
Smoking, 586
Smooth muscle:
  actin and myosin, 241–242
  of airways, 329, 331, 337
  in blood vessels, 285, 296
  excitation-contraction coupling, 242–246
  of GI tract, 403, 405
  length-tension relationship, 241–242

Smooth muscle:
  membrane potentials, 244–248
  multiunit muscles, 245, 247–249
  single-unit muscles, 247–248
  in small intestine, 433–434
  of stomach, 427
  structure, 240–242
  summary of characteristics, 248
Sneeze, 361
Sodium:
  absorption by intestinal tract, 414–416
  in activation, 154–155
  active transport, 111–112, 149–150
  channels, 155
  equilibrium potential, 149
  excretion of kidney, 374
  ions, 22–23, 25–28
  regulation of balance of, 378–391
  relation to extracellular volume, 380–381
  requirement for active transport of organic molecules, 412
  role in generation of action potential, 154–155
  role in hypertension, 317
  total body balance, 378
Soft palate, 421–422
Solubility, 27–28
Solutes, 27
Solvent, 27
Somatic, word defined, 558
Somatic nervous system, 184–186
Somatic sensation, 562–566
Somatomedins, 459
Somatosensory cortex, 558–559, 563
Somatostatin, 173, 177, 192, 205–206, 448
Sound, 574–576
Sound localization, 581–582
SPA, 565
Spatial summation, 171
Specific dynamic action, 464–465
Specific immune responses, 522, 533–544
Specific sensory pathways, 558, 562–563
Speech, 602, 629–630
Sperm, 479–486
  capacitation, 504
  transport, 504
Spermatids, 481–484
Spermatogenesis, 480–484, 487, 489–490
Spermatogonia, 481–484
Spermatozoa, 481–483

Sphincter of Oddi, 432
Sphygmomanometer, 284
Spinal cord, 179–180
Spinal nerves, 180–181, 184
Spindle apparatus, 65
Spindle fibers, 592–593
Spleen, 523–524
Standard 70-kg man, 7–8
Stapes, 576–577
Starch, 31, 91, 411
Starling's law of heart, 278
Starvation, 447
States of consciousness, 606–612
Steady state, 129
Stenosis of heart valve, 275
Sternum, 332
Steroid hormones, 199–200, 209–210
Steroids, 33, 35
Stimulus, 131
    intensity, 559–560
    localization, 560–561
    modality, 558–559
    response to, 163–165
    strength, 156, 165–166
Stimulus-produced analgesia, 565
Stomach:
    absorption, 406
    functions of, 403–406, 408
    hormones effecting, 419
    motility, 426–428
    mucus, 437
    pressure in, 422
    secretions of, 423–428
    ulcers, 436–437
Stress, 552–555
Stretch reflex, 593–597
Striated muscles, 212
Stroke, 188, 317, 319, 435
Stroke volume, 275–280, 282–283, 362–363
Subarachnoid space, 189
Subatomic particles, 21–22
Subclavian arteries, 355
Sublingual salivary gland, 403, 421
Submandibular salivary gland, 403, 421
Submucosa, 403, 405
Submucus plexus, 403, 405, 417
Substance P, 173, 177, 565
Substantia nigra, 602
Substrate phosphorylation, 81, 84–86

Substrates, 47, 74
Suckling, 511–512
Sucrose, 30–31
Sulfate, 97
Sulfhydryl group, 38–39
Sulfur, 22–23
Sulfur metabolism, 97
Summation:
    of contractile activity, 227–229
    of synaptic activity, 171
Superior sagittal sinus, 189
Superior vena cava, 255
Suppressor T cells, 534, 556
Suprathreshold stimuli, 156
Surface tension, 341–342
Surfactant, 342, 364
Swallowing, 179, 421–423
Sweat, 367, 389–390, 474–475, 477
Sylvian fissure, 563
Sympathetic nervous system, 185–189
    activity during exercise, 457
    effects on arterioles, 287–288
    effects on heart, 275–277, 279–280
    relation to adrenal medulla, 207
    response to stress, 552, 554–555
    role in regulation of organic metabolism, 454–455
Sympathetic trunks, 186
Sympathetic vasodilator fibers, 288, 306, 315
Synapse, 166–176
    chemical, 167
    electrical, 167
    excitatory, 168–171
    inhibitory, 169–171
    presynaptic inhibition, 176
Synaptic cleft, 167, 169
Synaptic delay, 168
Synaptic potentials, 152
Synaptic terminal, 167, 169
Synergistic muscles, 595
Systemic circulation, 254
Systole, 272–274
Systolic blood pressure, 283

T lymphocytes, 533, 534, 539, 543, 547
Tm, 376
t tubule (*see* Transverse tubule)
T wave, 271–273
Target cells, 194
Taste, 584–585

Taste buds, 558, 584
Tectorial membrane, 578–579
Teleology, 9–10
Temperature:
    effect on chemical reactions, 73
    effect on hemoglobin-oxygen saturation, 349
    effects of progesterone on, 503
    effects on ventilation, 361
    regulation of, 470–478
    sensation, 564
Temporal association cortex, 614
Temporal bone, 582
Temporal lobe, 182, 616–617, 630
Temporal summation, 171
Tendons, 3, 230–231
Tension of muscle, 225, 231
Termination code words, 51, 56
Tertiary structure of proteins, 37–39
Testes, 480, 485, 489
Testosterone, 192, 201–203, 209–210, 480
    effects of, 487–489
    in females, 488–489, 503
    neuromodulator, 177
    role in sexual differentiation, 516
    secretory changes with age, 517–520
    stimulation of erythropoietin, 262
    structure of, 33
Tetanus (disease), 175
Tetanus contraction, 228
Thalamus, 182–183, 558, 590, 607–608, 610–611
TH (*see* thyroid hormone)
Theca, 493
Thermoreceptors, 475–476, 564
Theta rhythm, 611
Thiamine, 76
Thick filaments, 214–217, 222, 230
Thin filaments, 214–219, 222, 230
Third ventricle of brain, 189
Thirst, 367, 389–391
Thoracic cage, 331, 335–336
Thoracic cavity, 264, 328
Thoracic nerves, 180
Thoracolumbar division of the autonomic nervous
    system, 185
Thoroughfare channels, 290
Threshold potential, 156
Thrombin, 320–325
Thrombosis, 325–326
Thromboxanes, 134

Thymine, 39–41, 51–52
Thymosin, 192, 524, 534
Thymus, 523–524, 533, 544
Thyroglobulin, 196–197
Thyroid, 192–193, 395, 461
Thyroid hormone, 461
    neuromodulator, 177
    effects on metabolic rate, 465
    effects on CNS, 625
    (*See also* Thyroxine)
Thyroid-stimulating hormone, 192, 196–197, 202–203,
    205
Thyrotropin-releasing hormone, 177
Thyroxine, 192, 196–197, 199–200, 205, 461
Tidal volume, 339–341
Tight junction:
    intestinal epithelium, 409, 411
    between Sertoli cells, 483
    structure of, 15, 122–123
Tissue repair, 528–529
Tissue thromboplastin, 324
Tissues, 3–4
Tolerance, 621
Tone, smooth muscle, 242–243
Tonsils, 523
Total peripheral resistance, 302–303
Touch, 563
Toxemia of pregnancy, 509
Trace elements, 22–23
Trachea, 328–329, 403, 421–422
Training, physical, 363
Transamination, 95–96
Transcriptase, 53–54
Transcription, 53–56, 60
Transepithelial transport, 375
Transfer RNA, 57–58
Transferrin, 260, 416
Transfusion of blood, 544–545
Translation, 53, 56–60
Transmural pressure, 278
Transplantation, 543
Trans-retinal, 569
Transverse tubule, 220–222
Tremor, 601
Triacylglycerols, 92–93, 98
    breakdown during postabsorptive state, 446–447
    digestion and absorption by GI tract, 412–415
    effect of insulin on, 449
    formation during absorptive state, 443–444

Triacylglycerols:
    structure of, 32–34
Triceps muscle, 232
Tricuspid valve, 264–265, 274–275
Tricyclic antidepressants, 619
Triglycerol (*see* Triacylglycerol)
Triiodothyronine, 192, 461
Trophoblasts, 504, 508
Tropomyosin, 218–219, 222, 242
Troponin, 218–220, 222, 242
Trypsin, 36, 412, 429–430
Trypsinogen, 429–430
TSH (*see* Thyroid stimulating hormone)
Tubal pregnancies, 504
Tuberculosis, 545
Tubular reabsorption, 370–371, 374–377
Tubular secretion, 370–371, 377
Tumor, 68
Twitch contraction of muscle, 226–227
Tympanic membrane, 576–578
Type II cells, 330, 342
Tyrosine, 174, 196–197

Ulcers, 436–437
Ultrafiltration across capillaries, 292–296, 299
Umbilical cord, 505–506
Underwater swimming, 357
Unsaturated fatty acid, 32, 34, 458
Upper esophageal sphincter, 421–422
Upright posture, 313–314
Uracil, 41, 53
Urea, 96–98, 441–442
    excretion by kidney, 374, 376
    excretion per day, 368
Uremia, 400
Ureters, 368–369, 377
Urethra, 480, 485, 492
Uric acid, 368
Urination, 377–378
Urine, 431
    formation in diabetes mellitus, 451
    osmolarity, 385–389
    volume, 367
Uterine smooth muscle, 198
Uterine tubes, 491–493
Uterus, 491–493
    anticlotting factors, 325
    contractions during parturition, 509–511
    during menstrual cycle, 501–502

Uterus:
    sperm transport, 504
Utricle, 582–583

Vaccine, 539
Vagina, 491–493, 501–503
Vagus nerve, 275
Valence, 23
Valves:
    of heart, 264–266, 273–275
    in lymphatics, 299
    in veins, 299
Vas deferens, 480, 484–485
Vascular resistance:
    autonomic control, 287–289
    autoregulation, 285–287
Vascular system, 280–299
Vasectomy, 487
Vasoactive intestinal peptide, 177
Vasopressin, 177, 207
    (*See also* Antidiuretic hormone)
Varicose veins, 314
Veins:
    pooling of blood in, 313–314
    pressure in, 277, 296–299
Velocity:
    of action potential propagation, 160
    of blood flow, 290–291
    of muscle shortening, 227, 233
Venous return, 278, 296–299
Ventilation of lungs, 333–342
    control of, 352–364
    during exercise, 342
    work of, 341–342
Ventral roots, 180–181
Ventricles:
    of brain, 188–190
    of heart, 264–265
Ventricular ejection, 273–274
Ventricular filling, 272–274
Venules, 255
Vertebral column, 180
Vestibular apparatus, 582–583, 590, 603–605
Vestibular glands, 491
Vestibular system, 582–584
Vestibule, 491, 493
Villi, 408, 410
Virilism, 489
Viruses, 522

Viruses:
    host response to, 542-543
    inhibition by interferon, 532-533
Viscosity, 282
Vision, 566-574
Visual association cortex, 586
Visual cortex, 558-559, 571, 624
Vital capacity, 339-340
Vital force, 9, 29
Vitalism, 9
Vitamin A, 76, 568
Vitamin B complex, 76
Vitamin B$_{12}$, 260-262, 414
Vitamin C, 76
Vitamin D, 192, 395, 416
Vitamin E, 556
Vitamin K, 323, 326
Vitamins, 76-77
    absorption: by large intestine, 435
    by small intestine, 414
Vitreous humor, 566, 570
Vocal cords, 328-329
Voltage, 146
Voluntary movement, 591
Vomiting, 428, 436
Vomiting center, 181
Vulva, 491

Walking, 605-606
Warm receptors, 475, 564

Water:
    absorption by intestinal tract, 414-416
    balance during pregnancy, 509
    concentration, 113-114
    excretion by kidney, 374
    regulation of balance, 378-391
    structure of, 26-28
    total body balance, 367
Water-soluble vitamins, 76-77
Wavelength, 566
Weight loss, 469-470
Weightlessness, 314
White blood cells (see Leukocytes)
White matter, 179, 181
White pulp, 524
Withdrawal symptoms, 621, 623
Wolffian ducts, 516
Womb (see Uterus)
Work, 462-463

X chromosome, 514-515
Xylocaine, 155

Y chromosome, 514-515
Yolk sac, 506-507

Z lines, 212, 214-215
Zona pellucida, 491, 493-494, 505
Zymogen granules, 121